AF784065

Geofluids

Developments in Microthermometry, Spectroscopy, Thermodynamics, and Stable Isotopes

Geofluids

Developments in Microthermometry, Spectroscopy, Thermodynamics, and Stable Isotopes

Vratislav Hurai

Monika Huraiová

Marek Slobodník

Rainer Thomas

ELSEVIER

AMSTERDAM • BOSTON • HEIDELBERG • LONDON • NEW YORK • OXFORD
PARIS • SAN DIEGO • SAN FRANCISCO • SINGAPORE • SYDNEY • TOKYO

Elsevier
Radarweg 29, PO Box 211, 1000 AE Amsterdam, Netherlands
The Boulevard, Langford Lane, Kidlington, Oxford OX5 1GB, UK
225 Wyman Street, Waltham, MA 02451, USA

ISBN: 978-0-12-803241-1

British Library Cataloguing in Publication Data
A catalogue record for this book is available from the British Library

Library of Congress Cataloging-in-Publication Data
A catalog record for this book is available from the Library of Congress

For information on all Elsevier publications visit our website at http://store.elsevier.com/

Contents

Preface

Fluids have played a principal role in the history of humankind. Hydrogen from liquid water in the oceans and carbon from CO_2-rich gas emanations from volcanic eruptions are the main constituents of the hydrocarbons produced by photosynthetic reactions. Fluids are responsible for the origin of raw materials, minerals, and specific metals, which stimulated revolutionary industrial changes and essential technological breakthroughs. In recent times, petroleum and methane have covered most of the energy demands of industrial civilization. The extension of knowledge on how, from which fluid, and in what conditions minerals originate help us to improve the effectiveness of exploration and exploitation of mineral deposits.

Geofluid research is divided into two major branches: One group of scientists studies modern free fluids in the hydrosphere, atmosphere, and exploratory drillings in the shallow lithosphere as well as the gases that emanate from the Earth's interior in volcanoes. The second group studies paleofluids, which participated in rock- and mineral-forming processes in the past. Paleofluids can be analyzed directly in tiny vacuoles hermetically sealed in minerals during their growth. Indirect methods are based on the reconstruction of fluid properties from fluid–mineral equilibria, and this technique has become increasingly important with the progress in thermodynamics and stable isotope research during the twentieth century.

Microscopic techniques that rapidly developed from the 16th century after the invention of the microscope lens provided a basis for the scientific investigation of fluids trapped as inclusions in minerals. The research blossomed with the introduction of modern laboratory techniques designed for the imaging and analysis of very small (micrometer- and nanometer-sized) objects using microbeam techniques in the second half of the twentieth century. The relatively short history, rapid development, and complexity of the interdisciplinary research are the main reasons that comprehensive textbooks focused on geofluids are rare. Newcomers can draw on several classical sources: the first textbook in the strict sense, written by Shepherd et al. (1985), the extensive monograph by Roedder (1984), and the short-course volume edited by Samson et al. (2003). There are also some elementary textbooks written in other languages, for example, Ukrainian (Voznyak, 2007), German (Leeder et al., 1987), Czech (Benešová and Ďurišová, 1989), and Russian (Lemmlein, 1973, Yermakov and Dolgov, 1979). Comprehensive state-of-the-art information about fluid inclusion research focused on specific geological environments can also be obtained from short-course volumes published by scientific societies (Hollister and Crawford, 1981, Goldstein and Reynolds, 1994), edited thematic journal issues (Lithos 2001, vol. 55, 322 pages), and paper selections from biennial European, Pan-American, and Asian Conferences on Current Research on Fluid Inclusions (ECROFI, PACROFI, ACROFI) published in regular scientific journals.

Fluid inclusion research alone only rarely yields the comprehensive information needed to decipher the origin of geofluids, and so must be supplemented by other methods. For instance, if a mineral precipitated from a homogeneous fluid, an independent constraint on the temperature or pressure must be obtained to fully characterize possible formation *P–T* conditions. This information can be derived indirectly, by analysis of a mineral followed by interpretation of the data in terms of mineral–fluid equilibria. Though this approach is commonly used in literally all scientific papers, relevant procedures are not treated in a comprehensive text, and beginners are forced to look for the specific information in textbooks dedicated to specific laboratory methods, or in innumerable research papers.

This book was written to illustrate how the direct and indirect methods of geofluid research can be combined to gain information on the origin of minerals and rocks. Fluid inclusion study is the central research tool, but it is supplemented by indirect methods employing chemical, thermodynamic, and stable isotope equilibria between minerals and fluids. Working examples with step-by-step explanations and calculations, and case studies help newcomers to learn and understand how the analytical data can be combined to decipher temperature, pressure and depth of mineral formation, composition, oxidation state and origin of mineral-forming fluid, and the dynamic character of a hydrothermal system. Much attention is also paid to fluids in magmatic systems. Each sample calculation can be performed by using simple mathematical functions involved either in an Excel spreadsheet or available computer programs.

The second objective of this book is to provide an all-in-one information source for students who are often beset with a limited time during elaboration of their MSc and PhD theses, and for practical geologists who are far from being specialists in geofluid research. Hence, extensive appendices with supporting data are also involved. For example, Raman tables containing 548 minerals and fluid species will be helpful in the identification of fluid inclusion phases and compounds, when their Raman signals are overlapped with that of the host mineral, thus making the identification with conventional search machines either equivocal or impossible.

This text is designed to provide an introduction rather than be an in-depth information source. Some simplifications can be encountered in places, owing to a number of methods treated. Hence, a detailed theory had to be omitted, and emphasis was instead placed on practical applications tested with our own original published and unpublished data and documentation. The limited available space resulted from the need to accommodate the content to geofluid-related courses lasting one or two semesters. Chapters 1–5 deal with specific subtopics of the fluid inclusion analysis and their interpretation and can thus serve as a text accompanying a one-semester course dedicated to fluid inclusion microthermometry. The remaining chapters can be either extracted as separate modules in independent economic geology, petrology, or isotope geochemistry courses, or used together with the preceding section in a two-semester course dealing with geofluids in a broader sense.

The text was completed for the 25th anniversary of the velvet revolution and the opening of the Iron Curtain in the former Czechoslovakia. It summarizes results achieved during this period, which would not be possible without close cooperation with more than 100 scientists from about 20 countries. We would like to express our gratitude to all of them. VH particularly thanks his mentors and colleagues who introduced him into the world of fluid inclusions and stable isotopes: Jana Ďurišová from the Czech Geological Survey (Prague, Czech Republic) at the outset of his scientific career, and Jochen Hoefs, Elfrun-Erika Horn, Fons van den Kerkhof, Uli Hein, Klaus Simon, and Uwe Wiechert during his research stay at at the Georg–August University of Göttingen in 1990–1992 and 1996. Special thanks from RT go to Wilhelm Heinrich from the GFZ (Potsdam, Germany) for his permanent interest in the inclusion and Raman work over many years, and to Paul Davidson for his unselfish cooperation and help in the melt inclusion research. Thanks are also due to Rastislav Milovský, who compiled the eutectic and peritectic points of water-salt systems listed in this textbook, to Martin C. Styan for linguistic corrections of major portion of the initial defective draft, and Libuše Plchová for helpful technical assistance. Financial funding for technical works connected with the textbook preparation was provided by the VEGA grant 02/0069/13. The laboratory of vibrational spectrometry and fluid inclusion microthermometry at the Geological Institute of Slovak Academy of Sciences was established due to financial support from the European Regional Development Fund under the project of the Centre of Excellence for Integrated Research of the Earth's Geosphere (ITMS-26220120064).

List of Abbreviations

A	Helmholtz energy
c	number of components
CP	critical point
C_P	isobaric heat capacity
C_V	isobaric heat capacity
d	diameter
D	density
E	eutectic point
EOS	equation of state
f	degree of freedom, variance
F_v	bubble filling degree of fluid inclusions
$G, \overline{G}$	Gibbs free-energy, molar Gibbs free energy
G	vapor or gas-rich phase (may be also liquid in gas-aqueous systems)
h	Mohs hardness
$H, \overline{H}$	enthalpy
K	equilibrium constant
L	liquid
$M, \overline{M}$	mass, molecular mass
P	peritectic point
P	pressure
p	number of phases
Q	heat
r	radius
S	entropy
S	solid
T	temperature (Kelvin)
t	temperature (° C)
τ	time
T_e	eutectic temperature (°C)
T_f	temperature of freezing (°C)
T_h	temperature of homogenization (°C)
T_i	temperature of initial melting (°C)
T_m	temperature of melting (°C)
T_n	temperature of bubble nucleation (°C)
TP	triple point
T_s	temperature of sublimation (°C)
U	internal energy
$V, \overline{V}$	volume, molar volume
w	work
μ_{JT}	Joule–Thomson coefficient

CONSTANTS:

c velocity of light in vacuum (299,792,458 m s^{-1})
h Planck's constant (6.626069×10^{-34} J s)
k_B Boltzmann's constant ($1.3806488 \times 10^{-23}$ J K^{-1})
g acceleration due to gravity at sea level (9.80665 m s^{-2})
N_A Avogadro's number ($6.02214129 \times 10^{23}$ mol^{-1})
R gas constant (8.3145 J K^{-1} mol^{-1}, 83.11665 cm^3 bar deg^{-1} mol^{-1})

CHAPTER

GENERAL CHARACTERISTICS OF GEOFLUIDS

1

CHAPTER OUTLINE

The subdivision of everyday matter into solid, liquid, and gas (vapor) is also maintained in chemical and physical terminology. An ideal gas is characterized by perfect disorder at the molecular level. Its counterpart is the crystal structure of a solid—the most ideally ordered package of atoms and molecules. In contrast, an exact and concise definition of liquid is sometimes ambiguous. Liquids can be most easily classified in terms of cohesive (attractive and repulsive) forces keeping them together. One can distinguish between ionic liquids or molten salts, metallic liquids composed of ions and moving electrons, molecular liquids fixed by van der Waals forces, and waterlike liquids, where hydrogen bonds are the dominant attractive forces. Glass represents a compromise between crystals and liquids. Quartz crystals as well as quartz glass are fixed by electrostatic bonds between silica and oxygen. Unlike in quartz, these bonds have uneven lengths in silicate melts, resulting in some physical properties reminiscent of those of typical liquids (Moore, 1979).

The subdivision of matter into solid, liquid, and vapor is sufficient for the specific and narrow range of thermodynamic conditions on the Earth's surface. Distinguishing between liquid and vapor may be rather problematic in the Earth's crust and mantle, which is at increased temperature and pressure. Hence, geologists prefer the term *fluid* to characterize the aggregation state of the matter with properties of liquids and gases. Silicate melt at high temperature (>700 °C) also has properties typical

Geofluids. http://dx.doi.org/10.1016/B978-0-12-803241-1.00001-0

of liquids, but this term cannot be applied to a solidified magma on the Earth's surface. Dynamic viscosity—the ability of matter to flow due to an oriented differential strain—is a limiting physical quantity that divides liquids from solids.

Having an active supply of atoms and molecules during the crystallization of solids is the second fundamental feature of a fluid. The main fluid compounds can be directly involved in the structure of the precipitating solid, for example, during crystallization of halite from NaCl-oversaturated aqueous fluid and crystallization of rock-forming minerals from cooling silicate melt. In most cases, however, the fluid serves only as a transporting medium for soluble mineral-forming compounds, as during precipitation of insoluble sulfides from aqueous solutions by decomposition of soluble thiocomplexes. Except for some metamorphic reactions exchanging ions in a solid state, almost all terrestrial and extraterrestrial solids originate in a fluid medium as indicated by fluid inclusions identified in literally all minerals, including diamonds and meteorites.

1.1 BRIEF HISTORY OF GEOFLUID OBSERVATION AND RESEARCH

The first reference to what were probably fluid inclusions in a mineral is found in the *Natural History* of Gaius Plinius Secundus (Pliny the Elder) written about 75 A.D. (Kesler et al., 2013). The first scientific document about fluid inclusions was written by Abu Reikhan al-Biruni (972–1048 A.D.). The Uzbek scholar described "inclusions in form of air bubbles and water droplets" in quartz, sapphire, and other minerals in his book *Precious Stones*. He was the first to attribute these inclusions to "sap of the Earth," from which minerals originated by "lithification." Al-Biruni thereby defined for the first time the primary focus of the study of fluid inclusion: deciphering the origin of minerals and rocks. Apart from this, al-Biruni attributed the spontaneous cracking of gemstones during cutting and polishing to fluid inclusions and recommended their removal from minerals using fine drilling (Lemmlein, 1950). The first systematic descriptions of fluid inclusions in precious stones were made by Ahmad al-Tifashi in Cairo and Albertus Magnus, Archbishop of Cologne, in the 13th century (Kesler et al., 2013).

Sporadic observations of "moving bubbles" in quartz were described by Boyle (1672, 1673) and Scheuchzer (1723), but true scientific interest in fluid inclusions is dated at the outset of the 19th century. Davy (1822) made the first attempts to determine the chemical composition of fluid inclusions in quartz crystals. He tried to open the inclusions by drilling under water, oil, and mercury, and described different behavior among the liberated gas bubbles. Brewster (1823) observed within some fluid inclusions two immiscible liquids (water and carbon dioxide), and first reported on halite "squares" in an aqueous phase. Moreover, Brewster isolated insoluble daughter crystals from inclusions and identified them as calcite. Brewster (1845) was the first to study the behavior of solid phases in fluid inclusions on heating. The observations performed by Davy and Brewster significantly supported the neptunistic hypothesis of mineral origin.

A factual revolution and turning point in the study of fluid inclusions is connected with Henry Clifton Sorby. In his classic paper, Sorby (1858) was the first to describe melt inclusions in volcanic rocks (Figure 1.1) and to prove experimentally that the liquid phase in most inclusions is represented by water. He specifically described samples from ore deposits and drew conclusions concerning ore formation that remained scientifically unfashionable for many years. Sorby heated minerals containing fluid inclusions in a sealed test tube, had the condensed vapor frozen out, and determined its crystallographic shapes and temperature of melting. During these experiments, he noticed precipitation of another

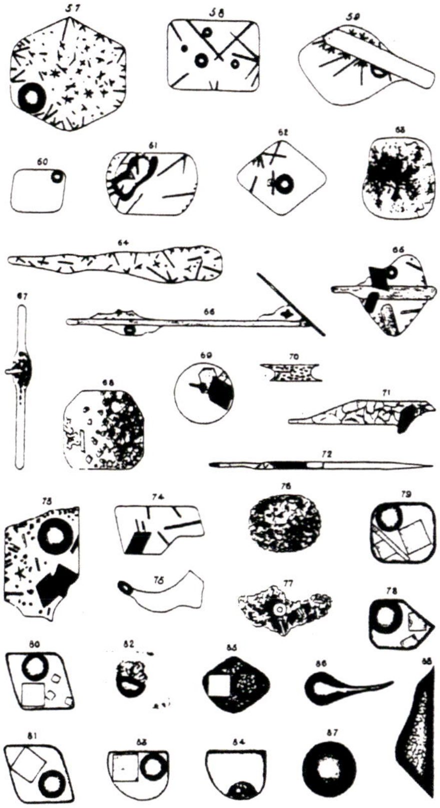

FIGURE 1.1

One of the first drawings of glassy and gaseous inclusions in feldspar phenocrysts in trachyte of Vesuvius, and those in pyroxenes of Scottish basalts (Sorby, 1858).

phase, provisionally identified as NaCl or KCl. Beyond that, he crushed quartz crystals and leached them in pure water to analyze soluble components in the inclusion fluids. He observed thermal expansion of salt-containing aqueous solutions in thin capillaries. These observations led him to conclude that the coefficient of thermal expansion of the sealed liquid must be one or two times greater than that of the host mineral, resulting in the formation of vapor bubbles in the aqueous inclusion on cooling. Consequently, he proposed a method of estimation of the crystallization temperature of minerals by heating the trapped fluid inclusions until the bubbles disappear. In such a way, Sorby defined the essential principle of the fluid inclusion thermometry. Sorby together with Buttler described in 1869 multiphase inclusions in emerald and spinel, and conducted experiments with homogenization of inclusions in sapphire. They found out that vapor bubbles in CO_2-containing inclusions always diminished at temperatures lower than 31 °C.

Substantial progress in knowledge on material composition of fluid inclusions was achieved at the end of the 19th century. Phillips (1875, cited in Hein, 1990) used a paraffin bath to observe changes in fluid inclusions on heating and compared his results with those obtained by Sorby. He concluded that an inclusion population in the same sample exhibits various liquid-to-vapor ratios and so also had different homogenization temperatures. According to this, he questioned Sorby's assumptions about the possibility to determine crystallization temperatures of minerals by measuring homogenization temperatures. Neither Phillips nor other scientists who challenged Sorby's hypotheses could know of the existence of various inclusion generations trapped at different times, which led to their reaching erroneous conclusions.

Sorby's hypotheses about fluid inclusions as "natural thermometers" seemed to be forgotten at the beginning of the 20th century. It was believed that only qualitative parameters, such as chemical composition, could be inferred from fluid inclusions. Könnigsberger (1901) and later Könnigsberger and Müller (1906, both cited in Hein, 1990) performed some measurements of fluid inclusion homogenization temperatures in quartz crystals from alpine-type fissures and admitted some restricted validity of Sorby's hypotheses. Nacken (1921) was the first to apply principles of physical chemistry to interpreting phase transitions in fluid inclusions and to estimate the formation temperature and pressure using phase diagrams for experimental systems. He concluded that CO_2-bearing inclusions cannot be employed for estimating temperature due to the influence of pressure and speculated that the homogenization temperature of aqueous inclusions could correspond to the formation temperature if it was considerably lower than that in the critical point of pure water.

Lemmlein (1929) introduced genetic aspects to the classification of fluid inclusions, and distinguished primary and secondary inclusions. He synthesized secondary inclusions within salt crystals, described the formation mechanism of such inclusions in healing cracks, and explained the process of inclusion wall recrystallization as necking down. A further breakthrough in fluid inclusion research occurred during the Second World War in connection with the use of piezoelectric properties of natural quartz in radiolocation and communication technology in the former Soviet Union. By this time, Western countries had already developed techniques for the industrial production of synthetic quartz. But systematic fluid inclusion research is connected with Yermakov (sometimes spelled "Ermakov") and his school. The first monograph dealing with fluid inclusions in minerals (Yermakov, 1950) summarized the scientific effort of his school during the Second World War. A completed and reworked English version of this monograph (Yermakov, 1965) was published in the United States during the period of enormously increased interest in fluid inclusions boosted by the effort to resolve the origin of strategic ore deposits. Early papers documented the range of temperatures and salinities in various types of magmatic intrusion-related ore deposits, and combined these data with leachate and stable isotope analysis (e.g., Hall and Friedman, 1963; Rye and O'Neil, 1968; Taylor, 1974). Somewhat later, fluid inclusions began to be investigated in metamorphic (Touret, 1971b, 1981) and sedimentary (Burruss, 1981a) minerals. Modern research is inseparably connected with Edwin W. Roedder (1919–2006), whose 644-paged monograph (Roedder, 1984) remains the landmark publication for all geofluid researchers.

The history of melt inclusions investigation as summarized by Lowenstern (2003) begins with descriptive observations in volcanic rocks undertaken by Sorby (1858) and Zirkel (1873). Experimental investigations, however, blossomed after a 100-year-long gap in the 1960s with the interest in lunar samples of magmatic rocks (Roedder and Weiblen, 1970) and terrestrial pegmatites and igneous rocks (Roedder and Coombs, 1967; Sobolev et al., 1967; Roedder, 1979, and others). The introduction of new analytical techniques in the second half of the 20th century triggered extensive research on major and trace element contents, stable and radiogenic isotopes at a μm-scale. In the early 1970s, an electron microprobe was first applied to the investigation of melt inclusions by Anderson (1974) and Clocchiatti (1975). Anderson et al. (1989) used infrared spectroscopic techniques to analyze melt inclusions in phenocrysts from the Bishop Tuff. Dunbar et al. (1989) was the first to determine the water content in melt inclusions using an ion microprobe. A method for determining water content in melt inclusions using confocal Raman microspectrometry was first presented by Thomas (2000). First analyses of trace elements using laser ablation and inductively coupled plasma mass spectrometry were performed by Kamenetsky et al. (1997) and Taylor et al. (1997). Simultaneous multielement analysis with a proton-induced X-ray emission was applied to unexposed melt inclusions by Kamenetsky et al. (2003). The synchrotron radiation X-ray fluorescence was used for the same purpose by Rickers et al. (2004). The melt inclusion research was summarized in a number of review papers (Frezzotti, 2001; Schiano, 2003; Audétat and Lowenstern, 2014), short course volumes (De Vivo and Bodnar, 2003; Webster, 2006), and thematic issues (Bodnar et al., 2013), which can be recommended for further in-depth study of this rapidly expanding research field.

1.2 ENTRAPMENT OF FLUIDS IN MINERALS

An inclusion can be rather scholastically defined as "a part of crystal, not corresponding to its regular structure, hermetically isolated during crystal growth, and creating a phase boundary with the host" (Yermakov and Dolgov, 1979). This definition is a modification of the definition of a chemical or thermodynamic system. Fluid inclusions represent a special type of inclusion composed of a fluid phase coexisting with the host mineral during crystallization or recrystallization. The term *fluid inclusion* also pertains to cases where the enclosed matter does not match the definition of fluid in terms of dynamic viscosity (e.g., solidified silicate or sulfide melt). The limiting factor here is the aggregation state of the fluid during entrapment and not that during observation (Roedder, 1984; Bodnar, 2003a).

Fluid inclusions in minerals visible to the naked eye are rare. Brewster (1823) observed multiphase fluid inclusions, 8.6 mm long, in topaz. Hidden (1882) described fluid-filled cavities in quartz, 6×0.4 cm in size. Zakharchenko (1950) discovered inclusions with volumes up to 1.5 cm^3 in quartz from the Pamir Mountains. One of the fluid inclusions occupied 40 vol.% of the quartz crystal, and vapor bubbles in these inclusions reached diameters up to 4 mm. Aqueous and gaseous fluid inclusions several millimeters long occur in quartz from secondary quartzite of the Šobov deposit of the Banská Štiavnica stratovolcano (Oružinský and Hurai, 1985). One of the most spectacular examples of exceptionally large petroleum inclusions was recently found in doubly terminated quartz crystals from the Baluchistan province of Pakistan (Figure 1.2).

Fluid inclusions only rarely exceed 0.1 mm (1000 μm), and such dimensions mostly refer to crystals. Most inclusions in massive mineral aggregates are generally 10–50 μm in size, thus requiring 500–1000 × magnification for comfortable observation under a petrographic microscope. Massive milky quartz usually contains a total of 10^9 fluid inclusions for each cubic centimeter, but their diameters are less than 1 μm (Roedder, 1984). The phase composition of such inclusions is indistinguishable under an optical microscope. Transmission electron and atomic force microscopy has revealed "bubbles" several nanometers in diameter in synthetic and natural minerals, and there is probably an uninterrupted continuum in inclusion sizes from those resolvable with the optical microscope down to those that contain only a few water molecules (Bodnar, 2003a).

In most crystals, small irregularities occur along growing surfaces due to structural failures—dendritic growth, spiral dislocations, fracturing, subparallel crystal growth, impurities on growing surfaces, and etching and partial dissolution triggered by an interrupted growth. Each irregularity of the

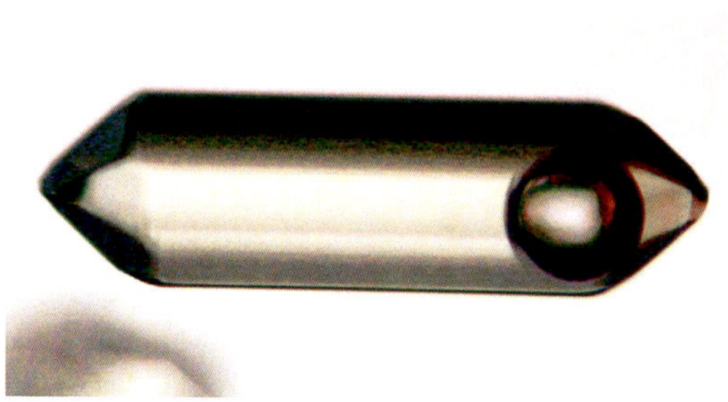

FIGURE 1.2

Photomicrograph of one of the smaller inclusions in quartz from the Baluchistan province (Pakistan) with oil and vapor bubble phases as seen in transmitted visible light. Note also a small rounded bitumen particle in the left tip of the inclusion, and a dark brown phase rimming the boundary between the vapor bubble and the blue fluorescent oil. Longer diameter of the inclusion is 0.563 mm. Other inclusions trapped in the quartz are up to 1 cm large and they are clearly discernible by the naked eye (front cover photo).

crystal surface serves as a potential fluid trapping site, which is hermetically sealed during the crystal growth to form an intracrystalline inclusion (Figure 1.3).

Because the intracrystalline inclusions originate during growth of crystals, they are called primary. Intracrystalline primary inclusions are either isolated or occur in three-dimensional clusters without preferred directional orientation (Figure 1.4). When they are arranged along incremental growth zones of the host crystal, the group of intracrystalline inclusions may acquire the two-dimensional planar orientation. Intercrystalline (intergranular) inclusions occur among crystals grown with a subparallel orientation. In contrast to intracrystalline inclusions, their shapes do not copy the crystal symmetry of the host mineral.

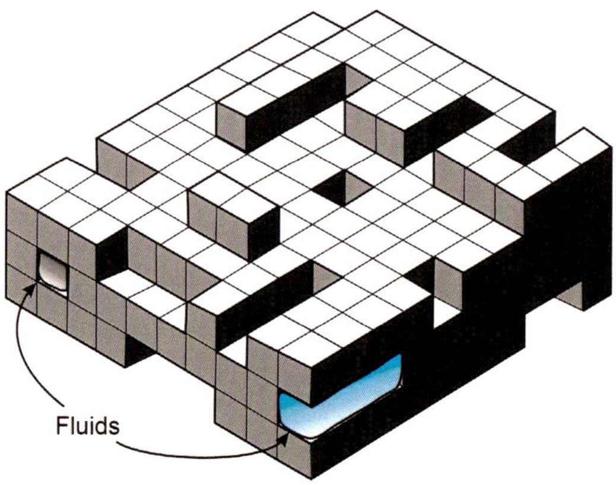

FIGURE 1.3

Schematic presentation of irregular crystal growth and origin of intracrystalline (primary) fluid inclusions.

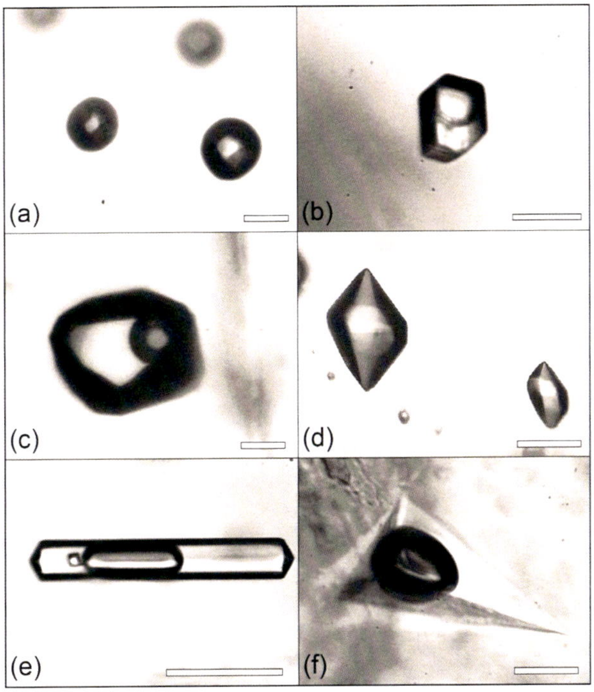

FIGURE 1.4

Several examples of intracrystalline, negative crystal-shaped (a–e) and intercrystalline (f) inclusions. (a) CO_2 liquid inclusions in Ca-plagioclase (igneous cumulate xenolith from Hawaii, courtesy of J. Lexa). (b) Aqueous fluid inclusion in retrograde metamorphic garnet (UHP gneiss of Pohorje Mts., Slovenia, Hurai et al., 2010). (c) Aqueous fluid inclusion in sphalerite (hydrothermal polymetallic veins near Soviansko, Nízke Tatry Mts., Slovakia, courtesy of J. Luptáková). (d) Methane liquid inclusions in quartz (veins in Eocene flysch sandstones, Vel'ký Lipník, Slovakia). (e) Tubular aqueous fluid inclusion in apatite (carbonatite of the Evate deposit, the Monapo ring structure, Mozambique, courtesy of P. Hvožd'ara). (f) Intercrystalline aqueous fluid inclusion in albite aggregate (Rožňava–Nadabula siderite deposit, Gemeric tectonic unit, Slovakia). Scale bars correspond to 20 μm.

Secondary inclusions occur along healed fractures whose formation is caused by brittle deformation and recrystallization subsequent to crystal growth. Pseudosecondary inclusions originate during deformations coeval with the crystal growth. Planar arrangement controlled by the shape of healed and recrystallized crack is diagnostic of both inclusion types (Figure 1.5).

Healed cracks with pseudosecondary inclusions terminate within a crystal or mineral grain. Each trail is coeval with the growth zone, where it terminates, and is older than the preceding zones (e.g., trails E and C in Figure 1.6). Healed crack with secondary inclusions in strict sense reaches outer crystal surface (e.g., trail F in Figure 1.6) or intersects grain boundaries in crystalline aggregates.

Shapes of inclusions range from irregular to negative crystal-shaped cavities, which copy the crystallographic form of the host mineral (Figure 1.4a–e). Individual secondary and pseudosecondary fluid inclusions mostly acquire regular or irregular two-dimensional shapes, but negative crystal shapes are

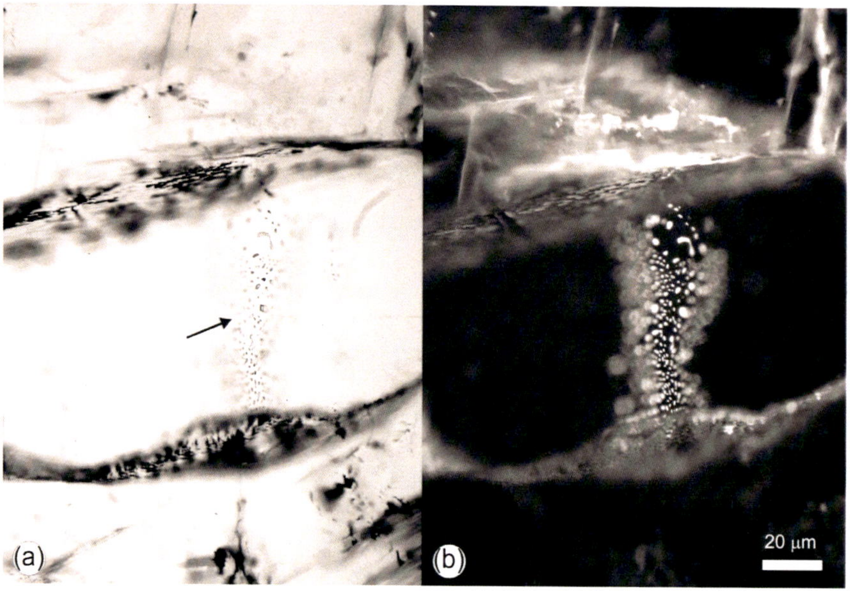

FIGURE 1.5

(a) Secondary crack in calcite veinlet in flysch sandstones with petroleum inclusions visualized by the fluorescence emission (b) induced by a 365 nm ultraviolet light.

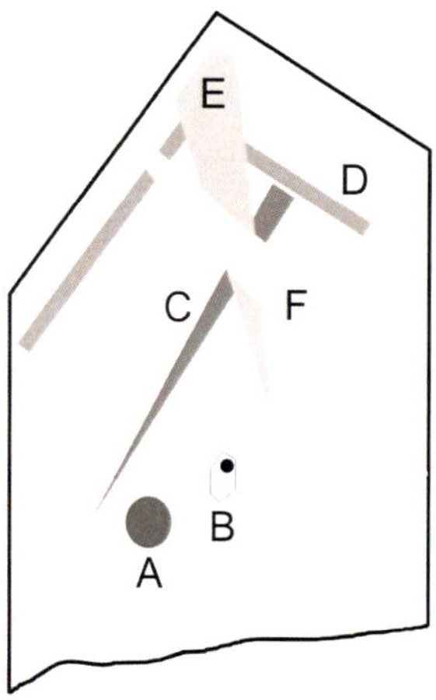

FIGURE 1.6

Schematic drawing of various genetic types of fluid inclusions: A depicts a three-dimensional cluster of primary inclusions, B is an isolated primary inclusion, D denotes primary inclusions arranged along a growth plane, C and E are planes of pseudosecondary inclusions terminated at growth planes, and F is a plane of secondary inclusions related to post-growth brittle deformation of the host crystal.

also common. The variability of fluid inclusion shapes is controlled by energetic stability. Large surfaces of irregular, planar-shaped inclusions are energetically unstable. For this reason, the large surface energy is minimized by dissolving inclusion walls in places of the increased energetic potential and depositing the dissolved material along energetically stable crystallographic axes of crystal lattice. This process results in the shortening of strongly elongated inclusions, rounding re-entrants of amoeboidal inclusions, and transforming the initially planar or irregular inclusions to a final, energetically most stable three-dimensional, negative crystal shape with the smallest surface for a given volume. Recrystallization of inclusion walls has been directly observed in "wet" synthetic quartz growing from NaOH–H_2O solutions and containing a significant amount of structurally bound water (Gratier and Jenatton, 1984).

Bodnar et al. (1985b) introduced the term *maturation* to describe the process of spontaneous fluid inclusion recrystallization taking place after fluid inclusion sealing. The maturation-related changes of fluid inclusion shapes occur at fixed volumes. Hence, the recrystallization does not modify the original volumetric properties of inclusions, except for cases where an elongated inclusion splits during the recrystallization into several smaller isolated inclusions. The process, called "necking down," can randomly redistribute various phases present in the original inclusion into newly formed inclusions, whose individual volumetric properties (phase composition, density) are no longer representative of the original fluid (Figure 1.7).

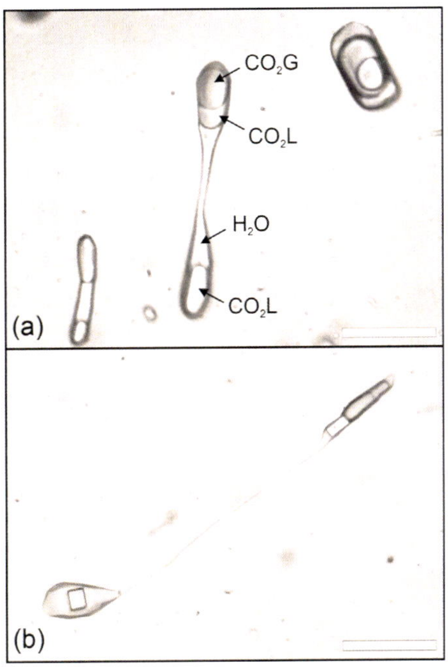

FIGURE 1.7

(a) Necking down of inclusions containing aqueous liquid (H_2O), CO_2-rich liquid (CO_2L), and CO_2-rich gas (CO_2G) phases. Apart from these phases, the inclusion in (b) also contains halite cubes. Note different phase ratios on both sides of elongated, necked-down inclusions, documenting a compositional and pressure gradient. Photomicrograph in (b) is merged from two frames to eliminate nonplanar focus. Scale bars equal to 10 μm.

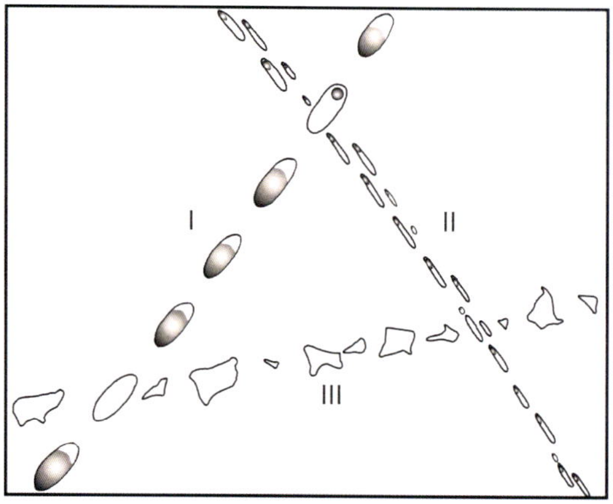

FIGURE 1.8

Spatial relationships among planes of secondary inclusions trapped consecutively in the order from system I to system III.

As a rule, negative crystal or regularly shaped inclusions are earlier than planar inclusions with irregular shapes. According to this concept, fluid inclusion shapes can serve as an auxiliary criterion for the estimation of relative ages of fracture-bound secondary/pseudosecondary fluid inclusion systems (Figure 1.8). However, the degree of maturation (shape factor) cannot be automatically extended as the universal criterion for distinguishing between primary and secondary inclusions, particularly in different minerals. An illustrative example was provided by Hurai et al. (2010), who described negative crystal-shaped aqueous inclusions in metamorphic garnet from a high-*P* metamorphic terrain of the Pohorje Mountains, Slovenia (Figure 1.4b). The randomly distributed fluid inclusion assemblage created a compact three-dimensional network diagnostic of a primary origin. However, the fluid-rich domains occurred only along cracks within the garnet and margins of garnet grains with an increased Mn content. Moreover, they were closely and spatially associated with inclusions of retrograde biotite and sillimanite. Hence, the apparently primary inclusions were in fact secondary inclusions resulting from a retrogression-related dissolution-reprecipitation of the high-pressure Mg-rich garnet.

Apart from primary, secondary, and pseudosecondary inclusions, exsolution inclusions are defined as a special group of the secondary inclusions originated mainly due to ductile deformation and recrystallization of high-pressure minerals accompanied by the release of structurally bound volatiles on decompression (Wilkins and Barkas, 1978). The process is accompanied by migration of fluid inclusions from the center toward grain margins, and their reentrapment in secondary inclusions along deformation lamellae and/or along grain boundaries.

1.3 BASIC STATE PROPERTIES OF GEOFLUIDS

As mentioned, fluid inclusions are considered thermodynamic and chemical *systems* separated from their surroundings by their host mineral. Regarding permeability of walls with respect to energy and matter, three kinds of thermodynamic systems can be distinguished (Chang, 1991):

1. *Isolated* systems—their walls prevent any exchange of energy and matter between the system and its surroundings.
2. *Closed* systems—they permit energy transfer but are impermeable to matter. Hence, the systems possess constant mass but variable energy content. Chemical reactions are possible within the system.
3. *Open* systems—energy and matter pass into or out of the system.

A *homogeneous* system is unified chemically and physically in the whole volume and consists only of a single phase (solid, liquid, or vapor). A system consisting of several phases is *heterogeneous*. Individual phases are separated by phase boundaries, along which properties (density, composition) change instantaneously. Chemical composition of a system or phase is expressed by independent chemical individuals called *components*. *Phases* have the same chemical composition in unary (one-component) systems. For instance, ice, aqueous liquid, and steam are different phases with the same chemical composition in the H_2O system. Phases in polycomponent (binary, ternary, etc.) systems always have different chemical compositions.

Fluid inclusions in minerals usually behave as closed systems because they exchange only energy with their surroundings. Hence, the supply of external energy (heat) may induce phase transformations and chemical reactions, which are directly observable under an optical microscope in transparent minerals and by using a special type of microscope in some opaque minerals. In the case of a closed system, these transitions are reproducible and reversible, which means that they occur at the same temperature when the experiment is repeated. Precise measurements of the temperatures of phase transitions serve as an analytical tool for determining the fluid composition and density. The method is usually referred to as *fluid inclusion microthermometry*.

In many cases, matter along with energy can also pass out or into fluid inclusions, which thus acquire properties of an open system, where phase transformations of the same kind (e.g., melting of a solid, vapor bubble disappearance) occur at various temperatures, when the experiment is repeated under the same conditions. This behavior is often used to check whether the measured fluid inclusion represents an open or closed system. In the case of the latter, temperatures of phase transformations cannot be used to interpret formation *P–T* conditions and must be rejected.

In some cases, individual fluid inclusions show reproducible phase transformations during freezing–heating experiments, but an unusual character of microthermometric data (e.g., variable temperatures of total homogenization) is observed in the cogenetic population of inclusions. If this behavior is associated with specific textures present around and within fluid inclusions (cracks around inclusions, wall recrystallization, satellite inclusions), this would indicate temporal opening of the inclusions and their resealing. These inclusions no longer reflect their formation *P–T* conditions, but they bear information on conditions of their reequilibration on the host's mineral way to the Earth's surface.

Thermodynamic and chemical systems, including fluid inclusions, are described using **state properties (functions, variables)**. *Macroscopic* properties are directly measurable state properties (e.g., pressure, volume, density, fugacity, composition) studied by mechanics and classical thermodynamics. In contrast, *microscopic* properties describe each particle of the system (e.g., interatomic distances, energy levels), and are of primary concern in statistical mechanics. A characteristic feature of state functions is that when the state of a system changes, the *magnitude of change of any state function depends only on the initial and final stages of the system and not on how the change is accomplished* (Chang, 1991).

Numerical values of *intensive* state properties do not depend on how much substance is present, and they are measurable at any place in the system (e.g., temperature, pressure, density). In contrast, *extensive* state properties are dependent on the substance quantity (number of moles) in the system. The numerical value of an extensive variable (e.g., volume) corresponds to the sum of the partial values in all homogeneous parts of the system. Extensive variables can be converted to intensive ones after dividing by the number of moles of the substance. Temperature, pressure, substance quantity, and molar volume are basic macroscopic properties conceivable from everyday experience. Apart from them, thermochemistry and thermodynamics use additional state properties defined by *thermodynamic laws* (see Chapter 6). Depending on the problem solved, appropriate state properties are selected to describe the system. Fluid inclusion analysis and interpretation are concerned primarily with temperature, pressure, volume (density), and concentration. Enthalpy and entropy are addressed when the behavior of free fluids and melts on cooling and heating is to be studied. Gibbs free energy is employed for the calculation of mineral–fluid equilibria.

1.3.1 TEMPERATURE

Temperature is a measure of the relative amount of heat added or taken away from a system. Heat is an amount of thermal energy that can be transferred from hot to cold systems.

Before the temperature of a system is determined, an empirical temperature scale has to be defined by measuring a state property at two reproducible fixed points. Since the 18th century, the volume of a liquid (e.g., mercury, ethanol) has been measured in a glass capillary and referred to the freezing and boiling points of water at atmospheric pressure. These values were arbitrarily defined as 0 and 100, and the scale was named the *Celsius temperature scale* after the Swedish astronomer Anders Celsius (1701–1744). Since the 19th century, the scientific community has referred to this scale as the centigrade scale and to the corresponding units as degrees centigrades or degrees Celsius (°C).

French balloon enthusiasts, the physicist Jacques Charles (1746–1823) and the chemist and physicist Joseph Gay-Lussac (1778–1850) observed that the volume of a gas expands when heated and contracts when cooled, and at any given (sufficiently low) constant pressure, the measured volume plotted against temperature creates straight lines with different slopes, which all intersect at zero volume at a temperature of −273.15 °C. In 1848, the Scottish mathematician and physicist William Thomson, Lord Kelvin (1824–1907), was the first to realize the significance of the phenomenon and identified the temperature of −273.15 °C as the lowest attainable temperature called the *absolute zero*. With this value as the starting point, he set up an absolute temperature scale, now called *the Kelvin temperature scale*. In thermodynamic calculations, only temperatures related to this temperature scale are used, having an advantage of being always positive.

By convention, we use the abbreviation T to denote absolute temperature in Kelvins, and t to indicate temperatures on the Celsius scale:

$$T[\mathrm{K}] = t[^\circ\mathrm{C}] + 273.15 \quad (1.1)$$

Since 1954, the Celsius scale has been defined by the absolute zero and the triple point of specially purified water determined precisely at 273.16 K and 0.01 °C (BIPM—the International Bureau of Weights and Measures, http://www1.bipm.org/en/si/si_brochure/chapter2/2-1/2-1-1/kelvin.html). Hence, Kelvin is the fraction 1/273.16 of the thermodynamic temperature of the triple point of water. Both temperature scales are accepted as international temperature scales approved by the BIPM (Preston-Thomas, 1990).

1.3.2 PRESSURE

Pressure (P) is defined in the international system of measures and weights (SI) as a force F per unit area

$$P = \frac{F}{d^2} \tag{1.2}$$

where d is distance (in meters), and the corresponding unit is N m^{-2} (Newtons per square meter) or Pa (Pascals). Because the units are very small, the measures of MPa (megapascal = 10^6 Pa) or GPa (gigapascal = 10^9 Pa) are used to avoid large numbers. In recent literature, pressure is often given as bar or kbar (10^3 bar). Numerous other obsolete pressure units have been used in the scientific literature. Workers in the oil and gas industry and engineers often refer to traditional units unrelated to the international SI system, such as the pound-force per square inch (psia). Common pressure units and their conversion formulas are summarized in Table 1.1.

1.3.3 QUANTITY AND CONCENTRATION

The amount of a substance is one of the seven base quantities of the SI system. The symbol for substance quantity is n and the corresponding unit is mole (mol), which is the amount of substance that contains in its ground state as many elementary entities as there are atoms in 0.012 kg of carbon 12 (http://www1.bipm.org/en/si/si_brochure/chapter2/2-1/2-1-1/mole.html). A fraction of n is called mole fraction.

In everyday use, the word *weight* is used to express mass, for which the basic SI unit is kilogram. However, in the SI system, weight is a force for which the unit is Newton. Hence, the concentration unit "weight percent" frequently used in literature should be replaced by "mass percent" (Bakker and Brown, 2003).

The majority of substances, including geofluids, are made of components whose relative concentrations must be specified. In aqueous fluids, H_2O—the dominant component—is called the *solvent*, whereas dissolved minor components (e.g., NaCl, $CaCl_2$, CO_2) are *solutes*. Concentrations of the solutes are most frequently expressed as mass percentages—m (mass of solute/mass of solvent × 100), molar fractions—$\bar{x}$ (number of moles of solute/sum of moles of solvent and the solute), or as molality—$\overline{m}$ (number of moles of solute in 1000 g of solvent). In binary systems, relationships between the concentration units are expressed by the following equations:

$$\bar{x}(m_i) = \frac{\dfrac{m_i}{\overline{M}_i}}{\dfrac{m_i}{\overline{M}_i} + \dfrac{100 - m_i}{\overline{M}_s}} \tag{1.3}$$

$$\overline{m}(m_i) = \frac{1000}{\left(\dfrac{100}{m_i} - 1\right)\overline{M}_i} \tag{1.4}$$

$$\overline{m}(\bar{x}_i) = \frac{1000\bar{x}_i}{\overline{M}_s(1 - \bar{x}_i)} \tag{1.5}$$

$$m(\overline{m}_i) = \frac{100\overline{m}_i\overline{M}_i}{1000 + \overline{m}_i\overline{M}_i} \tag{1.6}$$

Table 1.1 Pressure Units and their Conversions

Unit	Pa	Bar	mbar	Atm (760 torr)	kp/m^2	At (kp/cm^2)	Torr (mm Hg)	lbf/in^2 (psia)
1 Pascal (Pa)	1	10^{-5}	10^{-2}	9.86923×10^{-6}	1.01972×10^{-1}	1.01972×10^{-5}	7.50062×10^{-3}	1.45038×10^{-4}
1 Newton/m^2 (N m^{-2})	1	10^{-5}	10^{-2}	9.86923×10^{-6}	1.01972×10^{-1}	1.01972×10^{-5}	7.50062×10^{-3}	1.45038×10^{-4}
1 bar (bar)	10^5	1	10^3	9.86923×10^{-1}	1.01972×10^4	1.01972	7.50062×10^2	1.45038×10^1
1 millibar (mbar)	10^2	10^{-3}	1	9.86923×10^{-4}	1.01972×10^{-5}	1.01972×10^{-3}	7.50062×10^{-1}	$1.45038.10^{-2}$
1 physical atmosphere (atm)	1.01325×10^5	1.01325	1.01325×10^3	1	1.03323×10^4	1.03323	760	14.6959
1 kilopond/m^2 (kp/m^2)	0.80665	9.80665×10^{-5}	9.80665×10^{-2}	9.67841×10^{-5}	1	10^{-4}	735.559×10^{-4}	14.2233×10^{-4}
1 kilopond/cm^2 (kp/cm^2)	9.80665×10^4	9.80665×10^{-1}	9.80665×10^2	9.67841×10^{-1}	10^4	1	735.559	14.2233
1 millimeter Hg column (mm Hg)	133.322	133.322×10^{-5}	133.322×10^{-2}	1.31579×10^{-3}	13.5951	13.5951×10^{-4}	1	0.0193368
1 pound-force per square inch[a] (lbf/in^2)	6.89476×10^3	6.89476×10^{-2}	6.89476×10^1	0.0680460	7.0307×10^2	7.0307	51.7149	1

[a] *1 pound = 0.453 592 37 kg; 1 inch = 0.0254 meter.*

$$m(\bar{x}_i) = \frac{100\overline{M}_i}{\frac{\overline{M}_s}{\bar{x}_i} + \overline{M}_i - \overline{M}_s} \tag{1.7}$$

where the subscript i denotes the solute (e.g., NaCl, CO_2), and the subscript s corresponds to the solvent (e.g., H_2O), $\overline{M}$ is the molecular mass of solute or solvent.

Equation (1.7) can be extended for ternary systems with two solutes (i and j) to express the relative concentration of one solute in the solvent (s)

$$m(\bar{x}_i, \bar{x}_j) = \frac{100\overline{M}_{i,j}}{\frac{\overline{M}_s(1-\bar{x}_{j,i})}{\bar{x}_{i,j}} + \overline{M}_{i,j} - \overline{M}_s} \tag{1.8}$$

For example, to express the NaCl concentration relative to water in an H_2O–CO_2–NaCl mixture, Equation (1.8) reads

$$m\mathrm{NaCl} = \frac{100\overline{M}_{\mathrm{NaCl}}}{\frac{\overline{M}_{\mathrm{H_2O}}(1-\bar{x}\mathrm{CO_2})}{\bar{x}_{\mathrm{NaCl}}} + \overline{M}_{\mathrm{NaCl}} - \overline{M}_{\mathrm{H_2O}}} \tag{1.9}$$

Equation (1.3) can also be expanded to express the composition of a ternary fluid

$$\bar{x}(m_{i,j}) = \frac{\frac{m_i}{\overline{M}_i}}{\frac{m_i}{\overline{M}_i} + \frac{m_j}{\overline{M}_j} + \frac{100 - m_i - mj}{\overline{M}_s}} \tag{1.10}$$

Finally, Equation (1.6) rearranged for a ternary system reads

$$m(\overline{m}_i) = \frac{100\overline{m}_i\overline{M}_i}{1000 + \overline{m}_i\overline{M}_i + \overline{m}_j\overline{M}_j} \tag{1.11}$$

or

$$m(\overline{m}_j) = \frac{100\overline{m}_j\overline{M}_j}{1000 + \overline{m}_i\overline{M}_i + \overline{m}_j\overline{M}_j} \tag{1.12}$$

1.3.4 VOLUME AND DENSITY

Volume (V) is an important state property, because it defines the P–T trajectories of fluid inclusions and the direction of chemical reactions at changing pressure and temperature. The basic volume unit is m^3, but the derived unit cm^3 is also accepted. For thermodynamic calculations and gases, the unit to use is the *molar volume* ($\overline{V}$), defined as the volume of 1 mol of substance. The molar volume of an n-component system can be calculated as

$$\overline{V} = \sum_i^n \bar{x}_i \times \overline{V}_i \tag{1.13}$$

where $\bar{x}_i$ is the molar fraction, $\overline{V}_i$ is the molar volume [$cm^3\ mol^{-1}$ or kg mol^{-1}], and $\overline{M}_i$ is the molecular mass of the i-th component in the mixture (Burruss, 1981b).

Density (D) is defined as the mass of a volume unit of a substance

$$D = \frac{M}{V} = \frac{\overline{M}}{\overline{V}} \tag{1.14}$$

The basic density unit in the SI system is the kg m^{-3}, however, derived units (g cm^{-3}) are preferred in geological literature to avoid large numbers.

For a binary system, Equation (1.13) combined with (1.14) reduces to

$$\overline{V} = \frac{\overline{x}_i \overline{M}_i + \overline{x}_j \overline{M}_j}{D} \tag{1.15}$$

This equation is used to characterize the molar volume of natural gaseous inclusions, which contain two dominant gas components.

1.3.5 VISCOSITY

Dynamic viscosity—the ability of matter to flow due to an oriented differential strain—is a limiting physical quantity that divides liquids and solids. The basic physical unit of the dynamic viscosity, Pascal-second (Pa s), corresponds to viscosity of liquid with a laminar flow rate of 1/s and tangential strain of one Pascal (N m^{-2}) across the flow. Using cgs units (centimeter, gram, second), the viscosity is expressed in poises (P), what is dyn s cm^{-2} (1 dyn = 1 g cm/s^2 = 10^{-5} kg m/s^2 = 10^{-5} N). Dyn is in the cgs unit system; the measure of power (from Greek *dynamis*) and relationship between the two units is 1 P = 0.1 Pa s. This type of viscosity is designated as **dynamic viscosity**. The ratio between the dynamic viscosity and density is **kinematic viscosity** [m^2 s^{-1}].

Some values of dynamic viscosities of geological materials are as follows (Fyfe et al., 1978): water at room temperature −0.001 Pa s, molten basalt 10^2–10^3 Pa s, molten granite 10^9–10^{11} Pa s, dry rock >10^{14} Pa s. Although the exact dividing line between fluid and solid states is a matter of convention, the term *fluid* can be safely attributed to matter if its dynamic viscosity is 10^{10} times lower than that of a dry rock.

The viscosity is an important physical unit, which influences the production, transport, and eruption of magmas. Natural silicate magmas have viscosities ranging between 10^{-1} and 10^{14} Pa s (Giordano et al., 2008) depending on temperature, melt composition, and amount of dissolved volatiles. The most important species affecting the bulk magma viscosity is water (Giordano et al., 2004a) and fluorine (Giordano et al., 2004b), the contents of which determine the effusive or explosive behavior of volcanic eruptions (Dingwell, 1996). CO_2 and S only insignificantly influence the magma viscosity (Dingwell and Hess, 1998; Zimova and Webb, 2006; Morizet et al., 2007). However, in pegmatite-forming fluid and melt systems a large number of other components, including H_2O, OH^-, CO_2, HCO_3^-, CO_3^{2-}, SO_4^{2-}, PO_4^{3-}, H_3BO_3, F, and Cl, as well as the structure modifiers Li, Na, K, Rb, Cs, and Be, have a greater meaning for reducing the viscosity, sometimes to very low values (Thomas and Davidson, 2012b).

1.4 CHANGES IN FLUID INCLUSIONS

1.4.1 PHASE CHANGES

Fluid inclusions observed at room temperature usually consist of several solid, liquid, and vapor phases, thus representing heterogeneous systems. Some inclusions may be monophase, and they thus obey the properties of a homogeneous system. The phase composition of fluid inclusions at room temperature is thereby an important diagnostic and classification tool. The character of the dominant phase determines the description of the inclusion type (e.g., aqueous inclusions are dominated by aqueous liquid). The second adjective can be added to express the number of phases present (e.g., monophase, two-phase, polyphase aqueous inclusions). Several types of fluid and melt inclusions and their descriptive classification based on the room temperature phase composition are illustrated in Figure 1.9.

The determination of aggregation state is rather problematic in homogeneous, monophase inclusions. Some basic rules are helpful for the phase identification in fluid inclusions. Gases and liquids have variable refraction indices, depending on their density. The difference in the refraction index

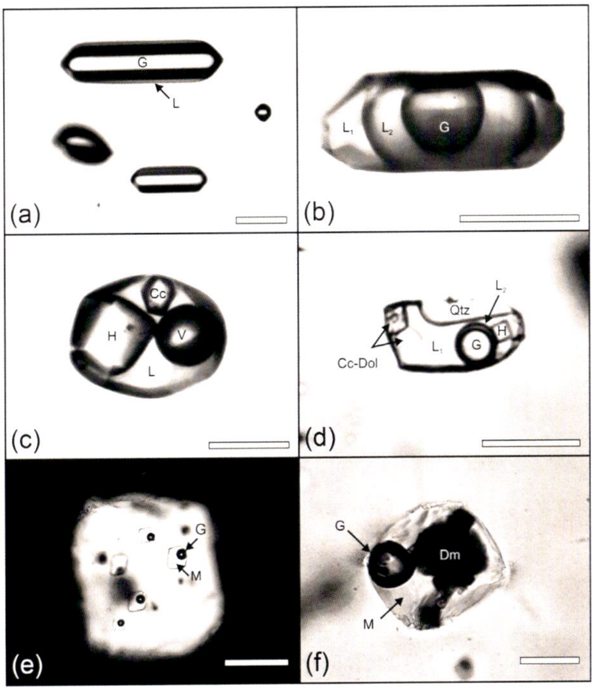

FIGURE 1.9

Polyphase, compositionally variable fluid inclusions, as observed at room temperature. (a) Gaseous inclusions in quartz of secondary quartzites of Šobov, Banská Štiavnica (Oružinský and Hurai, 1985), composed of CO_2 vapor (G) and aqueous liquid (L) phases. (b) Three-phase CO_2-H_2O inclusion in quartz from alpine fissure (Klenovec, Slovakia), composed of aqueous liquid (L_1), CO_2 liquid (L_2), and CO_2 vapor (G) phases. (c) Polyphase brine inclusion in quartz from Cristâlandia, Brasil (Marko et al., 2006), composed of aqueous liquid (L), vapor (V), halite (H), and calcite (Cc). (d) Polyphase aqueous inclusion in emerald from Brazil (courtesy of M. Dyda) composed of aqueous liquid (L_1), CO_2 liquid (L_2), CO_2 vapor (G), halite (H), quartz (Qtz), calcite-dolomite (Cc-Dol), and other unidentified phases. (e) Two-phase silicate glass inclusion in quartz phenocryst in rhyolite, consisting of quenched, solidified silicate melt (M) and gas bubbles (G). (f) Polyphase silicate melt inclusion in quartz xenocryst in rhyolite, consisting of silicate glass (M), vapor bubble (G), and numerous opaque (magnetite), brownish and green solid phases, represented by amphiboles and biotite (Dm). Scale bars in all photomicrographs correspond to 20 μm, except in (e) (50 μm).

between low-density vapor bubbles and the host mineral is so large that almost all light passing through the sample is reflected at the phase boundary. As a consequence, the low-density gaseous inclusions (Figure 1.9a) as well as vapor bubbles in silicate glass inclusion (Figure 1.9e and f) are almost opaque.

Aqueous inclusions containing liquid and vapor phases are the most frequent inclusion types in metamorphic and hydrothermal minerals. As a rule, the aqueous liquid always adheres to inclusion walls, being attracted by the charged silicate surface. Cubic hexahedral crystals in polyphase aqueous inclusions are usually halite, and crystals with high refractive index are usually carbonates (Figure 1.9c and d). If, at room temperature, an inclusion contains two colorless immiscible liquid phases and a vapor bubble, the aqueous liquid is in contact with the host mineral, and the second liquid phase adjacent to the vapor bubble is most likely composed of CO_2 (Figure 1.9b). The phase boundary between the CO_2 vapor and liquid phases disappears on slight warming to 31 °C (e.g., by the heat produced by an electric bulb).

Petroleum can be distinguished from aqueous solutions by its greenish, yellowish, and brown colors (Figure 1.2) and emission of variegated fluorescence under UV illumination. However, petroleum devoid of aromatic hydrocarbons will not fluoresce in the ultraviolet light.

Monophase inclusions are in most cases composed of mixtures of low-molecular-weight gases (e.g., nitrogen, methane) and carbon dioxide. They are practically optically indiscernible from monophase aqueous inclusions, whose presence, however, is limited to low temperature (<100 °C) hydrothermal minerals, or to late secondary inclusions.

The reason for the different phase compositions of fluid inclusions at room temperature stems from the phase rule discovered by Josiah Willard Gibbs (1839–1903) and published in 1875–1878. The number of phases (p) in a system is rigorously defined by the equation

$$p = c + 2 - f \tag{1.16}$$

where c is the number of components and f is the degree of freedom or variance, namely the number of state variables (temperature, pressure, composition), that can be independently changed without changing the number of phases present. According to the phase rule, the number of phases increases with the number of components and decreases with the increasing degree of freedom (variance). For instance, in a one component system ($c = 1$), a maximum of three phases (solid, liquid, and vapor) can be observed at a unique temperature and pressure, called *triple point*, where the degree of freedom equals zero. Hence, a small change in temperature and/or pressure leads to a phase transformation, which causes the disappearance of one or two of the three phases stable at the unique triple point.

The number of phases at the temperature of observation is an important indicator of fluid composition. For instance, four phases observed at room temperature (e.g., Figure 1.9c) means that the enclosed fluid system consists of at least two components. The same is valid if two immiscible liquids are observed (Figure 1.9b), because the number of liquid phases cannot exceed the number of components.

1.4.2 VOLUME CHANGES

All crystalline phases expand on heating at constant pressure due to enlargement of interatomic distances. A reversal process—contraction—occurs on cooling. The fractional increase of volume (V) with temperature (T) at constant pressure (P) is called the *coefficient of isobaric thermal expansion* α, and is defined as

$$\alpha = \frac{1}{V}\left(\frac{\partial V}{\partial T}\right)_P \tag{1.17}$$

Similarly, the *coefficient of isothermal compressibility* β is defined as the negative fractional change of volume due to pressure variations at constant temperature

$$\beta = -\frac{1}{V}\left(\frac{\partial V}{\partial P}\right)_T \tag{1.18}$$

The term $1/V$ is often replaced by V_0, namely the volume of a mineral or fluid at 25 °C and one bar.

An average thermal expansion of minerals with the exception of quartz is $\sim 3 \times 10^{-3}$ cm^3 K^{-1} at 1 bar confining pressure, and the compressibility at 20 °C corresponds to $\sim 5 \times 10^{-4}$ cm^3 bar^{-1} (Skinner, 1966; Birch, 1966); in other words, they differ by nearly one order of magnitude. Because the thermal expansion and compressibility act in opposite directions, mineral volumes at 300 °C and 2000 bar, 600 °C and 4000 bar, 900 °C and 6000 bar are roughly equal to those at atmospheric pressure and room temperature. Some minerals, originating at high pressures, can even increase their volumes on cooling. Generally, the overall volume change of solids except for quartz due to compressibility and thermal expansion is negligible, and it only rarely exceeds 1 vol.%. For thermodynamic calculations involving crustal processes, volume of minerals can be taken as a constant, independent of temperature and pressure.

Interestingly, low-temperature α-quartz expands on heating significantly more than other minerals, up to a maximum of about 4.5 vol.% at 573 °C and 1 bar, when its structure is reorganized in a high-temperature β polymorph, whose volume further slightly decreases with the increasing temperature. Isothermal (20 °C) compressibility of the α-quartz is also the largest compared to other minerals, reaching about 4.4 vol.% at 20 kbar pressure. Hence, even for this mineral, the total volume change only exceptionally exceeds 2% due to the counterbalancing effect of pressure.

As volume is the reciprocal function of density (Equation 1.14), the volume increase of a host mineral results in the density decrease of the enclosed fluid and vice versa. Hence, a 2% volume decrease on cooling of an aqueous fluid inclusion in quartz with the initial density of 1 g cm^{-3} results in the increase of the inclusion fluid density to 1.02 g cm^{-3} at ambient temperature and pressure. Such a difference can be ignored in the interpretation of fluid inclusion data, for which a larger scatter of densities is often observed in coeval inclusions. Therefore, fluid inclusions are usually considered **isochoric systems**, whose volume (density) remains fixed at changing *P*–*T* conditions. Their possible formation *P*–*T* conditions are defined by an **isochore**, which is the line connecting equal fluid volumes (densities) in the *P*–*T* coordinates.

With the development of the technique of synthetic fluid inclusions in quartz, the volume contraction and expansion of the host mineral due to changing temperatures and pressures are already involved in the experimental data, which should be corrected for this effect. In the case of synthetic fluid inclusions in quartz, d*P*/d*T* slopes of "isochores" are calculated as a function of liquid–vapor homogenization temperature (T_h) and are referred as iso-T_h lines in the phase diagram of the NaCl–H_2O system (Bodnar and Vityk, 1995; Bodnar, 2003c; Lecumberri-Sanchez et al., 2012) and the H_2O–CH_4 system (Lin and Bodnar, 2010). The iso-T_h lines are defined as those connecting the formation temperature and pressure with the homogenization temperature and pressure. The d*P*/d*T* slopes of the iso-T_h quasi-isochoric trajectories slightly differ from those in strictly isochoric systems, and should not replace each other. The iso-T_h trajectories, however, more precisely match the behavior of natural fluid inclusions in quartz and should preferably be used to constrain formation P–T conditions.

In contrast to solids, volume of fluids change considerably with temperature and pressure. At changing temperatures, the internal fluid inclusion pressure usually varies differently from that of the external fluid, which has no fixed volume and thereby behaves as an open system. The only exception is the rare case in which the pressure fluctuations of the surrounding fluid are similar to those within fluid inclusions. In the majority of cases, however, temperature variations during the crystal growth or the superimposed geological processes (metamorphism, uplift) impose a pressure differential between the quasi-isochoric, closed fluid inclusion system and the nonisochoric, free fluid surrounding the host mineral. The pressure differential, after reaching some critical value, is compensated by an irreversible (nonelastic) volume change of the fluid inclusion.

Three types of the nonelastic modifications of fluid inclusions occur in nature. Stretching—the irreversible volume change without optically visible mechanical failure (rupture) of inclusion walls—occurs at initial stages of fluid inclusion reequilibration. Gradually increasing overpressure results in the decrepitation (explosion), when the fluid inclusion walls instantaneously rupture and a portion of the inclusion fluid volume is released. In contrast, negative pressure gradient between fluid inclusion and its surroundings (internal underpressure) results in an implosion. Both processes are accompanied by cracking. The released inclusion fluid may be subsequently trapped within crystallographically oriented, healed cracks and microfissures. The reequilibration-related trails of secondary inclusions are called *decrepitation clusters*. Reequilibration is often accompanied by an extensive recrystallization of fluid inclusion walls, whose shapes in extreme cases revert from energetically stable regular forms (negative crystals) to unstable irregular or amoeboidal forms (Figure 1.10).

1.4.3 COMPOSITIONAL CHANGES

Apart from being closed, near-isochoric (near-constant volume) systems, fluid inclusions possess attributes of an isochemical (isoplethal) system, whose bulk mass and chemical compositions remain constant in spite of chemical reactions taking place due to supply or removal of external heat. In some cases, however, fluid inclusions can either release some compounds or gain others, thus behaving as an open system. Diffusion is the main exchange mechanism controlled by a chemical gradient between the inclusions and their surroundings. Because the chemical and pressure gradients occur simultaneously, the diffusion is often accompanied by irreversible volume changes.

The English metallurgist Roberts-Austen demonstrated in 1896 that gold migrates by diffusion in lead at 300 °C faster than sodium chloride in liquid water at 15 °C (Moore, 1979). This example shows how important the role of diffusion can be in nature. In the case of fluid inclusions, however, diffusion pertains mainly to small atoms and molecules of gaseous species. Pasteris and Wanamaker (1988) heated olivine-hosted inclusions of pure CO_2 for 1–161 h at temperature around 1400 °C in a slightly reducing atmosphere and showed that initially pure CO_2 inclusions were transformed into a CO_2–CO mixture. They explained this phenomenon by the diffusion of oxygen out of the inclusions, which meant toward the lower partial pressure of the oxygen. High reaction velocity ruled out diffusion along dislocations, which proceeds too slowly in these conditions. A diffusion caused by point dislocations well known in metallic compounds could be a more plausible mechanism to account for this phenomenon. Such diffusion is caused by the displacement of metallic ions from structural to interstitial positions and the concomitant formation of point defects composed of vacant and interstitial positions.

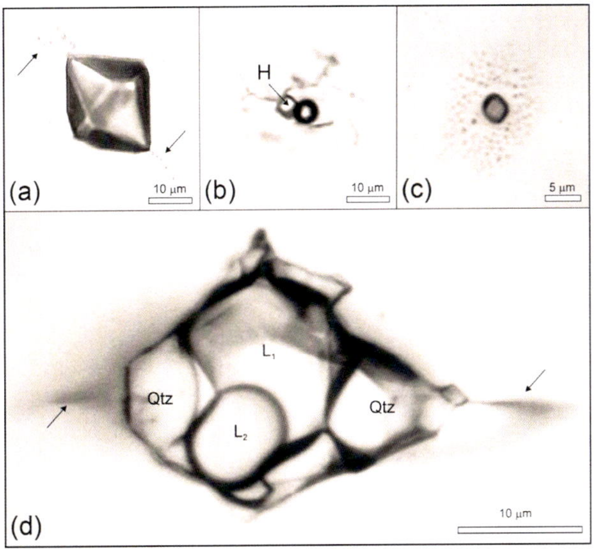

FIGURE 1.10

Examples of irreversible post-entrapment modifications of fluid inclusions. (a) Moderate reequilibration of a negative crystal-shaped methane inclusion in quartz documented by thin decrepitation cracks (indicated by arrows) emanating from inclusion tips along the c-axis. (b) Heavily reequilibrated brine inclusion in fluorite, showing strong recrystallization of inclusion walls. H designates halite. (c) Reequilibrated nitrogen inclusion in barite with planar cluster of secondary satellite inclusions (Hurai et al., 2008b). (d) Heavily reequilibrated CO_2–H_2O inclusion in quartz from an alpine fissure (Hurai and Horn, 1992), showing crystallographically oriented cracks with satellite inclusions and strongly recrystallized inclusion walls. L_1 denotes aqueous liquid, whereas L_2 is a CO_2 liquid phase deformed by the secondary quartz crystals (Qtz) growing inside the inclusion cavity.

The experiment with olivine demonstrated that fluid inclusions trapped in mantle xenoliths could have been oxidized during ascent in basaltic magma to the Earth's surface so that they are nonrepresentative of the original mantle fluid.

Diffusion of hydrogen into inclusions may cause reduction of CO_2 to CO_2+CH_4 mixture (Hall et al., 1991). In turn, a diffusive loss of hydrogen accounts for pure CO_2 inclusions in granulites, resulting from the reaction $CH_4+2H_2O \rightarrow CO_2+4H_2$ (Hollister and Burruss, 1976). Hall et al. (1991) recalculated that the addition of 12.5% of hydrogen at 400–550 °C to the H_2O–CO_2–CH_4 inclusions of the massive sulfide deposit at Ducktown (Tennessee) resulted in the increase in CH_4 concentration from the initial value of 10^{-7} to as much as 3 mol.%, that is by seven orders of magnitude. Massive diffusion of water in quartz has been experimentally proven by Sterner et al. (1995) in synthetic inclusions doped by D_2O. A diffusional loss of water molecules has been proposed to account for the correlation between the inclusion size and $\delta^{18}O$ values in hydrothermal quartz (Ohba et al., 1995).

A post-entrapment diffusion of hydrogen out of or into inclusions changes the fluid oxidation state. Consequently, some daughter minerals (calcite, chalcopyrite, hematite) fail to dissolve on heating the aqueous fluid inclusions, though they clearly represent daughter minerals precipitated after inclusion sealing. Apart from sluggish dissolution kinetics, the oxidation of inclusion fluid by outward-migrating

H^+ has been proposed as an alternative, and Mavrogenes and Bodnar (1994) experimentally corroborated this mechanism. They enabled the hydrogen to migrate into quartz-hosted fluid inclusions by heating them for several days in a reducing atmosphere at 600 °C. After the experiment, initially insoluble chalcopyrite daughter phases dissolved easily in the inclusions.

Though small hydrogen atoms migrate through a crystal lattice of silicates relatively easily, the hydrogen diffusion-induced changes of inclusion properties are in most cases negligible. This was documented by Kozlowski (1985) who performed a series of experiments with hydrous silicate melt inclusions. The water dissociated to H^+ at high temperature should diffuse out, thus resulting in the lowered density and an increasing homogenization temperature of the melt inclusions. However, though some inclusions were maintained at the temperature of 820 °C continuously over four years, no measurable changes in their homogenization temperatures were recorded.

Another example of negligible hydrogen diffusion comes from the Oklo uranium deposit (Gabon), where Dubessy et al. (1988) identified molecular hydrogen and oxygen in quartz-hosted fluid inclusions formed by radiolysis of water to H_2 and O_2 during spontaneous breakdown of radioactive elements. High hydrogen contents have been preserved in these inclusions for over 2 billion years in spite of the presumably high diffusivity of hydrogen in quartz. Hence, several lines of evidence suggest that fluid inclusion composition is influenced by the outward and inward diffusion only in exceptional cases.

CHAPTER

PHASE DIAGRAMS

2

CHAPTER OUTLINE

2.1 UNARY SYSTEMS

The state of a chemical system is described by intensive state properties, of which temperature (*T*), pressure (*P*), volume (*V*), and concentration (*X*) are essential for geologists. Because the variables are mutually dependent, only a limited set of them is needed to fully describe the state of the system. The number of variables (variance, degree of freedom) needed for the description of a system is defined by the phase rule (Equation (1.16)). For instance, two of three intensive variables, temperature, pressure, and volume, are needed for a complete description of a homogeneous one-component system with the number of components and phases each equal to one. The system is bivariant, because it has two degrees of freedom (*f*). However, if the same unary system consists of two phases (liquid + vapor, liquid + solid, solid + vapor), the degree of freedom decreases by unity, the system becomes univariant, so that only one of the three intensive variables needs to be defined to fully characterize the state of the system.

Phase equilibria are defined by an invariant point when $f=0$, by a univariant curve when $f=1$, a bivariant surface when $f=2$, and a three-dimensional space when $f=3$. Unary system ($c=1$) reaches the maximum degree of freedom $f=2$, if it consists of only of one of three (solid, liquid, or vapor) phases. Two-phase equilibria are univariant, projecting along curves. Liquid, solid, and vapor phases

Geofluids. http://dx.doi.org/10.1016/B978-0-12-803241-1.00002-2

can coexist in an invariant triple point. The critical point at a two-phase liquid–vapor curve, where the densities of the coexisting phases become identical, is also invariant (Figure 2.1).

Space in the three-dimensional phase diagram (Figure 2.1) is subdivided into three separated stability fields of a homogeneous solid, liquid, and vapor. The fields are bounded by two-phase curves. Liquid and vapor can coexist along a univariant curve called the boiling–condensation (= dew) curve, which extends from the triple point and ends at the critical point. The density of the vapor increases and that of the liquid decreases with the increasing temperature along this curve, until both phases have identical density and no phase boundary occurs in the critical point. Hence, pressure fluctuations at a temperature higher than that in the critical point permit a continuous change from liquid to vapor without any phase transformation, thus leaving the system homogeneous at any pressure (supercritical fluid). In contrast, the pressure fluctuations at a temperature below that at the critical point result in a discontinuous density change or heterogenization, when the homogeneous fluid instantaneously splits into liquid and vapor phases. The phase transformation is called boiling, if the homogeneous liquid liberates vapor bubbles, or condensation, if a liquid is condensed from the homogeneous vapor. In *P*–*T* coordinates, the boiling and the condensation curves overlap owing to identical saturation pressures (Figure 2.2).

The melting curve, where the solid coexists with the liquid, and the sublimation curve, where the solid coexists with the vapor, are the remaining two-phase boundaries in the phase diagram of a one-component system. The topology of other one-component systems is basically identical; however, the *P*–*T* parameters of critical and triple points are different and unique for each system (Table 2.1).

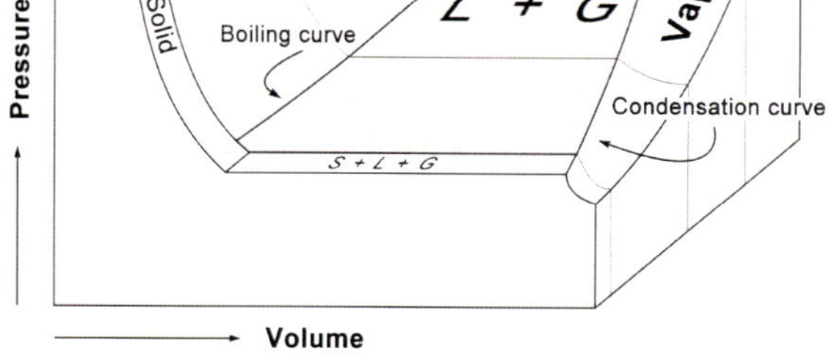

FIGURE 2.1

Schematic phase diagram of unary system (modified after Burruss (1981c)). S, L, and G denote solid, liquid, and vapor phases, respectively.

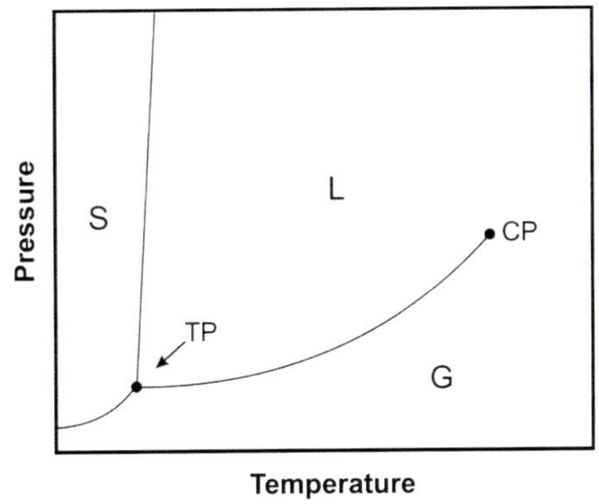

FIGURE 2.2

Schematic projection of two-phase curves of a one-component system in temperature–pressure coordinates.

2.2 PROPERTIES OF WATER

Water is the most important chemical substance on the Earth. There are 1.3×10^9 km^3 of water in oceans, 3.3×10^7 km^3 in the polar ice caps, 2×10^5 km^3 in glaciers, 2.2×10^3 km^3 in lakes and rivers (Debenedetti and Stanley, 2003). Nearly every aspect of our daily life is influenced or controlled by liquid water, which has specific properties resulting from its low molecular weight, asymmetric molecular structure, and the way individual water molecules group together.

The water molecule is polar, because one end is negative and the other is positive. Hence, water molecules attract each other like small magnets. Each positively charged atom of hydrogen connects with the negatively charged oxygen atom of a neighboring water molecule (Figure 2.3). In such a manner, so-called water clusters are formed, composed of four to nine molecules. Hydrogen bonding enables grouping of the clusters in more complex structures. The hydrogen bonds can have various lengths fluctuating between 1.73 and 3.94 Å. Temperature-dependent bending of the bonds gives rise to expanded and condensed clusters with various densities. A group of 280 water molecules represents the stable unit of the highest order. The stable clusters can further group into superclusters. Liquid water therefore consists of an infinite three-dimensional network of water molecules connected through hydrogen bonds. The outer perimeter of the stable water cluster composed of 280 molecules acquires a shape of rhombicosidodecahedron with 12 pentagons, 20 triangles, and 30 rectangles (Chaplin, 2000). The cluster thus has a pentagonal, fivefold symmetry, which does not exist in solid crystals but may exist in quasi-crystals. For liquids, the symmetry guarantees the tightest possible arrangement of molecules. Apart from water, a similar molecular arrangement is typical only of condensed precious gases and liquid metals. It is worth mentioning that the pentagonal symmetry of water was envisaged by Plato in the fourth century B.C., although he could not know which elements the water consisted of.

Water is unusual among ordinary natural fluids owing to a negative slope of the solid–liquid phase boundary, and the existence of about 20 stable and metastable crystalline and amorphous ice polymorphs at other than atmospheric pressures (Ludwig, 2001; Dunaeva et al., 2010). Consequently, the complete phase diagram of water has a great number of triple points and possibly two critical points (Figure 2.4). Metastable supercooled aqueous liquid can exist at temperatures between 0 and −42 °C.

Table 2.1 Triple and Critical Points of Simple Molecules

	Triple Point			Critical Point			
Species	***T*** **(°C)**	***P*** **(bar)**	***D*** **(g/cm³)**	***T*** **(°C)**	***P*** **(bar)**	***D*** **(g/cm³)**	**Source**
Ar	−189.2			−122.3	48.5		1
CO	−199.0			−140.0	38.4		2
				−140.0	35.0		3
COS	−138.2			104.8	65.7		2
				104.4	65.9		3
CO_2	−56.6	5.2	*0.0139 (G)*	31.06	73.8	***0.466055***	4, *5*, ***6***
					73.60		2
				31.0	73.87		3
	−56.6	5.17	*1.1796 (L)* *1.513 (S)*	30.98	73.75		7, *5*
CH_4	−182.47	0.12		−82.62	45.96		4
				−82.1	46.3		2
	−182.47	0.11696	0.45147 (L) 0.00025 (G)	−82.6	45.99	0.16266	8
				−82.1	46.41		3
C_2H_6	−183.3	1.1e−5		32.27	48.84		4
C_3H_8	−187.68	3.3e−9		96.7	42.5		4, 9
n-C_4H_{10}	−138.29	4.0e−6		152.03	37.96		4, 9
n-C_5H_{12}	−129.72			196.5	33.69		9
H_2				−239.9	12.97		3
H_2O	0.015	0.00611		374.12	221.1	0.32274	4
				374.1	221.19		3
			0.99979 (L) 4.8546e−6 (G)				10
	0.0098		0.917 (S)				11
				373.946	220.64		12
H_2S	−87.5	0.23		100.05	89.37		4
	−85.5			100.4	89.8		2
				100.4	90.08		3
N_2	−210.0	0.13		−146.95	34		4
NH_3				132.4	113		13
O_2				−118.4	50.76		3
				−118.55			14
SO_2	−72.7			157.8	78.5		2
				157.5	78.7		13

1, Moore (1979); 2, Murck et al. (1978); 3, Shi and Saxena (1992); 4, Allamagny (1976) cited in Mullis (1987); 5, Ražnjevic (1975); 6, Angus et al. (1976); 7, Ely et al. (1989); 8, Kleinrahm and Wagner (1986); 9, Hiza et al. (1982); 10, Wagner et al. (1994); 11, Mason (1971); 12, Wagner and Pruss (2002); 13, Takenouchi and Kennedy (1964); 14, Weast (1979). Multiple literature sources in some rows are linked with data by different letter styles.

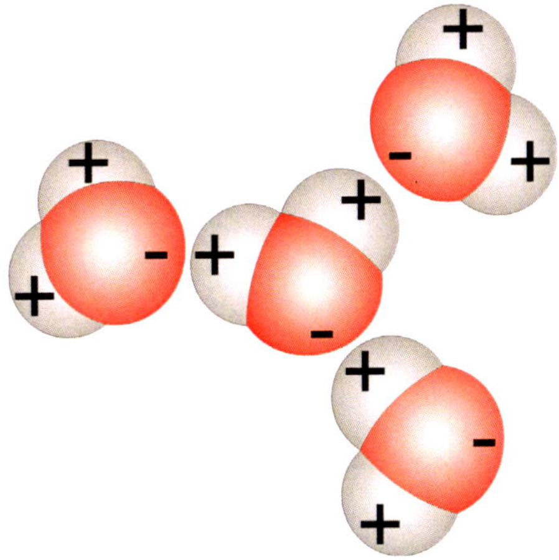

FIGURE 2.3

Water molecules consisting of two positively charged hydrogen (gray) atoms and one negatively charged oxygen (red) atom can form clusters by hydrogen bonding.

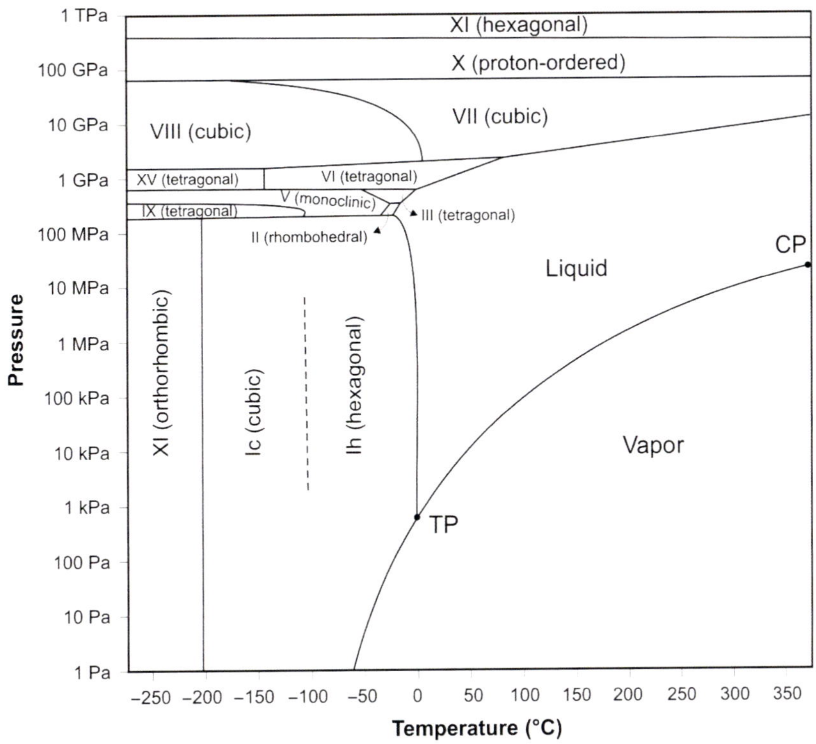

FIGURE 2.4

Phase stability diagram of water expanded to high pressures. Ice phases relevant to *P–T* conditions on Earth are as follows: Ih, normal hexagonal ice in biosphere; Ic, metastable cubic variant of the hexagonal ice occasionally present in upper atmosphere; XI, stable configuration of the hexagonal ice in ancient Antarctic ice; VI, tetragonal ice in aqueous inclusions under high residual pressure, trapped in diamonds. Other high-pressure ice modifications occur on icy moons orbiting around Jupiter, Saturn, and Uranus. http://www.lsbu.ac.uk/water/phase.html, published with permission from M. Chaplin.

When cooled fast enough, water, like other liquids, vitrifies. Unlike most other substances, though, glassy water can apparently exist in at least three distinct amorphous forms: low-density amorphous ice, high-density amorphous ice, and very-high-density amorphous ice below the glass transition temperature at ~−143 °C (Debenedetti, 2003). Immediately above this temperature, it transforms into a highly viscous fluid and ultimately crystallizes at ~−123 °C. The region between −42 and −123 °C is a "no-man's land," within which bulk liquid water cannot be studied (Debenedetti and Stanley, 2003).

Isochoric *P–T* projection of the phase diagram of water (Figure 2.5) is the most useful form of diagrams used in geofluid research. Other types of phase diagrams can also be constructed employing additional intensive variables defined by thermodynamic laws.

The ability of creating ordered molecular clusters combined with strong chemical bonding and low molecular weight result in unique water properties, which seem to be accommodated to biological processes. Water composes a total of 60 wt.% of terrestrial, and sometimes as much as 90 wt.% of marine organisms. Organisms could not originate and live without liquid water. As many as 64 counterintuitive "anomalous" properties of liquid water are known, which substantially differ from those of other ordinary liquids and most of which have not been adequately explained (Harvey and Friend, 2004). The best known of these is the density paradox: liquid water displays a density maximum at 4 °C,

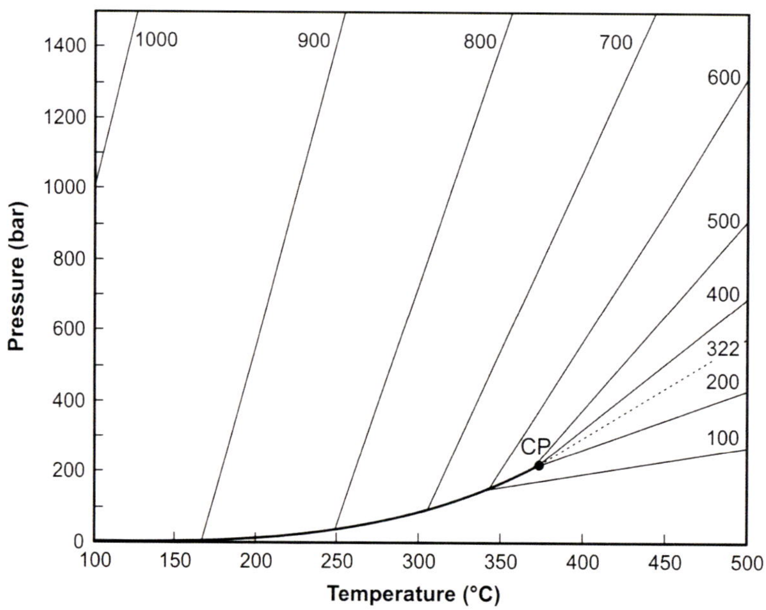

FIGURE 2.5

Isochoric *T–P* phase diagram of water with iso-density lines ranging between 100 and 1000 kg/m^3. Shown also is a two-phase (liquid–vapor) curve ending in the critical point, CP, at 373.946 °C, 220.64 bar, 322 kg/m^3 (Saul and Wagner, 1989; Wagner and Pruss, 2002) constructed using data from the NIST Chemistry Webbook. The critical density isochore (dashed line) divides the fields of liquid-like supercritical fluids at higher pressures and vapor-like supercritical fluids at lower pressures.

and becomes less dense rather than denser when it freezes. Most liquids become more viscous under pressure, but the reversal is true for cold liquid water (Ludwig, 2001; Mallamace et al., 2013).

Temperature and pressure are balanced on the Earth's surface in such a manner that water, as the only natural inorganic substance, can exist simultaneously in all three possible phases. Of course, there are other triple points in the phase diagram of water, but they permit the coexistence of three solid phases, or two solids with liquid. The temperature at the triple point of aqueous liquid, vapor, and solid, namely 0 °C, is extraordinarily high. All other substances with comparable or higher molecular weight exhibit much lower temperatures of melting. For instance, methane with molecular weight of 16 freezes below −182 °C and even carbon dioxide with the molecular weight of 44 solidifies at the temperatures below −56.6 °C.

The ability of water molecules to form various four-coordinated networks results in the extreme ice polymorphism. At atmospheric pressure and temperatures below 0 °C, liquid water transforms to hexagonal ice, which displays a structure with a sixfold symmetry, in which each H_2O molecule is linked with other four molecules. This ordering is less tight than that of the liquid water. Consequently, the density of hexagonal ice is lower than that of the coexisting liquid water. If the liquid water freezes, it expands by about 9% as a result of the density difference. In a closed space (e.g., in fluid inclusions or a sealed bottle), an overpressure 25 times higher than that of atmospheric pressure is generated as a consequence. This behavior of ice complicates human life, but is vitally important for organisms living in water. Owing to the low density, ice floats on water, so that oceans, lakes, and rivers freeze only on the surface. The frozen water serves as an insulator, which prevents freezing of deeper parts of the water column. Hence, the anomalous structure of the low-density hexagonal ice and the denser, pentagonally symmetric liquid water permits the survival of organisms in winters and recurrent ice ages.

Strong chemical bonds and the high ordering degree of liquid water prevent easy evaporation, and also cause an extremely high boiling temperature. With the critical temperature of 374 °C, boiling and condensation occur in a wide range of temperatures between the critical and triple points. The Earth's atmosphere exerts the pressure equivalent to 1 tonne per 0.1 m^2. This pressure allows for an abrupt conversion of the aqueous liquid to vapor (boiling) and vapor to aqueous liquid (condensation) up to 100 °C. Vapor condensates in the colder upper atmosphere and returns to the surface in the form of rainfall, which, together with the existence of large continents have a profound effect on the origin of intelligent life.

The high heat capacity of liquid water is the reason why the water can store a large amount of heat energy without increasing the temperature. Owing to low molecular weight, a fixed volume of liquid water contains many molecules, which can effectively absorb the heat energy. On heating, the heat energy is consumed for bending and breaking the strong hydrogen bonds. Along with the ability of evaporation and condensation, the high heat capacity of water is essential for living organisms. Liquid water in cells of living organisms serves as heat reservoirs preventing sudden temperature fluctuations and dehydration of organic tissues. Transformation of liquid water to vapor liberates excess heat and helps to maintain the constant temperature of organic cells. The large volume of liquid water in oceans maintains a stable temperature in spite of rotation and tilting of the Earth's axis, and irregularities of its orbital trajectory.

Only mercury has larger surface tension than water. This feature is also caused by extraordinary cohesion of water molecules and results in the seemingly banal fact of creating water tears instead of a thin film on a solid surface. The high surface tension enables movement of liquid water in thin capillaries against the direction of the gravitation field. This feature permits blood to circulate in veins, and liquid water together with dissolved minerals and nutrients to be transported from roots to leaves, which are in many cases located tens of meters above the groundwater table.

The high dissolving ability of water is probably the most important property from the viewpoint of geologists. Due to molecular polarity, water dissolves much more organic and inorganic species than any acid. Moreover, water creates ionic solutions with salts and complex metal compounds. This is particularly important for dissolution, transport, and deposition of minerals.

Water is present everywhere in the continental crust. High-salinity $CaCl_2$-rich brines have been pumped out from deep boreholes from depths of more than 8 km. Free liquid water can persist even in the upper mantle. This is shown by the presence of ice VI inclusions in eclogitic diamonds formed more than 100 km deep (Kagi et al., 2000). The preservation of ice VI was enabled by high residual pressure within the inclusions, as much as 1.9 GPa, which caused the conversion of the liquid water to the ice VI at room temperature, instead of to the hexagonal ice at temperatures below 0 °C and lower pressures. The trace amounts of water (up to 2000 ppm) stored in defects in nominally anhydrous minerals, such as pyroxene, olivine, and garnet, in the Earth's mantle is equivalent to the amount of water in the ocean and atmosphere combined (Bhattacharya, 2012).

The existence of a sufficient amount of water at temperatures and pressures close to 0 °C is an important factor but not the only prerequisite for the origin of life. Evolved organisms could not originate on a geologically inactive planet. Slow vertical movement, by approximately 1 cm per thousand years, of the partially plastic upper mantle is the engine of global tectonic processes. The upper mantle rocks partly melt during the ascent-related decompression and form less-dense basaltic melt at a temperature of around 1300 °C. The basalt is composed mainly of silicates of calcium, magnesium, and iron. The basaltic melt pours out along fractures in the sea floor and creates the oceanic crust. The oceanic crust can store a large amount of seawater in the form of OH minerals formed by submarine weathering. The hydrated oceanic crust may be transported to depths of 100–250 km (Abers, 2000). The OH-bound water together with free aqueous liquid can depress the temperature of partial melting of the oceanic crust from 1300 to as low as 700 °C. Such a low-temperature partial melting generates granitic melts rich in sodium, potassium, and calcium, which are much lighter than the basalt. The light granitic melts are then transported by volcanoes to the surface and deposited above the oceanic crust, thus creating the continental crust.

Organisms with a central nervous system could not originate without the continental crust and the ability of water to dissolve, condensate, and evaporate. In the course of billions of years, an enormous amount of salts have been dissolved in ocean, which now contains 32 g of salts in 1 L (3.2 mass%). The existence of ionic aqueous solutions is important for the origin of life and for the development of *Invertebrata* and organisms with outer carbonate, silica, or phosphate shells. In contrast, the origin of organisms with a central nervous system is directly connected with salt-free distilled water, which is produced in large quantities from the seawater by evaporation and condensation of water vapor above continents. The meteoric water mixes with the seawater in estuaries and creates a mineral-deficient environment. The origin of the vertebral column is a direct response of more primitive organisms to the lack of inorganic species. In such a manner, *Osteichthyes*—fish with a vertebral column—developed from *Placodermi* and as a successful species expanded in the oceans and later conquered the continents, where amphibians, reptiles, birds, mammals, and finally *Homo sapiens* have developed. It could be concluded that without the geological processes assisted by liquid water, the octopus would be the most intelligent creature living on the planet Earth.

2.3 AQUEOUS SYSTEMS WITH SALTS

Binary systems ($c=2$) have the maximum degree of freedom equal to 3, and the concentration (X) is the additional state variable needed to fully define the system. Compared with an unary system, triple point converts to quadruple point called the eutectic point, where two solids, liquid, and vapor coexist. Additional quadruple points (peritectic points) may occur if the fluid components react mutually to form solid phases with intermediate compositions. The invariable critical point of the unary system transforms to the critical curve composed of critical points for each particular concentration.

2.3.1 H_2O–NaCl SYSTEM

The system H_2O–NaCl is the reference system for aqueous fluids in nature. Its topology has been described by numerous authors, including Sourirajan and Kennedy (1962), Urusova (1975), Bodnar et al. (1985a), Gehrig et al. (1986), Pitzer and Pabalan (1986), Bischoff (1991), Bodnar (1994), and Lecumberri-Sanchez et al. (2012). Experimental works addressing selected portions of the system have been summarized in Bodnar (1992) and Bodnar and Vityk (1995).

Critical temperature and pressure rise proportionally with the total dissolved NaCl. For instance, the critical point of pure H_2O, 374 °C and 221 bar, shifts to 662 °C and 968 bar with 25 mass% NaCl dissolved (Knight and Bodnar, 1989). Furthermore, the saturated vapor pressure tends to decrease with the addition of NaCl. Consequently, at atmospheric pressure, pure water boils at 100 °C, compared with the boiling point of 107 °C diagnostic of a 25 mass% NaCl–H_2O solution (Haas, 1976).

The system NaCl–H_2O (Figure 2.6) is characterized by a continuous critical curve, connecting critical points of pure compounds ($CP_{H_2O}=374.1$ °C and 221.1 bar, $CP_{NaCl}=\sim 3600$ °C and ~258 bar). Pressures along the vapor-saturated halite liquidus are substantially lower; therefore the liquidus does not intersect the critical curve. This topological feature is consistent with some gas mixtures, such as CO_2–CH_4. The halite liquidus consists of two branches (G+L+H and L+G+H), extending from the hydrohalite–halite peritectic point (P) at 0.1 °C and 26.3 mass% NaCl, to the triple point of pure NaCl ($TP_{NaCl}=801$ °C, 0.0003 bar) and the critical point of halite—CP_{NaCl}. The vapor-saturated hydrohalite liquidus connects the eutectic and peritectic points (E–P), whereas a vapor-saturated ice liquidus extends from the eutectic point ($E=-21.1$ °C, 23.2 mass% NaCl) to the triple point of pure H_2O ($TP_{H_2O}=0.01$ °C, 0.006 bar).

A NaCl-undersaturated fluid is homogeneous at pressures above the critical curve and becomes heterogeneous along two-phase surfaces connecting the critical and three-phase curves. Along these surfaces, initially homogeneous fluid splits into liquid and vapor phases. Further decompression leads to the precipitation of halite in the L+G or G+L mixture along the three-phase curves, whereas the G+H, liquid-free assemblage is stable at very low pressures beneath the vapor-saturated halite liquidus.

The L+G region is subdivided into smaller parts by isoplethal loops corresponding to particular NaCl concentrations. Depending on salinity, the isopleths extend from ice, hydrohalite, or halite liquidi, pass through a critical point on the critical curve (e.g., CP_A and CP_B), and terminate on the G+L+H curve, thus encompassing the outer surface of the L+G region. The isoplethal loops are subdivided into boiling (L_B+G) and condensation (G+L_B) curves below and above the CP, unique for each NaCl concentration.

Halite liquidi are important components of the phase diagram of the H_2O–NaCl system at high temperatures and the concentrations >26.3 mass%. Halite can dissolve in the aqueous liquid in the presence of vapor along the vapor-saturated halite liquidus, corresponding to the L+G+H surface. The vast

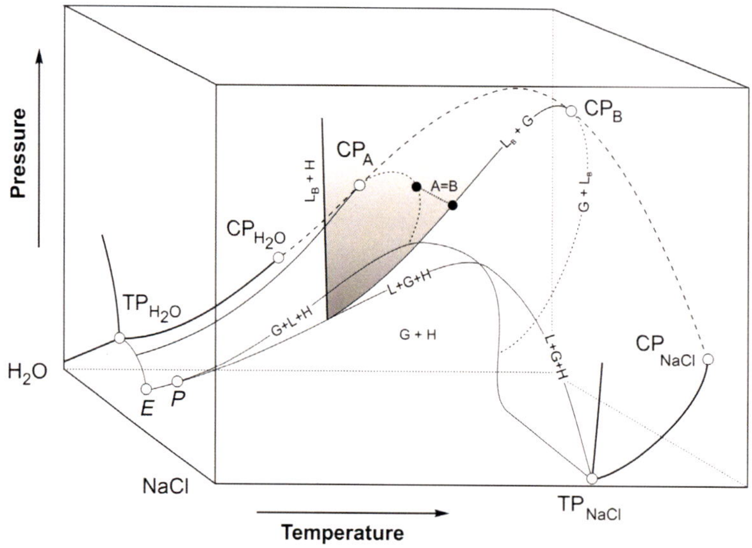

FIGURE 2.6

Distorted three-dimensional P–T–X projection of the NaCl–H_2O system modified after Bodnar et al. (1985a) and Gehrig et al. (1986). *E*, eutectic point (−21.1 °C, 23.2 mass% NaCl); *P*, peritectic point (0.1 °C and 26.3 mass% NaCl); TP_{H_2O}, triple point of pure H_2O (0.01–0.02 °C, 0.006 bar); CP_{H_2O}, critical point for pure H_2O (374.1 °C, 220.2–221.2 bar); CP_{NaCl}, critical point of pure NaCl (~3600 °C and ~258 bar); TP_{NaCl}, triple point of pure NaCl (801 °C, 0.0003 bar). L_B+H is vapor-unsaturated halite liquidus for the fluid of composition B. L_B+G and G+L_B are liquid+vapor curves for the composition B, extending from critical point CP_B of the particular fluid to the vapor-saturated halite dissolution curves (L+G+H, G+L+H). The fluid B remains homogeneous in the shaded region.

majority of natural brine inclusions correspond to the liquid-rich side of the L+G+H surface. Halite can also dissolve in the homogeneous liquid devoid of vapor bubbles. In this case, the temperature of the halite dissolution projects along an upright line extending from the L+G+H surface (Figure 2.7). Being slightly pressure-dependent, the line called vapor-unsaturated halite liquidus has a negative slope for the concentrations <50 mass% NaCl and is positively sloped for those exceeding 50 mass% NaCl. The corresponding dT/dP slope [°C/bar] can be expressed by the equation (Bodnar, 1994)

$$\frac{dT}{dP} = -38.38 + 0.9m - 0.0029m^2, \tag{2.1}$$

where m is the salinity in mass%.

A liquid with the composition B remains homogeneous at any P–T conditions within the shaded plane between the L_B+G and L_B+H curves, namely between the vapor-unsaturated halite liquidus and the boiling curve. In other words, the fluid with composition B precipitates halite along L_B+H or liberates a vapor bubble along L_B+G. The composition and density of the coexisting vapor along the boiling curve correspond to that at the same PT conditions on the opposite side of the L+G surface (e.g., the composition A in the point A=B), where a low-density, low-salinity vapor with the composition A is a stable phase. Moving along the L_B+G curve leads to changing density of the liquid B,

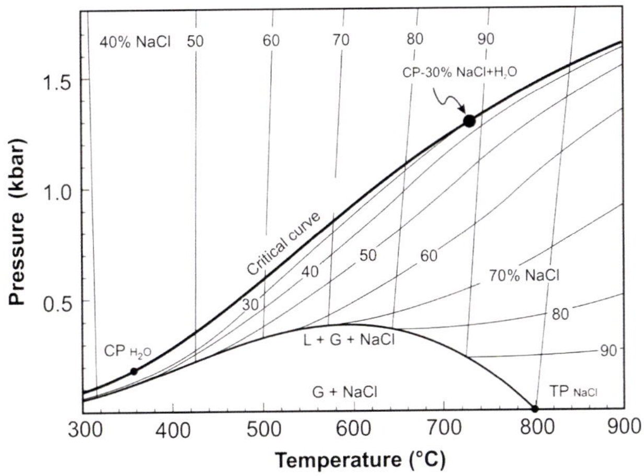

FIGURE 2.7

Isoplethal projection of a portion of the system H_2O–NaCl with vapor-unsaturated halite liquidi (Bodnar, 1994, steeply sloped lines), and isopleths of liquid phases inside the L+G field (Bodnar et al., 1985a, flat to moderately dipping curves). Reprinted from Bodnar, R.J., Burnham, C.W., Sterner, S.M., 1985. Synthetic fluid inclusions in natural quartz. III. Determination of phase equilibrium properties in the system H_2O–NaCl to 1000 °C and 1500 bars. Geochim. Cosmochim. Acta 56, 1861–1873, Copyright 1985, and Bodnar, R.J., 1994. Synthetic fluid inclusions: XII. The system H_2O–NaCl. Experimental determination of the halite liquidus and isochores for a 40 wt.% NaCl solution. Geochim. Cosmochim. Acta 58, 1053–1063, Copyright 1994, with permission from Elsevier.

whereas the vapor A changes both its density and its composition. Hence, each point within the two-phase L+G region has its own unique combination of the coexisting liquid and vapor phases.

The low-temperature portion of the phase diagram of the H_2O–NaCl system is depicted in detail in Figure 2.8. The three-phase L+V+H curve ends in the quadruple peritectic point P at 0.1 °C, where

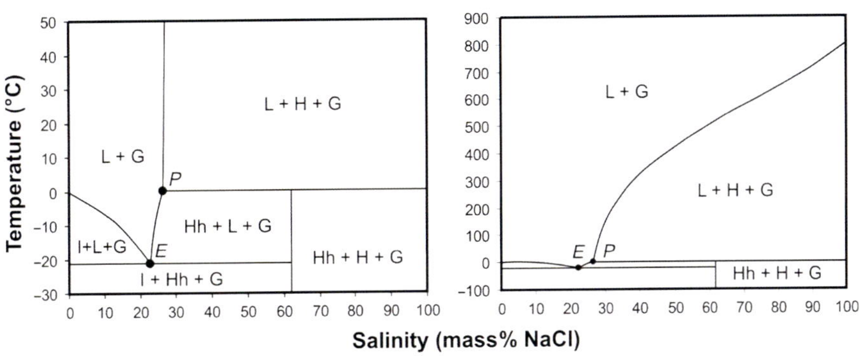

FIGURE 2.8

Low-temperature phase diagram of the H_2O–NaCl system constructed from experimental data (Sterner et al., 1988; Bodnar, 1993). L, G, H denote liquid, vapor, and halite, respectively. Hh is hydrohalite and I is ice. *E* and *P* denote eutectic and peritectic points, respectively.

halite reacts with water to form hydrohalite $NaCl{\cdot}2H_2O$, coexisting with a 26.3 mass% $NaCl-H_2O$ aqueous liquid and vapor. The eutectic point E at −21.1 °C is another quadruple point, where the hydrohalite coexists with ice, a 23.2 mass% $NaCl-H_2O$ aqueous liquid and vapor.

The hydrohalite contains 61.9 mass% NaCl. Hence, a totally frozen solution with the total concentrations lower than 61.9 mass% NaCl evolves at the eutectic point a variable proportion of an aqueous liquid with the concentration of 26.3 mass% NaCl. If the total concentration is less than 23.2 mass%, the eutectic solution is accompanied by ice, which dissolves along the vapor-saturated ice liquidus up to the temperature of 0 °C, corresponding to the triple point of pure water. If the total concentration is above 23.2 mass% NaCl, the eutectic solution coexists with hydrohalite, which either dissolves in the liquid below 0.1 °C and the concentrations below 26.3 mass%, or it converts to halite at the peritectic point P at higher total concentrations. If the total concentrations are above 61.9 mass% NaCl, an aqueous liquid liberates in the peritectic point P by the decomposition of hydrohalite.

2.3.2 MULTICOMPONENT H_2O–SALT SYSTEMS

Apart from the dominant NaCl, other salts can also be present in substantial amounts in natural aqueous solutions. The most abundant salts, apart from NaCl, are as follows (set up according to their decreasing abundance): KCl, $CaCl_2$, $MgCl_2$, $FeCl_2$, $FeCl_3$, sulfates, alkalic hydrocarbonates, carbonates. The properties of most natural aqueous fluids can be described in terms of two- or three-component systems. Each of these systems can be identified according to a specific eutectic temperature (T_e), that is, the minimum possible temperature of the stable existence of a liquid. High-salinity aqueous solutions frequently create hydrates, which dissolve to anhydrous compounds along peritectic curves. However, low-salinity aqueous solutions behave as simple systems and their salinity is referred to as the low-salinity region of the $H_2O-NaCl$ system, and expressed as total salinity in mass% NaCl equivalents.

The multicomponent aqueous systems with salts are rather less constrained by experimental data. As the number of phases increases with the number of components, the experiments and the visualization of phase diagrams become complicated. Of particular importance is the system $NaCl-CaCl_2-H_2O$—the reference system for natural brines, and the system $NaCl-KCl-H_2O$, featuring the majority of magmatogenic hydrothermal ore deposits (Figure 2.9). Other available experimental phase diagrams pertinent to natural fluid inclusions are available for $H_2O-NaCl-MgCl_2$ (Dubois and Marignac, 1997), $NaCl-LiCl-H_2O$ (Dubois et al., 2010), $H_2O-NaCl-NaHCO_3$, and $H_2O-NaCl-Na_2CO_3$ systems (Borisenko, 1977), and for the $H_2O-NaCl-FeCl_2$ and $H_2O-FeCl_2$ systems (Lecumberri-Sanchez et al., 2014; Steele-MacInnis et al., 2015) featuring porphyry copper deposits (Figures 2.10 and 2.11).

Several groups of aqueous solutions can be distinguished according to the temperatures of their eutectic points (see Appendix I). Chloride solutions with appreciable $CaCl_2$ exhibit T_e values around −50 °C. Some $CaCl_2$-rich aqueous solutions begin to melt at very low temperatures, below −80 °C, thus overlapping T_e values of LiCl and $CaBr_2$-rich solutions. Initial melting as low as −85 °C in the $CaCl_2$-dominated fluids is likely to be a manifestation of a metastable ice + halite + $CaCl_2{\cdot}nH_2O$ eutectic assemblage (Davis et al., 1990). Eutectic temperatures between −40 and −35 °C correspond to iron chloride-rich solutions, whereas those between −23 and −21 °C unambiguously identify the abundant chlorides of alkali metals. Higher eutectic temperatures are very rare, and they reveal sulfate-, carbonate-, and hydrocarbonate-rich solutions. Eutectic temperatures in carbonate and hydrocarbonate-rich solutions are very close to 0 °C and can thus often be erratically interpreted as very low-salinity chloride solutions.

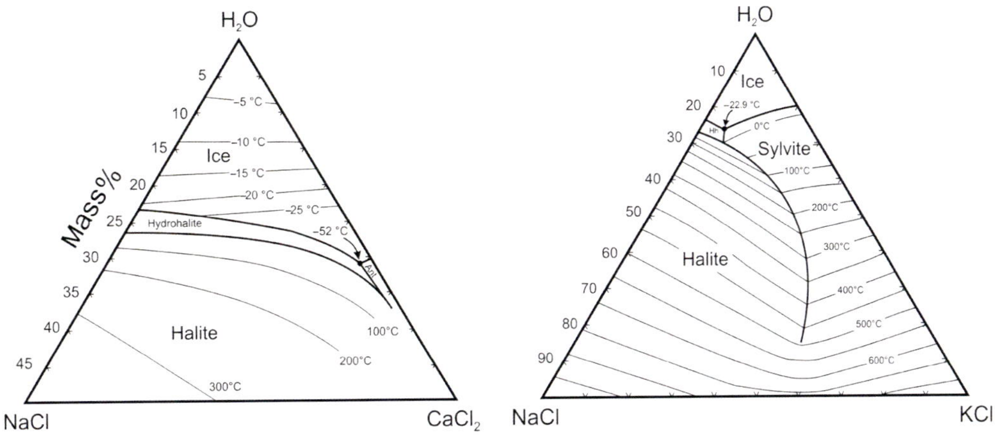

FIGURE 2.9

Isothermal phase diagrams (saturation solubility surfaces, cotectic curves, eutectic points) of the system $H_2O-NaCl-CaCl_2$ compiled from the experimental data of Vanko et al. (1988); Williams-Jones and Samson (1990)—the halite stability field; Oakes et al. (1990)—the ice stability field; and Yanatieva (1946)—the cotectic ice + hydrohalite curve. Phase diagram of the $H_2O-NaCl-KCl$ system was redrawn from Sterner et al. (1988). *Ant* denotes antarcticite—$CaCl_2{\cdot}6H_2O$, and solid circles with arrows correspond to eutectic (quintuple) points, where vapor and liquid coexist with other three solid phases. The right diagram reprinted from Sterner, S.M., Hall, D.L., Bodnar, R.J., 1988. Synthetic fluid inclusions. V. Solubility relations in the system NaCl-KCl-H_2O under vapour-saturated conditions. Geochim. Cosmochim. Acta 52, 989–1005, Copyright 1988, with permission from Elsevier.

Properties along boiling and condensation curves of the systems $CaCl_2-H_2O$, H_2O-KCl, and $H_2O-MgCl_2$ have been studied by Tkachenko and Shmulovich (1992), Bischoff et al. (1996), and Zhang and Frantz (1987). Isochores for halite-undersaturated $NaCl-CaCl_2-H_2O$ solutions in one-phase region can be interpolated from experimental data for binary subsystems (Zhang and Frantz, 1987; Oakes et al., 1990). The addition of $CaCl_2$ to $NaCl-H_2O$ increases the pressure along the critical curve and significantly expands the liquid + vapor region (Tkachenko and Shmulovich, 1992; Bischoff et al., 1996). Pressures in the two-phase regions and along three-phase curves may be substantially increased compared to the $H_2O-NaCl$ system by a $CaCl_2$ hydrolysis-related generation of HCl, and its preferential partitioning in the vapor phase (Bischoff et al., 1996). Pressures along the three-phase L + V + H curve, as well as those for vapor-unsaturated halite liquidi are unknown for the system $NaCl-CaCl_2-H_2O$ and more complex aqueous solutions.

2.4 AQUEOUS SYSTEMS WITH GASES

2.4.1 H_2O-CO_2 AND H_2O-CH_4 SYSTEMS

Experimental data on the two-phase region and the critical curve of the H_2O-CO_2 system were published by Tödheide and Franck (1963), Takenouchi and Kennedy (1964), Sterner and Bodnar (1991), and Blencoe et al. (2001) up to a pressure of 600 MPa. Experiments with H_2O-CH_4 mixtures were performed by Welsch (1973) and Christoforakos (1985). Synthetic H_2O-CH_4 fluid inclusions were

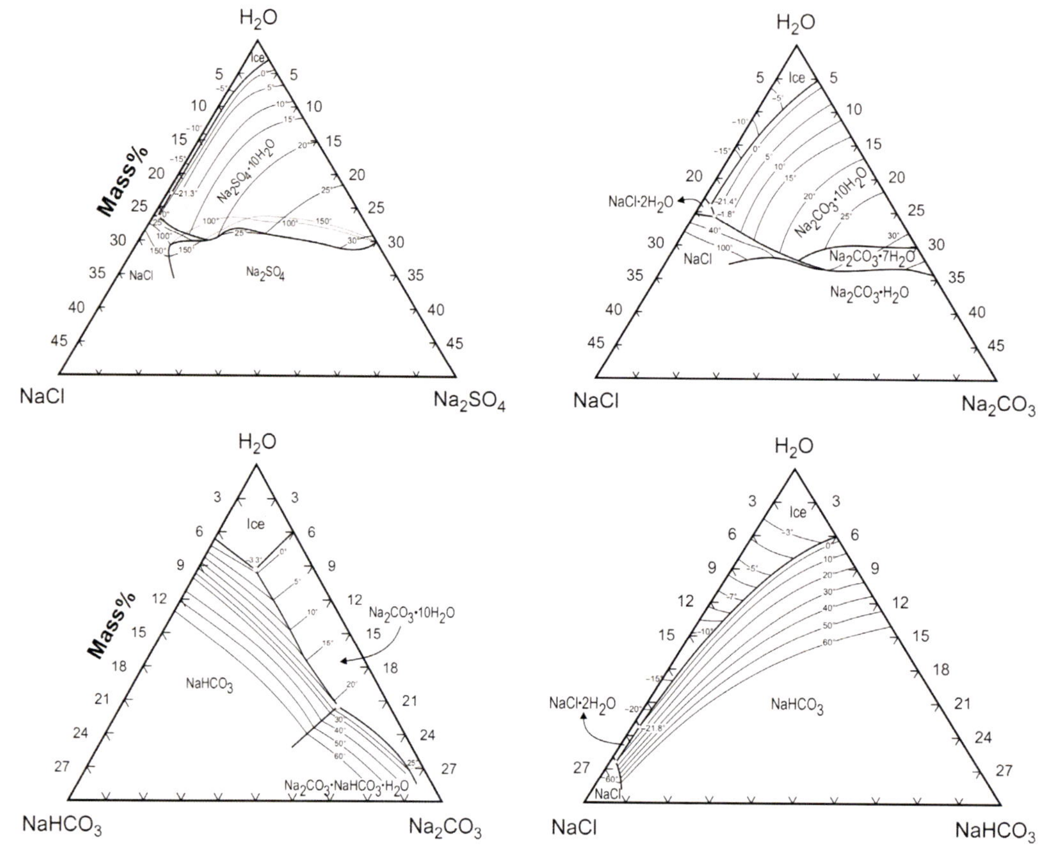

FIGURE 2.10

Isothermal phase diagrams of the $NaCl-Na_2SO_4-H_2O$, $NaCl-Na_2CO_3-H_2O$, $NaHCO_3-Na_2CO_3-H_2O$, and $NaCl-NaHCO_3-H_2O$ systems. Reprinted from Borisenko, A.S., 1977. Study of salt composition of fluid inclusions in minerals using cryometric technique. Geol. Geofiz. 8, 16–27, Copyright 1977, with permission from the Siberian Branch of the Russian Academy of Sciences, V.S. Sobolev Institute of Geology and Mineralogy and A.A. Trofimuk Institute of Petroleum Geology and Geophysics, Novosibirsk, Russia.

studied by Lin and Bodnar (2010). Equations of state describing the two-phase immiscibility regions of the H_2O-CO_2 and H_2O-CH_4 systems were proposed by Bowers and Helgeson (1983a), Christoforakos and Franck (1986), Heilig and Franck (1989), Duan et al. (1992b), and Churakov and Gottschalk (2003). Bakker and Diamond (2000) evaluated the ability of some equations of state to reproduce properties of conjugate immiscible phases. The topologies of geologically important gas–water systems, with particular emphasis on isochoric phase transitions applicable to fluid inclusion studies, were discussed in detail by Diamond (2001, 2003).

Compared to aqueous systems devoid of gases, saturation pressure dramatically increases with the addition of a very small amount of CO_2. For instance, at 250 °C, saturation vapor pressure in pure water

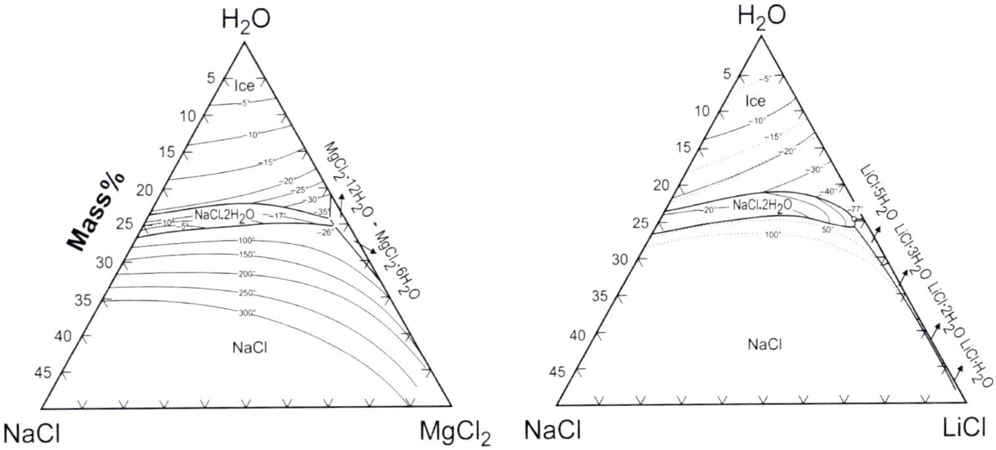

FIGURE 2.11

Isothermal phase diagrams of the NaCl–$MgCl_2$–H_2O (Dubois and Marignac, 1997) and the NaCl–LiCl–H_2O systems (redrawn for the triangular system in mass% coordinates according to data in Akopov (1963)—solid isotherms and two-phase cotectic boundaries and Dubois et al. (2010)—stippled isotherms). The diagram of the NaCl–$MgCl_2$–H_2O system reprinted with permission from the Society of Economic Geologists.

equals 39.5 bar, and increases up to 50 and 70 bar after the addition of 0.2 and 0.6 mol.% CO_2, respectively (Ellis and Golding, 1963).

Phase equilibria in the H_2O–CO_2 system are rather complicated due to contrasting properties of pure end-member molecules. CO_2 is a large, nonpolar, highly volatile molecule. Temperature and pressure in triple point of CO_2 are relatively low (−56.6 °C, 5 bar) compared to those at the critical point at 30.9782 °C, 7.3773 MPa (Span and Wagner, 1996). On the other side, the small and asymmetric (bipolar) water molecule is far less volatile due to strong hydrogen bonding and a dense, quasi-crystal structure. As a consequence, the triple (0.01 °C, 0.006 bar, see Table 2.1) and critical points of water (373.946 °C, 22.064 MPa, Wagner and Pruss, 2002) are considerably higher than expected considering a low molecular mass (18) comparable to that of methane (16) and other gases with substantially lower critical and triple points. Thus, the large immiscibility field of the H_2O–CO_2 system (Figure 2.12) is the direct consequence of the contrasting physical properties of the two mixing molecules.

The lowest variance elements in the phase diagram of H_2O–CO_2 system are quadruple lines, which project into points in the P–T space (Figure 2.13). Each quadruple point represents the equilibrium among four phases of fixed composition under the following unique P–T conditions (Diamond, 1995):

Q1 (−1.5 °C, 10 bar): CO_2-vapor, CO_2-clathrate, aqueous liquid, ice
Q2 (10 °C, 45 bar): CO_2-vapor, CO_2-liquid, CO_2-clathrate, aqueous liquid
Q3 (−56.6 °C, 5 bar): CO_2-solid, CO_2-vapor, CO_2-liquid, CO_2-clathrate

The critical curve of the H_2O–CO_2 system consists of two separate segments divided by a large immiscibility field. The shorter segment on the CO_2-rich side connects the critical point of pure CO_2 with the lower critical end point (LCEP) at 31.45 ± 0.3 °C, 7.4 ± 0.01 MPa, where 0.11–0.22 mol.% H_2O is dissolved in the CO_2-rich phase (Morrison, 1981; Song and Kobayashi, 1987; Wendland et al., 1999;

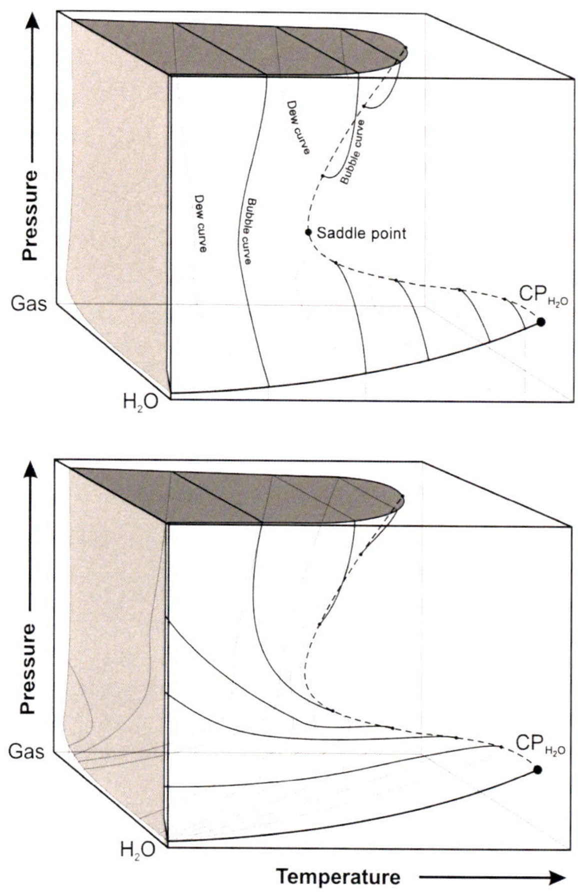

FIGURE 2.12

Schematic phase diagram of a gas–aqueous fluid system with a two-phase immiscibility field visualized by isotherms in the upper diagram and isopleths in the lower diagram. Modified after Tödheide and Franck (1963) and Diamond (2003).

Diamond and Akinfiev, 2003). The longer segment of the critical curve extends from the critical temperature and pressure of pure water (Figure 2.14) through a saddle point at 267 ± 2 °C, 200–245 MPa, and 31–41.5 mol.% CO_2 (Tödheide and Franck, 1963; Takenouchi and Kennedy, 1964; Sterner and Bodnar, 1991) to higher temperatures, pressures, and CO_2 concentrations, for example, 400 °C, 2 GPa, 45 mol.% CO_2 (Kerrick and Jacobs, 1981), up to ~5 GPa (Tödheide, 1982), where it terminates in an upper critical end point on a CO_2-clathrate dissociation surface (Diamond, 2001).

All gas–aqueous mixtures are homogeneous outside the space bounded by dew and bubble surfaces (see Figure 2.12). Inside the two-phase immiscibility space, fluids split into two conjugate carbonic and aqueous phases. An isothermal projection of the immiscibility field of the H_2O–CO_2 system (see Figure 2.14) enables rapid determination of the compositions of the coexisting CO_2-rich gaseous and H_2O-rich liquid phases at any combination of temperature and pressure.

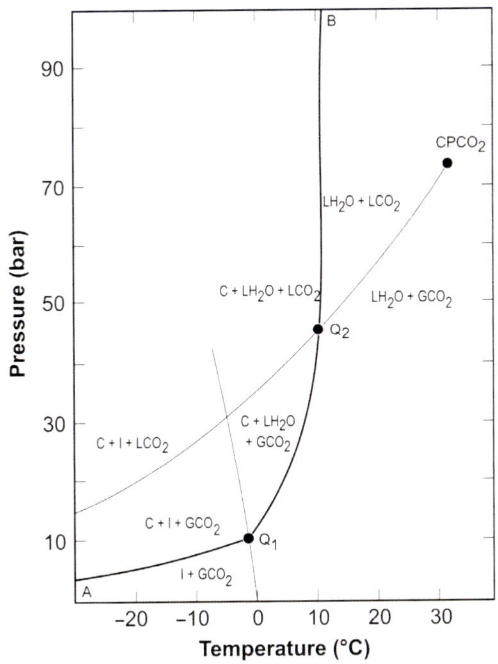

FIGURE 2.13

P–*T* projection of low-temperature portion of the phase diagram of the H_2O–CO_2 system (Collins, 1979). Abbreviations: C, CO_2-clathrate; I, ice. Q_1 and Q_2 are quadruple points.

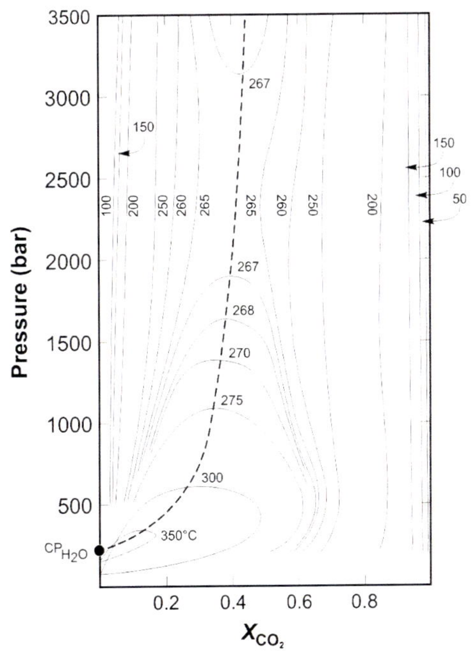

FIGURE 2.14

$P - X_{CO_2}$ (mole fraction) isothermal projection of the two-phase immiscibility surface of the system H_2O–CO_2 (Takenouchi and Kennedy 1964). Reprinted with permission from the American Journal of Science 262, 1055–1074.

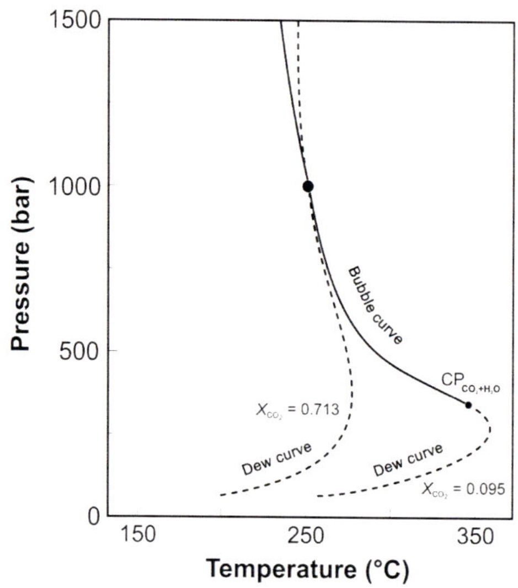

FIGURE 2.15

Isopleths (solvi) of carbonic and aqueous phases coexisting at 250 °C and 1000 bar in the H_2O-CO_2 system, as derived from the experimental data in Takenouchi and Kennedy (1964).

Solvus (dissolution curve) is the intersection of isopleths (constant composition surfaces) with the immiscibility surface. The solvus defines the gas solubility limit in an aqueous solution. The solvi for aqueous solutions with 9.5 and 71.3 mol.% CO_2 are displayed in P–T coordinates (Figure 2.15). These solutions are homogeneous at any P and T to the right of the corresponding solvus; however, they split into CO_2-rich gaseous phase and an aqueous liquid to the left of the solvus. It is noteworthy that the solvus of the 71.3 mol.% CO_2-H_2O fluid does not intersect the critical curve of the H_2O-CO_2 system. Regardless of pressure, homogeneous fluids with such composition possess the character of gas. In contrast, the solvus for 9.5 mol.% CO_2 intersects the critical curve of the system H_2O-CO_2 at a 350 bar pressure, when it passes from the gas field at a lower pressure into the liquid field at a higher pressure, where it intersects the solvus of the gas with 71.3 mol.% at 250 °C and 1000 bar, thus defining a unique combination of compositions of two coexisting phases.

The high-temperature portion of the phase diagram of the H_2O-CH_4 system is similar to that of the H_2O-CO_2 system, except for a less pronounced saddle of the two-phase surface and the inflection point of the critical curve located at 353 °C, 100 MPa, and 26 mol.% CH_4 (Welsch, 1973). The position of the upper critical point remains unknown. The two-phase immiscibility field is expanded to higher temperatures due to even more contrasting physical properties of water and methane and the corresponding very limited solubility of the two components. For instance, the H_2O-CO_2 mixture is miscible at any composition at temperatures above 269 °C and 1 kbar; however, the immiscibility is still possible at 355 °C if the H_2O-CH_4 mixture contains 28 mol.% CH_4 (Figure 2.16).

An additional consequence of the low solubility of CH_4 in H_2O and vice versa is almost pure end-member compositions of the conjugate immiscible phases at temperatures below 200 °C. The CH_4

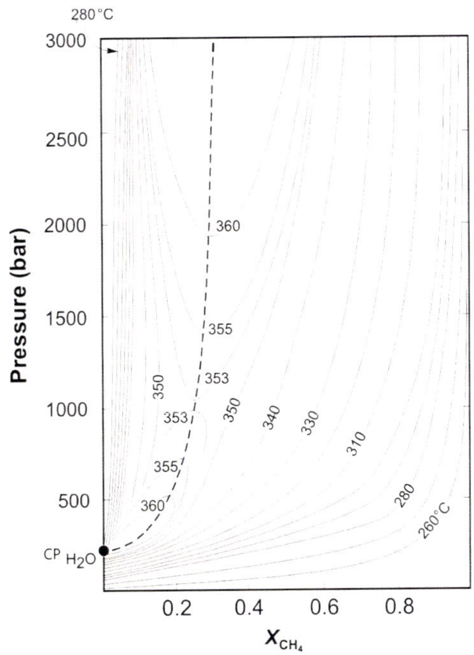

FIGURE 2.16

$P-X_{CH_4}$ (mole fraction) isothermal projection of the two-phase immiscibility surface of the system H_2O-CH_4. Reprinted from Krader, T., Franck, E.U., 1987. The ternary systems H_2O–CH_4–NaCl and H_2O–CH_4–$CaCl_2$ to 800 K and 250 MPa. Ber. Bunsenges. Phys. Chem. 91, 627–634. Copyright Wiley–VCH Verlag GmbH & Co. KGaA. Reproduced with permission.

content in the aqueous phase does not exceed 10 mol.%, and the coexisting CH_4-rich phase is even more genuine, containing less than 2 mol.% H_2O.

A change of the shape of the two-phase surface of the H_2O-CH_4 system after the addition of 8 mass% NaCl is illustrated in a phase diagram in Figure 2.17. Behavior of the system is essentially similar to other water–gas–electrolyte systems: addition of salts drastically decreases the capacity of the aqueous fluid to dissolve gas, thereby resulting in the expansion of the immiscibility field. In our case, the temperature minimum along the immiscibility surface is increased by 130 °C, and is now located at 486 °C and 1300 bar. In contrast to the binary H_2O-CH_4 system, isotherms of the immiscibility surface do not converge in the critical H_2O(+NaCl) point, but they are cut by the three-phase curve, along which the halite precipitates.

The low-temperature portion of the H_2O-CH_4 system differs from that of the H_2O-CO_2 system by the low critical point of pure CH_4, which is located within the CH_4-clathrate stability field. Consequently, the invariant Q2 point is not present and the three-phase CH_4-clathrate + aqueous liquid + CH_4-gas curve extends to very high pressures. Furthermore, the boiling and condensation of pure methane occur at considerably lower pressures than the CH_4-clathrate + CH_4-liquid + CH_4-vapor curve, and the LCEP lies above the critical point of pure CH_4 (Bakker, 1997).

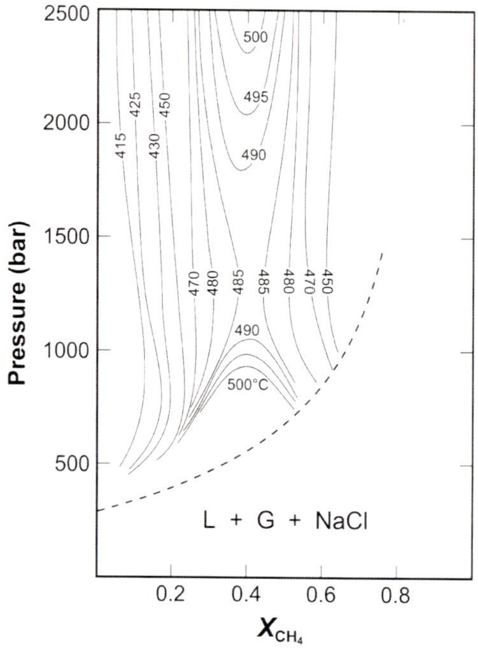

FIGURE 2.17

P–*X* projection of the H_2O–CH_4–8 mass% NaCl system. Reprinted from Krader, T., Franck, E.U., 1987. The ternary systems H_2O–CH_4–NaCl and H_2O–CH_4–$CaCl_2$ to 800 K and 250 MPa. Ber. Bunsenges. Phys. Chem. 91, 627–634. Copyright Wiley-VCH Verlag GmbH & Co. KGaA. Reproduced with permission.

2.4.2 H_2O–CO_2–NaCl AND H_2O–CH_4–NaCl SYSTEMS

The shapes of CO_2 dissolution curves in aqueous solutions with NaCl are similar to those in the H_2O–CO_2 system, but the addition of NaCl drastically reduces the capacity of the aqueous solution to dissolve CO_2. For instance, at a 250 °C temperature and a 500 bar pressure, pure water can dissolve 14.3 mass% CO_2, but only 4.1 mass% CO_2 can be dissolved in a 20 mass% NaCl–H_2O solution (Takenouchi and Kennedy, 1965). In other words, at given *P*–*T* conditions the solutions with the mentioned compositions are homogeneous, but they split into a gaseous phase depleted in NaCl and a CO_2-poor aqueous phase enriched in NaCl after increasing the concentration of one of the two solutes, namely either NaCl or CO_2.

The increasing CO_2 content in an NaCl aqueous solution increases the saturation pressure and also shifts the critical point to higher temperatures and pressures. For instance, the critical temperature of a 6 mass% NaCl–H_2O solution, 430 °C, increases to 480 °C after addition of 9.7 mol.% CO_2 (Roedder, 1984). Expansion of the immiscibility field is a logical consequence of this behavior. Consequently, the H_2O–NaCl–CO_2 solutions can exist in a heterogeneous state even at temperatures in excess of 400 °C and pressures greater than 2 kbar, for example, in metamorphic conditions.

Due to the increased number of variables and degree of freedom of a three-component system, the representation of the topology of the H_2O–CO_2–NaCl system demands an increased imagination, as

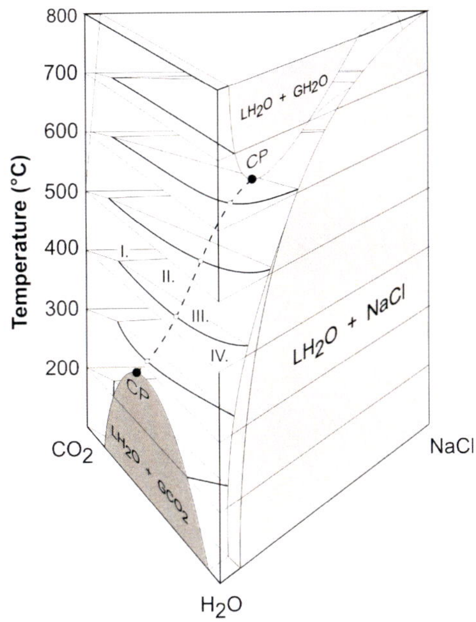

FIGURE 2.18

Phase diagram of the H_2O-CO_2-NaCl system at 1 kbar (modified after Gehrig, 1980; Duan et al., 1995).

seen on a 3D diagram depicted for the pressure of 1 kbar (Figure 2.18). The loop on the left side of the prism on the H_2O-CO_2 plane is the isobaric projection of a two-phase surface from Figure 2.12 and displays the limits of mutual solubility of the two components at 1 kbar. CP is the projection of the critical point of the H_2O-CO_2 system. Two shaded fields of two-phase equilibria occur on the $H_2O-NaCl$ plane. In the upper part, the two-phase L+G field intersects the critical curve of the $H_2O-NaCl$ system at the CP. The NaCl solubility curve occurs in the lower part of the $H_2O-NaCl$ plane. Halite coexists with saturated $H_2O-NaCl$ solutions below this curve, whereas homogeneous undersaturated NaCl solution are stable above this curve.

Isothermal sections of the prism displayed in Figure 2.18 are subdivided into several fields. Halite coexists with a CO_2-rich liquid in field I. Field II represents the region of three-phase equilibrium of NaCl, aqueous liquid (LH_2O), and CO_2 gas (GCO_2). Field III is the continuation of the two-phase LH_2O+GCO_2 equilibrium of the binary H_2O-CO_2 system in the ternary H_2O-CO_2-NaCl system and documents the influence of NaCl on the expansion of the immiscibility field. All components are perfectly miscible within field IV, which represents a homogeneous fluid.

It must be emphasized that binary planes H_2O-CO_2, $H_2O-NaCl$, and combinations of phase equilibria in the fields III and IV are the most representative of the fluids existing in the Earth's crust. Three-phase equilibria are extraordinarily rare and two-phase equilibria such as $NaCl+GCO_2$ do not exist at all. Some geologically relevant isobaric–isothermal cross sections of the $H_2O-NaCl-CO_2$ system are depicted in Figure 2.19.

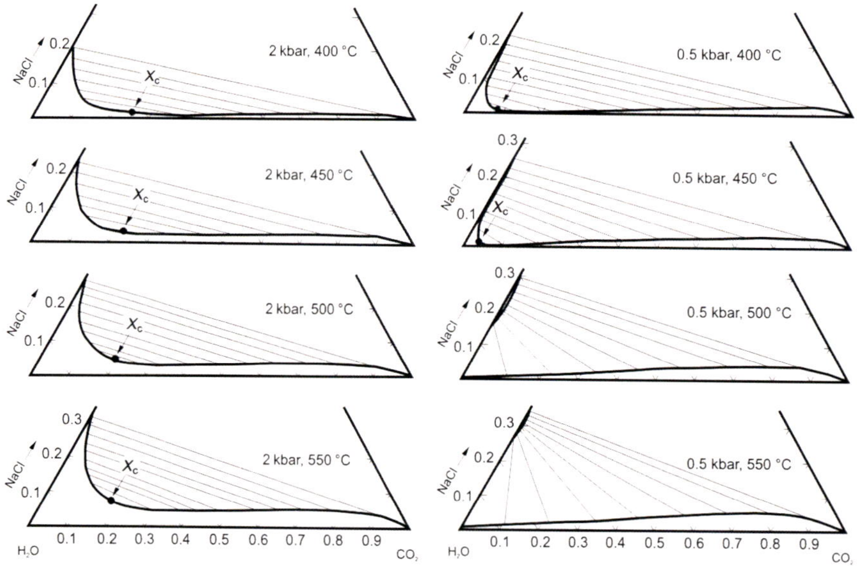

FIGURE 2.19

Isothermal and isobaric sections of the H_2O-CO_2-NaCl system (Bowers and Helgeson, 1983b; Bowers, personal comm.). Tie lines inside the immiscibility field connect compositions of the coexisting NaCl-rich liquid and CO_2-rich gaseous phases. X_c are critical compositions. Reprinted from Bowers, T.S., Helgeson, H.C., 1983. Calculation of the thermodynamic and geochemical consequences of nonideal mixing in the system H_2O–CO_2–NaCl on phase relations in geological systems: equation of state for H_2O–CO_2–NaCl fluids at high pressures and temperatures. Geochim. Cosmochim. Acta 47, 1247–1275, Copyright 1983, with permission from Elsevier.

2.5 GAS MIXTURES

Gas inclusions devoid of water are typical of metamorphic and magmatic minerals formed in the lower crust and upper mantle. Early studies showed essentially pure CO_2 composition of these fluids, but Raman spectroscopy introduced at the end of the 1970s revealed a spectrum of gas mixtures composed of CO_2, CH_4, and N_2 end-members. Other gas species (H_2S, SO_2, CO, H_2) occur occasionally in low amounts, and the fluid properties in most cases may be satisfactorily reproduced using experimental data on binary subsystems CO_2-N_2, CH_4-N_2, and CO_2-CH_4. Gases with three dominant components are rare and can be found mainly in low-grade metamorphic rocks, where they indicate a chemical nonequilibrium (e.g., van den Kerkhof, 1988; van den Kerkhof and Thiéry, 2001; Huizenga, 2005).

Gaseous mixtures identified in granulite-facies rocks are dominated by CO_2 (Touret 1971a,b), whereas those from eclogites are N_2-rich (Andersen et al., 1989, 1995; Stevens and Clemens, 1993; Janák et al., 1996). In the lower crustal lithologies, N_2-rich fluids dominate in metabasite migmatites, whereas carbon dioxide prevails in metapelite migmatites (Janák et al., 1999; Hurai et al., 2000). CO_2 originates by decarboxylation (oxidation of graphite and organic matter) in metapelitic rocks, decarbonatization reactions in calc-silicate rocks, or devolatilization of upper mantle or mantle-derived mafic intrusions. Methane is liberated mostly by retrograde metamorphic reactions at a lower partial pressure of oxygen, which radically decreases with decreasing temperature. Diagenesis and prograde anchimetamorphic reactions take place in reducing conditions and low partial pressure of oxygen, thus resulting in the liberation of methane and higher gaseous hydrocarbons by decomposition of kerogen and bitumens, and/or thermal cracking of petroleum at temperatures of >200 °C. Ammonium-rich micas crystallize at anchimetamorphic conditions, and the NH_3^+ ion can also isomorphically substitute potassium in other minerals (feldspars). Partial melting of micas and feldspars in amphibolite facies leads to massive liberation of nitrogen, which is trapped in newly formed minerals.

Carbon dioxide is the dominant volatile component in mafic intrusive rocks derived from the mantle. Carbon monoxide that is sometimes present in minor amounts together with the dominant CO_2 in fluid inclusions in mantle peridotites reflects a lower oxygen fugacity (Bergman and Dubessy, 1984). Rare nitrogen- (Andersen et al., 1995) and water-rich (Kumagai et al., 2014) inclusions in mantle peridotites are thought to be attributed to contamination from crustal fluids, or exsolved from sub-arc basaltic magmas formed by partial melting of mantle wedge triggered by slab-derived fluids. On the basis of abundant OH-bearing daughter minerals in CO_2-rich inclusions, Frezzotti et al. (2012a) argued for a much larger amount of H_2O in the shallow mantle, not only above subduction zones, but in intraplate and extensional settings as well. Small amounts of hydrogen sulfide, hydrogen, carbon monoxide, and methane have been found together with the dominant CO_2 in late-stage fractionation products of alkali basalts and the partial melts (anhydrous orthopyroxene granite) derived from the lower continental crust as a consequence of basalt underplating (Huraiová et al., 2005).

Strongly reducing fluids dominated by methane and higher gaseous hydrocarbons are typical of differentiated peralkaline intrusive rocks (Krumrei et al., 2007). The origin of the hydrocarbons is abiogenic, most likely connected with the Fischer–Tropsch synthesis during post-magmatic alteration assisted by Fe-oxides and silicates (Potter and Konnerup-Madsen, 2003).

Minerals from ancient natural nuclear reactors in Gabon and Canada contain appreciable amounts of oxygen and hydrogen originated by the radiation-induced hydrolysis of water (Dubessy et al., 1988). These occurrences are exceptional, because the oxygen concentrations in natural fluids are far below the detection limits of Raman spectrometry.

The topology of unary gaseous systems is essentially the same, except for unique conditions in triple and critical points diagnostic of each system. The topology of the $CO_2-CH_4-N_2$ system is influenced by the relatively high temperatures at the triple and critical points of CO_2 compared to methane and nitrogen. As a consequence, the topology along the CO_2-CH_4 join is basically identical to that of the $H_2O-NaCl$ system, where the three-phase curve does not intersect the critical curve of the system, whereas the critical curve of the CO_2-N_2 subsystem is divided by the three-phase curve into two separated parts (Figure 2.20). In contrast to the $H_2O-NaCl$ system, however, gas mixtures are fully miscible at any ratios, and they do not react with each other. Consequently, a larger variety of phase transformation occurs in gaseous systems, also involving the sublimation of solid into vapor. In contrast, sodium chloride is preferably concentrated in aqueous liquid of the $NaCl-H_2O$ system and except for the conditions near the critical curve, its concentration in vapor never exceeds 3 mass% (Bodnar et al., 1985a; Pitzer and Pabalan, 1986). Hence, natural fluid inclusions with halite do not homogenize by sublimation of the halite into the vapor phase, though this type of phase transformation is common in low-density CO_2-CH_4 inclusions.

Detailed analysis of binary subsystems of the $CO_2-CH_4-N_2$ system was done by Arai et al. (1971), Swanenberg (1979), Burruss (1981c), van den Kerkhof (1988), and others. Owing to the low temperatures and pressures of phase transitions, the binary systems are projected in $\overline{V}-X$ coordinates. The

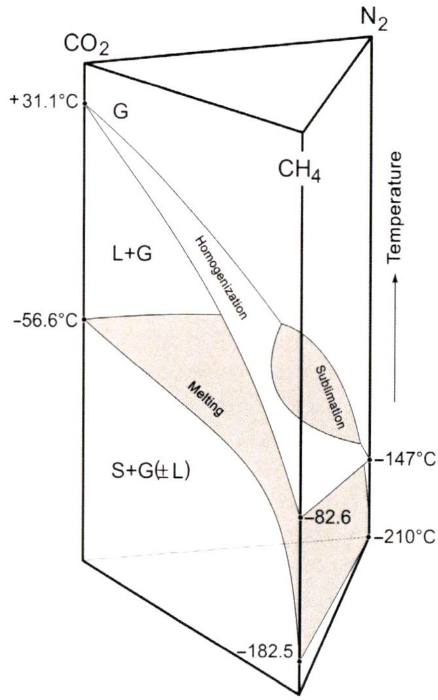

FIGURE 2.20

Schematic phase diagram of the CO_2–CH_4–N_2 system. The two-phase surface connecting critical points, and the three-phase surface connecting the triple points of pure end-members are shown. Reprinted from van den Kerkhof, A.M., 1990. Isochoric phase diagrams in the systems CO_2–CH_4 and CO_2–N_2: Application to fluid inclusions. Geochim. Cosmochim. Acta 54, 621–629, Copyright 1990, with permission from Elsevier.

$\overline{V}-X$ diagram for the system CO_2–CH_4 was introduced by Burruss (1981c) and later refined by Thiéry et al. (1994a). Similar diagrams published by van den Kerkhof and Thiéry (2001) appeared to be very practical, because they permit molar volumes and compositions to be derived simultaneously by plotting fluid inclusion microthermometry data. Bakker and Diamond (2000) constructed the $\overline{V}-X$ diagram also for the H_2O–CO_2 system.

Three-dimensional $\overline{V}-X$ projections (Figure 2.21) contain three-phase univariant curves, and two-phase divariant surfaces. Isothermal curves are projections of two-phase surfaces LS, LG, or GS separated by three-phase curves LGS and GLS. The three-phase curve of the CO_2–CH_4 system connects the triple points of CH_4 and CO_2 and has a pressure maximum at −69 °C and 49 bar, where the gas phase with ~89 mol.% CH_4 coexists with the liquid phase containing ~76 mol.% CH_4. The critical curve (L=G) connects the critical temperature and pressure of pure CO_2 and CH_4. Composition of the coexisting liquid, vapor, and solid phases along the three-phase curve is defined by a triangle, whose apices connect molar volumes and compositions of the coexistent L (CH_4+CO_2 mixture), G (CH_4+CO_2 mixture), and S (essentially pure CO_2) phases.

The critical curve of the CO_2–N_2 system is divided into two segments by a three-phase curve, along which the pressure is approximately five times greater than that along a three-phase curve of the

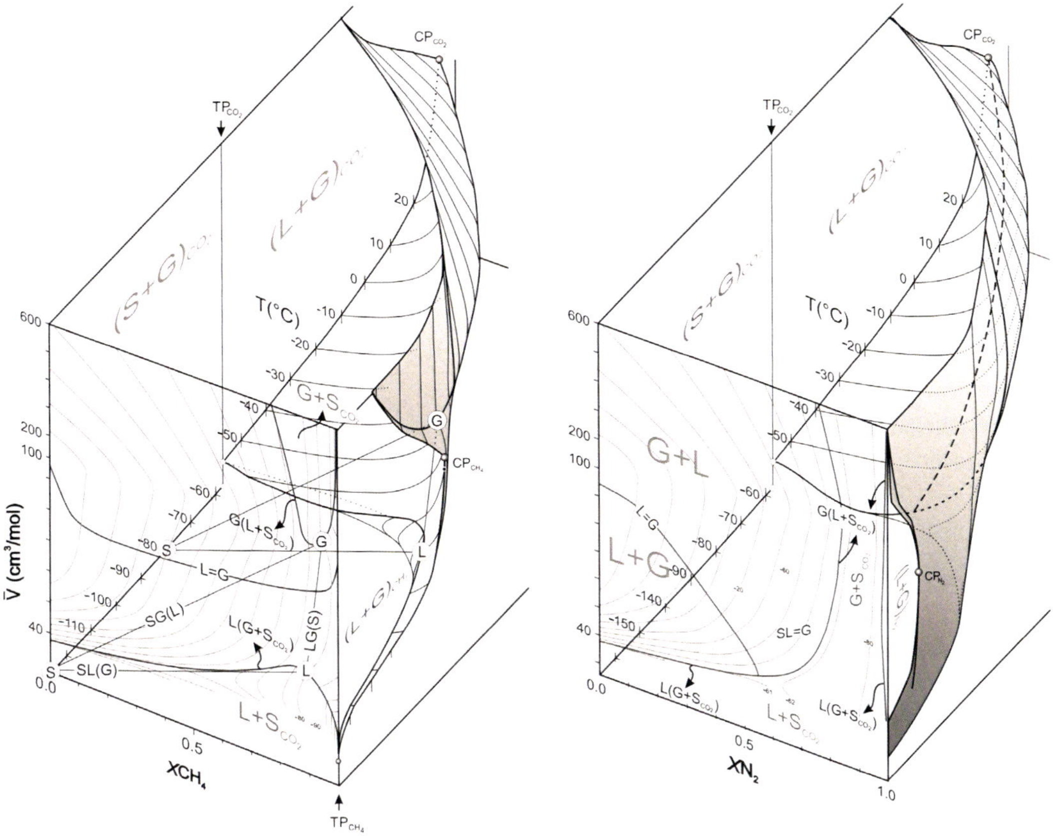

FIGURE 2.21

Three-dimensional $T-\overline{V}-X$ diagrams of the system CO_2–CH_4 (left) and CO_2–N_2 (right) constructed from isothermal projections (Thiéry et al., 1994a; van den Kerkhof and Thiéry, 2001). L = G is the critical curve. See text for more detailed explanation.

CO_2–CH_4 system. The lower intersection is located very near to the critical point of pure nitrogen, and this part of the LGS curve is in the $\overline{V}-X$ coordinates almost identical with the two-phase curve of pure N_2. The upper part of the three-phase curve extends from 235 bar, −61.2 °C, and $X_{N_2}=0.47$ (van den Kerkhof, 1990). In these conditions, critical homogenization and melting of CO_2 solid occur simultaneously. The CO_2 solid coexists with a supercritical CO_2–N_2 fluid between −61.2 and −147 °C, and sublimation of the solid in liquid or vapor occurs as a final phase transition.

The CH_4–N_2 system is rare in nature. Phase changes occur at temperatures below that of the critical point of pure methane (−82.6 °C). A limited miscibility of solid methane and solid nitrogen is possible, in contrast to the two previous systems. The mixed CH_4–N_2 crystals may contain up to 45 mol.% N_2. The CH_4-dominant crystals are cubic, whereas the N_2-rich crystals are hexagonal (van den Kerkhof, 1988). Vapor and liquid phases are miscible at any ratios. Three-phase equilibria occur at very low temperatures, between −211 and −183 °C. As a consequence, the CH_4–N_2 mixtures trapped in fluid inclusions exhibit only two-phase L+G equilibria, when cooled using liquid nitrogen (−196 °C), and the low-temperature equilibria can be observed only in freezing–heating stages cooled with liquid helium. The three-phase curve has a pressure maximum at −191 °C and 0.68 bar. The critical curve exhibits the maximum at −105 °C, 50 bar and ~44 mol.% N_2. The quadruple point is located at −210.6 °C and $X_{N_2} = 0.762$. Hence, temperatures of melting of the mixtures with 76–100 mol.% N_2 are essentially constant.

2.6 HYDROCARBONS

Fluid inclusions with higher hydrocarbons are typical of alkaline igneous intrusions and low-grade sedimentary rocks. With dominant gas, Mullis (1987) distinguished several zones in meta-sediments. The zone with higher hydrocarbons occurs in nonmetamorphosed and diagenetically modified sediments. The low- and medium-grade anchizone is typical of increased methane concentrations, whereas water and CO_2 tend to dominate with the increasing metamorphic grade. The composition of fluids correlates with vitrinite reflectance and illite crystallinity in country rocks and is buffered by the low oxygen fugacity and organic compounds. Generally, higher gaseous hydrocarbons (wet gas zone) only rarely occur in the sediments heated above 200 °C, whereas the methane (dry gas) zone is limited to temperatures between 200 and 270 °C. This zonation, observed in regionally metamorphosed nappes of the Alps may differ from that of sedimentary basins affected by focused, short-term heat flow associated with fluid fluxes (O'Reilly and Parnell, 1999). Exceptions can also be caused by a specific lithology. For instance, oil inclusions were observed at higher temperatures within the dry gas zone in proximity of bituminous sedimentary layers producing excess hydrocarbons on heating (Hurai et al., 2002b). In contrast, higher hydrocarbons are absent in carbon-deficient sandstones and clastic sediments even at lower temperatures.

The oil/gas ratio in hydrocarbon reservoirs decreases with temperature above 150 °C, when thermogenic methane originates by cracking of oil and/or kerogen. Essentially pure methane (thermogenic dry gas), with minor admixture of higher gaseous C_1–C_4 hydrocarbons (<5 mass%) is typically present at temperatures exceeding 200 °C. Mango and Hightower (1997) pointed out that composition of the CH_4-dominated dry gas does not fit that produced by experimental pyrolysis, which is considerably enriched in the C1–C4 hydrocarbons even at 200 °C. Obviously, apart from the temperature, the oil-to-gas ratio and the composition of natural hydrocarbons also depends on other factors, including the presence of metallic catalysts in country rocks.

Natural hydrocarbons are complex compounds, whose composition reflects the chemical diversity of the biosphere and variability of the hydrocarbon-producing processes. They can be subdivided into dry gas, gas-condensate, volatile oil and black oil, and their mixtures (Burruss, 1992; Munz, 2001). Some typical physical properties of hydrocarbon mixtures are listed in Table 2.2.

Main petroleum constituents can be subdivided into (Munz, 2001):

1. Paraffins (15–60%)—normal and branched alkanes with simple interatomic bonds and total number of carbon atoms between 20 and 40. Melting points vary between 50 and 90 °C. Alkanes are called saturated hydrocarbons, because all atoms are connected with single bonds.
2. Cycloalkanes—naphthenes (30–60%)—cyclic aliphatic hydrocarbons with the general formula C_nH_{2n}, containing one or more rings of carbon atoms in their structures. Their melting and boiling

Table 2.2 Physical Properties and Compositions (mol.%) of Some Typical Hydrocarbon Mixtures (Burruss, 1992)

Compound	Black Oil	Volatile Oil	Gas Condensate	Dry Gas
C1	48.9	64.4	87.1	86.7
C2	2.8	7.5	4.4	7.8
C3	1.9	4.7	2.3	3.0
C4	1.6	4.1	1.7	1.7
C5	1.2	3.0	0.8	0.9
C6	1.6	1.4	0.6	
C7+	42.2	14.9	3.8	
Density (°API)	34	50	61	
T_h (°C)	82	81	12	−51
State	Liquid	Liquid	Gas/liquid	Gas/liquid
Properties of the C7+ fraction				
Molecular weight	225	181	112	
Average C-number	16	13	8	
Boiling point at 1 atm	277 °C	227 °C	107 °C	
Density (g/cm^3)	0.84	0.80	0.75	

temperatures are higher than those of alkanes. The naphthenes are also saturated compounds, which contain only single bonds.

3. Aromatic hydrocarbons (3–30%) with alternating double and single bonds between carbon atoms forming benzene rings. The aromatic hydrocarbons can be monocyclic or polycyclic. The simplest aromatic compound is benzene, C_6H_6, consisting of a ring structure with six carbon atoms. The most aromatic and heaviest compounds of petroleum are asphaltenes.
4. Polar compounds containing sulfur, nitrogen, and oxygen.
5. Inorganic gases (N_2, CO_2, H_2S).
6. Up to 13 mass% water can be dissolved in hydrocarbons at temperatures up to 300 °C (Pironon et al., 2000).

Although there is a considerable variation between the ratios of organic molecules, the elemental bulk composition of petroleum is relatively constant: C (83–87%), H (10–14%), N (0.1–2%), O (0.05–1.5%).

Due to their chemical complexity, phase changes of hydrocarbon inclusions are more complicated than those in inorganic fluids. Quantitative characterization of composition and density based merely on temperatures of phase transitions is impossible, and the contribution of additional destructive (e.g., gas chromatography) and nondestructive analytical methods (UV-fluorescence, UV-Raman, microthermometry) is essential.

Methane is the simplest hydrocarbon molecule. Due to its low critical temperature, it occurs in the Earth's crust only as a homogeneous, supercritical (gas- or liquid-like) fluid. Increasing the number of carbon atoms leads to increasing critical temperature and decreasing critical pressure of more

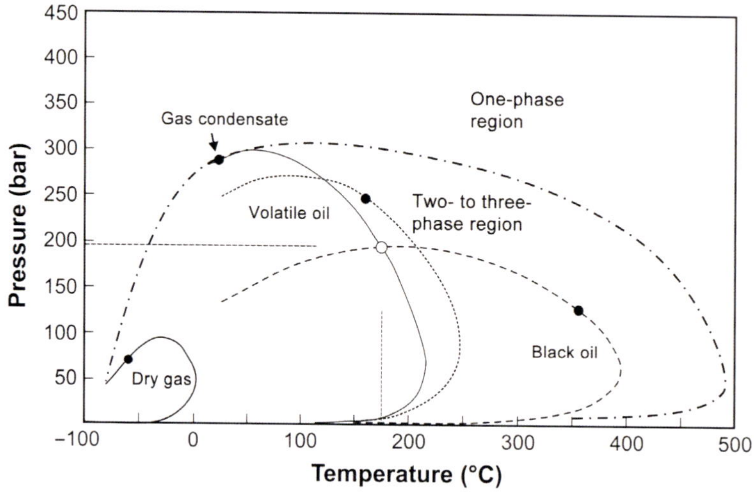

FIGURE 2.22

Two-phase immiscibility isoplethal surfaces for various petroleum reservoirs (Munz, 2001). Black circles represent critical points, dividing the bubble curve from the dew curve of the corresponding immiscibility loop. The dot-dashed line represents the maximum immiscibility limit, outside which hydrocarbons of any composition should be homogeneous at any pressure and temperature. The maximum immiscibility limit was calculated by the FIT-OIL freeware for reference oil compositions defined by the $\alpha-\beta$ model (Thiéry et al., 2002) and calculated using the equation of state of Peng and Robinson (1976). Open circle refers to $P-T$ conditions at which black oil and vaporlike gas condensate may coexist as two conjugate immiscible phases. Solid circles denote critical points for the unique hydrocarbon compositions.

complex hydrocarbon molecules (see Table 2.1). The complex chemistry of petroleum results in a complex phase behavior at low temperatures due to the infinite degree of freedom. At higher temperatures, when only two to three phases are present, the phase diagram of petroleum can be compared to that of aqueous solutions. For the interpretation of fluid inclusion data, two-phase curves, dividing the one-phase and liquid–vapor regions of individual petroleum types are important (Figure 2.22).

Note that the maximum temperature immiscibility limit (*cricondentherm*) is defined by the dew curve of black oil, whereas the maximum pressure limit (*cricondenbar*) is defined by the dew curve of gas condensate. The maximum temperature and pressure at which hydrocarbons may exist as heterogeneous fluid correspond to ~490 °C and 300 bar, respectively. Inside the isoplethal immiscibility loops, the hydrocarbon of a given composition splits into two or more coexisting phases with different densities. For instance, within the immiscibility region, the gas-condensate can coexist with black oil liquid at about 175 °C and 190 bar. Due to complex oil compositions with thousands of components, multiphase equilibria, such as liquid–liquid–gas, or liquid–liquid can also be possible within the immiscibility field, providing the phase density contrasts are sufficiently large. This behavior differs from that in binary fluid systems (e.g., $H_2O-NaCl$, H_2O-CO_2), where unique combinations of vapor and liquid, or gas-rich liquid and H_2O-rich liquid phases can exist at any $P-T$ point within the two-phase field.

2.7 SILICATE MELT

With 10 major oxides plus water as the most important volatile species, the complete phase diagram of a melt–H_2O system would have 13 dimensions. However, the composition of major oxides and some volatile species (fluorine, chlorine) can be easily determined using an electron microprobe. Hence, unlike for aqueous and gaseous systems, measurements of temperatures of phase transitions in multiphase melt inclusions have no practical meaning for the determination of the bulk system composition. However, the phase diagram of a silicate melt system saturated with a volatile component can be employed for the estimation of formation P–T conditions, because the water concentration in volcanic glass or melt inclusions can be directly measured using microspectrometric methods (Raman, SIMS, FT-IR).

The granite–H_2O system (Figure 2.23) represents the reference system for melt inclusion studies. The haplogranite with equal proportion of Na_2O and K_2O is chosen as the representative bulk

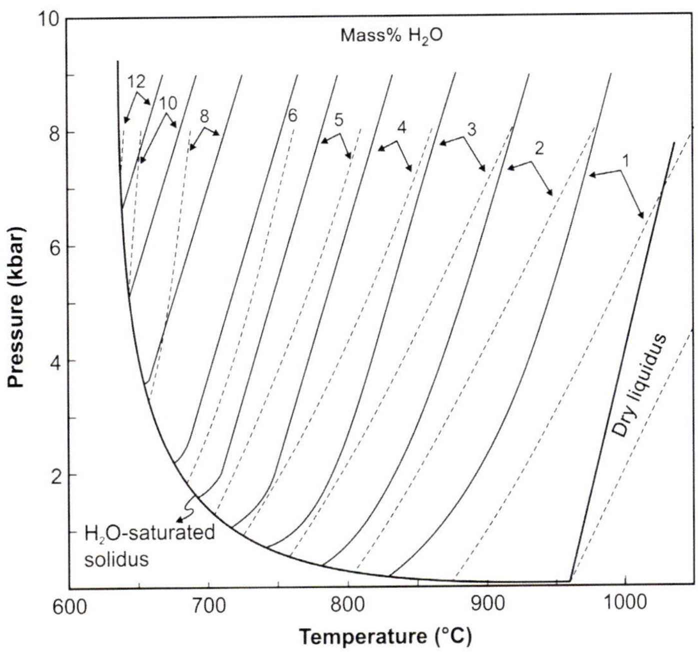

FIGURE 2.23

H_2O-saturated solidus in the system quartz–albite–orthoclase–H_2O with liquidi for given amounts of H_2O (solid lines) (Holtz and Johannes, 1994), and for natural subaluminous granitic and rhyolitic systems with eutectic compositions (Holtz et al., 2001). The system is fully crystallized at solidus. In contrast, no solid and vapor phases are present in liquidus. Reprinted from Holtz, F., Johannes, W., 1994. Maximum and minimum water contents of granitic melts: implications for chemical and physical properties of ascending magmas. Lithos 32, 149–159, Copyright 1994, and Holtz, F., Johannes, W., Tamic, N., Behrens, H., 2001. Maximum and minimum water contents of granitic melts generated in the crust: a reevaluation and implications. Lithos 56, 1–14, Copyright 2001, with permission from Elsevier.

composition for the granitic systems. The phase diagram is composed of the water-saturated solidus curve, along which the melt is fully crystallized. The solidus corresponds to eutectic curves in multi-component salt-water systems. All solids disappear along liquidus, which marks the *P–T* conditions at which neither gaseous nor solid phases are present in the system. Slopes of the haplogranite liquidi are always positive and constrain the minimum *P–T* conditions, at which the water-saturated melt with the given H_2O concentration becomes a single liquid phase. Positions of liquidi are strongly influenced by the amount of water dissolved.

CHAPTER

EQUATIONS OF STATE

3

CHAPTER OUTLINE

An equation expressing volume as a function of pressure and temperature is called the equation of state (EOS). Modern equations of state evolved from experimental studies that were carried out over centuries and resulted in several laws that represent important generalizations of macroscopic behavior of fluid systems. The series of gas laws involves Boyle's, Charles's, Gay-Lussac's, and Avogadro's laws, as well as the definition of the molar gas constant. The history of these discoveries with many practical examples is described in detail in chemistry textbooks (e.g., Moore, 1979; Chang, 1991).

3.1 BEHAVIOR OF IDEAL GAS

In the seventeenth century, the British chemist and philosopher Robert Boyle (1627–1691) investigated the pressure–volume relationships of gas and found out that as long as the temperature was held constant, the volume V of a fixed amount of gas was inversely proportional to the gas pressure P. In other words, the product of P and V was always constant. Introducing the proportionality constant k_1, Boyle's law can be written as

$$PV = k_1 \quad \text{or} \quad P = k_1 \times 1/V \tag{3.1}$$

For a given sample of gas under two different sets of conditions, we can write

$$P_1V_1 = P_2V_2 \tag{3.2}$$

where V_1 and V_2 are the volumes at pressures P_1 and P_2, respectively. Plotting P versus $1/V$ takes the form of a linear equation $y = ax + b$, where $a = k_1$ and $b = 0$.

The dependence of gas volume on temperature is given by

Geofluids. http://dx.doi.org/10.1016/B978-0-12-803241-1.00003-4

$$V/T = k_2 \tag{3.3}$$

and the equation is known as Charles and Gay-Lussac's law. Similarly to Equation (3.2), two different sets of conditions yield the relation

$$V_1/T_1 = V_2/T_2 \tag{3.4}$$

where V_1 and V_2 are the volumes of gases at temperatures T_1 and T_2, respectively.

The work of the Italian physicist and chemist Amedeo Avogadro (1776–1856) completed the studies of previous researchers. In 1811, he published the hypothesis that at the same P and T, equal volumes of different gases contain the same amount of molecules

$$V = nk_3 \tag{3.5}$$

where n represents the number of moles. In this way, he defined the Avogadro's number as the number of molecules in 1 mol of substance.

Equations (3.2) and (3.4) can now be combined into a master equation describing the behavior of gases

$$P\overline{V} = \frac{P_0\overline{V}_0 T}{T_0} = RT \tag{3.6}$$

The subscript 0 denotes standard conditions at 1 atm and 0 °C = 273.15 K, often abbreviated as Standard Temperature and Pressure (STP), when 1 mol of any gas occupies the volume of 22.41387 L. R is the proportionality value called *gas constant*.

The absolute value of the gas constant can be inferred by the substitution of properties of an ideal gas into Equation (3.6)

$$R = \frac{P_0\overline{V}_0}{T_0} = \frac{101{,}325\ \text{Pa} \times 22.41387 \times 10^{-3}\,\text{m}^3}{273.15\text{K}} = 8.314426\ \text{J}\,\text{mol}^{-1}\,\text{K}^{-1}$$

Using the pressure unit of 1 atm and liter as the volume unit, the numerical value of the gas constant is equal to 0.082057 L atm $\text{mol}^{-1}\ \text{K}^{-1}$.

3.2 EQUATIONS OF STATE FOR GEOFLUIDS

Generally, the EOSs for high temperatures and pressures describe properties of homogeneous, supercritical fluids, whereas those for low temperatures and pressures are mostly empirical equations describing the properties of coexisting phases in heterogeneous fluids. The EOSs for homogeneous fluids are used to calculate isochores of fluid inclusions from input parameters (density, composition) obtained by microthermometry, which means from the temperatures of phase transitions.

The simplest EOS valid for 1 mol of ideal gas is represented by Equation (3.6). For an optional amount of gas, the equation must be multiplied by the number of moles, and the state function of an ideal gas takes the final form

$$P\overline{V} = nRT \tag{3.7}$$

The ideal gas is a hypothetical gas, whose molecules do not attract or repel one another, and their volume is negligible compared to a gas container. Real gases can approach the ideality only at low temperatures and pressures (<5 atm). The repulsive forces of gas molecules tend to increase with the increasing temperature. With the increasing pressure, density of gas also increases, the molecules are closer to one another, and the intermolecular forces become significant to affect the motion of the molecules. In addition, the volume of real gas molecules also becomes a significant fraction of the total volume, thus causing deviations from the ideal gas behavior. For this reason, the Dutch physicist Johannes van der Waals (1837–1923) modified in 1873 the gas equation by involving corrections to the volume of gas molecules and their interactions

$$\left(P + a/\overline{V}^2\right)\left(\overline{V} - b\right) = RT \tag{3.8}$$

The term $\overline{V} - b$ is referred to as "free volume," namely the free space where molecules can move around. The parameter a is an expression of how strongly molecules of a given gas attract one another. The parameter b is linked with the volume of the gas molecules and their repulsive forces. There is a rough but not straightforward correlation between b and molecular size. Both constants are unique for each gas molecule, and are independent of temperature and pressure. Physical meaning of the equation can be read as follows (Cemič, 2005): external pressure P and attraction between molecules $\left(a/\overline{V}^2\right)$ act in the same direction, pushing molecules together. At equilibrium, this pressure is balanced by the thermal pressure $RT/\left(\overline{V} - b\right)$, holding the molecules apart. Hence

$$P + a/\overline{V}^2 = \frac{RT}{\overline{V} - b} \tag{3.9}$$

The introduction of two correction factors transformed the simple linear EOS of ideal gas into the cubic form (in $\overline{V}$).

The van der Waals equation satisfactorily describes the behavior of real gases at middle pressures. Inconsistencies occur at higher pressures, where repulsive forces prevail over attractive ones. The a, b parameters and gas constant R in the van der Waals equation and critical parameters (P_c, $\overline{V}_c$, T_c) of a given gas are directly correlated with the following equations:

$$T_c = 8a/27bR, \ \overline{V}_c = 3b, \ P_c = a/27b^2 \tag{3.10}$$

derived by solving Equation (3.9) together with its first and second derivatives (for details, see Anderson, 2005).

The EOS modified by Redlich and Kwong (1949), abbreviated hereafter as MRK, became the most popular version of the van der Waals equation. The equation was extended by empirical P- and T-dependent parameters, which significantly improve the ability of the equation to reproduce fluid parameters at higher temperatures and pressures. It retains the first term of the van der Waals equation, but modifies the second term

$$\left[P + \frac{a}{\overline{V}\left(\overline{V} + b\right)(T)^{0.5}}\right]\left(\overline{V} - b\right) = RT \tag{3.11}$$

Various adjustments of a and b parameters permit the MRK to be used universally for pure gases and their mixtures, as well as for H_2O–CO_2–NaCl fluids (Appendix II).

The a and b parameters have been originally defined as constants calculated from critical conditions and parameters. Soave (1972) and de Santis et al. (1974) were the first to derive correlation between the parameter a and temperature, leaving the parameter b fixed. Later, Holloway (1977, 1981) defined an empirical polynomial fit to the values calculated by de Santis et al. (1974) and rearranged the parameters a and b for pure H_2O. Kerrick and Jacobs (1981) expanded the MRK for H_2O and CO_2 in the supercritical region at higher PT and derived an EOS for H_2O–CO_2 mixtures. Further improvements for pure H_2O in sub- and supercritical regions were done by Halbach and Chatterjee (1982). Recently, the modifications of the MRK by Soave (1972) and Thiéry et al. (1994b) have been successfully applied to phase equilibria modeling in the CO_2–CH_4–N_2 system, and to construction of $\overline{V}-X$ diagrams of binary subsystems (Burruss, 1981c; van den Kerkhof and Thiéry, 1994, 2001).

The Peng–Robinson EOS (Peng and Robinson, 1976), abbreviated as PR EOS, is specifically focused on natural hydrocarbon gas and petroleum systems, and it has become the most popular in the oil industry. The EOS is similar to that of the Redlich–Kwong equation modified by Soave (SRK EOS), except for a slightly better performance of the PR EOS around the critical point, making this EOS better suited for gas/condensate systems. The PR EOS has the following form:

$$\left[P+\frac{a}{\overline{V}\left(\overline{V}+b\right)+b\left(\overline{V}-b\right)}\right]\left(\overline{V}-b\right)=RT \tag{3.12}$$

Peng and Robinson conserved the temperature dependency of the attractive term and the acentric factor introduced by Soave. However, they presented different fitting parameters to describe this dependency (α), and coefficients a and b were made functions of the critical properties by imposing the criticality conditions.

Natural aqueous fluid systems always contain dissolved salts. Bowers and Helgeson (1983a) implemented experimental data of Gehrig (1980) and rearranged the MRK for the description of phase properties in the H_2O–CO_2–NaCl system, including the two-phase liquid–vapor region. Bakker (1999) modified the equations of Bowers and Helgeson and adapted them to more complex fluid systems, involving H_2O, CO_2, NaCl, CH_4, and N_2.

Generally, the RK EOS is suitable for calculating gas phase properties (particularly fugacities), but not for liquids, when a correlation for the behavior of liquid phase must be implemented. The SRK EOS and the PR EOS require more parameters, although they serve a similar purpose to the RK EOS. The SRK EOS makes better predictions for polar systems. In contrast, the PR EOS obtains better liquid densities for nonpolar systems and is better tailored to hydrocarbon gas and condensate systems.

Empirical virial equations are additional types of EOSs used for the description of properties of real gases

$$\frac{P\overline{V}}{RT}=1+\frac{B(T)}{\overline{V}}+\frac{C(T)}{(\overline{V})^2}+\frac{D(T)}{(\overline{V})^3}+\cdots \tag{3.13}$$

where B, C, and D are parameters dependent on temperature and gas composition called the second, third, fourth, etc. virial coefficients. The equations are suitable for pure gases or their mixtures at high pressures or temperatures close to their condensation temperatures. The main disadvantage of the virial equations is the large number of constants, which must be fixed for each gas component. Their wider applicability for natural fluid systems is hindered by the lack of experimental data and problems with the modeling near-critical conditions.

Simplified functions of the Helmholtz energy are the last type of EOS (e.g., Haar et al., 1984; Setzmann and Wagner, 1991; Span and Wagner, 1996). This type of state function involves a set of differential equations correlating pressure, Gibbs free-energy, and isobaric heat capacity with density and temperature. The equations are extraordinary in their precise descriptions of properties of pure gases; however, their applicability to natural systems is limited by the lack of mathematical extensions for multicomponent gases.

The virial EOSs with polynomial formulation cannot be extrapolated beyond their applicability range. In contrast, the MRK employing molecular sizes and interaction *a*, *b* potentials (e.g., Shmonov and Shmulovich, 1974; Holloway, 1977, 1981; Saxena and Fei, 1987), as well as those formulated on the basis of statistical physics using the molecular dynamics and Monte Carlo simulation, can be extrapolated and used to calculate the volumes of literally all natural gases and their polycomponent mixtures (Shi and Saxena, 1992; Belonoshko et al., 1992).

These equations, however, do not satisfactorily reproduce the experimental data on water-rich fluids, which are in fact the most abundant fluids in the Earth's crust. Thus, the volumes of aqueous fluids with dissolved salts must be described using empirical fits to experimental data. Recently, Anderko and Pitzer (1993), Economou et al. (1995), Jiang and Pitzer (1996), and Liu et al. (1999) developed EOSs for calculating the properties of aqueous solutions with electrolytes based on the perturbation theory. The mathematical formulation of these EOSs is complex and stands beyond the scope of this textbook. The EOS of Anderko and Pitzer became more frequently employed in the study of fluid inclusions. It has been fitted to extensive experimental data and satisfactorily reproduces the vapor–liquid equilibria, fluid volumes, and NaCl solubilities in temperature range between 400 and 927 °C, and pressures up to 5 kbar.

PROBLEM 3.1 CALCULATING PRESSURE OF A MIXED-VOLATILE FLUID

Calculate pressure (bar) in a CO_2–CH_4–N_2 fluid with the molar volume of 50 $cm^3\ mol^{-1}$ at 400 °C. Composition of the fluid corresponds to $X_{CO_2} = 0.2,\ X_{CH_4} = 0.2,\ X_{N_2} = 0.6$.

Solution

Calculate the *b*-term of the CO_2–CH_4–N_2 mixture using Equations (II.28)–(II.30) in Appendix II

$$b = 29.7 \times 0.2 + 29.703 \times 0.2 + 26.8 \times 0.6 = 27.9606$$

Calculate the *a*-term for CO_2 in the CO_2–CH_4–N_2 mixture

$$a_{CO_2-i} = 46{,}000{,}000 \times 0.2 \times 0.2 + (46{,}000{,}000 \times 31{,}590{,}000)^{0.5} \times 0.2 \times 0.2 + (46{,}000{,}000 \times 15{,}382{,}000)^{0.5} \times 0.2 \times 0.6 = 6{,}556{,}830$$

Calculate the *a*-term for CH_4 in the CO_2–CH_4–N_2 mixture

$$a_{CH_4-i} = 31{,}590{,}000 \times 0.2 \times 0.2 + (31{,}590{,}000 \times 46{,}000{,}000)^{0.5} \times 0.2 \times 0.2 + (31{,}590{,}000 \times 15{,}382{,}000)^{0.5} \times 0.2 \times 0.6 = 5{,}433{,}627$$

Calculate the *a*-term for N_2 in the CO_2–CH_4–N_2 mixture

$$a_{N_2-i} = 15{,}382{,}000 \times 0.6 \times 0.6 + (15{,}382{,}000 \times 46{,}000{,}000)^{0.5} \times 0.6 \times 0.2 + (15{,}382{,}000 \times 31{,}590{,}000)^{0.5} \times 0.6 \times 0.2 = 11{,}374{,}771$$

Continued

PROBLEM 3.1 CALCULATING PRESSURE OF A MIXED-VOLATILE FLUID—CONT'D

Calculate the a-term for the CO_2–CH_4–N_2 mixture

$$a = a_{CO_2-i} + a_{CH_4-i} + a_{N_2-i} = 6,556,830 + 5,433,627 + 11,374,771 = 23,365,228$$

Because parameters a_i were adjusted to pressures in atm, the term a must be corrected to conform to the pressures in bars

$$a = 23,365,228/1.013 = 23,065,378$$

Insert the calculated a and b correction terms, and the molar volume in Equation (3.11) and calculate the pressure at 400 °C (673 K)

$$P = \frac{83.11665 \times 673}{50 - 27.9606} - \frac{23,065,378}{\sqrt{673} \times (50^2 + 50 \times 27.9606)} = 2310\,\mathrm{bar}$$

In case of the presence of the polar H_2O molecule, a T-dependent correction a-term for CO_2 should be used (Equation (II.27)). However, the calculation with the T-dependent term in anhydrous gas mixture has little influence on the resulting pressure, which is only slightly lower for the preceding example (2307 bar).

CHAPTER

FLUID AND MELT INCLUSION MICROTHERMOMETRY

4

CHAPTER OUTLINE

Microthermometry is a nondestructive analytical method based on the measurement of temperatures of phase transformations in fluid and melt inclusions. Chemical composition and density of the inclusions fluids can be derived by comparing the recorded temperatures with experimental phase diagrams of chemical systems or recalculated using equations of state (EOS). These data are then used to construct isochores (constant density lines in P–T coordinates) calculated from appropriate EOS. All these procedures are discussed in the following chapters.

Geofluids. http://dx.doi.org/10.1016/B978-0-12-803241-1.00004-6

4.1 OBSERVATION AND IMAGING

Initial information about the presence and phase composition of fluid inclusions can be obtained by the inspection of standard petrographic thin sections under an optical microscope equipped with microscope lenses with a higher magnification (40 × and more). Detailed fluid inclusion analysis, however, demands preparation of special wafers, usually 100–200 μm thick, polished on both sides and cleaned from glue, Canadian balsam, and other sticking agents and organic solvents that could evaporate on heating and contaminate the freezing–heating stage and the microscope lens. Thinner wafers are difficult to clean, handle, and store because they easily break into pieces. Thicker plates distort optical properties, making observations of fluid inclusions deep below the surface either difficult (mainly in birefrigent minerals) or impossible (using lenses with short working distance), and inducing thermal gradients on heating and cooling. Some soft minerals (fluorite, barite, calcite) can be heavily damaged both thermally and mechanically during sawing and polishing. Hence, the observation of fluid inclusions and microthermometric measurements are conducted on cleavage fragments gently separated from crystal aggregates using sharp mini-chisels.

Detailed descriptions of petrographic observations and fluid inclusion documentation are given in van den Kerkhof and Hein (2001). Microscopic observations at room temperature in visible polarized or unpolarized light permit optical properties of daughter minerals (Appendix III), distribution, shape, size, phase composition, and trapping chronology of fluid inclusions to be determined. This information is essential for the genetic classification of fluid inclusions. Either handy drawings (with the help of a camera lucida) or photomicrographs at various magnifications, starting with a general outline of the sample and followed by an outline of detailed positions of inclusion groups suitable for microthermometry, are produced (Figure 4.1). Pronounced scratches, fissures, and other markers enable rapid orientation in the sample (remember that the image is often seen upside-down in some microscopes). Relationships between inclusions should be documented at various magnifications. Drawings and photographs taken at larger magnifications should depict not only selected morphologically distinct fluid inclusions, but also important genetic phenomena, such as the intersection of inclusion trails, their relative ages, coexistence of inclusions with diverse phase volumes, and so on.

Detailed drawings of inclusion groups and individual inclusions are essential to avoid misorientation during accidental fragmentation of the sample. Important places with fluid inclusions should be marked using a water-resistant thin marker. Larger plates must be broken into smaller pieces owing to the limited area available on the movable holder of the heating–freezing stage. It is useful to draw the outer perimeter of the larger plate before breaking, and all obtained fragments, on a millimeter-squared paper for the reconstruction of exact positions of the measured fluid inclusions. In such a way, the core–rim relationships and fluid inclusion properties can also be reconstructed in larger crystals.

Near-infrared (IR) microscopy is used for observing and measuring phase transitions in fluid inclusions in some opaque minerals. The spectral range for the IR microscopic observations is limited by the detection capacity of commercially available IR-sensitive Charge-Coupled Device (CCD) cameras, ranging from 0.8 to 2.5 μm. The cameras are coupled with an optical microscope with special IR-transparent lenses and optical prism. For the IR observations, 80–180-μm-thick wafers polished on both sides are used.

Transparency of a mineral for a given spectral range is determined by the energy of IR-active vibrations of anions and molecules in crystal lattice. Trace elements with different charges can change the configuration of electrons in the host mineral, thus shifting the IR-transmittance gap. For this

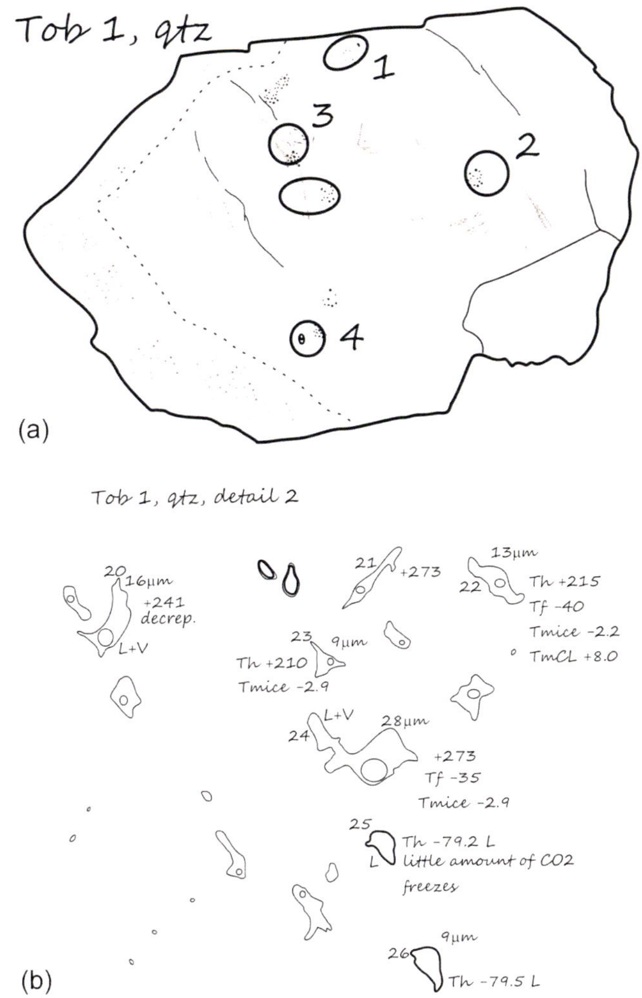

FIGURE 4.1

Sample documentation of a doubly polished wafer with fluid inclusions and measured microthermometric data.

reason, the IR-transparency can be very different even for the same mineral, depending on formation conditions and the contents of minor and trace elements. This feature is particularly typical of pyrite (Lüders and Ziemann, 1999). Galena exhibits an IR-absorption band that lies outside the spectral range of commercially available IR cameras with the highest resolution. Native metals and arsenopyrite are also opaque at any spectral range. Some minerals (enargite) are transparent at low temperatures at wavelengths below 1 μm, but they gradually lose their IR-transmittance on heating to temperatures above 160 °C, thus requiring the use of IR-cameras sensitive to 2.5 μm.

IR microscopy has been successfully applied to the study of fluid inclusions in pyrite (Richards and Kerrich, 1993; Lindaas et al., 2002; Kouzmanov et al., 2002), wolframite (Thomas, 1982; Campbell and

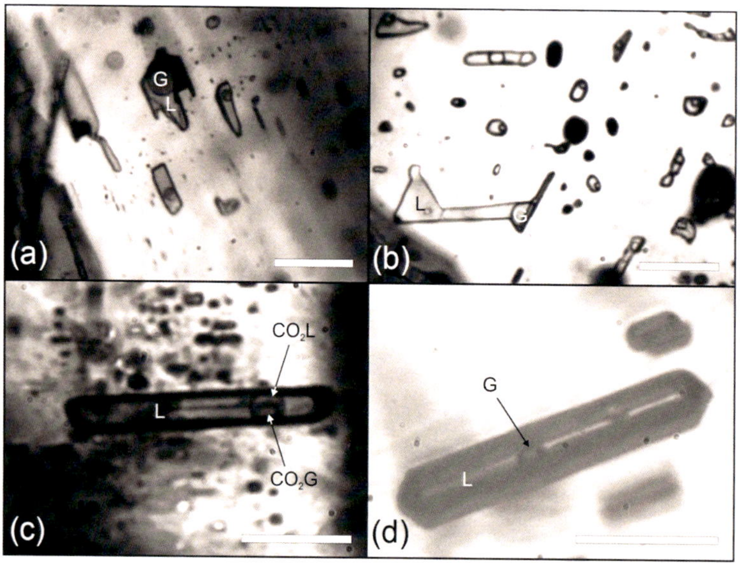

FIGURE 4.2

Transmitted infrared light photomicrographs of aqueous inclusions in (a) enargite (the Recsk–Lahóca porphyry-skarn deposit, Hungary, courtesy of E. Vágó and F. Molnár, ELTE Budapest, Hungary); (b) iron-rich, black sphalerite (the Tisovec skarn deposit, Slovakia); (c) stibnite (the Čučma hydrothermal deposit, Slovakia); (d) stibnite (the Dúbrava hydrothermal deposit, Nízke Tatry Mts., Slovakia). Scale bars 10 μm.

Robinson-Cook, 1987; Campbell et al., 1988; Campbell and Panter, 1990; Bailly et al., 2002), stibnite (Lüders, 1996; Chovan et al., 1999; Bailly et al., 2000; Germann et al., 2003; Hagemann and Lüders, 2003), enargite (Mancano and Campbell, 1995), hematite, hausmannite (Lüders et al., 1999, 2005), bournonite (Lüders, 1996), cassiterite (Campbell and Panter, 1990), and dark, Fe-rich sphalerite (Thomas, 1982; Hurai and Huraiová, 2011, and many others). Examples of aqueous fluid inclusions in enargite, stibnite, and sphalerite, as seen in the transmitted IR light, are shown in Figure 4.2.

Hydrocarbons illuminated by the ultraviolet light fluoresce in pale blue, green, yellow, orange, or red colors, depending on their composition and density (Figure 4.3). Ultraviolet illumination is a routine method commonly used for the rapid identification of hydrocarbons since the nineteenth century, when Reese (1898) was the first to observe inclusions of bitumens using this method.

An optical microscope for reflected light equipped with nonfluorescent objectives, a set of mirror filters for eliminating undesired wavelengths of the electromagnetic spectrum, and a mercury burner as the source of the UV is used for the observation of induced fluorescence. The maximum wavelength of the UV excitation usually corresponds to 365 nm. Wavelengths of the emitted visible light can be recorded and analyzed in a spectrograph. The excited fluorescence, which normally ranges between 400 and 700 nm, is characterized by maximum absorption (L_{max}) and quotient Q, expressing the intensity ratios of the 650 and 500 nm wavelengths (Stasiuk and Snowdon, 1997). The Q_{580} fluorescence emission integral quotient (the ratio of 580–700 nm to 400–580 nm spectral integrals) is used to scan cored wells for the presence of hydrocarbons (Kihle et al., 2012).

The main fluorescent compounds are aromatic hydrocarbons, whereas polar compounds with N, S, and O fluoresce less intensely at higher wave numbers in the red region of the electromagnetic

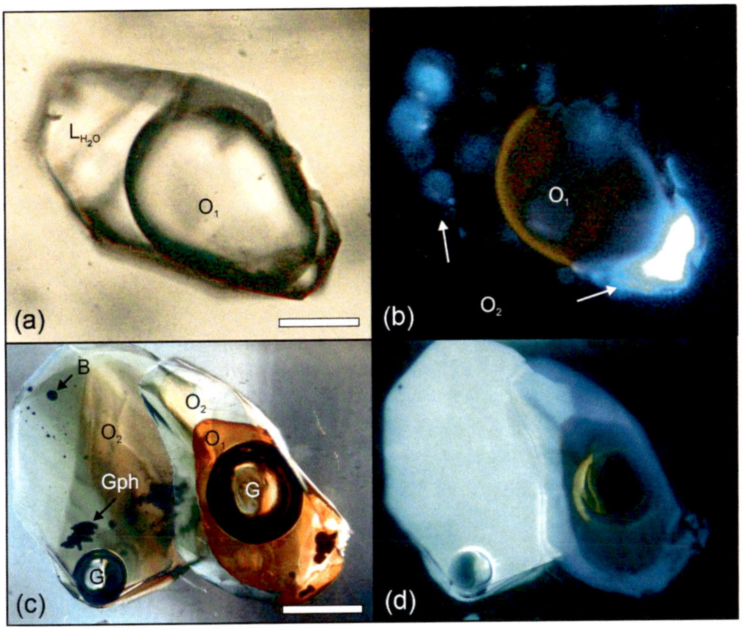

FIGURE 4.3

(a–b) Multiphase oil-bearing inclusion in quartz from mineralized joints of Paleogene sandstones of the Carpathian flysch basin (Vel'ký Lipník, Spišská Magura Mts.): (a) Transmitted light image in polarizing microscope, showing two colorless immiscible aqueous (LH_2O) and petroleum (O_1) phases. (b) The 365 nm UV-induced fluorescence emission reveals green-yellow fluorescent heavy oil (O_1) and numerous droplets of white-blue fluorescing, medium-to-light paraffinic oil adhering to the inclusion walls. (c–d) Two overlapping polyphase oil inclusions in quartz from Baluchistan, Pakistan, containing a vapor bubble (G), dark brown oil (O_1), yellow oil (O_2), bitumen globules (B), and clusters of graphitic material with a low degree of ordering (Gph). (d) UV-light-induced blue fluorescence emission of the yellow oil, contrasting with weak, yellowish emission of the brown oil. Bar scales correspond to 20 μm in (a) and 200 μm in (c).

spectrum. Saturated hydrocarbons fluoresce with maximum intensity in the blue-green region; however, this color may also be generated by contaminating aromatic hydrocarbons (Bertrand et al., 1986; Khavari, 1987; Burruss, 1991). The excited spectrum is influenced by several factors, and it does not necessarily reflect the hydrocarbon composition (Stasiuk and Snowdon, 1997; Munz, 2001).

Cathodoluminescence is employed for the observation of textural elements, which are invisible under a polarizing microscope. The cathodoluminescence enables visualization of oscillatory zoning or processes of recrystallization, thus permitting, among other things, an unambiguous identification of primary and secondary fluid inclusions. Microtextures connected with the fluid inclusions' re-equilibration can also be observed (Behr et al., 1989; van den Kerkhof and Hein, 2001).

Cathodoluminescence is excited by the bombardment of a polished conductive (carbon-coated) sample with an electron beam in a vacuum chamber. The bombardment stimulates energy level transitions in the mineral, thereby giving rise to a photon emission. The wavelengths and the intensities of the induced luminescence depend on the structural defects and the presence of trace elements—chromophores. For instance, the high-temperature magmatic quartz with high Al^{3+} contents fluoresces blue, and the low-temperature diagenetic quartz with a low Al^{3+} does not display any fluorescence. Wavelengths of the excited light vary with time.

Two cathodoluminescence systems are available. A microscope-mounted cold cathode operates with high gun currents and is applied to strongly fluorescing minerals, such as carbonate, feldspar, scheelite, and sodalite. A "hot-gun" system operating with lower currents is capable of inducing luminescence of quartz. CL detectors are also mounted on routinely used scanning electron microscope systems of modern electron microprobes. An example of cathodoluminescence-induced textures of carbonates formed as a result of hydraulic fracturing is shown in Figure 4.4.

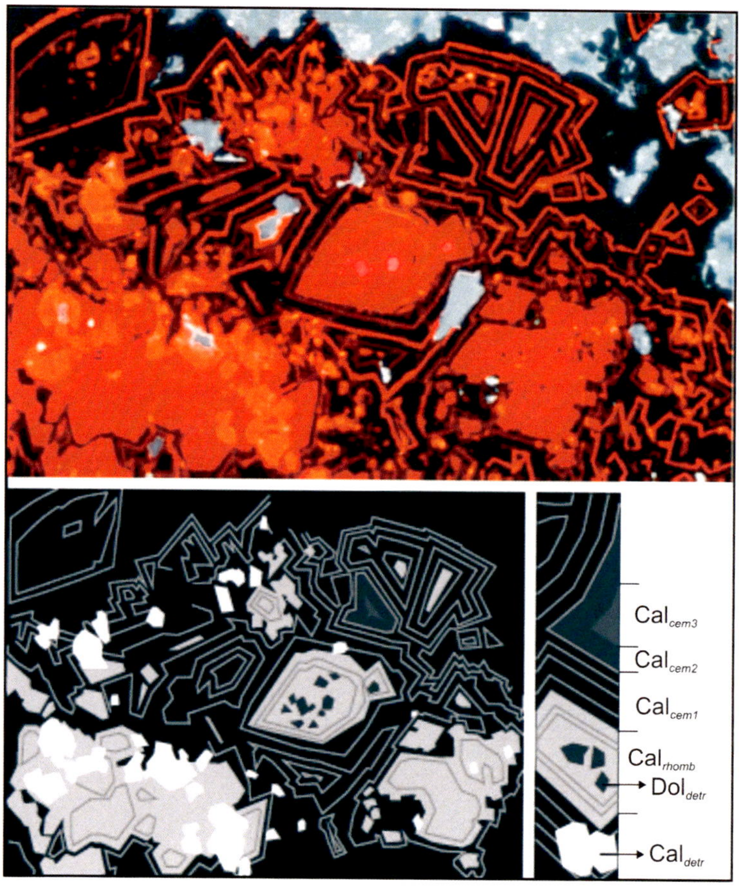

FIGURE 4.4

Example of typical cathodoluminescence texture of carbonate assemblage from hydraulic tectonic breccia from the base of a calcareous nappe (Milovský et al., 2012). The CL photomicrograph (top) is redrawn in the bottom simplified sketch to enhance CL textures. Cal, calcite; Dol, dolomite; *cem*, newly formed calcitic matrix; *rhomb*, newly formed crystals with rhythmic growth zoning; *detr*, calcite and dolomite detritic grains. Reprinted from Milovský, R., van den Kerkhof A., Hoefs, J., Hurai, V., Prochaska, W., 2012. International Journal of Earth Sciences (Geologische Rundschau) 101, 535–554. Fig. 4., with kind permission from Springer Science and Business Media.

4.2 SAMPLE HANDLING AND STORAGE

A syringe with a flat tip permits comfortable manipulation of very small fragments (1 mm^2 to 1 cm^2 in area). Wafer fragments are stuck on the tip of the syringe by an underpressure developed through flexible hosepipe either by mouth or via a handy vacuum pump (Figure 4.5).

The obtained fragments can be placed either within transparent plastic boxes, or packed within simple sealable bags made of a paper sheet, with dimensions fitting the firm cartoon boxes used for the storage and archiving of minerals (Figure 4.6).

4.3 CALCULATING VOLUMETRIC PHASE RATIOS

Inclusion phase ratios and total volumes of inclusions are the first quantitative information that can be gained from observations at room temperature. Basic relationships among volume (V), mass (M), and density (D) can also be employed to forecast phase volume ratios at room temperature in fluid inclusions with known compositions and densities.

Mass is an extensive state property that equals the sum of masses of all phases present in a fluid inclusion:

$$M = M_1 + M_2 + M_3 + \cdots + M_n \tag{4.1}$$

Combining Equations (1.14) and (4.1) we get

$$VD = V_1D_1 + V_2D_2 + V_3D_3 + \cdots + V_nD_n \tag{4.2}$$

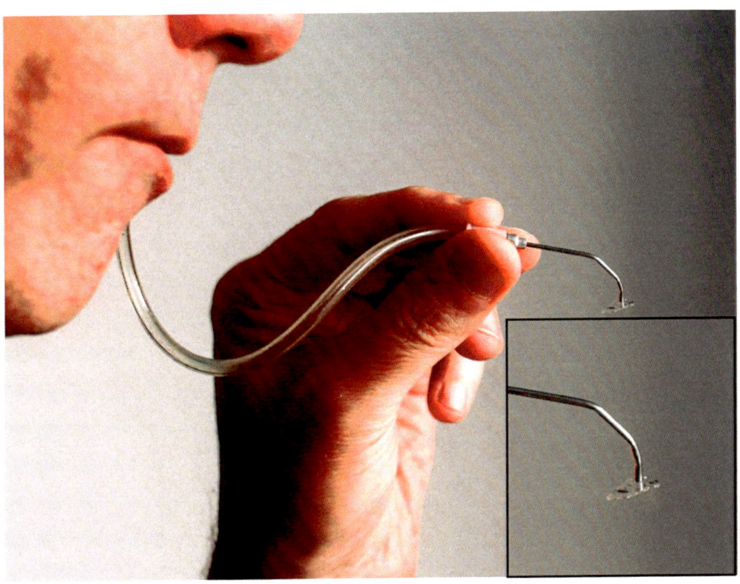

FIGURE 4.5

Syringe tip with flexible hosepipe and flat end for handling doubly polished wafers with fluid inclusions.

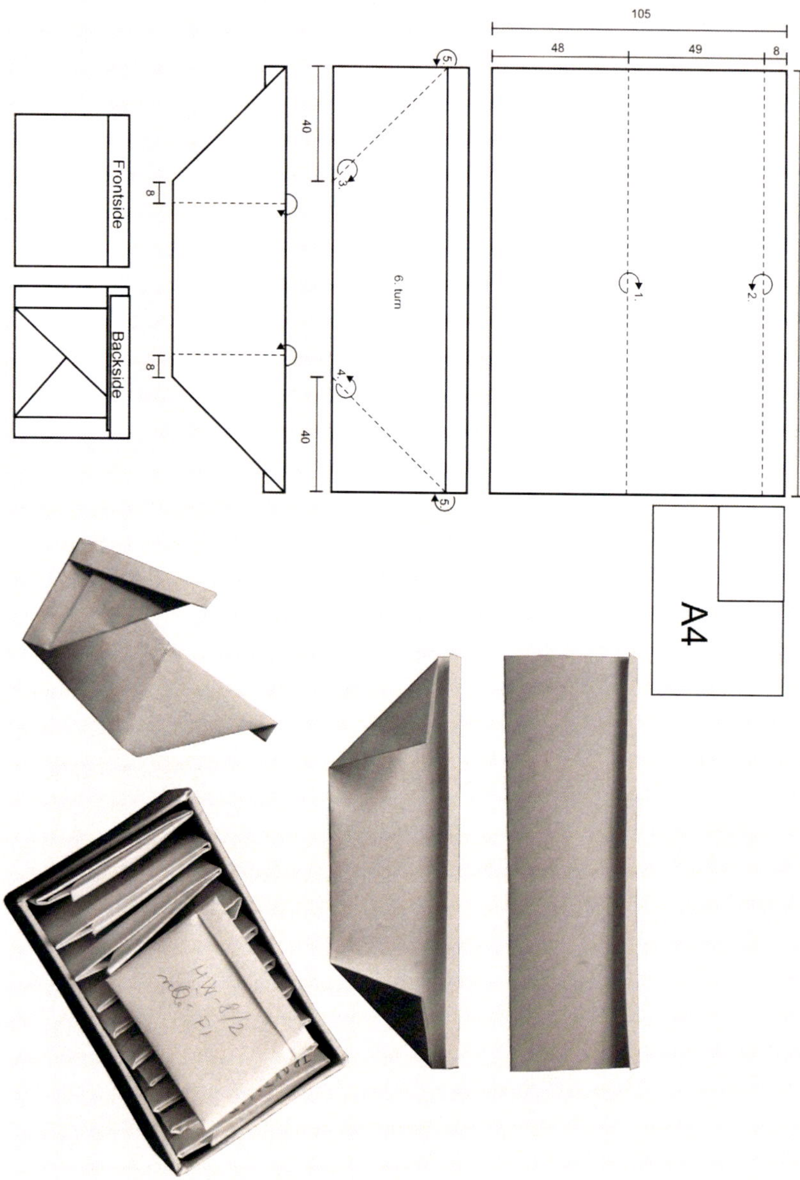

FIGURE 4.6

Making a sealable bag for archiving doubly polished wafers used in microthermometry.

For a fluid inclusion containing liquid and vapor phases, Equation (4.2) reduces to

$$VD = V_G D_G + (V - V_G) D_L \tag{4.3}$$

where subscripts G and L refer to vapor and liquid phases, respectively.

Equation (4.3) can be rearranged to obtain the volume V_G of the vapor bubble in a two-phase fluid inclusion

$$V_G = \frac{V(D - D_L)}{D_G - D_L} \tag{4.4}$$

and the volume fraction of the vapor bubble is

$$\bar{v}_G = \frac{D - D_L}{D_G - D_L} \tag{4.5}$$

Similarly, the volume fraction of aqueous liquid can be expressed as

$$\bar{v}_L = \frac{D - D_G}{D_L - D_G} \tag{4.6}$$

Masses and mass fractions of coexisting liquid and vapor phases can be expressed in a similar way. For example, the mass of the vapor bubble is equal to

$$M_G = \frac{M(m - m_L)}{m_G - m_L} \tag{4.7}$$

and its mass fraction

$$\frac{M_G}{M} = \frac{m - m_L}{m_G - m_L} \tag{4.8}$$

Similarly, the mass fraction of aqueous liquid is

$$\frac{M_L}{M} = 1 - \frac{M_G}{M} = \frac{m - m_G}{m_L - m_G} \tag{4.9}$$

where m is the total concentration of solute in the inclusion (mass%), m_L and m_G are concentrations of the solute in the liquid and vapor phases, respectively, and M is the weight of the inclusion.

Only densities of coexisting phases and total density of the inclusions at a given temperature are needed to calculate volume fractions of the liquid and vapor phases. For example, a vapor bubble occupies 8 vol.% in an inclusion of pure water with the total density $D = 0.917$ g cm^{-3} at 25 °C (Equation (4.5)), because $D_G = 0.000023$ g cm^{-3} and $D_L = 0.997047$ g cm^{-3}. This approach can be used to calculate volumetric phase ratios in NaCl–H_2O inclusions, where density at 25 °C is defined by Equation (II.15) in Appendix II, density of the vapor at this temperature equals to $D_G = 0.000023$ g cm^{-3}, and the total density of the inclusion can be calculated from homogenization temperature and concentration using Equation (II.14) (Appendix II).

To calculate the volumetric phase ratios in a three-phase inclusion composed of vapor, liquid, and solid phases, Equation (4.2) must be rearranged to

$$VD = V_G D_G + V_S D_S + (V - V_G - V_S) D_L \tag{4.10}$$

and the volume of the vapor phase corresponds to

$$V_G = \frac{V(D - D_L) - V_S(D_S - D_L)}{D_G - D_L} \tag{4.11}$$

PROBLEM 4.1 CALCULATING VAPOR VOLUME IN AQUEOUS INCLUSION

Calculate the vapor phase volume in a pure water inclusion at 25 °C, if the inclusion homogenizes to liquid at 150 °C.

Solution

The density of vapor-saturated aqueous liquid at 25 °C and 150 °C corresponds to 0.997 (Potter and Brown, 1977) and 0.924 g cm^{-3} (Haas, 1976), respectively, whereas that of vapor is essentially zero at 25 °C. After substituting for m_L, m, and m_G in Equation (4.8) and multiplying by 100, we get 7.3 vol.% of the vapor bubble at 25 °C.

PROBLEM 4.2 CALCULATING PHASE VOLUMES IN BRINE INCLUSION

Calculate phase ratios at 25 °C for an aqueous inclusion containing 40 mass% NaCl, when the inclusion homogenizes to liquid at 500 °C.

Solution

The density of the 40 mass% NaCl–H_2O solution at 500 °C corresponds to 0.9059 g cm^{-3} (Bodnar, 1983). Calculate molality $\overline{m}$ from Equation (1.4) [11.407 mol NaCl/kg H_2O]. The aqueous solution contains 666.66 g NaCl ($\overline{m}\cdot\overline{M}$—molecular mass of NaCl) per 1 kg H_2O. A total of 1666.66 g of the aqueous solution occupies the volume of 1839.8 cm^3 (Equation (1.14)).

The saturated NaCl–H_2O solution at 25 °C contains 26.46 mass% NaCl (Potter et al., 1977), that is, 6.1565 mol NaCl/kg H_2O, or 359.8 g NaCl/kg H_2O, and has a density of 1.1971 g cm^{-3}. The excess NaCl, that is, 306.9 g, precipitates as halite. Density of the halite at 25 °C is 2.1631 g cm^{-3} (Robie et al., 1979), and its volume (141.9 cm^3, 7.7 vol.%) can be calculated according to Equation (1.14). Assuming that the vapor phase above the halite-saturated aqueous solution at 25 °C does not contain any NaCl, the vapor bubble volume can be easily calculated from Equation (4.10):

$$V_G = (1839.8 \times (0.9059 - 1.1971) - 141.9 \times (2.1631 - 1.1971))/-1.1971$$
$$= 562\ \text{cm}^3\ (30.5\ \text{vol.\%})$$

PROBLEM 4.3 CALCULATING PHASE VOLUMES IN CARBONIC-AQUEOUS INCLUSION

Calculate phase volumes at 20 °C in a multiphase inclusion that contains 4 mol.% CO_2 and 15 mol.% NaCl, and its total density is $D = 0.95$ g cm^{-3}.

Solution

At the outset, one must transform mole percents into molar fractions (/100) and calculate the mass M_i of individual compounds in 1 mol of solution

$$M_i = \overline{x} \times \overline{M}_i \tag{4.12}$$

where $\overline{x}$ is the molar fraction and $\overline{M}_i$ is the molecular weight of the ith component ($M_{CO_2} = 1.7604$ g, $M_{NaCl} = 8.7664$ g, $M_{H_2O} = 14.5924$ g). Hence, a total of 120.65 g CO_2 and 600.75 g NaCl are dissolved in 1000 g of water. The solution mass of 1721.4 g occupies the volume of 1812 cm^3 (Equation (1.14)). However, only 358.7 g NaCl can be dissolved in 1000 g of H_2O at 20 °C. The rest, that is, 242.05 g, precipitates as halite, whose volume can be calculated from Equation (1.14) knowing the total density (2.163 g cm^{-3}). The volume of the NaCl-saturated

Continued

aqueous solution can be derived in the same way, because we know that its density at 20 °C is equal to 1.2105 g cm^{-3} (Equation (II.14), Appendix II). From the total inclusion volume of 1812 cm^3, the NaCl-saturated aqueous liquid occupies 1122.4 cm^3 (61.9 vol.%), halite 111.9 cm^3 (6.2 vol.%), and the rest, that is, 577.7 cm^3, is filled with CO_2, whose density is 0.2088 g cm^{-3} (Equation (1.14)). The CO_2 density is higher than that of vapor (0.1939 g cm^{-3}) and lower than that of liquid (0.7743 g cm^{-3}) coexisting along a two-phase curve of pure CO_2 at 20 °C (Equation (II.3), Appendix II). For this reason, CO_2 must split into the liquid and vapor phases whose volumes are calculated in the same way as outlined for the two-phase inclusions (Equation (4.3)). The CO_2 vapor and liquid phases will occupy 562.9 cm^3 (31.1 vol.%) and 14.8 cm^3 (0.8 vol.%), respectively. Consequently, the partial homogenization of CO_2 occurs to vapor, by the evaporation of the CO_2 liquid.

4.4 CALCULATING FLUID INCLUSION VOLUMES

Information about absolute fluid inclusion volume is needed when studying the effects of decrepitation, which is initiated at a certain threshold value. In the case of a prolonged regular shape, satisfactory results are achieved by a direct measurement of the inclusion diameter by using the calibrated micrometric eyepiece in two perpendicular directions. Flat irregular inclusions can be photographed, digitalized, and analyzed using image analysis computer programs. However, in most cases, these measurements are subject to large errors owing to the three-dimensional shape of inclusions. This is a particularly serious problem in the analysis of mixed gas–aqueous inclusions, in which inaccurate determinations of phase volumes at a certain temperature must be done to calculate the bulk fluid density and composition (Roedder, 1984; Bakker and Diamond, 2006). For petroleum inclusions, volumes can also be measured by laser scanning (Pironon et al., 1998).

Bodnar (1983) calculated inclusion volumes as an approximation to a simple, experimentally well-constrained system H_2O–NaCl. Analysis of Equation (4.4) shows that for the volume calculation of a two-phase liquid–vapor aqueous inclusion, we need only know the volume of a vapor bubble D_G at 25 °C (i.e., the temperature of observation), densities of the aqueous liquid D_L and the coexisting vapor D_G phases at this temperature, and the density D of the whole inclusion. The vapor bubble usually has a spherical shape so that its volume can be easily inferred from the bubble diameter (d) as $V_G = 1/6\pi d^3$. In flat inclusions, however, the equation for the circle area must be used: $V_G = \pi r^2$, where r is the bubble radius. Densities of aqueous liquid and vapor phases at 25 °C are as follows: 0.997047 g cm^{-3} (Potter and Brown, 1977) and 0.000023 g cm^{-3} (Keenan et al., 1978). The bulk density of the inclusion can be derived from the temperature of total homogenization and the EOS for the two-phase saturation curve (Appendix II).

If another phase is present, for instance halite, Equation (4.11) can be rearranged to

$$V = \frac{V_G(D_G - D_L) + V_S(D_S - D_L)}{D - D_L} \tag{4.13}$$

This situation does not pose a problem, because the halite usually crystallizes as a single hexahedron, whose volume V_S corresponds to cube of its edge, which is measurable using a micrometric eyepiece. The halite density at 25 °C corresponds to 2.1631 g cm^{-3} (Robie et al., 1979), and the density of the NaCl-saturated aqueous solution at 25 °C is 1.1971 g cm^{-3} (Potter and Brown, 1977). Assuming that the composition and density of vapor above the NaCl solution and pure water are roughly similar at 25 °C, the total density of the inclusion is calculated from the temperature of halite dissolution (Appendix II), and substituting the calculated parameters into Equation (4.13) yields the volume of the investigated inclusion.

4.5 EVALUATING TRAPPING MODES OF FLUID INCLUSIONS

The evaluation of trapping mode is essential information that determines the further strategy of measurements as well as the methods of fluid inclusion data reduction and interpretation.

Minerals usually precipitate from a homogeneous fluid. At higher pressures (such as P_1 in Figure 4.7), the fluid inclusion trapping conditions project within the field of homogeneous liquid, whereas the homogeneous vapor is trapped at a lower pressure P_3. After isolating the fluid inclusion in the growing crystal, further phase changes in the fluid inclusion are isochoric and isoplethal. In other words, volume and composition of the inclusion are fixed, unless specific P–T conditions are established, when the inclusion re-equilibrates due to high internal under- or overpressure. During ascent to the present-day surface, the inclusion fluid pressure follows the corresponding isochore until the two-phase curve is intersected. After certain undercooling is needed to overcome metastability, a small vapor bubble nucleates in the fluid inclusion trapped at higher pressure P_1, whereas a thin rim of liquid condensates in the inclusion trapped at a lower pressure P_3. On heating, homogenization of both inclusions to liquid (by diminishing of the vapor bubble) or vapor (by evaporation of the liquid rim) occurs exactly at the intersection of their isochores with the two-phase boiling or condensation curves, and this moment is recorded as the homogenization temperature (T_h). When the inclusions are cooled

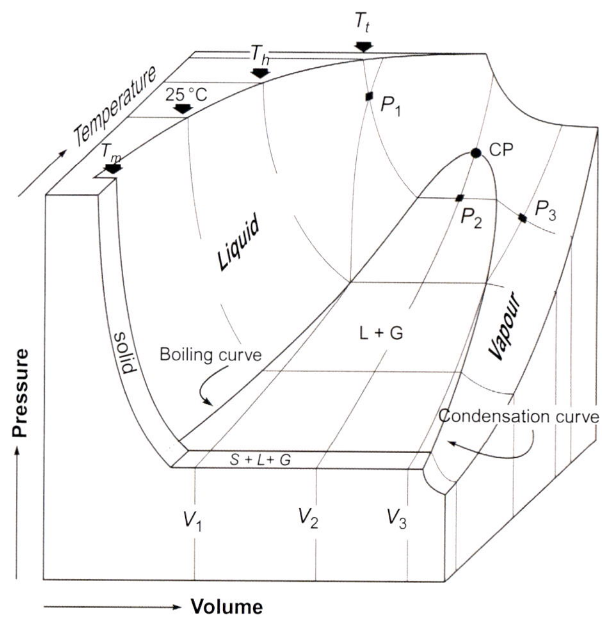

FIGURE 4.7

Schematic phase diagram of unary system (modified from Burruss, 1981c) with isochores of fluid inclusions trapping a homogeneous liquid with volume V_1, vapor with volume V_3, and a heterogeneous liquid + vapor mixture with bulk volume V_2 at the same temperature T_t, but different pressures (P_1–P_3). See Figure 4.8 for further explanation.

below their homogenization temperature, their isochores traverse across the liquid–vapor surface, until a three-phase boundary is intersected and this moment is recorded as the temperature of melting.

Densities of coexisting phases in a heterogeneous isochoric system are invariant. Hence, when the inclusion is cooled below its homogenization temperature, volumes of the coexisting phases must also change owing to changing densities of the coexisting liquid and vapor along boiling/condensation curves, and those of vapor and solid along the sublimation curve. At any moment, the phase ratios of all phases present in the inclusion obey the mass-balance rule, and the bulk inclusion density remains constant.

The homogeneous trapping mode can be easily discerned under an optical microscope, because phase ratios in a population of coeval fluid inclusions remain constant at any temperature below T_h (Figures 4.8 and 4.9a). This is, however, not the case for the inclusions trapping a heterogeneous fluid mixture, consisting of vapor and liquid, or even more phases. This situation is rather common in aqueous fluids with dissolved salts and gases, having high critical temperatures and pressures. The presence

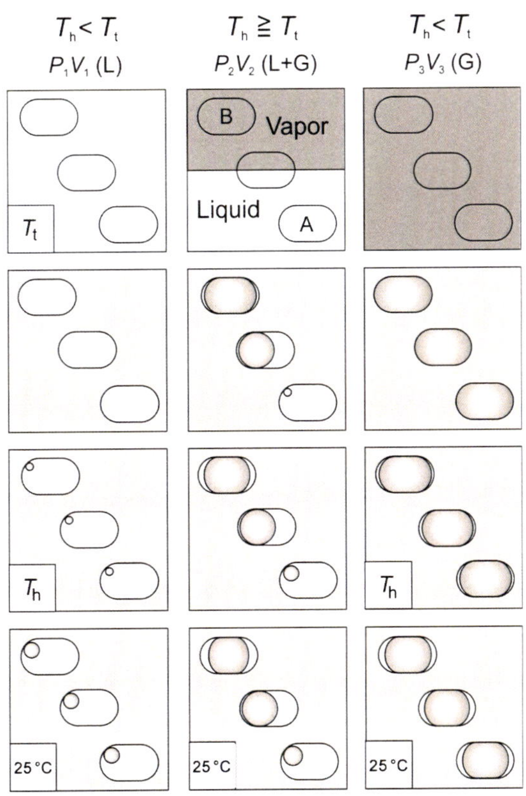

FIGURE 4.8

Schematic explanation of homogeneous and heterogeneous trapping at different pressures (P_1–P_3), within the fields of homogeneous liquid (V_1), homogeneous vapor (V_3), and heterogeneous liquid+vapor mixture (V_2), and the relations between trapping (T_t) and homogenization (T_h) temperatures. See Figure 4.7 for comparison.

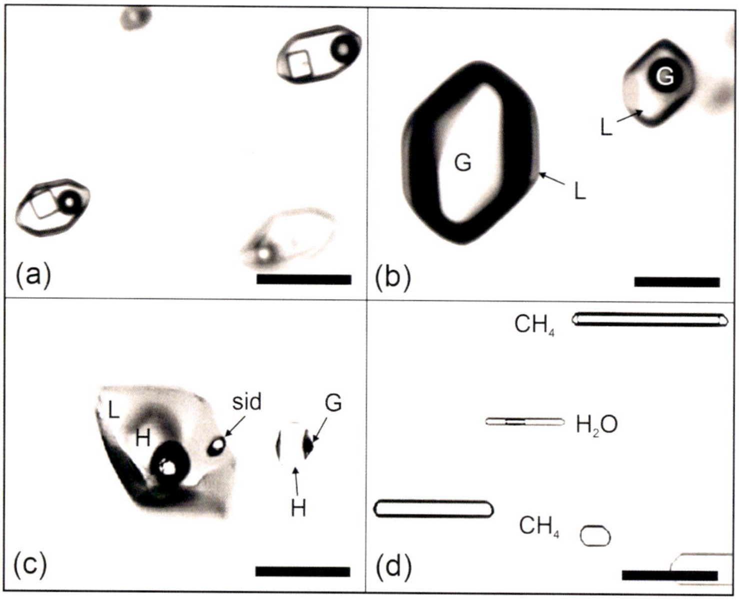

FIGURE 4.9

Homogeneous (a) and heterogeneous (b–d) trapping in various fluid systems: (a) high-salinity brines in quartz from tension gashes in Brasilian craton, Cristalândia, trapped at 280–400 °C and 3 kbar (Marko et al., 2006); (b) boiling low-salinity fluids in quartz from hydrothermal metasomatic silicites (secondary quartzite) formed at ~270 °C and 50 bar in a Miocene stratovolcano, Banská Štiavnica, Slovakia (Oružinský and Hurai, 1985); (c) halite-oversaturated brines in quartz from polymetallic siderite veins of low- to medium-grade Variscan basement, Rudňany deposit, Slovakia, trapped at 170–210 °C, 1.2–1.7 kbar (Hurai et al., 2002a); (d) heterogeneous CH_4 aqueous mixture in quartz from mineralized joints in Eocene flysch sandstones, Zálesie, Spišská Magura Mts., Slovakia, trapped at 140–190 °C and ~1 kbar (Hurai et al., 2006). Solid bar scales correspond to 20 μm.

of two phases decreases the variance of the system, thereby allowing the direct determination of the trapping *P–T* conditions.

If the trapping *P–T* conditions are lower than those in the critical point, a heterogeneous L + G mixture can be sealed within fluid inclusions. The coexisting phases can be trapped at random ratios. Hence, at the moment of trapping most inclusions contain liquid and vapor phases in variable proportions, and their homogenization temperatures are higher than those of trapping, reaching the maximum corresponding to critical temperature if the bulk density of the L + G mixture equals that at the critical point. Some inclusions, however, can trap homogeneous liquid (inclusion A in Figure 4.8) or homogeneous vapor (inclusion B), and they behave exactly as homogeneously trapped fluids. Their homogenization temperature, however, exactly matches that at the moment of the inclusion sealing because the inclusion fluid isochore is already within the L + G field and an unmeasurably small drop in temperature leads to boiling (bubble nucleation) in liquid inclusion or condensation of liquid rim in vapor inclusions.

The heterogeneous trapping mode can be easily discerned under the microscope according to variable phase proportions in a group of coeval inclusions, which homogenize to liquid and vapor at various temperatures (Figures 4.8 and 4.9b). For a statistically relevant number of measurements, the minimum homogenization temperature is taken as that corresponding to the actual crystallization temperature and the formation pressure can be derived from the EOS describing P-T conditions along the two-phase curve.

The trapping mode defines the strategy of further microthermometric measurements. In the case of a homogeneous trapping, consistent phase, and compositional properties, a minimum of 20 measurements (in the best case as many as 50) are needed for the construction of T_h and T_m frequency diagrams, and to identify limiting isochores of the isochoric envelope representative of the cogenetic fluid inclusion population. In the case of a heterogeneous trapping, only minimum T_h values of aqueous inclusions are measured to determine the local formation temperature. T_m values are needed for the characterization of the aqueous phase, but they are not further employed for the calculation of isochores. Pressures are derived from maximum and minimum densities of proximal gaseous inclusions, in which both T_h and T_m values must be determined to calculate the corresponding isochore. Hence, in the case of the heterogeneous trapping, quality of data that is the focus on most proximal aqueous and gaseous inclusions possibly sealed at the same time is preferred over quantity.

4.6 MEASURING TEMPERATURES OF PHASE TRANSITIONS

4.6.1 FREEZING–HEATING STAGES

Once the detailed fluid and melt inclusion documentation at room temperatures is made, temperatures of phase transitions can be measured. Commercially available microthermometric stages for the investigation of fluid inclusions are cooled by liquid nitrogen, enabling continuous observation within the temperature range from −196 to +600 °C. Freezing–heating stages specially designed for the study of fluid inclusions have been manufactured by CHAIXMECA (France), REYNOLDS (USA), and LINKAM (UK), of which only the last company remained on the market until recently.

In the last century, difficulty of recognition was the main limiting factor of melt inclusion studies. Magnifications of commercially available heating stages (such as from LEITZ-WETZLAR) were limited by the optical efficiency of 20 × or 40 × objectives, with working distances large enough to protect the lens from high temperatures in the heating cell. Small dimensions of fluid inclusions together with bad optics and the red light emitted from the glowing heater prevented scientists from obtaining high-quality microphotographs that documented the phase transitions in melt inclusions at high temperatures.

In more recent years, scientists have used modern heating stages from LINKAM to optically observe phase transitions in melt inclusions. The devices have a ceramic tube heater that completely encases the sample in a uniform temperature-controlled environment enabling high accuracy control of up to 1400 °C. The gas-sealed chamber enables observations in an inert atmosphere (nitrogen, argon), which prevents oxidation of Fe-bearing minerals. Lenses with higher magnification make it easier to observe and document phase changes. Chips and wafer fragments are mounted on a sapphire sample slide that can be moved in *XY* directions to better explore the sample. Scientists can use high-speed quench cooling by using a special manipulator that enables them to drop the investigated sample from the heater to a cold platform. By this method, cooling rates of up to 240 °C per second can be reached (Esposito et al., 2012). Quenching rates at the order of about 500 °C per second can be achieved by quenching the samples in liquid nitrogen (Thomas, 1994).

Microthermometric measurements must be preceded by the calibration of the microthermometric stage using synthetic or natural inclusions with known composition, such as inorganic and organic chemical substances with well-defined melting temperatures. Some of the most convenient and easily accessible substances are listed in Table 4.1. It is sometimes necessary to use a sapphire or quartz glass slide instead of normal glass to prevent reactions at high temperatures. A calibration curve consisting of several different points is constructed to correct the measured temperatures. A three-point calibration procedure is already involved in the software of the LINKAM stages, so that the corrected values are automatically displayed on-screen and need not be recalculated manually.

4.6.2 PHASE CHANGES

The solubility of salts and gases in aqueous solutions increases with temperature. Hence, a number of phases within inclusions decreases with the increasing temperature, and the opposite is true of cooling. If one is speaking about phase changes in fluid inclusions, he or she means the disintegration of initially

Table 4.1 Melting Temperatures of Various Chemical Substances Suitable for the Calibration of Microthermometric Stages (Selected from the Far More Extensive Compilation of Roedder (1984)). Melting Points of Elements are from Weast (1979).

Temperature (°C)	Compound	Formula
−112.0	Ethylalcohol	C_2H_5OH
−95.0	Toluene	$C_6H_5CH_3$
−56.8	n-Octane	$CH_3(CH_2)_6CH_3$
−56.6	Carbon dioxide	CO_2
−63.4	Trichloromethane	$CHCl_3$
−38.9	Mercury	Hg
0.1	Distilled water	H_2O
80.6	Naphthalene	$C_{10}H_8$
112.8	Sulfur	S
156.6	Indium	In
228.0	Saccharine	–
232.0	Tin	Sn
263.0	Phenolphthalein	–
306.8	Sodium nitrate	$NaNO_3$
327.4	Lead	Pb
330.3	Potassium nitrate	KNO_3
398.0	Potassium dichromate	$K_2Cr_2O_7$
419.6	Zinc	Zn
449.5	Tellurium	Te
630.5	Antimony	Sb
800.4	Halite	NaCl
848.0	Lithium fluoride	LiF
962.0	Silver	Ag
1064.0	Gold	Au

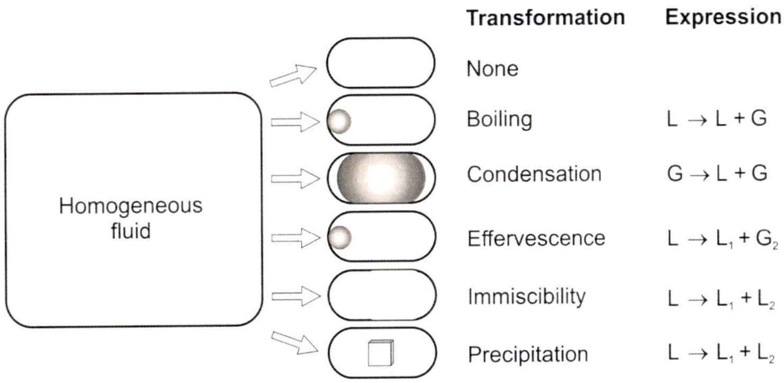

FIGURE 4.10

Phase transformations observed in fluid inclusions.

homogeneous systems into several phases (Figure 4.10). These changes—heterogenization—are a consequence of different temperatures of trapping and observation. Heterogenization, however, is not encountered in the inclusions of low-molecular-mass gases and gaseous mixtures. Due to critical points below 0 °C, these inclusions remain homogeneous at room temperature, and additional cooling is needed to achieve their heterogenization at very low temperatures.

Several kinds of phase transformations can be observed within fluid inclusions (Figure 4.10). Liberation of vapor from liquid ($L \rightarrow L+G$), and condensation of liquid from vapor ($G \rightarrow G+L$) are the most abundant phase changes. The chemical composition of the coexisting phases during boiling and condensation is the same, though concentrations of solutes may differ. Dissolved gases preferably concentrate into a vapor phase, giving rise to effervescence, when the composition of the gas bubbles significantly differs from the composition of the coexisting liquid. Higher partial pressure of the gas results in immiscible separation of a liquidlike gas and formation of the system consisting of two immiscible liquids ($L \rightarrow L_1+L_2$). This behavior is common in silicate melt systems, where silicate, carbonate, phosphate, and other liquids with contrasting densities and compositions may separate on cooling. Precipitation of solid phases from saturated solution ($L \rightarrow L+S$) is another type of heterogenization in natural fluid systems. The solubility of most salts is proportionate to temperature, so that an initially undersaturated solution can exceed its saturation limit on cooling to precipitate soluble components in the form of crystals. Aqueous solutions with electrolytes are the most typical systems, where the precipitation of soluble components occurs on cooling.

Natural fluids are in most cases complex systems, consisting mostly of water and dissolved gases and salts. Such systems exhibit a combined heterogenization on cooling, with various types of phase transformations in optional order, giving rise to multiphase fluid inclusions at room temperature. Complex brines can precipitate several solid phases—daughter minerals—on cooling, because the dissolved salts cannot form solid solutions at low temperatures. Characteristic of this behavior are the fluid inclusions trapped in pegmatite minerals, containing often up to several tens of daughter phases, some of them being stable only in hermetically sealed fluid inclusion.

4.6.3 PHASE TRANSITIONS IN A UNARY SYSTEM

Numerous sequences of phase transitions are theoretically possible in a one-component system, depending on the volume (density) of the trapped inclusion fluid (Figure 4.11). A sequence in which the isochoric sections directly hit the invariant points is very rare, whereas those intersecting the two-phase boundaries are substantially more common.

The isochores of medium-density fluids (D_3–D_5) intersect the boiling–condensation curve, triple point, and sublimation curve, thereby resulting in a sequence of phase transformations where homogenization is the last phase transition to occur on heating, thus defining the H-type phase transformation. If the fluid density exceeds that at the critical point (e.g., D_3 in Figure 4.11), the inclusion homogenizes to liquid by diminishing the vapor bubble, and the corresponding sequence of phase transformations on heating is $S+G \rightarrow S+L+G \rightarrow L+G \rightarrow L$ (Figure 4.12). Fluid inclusion densities lower than that at the critical point (D_5) result in the homogenization of vapor by the expansion of the vapor bubble, following the path $S+G \rightarrow S+L+G \rightarrow L+G \rightarrow G$.

Inclusions with densities exactly matching that at the critical point (D_4) represent special cases. In this case, volumes of the coexisting vapor and liquid phases do not change either on heating or cooling, and the inclusion homogenizes by the fading meniscus between the two phases giving rise to the critical homogenization.

If the fluid density is lower than that of vapor at the triple point (D_6 in Figure 4.11), the isochore intersects only the sublimation curve and the only phase transformation observed is the sublimation or nucleation of a solid phase on heating or cooling: $S+G \rightarrow G$, thereby defining the S-type phase transformation. The phase transformation is relatively common in gaseous inclusions in minerals precipitating at a low pressure in subvolcanic environments (e.g., Oružinský and Hurai, 1985), but it may often escape attention because the small volume of the nucleated CO_2 solid may be obscured by the dark,

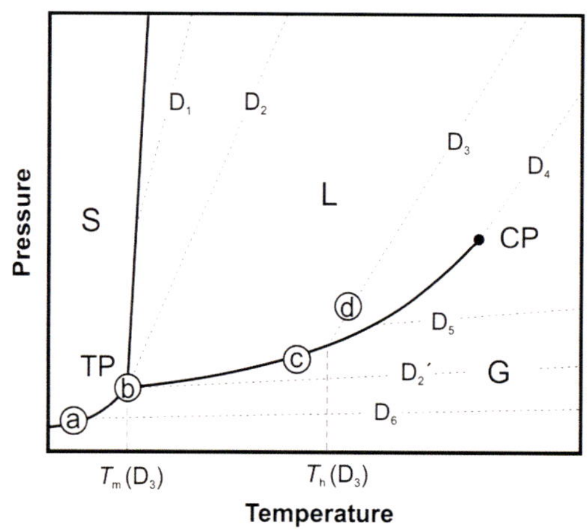

FIGURE 4.11

P–*T* projection of unary system with isochores D_1–D_6 intersecting various components of the phase diagram, that is, two-phase boundaries: D_1, D_3, D_5, D_6; invariant points: D_2, D_2', D_4.

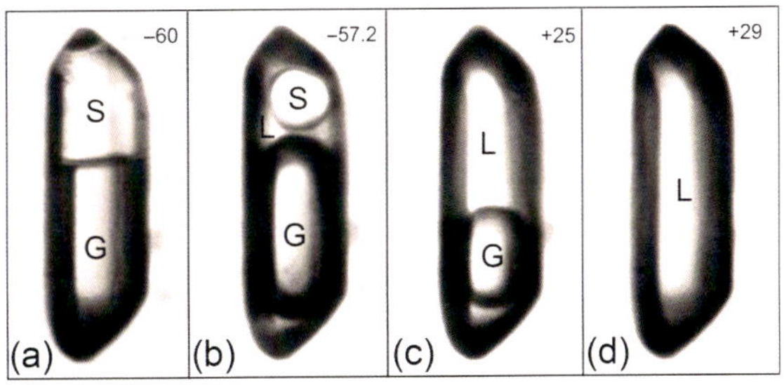

FIGURE 4.12

Phase assemblages in a medium-density CO_2 inclusion in quartz from an alpine fissure from the Spišsko-gemerské rudohorie Mts., Slovakia, at various temperatures (°C): (a) solid–vapor assemblage at sublimation curve below triple point; (b) the triple point and formation of liquid; (c) liquid and vapor phases at boiling curve; (d) homogeneous liquid.

optically opaque zone between the host mineral and the low-density vapor caused by a large difference in their refraction indices.

In a very special case, isochore intersects directly the triple point (D_2, D_2' in Figure 4.11), when the melting and the homogenization occur simultaneously. The corresponding sequence of phase transformations is $S+G \rightarrow S+L+G \rightarrow L$ or G depending on bulk density, which must exactly match that of the liquid or vapor, respectively, coexisting at the triple point of the unary system.

Figure 4.13 illustrates a very rare example of a superdense CO_2 fluid, the density of which is greater than that of liquid at the triple point, namely 1.18 g cm^{-3}. Note that the CO_2 solid occupies a substantial volume of the totally frozen inclusion (Figure 4.13a). The superdense inclusions homogenize by dissolution of solid in liquid at temperatures higher than that at the triple point, thus showing the S-type homogenization.

Superdense CO_2 can exist in a cold crust or mantle with thermal gradients lower than ~23 °C/km. Superdense CO_2 inclusions with $D > 1.2$ g cm^{-3} were described from upper-mantle xenoliths

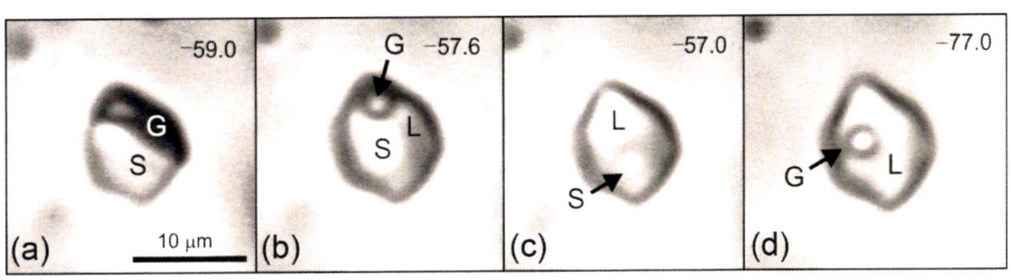

FIGURE 4.13

Serial of photomicrographs showing stable S-type (a–c) and metastable H-type (d) phase transitions in a superdense CO_2 inclusion in a quartz-stibnite vein hosted in low-grade metasediments (the Čučma deposit, Spišsko-gemerské rudohorie Mountains, Slovakia, Urban et al., 2006). S, L, and G denote solid, liquid, and vapor phases, respectively. The triple point observed at −57.3 °C is slightly below that of pure CO_2 due to admixture of nitrogen. Final melting occurred at −56.8 °C, and metastable homogenization to liquid at −58.9 °C. The corresponding fluid density was 1.187 g cm^{-3}.

(Solovova et al., 1990; Frezzotti et al., 1992). Densities of up to 1.21 g cm^{-3} recorded in amphibolite-facies migmatites of the Colorado Front Range are probably related to a secondary tectonic squeezing of the host quartz (van den Kerkhof and Olsen, 1990). Densities as high as 1.197 g cm^{-3} in hydrothermal sulfidic veins in low-grade volcanic-sedimentary rocks of Western Carpathians (Urban et al., 2006) are to the best of our knowledge probably the highest densities of tectonically unmodified CO_2 inclusions recorded so far in the Earth's crust.

In superdense CO_2 inclusions, a metastable homogenization by CO_2 vapor bubble disappearance (H type) occurs at $T < -56.6$ °C, and so below the triple point of pure CO_2, in the absence of a solid phase. The metastable H-type homogenization (Figure 4.13d) is enabled by the attenuation of phase transitions on cooling. This phenomenon is usually explained in terms of lack of nucleation centers and/or high surface tension of liquid that inhibits the nucleation of new phases. To overcome the metastability, an undercooling is needed to return the inclusion to the equilibrium state. For instance, the temperature of vapor bubble nucleation on cooling is lower than that of the bubble disappearance on heating, because the inclusion follows the metastable extension of the isochore from the field of the homogeneous liquid (Figure 4.14). In the same way, an undercooling as high as 50 °C is needed to shift the same inclusion from the metastable extension of the boiling curve to the sublimation curve and to nucleate a solid below the temperature at the triple point. If the undercooling is stopped prior to nucleation of the solid phase, the metastable homogenization to liquid can be observed not only in the superdense CO_2 inclusions, but also in gaseous mixtures with higher densities normally showing the S-type phase transformations (van den Kerkhof and Thiéry, 2001). The identical type of metastability is rather common in high-density NaCl–H_2O inclusions (Hurai et al., 2002a).

The density of hexagonal ice Ih ($D = 0.917$ g cm^{-3}) is lower than that of the associated aqueous liquid at the temperature and pressure of the triple point ($D = 0.9998$ g cm^{-3}). This unique feature

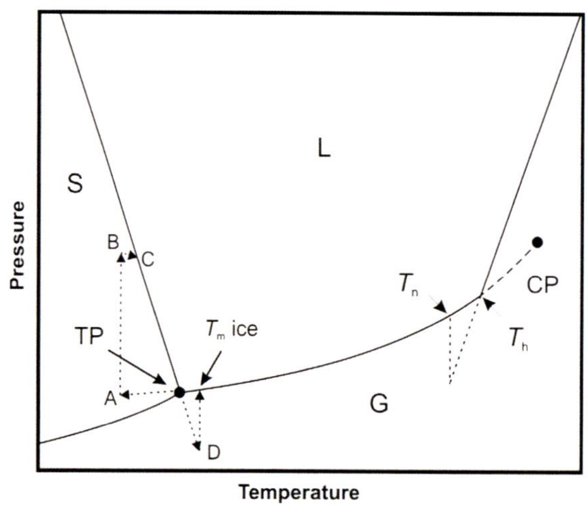

FIGURE 4.14

Schematic isochoric P–T projection of the H_2O system, with two-phase boundaries. S, L, and G denote solid, liquid, and vapor phases, respectively. $T_m$ice, metastable temperature of melting of ice; T_n, temperature of vapor bubble nucleation; T_h, temperature of homogenization.

of water maintains life in arctic regions and during cold climatic periods, because the icy cover of oceans, lakes, and rivers acts as an effective thermal insulator. On the other hand, the same effect causes serious problems in the microthermometric research of high-density aqueous inclusions, containing less than about 9 vol.% of vapor. This volume corresponds exactly to the relative volume expansion of the hexagonal ice during crystallization from the equivalent mass of the aqueous liquid. Due to the fixed fluid inclusion volume (isochoric system), the ice fills up the whole inclusion and triggers an internal overpressure, which normally results in the stretching and decrepitation of fluid inclusions sealed in soft minerals (calcite, barite). Small inclusions in silicates may withstand overpressures as high as several kilobars, however, the internal inclusion overpressure shifts the equilibrium from the metastable extension of the boiling curve (point A in Figure 4.14) into the field of homogeneous solid (point B). On warming, the ice melts along the melting curve along the C–D boundary, which has a negative slope in the H_2O system. Because nucleation of the vapor bubble is a metastable process, the triple point is crossed without any phase change and the ice resists in the absence of vapor to temperatures greater than 0 °C, until the equilibrium is restored at the point D, when the superheated ice disappears abruptly and simultaneously with the vapor bubble nucleation, thus returning the metastable equilibrium to the stable L+G boundary. The mechanism of the superheated ice nucleation, existing in aqueous fluid inclusions up to temperatures of 6.5 °C was first explained by Roedder (1967) who noticed that the difference between the stable and metastable transformation along the steeply dipping melting curve is so large, that a negative pressure should be developed on heating prior to return of the inclusion to the equilibrium state. The superheated ice cannot originate in other than aqueous fluid systems due to positive slopes of the melting curves and higher densities of solids compared to liquids. As a consequence, the volume of solid is smaller than that of the equivalent mass of liquid, and the stable G+S assemblage formed in the inclusions on freezing shifts the metastable L+G equilibrium onto the stable G+S sublimation curve.

Aqueous liquids in fluid inclusions cooled below the equilibrium L+G curve without vapor bubble nucleation are also put under the same metastable "superheated" state, when the matter undergoes a tensional stress without the loss of structural cohesion. The tensional stress, or "negative pressure," is also called the cavitation pressure. In synthetic fluid inclusions, cavitation pressures as low as −1.4 kbar have been estimated by the linear extrapolation of isochore from one-phase region below the boiling curve to the temperature of bubble nucleation (Shmulovich et al., 2009). It is worth remembering that stretched superheated metastable liquids under tension can also exist at the focus point of shock waves. Supercavitation is the use of the cavitation effect to create vapor bubbles in the liquid encompassing submarine projectiles or torpedoes. The bubbles greatly reduce friction, thereby enabling projectiles to reach extremely high speeds.

4.6.4 PHASE TRANSITIONS IN AQUEOUS INCLUSIONS

Compared to unary systems, one degree of freedom is added in binary systems together with additional possible phase transformations. For instance, the triple point converts to the quadruple point, where two solid phases coexist with the liquid and vapor. The solid phases melt within a certain temperature interval, thus giving rise to three-phase L+G+S curves. Peritectic quadruple points occur when one solid phase converts to another, for example, hydrohalite to halite or vice versa.

Low to moderately saline aqueous inclusions and low-density brines homogenize to liquid (total density is greater than that at the critical point) or to vapor (total density is lower than that at the critical

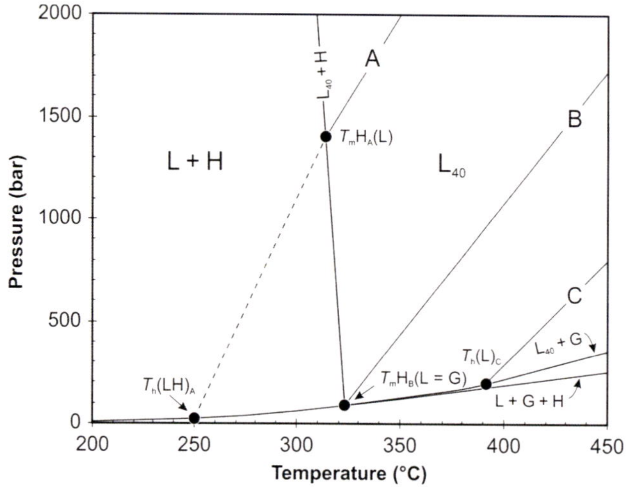

FIGURE 4.15

A portion of the H_2O–NaCl system with halite liquidus (L + H), vapor-saturated halite dissolution curve (L + G + H), and boiling curve (L + G) for a 40 mass% NaCl solution constructed from experimental data of Bodnar (1994). The isochores are shown for fluids A, B, and C with different densities and temperatures of phase transitions: T_mH(L), temperature of total homogenization by halite dissolution; T_h(LH), temperature of partial homogenization to liquid in the presence of halite; T_h(L), temperature of metastable total homogenization to liquid by bubble disappearance. The dashed line corresponds to the average isochoric slope in the L + H field, where the inclusion is composed of halite and aqueous liquid. Reprinted from Bodnar, R.J., 1994. Synthetic fluid inclusions: XII. The system H_2O–NaCl. Experimental determination of the halite liquidus and isochores for a 40 wt.% NaCl solution. Geochim. Cosmochim. Acta 58, 1053–1063, Copyright 1994, with permission from Elsevier.

point). The corresponding phase transformation sequences on heating are L+G→L or G, and L+G+S→L+G→L (high-salinity, low-density brines). However, a third important homogenization mode occurs in the high-salinity brines with a higher density, which may homogenize to liquid by the dissolution of halite according to the following scheme: L+G+S→L+S→L. This is because the isochoric path on heating leaves the three-phase L+G+S curve, traverses the two-phase L+S field, and finally enters into the field of the homogeneous liquid (the isochoric path A in Figure 4.15). In contrast, isochores for low-density, high-salinity fluid inclusions slide along the L+G+S curve until the halite dissolves in the presence of vapor, then follow the two-phase L+G surface and finally enter the homogeneous liquid field (the isochoric path C in Figure 4.15). In a very special case, the total homogenization to liquid may occur by simultaneous dissolution of vapor bubble and halite, at the intersection point of the three-phase L+G+S curve and two-phase L+G surface (the isochoric path B in Figure 4.15).

Another possible sequence of phase transitions, L+G+S→G+S→G, has not been recorded in natural aqueous fluid inclusions, because the dissolved salts are strongly partitioned in the high-density liquid phase. This phenomenon leads to the formation of extremely high-salinity aqueous liquids that originate by the fluid exsolution from felsic silicate melts at a low pressure, which is called second or retrograde boiling (Burnham and Ohmoto, 1980).

A total of three solids—ice, hydrohalite, and halite—may exist in the system H_2O–NaCl. Ice and hydrohalite coexist in the eutectic (quadruple) point *E* at −21.2 °C, whereas hydrohalite and halite are

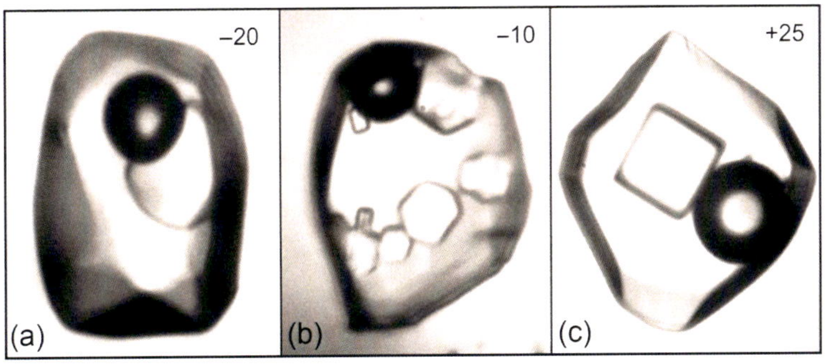

FIGURE 4.16

Aqueous fluid inclusions with various NaCl concentrations, showing three-phase assemblages of the aqueous liquid and vapor with ice (a), hydrohalite (b), and halite (c). Numbers in upper-right corners are temperatures in °C.

stable at the quadruple peritectic point P at 0.1 °C (Figure 2.8). The halite + hydrohalite assemblage often observed in frozen inclusions is metastable.

The temperature of melting of the last solid phase determines the total salinity of the system. For instance, ice is the last solid to melt at −15 °C in the aqueous inclusion in Figure 4.16a and the total salinity derived from the vapor-saturated ice liquidus equals 18.6 mass% (Table 4.2, or Equation (II.11) in Appendix II). In contrast, halite is the last solid phase to melt at 275 °C in the aqueous inclusion displayed in Figure 4.16c, and the total salinity derived from the vapor-saturated halite liquidus corresponds to 39.3 mass% (Equation (II.11), Appendix II). The exact determination of the salinity of the inclusion displayed in Figure 4.16b is not possible, because the hydrohalite melts metastably within a wide range of temperatures far above the stable peritectic temperature of 0.1 °C.

All three inclusions displayed in Figure 4.16 contain variable volume of vapor in the moment of the solid phase dissolution. Hence, the inferred salinity does not match exactly the three-phase curve, and the salinity and the volume of the vapor phase should be theoretically involved in the total salinity calculation.

Differences between the calculated and the actual salinity in NaCl-unsaturated aqueous inclusions can be inferred from Equations (4.4) and (1.14). After combination and rearrangement we get

$$m = m_L - \overline{V}_G D_G (m_L + m_G), \tag{4.14}$$

where m, m_L, and m_G are mass% NaCl concentrations in whole inclusion, aqueous liquid, and vapor bubble, respectively, $\overline{V}_G$ is the volume fraction of a vapor bubble, and D_G is the density of a vapor bubble. Further rearrangement results in the equation correlating the bulk salinity (m) with that of the liquid phase (m_L)

$$m_L = \frac{m + \overline{V}_G D_G m_G}{1 - \overline{V}_G D_G} \tag{4.15}$$

Because the salinity and density of vapor phase along ice liquidus is essentially zero, Equation (4.15) reduces to $m_L = m$. In other words, the total salinity of halite-unsaturated aqueous inclusions is equal to that of the liquid phase, regardless of the vapor bubble volume.

Table 4.2 Salinities Corresponding to Measured Freezing Point Depression (FDP) Below 0 °C, Calculated According to Equation (II.11) in Appendix II (Bodnar, 1993)

FDP (°C)	0.0	0.1	0.2	0.3	0.4	0.5	0.6	0.7	0.8	0.9
0.	0.00	0.18	0.35	0.53	0.70	0.88	1.05	1.22	1.40	1.57
1.	1.74	1.91	2.07	2.24	2.41	2.57	2.74	2.90	3.06	3.23
2.	3.39	3.55	3.71	3.87	4.03	4.18	4.34	4.49	4.65	4.80
3.	4.96	5.11	5.26	5.41	5.56	5.71	5.86	6.01	6.16	6.30
4.	6.45	6.59	6.74	6.88	7.02	7.17	7.31	7.45	7.59	7.73
5.	7.86	8.00	8.14	8.28	8.41	8.55	8.68	8.81	8.95	9.08
6.	9.21	9.34	9.47	9.60	9.73	9.86	9.98	10.11	10.24	10.36
7.	10.49	10.61	10.73	10.86	10.98	11.10	11.22	11.34	11.46	11.58
8.	11.70	11.81	11.93	12.05	12.16	12.28	12.39	12.51	12.62	12.73
9.	12.85	12.96	13.07	13.18	13.29	13.40	13.51	13.62	13.72	13.83
10.	13.94	14.04	14.15	14.25	14.36	14.46	14.57	14.67	14.77	14.87
11.	14.97	15.07	15.17	15.27	15.37	15.47	15.57	15.67	15.76	15.86
12.	15.96	16.05	16.15	16.24	16.34	16.43	16.53	16.62	16.71	16.80
13.	16.89	16.99	17.08	17.17	17.26	17.34	17.43	17.52	17.61	17.70
14.	17.79	17.87	17.96	18.04	18.13	18.22	18.30	18.38	18.47	18.55
15.	18.63	18.72	18.80	18.88	18.96	19.05	19.13	19.21	19.29	19.37
16.	19.45	19.53	19.60	19.68	19.76	19.84	19.92	19.99	20.07	20.15
17.	20.22	20.30	20.37	20.45	20.52	20.60	20.67	20.75	20.82	20.89
18.	20.97	21.04	21.11	21.19	21.26	21.33	21.40	21.47	21.54	21.61
19.	21.68	21.75	21.82	21.89	21.96	22.03	22.10	22.17	22.24	22.31
20.	22.38	22.44	22.51	22.58	22.65	22.71	22.78	22.85	22.91	22.98
21.	23.05	23.11	23.18							

Chou (1987) analyzed m and m_L along the vapor-saturated halite liquidus and concluded that the largest discrepancies occur when halite dissolves at ~550 °C in the presence of a vapor bubble. The inaccuracy of the salinity determination increases with the temperature and the volume of the associated vapor bubble, because the density and the salinity of the vapor also increase with temperature. The discrepancies between the actual and measured salinity can attain several mass% and contribute to uncertainties resulting from the inaccuracy of the measured temperature, and deviations of natural multicomponent fluid inclusion systems from the binary H_2O–NaCl system.

The vapor-unsaturated melting of halite can be affected by pressure, which steeply rises in fluid inclusions heated above the temperature of vapor bubble disappearance. This is not the case with the inclusions containing 51 mass% NaCl (Equation (2.1)), in which the vapor-unsaturated halite liquidus is perpendicular to the temperature axis, and there is no difference between the vapor-saturated and vapor-unsaturated halite dissolution temperatures. However, due to variable slopes of other halite liquidi, the total salinity will be underestimated at concentrations of less than 51 mass% NaCl, and overestimated at greater concentrations, when recalculated in terms of the vapor-saturated halite liquidus (Equation (II.10), Appendix II).

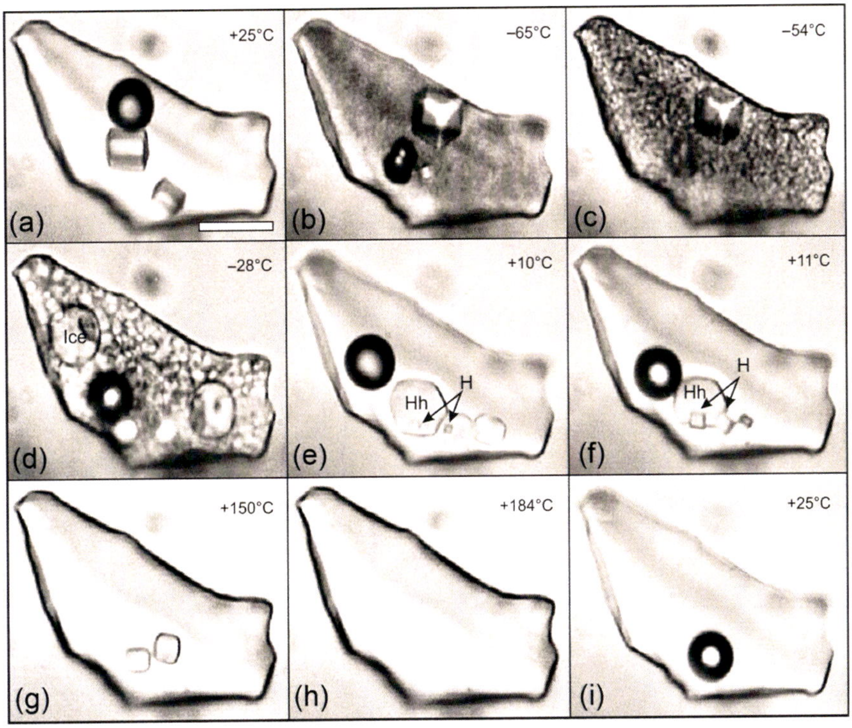

FIGURE 4.17

Phase transformations in a brine inclusion in quartz from hydrothermal polymetallic siderite veins of the Western Carpathians (Hurai et al., 2002a). The halite-bearing inclusion composed of two halite crystals, vapor bubble, and aqueous liquid at room temperature (a) becomes completely solid (b) after double freezing at −80 and −65 °C. Initial melting of the completely frozen inclusion is indicated by the coarsening at −54 °C (c) near the eutectic point. Halite is resorbed by growing hydrohalite crystals after several cooling/warming cycles and the ice melts in the presence of hydrohalite, aqueous liquid, and vapor bubble at −25.4 °C (d). Hydrohalite instantaneously transforms metastably into halite (peritectic melting) at temperatures above 10 °C (e, f). The vapor bubble diminishes at +150 °C prior to halite dissolution (g) and the inclusion becomes homogenized at +184 °C after dissolution of halite crystals (h). On cooling, the halite fails to nucleate, giving rise to a metastable L+G assemblage (i). The numbers in upper-right corners refer to the temperature of the cooling/heating stage. The bar scale in (a) corresponds to 10 μm.

Inclusions exhibiting stable total homogenization by halite dissolution may homogenize also metastably, by bubble disappearance, if the halite fails to precipitate on cooling (Figure 4.17i). This behavior is observed in the inclusions with total salt concentrations of less than 35 mass% (Hurai et al., 2002a). The metastable homogenization—T_h(L)—occurs in the moment when the one-phase-field isochore intersects a metastable extension of the L+G curve in the L+H field (Figure 4.18).

Differences among the T_h(LH), T_h(L), and T_mH(L) values tend to diminish with the decreasing fluid density. For instance, a 40 mass% NaCl–H_2O fluid inclusion labeled A in Figure 4.18, should yield T_h(LH) = 250 °C, T_h(L) = ~230 °C and T_mH(L) = 313 °C. The inclusion B with the same composition

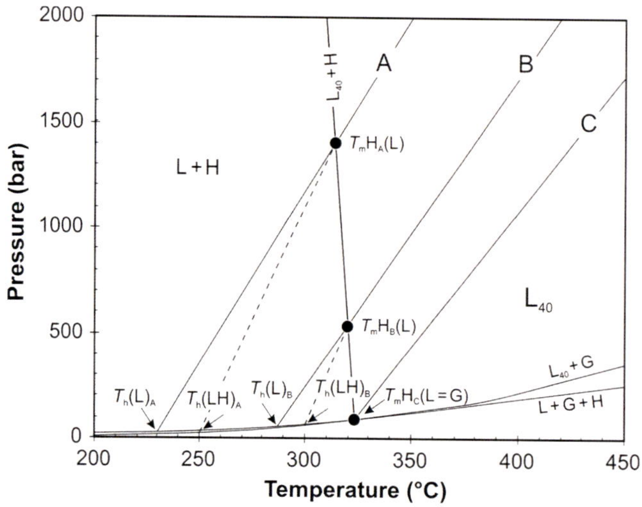

FIGURE 4.18

A portion of the H_2O–NaCl system with halite liquidus (L+H), vapor-saturated halite dissolution curve (L+G+H) and boiling curve (L+G) for a 40 mass% NaCl solution constructed from experimental data of Bodnar (1992, 1994). Isochores are shown for fluids A, B, and C with different densities, and corresponding temperatures of phase transitions: T_mH(L), temperature of total homogenization by halite dissolution; T_h(LH), temperature of partial homogenization to liquid in the presence of halite; T_h(L), temperature of metastable total homogenization to liquid by bubble disappearance. Dashed lines are average isochoric slopes in the L+H field, where the inclusions are composed of halite and aqueous liquid. The isochoric slopes within the L+H field are in fact curves, because the salinity increases on heating due to halite dissolution (Lecumberri-Sanchez et al., 2012).

but a lower density would exhibit the respective phase transitions at 300, ~297, and 320 °C. Finally, all three temperatures of phase transitions converge at 323 °C for the inclusion labeled C, whose one-phase-field isochore extends directly from the vapor-saturated halite liquidus, without intersecting the L+H field. This temperature exactly corresponds to the vapor-saturated halite liquidus of the particular inclusion—T_mH(L=G).

The T_mH(L=G) value can be determined using microthermometric data from a population of coeval brine inclusions with the same concentration and different density. In this case, the T_h(LH), T_h(L), and T_mH(L) values and their differences plot along straight lines, which converge at T_mH(L=G). Alternatively, the vapor-saturated halite liquidus can also be determined in a single inclusion using a method of sequential re-equilibration (Hurai et al., 2002a). The investigated brine inclusion is stepwise overheated above its total homogenization temperature and the temperatures of phase transitions are recorded during each cycle, until the inclusion decrepitates. The T_mH(L=G) value can then be located for the inclusions whose composition remains unmodified by leakage, which means that they follow the isoplethal path. However, the leakage can be identified easily either optically (creation of cracks emanating from tips of the inclusion, often accompanied by instantaneous bubble nucleation), or by a change of the slope of the correlation line connecting the T_h(LH), T_h(L), and T_mH(L) values with their differences recorded during individual heating–cooling cycles (Figure 4.19).

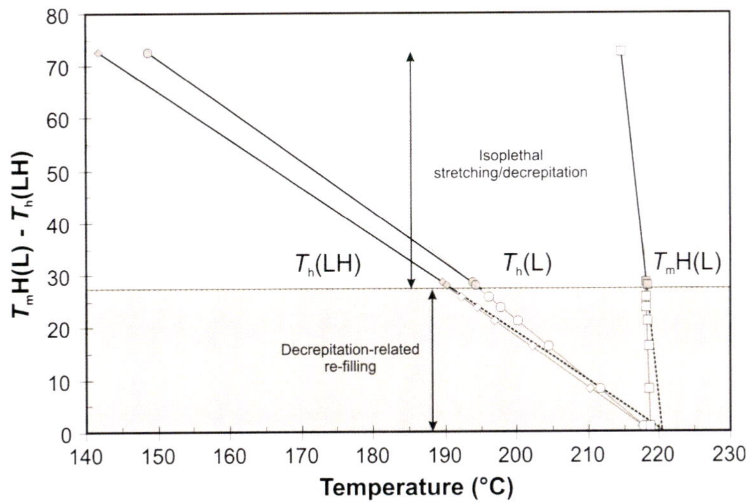

FIGURE 4.19

Plot of the microthermometric data from a stepwise re-equilibration experiment, illustrating the isoplethal re-equilibration followed by leakage reflected in changed slope of regression lines. A total of eight overheating cycles was made until the inclusion reached the L+G+H boundary. Note that the original vapor-saturated halite liquidus was downshifted from ~221 to ~217 °C due to the leakage occurring during the third cycle (Hurai et al., 2002a).

Most natural aqueous solutions are dominated by NaCl accompanied by another salt, which increases the degree of freedom of the fluid inclusion system. Accordingly, the temperatures to melt two solids are needed to derive the composition in ternary systems. Although numerous solubility phase diagrams exist (e.g., Borisenko, 1977, 1982), most practical problems can be solved using experimental phase diagrams of the H_2O–NaCl–$CaCl_2$ and H_2O–NaCl–KCl systems.

The system H_2O–NaCl–$CaCl_2$ is typical of $CaCl_2$-dominated eutectic solution, with the composition projecting close to the $CaCl_2$–H_2O join (Figure 4.20). Consequently, NaCl-rich compounds, hydrohalite and halite associated with ice, will be the last solid phases to melt in the vast majority of natural fluid inclusions. Schiffries (1990), however, reported on extremely high $CaCl_2$ concentrations and dissolution of antarcticite ($CaCl_2{\cdot}6H_2O$) in brine inclusions of magmatic origin trapped in basic intrusions.

Figure 4.17 illustrates phase transformations in a brine inclusion with T_e value around −54 °C diagnostic for the H_2O–NaCl–$CaCl_2$ system. On heating above T_e, the inclusion contains a vapor bubble, aqueous liquid, hydrohalite, and ice, thus moving along the ice+hydrohalite cotectic curve (the melting path extending from the eutectic point E in Figure 4.20). Total dissolution of the ice occurs at −25.4 °C, thus leaving the hydrohalite as the only stable solid in the fluid inclusion. At this moment, the fluid isochore leaves the ice+hydrohalite cotectic curve and enters the hydrohalite stability field. Further temperature increase leads to sluggish dissolution of the hydrohalite, and composition of the associated solutions changing along the tie line connecting the temperature of ice dissolution along the ice+hydrohalite cotectic curve with the hydrohalite composition (61.9 mass% NaCl) on the NaCl–H_2O join (the melting path 1). The hydrohalite should break down to halite and release water after

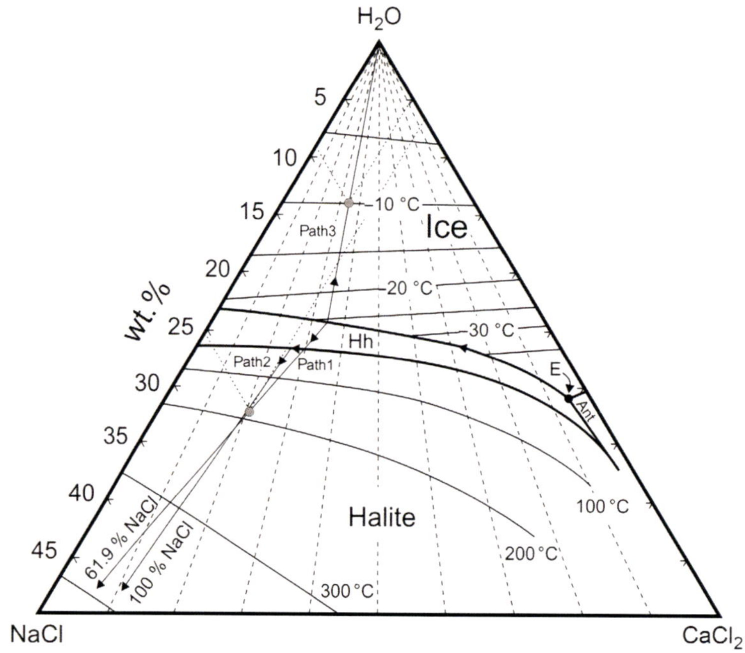

FIGURE 4.20

Phase diagram of the H_2O–NaCl–$CaCl_2$ system (compiled from Vanko et al., 1988; Williams-Jones and Samson, 1990; Oakes et al., 1990; Yanatieva, 1946). See the text and Figure 4.17 for an explanation of melting paths 1 and 2 designated by arrows.

hitting the hydrohalite + halite cotectic curve. In this case, the melting path should follow for a short distance the halite–hydrohalite cotectic curve until all hydrohalite is converted to halite. Afterward, the composition of residual liquid should move along the tie line connecting the temperature of hydrohalite breakdown on the H + Hh cotectic curve with the NaCl apex of the phase diagram, thus following the melting path 2. In real inclusions, however, the halite-hydrohalite transition is metastable and so cannot be used for the determination of salinity. This problem can be circumvented by using the intersection of the temperature of total halite dissolution (185 °C) with the melting path 1, which defines the final composition of the fluid devoid of any solid phases (25.1 mass% NaCl, 6.5 mass% $CaCl_2$, ~0.8 NaCl/NaCl + $CaCl_2$ mass ratio). However, the halite did not dissolve in the presence of the vapor bubble, and the actual composition is somewhat different owing to an increased pressure. The differences between the temperatures of vapor-saturated and vapor-unsaturated halite melting attain 4–7 °C in natural brines with the total salt concentration of 30–35 mass% (Hurai et al., 2002a), thus resulting in the underestimation of the actual total salinity by less than 1 mass% NaCl equation.

Either hydrohalite or ice is the last solid to melt in the halite-undersaturated aqueous inclusions. In the first case, the hydrohalite melting is also metastable, except for the temperatures very close to the ice + hydrohalite cotectic curve, and cannot be directly used for the determination of the fluid composition. However, due to a high density of isotherms within the hydrohalite stability field, the fluid composition can be estimated sufficiently precisely, using just the compositional limits defined by two intersection points of the corresponding tie line with the ice + hydrohalite and hydrohalite + halite cotectic boundaries.

Several problems can be faced during attempts to freeze natural brine inclusions. Some inclusions fail to freeze out even in liquid nitrogen, namely at −196 °C, but sometimes they do this on slow warming between −80 and −50 °C. The failure to nucleate clathrates, halite, or vapor bubble on cooling fully homogenized inclusions may lead to various metastable freezing paths (M1–M3), schematically depicted in Figure 4.21. Double freezing indicates nucleation of ice and hydrohalite, and such inclusions usually exhibit stable phase transition along the S-path pronounced by a thick solid line. If a small amount of CO_2 is present, a triple freezing is observed. Other complications may arise from the failure of halite to convert to hydrohalite. In this case, several freezing–heating cycles must be made. If the

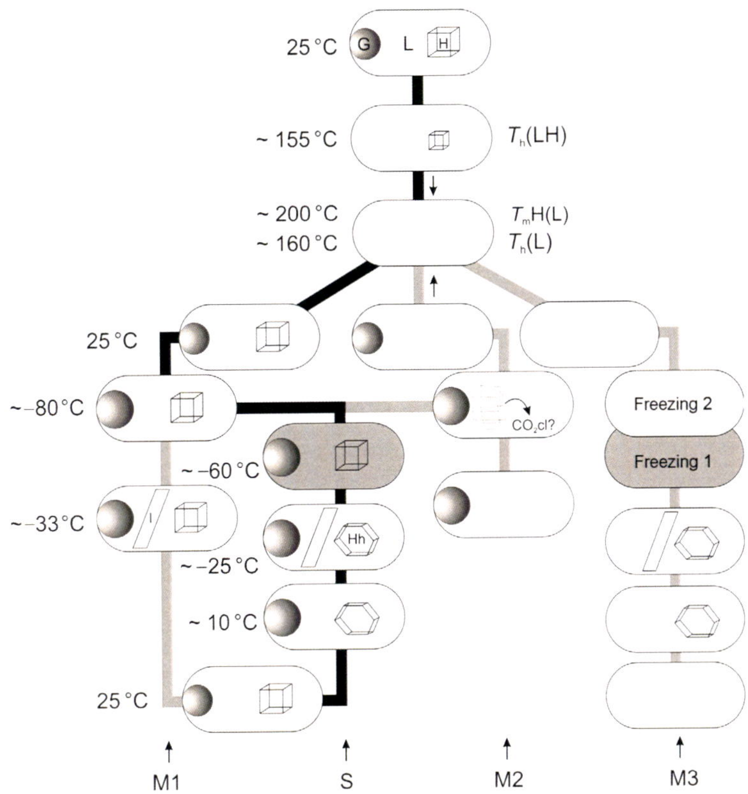

FIGURE 4.21

Systematics of phase transitions in brine inclusions, with approximate temperature ranges typical of each phase change as observed in natural inclusions in quartz from siderite veins of the Western Carpathians (Hurai et al., 2002a). The most frequently observed sequence is indicated by the solid line (see also Figure 4.17 for comparison). Other paths are metastable sequences, originating due to failure to nucleate solid phases and/or vapor bubble on cooling and freezing. H, halite; Hh, hydrohalite; I, ice; CO_2cl, CO_2 clathrate.

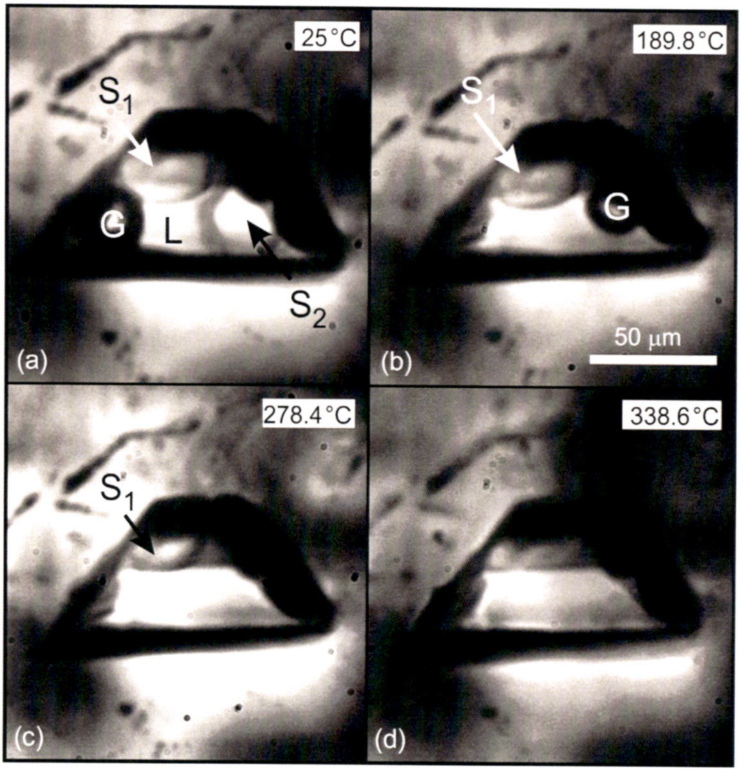

FIGURE 4.22

Infrared images of primary fluid inclusion in Fe-sphalerite from the Tisovec skarn deposit genetically related to a Neogene diorite intrusion (Hurai and Huraiová, 2011). The shown sequence of phase transitions includes: (a) the appearance of the inclusion at room temperature; (b) the dissolution of sylvite; (c) the partial homogenization of vapor bubble; (d) the total homogenization by halite dissolution. S_1, solid phase, probably halite; S_2, solid phase, probably sylvite; G, vapor phase.

conversion is incomplete, the calculated salinities are overestimated, because ice melts at a lower temperature in the presence of halite than it does in the presence of hydrohalite.

Phase transitions in a high-salinity H_2O–NaCl–KCl inclusion are illustrated in Figure 4.22. At room temperature, the inclusion contains halite (S_1) plus another solid (S_2), presumably sylvite, which always dissolved at a lower temperature, between 60 and 190 °C. The halite dissolved at 339 °C, and final composition inferred from the phase diagram in Figure 4.23 reads 29 mass% NaCl and 28 mass% KCl. The resulting total solute concentration (57 mass%) may slightly differ from the actual value because of the absence of vapor during the melting of halite.

Density (molar volume) of aqueous inclusions is routinely calculated from that in the reference H_2O–NaCl system, using empirical EOSs calibrated to experimental data for two-phase L+G curves

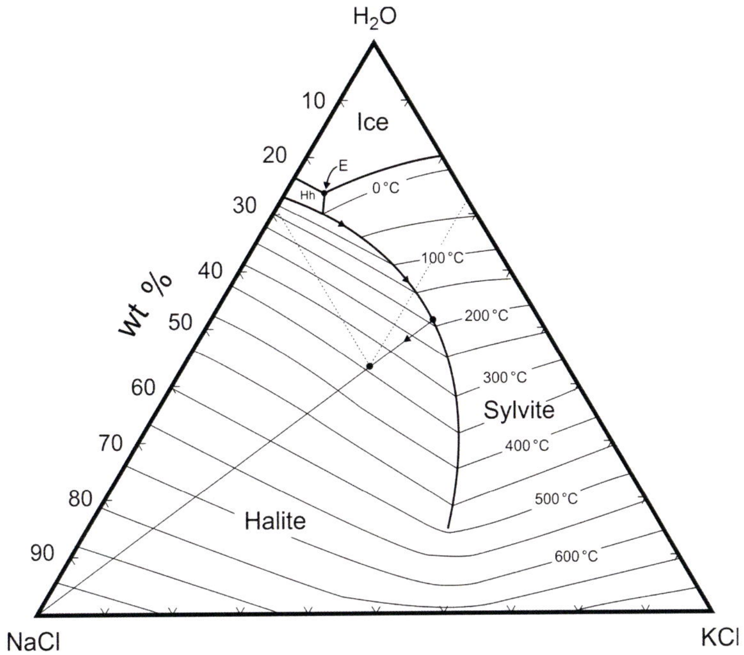

FIGURE 4.23

Vapor-saturated solubility surfaces and cotectic curves in the system H_2O–NaCl–KCl (Sterner et al., 1988) with a melting path (arrows) defined by phase transitions in the aqueous inclusion displayed in Figure 4.22.

(Haas, 1976; Bodnar, 1983), or three-phase L+G+H curves (Bodnar, 1983; Bischoff, 1991). All these equations are listed in Appendix II. Zhang and Frantz (1987) published equations correlating the density and homogenization temperatures of fluid inclusions in the H_2O–KCl and H_2O–$CaCl_2$ systems; however, these equations overestimate the calculated densities.

PROBLEM 4.4 CALCULATING VAPOR-SATURATED HALITE LIQUIDUS

Calculate the vapor-saturated halite liquidus in artificially re-equilibrated halite-bearing fluid inclusion using the microthermometric data in Table 4.3. Calculate the fluid composition using the vapor-saturated solubility diagram of the system H_2O–NaCl–$CaCl_2$ (Figure 4.20).

Solution

Construct the diagram correlating temperatures of phase transitions and their differences, create a linear data fit, and derive the vapor-saturated halite liquidus at T_h(LH) = T_h(L) = T_mH(L) (Figure 4.24).

The calculated vapor-saturated halite liquidus is located at 216 °C, and combined with $T_m$ice = −25.8 °C it yields the bulk composition corresponding to 27.4 mass% NaCl and 6.4 mass% $CaCl_2$. Note that the difference between the vapor-unsaturated and the vapor-saturated halite liquidi is 5.2 °C, and the total salinity derived from the T_mH(L) value is underestimated by 0.3 mass%.

Table 4.3 Temperatures of Phase Transitions in Artificially Re-equilibrated Brine Inclusion FI-5, 40 × 20 μm, T_e = −58 °C

Cycle	$T_m$ice	T_h(LH)	T_mH(L)	T_h(L)	Overheat
1.	−25.8	145.4	210.8	152.6	225
2.	nd	186.5	213.9	190.1	228
3.	nd	189.8	214.2	–	230
4.	−25.8	191.8	214.3	–	233
5.	nd	195.4	214.5	–	237.5
T_mH(L=V)		***216.0***	***216.0***	***216.0***	–

Composition: 27.4 mass% NaCl + 6.4 mass% CaCl$_2$.
Composition derived from T_mH(L): 26.9 mass % NaCl + 6.6 mass % CaCl$_2$.

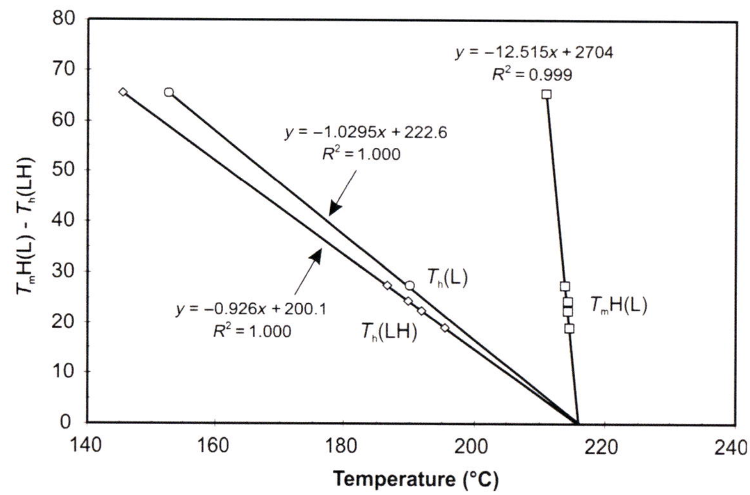

FIGURE 4.24

Linear fits to data in Table 4.3.

PROBLEM 4.5 CALCULATING MOLAR VOLUME OF AN H_2O–NaCl–$CaCl_2$ INCLUSION

Calculate the molar volume of an H_2O–NaCl–$CaCl_2$ inclusion using microthermometric data in Table 4.3.

Solution

Recalculate the bulk composition of the inclusion (27.4 mass% NaCl, 6.4 mass% $CaCl_2$) to molar fractions. Derive the composition of the aqueous liquid coexisting with halite at 145 °C in the following way:

1. Connect the NaCl apex in the phase diagram of the H_2O–NaCl–$CaCl_2$ system (Figure 4.20) with the temperature of halite dissolution (216 °C) on the tie line connecting the hydrohalite composition (61.9 mass% NaCl) and the temperature of ice melting (−25.8 °C) on the ice-hydrohalite cotectic curve.

Continued

2. Derive the composition of the NaCl-saturated liquid at 145 °C along the tie line (23.5 mass% NaCl, 6.7 mass%$CaCl_2$).
3. Calculate the molar fraction of NaCl (0.0927).

Molar fraction of NaCl in halite ($\overline{x}_H$) coexisting with the aqueous liquid at 145 °C can be inferred from the **mass-balance equation**

$$\overline{x}_H = (\overline{x}\text{NaCl} - \overline{x}\text{NaCl}_L)/(\overline{x}\text{NaCl}_H - \overline{x}\text{NaCl}_L) = 0.02083 \quad (4.16)$$

where $\overline{x}$NaCl is the bulk NaCl molar fraction at 216 °C (0.1116), $\overline{x}\text{NaCl}_L$ is the NaCl molar fraction in the liquid phase at 145 °C (0.0927), and $\overline{x}\text{NaCl}_H$ is that in halite (1.0).

Density [g cm^{-3}] of the aqueous liquid at 145 °C can be calculated from the empirical fit to a three-phase curve of the H_2O–NaCl system (Equation (II.13) or (II.14), Appendix II), whereas that of halite is defined by the linear fit to thermal expansion data of Skinner (1966) valid for temperatures from 20 to 600 °C

$$D_H = 2.1737 - 0.0003 \times T \quad (4.17)$$

Express halite and aqueous liquid densities at 145 °C (2.13 and 1.158 g cm^{-3}, respectively) as molar volumes (27.4355 and 19.9111, respectively). Enter molar fractions and the molar volumes of the halite and the coexisting aqueous liquid in Equation (4.16) and calculate the total molar volume of the inclusion

$$\overline{V}_{\text{tot.}} = \overline{x}_H \overline{V}_H + \overline{x}_L \overline{V}_L \quad (4.18)$$

$$\overline{V}_{\text{tot.}} = 0.02083 \times 27.4355 + (1 - 0.02083) \times 19.9111 = 20.07 \text{cm}^3/\text{mol}$$

Calculation with the temperature of vapor-unsaturated halite melting (210.8 °C) and the salinity of 32.4 mass% NaCl equation derived from Equation (II.11), Appendix II, yields the molar volume of the same inclusion equal to 20.74 cm^3 mol^{-1}.

4.6.5 PHASE TRANSITIONS IN GASEOUS INCLUSIONS

Inclusions of gaseous mixtures can be divided into two major groups according to the homogenization mode, which means the final phase transformation observed on heating (van den Kerkhof, 1988, 1990; Thiéry et al., 1994a; van den Kerkhof and Thiéry, 2001). The inclusions that homogenize either to liquid by the vapor bubble disappearance, or to vapor by the evaporation of liquid, belong to the H group. Inclusions that homogenize by melting or sublimation of a solid create the S group. The number of observed phase transformations from –160 °C to the total homogenization defines the individual subtypes numbered 1–4 (Figure 4.25).

The presence of CH_4 or N_2 in the CO_2-dominated inclusions is reflected in a shift of the temperature of melting of CO_2 solid below that at the triple point of pure CO_2, namely −56.6 °C. In contrast, minor CO_2 in CH_4- or N_2-dominated fluids would precipitate as CO_2 solid at temperatures below −100 °C, thus leaving behind an essentially pure N_2 or CH_4 residual fluid. A solid phase does not form in CH_4–N_2 inclusions cooled in liquid nitrogen, and the only observable phase transition is the homogenization to liquid or vapor. In the absence of critical homogenization, the composition of such inclusions can be determined only by means of Raman spectroscopy.

Phase transitions in an inclusion dominated by CH_4 are illustrated in Figure 4.26. The sequence of observed phase transitions corresponds to the H2 type: L+G+S→L+G→L. The inclusion consists

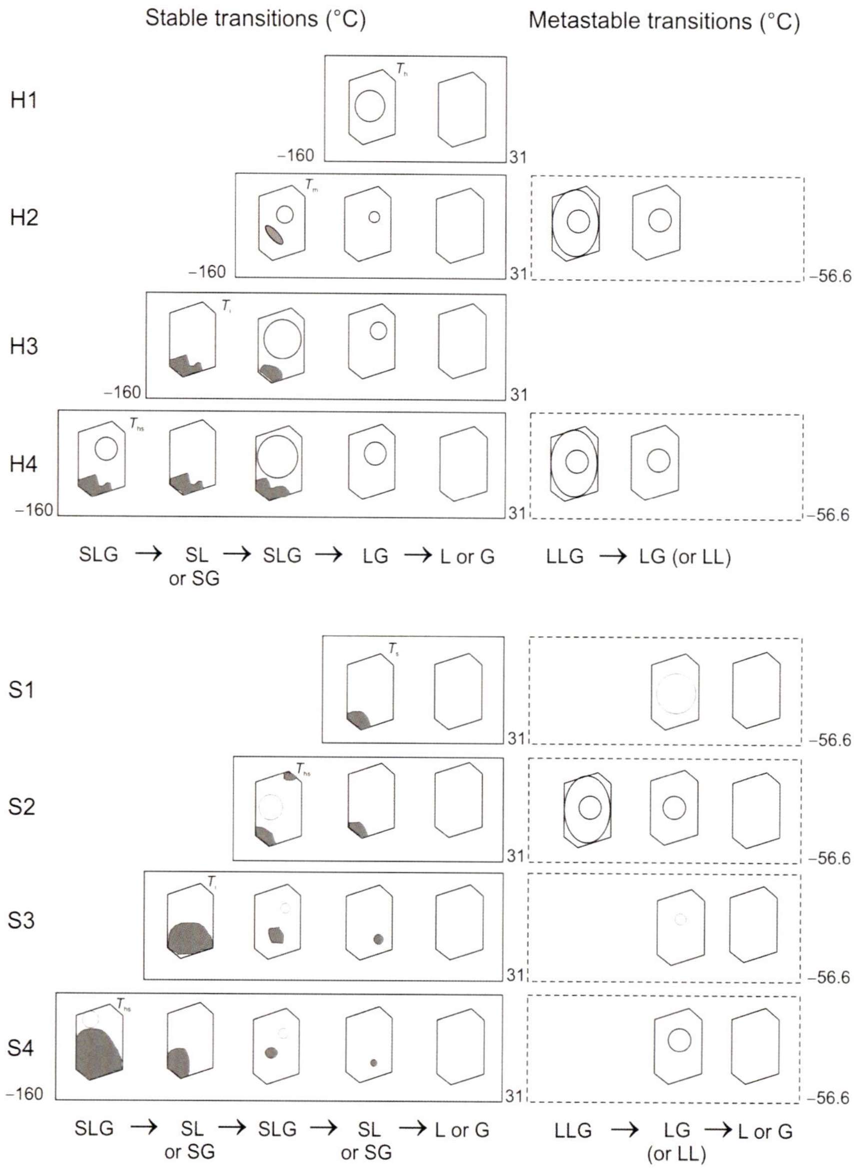

FIGURE 4.25

Classification of phase transitions in CO_2–CH_4–N_2 fluid inclusions. Reprinted from van den Kerkhof, A.M., Thiéry, R., 2001. Carbonic inclusions. Lithos 55, 49–68, Copyright 2001, with permission from Elsevier.

of two solid phases and a vapor bubble at a temperature of −196 °C. The temperature of initial melting (−182.6 °C) is close to that at the triple point of pure methane. Octahedral shape of dendritic crystals and melting at −110 °C reveal CO_2, although hydrogen sulfide may also form similar crystals (van den Kerkhof and Hein, 1992). Therefore the presence of CO_2 must be verified by Raman spectrometry.

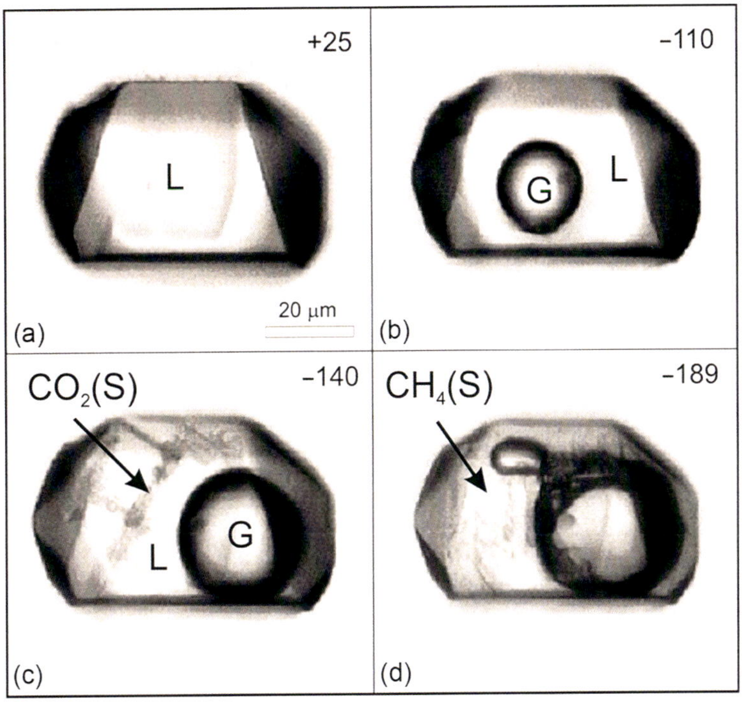

FIGURE 4.26

Phase transitions in a CH_4-dominated fluid inclusion in quartz from mineralized joints of Paleogene sandstones, Vel'ký Lipník, Spišská Magura Mts. (Hurai et al., 2002b). Homogeneous inclusion (a) splits to liquid and vapor phases on cooling below −100 °C (b). Dendritic crystals precipitate instantaneously at −120 °C (c), and the residual liquid freezes out at −194 °C (d). Stable phase transitions involve initial melting at −182.6 °C, melting of the dendritic crystals at −110 °C, and homogenization to liquid at −90 °C. The observed sequence of phase transformations from −160 °C, that is SLG→LG→G, corresponds to the H2-type diagnostic of CO_2–CH_4 mixtures.

Composition and molar volume of the illustrated inclusion can be easily inferred from the isothermal $\overline{V}-X$ phase diagram of the CO_2–CH_4 system (Figure 4.27). The main advantage of this projection is the direct correlation of microthermometric (temperature of phase transition) and Raman (composition) data without the need for detailed knowledge of the system topology. The obtained parameters can be directly used as input data to calculate isochores from the corresponding EOS.

A unique sequence of phase transitions is possible in the CO_2–CH_4 system due to a saddle-shaped three-phase curve, which is intersected twice by specific SL(G) tie lines (Figure 4.27c). This behavior results in additional phase transitions in the inclusions containing 15–86 mol.% CO_2, with molar volumes between 38 and 63 $cm^3\ mol^{-1}$. For instance, some gaseous inclusions in quartz from Paleogene sandstones (the Dukla unit, Outer Carpathians) exhibit the S+L+G→S+L phase transition at −77.6 °C, bubble renucleation S+L→S+L+G at −68.4 °C, melting of the CO_2 solid (S+L+G→L+G) at −66.5 °C, and total homogenization to liquid (L+G→L) at −65 °C. The sequence of phase transitions corresponds to the H4 subtype. The inclusion contains 38 mol.%

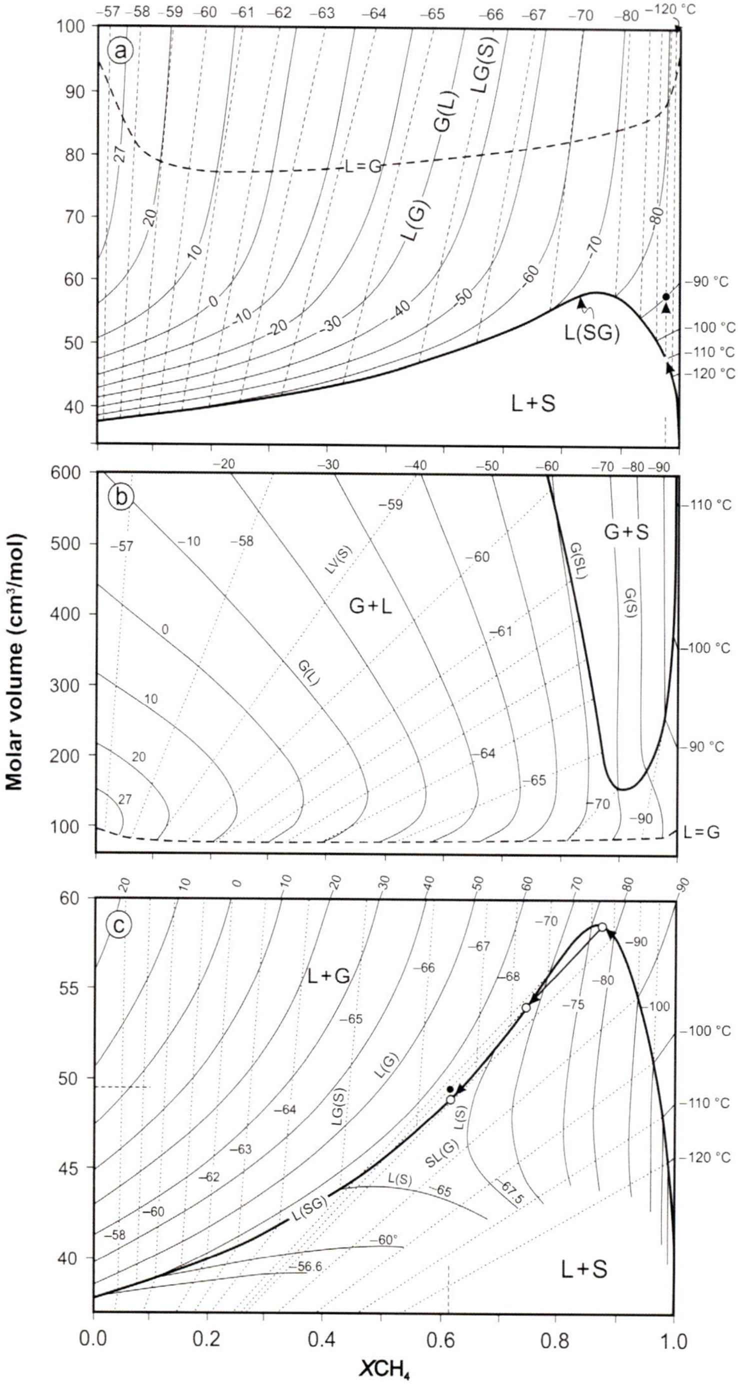

FIGURE 4.27

Isothermal $\overline{V}-X$ projections of the CO_2–CH_4 system (van den Kerkhof and Thiéry, 2001, van den Kerkhof, personal comm.). The dashed L=V curve is a projection of the critical curve; thick LGS curves are projections of the three-phase curve; isotherms L(G), G(L), and L(S) are projections of the two-phase boiling, condensation, and

Continued

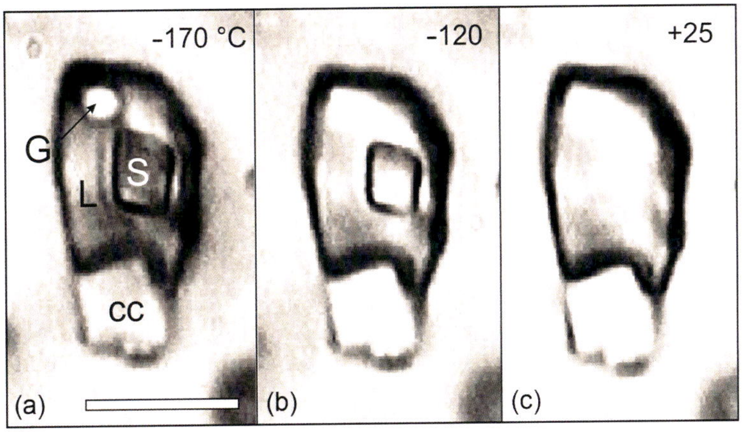

FIGURE 4.28

Phase changes in a N_2–CO_2 inclusion from leucosome of metabasite migmatite from the Western Tatra Mountains (Hurai et al., 2000). Partial homogenization to liquid took place at −159 °C and the inclusion homogenized by dissolution of CO_2 solid at −75 °C. The sequence of phase transformations corresponds to the S2 type. cc is calcite. Scale bar represents 10 μm.

CO_2 and its molar volume is 49 cm^3 mol^{-1}, as inferred from the phase diagram in Figure 4.27c. Similar H4-type CO_2–CH_4 inclusions have been observed in quartz from a migmatized gneiss of Norway (van den Kerkhof et al., 1991).

The most common phase transition sequence in N_2–CO_2 inclusions corresponds to the S2 type (Figure 4.28). In this particular case, the partial homogenization to liquid occurred at −159 °C (S+L+G→S+L) and the CO_2 solid dissolved in the liquid phase at −75 °C. According to the phase diagram in Figure 4.29, the inclusion contains 13 mol.% CO_2 together with the dominant N_2, and the bulk molar volume corresponds to 45 cm^3 mol^{-1}.

More complicated sequences of phase transitions are possible at higher CO_2 concentrations, when SL(G) or SG(L) tie lines traverse across the whole two-phase LS field and crosscut the upper limb of the three-phase LGS curve. In this case, a renucleation of gas bubble or condensation of liquid occurs during the rapid melting of the CO_2 solid phase. Final homogenization to liquid or vapor occurs within the LG field. Such an inclusion is illustrated in Figure 4.30, where partial homogenization to liquid of the N_2-rich phase occurred at −149 °C. The gas bubble renucleated immediately before the CO_2 solid dissolution at −61.4 °C, so that the L+S+G assemblage existed over a very short interval of 0.2–0.3 °C.

FIGURE 4.27—Cont'd sublimation surfaces, respectively. Dashed tie lines LG(S) connect compositions and molar volumes of liquid phases along the three-phase LSG curve with those of gaseous phases along the GSL curve. Dashed tie lines SL(G) connect the composition and molar volumes of the liquid phase along the LGS curve with those of the coexisting CO_2 solid. The solid circle in (b) is the projection of the microthermometric data from the inclusion illustrated in Figure 4.26. The solid circle in (c) is the projection of the H4-type inclusion discussed in the text. Slopes of tie lines within the L+S field reflect molar volumes of CO_2 solid at different temperatures, as inferred from data along the sublimation curve of pure CO_2 (Ražnjevic, 1975). Phase diagram reprinted from van den Kerkhof, A.M., Thiéry, R., 2001. Carbonic inclusions. Lithos 55, 49–68, Copyright 2001, with permission from Elsevier.

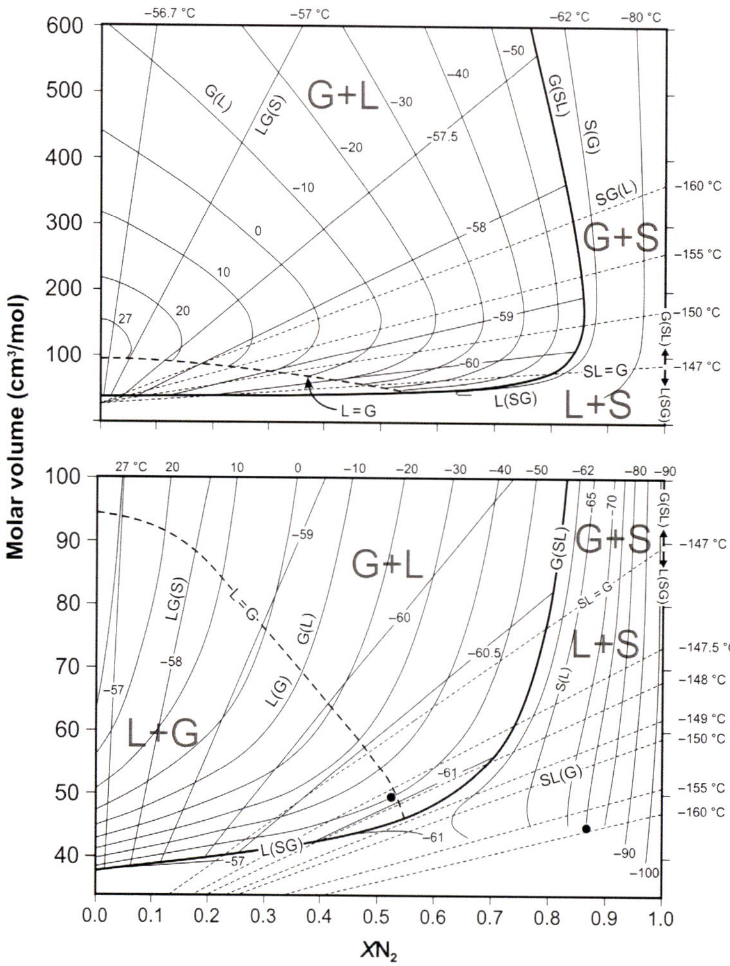

FIGURE 4.29

Isothermal $\overline{V}-X$ projections of the CO_2–N_2 system (van den Kerkhof and Thiéry, 2001, van den Kerkhof, personal comm.). The dashed L=V curve is the projection of the critical curve; thick curves are projections of the three-phase LGS curve; isotherms L(G) and G(L) are projections of the two-phase boiling and condensation surfaces, respectively. Dashed tie lines SL(G) and SG(L) connect compositions and molar volumes of the liquid phase along the lower three-phase LSG curve with those of the coexisting CO_2 solid. Solid circles are projections of phase transformations in the fluid inclusions documented in Figures 4.28 and 4.30. Phase diagram reprinted from van den Kerkhof, A.M., Thiéry, R., 2001. Carbonic inclusions. Lithos 55, 49–68, Copyright 2001, with permission from Elsevier.

Shortly after the CO_2 solid dissolution, the total homogenization to liquid was observed at −61 °C, but in fact, the final homogenization was close to the critical point, because the meniscus between the solid and liquid phases was almost invisible. The observed phase changes document that a very low heating rate (max. 0.5 °C/min.) must be maintained to record rapid, near-critical phase transformations.

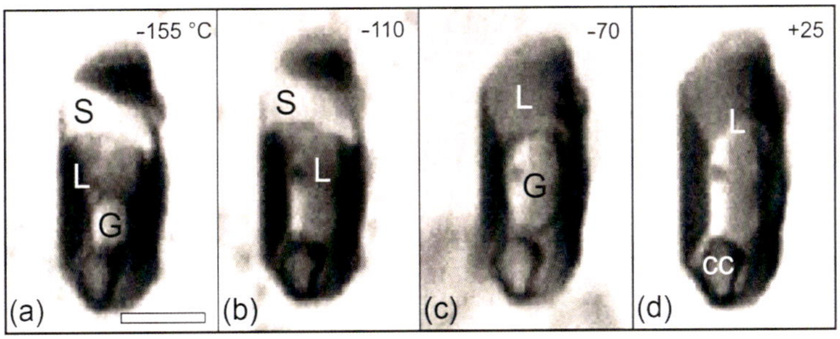

FIGURE 4.30

Phase changes in a CO_2–N_2 inclusion from leucosome of metabasite migmatite from the Western Tatra Mountains (Hurai et al., 2000). Partial homogenization to liquid took place at −149 °C, but the vapor bubble nucleated again at −61.4 °C shortly before the dissolution of the CO_2 solid. Total homogenization to liquid occurred at −61 °C. The LG assemblage displayed in (c) is metastable. The sequence of phase transformations, S+L+G→S+L→S+L+G→L+G→L corresponds to the H4 type. cc is calcite. Scale bar in (a) represents 5 μm.

It is worth noting that the CO_2 melting in the presence of liquid and vapor took place at a lower temperature (−61.4 °C) than the minimum along the upper limb of the three-phase curve (−61.1 °C). Moreover, the SL(G) tie line of −149 °C did not intersect the upper limb of the three-phase curve, so that the inclusion could not homogenize within the L+G field. Hence, either the phase diagram of the CO_2–N_2 system around the critical curve is imprecise, or additional gaseous compounds are present. Indeed, Raman spectroscopy revealed trace amounts of methane and H_2S, together with 55 mol.% CO_2 and 45 mol.% N_2, whereas the phase transitions projected in the CO_2–N_2 phase diagram indicate 47 mol.% CO_2 and molar volume of 49 cm^3 mol^{-1}. It must be emphasized; however, that the uncertainty in the determination of the temperature of partial homogenization is about ±0.5 °C and that of the CO_2 melting around ±0.2 °C. Hence, owing to the high density of isotherms close to the intersection of the critical curve with the upper three-phase curve, the composition of the inclusion in Figure 4.30 cannot be determined with precision better than ±5 mol.%.

Using microthermometric data only, nitrogen or methane admixture in CO_2-rich inclusions cannot be revealed if the sequence of phase transformations corresponds to the H1–H3 subtypes. A minor amount of nitrogen only slightly decreases the temperature of melting of the CO_2 solid below −56.6 °C; however, its effect on the lowering the final homogenization temperature is considerably greater. Thus, geologically unreasonably high apparent densities of the CO_2-rich inclusions can be determined without a control from Raman spectroscopic data. In contrast, the CH_4 admixture has a minor effect on lowering the final homogenization temperature of CO_2-rich inclusions.

The presence of methane in ternary gaseous mixtures can be revealed according to the temperature of melting of CO_2 solid below −61.2 °C—the temperature minimum in the CO_2–N_2 binary system. Further evidence is provided by the metastable existence of two immiscible liquids on cooling within the interval between the critical temperature of methane (−82.4 °C) and that of the precipitation of the CO_2 solid occurring in most cases between −90 and −110 °C (Figure 4.31). Except for some cases exemplified hereafter, the composition of ternary gaseous mixtures can be determined only using Raman spectrometric data.

Molar volumes of ternary gaseous inclusions can be determined only if their composition had been inferred from Raman spectroscopic data, or a critical phase transition occurs. For instance, the

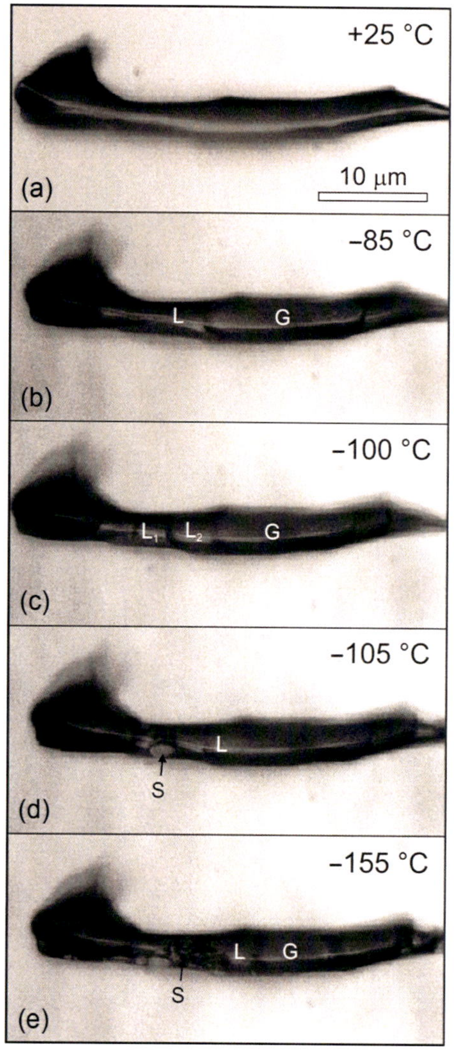

FIGURE 4.31

Phase transformations in CO_2–CH_4–N_2 inclusion in quartz from tectonic breccia at the base of the Magura nappe, Tertiary accretion wedge of the Outer Carpathians (Zborov-1 borehole, 3409 m). The inclusion is homogeneous at room temperature (a), and a vapor bubble liberates on cooling below −82 °C (b). Metastable coexistence of two liquids between −85 and −100 °C (c) reveals the existence of methane and CO_2. The CO_2 solid precipitates at −100 °C (d) and liberation of the vapor bubble below −105 °C reveals the N_2+CH_4 mixture (e). The stable sequence of phase transformations on heating (L+G+S→L+S→L+G+S→L+G→L) corresponds to the H4 subtype. Partial homogenization to liquid occurs at −105 °C, the CO_2 solid melts at −64.2 °C, and final homogenization (near-critical) to liquid takes place at −52.6 °C. Another partial, metastable homogenization by expansion of a vapor bubble (L_1+L_2+G→L_1+G) was observed at −86.7 °C. Raman spectroscopy proved the following bulk composition: 51 mol.% CH_4, 30 mol.% CO_2, and 19 mol.% N_2.

near-critical total homogenization illustrated in Figure 4.31 permits the composition and molar volume to be derived from the critical surface of the ternary system using the EOS for gaseous mixtures (Thiéry et al., 1994a,b). The calculation with the HOMOGEN computer program (Bakker, 2003) locates the position of the critical point of the CO_2–N_2–CH_4 system along the pseudosection with 51 mol.% CH_4, 30 mol.% CO_2, and 19 mol.% N_2 (determined by Raman spectrometry) at −52.6 °C, 190 bar, and 52 cm^3 mol^{-1}.

The bulk molar volume of the same inclusion can also be calculated as the sum of molar volumes of individual phases multiplied by their mole fractions. In our case, it is convenient to calculate the molar volume at −105 °C, when the CH_4+N_2 mixture homogenized in the presence of CO_2 solid phase. Since the solubility of CO_2 solid in the gaseous mixture is negligible at low temperatures (van den Kerkhof, 1988), the bulk molar volume can be calculated as the sum of the coexisting CO_2 solid with 100 mol.% CO_2 and the CH_4–N_2 liquid with 27 mol.% N_2 and 73 mol.% CH_4. The molar volume of CO_2 solid at −105 °C corresponds to 27.5 cm^3 mol^{-1} (Ražnjevic, 1975), and the molar volume of the CH_4–N_2 liquid (61 cm^3 mol^{-1}) can be inferred from the diagram correlating homogenization temperatures and molar volumes in CH_4–N_2 inclusions (Figure 4.32). Substitution of the obtained data in Equation (1.13) yields

$$\overline{V} = \overline{x}_{CO_2}\overline{V}_{CO_2} + \overline{x}_{(CH_4+N_2)}\overline{V}_{(CH_4+N_2)} = 0.3 \times 27.5 + 0.7 \times 61 = 51\ cm^3/mol.$$

The method can also be used to calculate molar volumes of other CO_2-rich inclusions of the H4 or S2–S4 subtypes. For instance, Raman analysis detected only traces of CH_4 and H_2S together with the dominant N_2 (45 mol.%) and CO_2 (55 mol.%) in the inclusion illustrated in Figure 4.30. At −149 °C, the inclusion consists of essentially pure CO_2 solid (26.4 mol.%, Ražnjevic, 1975) and almost pure N_2 liquid with the molar volume of 62.4 cm^3 mol^{-1} inferred from the boiling curve of pure nitrogen (Jacobsen et al., 1986). Substitution of these data in the above equation yields the bulk molar volume of 43 cm^3 mol^{-1}, which is somewhat lower than the value of 49 cm^3 mol^{-1} determined using the

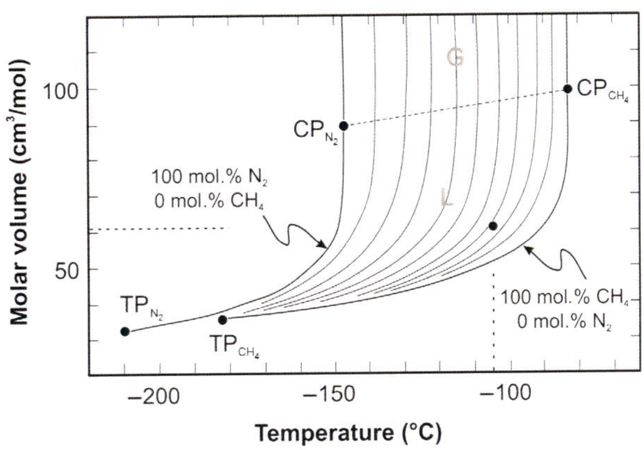

FIGURE 4.32

The relationship between temperature and molar volume in the CH_4–N_2 system (van den Kerkhof, 1988), compiled using EOSs proposed by Angus et al. (1976, 1979), and data from Swanenberg (1980). The solid circle is the projection point of the CH_4–N_2 liquid from Figure 4.31d. Phase diagram reprinted with kind permission from A.M. van den Kerkhof.

projection of the phase transitions onto the phase diagram of the CO_2–N_2 system. The inconsistency is most probably caused by the inaccuracy of the experimental data near critical conditions, and by a minor admixture of additional gas species, particularly H_2S.

Molar volume of a ternary CH_4–CO_2–N_2 mixture with $XCO_2 > 0.6$ can also be estimated using two-component subsystems, because isochores for liquid and vapor phases along the L–G curve form almost parallel lines (Thiéry et al., 1994b). Hence, the bulk molar volume of the CO_2–CH_4–N_2 inclusion can be inferred from $\overline{V}-X$ projections of the CO_2–CH_4 and CO_2–N_2 subsystems by using a pseudocomposition X^*. The pseudocompositions calculated from the appropriate equation are then combined with the temperature of phase transition to estimate the molar volume of the three-component mixture using the $\overline{V}-X$ diagrams of two-component mixtures (Problem 4.6).

PROBLEM 4.6 CALCULATING MOLAR VOLUME OF A GASEOUS INCLUSION

Calculate the molar volume of the gaseous inclusion containing 70 mol.% CO_2, 14 mol.% N_2, and 16 mol.% CH_4. Total homogenization of the inclusion to liquid occurs at −13 °C, and CO_2 solid melts at −61 °C (compare with data in Figure 7.9 and Problem 7.2). Calculate the molar volume and the composition of the inclusion in terms of binary CO_2–CH_4 and CO_2–N_2 systems and compare the obtained molar volumes with those recalculated in terms of the three-component CO_2–CH_4–N_2 system.

Solution

For the projection in the CO_2–CH_4 system, $X^*_{CH_4} = 1.1(X_{CH_4} + X_{N_2})$ for the homogenization to liquid and $X^*_{CH_4} = 0.9(X_{CH_4} + X_{N_2})$ for the homogenization to vapor. For the projection in the CO_2–N_2 system, $X^*_{N_2} = 0.9(X_{CH_4} + X_{N_2})$ for the homogenization to liquid and $X^*_{N_2} = 1.1(X_{CH_4} + X_{N_2})$ for the homogenization to vapor (Thiéry et al., 1994b). Pseudocompositions calculated according these equations correspond to $X^*_{CH_4} = 0.33$, $X^*_{N_2} = 0.27$, respectively. Molar volumes estimated using $\overline{V}-X$ diagrams of the CO_2–CH_4 and CO_2–N_2 subsystems (Figures 4.28 and 4.29) are 57.5 and 55 $cm^3\ mol^{-1}$, respectively. The difference in the molar volume estimations stems from the fact that the inclusion contains a small amount of H_2S (<0.5 mol.%), which affects its total homogenization temperature.

4.6.6 PHASE TRANSITIONS IN H_2O–CO_2 AND H_2O–CO_2–NaCl INCLUSIONS

Carbon dioxide reacts with water to CO_2 clathrate at temperatures below 10 °C, and the assemblage of additional solid phases is controlled by pressure (Figure 4.33). The CO_2 clathrate cannot precipitate in low-density H_2O–CO_2 inclusions with the CO_2 concentration less than 3.6 mass% (isochore 1), because the isochore intersects the ice liquidus below the Q1 quadruple point. In such inclusions, the temperature of ice melting shifts from 0 to −1.5 °C proportionally with the increasing CO_2 concentration. Similar fluid inclusions are trapped at a low pressure in subvolcanic systems or geothermal fields (Hedenquist and Henley, 1985).

The CO_2 clathrate is a stable phase only at CO_2 concentrations greater than 3.6 mass%. The isochores of fluid inclusions with 3.6–8.8 mass% CO_2 intersect the CO_2 clathrate liquidus between the quadruple points Q1 and Q2 (isochore 2), and the clathrate coexists with aqueous liquid and CO_2-rich vapor. In this case, the internal fluid inclusion pressure is too low to condensate the CO_2 liquid phase, which forms only at CO_2 concentrations greater than 8.8 mass%, when the isochores 3A and 3B intersect the three-phase ($LCO_2 - LH_2O - GCO_2$) curve. On cooling, the isochores 3A and 3B pass

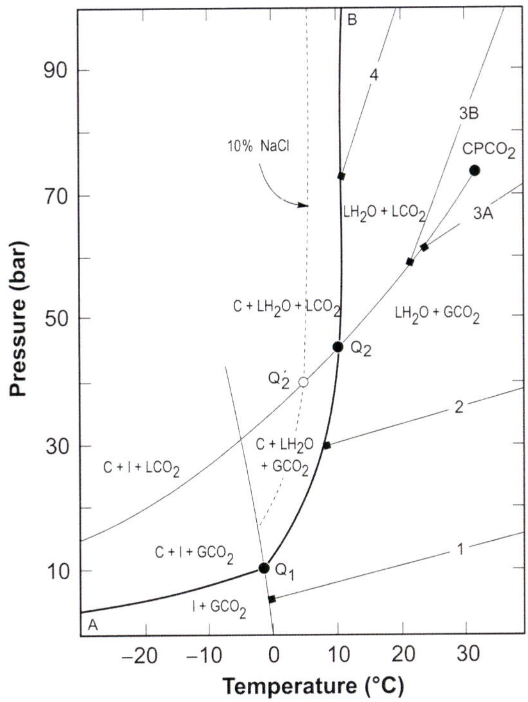

FIGURE 4.33

Low-temperature portion of the system H_2O–CO_2 (Collins, 1979) with isochores (1–4) of fluid inclusions with variable densities and CO_2 concentrations. C, CO_2 clathrate; I, ice. Phase diagram reprinted with kind permission from Society of Economic Geologists.

through the Q2 quadruple point at 10 °C, where the CO_2 liquid, CO_2 gas, and H_2O liquid phases coexist with the CO_2 clathrate. The isochore 4 of the densest H_2O–CO_2 inclusions passes out of the three-phase curve, and hits the CO_2 vapor unsaturated CO_2 clathrate liquidus between the Q2 and B points. In this rare case, the CO_2-rich phase breaks down to CO_2 liquid and CO_2 vapor phases along the metastable extension of the three-phase (LH_2O – LCO_2 – GCO_2) curve in the clathrate stability field below Q2. On heating, the CO_2-rich phase homogenizes to liquid prior to the CO_2 clathrate dissolution along the Q1-B curve in the presence of CO_2 and H_2O liquid phases. Owing to high pressure, the temperature of CO_2 clathrate melting is slightly shifted above 10 °C.

Salts do not enter the CO_2 clathrate structure, thus increasing the concentration of the residual aqueous phase once the clathrate crystallizes. Hence, the temperature of ice melting yields erratic (overestimated) total salinities, if the CO_2 clathrate is present in the inclusion. The CO_2 clathrate can be easily overlooked in low-salinity aqueous inclusions, owing to its index of refraction similar to that of the associated aqueous liquid. Addition of NaCl in the H_2O–CO_2 fluid shifts the Q2 quadruple point to lower temperatures, thus allowing for determination of total salinity (in mass% NaCl equivalents) using the empirical equation (Darling, 1991) valid for 0–24 mass% NaCl equivalents

$$\text{Mass\% NaCl equivalents} = 0.00098241(10 - T_m)(T_m^2 + 45.385T_m + 1588.75), \tag{4.19}$$

where T_m is the temperature of CO_2 clathrate melting in the presence of CO_2 liquid, CO_2 vapor, and H_2O liquid phases. Another equation derived by Bozzo et al. (1973) has also been frequently used.

The Q2 assemblage in H_2O–NaCl–KCl–CO_2 solutions can be described by the following equation (Diamond, 1992a)

$$\text{Mass\% (NaCl+KCl)} = 15.5296 + 4.2947X_{KCl} + 0.5061X_{KCl}^2 - 1.0709T_m - 0.4751X_{KCl}T_m - 0.048T_m^2 \quad (4.20)$$

where T_m is the temperature of CO_2 clathrate dissociation (°C) and X_{KCl} denotes the relative mass fraction of KCl calculated as mass% KCl/(mass% KCl + mass% NaCl). Estimation of the NaCl and KCl mass fractions is, however, problematic in low-salinity aqueous inclusions and must be determined indirectly, for instance, from the leachate chemistry.

On cooling, the CO_2 clathrate nucleates in carbo-aqueous fluid inclusions far below the stable liquidus temperatures. Even at low CO_2 concentrations, the clathrate can be identified by double freezing, when the CO_2 clathrate nucleates instantaneously between −35 and −40 °C, followed by the crystallization of the ice + salt hydrate mixture at lower temperatures. In some cases, homogenization of the CO_2-rich phase (to liquid or vapor) in the metastable absence of the CO_2 clathrate can be measured and used for the salinity calculations.

Diamond (1992a) inferred additional empirical equations (Appendix II) that allow for determination of the salinity along the univariant curves of the H_2O–NaCl–CO_2 system, provided that the inclusion pressure is fixed by the temperature of metastable partial homogenization of the CO_2-rich phase (in the metastable absence of the CO_2 clathrate).

One of the most common sequences of phase transitions, corresponding to the 3B isochoric trajectory in Figure 4.33, is documented in Figure 4.34. Melting of CO_2 solid at −56.6 °C indicates nearly pure CO_2. Salinity calculated from the CO_2 clathrate melting temperature using Equation (4.19) corresponds to 1.24 mass% NaCl equation. Since the inclusion decrepitates prior to total homogenization, bulk density can be calculated only as the sum of densities of CO_2- and H_2O-rich phases present in the inclusion at the moment of partial homogenization of the CO_2-rich phase. The calculation procedure further demands volume proportions of the two coexisting phases to be derived using digital image analysis. The volume percentages of the H_2O- (L_2) and CO_2-rich (L_1) phases correspond to 32 and 68 vol.%, respectively. Densities of the respective H_2O and CO_2 phases, 1.004 and 0.531 g cm^{-3}, can be calculated from Equations (II.15) and (II.3) in Appendix II. The total density of the inclusion is 0.682 g cm^{-3}.

Bakker and Diamond (2000) used the experimental data from Sterner and Bodnar (1991) for the construction of a $\overline{V} - \overline{x}$ diagram of the H_2O–CO_2 system (Figure 4.35). This type of diagram is commonly employed for the interpretation of microthermometric data from gaseous mixtures. If the temperature of total homogenization cannot be measured, the volume fraction of the CO_2-rich phase can also be used to infer the bulk molar volume.

Figure 4.35 also includes the correction for CO_2 dissolved in aqueous liquid. The mutual solubility of CO_2 in H_2O-rich phase and H_2O in CO_2-rich phase should be taken into account when calculating phase proportions at room temperature. Roedder (1984) reports on a maximum of 0.1 mass% H_2O dissolved in the CO_2-rich phase at +22.6 °C. However, the CO_2 content in the H_2O-rich phase is considerably higher, reaching about 5.5 mass% near the lower critical end point of the H_2O–CO_2 system. The empirical least-squares fit to data of Dolgov et al. (1968) enables the calculation of the CO_2 content in the H_2O-rich phase at temperatures between 0 and 31 °C:

$$\text{Mass\% } CO_2 = 6.9956 - 0.06455T + 4.83476 \times 10^{-4}T^2 \quad (4.21)$$

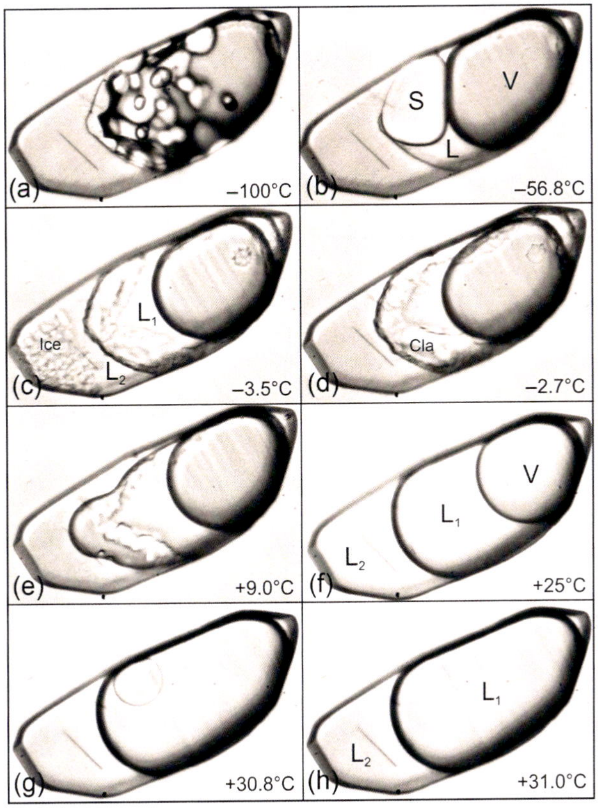

FIGURE 4.34

Phase transformations in an H_2O–CO_2–NaCl inclusion in quartz from an alpine fissure (Klenovec, Slovakia). On heating totally frozen inclusion (a), CO_2 solid (S) melts at −56.6 °C, that is, close to the triple point of pure CO_2 (b). Further heating leads to phase transformations in the aqueous phase, which are hardly discernible, except for rapid melting of ice (c–d) in the presence of aqueous liquid (L_2), CO_2 liquid (L_1), and CO_2 clathrate (Cla). The CO_2 clathrate melts at 9.4 °C (e–f), and partial homogenization of the CO_2-rich phase to liquid occurs at 30.9 °C. V denotes CO_2-vapor phase. Further heating led to decrepitation of the inclusion prior to total homogenization due to high internal overpressure. The longer diameter of the inclusion corresponds to 40 μm.

where T is the temperature (°C) along the three-phase ($LH_2O - LCO_2 - GCO_2$) curve. The solubility curve is applicable only for very low salinity aqueous fluids. The computer program Q2 (Bakker, 1997) can be used for calculations with the H_2O–NaCl–CO_2 system and increased NaCl contents.

Phase changes in an inclusion with a lower CO_2 concentration, corresponding to the isochore 2 in Figure 4.33, are documented in Figure 4.36. Composition and density of the inclusion can be calculated using the ICE computer program (Bakker, 1997), where input parameters are as follows: volume of vapor bubble at the moment of CO_2 clathrate dissolution (12.5 vol.%), salt composition and concentration (mol.% relative to water), and temperature of ice dissolution. The salinity

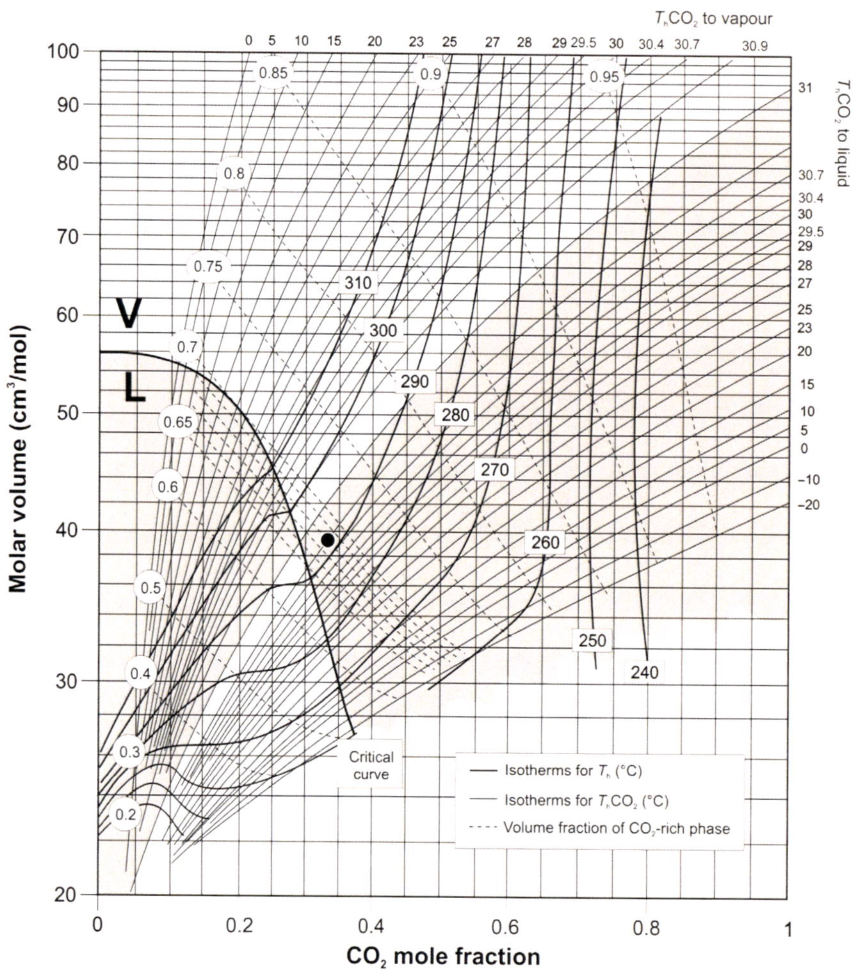

FIGURE 4.35

$\overline{V}-\overline{x}$ diagram of the H_2O–CO_2 system (Bakker and Diamond, 2000). The numbers in rectangles are temperatures of total homogenization (to liquid or vapor). Numbers in circles and ellipses denote volume fractions of CO_2-rich phase at the moment of partial homogenization. The solid circle is the projection point of the inclusion displayed in Figure 4.34 and discussed in the text. Considering the pure H_2O–CO_2 system, the inclusion in Figure 4.34 should homogenize to vapor at ~292 °C. Phase diagram reprinted from Bakker, R.J., Diamond, L.W., 2000. Determination of the composition and molar volume of H_2O–CO_2 fluid inclusions by microthermometry. Geochim. Cosmochim. Acta 64, 1753–1764, Copyright 2000, with permission from Elsevier.

determination according to Equation (4.19) is not possible, because the metastable CO_2 liquid did not condensate in the absence of the CO_2 clathrate. The eutectic temperature is close to the H_2O–NaCl–$CaCl_2$ system, and the NaCl mass fraction of 0.896 is defined from the simultaneous dissolution of ice and hydrohalite along the cotectic curve at −22.4 °C (Table 4.4). We convert the mass

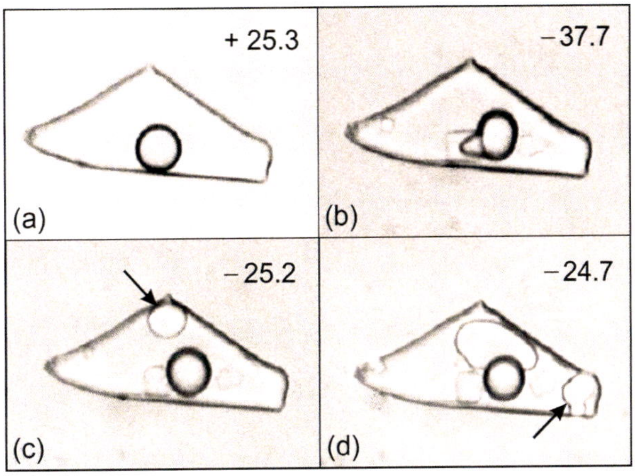

FIGURE 4.36

Phase changes in an aqueous inclusion with a lower CO_2 content in quartz from an alpine fissure (Hurai et al., 1997). The sequence of phase transformations corresponds to the isochore 2 in Figure 4.33. The inclusion contains only aqueous liquid and CO_2 vapor phases at 25 °C (a). Metastable cubic CO_2 clathrate crystals grown during sequential freezing coexist with the aqueous liquid down to −40 °C (b). The CO_2 clathrate crystallizes around the CO_2 vapor bubble, but smaller isolated crystals are also discernible within the aqueous liquid phase. The diagnostic features of the CO_2 hydrate are its low index of refraction, sluggish volume change on cooling and heating, and production of gas bubbles during fast heating. (c) Illustrates the metastable assemblage of ice (arrow), CO_2 clathrate, aqueous liquid, and CO_2 vapor immediately before the ice dissolution. (d) Illustrates the stable assemblage of hydrohalite (arrow), CO_2 clathrate, ice, aqueous liquid, and CO_2 vapor. Onset of melting (eutectic point) was observed at −49 °C, thus proving an increased concentration of calcium chloride. Ice and hydrohalite dissolved simultaneously at −22.4 °C, whereas the CO_2 clathrate dissolved at −5 °C. Total homogenization to liquid occurred at 223 °C. The longer diameter of the inclusion is 28 μm. The numbers in the upper-right corners are temperatures (°C).

Table 4.4 NaCl/(NaCl + $CaCl_2$) Weight Fractions Along the Ice–Hydrohalite Cotectic Curve of the System H_2O–NaCl–$CaCl_2$ (Yanatieva, 1946; Oakes et al., 1990)

T (°C)	0.0	0.1	0.2	0.3	0.4	0.5	0.6	0.7	0.8	0.9
−21.	–	–	1.000	0.992	0.983	0.974	0.966	0.957	0.948	0.939
−22.	0.930	0.921	0.913	0.905	0.896	0.888	0.881	0.873	0.865	0.857
−23.	0.848	0.840	0.831	0.823	0.814	0.806	0.798	0.790	0.782	0.775
−24.	0.767	0.760	0.753	0.747	0.740	0.733	0.726	0.719	0.712	0.705
−25.	0.697	0.691	0.686	0.679	0.673	0.667	0.660	0.654	0.649	0.643
−26.	0.637	0.630	0.625	0.620	0.615	0.608	0.601	0.595	0.590	0.584
−27.	0.679	0.573	0.567	0.562	0.556	0.552	0.547	0.542	0.536	0.531
−28.	0.525	0.521	0.516	0.511	0.505	0.500	0.494	0.489	0.484	0.478
−29.	0.472	0.466	0.461	0.456	0.450	0.445	0.439	0.434	0.429	0.423
−30.	0.418	0.413	0.408	0.403	0.399	0.395	0.390	0.385	0.380	0.375
−31.	0.370	0.366	0.362	0.358	0.353	0.349	0.345	0.341	0.336	0.332
−32.	0.328	0.324	0.320	0.316	0.312	0.308	0.304	0.300	0.297	0.293
−33.	0.289	0.285	0.281	0.277	0.274	0.270	0.266	0.263	0.260	0.256
−34.	0.252	0.249	0.245	0.242	0.238	0.235	0.232	0.228	0.224	0.221

fractions of salts to molar fractions dividing by their molecular mass and substitute all the parameters into the ICE computer program to obtain the total density (0.7643 g cm^{-3}) and bulk composition ($X_{H_2O} = 0.8446, X_{CO_2} = 0.024, X_{NaCl} = 0.1204, X_{CaCl_2} = 0.0111$) assuming that the vapor bubble is composed of pure CO_2 gas. This procedure can be used for complex gas mixtures (CO_2–CH_4–N_2) with composition determined by Raman spectrometry, and also for frequent cases, when ice dissolves after hydrohalite (i.e., lower salinities).

4.6.7 PHASE TRANSITIONS IN H_2O–CH_4 AND H_2O–CH_4–NaCl INCLUSIONS

Phase changes in an aqueous inclusion with a high methane concentration are documented in Figure 4.37. At room temperature, the inclusion is composed of aqueous and methane liquids. Formation of CH_4-clathrate in the aqueous phase is indicated by double freezing. A CH_4 vapor bubble is liberated at about −100 °C; however, the CH_4-liquid does not freeze until −196 °C.

The calculation of the total salinity is complicated by the lack of experimental data on the CH_4-clathrate dissolution, which is limited by the pressure of 200 bar, corresponding to CH_4 molar volumes greater than 38.7 cm^3 mol^{-1} (Figure 4.38). The maximum limit of salinity is provided by the temperature of ice melting (−1.8 °C), which would correspond to 3 mass% NaCl equation. This value is, however, overestimated, because the substantial portion of pure H_2O is bound in the CH_4-clathrate

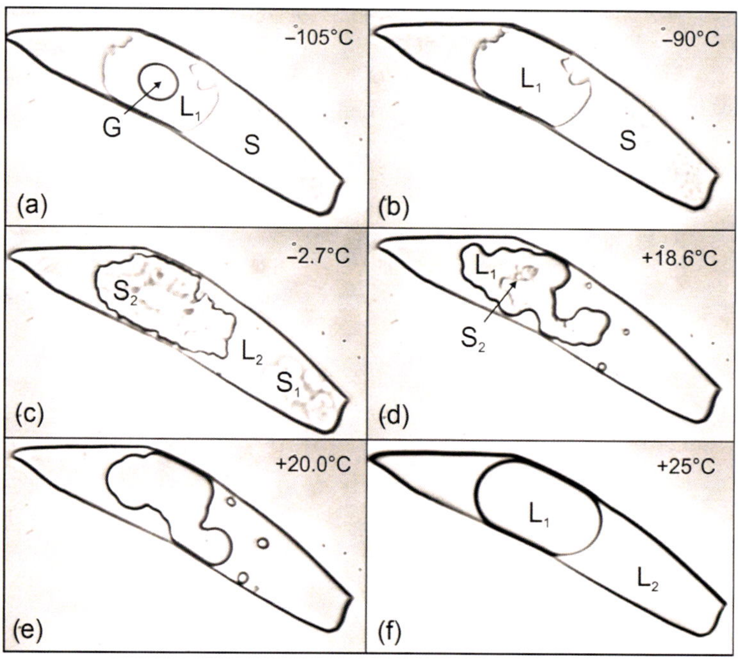

FIGURE 4.37

Phase transformations in an aqueous inclusion with methane trapped in quartz from calcite veins in Paleogene sandstones (Zálesie, Spišská Magura, Slovakia). Partial homogenization of the CH_4-rich phase (L_1) to liquid occurred at −90.1 °C ((a)-(b)), ice (S_1) melted completely at −1.9 °C (c), and CH_4-clathrate (S_2) dissolved at 22.8 °C ((e)-(f)). L_2 designates the aqueous liquid and G is the methane-rich vapor.

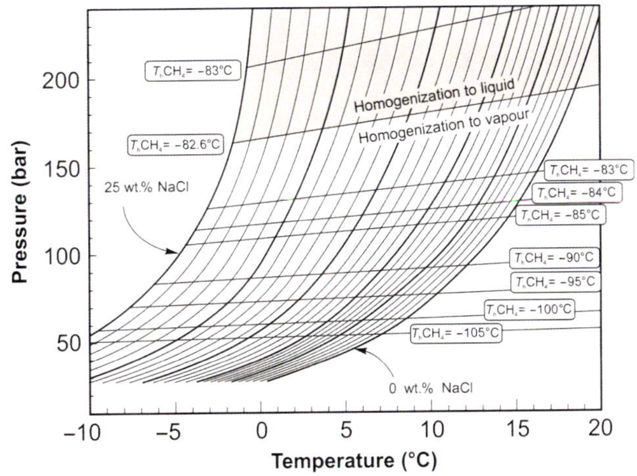

FIGURE 4.38

Dissociation curves of CH_4-clathrate (Diamond, 1992b, 1994) in aqueous solutions with variable salinity plotted onto isochores of pure methane (Jacobs and Kerrick, 1981).

structure. Hence, the total salinity of 1 mass% will be further considered, as measured in neighboring inclusions with a lower methane concentration.

The density of the CH_4-rich phase corresponds to 0.262 g cm^{-3}, and it is inferred from the temperature of partial homogenization (−90.1 °C) and the EOS for boiling curve of pure methane (Equation (II.7), Appendix II). The CH_4-rich phase occupies 37 vol.% of the inclusion at the moment of the CH_4-clathrate dissolution. The density of 1 mass% NaCl–H_2O solution at +22.8 °C corresponds to 1.0046 g cm^{-3} (Equation (II.15), Appendix II), and after substituting volumes and densities into Equation (4.2), we get the bulk inclusion fluid density (0.73 g cm^{-3}). The composition of the inclusion expressed in molar fractions can be calculated using the following procedure:

1. The number of moles (m) of each compound (i) can be expressed as

$$\overline{m}_{(i)} = \frac{DVm_i}{100\overline{M}_i}, \tag{4.22}$$

where D is the density of CH_4- or H_2O-rich phase, V denotes the phase volume (vol.%), $\overline{M}$ is the molecular mass, and m is the concentration (mass%) of i-th compound in L_1 or L_2.

2. Molar fraction of the i-th compound in the whole inclusion corresponds to

$$\overline{x}_i = \overline{m}_i / \sum \overline{m}, \tag{4.23}$$

where $\sum \overline{m}$ is the sum of moles of the solvent (H_2O) and solutes (CH_4, NaCl) in the inclusion. Of course, this calculation procedure can be used for any gas or salt composition. The composition and molar volume of the CH_4-rich inclusion displayed in Figure 4.37 correspond to $X_{CH_4} = 0.148, X_{H_2O} = 0.849, X_{NaCl} = 0.003$, and 24.4 cm^3 mol^{-1}. Similar parameters ($X_{CH_4} = 0.149, X_{H_2O} = 0.846, X_{NaCl} = 0.005$), and 24 cm^3 mol^{-1}) are obtained using the HOMOGEN computer program (Bakker, 2003). The same calculation procedure is also involved in the FLINCOR computer program (Brown, 1989).

4.6.8 PHASE TRANSITIONS IN PETROLEUM INCLUSIONS

Phase changes in a gas-condensate inclusion are illustrated in Figure 4.39. Such inclusions usually exhibit only one reproducible phase change—the total homogenization to liquid or vapor. A large expansion rate of gas bubble is the diagnostic feature of this type of fluid inclusions. Some higher hydrocarbons occur as solid phases or immiscible liquids at room temperature and additional ones are gradually separated from the residual hydrocarbon liquid on cooling. Refractive indices of the separated solid hydrocarbons are close to that of the residual liquid; therefore, they are hardly discernible and their melting points cannot be determined unequivocally. The gas-condensate inclusions do not freeze out completely even at −196 °C, mainly due to the presence of propane, which has the lowest temperature of the triple point among the C1–C4 hydrocarbons (Table 2.1). Total homogenization temperatures of gas-condensate inclusions roughly indicate their composition, because critical temperatures of hydrocarbons increase with their carbon number. The inclusion in Figure 4.39 homogenizes at +55 °C, thus providing the evidence for C3 and higher hydrocarbons.

The inclusion illustrated in Figure 4.40 contains a mixture of higher liquid hydrocarbons. At room temperature, the inclusion is composed of light-brown, yellow-green fluorescing oil and black asphalt/bitumen globules adhering to the inclusion's walls (S_1). On cooling, a portion of oil solidifies to waxy

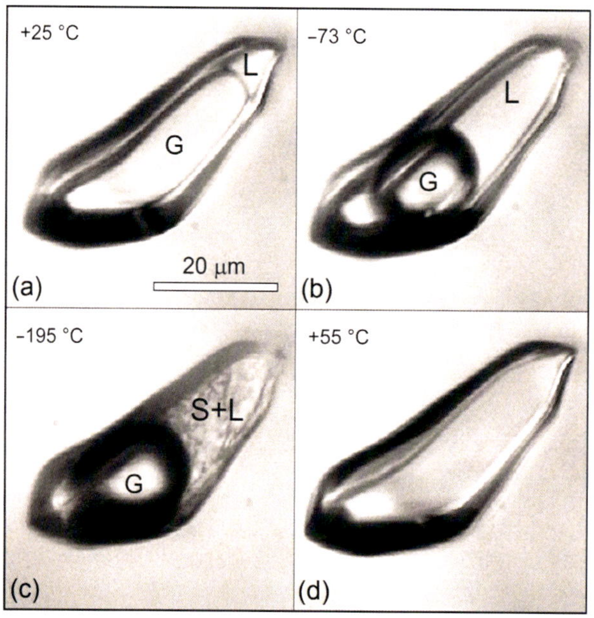

FIGURE 4.39

Primary gas-condensate inclusion in quartz from mineralized joints in Paleogene sandstones (Lipany-1 borehole, Levočské vrchy Mts.). The inclusion consists of vapor and liquid at room temperature (a). The liquid rapidly expands on cooling and occupies the largest volume of the inclusion at −73 °C (b). Solid hydrocarbons (S) begin to precipitate below −80 °C; however, a portion of the inclusion content remains unfrozen even at −195 °C, as documented by the rounded gas bubble (c). Total homogenization to vapor occurs at +55 °C (d). Temperatures of melting are not clearly discernible.

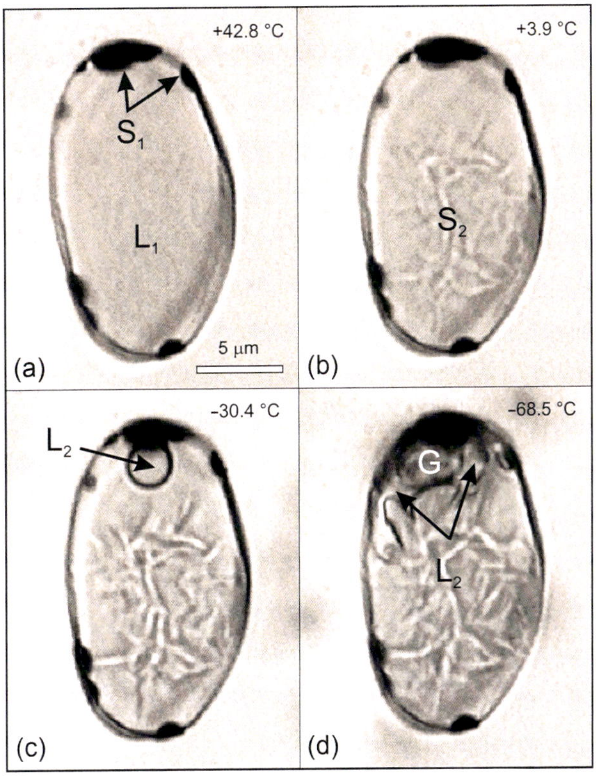

FIGURE 4.40

Phase transformations in gas-saturated heavy oil inclusion in quartz from flysch sediments (Hurai et al., 2002b). At room temperature, the inclusion is filled with light-brown oil (L_1) and black insoluble asphaltic blebs (S_1) stuck to the inclusion's walls (a). On cooling, waxy paraffinic substances (S_2) precipitate continuously below +5 °C (b), and immiscible gaseous CH_4-rich liquid (L_2) separates instantaneously below −30 °C (c). At temperatures below −60 °C, the methane-rich liquid liberates a vapor bubble (d). The methane-rich bubble homogenizes at −26 °C, thus indicating the presence of higher gaseous hydrocarbons.

substances. A methane-rich liquid separates on further cooling and splits later into vapor and liquid phases. Simultaneously, a plethora of solid phases precipitate from the residual liquid until they occupy almost the whole volume of the inclusion and further phase changes are no longer discernible.

The gas chromatogram obtained by leaching of crushed quartz revealed the predominance of C17–C22 n-alkanes in the saturated fraction. The inclusion hydrocarbons can thus be classified as a mixture of volatile naphthenic and paraffinic oil.

Phase transitions in five different types of gas-condensate and oil inclusions have been documented by Grimmer et al. (2003). They used confocal-scanning laser microscopy to measure the liquid-to-vapor ratio, modeled the fluid composition by the PIT software (Thiéry et al., 2000), and validated the modeled compositions by the Fourier-transform infrared spectroscopy.

Densities of petroleum inclusions can be calculated in several ways. The combination of microthermometric data (total homogenization temperature or vapor bubble volume at given T) with the composition of present-day hydrocarbons (e.g., Bodnar, 1990; Munz et al., 1999) can only be applied for recent reservoir fillings, when the inclusion fluid composition is approximately the same as that of the reservoir fluid (Munz, 2001). In other cases, the chemical information about hydrocarbon fluids can be obtained using gas chromatography–mass spectrometry (e.g., George et al., 1997), liquid chromatography (Pang et al., 1998), or nuclear magnetic resonance (Dereppe et al., 1994). These techniques, however, do not allow discrimination between multiple generations of fluid inclusions (Bourdet et al., 2008), and moreover, they are sensitive to sample contamination (Thiéry et al., 2002).

Individual fluid inclusions can be analyzed by the Fourier-transform infrared spectroscopy (e.g., Guilhaumou et al., 1990; Grimmer et al., 2003), however, the inclusions must be larger than 10 μm in diameter. Stasiuk and Snowdon (1997) correlated the UV-induced fluorescence of oil with its composition and density; this relationship, however, was proven to be not straightforward, and Munz (2001) concluded that the fluorescence can be used as a discrimination tool only for genetically related petroleum inclusions. Volumes of coexisting liquid and vapor phases in fluorescent oil inclusions can be precisely measured using the confocal scanning laser microscopy (e.g., Pironon et al., 1998; Aplin et al., 1999).

4.6.9 PHASE TRANSITIONS IN MELT INCLUSIONS

The appearance of melt inclusions is different in extrusive and intrusive rocks, mainly due to the contrasting cooling rates of these two rock types. Melt inclusions in extrusive and effusive rocks are generally composed predominantly of homogeneous, sometimes partly devitrified silicate glass, and small spherical shrinkage bubbles. In the case of more acidic silicate melts, the shrinkage bubble may also contain aqueous liquid and vapor phases (Figure 4.41a). If the cooling rate is very rapid, formation of the vapor bubble may be suppressed, and the inclusions consist only of a homogeneous silicate glass phase. The situation is essentially the same as observed in aqueous inclusions, where the formation of vapor bubble is attenuated on rapid cooling. The supercooled silicate melt inclusions devoid of vapor bubbles are diagnostic of Plinian-type eruptions and viscous acidic melts (Figure 4.41b). In contrast, melt inclusions in phenocrysts in rhyolites and basalts usually contain two-phase inclusions resembling in terms of the phase composition and appearance the aqueous inclusions occurring in literally all hydrothermal minerals (Figure 4.41c). Silicate melt inclusions in intrusive and plutonic rocks are completely crystallized due to small cooling rates. The vapor bubble is strongly deformed, and it frequently occurs as a thin film coating the surfaces of the crystallized minerals (Figure 4.41d). The fully recrystallized inclusions are often opaque in transmitted light, and they can be easily overlooked or erroneously interpreted as accidentally trapped clusters of mineral aggregates (Esposito et al., 2012; Bartoli et al., 2013).

Microthermometric measurement of melt inclusions and their successful application as geothermometers is subject to preconditions defined by Roedder's rules. From these criteria it follows that the natural cooling process should be reversible, which means that heating the melt inclusions from room temperature to trapping conditions should lead to phase transitions in a reverse order. However, the sluggish kinetics of the crystallization and melting processes leads to serious problems, particularly in water-depleted acidic melts. In contrast, observation of phase transitions in mafic melts is less influenced by kinetics, but it is often hampered by oxidation of the host minerals, mainly pyroxenes and

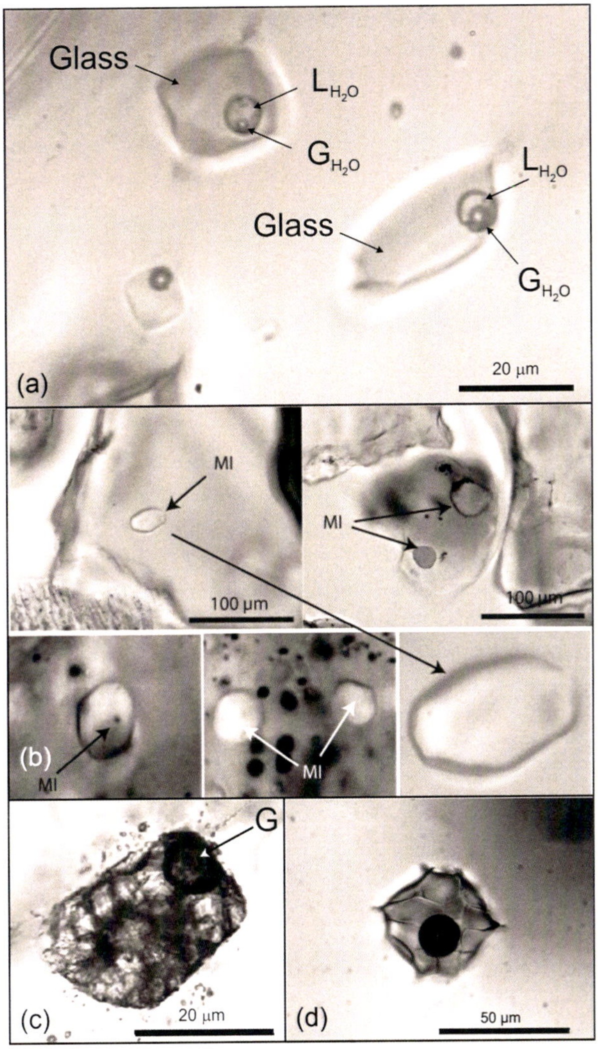

FIGURE 4.41

Various types of melt inclusions trapped in intrusive and effusive magmatic rocks. (a) Melt inclusions in quartz from pincinite xenoliths in alkali basalt (Pinciná maar, Slovakia) with aqueous liquid (LH_2O) and vapor (GH_2O) phases in the shrinkage bubble surrounded by homogeneous glass of rhyolitic composition. Note also decrepitation cracks emanating from tips of the inclusions. (b) Series of photomicrographs documenting rapidly quenched metastable homogeneous melt inclusions (MI) devoid of shrinkage bubbles in quartz grains from the Råda sand of Sweden deposited from Plinian eruption-related clouds extending from Iceland over Sweden at the outset of the post-glacial period. (c) Multiphase, fully recrystallized melt inclusion in quartz from the Eibenstock granite (W. Erzgebirge, Germany), showing vapor bubble (G), numerous solid phases (quartz, feldspar, mica), and also decrepitation crack emanating from the upper tip of the inclusion. (d) Melt inclusion in olivine phenocryst from the Etna volcano (Italy), containing shrinkage vapor bubble and homogeneous glass.

olivine. The fact that fluid and melt inclusion volume can be affected by irreversible lattice changes, such as the α–β transition in magmatic quartz, is also generally not considered.

For accurate reconstruction of formation *P–T* conditions of melt inclusions, the following partial problems must be solved: 1. Selection of suitable inclusions and their proper genetic classification. 2. Measurement or estimation of the degree of fill. 3. Measurement of **solidus**, beginning of melting. 4. Measurement of **liquidus**, temperature of melting of the last solid phase. 5. Measurement of total homogenization, namely the temperature of bubble disappearance in melt, **fluid-saturated liquidus**. 6. Comparison of results with independent temperature and/or pressure constraints. This procedure is equally valid for fluid and melt inclusions, although calculation of isochores is not necessary in the melt inclusion research.

Using "standard" melt inclusions in granitic quartz, the following phase changes during heating (=re-homogenization) can be observed and measured. For the sake of simplicity we separate this process into two different tasks:

1. Investigation of the volatile subsystem present in the shrinkage bubble.
2. Investigation of the bulk melt–fluid system.

Granitic melt inclusions usually contain a volatile subphase (aqueous liquid + vapor bubble, Figure 4.41a). Typical granite minerals (quartz, feldspars, mica) crystallize from the silicate glass phase (Figure 4.41c). If the inclusions are large enough (~100 μm), some minor components can be detected. Studying the volatile subphase permits the estimation of water content in the system (which must be corrected for the amount of water bound in micas), the detection of soluble components dissolved in the aqueous phase (cations, anions, and trace elements), and the composition of dissolved gases. The solid phase and the gas components can be effectively studied using micro-Raman spectroscopy. The bulk system can be chemically characterized after total homogenization and quenching with different analytical methods, such as electron microprobe, secondary ion mass spectrometry (SIMS), Raman spectroscopy, laser ablation ICP-MS, FT-IR spectroscopy, and the like.

Total homogenization of volatile silicate melt inclusions with larger vapor bubbles or higher density of the volatile subsystem is often impossible, because the internal pressure in the inclusion increases after partial homogenization very rapidly, up to the point of partial or total decrepitation. Water, CO_2, and sometimes HF escape through dislocations and decrepitation cracks, thus leaving behind a nearly dry silicate melt system with an artificially high homogenization temperature commonly exceeding by hundreds of degrees Celsius the true homogenization temperatures. In addition, fluid inclusions accompanying the melt inclusions also decrepitate on heating, thus fracturing the host mineral and opening the neighboring melt inclusions (Webster et al., 1997). These problems can be circumvented by heating within cold-seal high-pressure vessels (Student and Bodnar, 2004) or in a hydrothermal diamond anvil cell (Li and Li, 2014). As a rule, melt inclusions in granitic rocks cannot be homogenized using the conventional heating stage, although this method remains useful for water-poor andesitic or basaltic melts. Student and Bodnar (1996, 1999) discussed the behavior of crystalline melt inclusions in quartz during heating and cooling, and presented a model that predicts the *P–V–T–X* evolution of such inclusions during crystallization of a hydrous haplogranite melt.

Figure 4.42 schematically depicts phase changes in a granitic melt inclusion. At point 1, the fluid subphase in the melt inclusion is composed of aqueous liquid, vapor bubble, crystallized solid phases, and glass. With the increasing temperature, the *P–T* trajectory follows the liquid–vapor curve, and the vapor phase shrinks or expands and finally disappears at point 2, which corresponds to the temperature

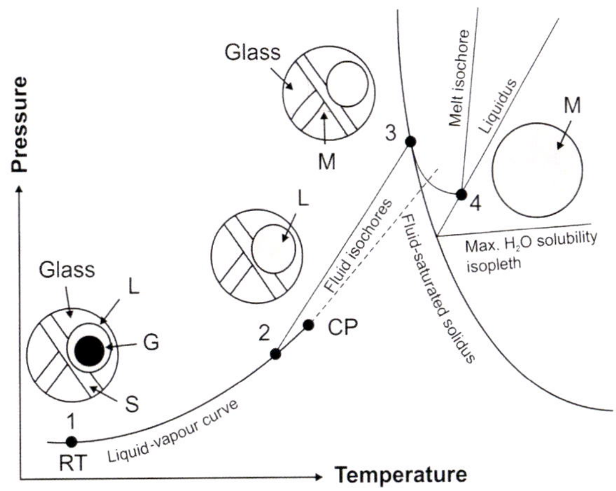

FIGURE 4.42

Schematic outline of phase changes in crystallized silicate melt inclusions containing aqueous liquid (L) and vapor (G) phases in the shrinkage bubble at room temperature (RT). Partial homogenization of the fluid subsystem to liquid occurs at point 2, whereas the first melt (M) appears at point 3, where the fluid isochore intersects the fluid-saturated solidus. Transformation of aqueous liquid to vapor occurs between points 3 and 4 after intersecting the critical isochore (dashed line) of the fluid subsystem. The water continually dissolves in the melt between points 3 and 4. Total homogenization by total dissolution of the vapor bubble and conversion of all solid phases (S) to melt occurs at point 4, when the silicate melt occupies the whole inclusion volume. On further heating, the inclusion pressure follows the one-phase melt isochore. CP denotes the critical point of the fluid subsystem.

of partial homogenization to liquid or vapor. On further heating, the melt inclusion system moves along the fluid isochore and the incipient melting of solid phases occurs at point 3 after hitting the fluid-saturated solidus of the melt–H_2O system. With the onset of melting, the water from the bubble is gradually dissolved in the melt whose volume increases at the expense of the molten solid phases. During this process, the bubble volume decreases and aqueous liquid may convert to vapor without phase transition, if the internal inclusion pressure drops below the critical isochore depicted by the dashed line. Finally, water is totally exhausted and solid phases are completely molten at point 4, which corresponds to the temperature of total homogenization on the fluid-saturated liquidus. On further heating, the P–T trajectory follows the one-phase melt isochore, whose slope is almost subvertical, because the compressibility of the silicate melt is negligible. For this reason, the total homogenization temperature, if properly determined, is nearly equal to that of trapping, and the intersection of the T_h value with the fluid-saturated liquidus also defines the lowest possible trapping pressure. Hence, for the sake of thermobarometry, proper determination of water content in melt inclusions is essential.

It must also be emphasized that the conventional microthermometry with direct observation of phase transitions commonly leads to overestimation of the total homogenization temperature in melt inclusions in granitic rocks. The homogenization is slow due to high melt viscosity and the correspondingly low diffusivity, if contents of volatiles are decreased. At ambient pressures, the inclusion volatiles (mainly H_2O, HF) escape either through microcracks or by diffusion. Hence, it is often possible to measure only "apparent" homogenization temperatures, which depends on the rate and duration of the heating experiment, and also on inclusion diameters.

The quenching technique proposed by Thomas (1994) and further refined in Thomas and Klemm (1997) can provide an answer to the problem of how to cope with the kinetic barriers of phase transformations in viscous acidic melt inclusions. Doubly polished chips are heated in a furnace to a constant temperature that is higher than the solidus for the given melt composition. Samples are then quenched and examined under the microscope for the evidence of homogenization. After quenching, the smaller inclusions will be completely homogenized, but the larger ones will not. As a criterion for the establishment of the equilibrium in the melt inclusion systems, only inclusions with critical diameters d_K are considered. In homogenization experiments, the critical diameter is taken as the diameter of the inclusions in which the vapor bubble is only just visible under the microscope. Hence, diameters of all inclusions in the sample are measured using a calibrated micrometric ocular. Standard duration of the experiment at various temperatures is maintained. The method of calculating water content from the critical diameter is discussed in Section 5.3.5.

If the critical inclusion diameters (d_K) are measured at different temperatures with the same duration of each heating experiment, the data can be extrapolated linearly to determine the homogenization temperature of a hypothetical, infinitesimally small melt inclusion, in which no kinetic barriers due to diffusion exist. This temperature would correspond to the stable fluid-saturated liquidus for melt inclusions of the given composition. For instance, stable fluid-saturated liquidus of 738 ± 6 °C for the Kirchberg granite was obtained by the linear extrapolation of critical diameters measured at various temperatures to $d_K = 0$, as exemplified in Figure 4.43 for two data subsets from Table 4.5. This method is, however, extremely time-consuming and can be applied only in samples with a large amount of

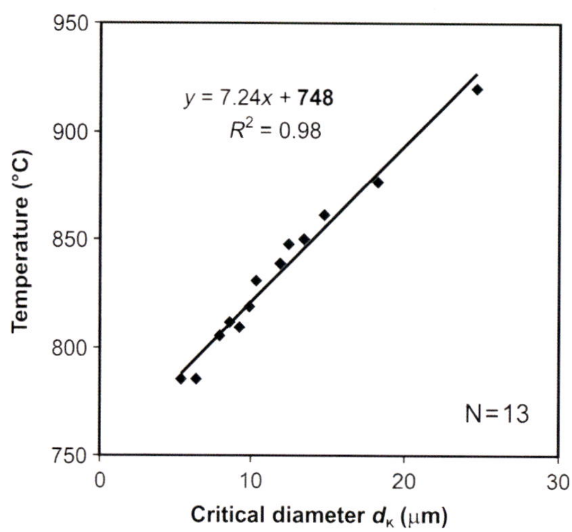

FIGURE 4.43

Estimation of stable homogenization temperature (fluid-saturated liquidus) for melt inclusions from Kirchberg granite. The confidence value R^2 and a linear equation fitted to the dataset in Table 4.5 are shown. The temperature intercept of the linear equation (the number in bold) represents the temperature of homogenization of a hypothetical, infinitesimally small melt inclusion free of kinetic effects.

Table 4.5 Critical Diameters for Melt Inclusions in Granite from Kirchberg, W. Erzgebirge, Germany, Measured at Various Temperatures with a Constant 20-H Duration of Each Heating Experiment (Thomas, 1994)

Run	Temperature (°C)	d_K (μm)
1	785	6.4
2	785	5.4
3	805	7.9
4	809	9.2
5	812	8.6
6	819	9.9
7	831	10.3
8	839	11.9
9	848	12.4
10	850	13.4
11	861	14.7
12	877	18.2
13	920	24.6

primary inclusions with uniform chemical and phase composition, and water content, originating at the same temperature. Decrepitation cracks and entrapment of a heterogeneous melt–fluid mixture make the use of this method impossible. It has recently become possible to determine the water content of homogenized melt inclusions using microbeam analytical techniques (Raman, SIMS).

In extremely hydrated silicate-melt systems, for example, in granite pegmatites, phase changes in melt inclusions are more complicated, because melt–fluid or even melt–melt–fluid immiscibility is typical. The second immiscible liquid can be represented in all magmatic rocks, regardless of composition, by aluminosilicate melt (Thomas and Davidson, 2012b; Charlier and Grove, 2012), iron-rich silicate melt (e.g., Jakobsen et al., 2005; Veksler, 2009), fluoride melt (Klemme, 2004), complex chloride salt melt (Kamenetsky et al., 2003), chloride-carbonatite melt in kimberlites (Kamenetsky et al., 2007), iron- and titanium-rich oxide melt (Hurai et al., 1998), and exotic melts close to the magmatic-hydrothermal transition (Veksler, 2004; Borisenko et al., 2011). The list of papers documenting magmatic melt immiscibility is incomplete. The most recent overview has been provided by Kamenetsky and Kamenetsky (2011).

SUGGESTED FURTHER READING

van den Kerkhof, A.M., Hein, U., 2001. Fluid inclusion petrography. Lithos 55, 27–47.

CHAPTER

INTERPRETATION OF MICROTHERMOMETRIC DATA

5

CHAPTER OUTLINE

5.1 CALCULATING DENSITIES AND ISOCHORES

Constructing isochores for fluid inclusions involves two stages (Wilkinson et al., 1998): (1) the position of the boiling or condensation curve must be ascertained; (2) orientations of the isochores originating over the range of homogenization temperatures must be determined. In such a way, an isochoric envelope is constructed for individual fluid systems or populations of coexisting inclusions, outlined by

Geofluids. http://dx.doi.org/10.1016/B978-0-12-803241-1.00005-8

the isochores of the least dense and the densest inclusions and by two phase curves (L+V or L+H). This approach is valid for essentially all types of fluids, whose composition and density is deciphered from temperatures of phase transitions or Raman spectroscopy.

5.1.1 AQUEOUS INCLUSIONS

Fluid inclusion microthermometric data can be reduced using various stand-alone models and equations of state designed primarily for the calculation of *P–V–T–X* parameters of individual fluid systems. No single model, however, covers the complete range of *P–V–T–X* conditions of interest in most fluid inclusion studies. In addition, further calculations must be performed to convert the measured temperatures of phase transitions into common state properties (composition, molar volume, density) used as input variables in equations of state. Hence, interpretation of microthermometric data can be tedious and often requires switching between various models and correlation equations to obtain the desired information.

To facilitate tedious calculations, several computer programs have been written that directly use the measurable microthermometric quantities as independent variables (Appendix VII). Some of the listed programs are either obsolete or do not run on recent computer platforms. FLINCOR was the first broad-based attempt to provide a computer program for calculating fluid inclusion properties from microthermometric data (Brown, 1989). The Windows version of this program still runs in recent operating environments, but the program lacks upkeep and is no longer recommended for further use. A MACFLINCOR version designed for the Apple Macintosh platform is a newer version, which has a more user-friendly interface that is conducive to use while making microthermometric measurements (Bakker and Brown, 2003).

The CLATHRATE and FLUIDS program packages (Bakker, 1997, 2003, 2009, 2012) are alternatives to the MACFLINCOR, offering solutions tailored for any specific problem associated with fluid inclusion data interpretation. The CLATHRATE package is subdivided into DENSITY, ICE, Q2, NOSALT, and CURVES programs. The first three treat specific types of clathrate melting behavior in the H_2O–CO_2–CH_4–N_2–C_2H_6–salt system, whereas the NOSALT calculates maximum clathrate melting temperatures in binary H_2O–gas mixtures. The CURVES program is designed for the construction of *P–V–T–X* diagrams with clathrate stability curves. The FLUIDS package is aimed at modeling fluid inclusions, in which no clathrate nucleates. The BULK program calculates bulk densities, molar volumes, and compositions using information gathered from microthermometry, Raman spectroscopy, and other methods. The ISOC program calculates isochores for user-specified fluid compositions and densities. The LONERS is a group of programs, each calculating *P–V–T–X* and additional state properties from state equations of individual fluid systems. The AqSo is a module for the calculation of the solubility of gases and salts in aqueous solutions.

HokieFlincs_H2O-NaCl (Steele-MacInnis et al., 2012) is an Excel spreadsheet for interpreting microthermometric data from H_2O–NaCl inclusions, using measured temperatures of ice, hydrohalite, and halite dissolution, and total homogenization temperatures as input variables. In addition to previous programs, it also enables the calculation of isochores for very high-density H_2O–NaCl inclusions, which homogenize by halite disappearance, namely $T_mH > T_h$. Salinity, pressure of homogenization, density, and the dP/dT slope of the corresponding isochore are calculated using empirical equations fitted to experimental data on synthetic inclusions (Lecumberri-Sanchez et al., 2012). The Excel spreadsheet has an advantage of processing and simultaneously comparing a great number of individual measurements (up to 300). The program does not calculate isochores but T_h–isolines uncorrected to compressibility and expansion of the host quartz.

Differences between the microthermometric data from synthetic fluid inclusions and experimental data obtained using a volume-calibrated autoclave are negligible at low to moderate salinities, where discrepancy between the two datasets is not greater than 100 bar and pressures less than 5 kbar. Larger discrepancies exist at higher salinities and densities, where the isochores derived from the autoclave experiments are underestimated by more than 1 kbar, for example, at 300 °C and around 5 kbar (Figure 5.1). Even larger discrepancies exist between various empirical fits to experimental data and equations of state for the H_2O–NaCl system. This problem primarily concerns interpretations of fluid inclusion data from metamorphic settings, where discrepancies arising from various equations and datasets may attain several kbars, thus significantly affecting interpretations of peak or retrograde conditions (e.g., Hurai, 1989). In turn, the discrepancies tend to diminish with the decreasing pressure, and they can be almost ignored in most hydrothermal settings with fluid pressures below 2 kbar.

Isochores for high-density, halite-bearing fluid inclusions that homogenize by halite dissolution can also be derived from metastable fluid inclusion behavior (Hurai et al., 2002a). The method, however, can only be applied to inclusions that can be cooled after the total homogenization to renucleate the vapor bubble without halite, and which can then be reheated to the temperature of the metastable vapor dissolution in the absence of halite (for details, refer to Figure 4.17 and related text). In that case, the line connecting the temperature of metastable total homogenization by bubble disappearance—Th(L)—on the liquid+vapor curve with that of stable total homogenization temperature by halite dissolution on the vapor-unsaturated halite liquidus defines the dP/dT slope of the isochore in the liquid-only region. The method is, however, applicable only to aqueous fluid inclusions devoid of dissolved gases, and salinities less than ~35 mass% NaCl eq. Remember, that if $T_m > T_h$, halite melts in vapor-unsaturated conditions, and the actual salinity is not exactly known. To infer the actual salinity from vapor-saturated halite liquidus, an artificial re-equilibration of the inclusion must be done. The procedure is treated in detail as Problem 4.4.

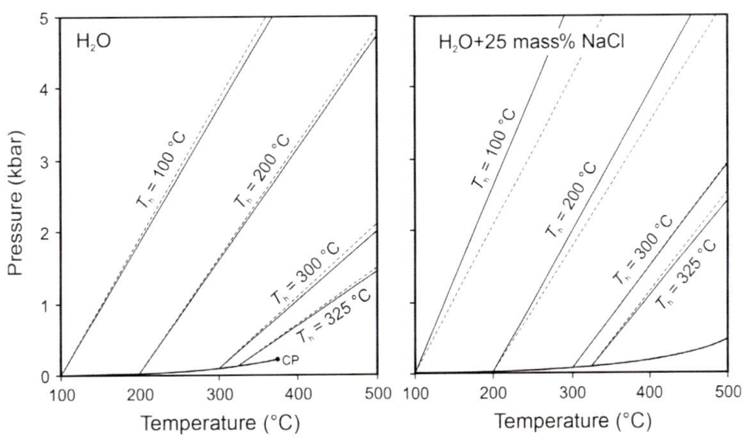

FIGURE 5.1

Comparison of T_h–isolines (solid lines) and isochores (dashed lines) of pure water and 25 mass% NaCl–H_2O solutions calculated using data from synthetic fluid inclusions in quartz (Lecumberri-Sanchez et al., 2012) and those gathered from a volume-calibrated autoclave (Gehrig, 1980), respectively.

Let us now calculate the isochoric slope for the inclusion FI-5 (Problem 4.4) and compare it with the experimental data of Lecumberri-Sanchez et al. (2012). For $T_h(H) = 145.4$ °C and $T_mH(L) = 210.8$ °C, the HokieFlincs_H2O-NaCl program yields the pressure at halite dissolution equal to 1 358 bar and a dP/dT isochoric slope in the liquid-only region corresponding to 25.05 bar/°C. However, the vapor unsaturated halite liquidus for the calculated salinity, 33.2 mass% NaCl, does not exactly match the starting point of the isochore. Hence, the vapor-unsaturated halite liquidus must be adjusted, and the corrected salinity is now 33.28 mass% NaCl, corresponding to a 226.5 °C temperature of vapor-saturated liquidus on the L+G+H curve (Figure 5.2). Linear extrapolation of the calculated isochore to the L+G curve defines the metastable homogenization in the absence of halite at 157.5 °C. All these data are, however, offset from the values inferred from the inclusion FI-5: $T_h(L)$ is overestimated by 4.9 °C, and $T_mH(L)$ by 10.5 °C. In contrast, the inferred isochoric slope corresponds to 22.6 bar/°C, and it does not differ too much from that calculated by the HokieFlincs_H2O-NaCl program. Discrepancies between the measured and calculated temperatures of phase transitions most likely result from the different slope of the vapor-unsaturated halite liquidus of polycomponent natural brines compared to that in the experimental binary H_2O–NaCl system and to a lesser extent also from slightly different isochoric slopes (Hurai et al., 2002a).

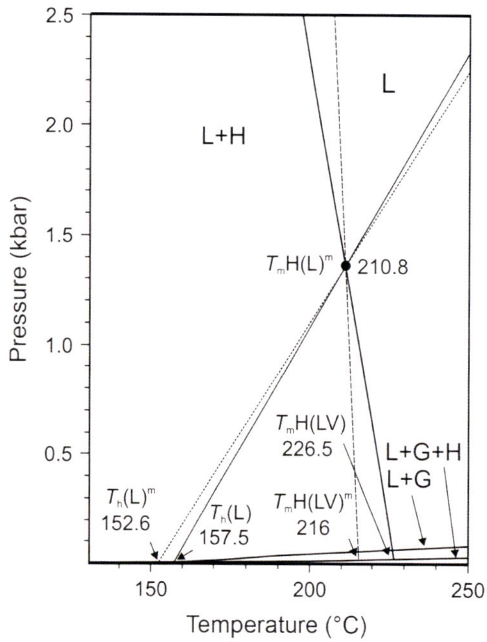

FIGURE 5.2

Comparison of isochore and vapor-unsaturated halite liquidus (stippled and dashed lines, respectively) derived from microthermometric data on the inclusion FI-5 (Problem 4.4) with those calculated from the same dataset using the HokieFlincs_H2O-NaCl program (Steele-MacInnis et al., 2012) based on experimental data from Lecumberri-Sanchez et al. (2012). Superscript ^{m} refers to measured temperatures; other values are calculated. L+H and L denote stability fields of liquid+halite and liquid, respectively. L+G+H and L+G are projections of the vapor-saturated three-phase curve and the two-phase surface of the H_2O–NaCl system.

5.1.2 HYDROCARBON INCLUSIONS

The UV-induced fluorescence color of hydrocarbon inclusions can be roughly approximated to the American Institute Petroleum (API) gravity, which relates to the density D (g cm^{-3}) by the equation (Bodnar, 1990)

$$\mathrm{API} = 141.5/D - 131.5 \tag{5.1}$$

If the API gravity is greater than 10, the oil floats on freshwater; if less than 10, it is heavier and sinks in freshwater.

A shift of the excited wavenumber toward the blue segment of the visible electromagnetic spectrum is caused by an increasing concentration of saturated hydrocarbons. Blue fluorescence colors (410–440 nm) correspond to mature paraffinic light oil with the density of 45–50 API. In contrast, a shift toward the red and orange parts of electromagnetic spectrum (530–550 nm) is caused by aromatic hydrocarbons, resins, and asphaltenes, which are important constituents of naphthenic or biologically degraded oil with the density of 10–15 API (Stasiuk and Snowdon, 1997).

The Q_{580} fluorescence emission integral quotient (the ratio of 580–700 nm to 400–580 nm spectral integrals) shows excellent negative correlation ($R^2 = 0.95$) with the API gravity (Figure 5.3), and can be used as a quantitative indicator of oil composition and density. However, though the API is assumed to increase with increasing maturity, this is not true for a migrated oil, whose density is modified by degassing, water washing, and biodegradation (Kihle et al., 2012). Hence, the florescence color can be used only for discrimination of genetically related oils (Munz, 2001).

The homogenization temperature combined with information about the chemical composition of the inclusion fluid serves as input data to calculate two-phase envelopes and isochores with commercially available software based on the Soave–Redlich–Kwong and the Peng–Robinson equations of state. A VTFLINC program developed by Aplin et al. (1999) requires an initial oil composition to launch the calculation of the total homogenization temperature (T_h) and the degree of filling with vapor bubble (F_v) at a certain temperature below T_h. These two calculated parameters are then correlated with those measured by microthermometry. The most problematic point, however, is the characterization of the oil composition. The present-day reservoir oil composition is used as an initial regression parameter. Then, C1–C5 gas components are titrated out or into the fluid inclusion composition until a match is obtained with the measured T_h and F_v values. Obviously, the accuracy of the method depends on the estimate of the initial oil composition.

A procedure for compositional characterization of inclusion oil based on online gas chromatography was mentioned by Munz (2001). C1–C5 compounds are treated as individual components, whereas the range from C6 to C30+ is subdivided into pseudocomponents by integrating ranges defined as single-carbon-number groups. Then, average values for critical properties of the pseudocomponents known from literature are used. The properties of the residue are estimated by extrapolation of the distribution of components in the range of C8–C30.

Thiéry et al. (2000, 2002) used the model from Montel (1993) to characterize the complex oil composition by using only two parameters. The parameter α sets the amount and distribution of high molecular components (C10+), whereas β is related to the content of volatile molecules like methane and C2–C6. Density and liquid-vapor boundaries are calculated with the modified Peng–Robinson equation of state using the PIT software and T_h and F_v values as input variables (Thiéry et al., 2000, 2002). This procedure is possible owing to the fact that the shape of the F_v–T curve, as well as the total homogenization temperature, is unique for any given composition and density of the inclusion oil.

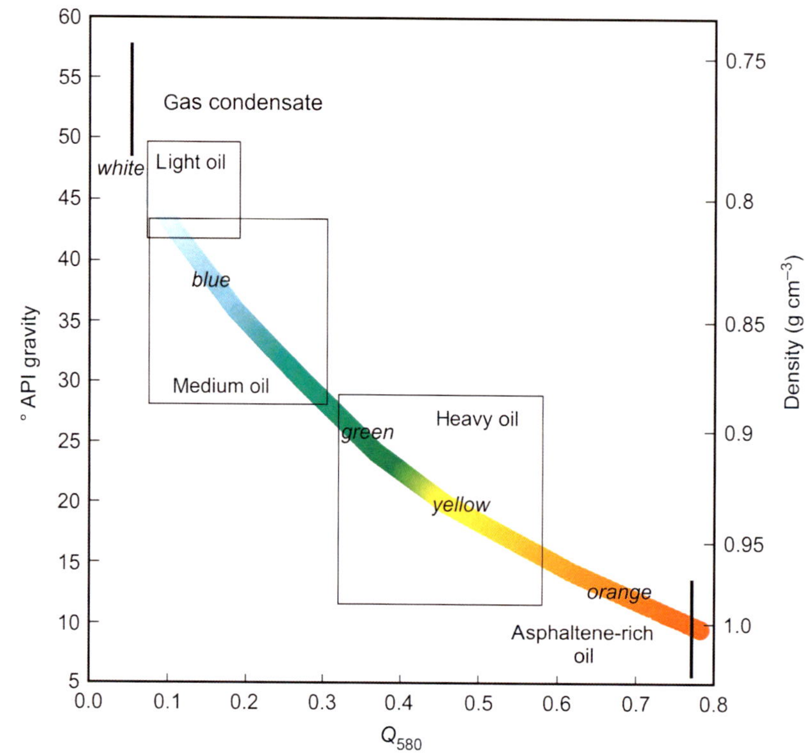

FIGURE 5.3

Correlation plot of density (API, g cm^{-3}), Q_{580} emission integral, and UV-induced fluorescence colors of petroleum standards (Skaare et al., 2011, Kihle et al., 2012). Reprinted from Kihle, J., Hurum, J.H., Liebe, L., 2012. Preliminary results on liquid petroleum occurring as fluid inclusions in intracellular mineral precipitates in the vertebrate of Pliosaurus funkei. Norsk Geol. Tidsskr. 92, 341–352, Copyright 2012, with permission.

The α–β compositional model is a simple way to characterize graphically the composition of any hydrocarbon fluid (Figure 5.4). The model fits the range of possible α–β values to the measured T_h and F_v values. Further refinement can be achieved by correlation of the microthermometric data with UV-induced fluorescence color, API gravity, gas chromatography, volume expansion rate of gas bubble, and more. The PIT software is no longer publicly available, but the surrogate FIT-Oil program was developed with specific algorithms and calculation formulas. The program calculates and graphically displays α–β curves, immiscibility loops, and isochores for measured inclusion fluid parameters.

Attempts were recently made to develop even more simplified models capable of predicting saturation pressures of crude oils with the minimum number of input variables. Ping et al. (2011) showed that the saturation pressure and density of a petroleum inclusion can be defined by T_h, F_v, and a hypothetical compositional distribution of C1, C2, C3, C4, C5, C6 components together with the C7+ fraction by simple tuning of the methane molar fraction. Moreover, the C7+ fraction and the petroleum type can be satisfactorily inferred from the slope of the correlation curve in the F_v–T diagram in artificially stretched oil inclusions (Bourdet et al., 2008). Black oil shows an almost linear correlation

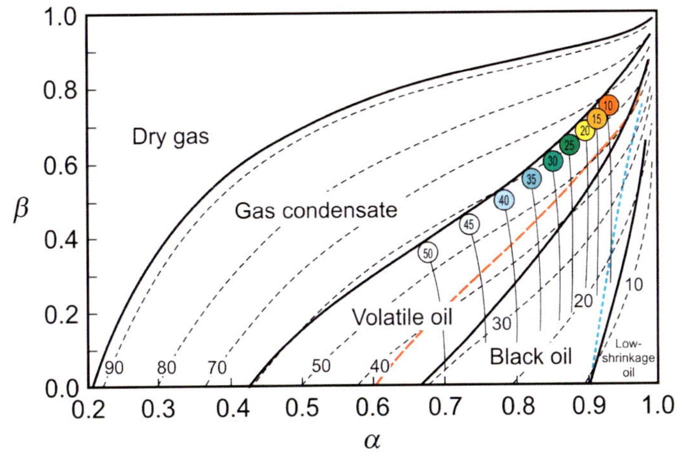

FIGURE 5.4

The α–β space (Ping et al., 2011) with thick solid lines delineating fields for different hydrocarbon–petroleum reservoirs according to the classification of McCain (1994). Thin dashed lines correspond to methane contents (mol.%), and thin subvertical curves labeled by numbers in circles are API iso-gravity curves calculated using the FIT-Oil program. The color roughly corresponds to that presumably induced by the 365 nm UV-light. Oils with less than 10°API gravity do not fluoresce under UV. Two α–β dashed curves are those calculated for oil inclusions in quartz from Pučov (orange curve) and Radotín (blue curve), documented in Figure 5.5 using the bubble diameters and total homogenization temperatures summarized in Table 5.1. The α–β space reprinted from Ping, H., Thiéry, R., Chen, H., 2011. Thermodynamic modelling of petroleum inclusions: the prediction of the saturation pressure of crude oils. Geofluids 11, 328–340, 2011, with permission from John Wiley & Sons, Inc.

of vapor volume fraction at 20 °C correlated with the total homogenization temperature, whereas curvature of the correlation curve increases toward the light oil and gas–condensate compositions.

Sample calculations of two-phase envelopes and isochores were made for the two oil inclusions displayed in Figure 5.5 using the FIT-Oil software. The relationship between bubble volumes and temperatures was measured using the object analysis included in the UTHSCSA Image Tool (http://compdent.uthscsa.edu/dig/itdesc.html), and the results are summarized in Table 5.1. The calculated α–β curves together with intensive bright blue UV-induced fluorescence emission show that the inclusion from Pučov contains light volatile oil with an API gravity around 45 (Figure 5.4). The α–β curve for the inclusion from Radotín plots inside the field of black oil, inconsistently with the dark blue UV-induced luminescence, and indicating API gravity values around 35. Hence, it is likely that the inclusion has leaked, which caused an increase in the total homogenization temperature. Indeed, the inclusion exhibits a total of homogenization temperature that is ~70 °C higher than those of other smaller oil inclusions in the same quartz.

Assuming an API gravity of 45°, the F_V–T correlation measured on the oil inclusion from Pučov corresponds to α and β values of 0.742 and 0.376, respectively, and the bulk composition calculated using the FIT-Oil program is as follows (mol.%): methane 50.03, ethane 10.38, propane 7.69, iso-butane 1.50, *n*-butane 3.36, iso-pentane 1.94, *n*-pentane 3.11, hexane 4.64, heptane 5.00, octane 3.76, nonane 2.84, decane 2.15, high-cetane number hydrocarbons 3.56, low-cetane number

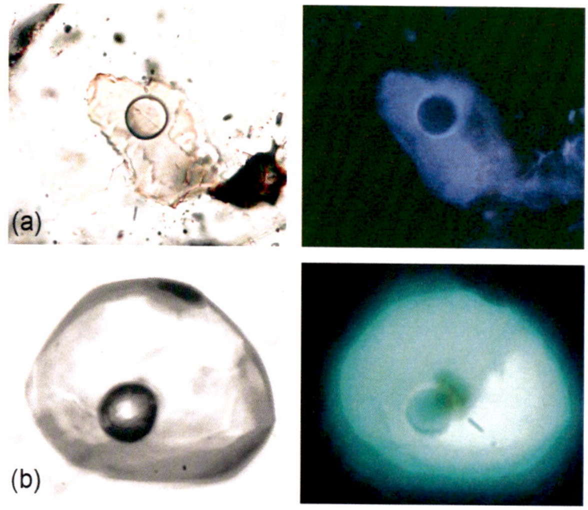

FIGURE 5.5

Transmitted light and UV-luminescence images of two oil inclusions used for calculation of two-phase envelopes and isochores using the FIT-Oil software (Thiéry et al., 2000, 2002). (a) Light-brown oil inclusion in quartz from Radotín. (b) Transparent colorless oil inclusion in quartz from Paleocene–Eocene sandstones of the intra-Carpathian fore-arc flysch basin (Pučov, Slovakia). All photomicrographs were taken at 22.7 °C. Luminescence was induced by a 365-nm reflected UV light. The longer diameter of both inclusions is about 50 μm.

Table 5.1 Bubble Volume (vol.%)—Temperature Dependence in Oil Inclusions Displayed in Figure 5.5

Pučov		Radotín	
Temperature (°C)	**Vol.% Gas**	**Temperature (°C)**	**Vol.% Gas**
−2.2	14.00	22.7	13.52
7.3	11.56	40.2	13.00
19.6	8.90	60.3	12.32
23.7	7.30	80.1	10.94
30.2	4.97	100.2	9.11
35.2	2.50	120.2	6.53
38.5	0.00	130.4	4.81
		140.2	3.00
		150.7	0.00

hydrocarbons 0.04. Now, the corresponding two-phase envelope and isochore can be calculated easily using the equation of state included in the FIT-Oil program (Figure 5.6).

It should be noted that isochores cannot be calculated using the F_v–T relationships in oil inclusions, which homogenize at very low temperature. The same pertains to high-density volatile oil inclusions similar to those documented in Figure 4.40. The problem can be circumvented by artificial inclusion

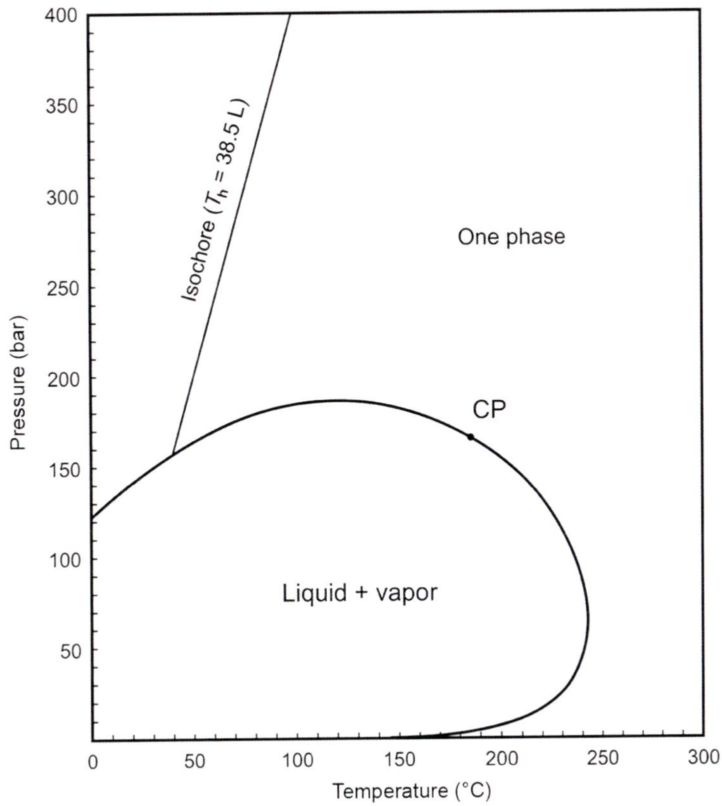

FIGURE 5.6

Two-phase envelope and isochore of an oil inclusion documented in Figure 5.5b calculated using the FIT-Oil program and data from Table 5.1.

stretching or decrepitation and subsequent measurement of the F_V–T relationship in the stretched inclusion that homogenizes at a higher temperature. The L+G boundary can then be reconstructed using the FIT-Oil, which enables the calculation of isochores also for lower T_h values.

5.1.3 GASEOUS INCLUSIONS

The procedure of calculating isochores of gaseous inclusions is essentially the same as described for aqueous inclusions, but one need not worry about the position of the two-phase boundary, which lies far below the temperatures and pressures relevant to geological processes. Hence, the construction of isochores can be done using an appropriate equation of state, providing the density and composition is known from microthermometry and/or Raman spectroscopy. For pure gases, the web calculator (Appendix VII) maintained and regularly updated by the National Institute of Standards and Technology (NIST) is the best choice for calculating fluid properties in saturation conditions and various isothermal and isobaric projections permitting the construction of isochores.

The bulk density (or molar volume) of a gaseous mixture is a critical parameter, which must be derived prior to calculation of an isochore. This is possible from microthermometric data only for mixtures, which can be approximated to a binary system or two binary subsystems (Thiéry et al., 1994b). Once the bulk composition and density (molar volume) are known, isochores can be calculated using the MRK EOS of Holloway (1977, 1981), which is included in all versions of programs for fluid inclusion data reduction and interpretation (FLINCOR, MACFLINCOR). For extremely high-pressure conditions, the virial-type equation of state of Zhang and Duan (2009) for C–O–H gaseous mixtures involved in the GFluid program available as an Excel spreadsheet (Zhang and Duan, 2010) can be recommended. The program is no longer available at the website quoted in the original paper, but it is still downloadable together with other programs from the Duan's research group at the website cited in Appendix VII. The GFluid program does not calculate isochores, but the molar volume of specified gas mixture can be determined at given *P*–*T* coordinates. Hence, the desired molar volume can be located at fixed *P* by changing *T* and vice versa using the SOLVER tool in an Excel spreadsheet.

5.2 CALCULATING FLUID AND MELT VISCOSITIES

Viscosity, rather than density, is the critical parameter that concerns melt inclusionists. Though the density of magma is an important parameter for estimating depths of burial and other derived melt parameters (Bottinga and Weill, 1972), it is less useful for the estimation of formation *P*–*T* conditions. Unlike fluids, silicate melts are almost incompressible, their isochores are almost vertical: hence, the difference between the temperature of total homogenization and that of trapping is negligible even for the melts formed at high pressures. We therefore do not need the densities and isochoric phase trajectories to be determined for the homogeneous melt-fluid system from a corresponding equation of state. Moreover, optical determination of the homogenization temperature is imprecise, because the gas bubble dissolution in the melt is controlled by diffusion, and it is thus strongly influenced by the inclusion diameter and the duration of the heating experiment (Thomas, 1994). In contrast, however, the viscosity is employed for the calculation of the amount of water dissolved in the silicate melt, and ultimately also for the calculation of formation *P*–*T* conditions (Section 5.3.5).

For melt compositions determined using either electron microprobe analysis of rehomogenized melt inclusions or bulk chemical analysis of host magmatic rock, two viscosity models can be considered. Earlier models (Shaw, 1972; Bottinga and Weill, 1972) employ the Arrhenius-type linear correlation between the natural logarithm of viscosity and reciprocal temperature, assuming the silicate melt is a Newtonian fluid. The model is suitable for the calculation of viscosity of water-bearing melts for temperatures close to liquidus, which means at higher temperatures. However, the Arrhenian-type models fail to reproduce melt properties at temperatures close to solidus.

Later experiments showed that magmas could behave as non-Newtonian fluids, showing a nonlinear viscosity dependence on temperature (Dingwell et al., 1993; Richet and Bottinga, 1995). Hence, the models involve a greater number of correction parameters, and they also consider the influence of fluorine—the volatile component most intensely affecting the melt viscosity besides water (Hui and Zhang, 2007; Giordano et al., 2008). The estimation of viscosity of water-rich melts and supercritical fluids with 10–80 mass% of dissolved silicates can be performed in situ by the observation of falling spheres in a diamond anvil cell (Audétat and Keppler, 2004).

Calculation of the melt viscosity after Shaw (1972) employs partial binary mixtures of oxides with SiO_2, assuming that the viscosity is controlled by the degree of Si–O bonds polymerization. The method can be applied to natural silicate melts containing SiO_2, Al_2O_3, TiO_2, Fe_2O_3, FeO, MnO, MgO, CaO, Na_2O, K_2O, and H_2O within the temperature interval of 600–1700 °C and the viscosity range of $10^{-0.2}$–10^{12} Pa s. An average slope of the viscosity versus temperature correlation curves is calculated from individual oxides (a total of 5 parameters), excluding MnO and P_2O_5. However, MnO can be added to FeO (Winter, 2010). The resulting equation reads

$$\ln \eta = s\cdot(1000/T) - c_T\cdot s + c_\eta \tag{5.2}$$

where η is the viscosity in poises, T is temperature [K], s is the average slope of $\ln \eta - 10^4/T$ correlation lines, c_T and c_η are coordinates of the intersection point of various Arrhenian lines estimated as $c_T = 1.5$ and $c_\eta = -6.4$.

As an alternative, the model of Giordano et al. (2008) can be applied to natural magmas containing SiO_2, Al_2O_3, TiO_2, FeO_{tot}, MnO, MgO, CaO, Na_2O, K_2O, P_2O_5, H_2O, and F. The temperature dependence of viscosity at atmospheric pressure is expressed using the Vogel–Fulcher–Tamann (VFT) relationship

$$\log\eta = A + \frac{B}{T-C} \tag{5.3}$$

where T is temperature [K], A is the constant [-4.55 ± 0.21], B and C are parameters calculated as follows

$$B = \sum_{i=1}^{7}[b_i M_i] + \sum_{j=1}^{3}[b_{1j}(M1_{1j} - M2_{1j})]$$
$$C = \sum_{i=1}^{6}[c_i N_i] + [c_{11}(N1_{11} - N2_{11})] \tag{5.4}$$

M and N are combinations of mol.% oxides from Table 5.2. The model involves a total of 18 parameters: the constant A, 10 coefficients for the calculation of B, and 7 coefficients for the calculation of C.

Table 5.2 Coefficients for the Calculation of VFT Parameters B and C from the Composition of a Melt Expressed in mol.% (Giordano et al., 2008)

Oxides		Values		Oxides		Values	
b_1	SiO_2+TiO_2	159.6	(7)[a]	c_1	SiO_2	2.75	(0.4)
b_2	Al_2O_3	−173.3	(22)	c_2	TA[c]	15.7	(1.6)
b_3	$FeO_T+MnO+P_2O_5$	72.1	(14)	c_3	FM[d]	8.3	(0.5)
b_4	MgO	75.7	(13)	c_4	CaO	10.2	(0.7)
b_5	CaO	−39.0	(9)	c_5	NK[e]	−12.3	(1.3)
b_6	Na_2O+V[b]	−84.1	(13)	c_6	ln(1+V)	−99.5	(4)
b_7	$V+\ln(1+H_2O)$	141.5	(19)	c_{11}	$(Al_2O_3+FM+CaO-P_2O_5)(NK+V)$	0.3	(0.04)
b_{11}	(SiO_2+TiO_2)(FM[d])	−2.43	(0.3)				
b_{12}	$(SiO_2+$TA[c]$+P_2O_5)$ (NK[e]$+H_2O$)	−0.91	(0.3)				
b_{13}	(Al_2O_3)(NK)	17.6	(1.8)				

[a]*Numbers in brackets represent 95% confidence limits.*
[b]$H_2O+F_2O_{-1}$
[c]$TiO_2+Al_2O_3$
[d]*FeO(tot)+MnO+MgO*
[e]Na_2O+K_2O

PROBLEM 5.1 CALCULATING VISCOSITY OF SILICATE MELT I

Calculate the viscosity of silicate melt using the method of Shaw (1972) at 800 °C for the peraluminous haplogranite (bulk chemical analysis in mass% from Thomas and Klemm (1997)).

Solution

The calculation can be done using Excel spreadsheet in only a few columns (Table 5.3). The molar proportion is calculated from mass% oxides by dividing by the molecular mass (column 4). If Fe_2O_3 is present, its molar proportion is obtained by dividing its mass by molecular mass and multiplying by 2. The molar proportion of Al_2O_3 is calculated similarly (but not those of Na_2O, K_2O, and H_2O). Molar proportions are recalculated to molar fractions in column 5. Column 6 shows Arrhenian slopes of binary mixtures of oxides with SiO_2 extrapolated to pure SiO_2. The slopes are multiplied by SiO_2 molar fraction in column 7. All values from column 7 (except for SiO_2) are multiplied by molar fraction of the corresponding oxide in column 8. The average slope (s) of the mixture of oxides including water is calculated as the total of column 8 divided by $1-X_{SiO_2}$. This value is inserted in Equation (5.2) to obtain the viscosity. The chemical analysis normalized to 100 mass% is given in column 9.

Table 5.3 Calculation of Viscosity of Peraluminous Haplogranitic Melt According to Shaw (1972)

1	2	3	4	5	6	7	8	9
Oxide		**Mol. Mass**	**Mol. Prop.**	**Mol. Fract.**	s_i^o	$s_i^o X_{SiO_2}$	$X_i (s_i^o X_{SiO_2})$	**Normalized**
Units	Mass%	g/mol	mol/100 g	mol/1 mol				Mass%
SiO_2	71.45	60.09	1.189	0.666				70.52
TiO_2	0.32	79.88	0.004	0.002	4.5	3.00	0.01	0.32
Al_2O_3	14.76	101.96	0.290	0.162	6.7	4.46	0.72	14.57
Fe_2O_3		159.70	0.000	0.000	3.4	2.26	0.00	0.00
FeO	2.49	71.85	0.036	0.020	3.4	2.26	0.05	2.46
MnO	0.13	70.94	0.002	0.001	3.4	2.26	0.00	0.13
MgO	0.78	40.31	0.019	0.011	3.4	2.26	0.02	0.77
CaO	2.01	56.08	0.036	0.020	4.5	3.00	0.06	1.98
Na_2O	3.72	61.98	0.060	0.034	2.8	1.86	0.06	3.67
K_2O	3.52	94.20	0.037	0.021	2.8	1.86	0.04	3.47
P_2O_5	0.14	141.94	0.001	0.001				0.14
H_2O	2.00	18.02	0.111	0.062	2.0	1.33	0.08	1.97
Total	101.32		1.79	1.00			1.05	100.00
Mean slope $(s)=\Sigma X_i\,(s_i^o X_{SiO_2})/(1-X_{SiO_2}) = 3.138$								
ln η				18.135	$\eta = 7.51 \times 10^7$ Poise			
ln η				15.832	$\eta = 7.51 \times 10^6$ Pa s			

PROBLEM 5.2 CALCULATING VISCOSITY OF SILICATE MELT II

Calculate the viscosity of silicate melt using the method of Giordano et al. (2008) at 800 °C for peraluminous haplogranite, and compare the result with that obtained from the previous model (Problem 5.1).

Solution

Only FeO_T is needed for calculation. Hence, if Fe_2O_3 is included in the chemical analysis, it must be recalculated to FeO and added to FeO. The first three columns in Table 5.4 are recalculated identically with the previous example. Molar percentages instead of molar fractions are listed in column 4. The *B* and *C* parameters are calculated according to the scheme in Equation (5.4) and Table 5.2. For instance, $b_1 = 159.6$ (coefficient b_1) × (72.55 + 0.24 mol.% $SiO_2 + TiO_2$) = 11,618.64. The final *B* and *C* values represent totals of all particular b_iM_i and c_iM_i parameters.

Table 5.4 Calculation of Viscosity of Peraluminous Haplogranitic Melt According to Giordano et al. (2008)

	1	2	3	4	5		6		7
	Mass%	Mol. Mass	Mol. Prop.	Mol.%	A		B		C
SiO_2	71.45	60.08	1.189	72.55		b_1M_1	11,618.64	c_1M_1	199.5
TiO_2	0.32	79.90	0.004	0.24		b_2M_2	−1530.5	c_2M_2	142.5
Al_2O_3	14.76	101.96	0.145	8.83		b_3M_3	164.839	c_3M_3	28.3
Fe_2O_3		159.69	0.000	0.00		b_4M_4	89.4	c_4M_4	22.3
FeO_T	2.49	71.85	0.035	2.11		b_5M_5	−85.3	c_5M_5	−73.1
MnO	0.13	70.94	0.002	0.11		b_6M_6	−877.7	c_6M_6	−204.1
MgO	0.78	40.31	0.019	1.18		b_7M_7	1248.9	$c_{11}M_{11}$	54.8
CaO	2.01	56.08	0.036	2.19		$b_{11}M_{11}$	−602.6		
Na_2O	3.72	61.98	0.060	3.66		b_{12}	−945.3		
K_2O	3.52	94.20	0.037	2.28		b_{13}	923.52		
P_2O_5	0.14	141.94	0.001	0.06					
H_2O+	2.00	18.01	0.111	6.77					
Sum	101.32		1.639	100.00	−4.55		10,004		170.3
						log η	6.53	Pa s	
						ln η	15.04	Pa s	
						η	3.40×10^6	Pa s	

Audétat and Keppler (2004) experimentally determined the viscosities of supercritical fluid-melt mixtures and showed that the effect of H_2O on viscosity of silicate melts is different at low and high water contents. The viscosities decreased exponentially during the first 20 mass% H_2O, and linearly at higher H_2O contents. They derived the equation

$$\log \eta = -4.2 + [0.008c_s] \times [(10{,}000/T) - 2] \tag{5.5}$$

where the viscosity η is given in Pa s, T in Kelvins, and c_s is the concentration of the dissolved silicate in mass%. The equation is applicable to aqueous fluids with up to 80 mass% dissolved silicate in the temperature range of 600–1200 °C. At high H_2O concentrations, the influence of dissolved silicate (Ca- versus Na-rich) is negligible, however, for albite–H_2O melts with low H_2O contents an extended cubic fit to experimental data should be used.

The authors demonstrated that the viscosity of a fluid with 20 mass% of dissolved silicate at 800 °C is equal to that of pure water at room temperature (10^{-3} Pa s) and that of a fluid with 50 mass% silicate in the same condition is comparable to olive oil at room temperature (10^{-1} Pa s). The contrasting effect of H_2O on viscosity of silicate melts at low and high water contents can be explained in terms of the contrasting mode of H_2O dissolution: whereas the low-H_2O melts dissolve water mainly as OH, which causes the breakage of tetrahedra links (Burnham, 1975), the mainly molecular H_2O at high water contents does not have a further effect upon the melt depolymerization (Stolper, 1982).

5.3 *P–T* ESTIMATES

5.3.1 HOMOGENEOUS FLUIDS

Fluid inclusion barometry is more precise than other conventional empirical barometers based on mineral equilibria. However, the possibility of simultaneous determination of temperature and pressure is restricted only to relatively rare heterogeneously trapped fluid inclusions. More abundant homogeneously trapped fluid inclusions permit either P or T to be calculated by fixing one of the two variables by an independent thermometer or barometer.

Before doing this, the inclusions must be tested whether or not they obey rules formulated by Roedder (1984). Roedder's rules state that the properties of fluid inclusions reflect the trapping conditions if (1) the inclusions trapped a single-phase fluid (criterion of homogeneity), (2) nothing has been added to or removed from the inclusions after trapping (criterion of constant mass), and (3) the inclusion volume remained constant following trapping (criterion of constant volume), (4) the effect of pressure is either known or it can be neglected or estimated independently (criterion of pressure), and (5) the origin and genetic significance is known (criterion of genesis). If these criteria are fulfilled, formation P–T conditions can be inferred from the isochoric envelope constructed from densities and compositions of the homogeneously trapped fluid inclusions.

Homogeneous fluids can be trapped everywhere along the isochore extending from the two-phase boundary and traversing across the fields of homogeneous liquid or vapor. The significance of the total homogenization temperature (T_h) was often overestimated in early fluid inclusion investigations, where it was directly correlated with the temperature of crystallization. In fact, the T_h value is a function of fluid density, which depends not only on temperature, but on pressure and concentration (in polycomponent systems) as well. Let us suppose, for instance, that homogeneous freshwater inclusions were sealed in quartz precipitating in a subvolcanic hydrothermal vein at 300 °C and 730 bar. Such inclusions would homogenize to liquid at 250 °C, and they would be indiscernible from those trapped in metamorphic quartz at 500 °C and 3500 bar, because their homogenization temperatures are identical. To infer actual trapping P–T parameters, an independent thermometer or barometer must be applied, and the missing variable, either temperature or pressure, can then be derived by intersection with the isochore calculated by using an appropriate equation of state. Thermometers are preferred owing to much better precision and uncertainty limits compared to barometers.

Four major groups of geothermometers involve (1) element exchange reactions, (2) equilibrium stable isotope fractionation between coexisting minerals, (3) solvus thermometers, and (4) thermometers based on structural ordering of minerals. A review of the available thermometers and barometers is beyond the scope of this textbook, but this topic has been addressed in several other extraordinary publications (e.g., Essene, 1982, 1989; Bohlen and Lindsley, 1987; Spear, 1993; Misra, 2012). Equilibrium cation-exchange thermometers developed for hot mineralized groundwater of geothermal fields (Section 8.4.1) are also useful, because they allow temperatures to be determined directly from ratios of major cations in aqueous fluid inclusions in hydrothermal minerals. For magmatic and high-grade metamorphic rocks, trace element contents measured using LA-ICPMS or electron microprobe (EPMA) in quartz can be used as an independent thermometer and combined with fluid isochores in the same quartz host in which they define the formation pressure. The Ti-in-quartz thermometer (Wark and Watson, 2006) seems to be most reliable, with the provision that the Ti-saturation during quartz crystallization is proven by inclusions of rutile or another Fe–Ti oxide. Detection limits for measuring Ti contents in quartz are 18–43 ppm for EPMA and about 3 ppm for LA-ICPMS (van den Kerkhof et al., 2004), which corresponds to the lowermost temperature limits of 575–650 °C and 450 °C, respectively (van den Kerkhof et al., 2014). An alternative to conventional thermometers and barometers is provided by multicomponent mineral equilibria calculated at a constrained bulk composition of metamorphic, magmatic, and (rarely) hydrothermal assemblages (Section 6.5). Methods of determining crystallization temperatures using the stable isotope composition of coexisting minerals crystallized at a thermodynamic equilibrium are treated in Section 9.4.

Determination of homogenization pressure has meaning if the pressure correction is small or negligible. In this case, the homogenization temperature also approximates that of trapping. This approach can be employed, for instance, in low- to medium-temperature subvolcanic hydrothermal systems. In high-temperature subvolcanic systems, such as Cu-porphyry deposits, and metamorphic environments, pressure also becomes an important parameter. With the increasing temperature, the solubility potential of aqueous solutions increases dramatically, resulting in the formation of high-density, halite-bearing inclusions, in which the homogenization of vapor bubbles occurs in the presence of halite, and halite dissolution is the last observed phase transformation. Owing to the steep slope of vapor-unsaturated halite liquidus, the fluid pressure at the moment of total homogenization in the brine inclusions that homogenize by halite dissolution is high, and it can be taken as a rough approximation of the minimum possible trapping pressure (Figure 5.7).

The high-density, high-salinity portion of the H_2O–NaCl system was covered by experimental data only recently (Lecumberri-Sanchez et al., 2012). The equations are also available in the form of an Excel spreadsheet named HOKIEFLINCS_H2O-NACL (Steele-MacInnis et al., 2012), which allows density, homogenization pressure, and salinity to be derived directly from the temperatures of bubble disappearance and halite dissolution for both homogenization modes, namely homogenization by bubble disappearance $T_h > T_m$, and by halite dissolution $T_h < T_m$ (Figure 5.8). The experimental data have not been corrected to thermal expansion and compressibility of host quartz, and so should not be automatically used to interpret fluid inclusion microthermometry data from other minerals.

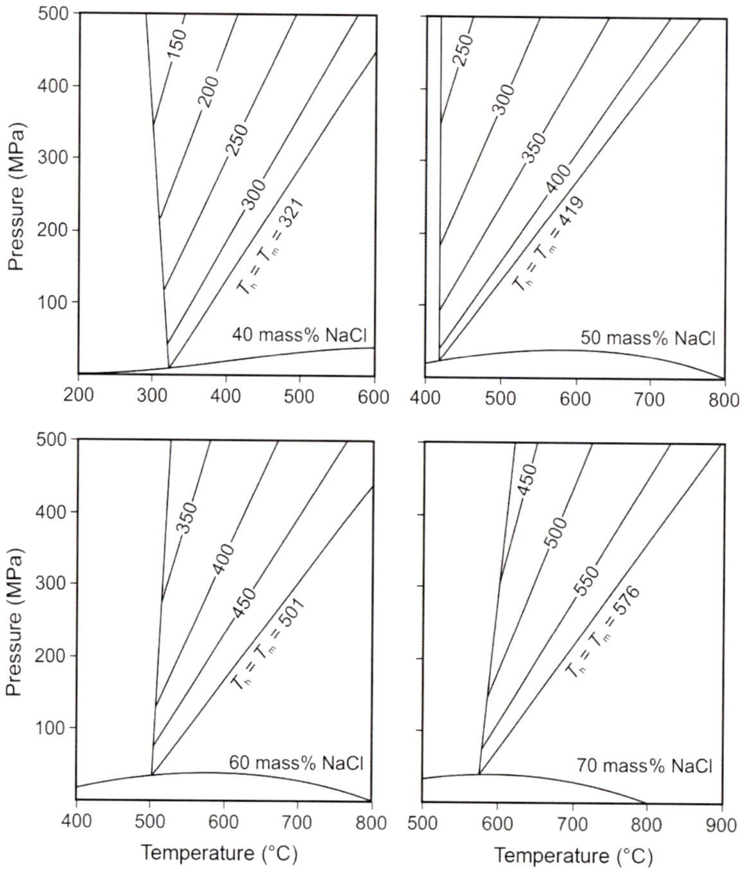

FIGURE 5.7

Isochores of high-density NaCl–H_2O brines. Reprinted from Lecumberri-Sanchez, P., Steele-MacInnis, M., Bodnar, R.J., 2012. A numerical model to estimate trapping conditions of fluid inclusions that homogenize by halite disappearance. Geochimica et Cosmochimica Acta 92, 14–22, Copyright 2012, with permission from Elsevier.

5.3.2 HETEROGENEOUS AQUEOUS FLUIDS

The H_2O–NaCl system is the reference system for all natural aqueous fluids with dissolved salts. Critical temperatures in the system increase with the increasing salt concentration, and a heterogeneous L+G equilibrium is possible in a wide range of P and T, including magmatic conditions (Figure 5.9). Apart from the L+G equilibrium, heterogeneous fluids composed of the aqueous liquid with dispersed halite crystals also occur at high NaCl concentration, and a boiling fluid with suspended halite crystals is stable along the three-phase L+G+H surface. The various types of heterogeneous trapping modes must be carefully distinguished using microthermometry data. This is particularly important for the high-salinity fluids trapped within the L+G field and along the L+G+H curve. In the first case, the inclusions must homogenize to liquid and to vapor, but they cannot homogenize by halite dissolution. In contrast, all three homogenization modes occur when the fluids are trapped along

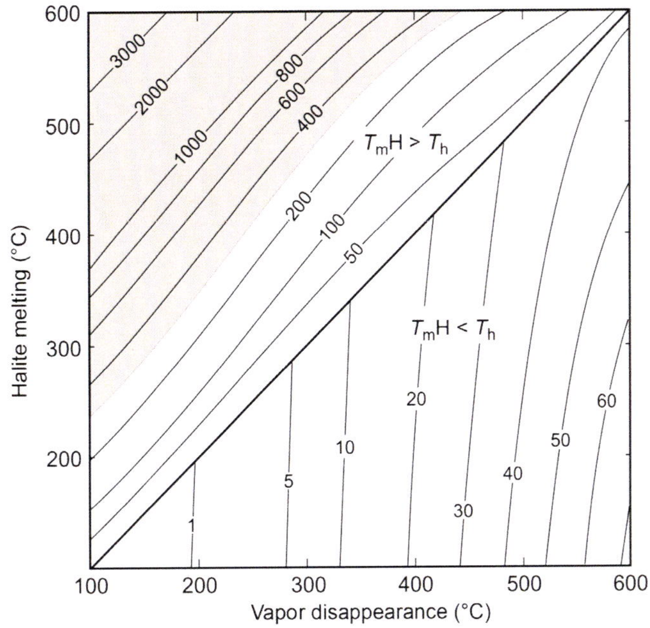

FIGURE 5.8

Homogenization pressure (MPa, to obtain pressure in bars, numbers must be multiplied by 10) as a function of the temperature of halite dissolution (T_mH) and that of vapor bubble homogenization (T_h). Total homogenization occurs by halite dissolution $T_mH(L)$, or vapor bubble disappearance $T_h(L)$. Partial homogenization occurs by halite dissolution in the presence of a vapor bubble $T_mH(LG)$, or by bubble disappearance in the presence of halite $T_h(LH)$. The gray area denotes the range outside the validity of the experimental data, where isobars have been constructed by graphical extrapolation from lower pressures. Reprinted from Lecumberri-Sanchez, P., Steele-MacInnis, M., Bodnar, R.J., 2012. A numerical model to estimate trapping conditions of fluid inclusions that homogenize by halite disappearance. Geochimica et Cosmochimica Acta 92, 14–22, Copyright 2012, with permission from Elsevier.

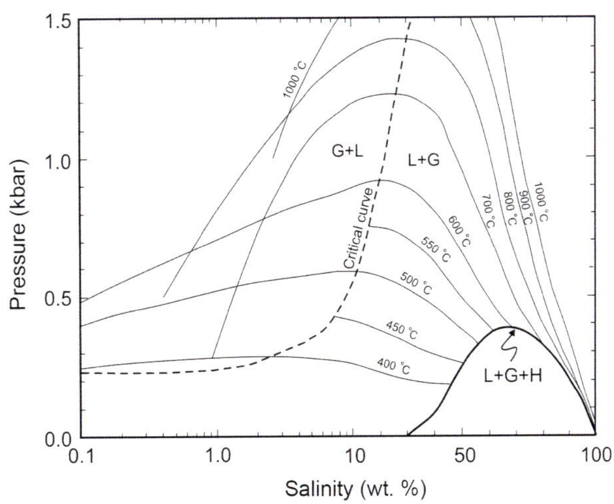

FIGURE 5.9

Isothermal *P*–*X* projection of the H_2O–NaCl system. Reprinted from Bodnar, R.J., Burnham, C.W., Sterner, S.M., 1985. Synthetic fluid inclusions in natural quartz. III. Determination of phase equilibrium properties in the system H_2O–NaCl to 1000°C and 1500 bars. Geochimica et Cosmochimica Acta 49, 1861–1873, Copyright 1985, with permission from Elsevier.

the three-phase curve, and the halite-dominated fluid inclusions should also be present. Improper determination of the trapping mode may lead to erratic pressure estimates.

Boiling fluids are trapped within the L+G field, to the right of the critical curve in the *P–X* isothermal projection of the H_2O–NaCl system (Figure 5.9). In this case, a small volume of a low-salinity vapor is liberated. In contrast, condensation occurs when a low-salinity fluid is trapped in the G+L field, to the left of the critical curve. In this case, a small volume of liquid with extremely high salinity originates due to the strong affinity of electrolytes to the aqueous liquid, and their comparably low solubility in the vapor phase. This phenomenon is frequently observed in porphyry-related igneous intrusions (Cline and Bodnar, 1991; Beane and Bodnar, 1995; Bodnar, 1995) and sea-floor hydrothermal systems (Bischoff and Pitzer, 1985; Bischoff, 1991).

The appearance of fluid inclusions trapping a heterogeneous H_2O–NaCl fluid with subcritical densities is illustrated in Figure 5.10. Inclusions that sealed coexisting, mutually uncontaminated phases, should homogenize at an equal temperature, corresponding to that of trapping. Simultaneously, their salinities must intersect the inferred isotherm at the same pressure in the *P–X* projection (Figure 5.9). The pressure

T = 450 °C

P (bar)	*D* (g/cm³)	mass % NaCl	Volume % G	Volume % H	$T_{m NaCl}$
300	0.935	38.0	27.4	6.8	301
	0.14	0.1	86.0		
400	0.635	14.0	42.2		
	0.33	2.5	67.5		

T = 500 °C

P (bar)	*D* (g/cm³)	Mass % NaCl	Volume % G	Volume % H	$T_{m NaCl}$
400	0.982	47.0	28.2	12.7	410
	0.15	0.3	85.0		
500	0.843	34.0	32.8	4.0	239
	0.24	2.5	76.3		

T = 550 °C

P (bar)	*D* (g/cm³)	mass % NaCl	Volume % G	Volume % H	$T_{m NaCl}$
400	1.026	54.0	28.6	17.8	479
	0.14	0.2	86.0		
500	0.972	50.0	30.4	14.4	441
	0.18	0.6	82.0		
600	0.879	42.0	33.5	8.6	353
	0.24	2.0	76.3		
700	0.756	31.0	38.6	2.2	181
	0.32	4.8	69.0		

T = 600 °C

P (bar)	*D* (g/cm³)	Mass % NaCl	Volume % G	Volume % H	$T_{m NaCl}$
400	1.183	66.0	24.9	29.4	580
	0.12	0.1	88.0		
500	1.059	59.0	29.0	21.7	523
	0.15	0.4	85.0		
600	0.967	53.0	32.2	16.1	470
	0.18	1.0	81.8		
700	0.889	47.0	35.0	11.5	410
	0.22	2.0	78.2		

FIGURE 5.10

The room temperature (25 °C) appearance of aqueous inclusions that trapped coexisting, mutually uncontaminated phases within the L+G region of the H_2O–NaCl system at 450–600 °C and various pressures. The inclusions are planar projections of those having a cylindrical shape with semispherical ends on both sides. G, vapor; H, halite; *m*NaCl, mass% NaCl; *D*, bulk density [g cm^{-3}]; T_mNaCl, temperature of halite dissolution [°C]. Volumetric phase ratios were calculated using principles outlined in Section 4.3 and the following experimental data: NaCl concentration in liquid (Bodnar et al., 1985a), NaCl concentration in vapor (Pitzer and Pabalan, 1986), liquid density (Bodnar, 1983), vapor density (Gehrig, 1980), temperature of halite dissolution (Potter et al., 1977).

is equivalent to that at the moment the inclusion was sealed. Hence, both the salinity and the homogenization temperature can be used to determine formation *P–T* conditions. However, measurements of the homogenization temperatures of vapor-dominated inclusions are unreliable, and apparent T_h values are often 150–300 °C lower than actual ones due to the incapability of human eyes to discern a thin liquid film along walls of a regular inclusion in optical microscope (Bodnar et al., 1985a; Sterner, 1992). The true temperature of homogenization to vapor can only be observed in the tips of flat, irregular inclusions.

A special type of heterogeneous trapping occurs along the L+G+H curve of the H_2O–NaCl system. In this case, the inclusions seal random mixture of three coexisting phases, and several homogenization modes occur in a population of coeval inclusions: to liquid by bubble disappearance and/or halite dissolution, or to vapor by evaporation of liquid. Halite-dominated or pure halite inclusions unambiguously indicate the L+G+H trapping mode. The inclusions in which the halite and the vapor bubble disappear simultaneously most likely trapped a homogeneous liquid, and their total homogenization temperature corresponds to that of trapping. Trapping pressure can then be derived from Equation II-12 (in Appendix II). In the absence of inclusions showing the simultaneous bubble and halite dissolution, the minimal temperature of homogenization by vapor bubble disappearance determined in a statistically relevant group of coeval inclusions can be taken as that approximately corresponding to the trapping temperature.

Figure 4.9c illustrates the heterogeneous trapping within the L+H field, which is characterized by the following features: (1) all inclusions contain halite, (2) halite volumes are variable (pure halite inclusions could be present), (3) liquid-dominated inclusions homogenize by halite disappearance, whereas the vapor bubble dissolves prior to halite melting. Minimal total homogenization temperature by halite dissolution corresponds to that of trapping, whereas the trapping pressure can be located along the vapor-unsaturated halite liquidus according to the scheme outlined in Figure 5.11.

Unknown NaCl concentration is the main problem faced when locating the pressure along the vapor-unsaturated halite liquidus, whose slope can be negative or positive depending on the NaCl concentration (Bodnar, 1994). The method of sequential re-equilibration (Problem 4.4) offers only a partial

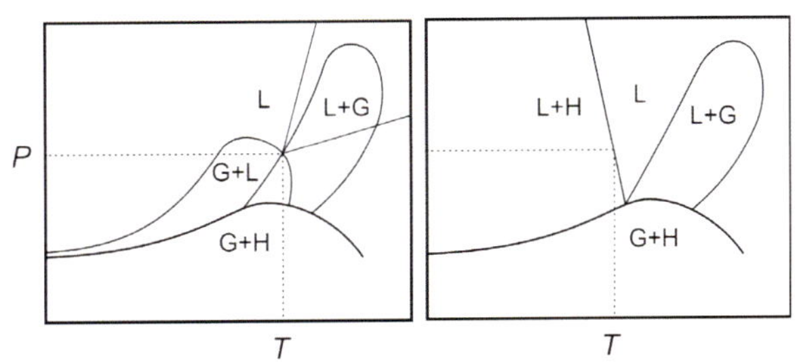

FIGURE 5.11

Schematic outline of principles for determining the formation *P–T* conditions of fluid inclusions that trapped a heterogeneous fluid with a subcritical density within the L+G field (left plot) and L+H field (right plot) of a salt-aqueous system. In the first case, the temperature and pressure are located at the intersection of isopleths for the coexisting liquid and vapor phases. The oblique lines are isochores of the homogeneous fluids that coexist as two immiscible phases at given *P* and *T*. In the second case, the trapping *P–T* conditions are located along vapor-unsaturated halite liquidus (negatively sloped line).

solution, because slopes of the vapor-unsaturated halite liquidi in natural multicomponent fluid systems probably differ from those in the experimental H_2O–NaCl system (Figure 5.2). For instance, Hurai et al., (2002a) estimated the slope of the vapor-unsaturated halite liquidus in natural 30–35 mass% NaCl–$CaCl_2$–H_2O fluids to be 3–3.5 × steeper than that of the H_2O–NaCl system with equivalent bulk salt concentrations. Accordingly, the salinity inferred from the temperature of vapor-unsaturated halite dissolution should not differ much from that of the vapor-saturated state, but the use of experimentally determined slopes of halite liquidi in the synthetic H_2O–NaCl fluid inclusions probably results in significant, perhaps 60–80% underestimation of the calculated fluid pressure.

The heterogeneous trapping of subcritical fluid within the L+G field is known as the secondary (retrograde) boiling taking place in igneous intrusions and producing economic Cu–Mo porphyry ores (Burnham, 1979). The residual silicate melt within a cooling and fractionating granodiorite intrusion tends to be enriched in volatiles. When the pressure-dependent saturation limit is overstepped, the free fluid phase is abruptly liberated from the residual melt, resulting in a drastic volume increase of the fluid-melt mixture. At a certain moment the expansion cannot be counterbalanced by plastic deformation of the magmatic reservoir and the surrounding rocks, and a system of cracks and explosive breccias is created above the intrusion. The fluid overpressure within the magmatic reservoir is compensated at the onset by the increase of the available volume by propagating cracks, and also by escape of volatiles from the system. The pressure changes initiated primarily by the cooling intrusion are recurrent—the pressure release coincidental with the brittle deformation initiates crystallization of minerals in open cracks, thus leading to the crack sealing and the pressure increase, until the brittle deformation is triggered again (Ramsay, 1980). This process is repeated until the silicate melt is fully solidified. If the recurrent pressure variations occur within the condensation region of the H_2O–NaCl system, a small volume of high-salinity brine is liberated by condensation from the low-salinity homogeneous fluid exsolved from the residual silicate melt. Both processes, namely fluid exsolution and breakdown to liquid and vapor phases, occur either simultaneously in shallow intrusions or are attenuated in the intrusions emplaced at medium depths. At higher pressures, the exsolved supercritical fluid liberates only low-salinity vapor bubbles by boiling.

Volumes of the coexisting vapor and liquid phases can be easily calculated using mass and volume balance Equations (4.2)–(4.6). For instance at 600 °C and 400 bar, a 5 mass% NaCl–H_2O fluid would split to 0.8 vol.% (7.5 mass% NaCl) of liquid with 66 mass% NaCl, and vapor with only 0.05–0.1 mass% NaCl (Figures 5.9 and 5.10). The low-density vapor phase preferably concentrates CO_2, HCl, HF, and other magmatic volatiles, which causes extensive acid alteration around a crack system. In contrast, the dense brine remains in the system, and leaches out metals from the parental intrusion and the surrounding rocks. These metals later precipitate in the form of sulfides mainly between the potassic and sericitic alteration zones, overlapping the boundary between the magmatic and meteoric hydrothermal circulation (e.g., Lowell and Guilbert, 1970; Reynolds and Beane, 1985).

Bodnar (1992) analyzed the experimental albite–H_2O–NaCl system and concluded that the salinity of the fluid exsolved during the fractional crystallization depends on depth and pressure. At pressures greater than 1.3 kbar, the exsolved fluid has a medium-to-low salinity (<12 mass%), which approaches zero with the progress of crystallization. In contrast, at pressures of less than 1.3 kbar, the fluid phase exsolves at 50–80% fractionation and contains less than 5 mass% NaCl, but this value rapidly increases up to 60 mass% by the end of fractionation. Salinity of the exsolved fluid remains nearly constant, at ~12 mass%, when the intrusion fractionates at 1.3 kbar.

Šulgan and Hurai (1989) modeled phase changes during crystallization of a quartz-monzonite intrusion with 4 mass% of fluid at different depths to elucidate the character of mineralization of porphyry systems. If the monzonite intrusion cooled at a depth greater than 6 km ($P > 1.7$ kbar), the fluid phase exsolved shortly before the total solidification of the residual melt at 700 °C. The exsolved fluid was enriched in molybdenum, because copper as a compatible element had entered the structure of mafic minerals. Because of high ambient pressure, the exsolved fluid with 5 mass% NaCl was homogeneous. Emplacement of the monzonite intrusion at 4 km ($P = 1$ kbar) resulted in the fluid oversaturation of the residual melt at 900 °C. The exsolved fluid with 5 mass% immediately disintegrated into vapor with 2.6 mass% NaCl (96 mass% of total fluid mass) and liquid with 63 mass% NaCl, accumulating copper in the form of chloride complexes. Shallow emplacement (<1 km) was not favorable for the formation of a Cu-porphyry deposit, because the fluid escaped from the system at early stages of the magmatic differentiation.

The water-saturated solidus in the albite–H_2O–NaCl system is accompanied by fluid immiscibility at pressures < 1.6 kbar (Bodnar, 1992). For instance, a fluid phase with subcritical density and the salinity of ~30 mass% NaCl exsolved at 70% albite crystallization at 850 °C and 1.5 kbar immediately split into a vapor with 20 mass% NaCl and liquid with 44 mass% NaCl. The continued crystallization produced the fluid with gradually decreasing salinity, down to 20 mass% NaCl at a 90% crystallization. Fluid inclusions trapped at this late-magmatic stage are characterized by relatively low homogenization temperatures (~300 °C) and they are indistinguishable from those trapped in hydrothermal conditions. Halite-bearing inclusions trapped in deep-seated intrusions (>1.6 kbar, 8 km) represent an early magmatic fluid exsolved during initial stages of the fractional crystallization. The same type of fluid inclusions in shallow intrusions (e.g., 0.5 kbar, 2 km) represent either products of secondary boiling and splitting of the magmatic fluids with low salinity and subcritical density, or they correspond to late-magmatic fluids liberated during advanced (>98%) fractional crystallization of the parental intrusion.

5.3.3 HETEROGENEOUS GAS-AQUEOUS FLUIDS

In contrast to salts, dissolved gases are preferably partitioned in the vapor phase during fluid heterogenization. The gas solubilities in aqueous solutions are variable: the most soluble gas is carbon dioxide, whereas nitrogen is the least soluble. The immiscibility field shrinks with the increasing pressure and expands with the increasing salt concentration (Figure 5.12).

Several examples of the room temperature (20 °C) appearance of fluid inclusions trapped within the two-phase immiscibility field of the H_2O–CO_2 system along the 250° and 300 °C isotherms are illustrated in Figure 5.13. Note, that a phase density inversion occurs at 570 bar and 300 °C, and at 2000 bar and 250 °C. At higher pressures, the density of the CO_2-rich gaseous phase is greater than that of the associated H_2O-rich liquid phase. This phenomenon occurs because the molecular mass of CO_2 is higher than that of H_2O.

Immiscibility in the H_2O–CO_2–NaCl system may result in the production of high-salinity brine even if the bulk fluid salinity is low (Figure 5.14). This phenomenon is qualitatively similar to that observed in H_2O–NaCl fluids. In contrast, however, the brine exsolution may occur at substantially higher pressures, above 2 kbar, in regional metamorphic conditions (Bowers and Helgeson, 1983b; Trommsdorf and Skippen, 1986). This is a consequence of the effect of dissolved gases, which

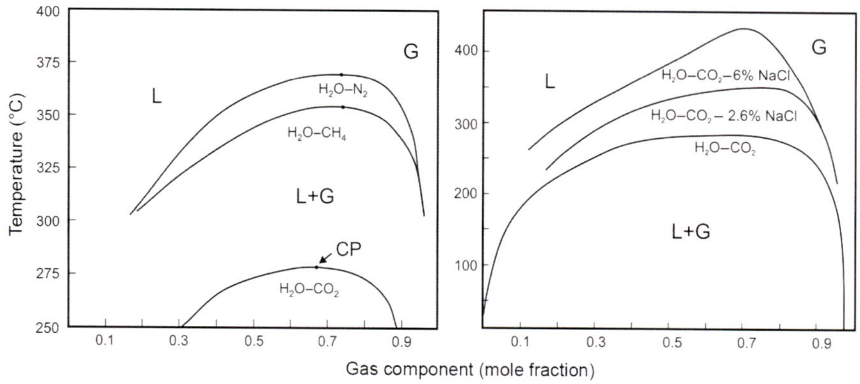

FIGURE 5.12

The left diagram depicts limits of dissolution (solvi) for H_2O–N_2, H_2O–CH_4, and H_2O–CO_2 systems at 1 kbar pressure, constructed using various EOSs (adapted from Diamond, 2003). The right diagram portrays solvi of the H_2O–CO_2–NaCl system at 1 kbar pressure and different salt concentrations (Tödheide and Franck, 1963; Gehrig, 1980; Hendel and Hollister, 1981).

increases the saturation vapor pressure, shifts critical points to higher P and T, and expands the immiscibility field.

Compositions and densities of coexisting phases in binary systems do not change with changing bulk fluid composition at fixed P and T, unless the solubility surfaces are intersected. This is, however, not the case for ternary systems, where compositions of the coexisting liquid and vapor phases may vary at constant P and T by changing the concentration of one solute. In other words, the salinity of the liquid phase may increase only with the increasing CO_2 concentration.

Immiscible H_2O–CO_2–NaCl fluids are responsible for metamorphic Au-bearing quartz veins (e.g., Craw, 1992; Craw et al., 1993; Jia et al., 2000; Lüders et al., 2005). Urban et al. (2006) and Hurai et al. (2008a) described the immiscibility of high-density to superdense CO_2 or CO_2–N_2 gases and high-salinity brines during various stages of hydrothermal quartz-stibnite and siderite vein formation. Metamorphic veins with CO_2–aqueous immiscible fluids also originated during tectonic compression (Huff and Nabelek, 2007) and post-metamorphic decompression (Diamond, 1990), and they are therefore particularly useful for constraining the metamorphic P–T trajectories. Immiscible methane-rich aqueous fluids in crack-seal veins and alpine fissures recorded peak- and post-metamorphic conditions in accretion wedges (Mullis, 1979; Evans, 1995; Evans and Battles, 1999; Mullis et al., 1994; Vityk et al., 1996; Lewis et al., 2000; Matsumura et al., 2003; Hurai et al., 2002b, 2006). The transition from CH_4-rich immiscible to H_2O-rich ($\pm CO_2$) homogeneous fluid zones occurred at 270 °C within a low-grade metamorphic complex of the Swiss Alps (Tarantola et al., 2007). Immiscibility significantly affects metamorphic reactions and mineral equilibria (Bowers and Helgeson, 1983b; Heinrich, 2007).

Fluid inclusions sealed within the two-phase immiscibility field usually contain random mixtures of two coexisting phases, that is, they are heterogeneous at the moment of entrapment. However, some inclusions may have trapped mutually uncontaminated aqueous or gaseous endmember phases, thus being homogeneous at the moment of entrapment. On cooling below the trapping temperature T_t, isochores of the carbonic and aqueous phases traverse along respective dew

P (bar)	D (g/cm^3)	X_{CO_2}	T = 250 °C	CO_2-rich phase		Volume at 20 °C (%)		
				D (g/cm^3)	T_h (°C)	H_2O	CO_2 L	CO_2 G
500	0.871	0.070		0.515	31.0	73.8	14.5	11.7
	0.539	0.720		0.502	31.1	7.4	49.2	43.4
1000	0.899	0.120		0.692	26.2	67.6	27.8	4.6
	0.772	0.695		0.741	22.8	11.8	83.2	5.0
1500	0.929	0.149		0.801	17.4	65.2	34.8	
	0.883	0.685		0.864	9.7	14.0	86.0	
2000	0.957	0.166		0.884	7.0	64.5	35.5	
	0.955	0.675		0.946	−3.1	15.7	84.3	
2500	0.981	0.170		0.950	−3.7	65.5	34.5	
	1.006	0.675		1.008	−14.8	16.6	83.4	

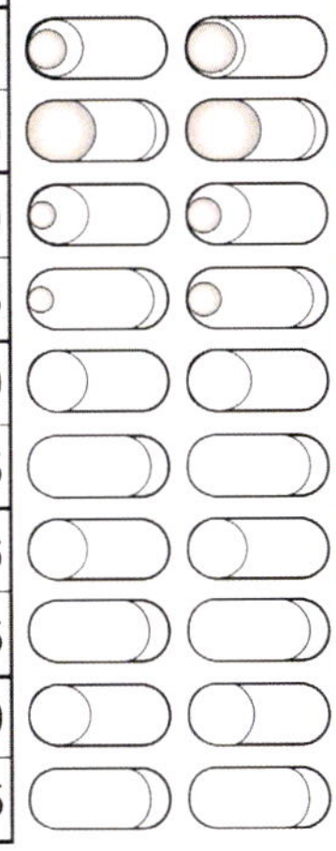

P (bar)	D (g/cm^3)	X_{CO_2}	T = 300 °C	CO_2-rich phase		Volume at 20 °C (%)		
				D (g/cm^3)	T_h (°C)	H_2O	CO_2L	CO_2G
200	0.773	0.023		0.157	14.3	73.2		26.8
	0.167	0.352		0.103	1.5	7.2		92.8
300	0.769	0.049		0.273	27.1	68.5	4.3	27.2
	0.298	0.454		0.221	23.1	9.8	4.3	85.9
400	0.765	0.079		0.362	30.4	63.4	10.6	26.0
	0.414	0.480		0.329	29.6	12.7	20.3	67.0
500	0.753	0.125		0.443	31.1	55.9	18.6	25.2
	0.514	0.460		0.417	31.0	16.7	32.0	51.3
600	0.225	0.225		0.508	31.1	41.7	31.5	26.8
	0.335	0.335		0.499	31.1	28.9	37.4	33.7

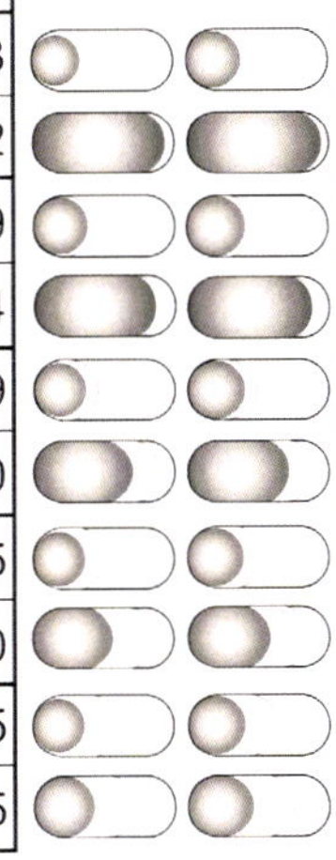

FIGURE 5.13

Appearance of fluid inclusions (at 20 °C) that trapped coexisting immiscible phases in the H_2O–CO_2 system at 250° and 300 °C, and various pressures. The inclusions in the left column are flat, whereas those in the right column are planar projections of cylindrical inclusions with hemispherical ends. *D* is the bulk density in the left column and that of CO_2-rich phase in the right column. X_{CO_2} is the bulk CO_2 molar fraction, T_h is the homogenization temperature of the CO_2-rich phase. The last three columns on the right side contain volume percentages of aqueous liquid, CO_2 liquid, and CO_2 vapor phases at 20 °C. The phase volumes were recalculated according to principles outlined in Section 4.3 using the following experimental data: density and homogenization temperature of CO_2-rich phase (Angus et al., 1976), density of water at 20 °C (Elsner et al. 1982), density of H_2O–CO_2 mixture (Bowers and Helgeson, 1983a), composition of coexisting L and G phases (Tödheide and Franck, 1963). Mutual solubility of H_2O in CO_2 and vice versa at room temperature was ignored.

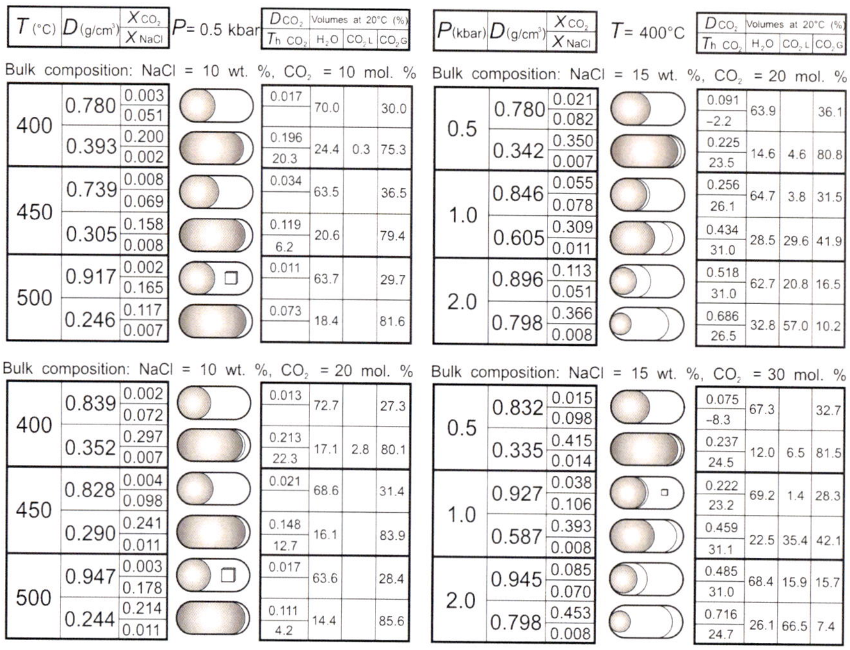

FIGURE 5.14

Appearance of fluid inclusions (at 20 °C) that trapped coexisting immiscible phases in the H_2O–CO_2–NaCl system at various temperatures and pressures. The drawings represent planar projections of cylindrical inclusions with hemispherical ends. Explanation of abbreviations is similar to that in Figure 5.13. Phase volumes were recalculated according to the principles outlined in Section 4.3 using the following experimental data: CO_2 density and homogenization temperature (Angus et al., 1976), density of aqueous liquid at 20 °C (Potter and Brown, 1977), density and composition of coexisting gaseous and aqueous phases in the H_2O–CO_2–NaCl mixture (Bowers and Helgeson, 1983a). Mutual solubility of CO_2 and H_2O at 20 °C was ignored.

(condensation) and bubble (boiling) surfaces, respectively (Figure 5.15). According to the lever rule, the isochoric cooling within the two-phase field causes nucleation and enlargement of the carbonic bubble at the expense of the aqueous phase in the aqueous inclusion to maintain the bulk composition fixed at T_t. Simultaneously, an aqueous rim condenses and enlarges at the expense of the carbonic phase in the carbonic inclusion. Upon heating, both inclusions homogenize simultaneously at T_t and remain homogeneous at higher temperatures, following the isochoric P–T paths of monophase supercritical fluids. Inclusions with transitional compositions change in a similar manner, but they homogenize at temperatures higher than T_t. This inclusion behavior is basically identical with that observed in fluid inclusions that trapped boiling aqueous fluids.

Individual assemblages of coeval, heterogeneously trapped inclusions should possess the following characteristics (Roedder and Bodnar, 1980; Pichavant et al., 1982; Ramboz et al., 1982; Diamond, 1990; Hurai, 1992, 2010):

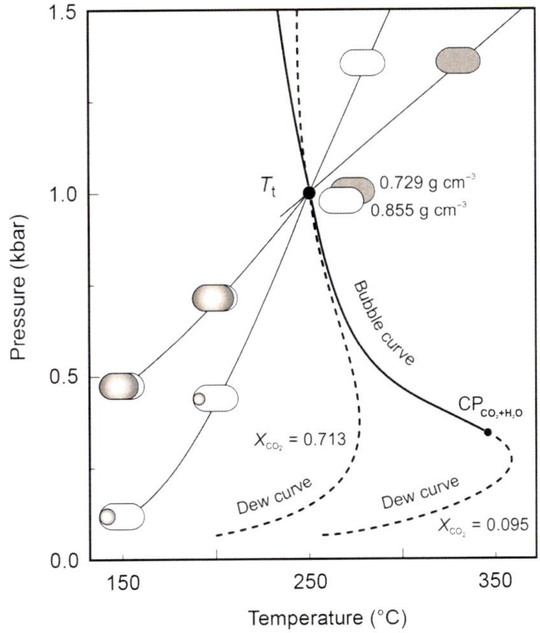

FIGURE 5.15

Isopleths and isochores of carbonic and aqueous phases coexisting at 250 °C and 1 kbar in the H_2O–CO_2 system. The changing slopes and curvatures of two isochores are worth noting when crossing the two-phase immiscibility surface outlined by isopleths of the two conjugate phases. The isochoric trajectories and phase compositions were calculated using the EOS of Duan et al. (1992b); the isopleths were redrawn from phase diagrams in Takenouchi and Kennedy (1964).

1. Variable phase ratios at room temperature, ranging from gas-dominated to water-dominated compositions.
2. Gas-dominated inclusions should homogenize to vapor, and the water-dominated inclusions to liquid. Homogenization temperatures range from T_t to a maximum temperature limit defined by "cricondentherm" (Mullis, 1979; Diamond, 1990, 2001).
3. Isochores and isopleths of inclusions with endmember compositions intersect at a single point corresponding to T_t and P_t.

Simultaneous application of all three criteria to natural fluid inclusions is rare (Mullis, 1979). In real inclusions, they are overlapped by fluctuating pressures in crack-seal veins (Hurai et al., 2006), post-entrapment diffusion of hydrogen (Hall et al., 1991) and water (Bakker and Jansen, 1991, 1994), graphite precipitation in fluid inclusions (Bergman and Dubessy, 1984; Cesare, 1995), and the necking down. The dew-point temperature, namely the total homogenization to vapor- or liquidlike gas, is often underestimated owing to the inability to discern visually the moment when the gas phase fills the whole inclusion cavity (Bodnar et al., 1985a; Sterner, 1992), or it is impossible to measure due to decrepitation and/or stretching prior the total homogenization of the high-density gaseous inclusions (Diamond, 1990).

The minimum homogenization temperature in a population of coeval aqueous inclusions is thought to approach the trapping temperature in the case of a heterogeneous trapping. Loucks (2000) proposed a statistical treatment of homogenization temperatures to discriminate between necked-down inclusions and those containing an immiscible fluid. When the trapping temperature has been correctly determined, the pressure can be located at the intersection with the isochore of the coexisting gaseous inclusion. This approach is known as the T_h–isochore intersection method (Mullis, 1987) and is widely used to constrain the fluid pressure.

The two-isochore intersection method is also applied for fluid inclusions with contrasting densities and compositions sealed at different times, but presumably in the same P–T conditions. This method is, however, invalid both theoretically and practically. If the homogeneous inclusion assemblages were trapped at different times in the same place, intersection of their isochores is meaningless without any other evidence as to how P–T conditions evolved during the time interval separating the inclusion entrapment. If the contrasting homogeneous fluid inclusion assemblages were trapped at different places but were coeval, intersection of their isochores could define their trapping P–T conditions, but the petrographic evidence for contemporaneity in different samples is very rare. The fact that isochores of compositionally contrasting homogeneous fluids almost always intersect somewhere in their P–T coordinates has been widely misused in geological literature. At any rate, calculation of the gas density and its isochore is essential for both methods.

Figure 5.15 illustrates that the shape of isochoric trajectory changes from linear to curved, when traversing along two-phase dew or bubble surfaces below T_t. Upon further cooling, the two-phase isochoric trajectories of gaseous inclusions kink after hitting one of the following 3-phase univariant curves:

1. CO_2 vapor, CO_2 liquid, and water coexistence curve in the H_2O–CO_2 system—the LLG curve.
2. Clathrate liquidi in the H_2O–CO_2 and H_2O–CH_4 systems.

For instance, the isochore of the CO_2-rich phase trapped at 1 kbar and 180 °C intersects the upper segment of the CO_2-clathrate liquidus, along which water and CO_2 liquid are stable coexistent phases (Figure 5.16). In reality, however, no phase change occurs at this moment, and the assemblage of the carbonic and aqueous liquids persists until a CO_2-vapor bubble nucleates at the intersection of the two-phase isochoric trajectory with the metastable extension of the LLG curve within the clathrate stability field (Diamond, 1992a). The clathrate crystallizes abruptly after surpassing a metastable barrier at around −35 °C, and the remaining carbonic liquid freezes out at around −90 °C. Progressive heating of the frozen inclusion first leads to CO_2 solid melting in the presence of carbonic liquid, clathrate, and CO_2 vapor at the quadruple point Q3 at −56.6 °C followed by the homogenization of the CO_2 vapor bubble at the quadruple point Q2 at 10 °C, and the dissociation of the CO_2 clathrate above 10 °C in the presence of CO_2 liquid and water. If the same inclusion is heated immediately after renucleation of the CO_2-vapor bubble, but prior to the clathrate crystallization, a metastable homogenization of the carbonic phase to liquid takes place at 6.4 °C along the metastable extension of the LLG curve below Q2.

Since the Q3 quadruple element of the H_2O–CO_2 phase diagram lies only 0.04 °C below the triple point of pure CO_2, and the LCEP is also located very close to the critical point of pure CO_2, the effect of the dissolved water on the melting temperature of the CO_2 solid, as well as on the partial homogenization of the carbonic phase, either stable or metastable, is immeasurable by microthermometry. The same is valid for the CH_4–H_2O system, where the three-phase clathrate-carbonic liquid-carbonic vapor

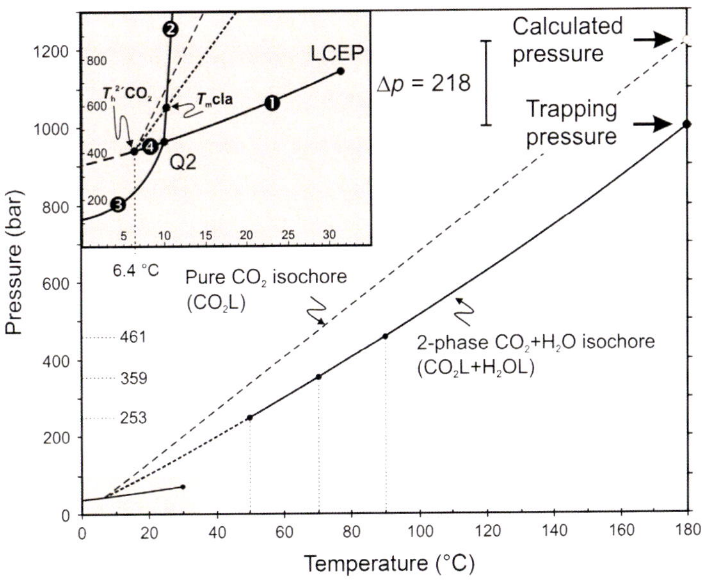

FIGURE 5.16

The low-temperature portion of the phase diagram of the H_2O–CO_2 system with the lower critical end point (LCEP), the quadruple point Q2 (clathrate, water, CO_2 vapor, and CO_2 liquid) and the three-phase curves: (1) water, CO_2 liquid, CO_2 vapor (LLG curve); (2) clathrate, CO_2 liquid, water; (3) clathrate, CO_2 vapor, water. The isochoric trajectory of the gaseous phase trapped within the two-phase immiscibility region at 180 °C and 1 kbar intersects a metastable extension of the LLG curve at 6.4 °C. The intersection temperature corresponds to a metastable homogenization of the CO_2-vapor bubble in the presence of carbonic liquid and water—$T_h^{2'}CO_2$. Assuming a pure CO_2 system, the equivalent density corresponds to 0.887 g cm^{-3} (Span and Wagner, 1996). Stable temperature of the clathrate dissolution in the presence of carbonic liquid and water is marked as $T_m$cla. The one-phase (pure CO_2) isochore intersects the 180 °C isotherm at 1218 bar, thereby resulting in the overestimation of the trapping pressure by 218 bar. See Problem 5.3 for the sample calculation of the density of the "equivalent pure CO_2 gas" and Figure 5.17 for properties of conjugate phases within the fluid inclusion at temperatures of 50, 70, and 90 °C. Both isochoric trajectories and properties of the conjugate phases were calculated according to Duan et al. (1992b). Other available EOSs for pure CO_2 overestimate the trapping pressure by 142–173 bar (Hurai, 2010).

curve is located close to the LG curve of pure methane (Diamond, 2003). For this reason, bulk densities and isochores of gaseous inclusions are calculated from EOSs of anhydrous gases or their mixtures. Most applications (e.g., Mullis, 1979; Evans, 1995; Evans and Battles, 1999; Mullis et al., 1994; Vityk et al., 1996; Lewis et al., 2000; Matsumura et al., 2003; Hurai et al., 2002b, 2006) pertain to the H_2O–CH_4 system, where solubility of H_2O in the gas phase is negligible even at very low salt concentrations. However, some applications also include the H_2O–CO_2 (Urban et al., 2006) or H_2O–N_2 systems (Hurai et al., 2008b) with increased salt concentrations.

The isochore calculation using anhydrous gas is not as precise as has been widely assumed. For instance, depending on the equation of state used, the described method overestimates the trapping pressure by 218 bar (Figure 5.16) for the CO_2–H_2O inclusion sealed at 180 °C and 1 kbar within the two-phase immiscibility field. This discrepancy is caused by differing slopes of the curved isochoric trajectory of the two-phase CO_2–H_2O fluid and the linear isochoric trajectory of the monophase pure CO_2 fluid. Depending on density and composition, water or clathrates with variable stoichiometries can be present during the partial homogenization of the carbonic phase in the H_2O–CO_2 inclusions. Moreover, a metastable partial homogenization may also occur in the presence of supercooled water between 0° and ~−35 °C, or a clathrate–ice mixture below −35 °C.

PROBLEM 5.3 CALCULATING PRESSURE CORRECTION FOR A CO_2–H_2O INCLUSION

Calculate the pressure correction for a CO_2–H_2O gaseous inclusion with partial homogenization of the carbonic phase in the presence of water. The inclusion was trapped at 180 °C and 100 MPa (Figure 5.16).

Solution

Calculations can be done using the highly accurate EOS for the H_2O–CO_2–CH_4 system (Duan et al., 1992b), which applies to a temperature range of 50–1000 °C and pressures from 0 to 3 kbar.

1. Calculate phase equilibrium at 180 °C and 1000 bar for optional fluid composition inside the immiscibility region. Save the composition and density of the carbonic phase ($X_{CO_2}=0.89959$, $X_{H_2O}=0.10041$, $\rho_{bulk}=0.83163$ g cm^{-3}).
2. Calculate the phase composition of the carbonic inclusion at various temperatures below that of trapping (T_t). Start with 90 °C. Find the bulk density for the inferred bulk composition ($X_{CO_2}=0.89959$, $X_{H_2O}=0.10041$) by iteration: calculate the properties of the coexisting vapor and liquid phases at various pressures; construct a regression line correlating the pressure and the bulk density, and find by interpolation the pressure at which the bulk density corresponds to that calculated in step 1. The bulk inclusion density at $T<T_t$ can be expressed as $D_{bulk}=\varphi_G D_G+\varphi_L D_L$, where φ and D are respective volume fractions and densities (g cm^{-3}) of the conjugate gas (G) and liquid (L) phases. At 90 °C, the gaseous inclusion trapped at 180 °C and 1 kbar contains 91.058 vol.% of CO_2-rich liquid with a density of 0.81693 g cm^{-3}, and 8.942 vol.% of water with a density of 0.98127 g cm^{-3} under a pressure of 43.69 MPa.
3. Select another temperature and repeat the procedure in step 2. Complete compositions (X_{CO_2}) and volume fractions of the gaseous phase in the carbonic inclusion at 50°, 70°, and 90 °C.
4. Calculate the volume fraction of the hypothetical pure CO_2 phase by linear extrapolation of the regression line passing through the data obtained in step 3 to $X_{CO_2}=1$ ($\varphi_{CO_2}=0.89767$, Figure 5.17), and calculate the density of pure CO_2:
   (a) Convert the bulk CO_2 mole fraction to a mass fraction
   $$m_{CO_2}=\overline{M}_{CO_2}/((\overline{M}_{H_2O}/X_{CO_2})+\overline{M}_{CO_2}-\overline{M}_{H_2O}),$$
   where $\overline{M}$ are molecular masses ($\overline{M}_{CO_2}=44.009$, $\overline{M}_{H_2O}=18.0148$), and $X_{CO_2}=0.89959$ (result: 0.956306).
   (b) Calculate the total mass of CO_2 as $M_{CO_2}=m_{CO_2}D_{bulk}$ (result: 0.79529).
   (c) Calculate the density of pure CO_2 as $D_{CO_2}=M_{CO_2}/\varphi_{CO_2}$ (result: 0.88595 g cm^{-3}).
5. Calculate the pressure at 180 °C from the equation state for pure CO_2 using D_{CO_2} and quantify the pressure correction Δp as the difference between the trapping and calculated pressures.

The initial bulk composition of the carbonic phase ($X_{CO_2}=0.89959$, CO_2 volume fraction $\varphi_{CO_2}=1$) or an optional number of points located between $T_h=T_t$ and $T_h^{2'}$ CO_2 can also be included in the data regression in step 4, but the obtained pressure correction will be slightly different, owing to a nonlinearity of the data along the two-phase isochoric trajectory. For instance, data for 180°, 90°, 70°, and 50 °C yield very good Pearson's correlation coefficient ($R^2=0.9998$), but the resultant density of the pure CO_2 gas is slightly higher (0.88700 g cm^{-3}) than that calculated using only the 90–50 °C segment of the two-phase trajectory.

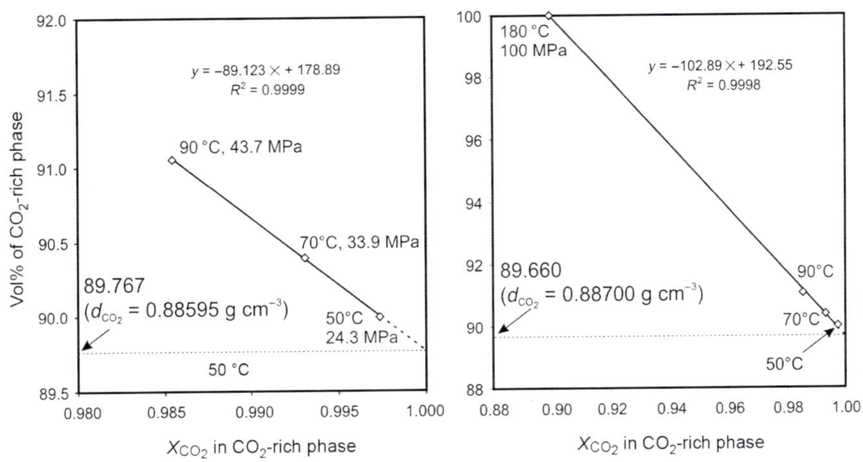

FIGURE 5.17

CO_2 mole fraction in CO_2-rich phase and its volume percentage in the gaseous inclusion trapped at 180 °C and 1 kbar within the immiscibility field of the H_2O–CO_2 system (see Figure 5.16 for comparison). Densities are also shown for the "equivalent pure CO_2 gas" calculated in Problem 5.3 by the extrapolation of a regression line constructed from three points calculated for temperatures of 50, 70, and 90 °C along the isochoric two-phase trajectory (left diagram) and from four points of the same trajectory (right diagram), including the starting point at 180 °C and 1 kbar, when the trapped CO_2-rich fluid becomes homogeneous. All calculations are based on the EOS of Duan et al. (1992b).

Hurai (2010) published several diagrams and computation protocols to evaluate the effects of small amounts of normal or supercooled aqueous liquid, clathrates, and hexagonal ice on the homogenization temperature of the associated CH_4 or CO_2 gas, and their isochoric slopes calculated assuming pure anhydrous gases. The pressure corrections were significant for the CO_2–H_2O inclusions with partial homogenization of the carbonic phase in the presence of water. The total pressure correction for water dissolved in the methane and for expansion of the host quartz was lower than ± 40 bar in the temperature range of 160–240 °C and pressures up to 3 kbar. All gaseous inclusions trapped close to their critical conditions (i.e., with a higher proportion of H_2O) yielded erratic results when interpreted in terms of an anhydrous supercritical gas. Inconsistencies existed regarding the volume effects of the supercooled water in the binary H_2O–CO_2 system down to a temperature of ~ -30 °C. The effect of dissolved salts was also difficult to quantify, owing to the nonexistence of an EOS that describes properties of conjugate phases to sufficiently low temperatures. The actual cage occupancy of clathrates at low temperatures still remains unclear.

The T_h–isochore intersection method (Mullis, 1987) remains the most versatile technique for the constraining formation P–T conditions of natural immiscible fluids. Its use is made advantageous by the fact that exact bulk composition of the fluid inclusions need not be known and that temperatures of homogenization, either partial in gaseous endmembers, or total in aqueous endmembers, are measurable regardless of the size and shape of inclusions. At higher CO_2 densities, metastable partial homogenization temperatures (in the absence of CO_2 clathrate and the presence of supercooled water) better reproduce the trapping conditions than those derived from stable assemblage with CO_2 clathrate.

The T_hCO_2 and $T_h^{2\prime}CO_2$ values (Figure 5.16) can be used for the detection of water in apparently pure CO_2 inclusions. A difference of up to 5 °C between these values was observed in seemingly monophase, negative crystal-shaped CO_2 inclusions, in which the aqueous liquid rim was optically indiscernible at room temperature. In turn, many natural CO_2-rich inclusions did not exhibit a measurable difference between the T_hCO_2 and $T_h^{2\prime}CO_2$ values. This indicates the presence of pure anhydrous gas. Though closely spatially associated with aqueous inclusions, such water-free gaseous inclusions cannot represent the immiscible gaseous phase exsolved from a heterogeneous fluid and so cannot be used for the calculation of trapping pressure using the T_h–isochore intersection method.

5.3.4 HETEROGENEOUS HYDROCARBONS AND WATER–PETROLEUM MIXTURES

The T_h–isochore intersection method is often used to constrain formation P–T conditions of coexisting aqueous and petroleum inclusions formed from an immiscible water–oil mixture. Here, the minimum homogenization temperature of aqueous inclusions is thought to approach the true trapping temperature, and its intersection with isochores of coeval petroleum inclusions brackets the trapping pressure (Figure 5.18). Because the thin rim of immiscible aqueous phase adheres to the walls of petroleum inclusions, their total homogenization temperature cannot be observed and determined. Thus, the apparent total homogenization temperature, either to vapor or to liquid, represents in fact the partial homogenization temperature, which corresponds to the intersection point of the isochore of the hydrocarbon-rich phase with the two-phase envelope ($T_h$oil). For successful application, total homogenization temperatures of aqueous inclusions (to liquid) must exceed those of the coexisting petroleum inclusion. This approach has been used in numerous studies of petroleum reservoirs (Nedkvitne et al., 1993; Munz et al., 1998, 1999).

Polyphase immiscibility is also possible in hydrocarbon systems at low pressures. An example is shown in Figure 5.19, which documents a trail of secondary inclusions in quartz from bituminous black shales filled with nonfluorescing black oil (L_1), gas-condensate liquid (L_2) with white fluorescence emission under 365 nm UV light, and gas-condensate vapor (G). The gas-condensate phase homogenizes either into vapor or to liquid over a wide range of temperatures, documenting a heterogeneous liquid–liquid–vapor mixture. The minimum temperature of homogenization (~60 °C) corresponds to the true trapping temperature. Because the composition of the condensate is not exactly known, the trapping pressure (<280 bar) can only roughly be estimated from the position of the two-phase immiscibility curve (Figure 2.22).

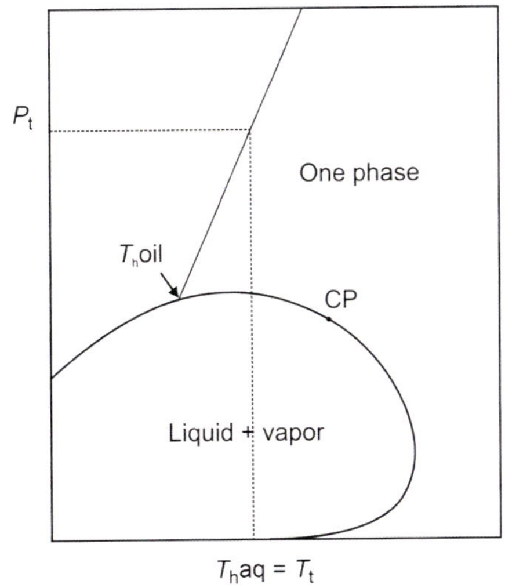

FIGURE 5.18

Schematic diagram showing the trapping pressure determination for immiscible oil–water mixtures using the T_h–isochore intersection method, if $T_h$aq > $T_h$oil. The two-phase boundary corresponds to that of oil, whereas the T_h value determined from aqueous inclusions is used to constrain the trapping pressure along the oil isochore. The method is essentially identical to that used to constrain formation P–T conditions in gas-aqueous heterogeneous mixtures.

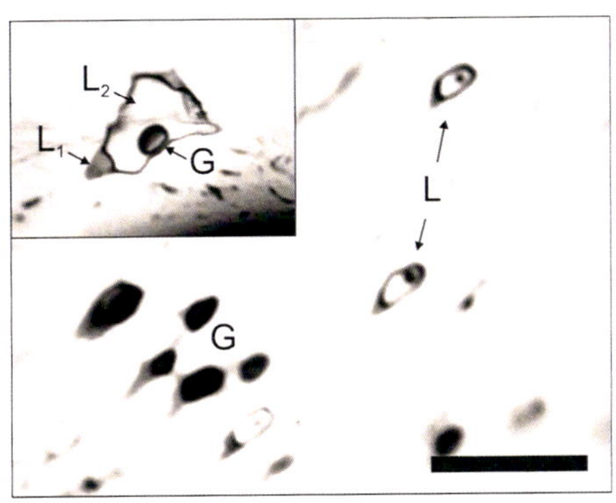

FIGURE 5.19

Heterogeneous (3-phase) black oil (L_1), condensate liquid (L_2), condensate vapor (G) mixture trapped in secondary healed cracks in quartz from fissures in Early Oligocene black shales near Pučov, central Carpathian flysch basin, Slovakia (courtesy of J. Soták). Bar scale corresponds to 20 μm. Note that the condensate phase exhibits variable liquid-to-vapor ratios, thus indicating the three-phase liquid–liquid-vapor immiscibility.

5.3.5 *P–T* CONSTRAINTS FROM MELT INCLUSIONS

Using melt and fluid inclusions in quartz, Thomas et al. (2000) demonstrated that two different coexisting melt inclusion types define a solvus boundary in an X_{H_2O}–T pseudobinary system. In the pegmatite–granite from Ehrenfriedersdorf (Erzgebirge, Germany), two melt inclusion types were classified as type A and type B, where type A includes relatively water-poor, peraluminous silicate melts, while type B comprises water-rich peralkaline inclusions. In subsequent years, they revealed melt inclusions trapped near the solvus crest as near-critical melt inclusions, and they designated the coexisting primary magmatic fluid inclusions as the type-C (Thomas and Davidson, 2012b).The melt and fluid inclusions related to the pseudobinary solvus form a complex assemblage, which means that all melt inclusions that lie on the solvus form a uniform paragenesis.

Figure 5.20 shows the relationship of the type A and B melt inclusions in a schematic X_{H_2O}–T plot of the pseudobinary silicate melt–H_2O system. The data come from the Variscan Ehrenfriedersdorf complex, the critical point in this case corresponds to 720 °C (Thomas et al., 2000; Thomas and Davidson, 2013). Bulk density of the supercritical phase of about 1.2 g/cm^3 can be estimated from the near-critical melt inclusions. Note that in trapping conditions, the three inclusion types illustrated were

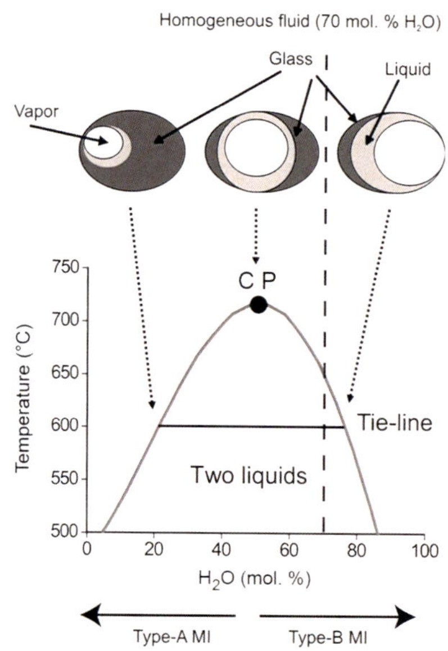

FIGURE 5.20

The relationship of type A and type B melts plotted in the pseudobinary phase diagram of the silicate melt–H_2O system. The diagram is derived from data from the Variscan Ehrenfriedersdorf granite–pegmatite system. CP is the critical point, in this case 720 °C. Melts above the solvus are homogeneous or supercritical, and those inside the solvus curve are heterogeneous. A total of three different types of melt inclusions (type A, near critical, and type B) are shown schematically in the upper portion as shown at room temperature after rehomogenization. Note that in trapping conditions, all the inclusions would have been homogeneous: (1) water-rich silicate melts (<20 mass% H_2O), (2) near-critical, water-rich silicate melts (20–35 mass% H_2O), and (3) silicate-poor fluids (>35 mass% H_2O).

homogeneous, and the schematic phase separations are post-trapping (Thomas et al. (2006b), and discussion in Thomas and Davidson (2012b)). An analysis of the inclusion diversity in the pegmatite quartz, however, showed that a vapor phase can also coexist under most conditions, implying the existence of a three phase equilibrium vapor+melt$_1$+melt$_2$. This immiscibility type was proven experimentally by Veksler and Thomas (2002). Due to pressure fluctuations, composition of the vapor phase varied over broad limits depending on the position of the vaporus curve relative to the solvus curve. The general case can be described in terms of the coexistence of the type-A and type-B melt inclusions, coexisting with the type-C fluid inclusions.

An important ultimate cautionary note is as follows: **the evolved granite is not a pegmatite**! In contrast to granites, processes like melt–melt–fluid immiscibility and the existence of supercritical fluids with unusual properties (low density, viscosity and surface tension, high diffusivity, reactivity and mobility) are essential for the formation of pegmatites. Thomas and Webster (2000) and Thomas and Davidson (2015) have shown that the shapes of pseudobinary solvi obtained from melt inclusions in quartz from Ehrenfriedersdorf, Zinnwald, Malkhan, Königshain, and Tanco pegmatites are similar, thus, the near-critical pegmatite-forming melt can be described as a supercritical fluid, and an uninterrupted continuum exists between silicate melts and fluids in supercritical conditions (Figure 5.21).

The concept of critical diameter of silicate melt inclusions introduced by Thomas (1994) and refined by Thomas and Klemm (1997) enables the determination of melt water concentration, which in combination with the phase diagram for the melt–H_2O system defines formation P–T conditions. The procedure is based on the assumption that the homogenization of fluid-filled bubble in viscous silicate melt inclusions is a diffusion-controlled kinetic process. Diffusion rates can be expressed in terms of the Einstein–Stokes function. For free diffusion of spherical molecules moving through a viscous fluid

$$D = \frac{k_B T}{6\pi\, r_0 \eta} \tag{5.6}$$

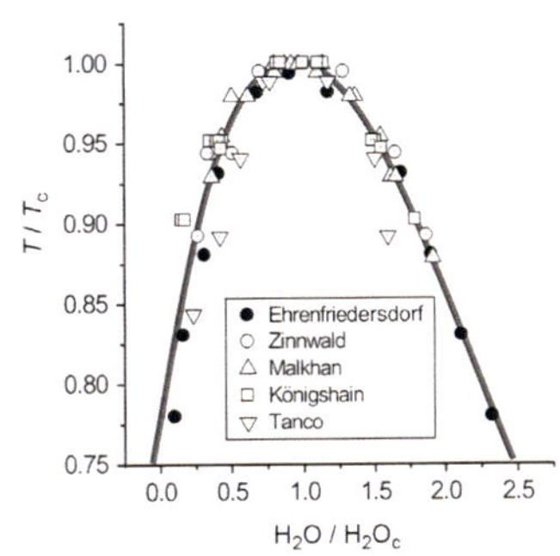

FIGURE 5.21

Pseudobinary solvus (reduced coordinates according to the principle of corresponding states) derived from melt inclusions from Ehrenfriedersdorf, Zinnwald, Malkhan, Königshain, and Tanco pegmatites. To show the five different systems on one plot, data from each system have been normalized to their critical temperatures (Ehrenfriedersdorf = 718 °C, Zinnwald = 705 °C, Malkhan = 721 °C, Königshain = 750 °C, and Tanco = 762 °C) and water concentrations. T_c is critical temperature, H_2O_c is the water concentration at the critical temperature (Thomas and Davidson, 2015).

where D is the diffusion coefficient [$m^2\ s^{-1}$], T is temperature [K], k_B is the Boltzmann's constant, η is the viscosity [Pa s], and r_0 is the radius of the spherical diffusion particle [m].

For diffusion over distance, the following relationship can be employed

$$r^2 = 4Dz \tag{5.7}$$

where z is time [seconds], r is the mean distance from the starting point, q is numerical constant dependent on dimensionality.

Solving Equation (5.7) for D we get

$$D = \frac{r^2}{4z} \tag{5.8}$$

and the combination of Equations (5.6) and (5.8) yields the general expression relating the critical melt inclusion radius r_K [in meters] with the dynamic viscosity η [Pa s]

$$\eta = f\frac{k_B T}{r_K{}^2} \tag{5.9}$$

where f is a function of experiment duration z [seconds], and the radius of a diffusing particle is r_0

$$f = 2z/3\pi r_0 \tag{5.10}$$

It should be emphasized that the frictional force coefficient $6\pi r\eta$ (Stoke's drag) in Equation (5.6) is a minimal value valid for spherical molecules. Asymmetric shapes, non-elastic interaction with the solvent (hydration shell) will increase the radius of the diffusing particle (Reynold's number). In the silicate melt–water systems, additional factors also become effective (Thomas, 1994), hence the real dimensions of the diffusing particles are difficult to estimate, though it is dependent on molar volume and molecular weight. For the purpose of this model, the diffusing particle radius r_0 was set arbitrarily to 1.1×10^{-9} m, and it serves merely as an auxiliary parameter without any special physical meaning, which is fitted to the viscosity model of Shaw (1972) and water contents in natural silicate melt systems determined by independent methods. The new particle diameter radius replaces the correction parameter k defined by Thomas and Klemm (1997) in Equation (5.8) for the same purpose. For a 20-h duration of heating, the f value becomes 1.389×10^{13}. For other heating durations, f must be recalculated according to Equation (5.10).

For the determination of viscosity and water content, heating experiments should be carried out with groups of variably sized melt inclusions at different temperatures, such as 700, 800, and 900 °C. This serves as a control for a possible diffusive loss of water, and for the discrimination of not coeval groups of melt inclusions with different water contents. If no water diffusion occurred, the group of inclusions should plot along a single isopleth with constant H_2O content in the $\ln r_K$ versus $1/T$ or $\ln \eta$ versus $1/T$ correlation diagram, thus also defining the water content in the magmatic system. Moreover, the measurement of critical inclusion diameters at various temperatures permits determination of the temperature of homogenization of hypothetical, infinitesimally small inclusions, in which no diffusion-related kinetic effects occur. This temperature corresponds to that of the vapor-saturated liquidus of the specific melt inclusion system, and it serves for the estimation of P–T conditions once the water content is determined from the viscosity or by Raman spectroscopy.

Table 5.5 summarizes data from the Kirchberg granite (Thomas, 1994) recalculated using the refined model with the kg m s system of physical units. The measured inclusions belong to a coherent cogenetic and coeval group of inclusions with the uniform water content corresponding to 6.7 ± 0.1

Table 5.5 Determination of Water Content in Melt Inclusions from the Kirchberg Granite (Sample OG3) Using the Quenching Method (Thomas, 1994)

Run	T (°C)	$10^4/T$ (K)	d_K (μm)	ln η* (Pa s)	Mass% H_2O
1	785	9.450	6.4	9.8943	6.64
2	785	9.450	5.4	10.2341	6.34
3	805	9.275	7.9	9.4919	6.63
4	809	9.241	9.2	9.1909	6.84
5	812	9.215	8.6	9.3285	6.65
6	819	9.156	9.9	9.0534	6.78
7	831	9.057	10.3	8.9851	6.63
8	839	8.992	11.9	8.7036	6.75
9	848	8.919	12.4	8.6293	6.66
10	850	8.904	13.4	8.4760	6.77
11	861	8.817	14.7	8.3005	6.75
12	877	8.695	18.2	7.8874	6.88
13	920	8.381	24.6	7.3215	6.71
14	998	7.867	42.4	6.2960	6.52
				Average	6.68
				$\pm 1\sigma$	0.10

Viscosity calculated after Shaw (1972) using the following chemical composition (mass% oxides): SiO_2 76.43, TiO_2 0.41, Al_2O_3 12.97, Fe_2O_3 0.50, FeO 0.44, MnO 0.19, MgO 0.33, CaO 0.78, Na_2O 3.44, K_2O 4.72, P_2O_5 0.06 combined with variable water contents.

mass% H_2O. With less confidence, the water content can also be calculated from one heating experiment. Once the water content of the silicate melt system is determined, minimum formation P–T conditions can be located along liquidus for the given H_2O concentration (Figure 5.22) intersected by the total homogenization temperature of the melt inclusion.

The study of coeval silicate melt and aqueous inclusions produced experimentally in the haplogranite–H_2O–NaCl–KCl system showed that isochores of the aqueous inclusions passed through the experimental conditions (Student and Bodnar, 1999). Homogenization temperatures of the coexisting melt inclusions were dependent on the homogenization technique and the heating rate, but minimum T_h values determined in the smallest inclusions were consistent with the experimental temperatures. The T_h values determined through incremental heating in a tube furnace using the rate of 10 °C/day were consistent in all inclusions regardless of their size, and they were also in agreement with the run conditions. Hence, the T_h–isochore intersection method is feasible in water-oversaturated granitic systems, as soon as the homogenization temperature of melt inclusions is correctly determined in variably-sized inclusions at constant run conditions.

Some melt inclusionists locate the formation P–T conditions at the intersection of the total homogenization temperature and the isochore of the fluid subsystem, mostly water, present inside the shrinkage bubble of melt inclusions below solidus. Contrary to the previous situation, this approach leads to overestimation of the fluid pressure, because the fluid density decreases between solidus and liquidus, as the free fluid dissolves in the increasing melt volume (Figure 4.42). Hence, extrapolation of the sub-solidus

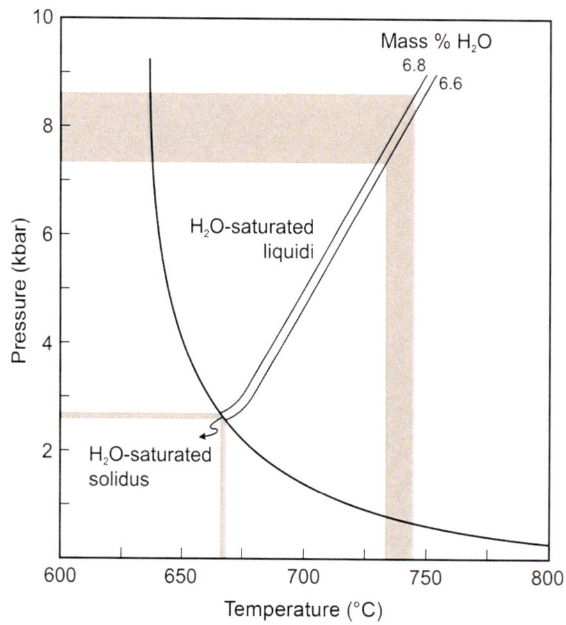

FIGURE 5.22

Determination of formation *P*–*T* conditions of the Kirchberg granite, Germany, using the quenching method and viscosity data. About 2.7 kbar of formation pressure and temperature of ~670 °C is obtained at the intersection of water-saturated liquidi for 6.7 ± 0.1 mass% H_2O with the haplogranite solidus (Holtz and Johannes, 1994), whereas the pressure of ~8 kbar is constrained by the intersection of the H_2O-saturated liquidi with the actual formation temperature of 739 ± 6 °C derived from critical diameters of melt inclusions determined at various temperatures and extrapolated to an infinitesimally small melt inclusion devoid of kinetic barriers triggered by the melt viscosity (see Tables 4.5 and 5.5 for data).

fluid isochore to the liquidus, or intersection of the fluid isochore with optically determined homogenization temperature significantly overestimates the estimated fluid pressures. For instance, the aqueous fluid subsystem of melt inclusions in metaluminous pincinite displayed in Figure 4.41a homogenized to liquid at a minimum temperature of 133 °C, thus corresponding to a density of 0.978 g cm^{-3} (salinity of 6.3 mass% NaCl equation determined from the temperature of melting of ice). Assuming the formation temperature of ~750 °C indicated by the orthopyroxene–ilmenite–melt thermometer (Section 6.5.4) or ~820 °C inferred from pseudosections (Section 6.5.6), unrealistically high fluid pressures between 11 and 12.3 kbar are obtained, which are inconsistent with the maximum lithostatic load of about 8 kbar provided by a 28–30 km thick continental crust during Pliocene. Similarly, Naumov et al. (1992, 1996) determined trapping pressures of up to 17 kbar at 800–900 °C from densities of aqueous liquid preserved in shrinkage bubbles of some melt inclusions in plagioclase phenocrysts of Miocene andesites of the Štiavnica stratovolcano (central Slovakia), corresponding to a formation depth of ~60 km. The abnormal character of the pressure estimates is underlined by the presence of low-density (0.19–0.35 g cm^{-3}) CO_2 inclusions in the centers of the same phenocrysts. The situation described is analogous to that of gas-aqueous fluid inclusions (Hurai, 2010), where the one-phase (liquid-only) isochore of the anhydrous CO_2 fluid projects above the two-phase (liquid–liquid) isochore of the CO_2–H_2O mixture.

PROBLEM 5.4 CALCULATING BEHAVIOR OF A MELT INCLUSION ON HEATING

Estimate whether melt inclusions of peraluminous haplogranite with 2 mass% dissolved water (Problem 5.1) become homogenized after heating for 20 hours at 800 °C.

Solution

Solve Equation (5.9) for r_K and insert the melt viscosity at a given temperature and water content (ln $\eta = 15.832$)

$$r_K = \sqrt{\frac{f k_B T}{\eta}} \quad (5.11)$$

$$d_K = 2r_K = 2 \times \left[1.389 \times 10^{13} \times 1.3806488 \times 10^{-23} \times 1073.15/e^{15.832}\right]^{0.5} = 3.3 \times 10^{-7}\ \text{m} = 0.33\ \mu\text{m}$$

where e is the Euler's number—the basis of decadic logarithm. This calculation shows that because of very high viscosity, only inclusions with diameters smaller than the calculated d_K value will be homogenized and hence, either duration or temperature of heating or both must be increased to homogenize the inclusions discernible under the optical microscope.

5.3.6 *PRESSURE* ESTIMATES FROM RE-EQUILIBRATED FLUID INCLUSIONS

Re-equilibration textures look different around inclusions irreversibly modified under internal over- and underpressure. Larger inclusions tend to re-equilibrate first under internal overpressure, whereas the internal underpressure causes preferential re-equilibration of the smallest inclusions (Vityk et al., 1994). Each *P–T* trajectory is typical of its own type of re-equilibration texture that can be used as a tectonothermobarometer (Crawford, 1992; Vityk et al., 1994, 1996; Vityk and Bodnar, 1995; Boullier, 1999; Bodnar, 2003b). Aqueous fluid inclusions in quartz also re-equilibrate under a deviatoric stress, and can thus serve as monitors of both the orientation and the magnitude of the tectonic stresses during a low-strain, ductile deformation of quartz-bearing rocks (Tarantola et al., 2010a, 2012). Apart from the volume changes, the re-equilibrated aqueous inclusions lose water by diffusion, thereby passively increasing their salt and gas contents and triggering the plastic deformation of the surrounding quartz (Tarantola et al., 2010b) by the mechanism of hydrolytic weakening (Gratier and Jenatton, 1984).

The magnitude of the critical pressure gradient needed to initiate irreversible volume changes of fluid inclusions is controlled by several factors, of which the most important are structural and physical properties of the host mineral (hardness, cleavage) and absolute volume (size) of fluid inclusions. Fluid inclusions in soft minerals (barite) tend to stretch upon internal overpressure, whereas those in harder (quartz) or cleavable minerals (calcite, fluorite) decrepitate almost immediately after overstepping some critical pressure gradient. Naumov et al. (1966) and Tugarinov and Naumov (1970) estimated the critical pressure gradients for barite at 250 bar, calcite at 350 bar, fluorite at 350 bar, quartz at 850 ± 50 bar, and beryl at 870 bar. According to these data, there is a passable correlation between the Mohs hardness (h) and the threshold value for the decrepitation pressure ($R^2 = 0.93$)

$$P(\text{bar}) = 122h - 45 \quad (5.12)$$

This relationship, however, is valid only for inclusions larger than several tens of micrometers. The smaller is the inclusion, the larger pressure gradient is needed to trigger the decrepitation. Leroy

(1979) found that quartz-hosted inclusions with the volume of 100 μm^3 can withstand an internal overpressure of 1400 bar. Similarly, very small inclusions in olivine survive without decrepitation an internal overpressure as much as 7 kbar at 1200 °C (Roedder, 1965). Apart from quartz (Bodnar et al., 1989a), fluid inclusion re-equilibration has been experimentally studied and the inclusion diameter-controlled critical pressure gradients have been determined in barite (Ulrich and Bodnar, 1988), sphalerite, fluorite (Bodnar and Bethke, 1984), and calcite (Prezbindowski and Larese, 1987; Ujiie et al., 2008).

Bodnar et al. (1989a) approximated the internal pressure needed to induce a nonelastic deformation (decrepitation) in quartz by the empirical exponential equation

$$P(\text{kbar}) = 4.26d^{-0.423} \tag{5.13}$$

where d is the inclusion diameter in μm. The equation is valid for regular, three-dimensional inclusions, showing negative crystal shapes.

Bodnar and Bethke (1984) derived a similar relationship between the inclusion size and the minimum pressure required to initiate stretching of fluid inclusions in fluorite

$$P(\text{bar}) = 900 - 147 \log V \tag{5.14}$$

where V is the inclusion volume in μm^3. Hence, the determination of the critical inclusion diameter/volume dividing the decrepitated and nondecrepitated inclusions helps us to estimate the critical pressure differential, and thereby constrain the post-entrapment P–T history.

Much attention has been paid to the interpretation of fluid-inclusion microthermometric data from calcite cements in sedimentary rocks (Goldstein, 1986, 2001; Goldstein and Reynolds, 1994; Tobin and Claxton, 2000). The character of the distribution of fluid inclusion homogenization temperatures in calcite cements is an important indicator of the basin's history. Cements precipitated in vadose environments have low salinities and variable liquid-to-vapor ratios due to entrapment of various proportions of near-surface freshwater and air. Liquid-only aqueous inclusions are common. Carbonate cements precipitated in low-temperature phreatic environments contain low-salinity aqueous inclusions without vapor bubbles, or with very small vapor bubbles, homogenizing at a narrow range of temperatures. With progressive burial, cements precipitated in low-temperature phreatic environments re-equilibrate due to high internal ovepressure to a fluid inclusion population with variable homogenization temperatures and salinities (Goldstein, 1986). Experimental short-termed thermal stretching leads initially to right-skewed T_h distribution, which evolves toward the left-skewed distribution, as the median is reset to the increased temperature. With advanced stretching, homogenization temperatures of most inclusions are accommodated to new P–T conditions, but some inclusions still preserve homogenization temperatures consistent with the initial low-temperature phreatic conditions (Prezbindowski and Larese, 1987; Tobin and Claxton, 2000). The opposite evolution of distribution shape was observed during isothermal decompression, when some inclusions survived isothermal decompression back to ambient pressures (Bourdet and Pironon, 2008). High-temperature unimodal, bimodal, and multimodal T_h distributions are not consistent with the thermally induced stretching, and are more likely to reflect multiple hydrothermal events, and leaked and refilled inclusions (Tobin and Claxton, 2000).

Barker and Goldstein (1990) found a linear correlation between the mean vitrinite reflectance and the maximum temperature recorded by total homogenization temperatures of stretched aqueous inclusions in calcite from diverse geological systems associated with burial diagenesis, hydrothermal alteration, and low-grade metamorphism. The reduced regression equation reads

$$\ln R_m = 0.00811\ T_h - 1.26 \tag{5.15}$$

where R_m is the mean vitrinite reflectance and T_h is the mean or mode homogenization temperature of aqueous fluid inclusions in calcite cement (in °C) assumed to approach the peak thermal overprint of the organic matter in the surrounding sediments.

Tobin and Claxton (2000) showed that median T_h values determined from fluid inclusions in calcite generally slightly underestimate T_{peak} values by 15–25 °C. They proposed special selection criteria for identifying the samples with predominantly stretched fluid inclusions that are most suitable to the determination of maximum thermal exposure in sedimentary basins. In comparison to previous studies, they provided a more complete set of data that satisfied the sample selection requirements and established another empirical relationship between calcite-hosted fluid inclusion T_h values and vitrinite reflectance data

$$\%R_m = 1.9532\ \log T_h - 2.9428. \tag{5.16}$$

The correlation between the homogenization temperature of fluid inclusions and the vitrinite reflectance provides a tool for the assessment of the thermal maturity of samples, when conventional organic maturity techniques fail to provide reasonable data.

5.4 CALCULATING DEPTHS AND GEOTHERMAL GRADIENTS

Fluid inclusions are sensitive geobarometers, from which depths of burial and geothermal gradients can be derived with a high precision. The fluid pressure is directly correlated with depths according to Pascal's law stating that the pressure exerted in a confined incompressible fluid is transmitted equally in all directions

$$\Delta P = \Delta h D g \tag{5.17}$$

where ΔP is the hydrostatic pressure [Pa], or the difference at two points within a fluid column due to mass of the fluid, D is the fluid density [kg m^{-3}], g is the sea-level acceleration due to gravity [9.81 m/s^2], Δh is the height of the fluid column.

However, the relationship between fluid-inclusion trapping pressure and depth is not straightforward. Figure 5.23 depicts various situations that may occur during crystallization of a mineral. Drawing A depicts the most simple situation, when a static homogeneous fluid occupies the whole volume of a fissure reaching the Earth's surface. The fluid pressure in this case is directly correlated with the depth. Complications may arise from the fact that hot fluids expand upward with the decreasing pressure. The fluid rising up within the fissure partially cools down, and transmits heat to the surrounding rocks by convective heating. Hence, temperature of the surficial fluid must in fact be lower than that of the deep-seated fluid. The actual integrated density of the fluid column will be higher than that observed within fluid inclusions, because the density of the aqueous liquid is inversely correlated with temperature. Consequently, the apparent depth derived from the density of the inclusion fluid will be greater than the actual depth. High-salinity brines can be heated during adiabatic decompression (Wood and Spera, 1984), thus counterbalancing the loss of energy due to the convective heating of country rocks.

Drawing B depicts another complication, when the upper part of the fluid column is fluxed by low-salinity meteoric water. The dilution with freshwater may also partly eliminate the effect of the heat transfer to surrounding rocks on the bulk density of the fluid column.

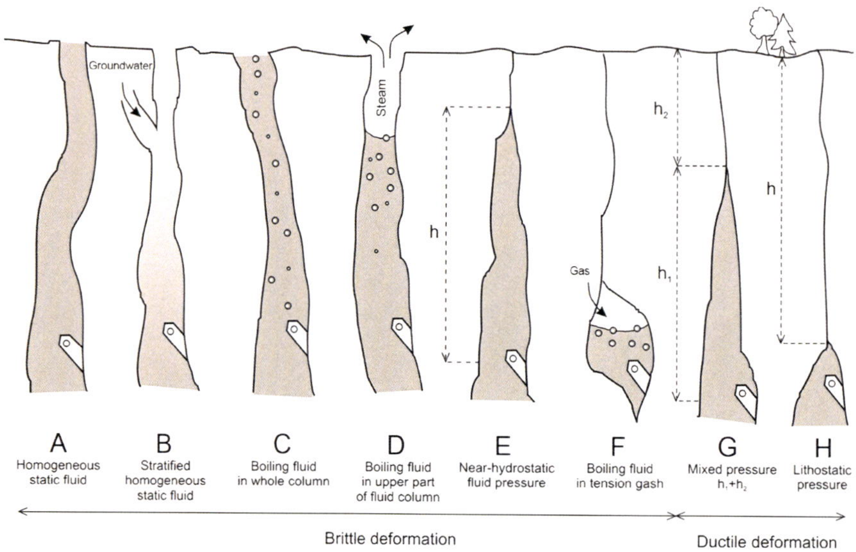

FIGURE 5.23

Relationships between hydrostatic and lithostatic pressure during crystallization of mineral (Roedder and Bodnar, 1980). Reprinted with permission from Annual Review of Earth and Planetary Sciences™ 8, 263–301.

Drawings C and D correspond to situations, where the fissure is filled with a boiling aqueous fluid. In this case, the actual pressure is defined by the two-phase boiling curve (e.g., Haas, 1976), however, the bulk density of the fluid column is unknown, because the fluid is composed of aqueous liquid and vapor phases, whose proportions vary with depth. Moreover, the fissure is often only partly filled with boiling aqueous fluid, whereas steam accumulates in the uppermost part. Another complication arises from the fact that hydrostatic pressure from the bottom of the fluid column is transferred to the top by expanding gas bubbles (Sahagian and Proussevitch, 1992). In submarine hydrothermal vents or fissures in accretion wedges, the actual hydrostatic pressure is increased by the hydrostatic load of the overlying ocean.

Drawings E through H depict closed systems, where a fluid-filled fissure does not communicate with the surface. If the fissure is sealed close to the surface (E), the static fissure fluid is nearly under hydrostatic pressure. Drawing F illustrates an extreme case of tension gashes, when tectonic strain enlarges the volume of a cavity, and this triggers boiling of the fluid. In this case, the apparent depth derived from the fluid inclusion density will be much lower than the actual depth.

Fluid pressure in a vein or fissure sealed at a greater depth would correspond to the sum of hydrostatic and lithostatic loads (drawing G). Deep-seated cavities would be under lithostatic pressure (drawing H) in ductilely deformed rocks. The same is valid for perfectly sealed fissures with a limited vertical extent, or those oriented horizontally (sills).

According to models of rock rheology (e.g., Sibson, 1990, 1992; Holbrook, 1999) and leak-off data from sedimentary basins (Gaarenstroom et al., 1993), fracture fluid pressure cannot exceed the lithostatic load plus cohesive (tensile) strength of the rock without hydraulic fracturing. The fracture pressure is generally lower than that of overburden (lithostatic) pressure, typically reaching 70–90% of overburden at depths of up to 5 km. The pressure needed to initiate the fracture opening increases with depth, approaching the lithostatic pressure gradient at around 5 km (Figure 5.24). Minimal fracture pressure approaches that of the hydrostatic pressure gradient in the case of static fluid. In a dynamic fluid column, an overpressure (supra-hydrostatic pressure) is established.

Hydrothermal vein systems may originate by several mechanisms differing in terms of the fluid overpressure and tectonic regime. Mesothermal veins are created at depths between 5 and 15 km along active faults under fluid pressure corresponding to that of the lithostatic load plus tensile strength of the surrounding rocks. The source of the supra-lithostatic fluid overpressure, whose magnitude probably does not exceed 100 bar, is a seismic zone located in an impermeable rock environment strained by horizontal tectonic compression. The associated fluid overpressure causes hydraulic reactivation of overthrust faults and shear zones, which serve as escape channels for hydrothermal fluids. This fault-valve process is characterized by fluctuating fluid pressures oscillating between the supra-lithostatic and hydrostatic limits (Sibson et al., 1988; Sibson, 1990; Boullier and Robert, 1992; and

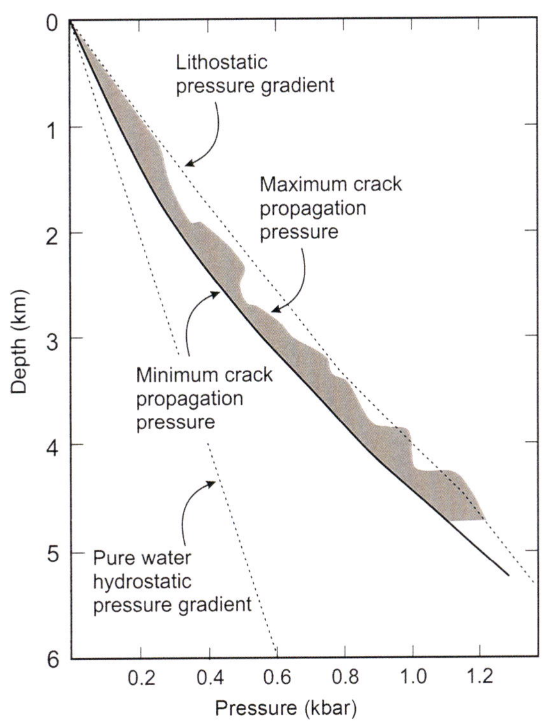

FIGURE 5.24

Leak-off experiments and measurements of formation (pore) pressure in oil drillholes in Jurassic sediments of the North Sea (Holbrook, 1999) showing values of minimal and maximal fluid pressures during propagation of fractures in water-saturated rocks. Reprinted from Holbrook, P., 1999. A simple closed form force balanced solution for pore pressure, overburden and the principal effective stresses in the Earth. Marine and Petroleum Geology 16, 303–319, Copyright 1999, with permission from Elsevier.

others). The vein systems originating in this way are vertically aligned, but sometimes they are combined with horizontal sills that are active during overpressurization.

A suction-pump effect is the reverse of the previous mechanism, producing dominantly epithermal, shallow (<2 km) vertical veins during extension or transtension (Sibson, 1987). The mineralized veins are originally strain-created, open dilatation cracks that sucked fluids at a sub-hydrostatic pressure by seismic (Sibson et al., 1975) or tectonic (Fyfe et al., 1978) pumping.

Subhorizontal veins originate from a water sill (Fyfe et al., 1978), which forms during dewatering and compaction of sediments triggered by the overburden pressure. The liberated fluids ascend and accumulate beneath an impermeable barrier. The fluid pressure beneath the barrier is gradually equilibrated with that in depths, thus generating a local supra-lithostatic overpressure. Breaking down of the coherent impermeable horizon liberates the accumulated overpressure and gives rise to another water sill located higher. The fluid pressure fluctuates episodically from lithostatic/supra-lithostatic during pressurization of the water sill, to near-hydrostatic after breaking the impermeable barrier.

A self-induced pressure equilibration (Gaboury and Daigneault, 2000) gives rise to regular flat veins arranged along intrusive dykes intersecting stratified sediments and volcanoclastics interlaced with impermeable horizons. A deep magmatic intrusion is the principal source of the fluid overpressure. The volcanic dykes serve as conduits for percolating fluids. The subhorizontal veins are opened gradually from top to bottom, when the fluid pressure in the feeding channel—the intrusive dyke—overcomes the local lithostatic load approximately by one-third of the tension strength of the surrounding rock. The fluid pressure within the sill and the feeding channel equilibrate at a hydrostatic when the system opens. Instantaneous pressure pulses exceeding the rock tension strength give rise to explosive or hydraulic breccias (Phillip, 1972).

An instantaneous fall in pressure coinciding with hydraulic fracturing often triggers fluid boiling and precipitation of minerals. The newly formed minerals fill open pores and eventually seal the cracks. This phenomenon, called a crack-seal mechanism (Ramsay, 1980), repeats periodically over a period of time. The precipitating minerals trap fluids with highly variable densities, reflecting recurrent (near-)hydrostatic and (near-)lithostatic pressures. In the crack-seal veins, maximal fluid pressure corresponding to that of the lithostatic load is approximately 3 times greater than the minimal one, reflecting the hydrostatic load. The pressure fluctuation results from the density differences of the surrounding rocks and that of the static or boiling fluid in an open fracture. The total fluid pressure recorded by fluid densities at the bottom of periodically opened and sealed cracks in accretion wedges or submarine hydrothermal vents must be corrected to the effects of seawater column and an advective overpressure generated by rising and expanding gas bubbles.

The fluid overpressure exceeding that of the static fluid column is generated when bubbles in a heterogeneous fluid mixture try to expand when rising upward, but cannot increase their volume due to incompressibility of the surrounding liquid (Sahagian and Proussevitch, 1992; Osborne and Swarbrick, 1997). For instance, densities of the conjugate methane and aqueous phases calculated from the equation of state for the binary H_2O–CH_4 system (Duan et al., 1992b) change from 0.385 g/cm^3 and 0.964 g/cm^3, respectively, at 220 °C and 3 kbar, to 0.174 g/cm^3 and 0.889 g/cm^3, respectively, at 205 °C and 0.5 kbar, thus corresponding to about 9% expansion of the aqueous liquid contrasting with a 122% expansion of the associated methane bubbles. As a consequence of the bubble expansion, the hydrostatic pressure is transported from bottom to top of the buoyant fluid column, thus redoubling the hydrostatic pressure of the fixed height of the fluid column (Figure 5.25).

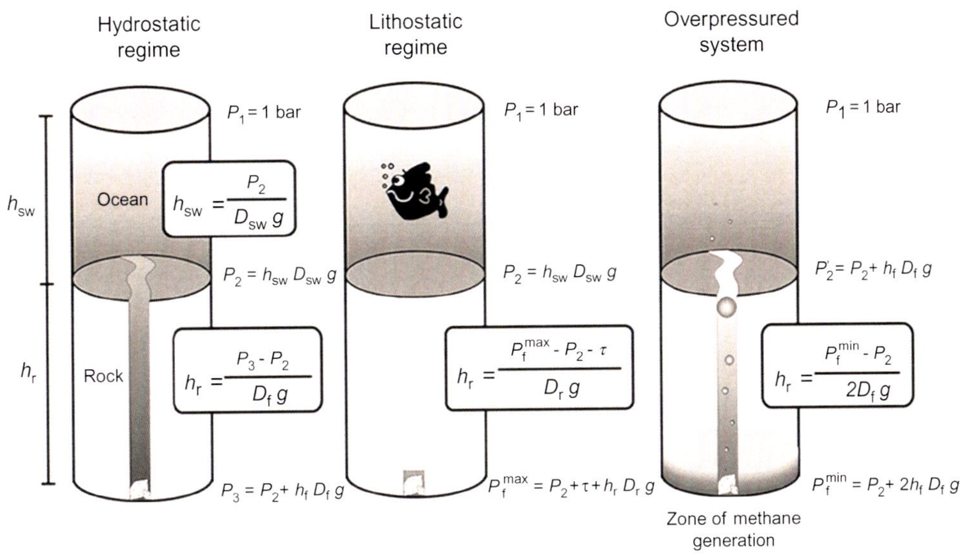

FIGURE 5.25

Cartoon depicting pressure components at the bottom of crack-seal vein in a sedimentary accretion wedge, where the methane generation zone liberates gas bubbles during hydraulic fracturing. *h*, *D*, τ, and g denote height, density, tensile strength of rocks, and the gravitation constant, respectively. P_1 is the atmospheric pressure, P_2 is the hydrostatic load of seawater column, P_2' is the supra-hydrostatic overpressure generated by rising and expanding gas bubbles, P_3 is the sum of P_2 and that of the fracture fluid. Subscripts f, sw, and r denote fluid, seawater, and rock.

Assuming an open fracture under a seawater column, the pressure at the bottom of the fracture filled with the buoyant fluid is

$$P_f^{min} = P_{sw} + 2P_f \tag{5.18}$$

where P_{sw} is the load of the seawater column, and P_f is the load of the fracture fluid column (atmospheric pressure is negligibly low). In contrast, the total fluid pressure at the bottom of a cemented and sealed fracture corresponds to

$$P_f^{max} = P_{sw} + P_r + \chi \tag{5.19}$$

where P_r stands for the lithostatic load of the rock column, and χ is the cohesive (tensile) strength of the fracture infilling [Pa]. Assuming the crack-seal mechanism and recurrent hydro- and lithostatic fluid regimes, the height of rock column (h_r) can be inferred from the minimal and maximal fluid inclusion trapping pressures (P_f^{min}, P_f^{max} in Pa) and Pascal's principle (Equation (5.17)) by solving the following equation (Hurai et al., 2006)

$$h_r[m] = \frac{P_f^{min} - (h_{sw}D_{sw}g)}{2D_f g} = \frac{P_f^{max} - \chi - (h_{sw}D_{sw}g)}{D_r g} \tag{5.20}$$

where D_r (2600–2750 kg m^{-3}) and D_{sw} (1028 kg m^{-3}) are average densities of typical sedimentary crust and seawater, respectively, g is the acceleration due to gravitation (9.81 m s^{-2}), and h_{sw} denotes the height of the seawater column. 100 bar is a typical χ value of fresh, nonfractured rock (Lockner, 1995) that can be set as a constant. Maximal bulk fluid density (D_f) and depth of burial equivalent to the height of the rock column h_r can be calculated by setting h_{sw} to zero. In turn, the minimal density of the gaseous phase in a heterogeneous fluid set as bulk D_f defines the maximal h_{sw} and minimal h_r. Combining the inferred depths with the minimum and maximum temperatures indicated by fluid inclusions, thermal gradients resulting from conductive (i.e., that induced by heat diffusion in solid rock), and convective (i.e., that induced by heat transfer in a fluid medium) heatings can also be calculated (Hurai et al., 2006).

Supra-lithostatic fluid overpressure can be generated during fracture propagation at depths greater than 4 km (Figure 5.24) in water-unsaturated rocks and sedimentary basins with heated formation water sealed under an impermeable barrier (Roedder and Bodnar, 1980). Fluid inclusions in quartz from sub-vertical cracks in flysch sediments of the Tertiary accretion wedge of the Western Carpathians recorded local fluid pressures exceeding the lithostatic load by 15–20%. The overpressure was generated here due to the thermal cracking of organic-rich sediments and/or oil, and it occurs only at depths below 9 km and temperatures above 200 °C (Hurai et al., 2006). Milovský et al. (2002, 2012) reported on significant supra-lithostatic fluid overpressures induced by the friction heat along the subhorizontal base of a moving superficial nappe.

PROBLEM 5.5 CALCULATING DEPTH OF BURIAL

Calculate the depth at a pressure of 2100 bar, assuming hydrostatic and lithostatic loads, and densities of the fluid and the surrounding rocks equal to 1.0 and 2.75 g/cm^3, respectively.

Solution

Convert the pressure and the density to SI units (2100×10^5 Pa, 2750 kg/m^3). Calculate the depth of crystallization assuming the lithostatic load using Equation (5.17)

$$h = 210{,}000{,}000/(2750 \times 9.81) = 7784 \text{ m}$$

Calculate the depth of crystallization assuming the hydrostatic load

$$h = 210{,}000/(1000 \times 9.81) = 21{,}407 \text{ m}$$

It is convenient to remember approximate lithostatic pressure 1 km deep for quick conversion of thermobarometric data. Rounding the gravitation constant to 10 m/s^2, the pressure in bars will correspond to the density in g/cm^3 multiplied by 100, for instance

$$P_{1\text{ km}} = 1000 \text{ m} \times 2750 \text{ kg/m}^3 \times 10 \text{ m/s}^2 = 27{,}500{,}000 \text{ Pa} = 275 \text{ bar, i.e., } 2.75 \text{ g/cm}^3 \times 100$$

Depths of burial, conductive, and convective geothermal gradients across the Tertiary accretion wedge of the Western Carpathians were calculated from microthermometric data plotted in Figure 5.26 using Equation (5.20). The results of these calculations are summarized in Table 5.6.

An independent check of validity for Equation (5.20) is provided by thermal maturity data from deep boreholes. The maximum temperature exposure (T_{max} in °C) and geothermal gradients calculated from the mean vitrinite reflectance (R_r%) according to Barker (1983) and Barker and Pawlewicz (1986)

$$T_{max} = \frac{\ln R_r\% + 1.2}{0.0078} \tag{5.21}$$

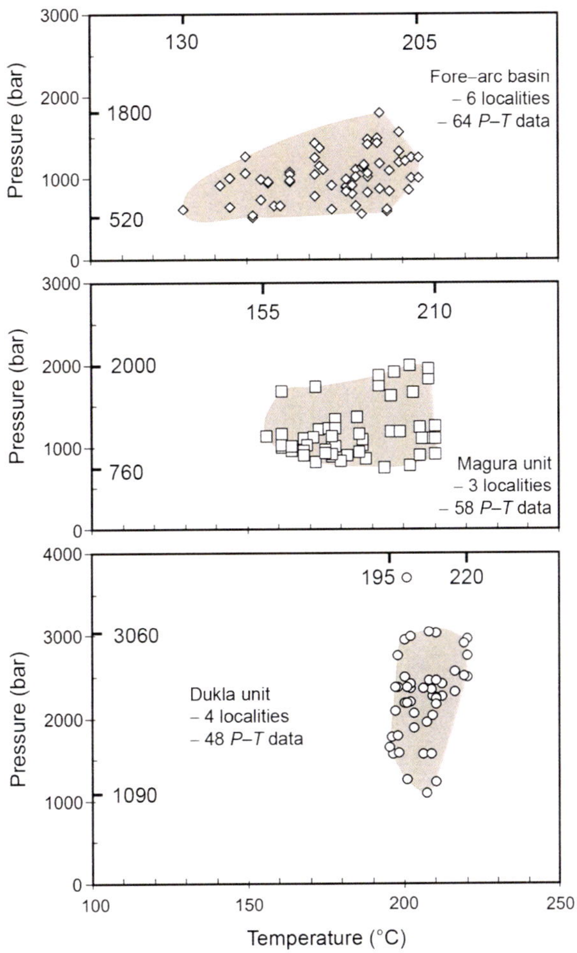

FIGURE 5.26

Trapping temperatures and pressures inferred from microthermometric data on coeval methane-rich and aqueous inclusions in blocky and drusy quartz from mineralized joints of flysch sandstones of Tertiary accretion wedge and fore-arc basin of Western Carpathians (Hurai et al., 2006) calculated using a T_h–isochore intersection method and EOSs for pure CH_4 (Kleinrahm and Wagner, 1986), and binary CO_2–CH_4 systems (Holloway, 1981; Thiéry et al., 1994a). All datapoints represent samples collected from surface outcrops. *P–T* data obtained from fluid inclusions in drillholes in a similar way are presented in Figure 5.27.

are consistent with the temperatures derived by the T_h–isochore intersection method from CH_4- and CH_4–CO_2-rich aqueous inclusions. For instance, a linear regression to mean vitrinite reflectance from the Hanušovce borehole (HAN-1) in the area covered by a fore-arc basin yielded an average geothermal conductive gradient of 21 °C/km and intercept of 73 °C on the recent surface, corresponding to a 3.5-km column of rocks removed by erosion (Figure 5.27). The derived gradient is within the range of 20–24 °C/km inferred from the fluid model. Moreover, *P–T* data from quartz crystals sampled from

Table 5.6 Summary of Burial Parameters Calculated Using Equation (5.20) and *P–T* Data in Figure 5.26

Unit	Rock *T* [°C]	Fluid *T* [°C]	Min.-Max. Sea Column (km)	Max.-Min.* Rock Column (km)	Gradients (°C km^{-1}) Conductive	Gradients (°C km^{-1}) Convective	Bulk Fluid *D* (g cm^{-1}) at Hydrostatic *P*
Fore-arc basin	130	205	0.0	6.6	20	31	0.40
			3.1	5.4	24	38	0.20
Magura nappes	155	210	0.0	7.4	21	28	0.53
			6.0	5.4	29	39	0.24
Dukla nappes	195	220	0.0	11.4	17	19	0.49
			6.1	9.0	22	24	0.27

Burial depths equivalent to the rock column eroded from recent surface.

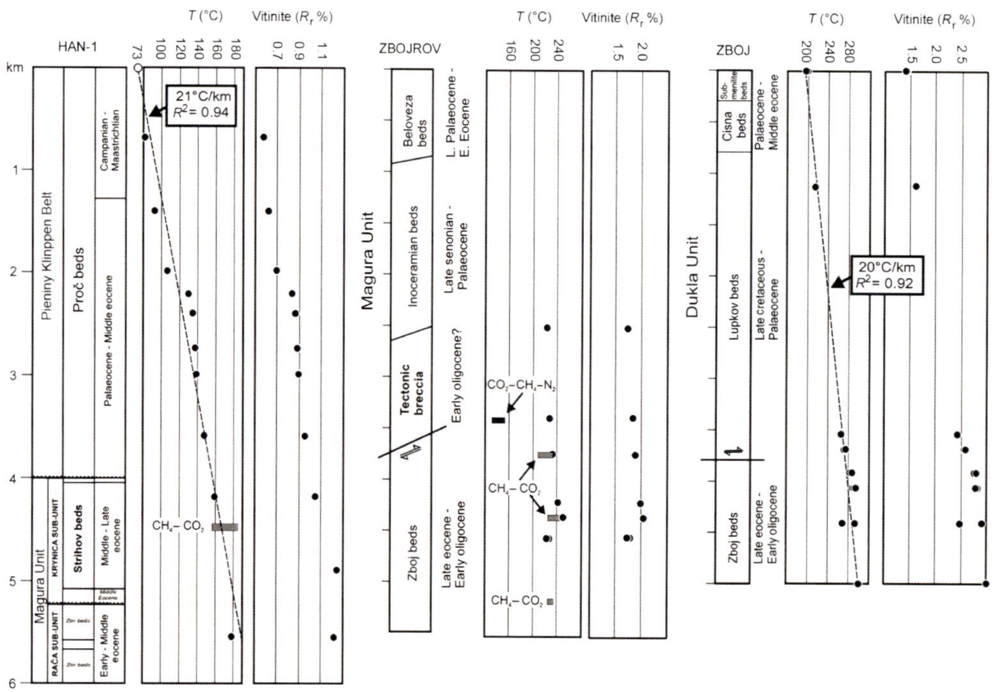

FIGURE 5.27

Stratigraphic logs and thermal maturity data from the vitrinite reflectance overlapped by fluid inclusion data for deep drillings in flysch sediments of the Tertiary accretion wedge of the Outer Carpathians (Zborov, Zboj) and the fore-arc basin (Hanušovce—HAN-1). Geological background: Leško et al. (1985), Ďurkovič et al. (1982), Wunder et al. (1991).

the depth of 4469 m overlap *P–T* conditions along the maximum thermal maturity trendline, yielding additional independent pressure control on the *P–T* parameters derived from fluid inclusions. A complete thermal maturity profile is not available from the Zborov drillhole located in the area with the Magura nappes on the recent surface; however, data from CH_4–CO_2-bearing aqueous inclusions from different depths are consistent with the vitrinite mean reflectance in surrounding flysch sediments. In contrast, *P–T* data from CO_2–CH_4–N_2-rich aqueous fluids percolating through a brecciated tectonic boundary between two tectonic superunits are inconsistent with the maximum thermal overprint, thus indicating a separate hydrothermal event non-coeval with the CH_4-bearing fluids. Fluid inclusion data are not available from the deep drillhole (Zboj) in the area covered by the Dukla nappes. However, the well-defined thermal maturity trend ends on the recent surface at a temperature of 199 °C, which indicates a pile of now missing overburden ~10 km thick. The inferred conductive gradient, 20 °C/km, is also consistent with the fluid model based on quartz samples collected from surface outcrops. It should be noted that the thermal maturity did not change across tectonic boundaries. This indicates that the maximum thermal overprint coincidental with the CH_4-rich fluids is genetically related to a post-nappe tectonics. The broadening of the fluid inclusion-derived temperature ranges from the lowermost tectonic units (Dukla nappes and subjacent Zboj units) to the uppermost units of the Magura nappe and adjacent fore-arc basin; in other words, the increasing divergence of the conductive and convective gradients (Figure 5.26) suggests uplift- and extension-related lateral and vertical fluxes of hot deep-seated fluids into colder sediments (Hurai et al., 2006).

Microthermometric data provide constraints on minimum formation depths and possible limits for the conductive and convective (adiabatic) flows also in metamorphosed accretion wedges, where homogeneous fluids percolated. For instance, superdense CO_2 fluid inclusions in quartz-tourmaline and quartz-stibnite hydrothermal veins of the Gemeric tectonic unit indicate that the geothermal gradient in the Variscan basement subjacent to an Early Cretaceous accretion wedge of Western Carpathians could not exceed 23 °C/km (Urban et al., 2006); otherwise the superdense CO_2 could not exist. Similarly, very high-density nitrogen inclusions in barite from siderite-barite veins from the same tectonic unit indicate possible depths of formation between 6 and 14 km, and geothermal gradients between 13 and 38 °C/km at reasonable temperatures (the shaded area in Figure 5.28).

5.5 TREND ANALYSIS

5.5.1 T_h–SALINITY PLOTS FROM AQUEOUS INCLUSIONS

Correlation of homogenization temperature and salinity is often used to decipher the origin and precipitation mechanism of hydrothermal fluids. In geothermal and subvolcanic epi- to mesothermal hydrothermal systems, the homogenization temperature is regarded as true trapping temperature whereas the salinity is calculated from the temperature of melting of the last solid phase (ice, hydrohalite, halite). Care must be exercised when using this approach in the hydrothermal systems under a higher pressure, as well as in those containing appreciable amounts of divalent cations and/or dissolved gases, mainly CO_2. In the first case, the actual trapping temperature may be significantly higher than that of the fluid inclusion homogenization in the case of the homogeneous trapping. In the second case, the apparent salinity expressed in terms of NaCl mass% equivalents is overestimated owing to different positions of the vapor-saturated ice and halite liquidi in the NaCl-bearing systems with chlorides of divalent cations. Moreover, dissolved gases also contribute to the depression of the melting point of

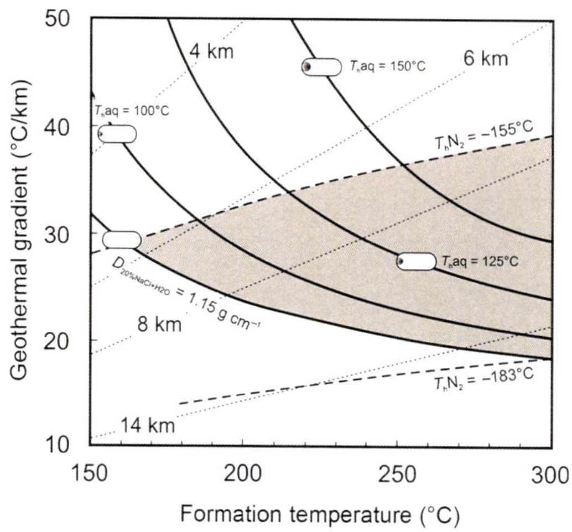

FIGURE 5.28

Formation P–T parameters and geothermal gradients during barite crystallization in the Droždiak siderite-barite polymetallic vein in the Gemeric tectonic unit (Hurai et al., 2008b). Isochores of primary nondecrepitated nitrogen inclusions (long-dash curves) were calculated from homogenization temperatures (T_h) using the EOS for pure N_2 (Span et al., 2000). Isochores for 20 mass% NaCl–H_2O fluid inclusions for the indicated homogenization temperature (solid curves) were derived from experimental data of Gehrig (1980). The shaded field encompasses possible formation P–T conditions of the barite. Dotted lines are depths inferred from the average crust density (2.75 g cm^{-3}), assuming the lithostatic load. The 1.15 g cm^{-3} 20% NaCl–H_2O isochore defines the upper stability limit for aqueous inclusions that remain homogeneous on cooling to room temperature. Such primary inclusions occurred in barite along with two-phase aqueous and high-density nitrogen inclusions, and provide another limit for the depths and geothermal gradients during the barite crystallization.

ice in low-salinity aqueous solutions (Hedenquist and Henley, 1985). In spite of this, the simple correlation of homogenization temperature and salinity yields useful first insight into the possible origin of hydrothermal fluids, although the fields' diagnostic of individual deposit types is not definitive, and compositions may exist outside the ranges depicted in Figure 5.29.

Dubessy et al. (2003) have considered the following precipitation mechanisms in gas-free, low-salinity hydrothermal systems triggered by the mixing of fluids with temperatures T_1 and T_2: (1) the isothermal mixing, with the fluid temperature controlled by the conductive heating from host rocks; (2) the adiabatic mixing whereby the temperature of the mixed fluid results merely from the convective heating and the proportion of the two mixing fluids, without heat exchange with surrounding rocks; (3) the mixing of fluids accompanied by the partial heat exchange between the mixing fluids and the surrounding rock. Each type results in a specific geometric distribution of microthermometric data in T_h–T_m diagrams, where T_m represents the temperature of ice melting, which is the last solid phase to melt (Figure 5.30).

Dissolved CO_2 gas plays an important role in the aqueous solutions of active and fossil geothermal and epithermal volcanic systems. The dissolved carbon dioxide causes an extra shift of the ice melting temperature, in addition to that generated by dissolved salts (Hedenquist and Henley, 1985). As shown

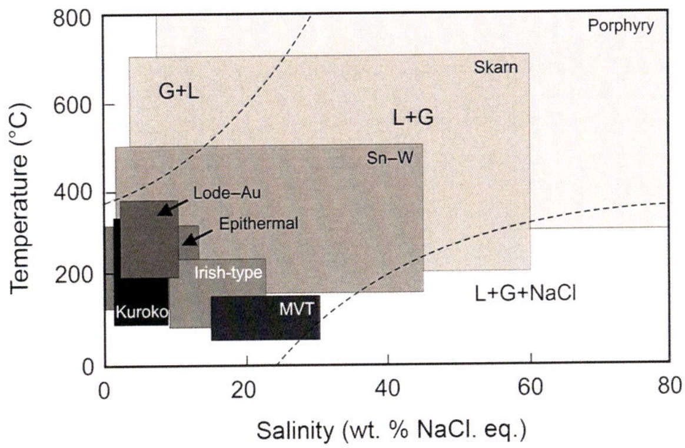

FIGURE 5.29

Typical ranges of ore-precipitating fluids projected in temperature-salinity coordinates (Wilkinson, 2001). Critical curve (upper-left dashed curve) and vapor-saturated halite liquidus of the NaCl–H_2O system (lower-right dashed curve) separate the fields of vapor-dominated (G + L), aqueous liquid-dominated, halite-undersaturated (L + G), and halite-oversaturated (L + G + NaCl) fluids. MVT, Mississippi Valley-type (telethermal) deposits; Sn–W, metasomatic granite-related (tin-wolfram) greisen deposits. Reprinted from Wilkinson, J.J., 2001. Fluid inclusions in hydrothermal ore deposits. Lithos 55, 229–272, Copyright 2001, with permission from Elsevier.

in Figure 4.33, the maximum shift due to the dissolved CO_2 corresponds to −1.5 °C assuming a pure H_2O–CO_2 system. The T_m–T_h diagram allows the discrimination between dilution and boiling in gas-free hydrothermal systems, and degassing in cooled gas-rich hydrothermal systems (Figure 5.31).

Mixing of cold, near-surface freshwaters with hot, higher saline, deeply circulated waters leads to a typical dilution trend, the slope and shape of which depend on the ratio of the mixing fluids. Extrapolation of the dilution trend to zero salinity ($T_m = 0$ °C) indicates whether the dilutant was a cold groundwater ($T_h < 100$ °C) or a steam-heated water ($T_h > 100$ °C). Boiling in a convective upflow also leads to a systematic, but relatively small, increase in the salinity accompanied by a loss of CO_2 and other gases. In a fluid where CO_2 dominates over dissolved salts, the degassing would result in a near-vertical trend in T_m values. In contrast, the boiling of a gas-poor fluid results in a slight salt enrichment of the residual fluid caused by the escape of essentially salt-free steam. However, the extreme adiabatic boiling and steam loss from 300 to 100 °C would increase the original salinity only by a factor of 1.7, for instance, from 0.2 to 0.34 mass% (Hedenquist and Henley, 1985).

High-salinity, high-density, NaCl-dominated brine inclusions, sometimes enriched in chlorides of divalent cations, are common in carbonate vein and metasomatic deposits (Hurai et al., 2002a, 2008a), porphyry-copper deposits (Bodnar and Beane, 1980), skarns (Baker and Lang, 2003; Hurai and Huraiová, 2011), iron oxide–gold deposits (Niiranen et al., 2007), in medium- to high-grade metamorphic environments (Scambelluri et al., 1997), and as fluid lubricants facilitating the nappe movement (Milovský et al., 2002, 2012). Trend analysis of such fluids using T_h–T_m diagrams faces numerous problems. The salinity of halite-saturated fluid at a given temperature is a function of pressure, which is unknown if the vapor phase is absent in the moment of total homogenization; that means the inclusion homogenizes by the halite disappearance (Figure 4.17). Hence, the common procedure of direct

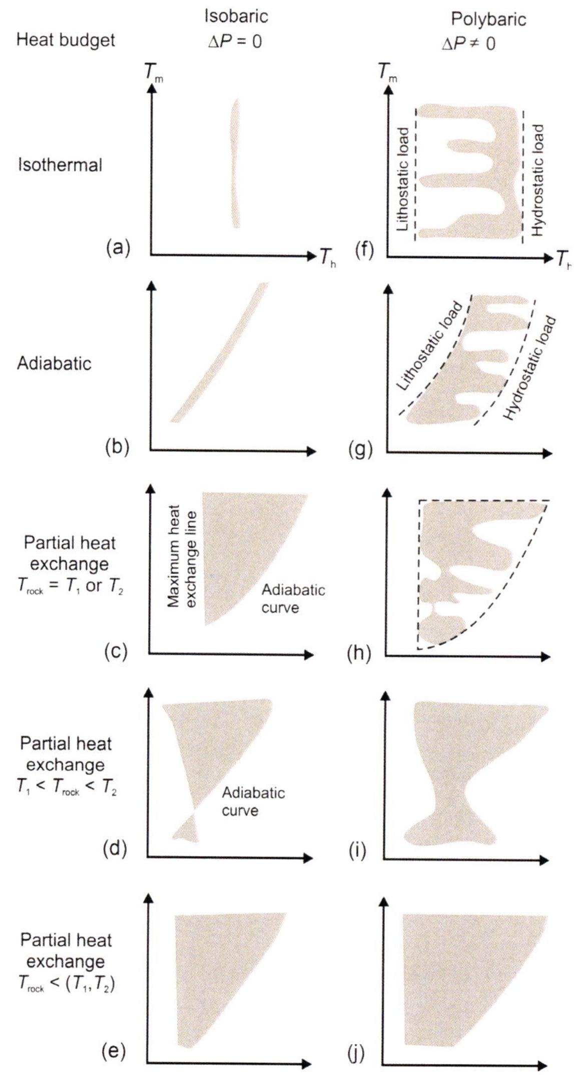

FIGURE 5.30

T_h–T_m diagrams for fluid mixing between a hot low-salinity and cold high-salinity fluids and different heat budgets: (a) isothermal; (b) adiabatic; (c–e) partial heat exchange between the mixed fluid and adjacent rocks, (a–e) isobaric mixing; and (f–j) nonisobaric mixing. At nonisobaric conditions, the fluid pressure is supposed to vary between those of lithostatic and hydrostatic loads (crack-seal veins), and the corresponding datapoints are distributed over the whole envelope. If the pressure fluctuations affect only some parts of the mixing process, the datapoints are distributed as complex patterns, with main trends parallel to the T_h axis. Reprinted from Dubessy, J., Derome, D., Sausse, J., 2003. Numerical modelling of fluid mixings in the H_2O–NaCl system: application to the north Caramal U prospect (Australia). Chemical Geology 194, 25–39, Copyright 2003, with permission from Elsevier.

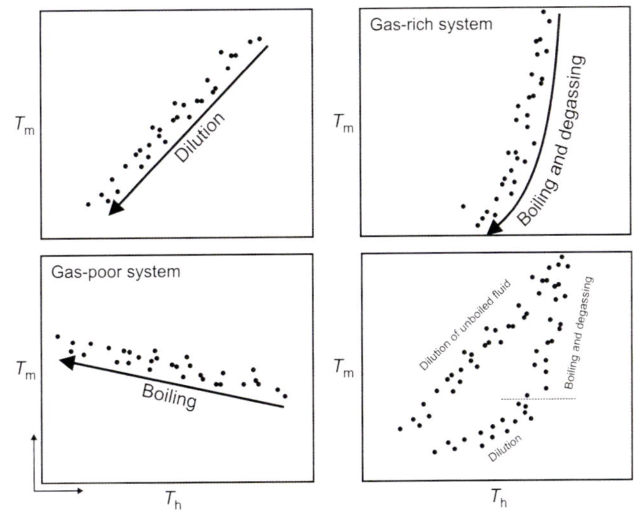

FIGURE 5.31

Schematic diagrams showing the dilution and boiling trends indicated by T_m and T_h values in aqueous fluid inclusions in minerals precipitated from gas-rich (steam-heated) and gas-poor (liquid-dominated) hydrothermal systems (Hedenquist and Henley, 1985). Arrows designate evolution trends. The lower-right diagram corresponds to the data distribution observed in the Broadlands geothermal field, New Zealand, reflecting the combined effect of the degassing and dilution on boiled and unboiled gas-rich fluids. Reprinted with permission from the Society of Economic Geologists.

calculation of the fluid salinity from the halite dissolution temperature cannot be applied because the empirical equations relating the halite dissolution temperature to salinity are valid only under vapor-saturated conditions along the univariant L+G+H curve. Some problems inherent in the interpretation of microthermometric data on high-density brines can, however, be solved using temperatures of vapor-absent halite dissolution (T_mH) and that of metastable (halite-absent) homogenization (T_h), which can be measured in brines with less than 40 mass% of dissolved salts. Plot of T_h and T_mH values against their difference (T_m–T_h) enables not only the true salinity to be inferred from a group of coeval inclusions or from an individual, artificially stretched inclusion (Problem 4.4) but reveals also the possible crystallization *P–T* trajectory (Milovský et al., 2012).

Possible crystallization *P–T* trajectories projected through T_h and T_mH values using the experimental phase diagram of the 40% NaCl–H_2O system are shown in Figure 5.32. The A–B curve represents a decompression trend, whose concave shape mimics the self-heating of the adiabatically expanding aqueous solution (Wood and Spera, 1984). In T_h–T_mH coordinates, the adiabatic decompression of homogeneous fluid results in a V-shaped trend that converges on the vapor-saturated halite liquidus, thus defining the true salt concentration. The adiabatic decompression trend may potentially intersect the halite+liquid field when the halite precipitation is triggered. This process is recorded in the T_h–T_mH plot by the T_mH scatter at an angle of ~30° from the vapor-unsaturated halite liquidus, whereas T_h values remain constant.

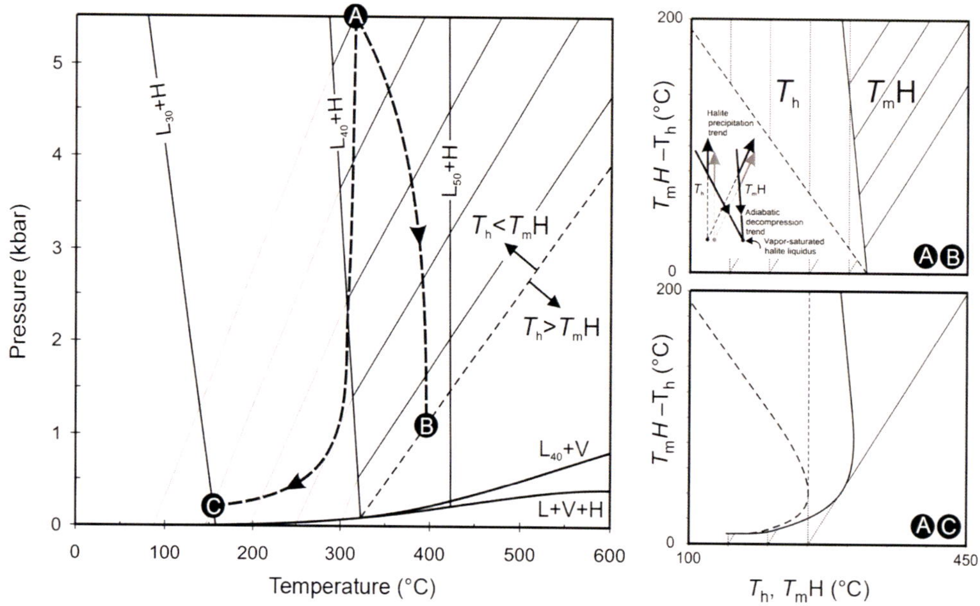

FIGURE 5.32

Possible *P–T* trajectories plotted on the phase diagram of the 40 mass% NaCl–H_2O system with vapor-unsaturated halite liquidi for 30, 40, and 50 mass% NaCl (Bodnar, 1994). Isochoric slopes in L-only (homogeneous, to the right of the L_{40}+H line) and L+H (heterogeneous, to the left of the L_{40}+H line) fields were estimated according to stable and metastable homogenization temperatures in experimentally re-equilibrated ~30 mass% natural brine inclusions in quartz (Hurai et al., 2002a). Only high-density brine inclusions with final homogenization by halite dissolution ($T_h < T_m$H) were considered. Two diagrams on the right side show possible distribution patterns of T_h and T_mH values plotted against their difference as produced by following *P–T* trajectories: the simple V-shaped pattern—adiabatic compression/decompression of a homogeneous (L-only) fluid with a fixed salinity, mirrored double V-shaped pattern—adiabatic compression/decompression in L+H field with fixed salinity, the curvilinear pattern—cooling accompanied by the precipitation of halite; and the decrease in the total fluid salinity in final stage (Milovský et al., 2012).

5.5.2 T_h–F_v PLOTS OF PETROLEUM INCLUSIONS

During the evolution of a sedimentary basin, oil may be altered during several processes (Thiéry et al., 2002): phase separation due to unmixing (immiscibility); leaching by incoming gas or aqueous solution; biodegradation; and cracking. Some of these processes can be revealed by the comparison of the total homogenization temperature (T_h) and the relative bubble volume (F_v) determined at $T < T_h$ in petroleum inclusions (Figure 5.33).

The positive correlation between F_v and T_h may be attributed to *P–T* changes. The trend 1 in Figure 5.33 illustrates the isothermal decompression (100 °C, 300–200 bar) of heavy black oil. The decompression leads to an increase in both T_h and F_v values, and a similar trend is produced by the isobaric heating and cooling. Since the fluid composition does not change during this process, various generations of fluid inclusions must coincide in the α–β compositional space (Figure 5.4).

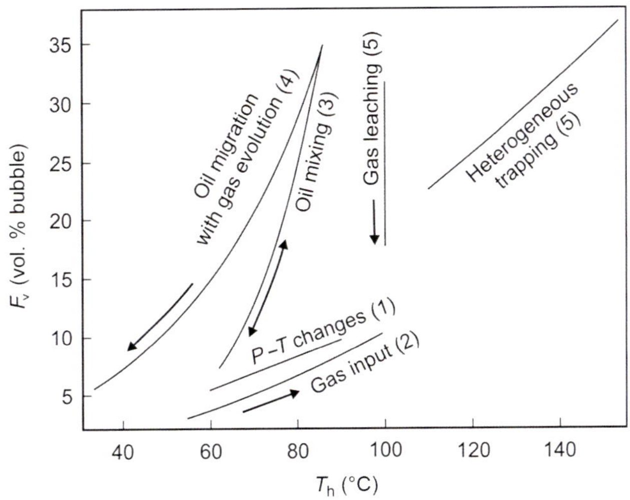

FIGURE 5.33

Effects of different processes on the total homogenization temperature (T_h) and the bubble volume (F_v) determined at 20 °C as modeled using the VTFlinc computer program. Reprinted from Thiéry, R., Pironon, J., Walgenwitz, F., Montel, F., 2002. Individual characterization of petroleum fluid inclusions (composition and P–T trapping conditions) by microthermometry and confocal laser scanning microscopy: inferences from applied thermodynamics of oil. Marine and Petroleum Geology 19, 847–859, Copyright 2002, with permission from Elsevier.

Petroleum reservoirs can be replenished by biogenic or thermogenic gas. Trend 2 simulates a sealed reservoir filled with a black oil, which is continually fluxed by gaseous methane at a constant temperature (100 °C) and pressure (300 bar). The resulting T_h–F_v curve exhibits a trend similar to that diagnostic of P–T fluctuations; however, in the α–β compositional space the gas input is discerned by a strong increase in the β parameter contrasting with almost constant α value.

The mixing is probably common in most petroleum reservoirs. Trend 3 represents the case, when light ($\alpha=0.84$, $\beta=0.6$) and heavier oil ($\alpha=0.9$, $\beta=0.39$) mix at random proportions at constant temperature (100 °C) and pressure (300 bar). The resulting T_h and F_v values of light oil inclusions decrease rapidly with small amounts of the heavy oil added. In contrast, the addition of a small amount of light oil has less dramatic effect on the properties of heavy oil inclusions. A curved mixing trend connecting compositions of the two mixed endmembers is visible in the α–β plot.

The liquid–gas unmixing can occur in petroleum migrating upward due to decreasing temperature and pressure. The less dense gaseous phase escapes continually, thus leaving behind residual petroleum fractions depleted in light molecular components. This process is exemplified by trend 4, reflecting the devolatilization of light oil ($\alpha=0.84$, $\beta=0.6$) along a 33 °C/km geotherm, starting at 100 °C and 300 bar. The oil became oversaturated with gas at 86 °C and 258 bar. From this point, the gas begins to escape. The first liberated gas is the gas-condensate progressively evolving to wet and dry gas at final stage. In the α–β space, a dramatic decrease in β-values of the residual oil fraction is accompanied by the moderate increase in α at the outset and decrease at the end of the process, thus resulting in an almost vertical general trend.

The fluid unmixing may occur also in situ at appropriate P–T conditions projecting within the immiscibility loop. In this case, fluid inclusions trap random mixtures of two or three coexisting phases, as documented in Figure 5.19. Let us consider the volatile light oil (α=0.9, β=0.6). At 90 °C and 248 bar, the petroleum splits into a mixture of gas (5 vol.%) and oil (95 vol.%), the composition of which is similar to that of the precursor oil. The immiscible gas phase corresponds to gas-condensate with α=0.77 and β=0.8. Inclusions trapping the random mixture of two coexisting phases exhibit increased and positively correlated T_h and F_v values (trend 5). The heterogeneous trapping trend can be distinguished from similar trends 1 and 2 by a steeper slope. In the α–β space, the heterogeneous trapping is reflected by nearly constant α and strongly changing β parameter.

Gas leaching is another mechanism, possibly affecting open-system petroleum reservoirs. Fluxes of thermogenic dry gas, predominantly methane, oversaturate the reservoir petroleum, producing a buoyant oil–gas fluid with the emulsion of gas droplets. Light molecular volatile components (C2+) dissolved in the petroleum are preferentially partitioned in the gas phase and are subsequently extracted from the reservoir in the escaping gas. The process is featured by the gas-leaching trend 5, representing a gas-saturated oil (α=0.847, β=0.559), to which a small amount of methane was added at 100 °C and 240 bar. After equilibration, the surplus gas was removed and new parameters of the residual petroleum were recalculated. In contrast to the fluid unmixing process, the gas-leaching trend is almost vertical, as the T_h remains constant, while the F_v decreases dramatically. The gas leaching produces a horizontal trend in the α–β space, as α increases at fixed β.

Post-entrapment stretching of oil inclusions produces trends similar to the P–T variation trend, due to changing density at fixed composition (Thiéry et al., 2002; Bourdet et al., 2008). If the stretching takes place, minimum T_h and F_v values are the most representative of the initial petroleum composition. In contrast, leakage is accompanied by the modification of both density and composition of petroleum inclusions. As a consequence, negative correlation between T_h and F_v values is established. Compositions of all investigated inclusions are different and their α–β curves do not coincide in the α–β compositional space. The dispersion of points results from the combination of stretching and the loss of light molecular gas components.

SUGGESTED FURTHER READING

Bodnar, R.J., 2003b. Reequilibration of fluid inclusions. In: Samson, I., Anderson, A., Marshall, D. (Eds.), Fluid Inclusions: Analysis and Interpretation. Miner. Assoc. Canada, Short Course 32, 213–230.

Goldstein, R.H., 2001. Fluid inclusions in sedimentary and diagenetic systems. Lithos 55, 159–193.

Tobin, R.C., Claxton, B.L., 2000. Multidisciplinary thermal maturity studies using vitrinite reflectance and fluid inclusion microthermometry: A new calibration of old techniques. AAPG Bulletin 84, 1647–1665.

CHAPTER

FLUID THERMODYNAMICS

6

CHAPTER OUTLINE

In most cases, geological processes cannot be observed directly for several reasons: They do not progress in a real time, they are beyond the reach of human activity, they have occurred in the past, or they are too complex to be experimentally modeled. However, the thermodynamic parameters of geological processes can be modeled by mineral–fluid equilibria. In the same way, it is possible to calculate the

Geofluids. http://dx.doi.org/10.1016/B978-0-12-803241-1.00006-X

composition of fluids in equilibrium with these minerals and to compare the results of theoretical calculations with the actual fluid compositions observed in fluid inclusions.

Problems inherent to the reconstruction of P–T conditions of mineral–fluid equilibria are solved by thermodynamics. We address only some fundamental concepts of thermodynamics here. More exhaustive coverage of the topic at the introductory level is given in applied thermodynamics texts that focus on minerals and rocks (e.g., Chatterjee, 1991; Fletcher, 1993; Anderson, 2005; Cemič, 2005; Philpotts and Ague, 2009) and chemistry textbooks (Chang, 1991).

6.1 THERMODYNAMIC LAWS

Thermodynamics deal with interconversions of heat and other forms of energy (Chang, 1991). The fact that heat is a kind of energy related to the motion of molecules was experimentally proven by James Prescott Joule (1818–1889) in the 1840s. The **internal energy** (U) of a system, however, consists not only of the kinetic energy of molecules but also other forms, such as kinetic and potential energy of atomic nuclei and their shells, and potential energy of intermolecular interactions.

The first law of thermodynamics states that energy can be converted from one form to another, but it cannot be destroyed or created. The total amount of energy in the universe remains constant. Thus, an energy gain in a system must be compensated by the energy loss in its surroundings. Furthermore, because the energy changes from one form to another, the energy lost by one system can be gained by another system in a different form.

There are three basic kinds of thermodynamic processes: heat transfer, mechanical work (volume transfer), and mass transfer. We are not able to express the magnitude of U in absolute numbers, but we can express the change of the internal energy of a closed system from initial (U_1) to final (U_2) conditions in terms of the total amount of **heat** (Q) added and **work** (w) done by the surroundings

$$U_2 - U_1 = \Delta U = Q - w \tag{6.1}$$

where ΔU is in J [kg m s^{-2}]. By convention, the energy put into the system in the form of heat has a positive sign, whereas that in the form of work is negative, as the positive work is done on the surroundings.

A useful example of mechanical work is the expansion of gas in a movable piston cylinder. As it expands, the gas pushes the piston upward against an opposing constant external pressure P. Because the pressure is a force per area unit (Equation (1.2)) and volume is d^3

$$Fd = PV \tag{6.2}$$

and the work done by the gas on the surroundings is equivalent to the change in the gas volume

$$w = P(V_2 - V_1) \tag{6.3}$$

Substituting (6.3) in (6.1) results in

$$U_2 - U_1 = \Delta U = Q - P(V_2 - V_1) \tag{6.4}$$

and after rearranging

$$Q = (U_2 + PV_2) - (U_1 + PV_1) \tag{6.5}$$

where Q is the heat involved in the system at a constant pressure.

Temperature, pressure, mass, and volume are macroscopic state properties conceivable from an everyday experience. Energy is also a state property because its magnitude is independent on the path followed by the system from the initial to the final state. In contrast, work is not a state property because its magnitude does not depend only on initial and final stages (ΔV) but on the external opposing pressure as well (Equation (6.3)). Heat is also not a state property because it is dependent on work (Equation (6.1)). Together with energy, the PV term is also a state property because in the case of an ideal gas, it is equal to RT (Equation (3.6)). Consequently, the change in $U+PV$ must be also a state property, which is given a special name—**enthalpy** (H)

$$H = U + PV \tag{6.6}$$

Enthalpy is addressed in geofluid research, for instance, when describing a cooling process, whereby a fluid is pushed through an orifice into a chamber at a lower pressure. If the pressures on two sides of the orifice remain constant, the process will be *isenthalpic*, which means that the enthalpy of the system does not change.

If heat is given off, a chemical reaction or a process is said to be *exothermic*, and the corresponding change in enthalpy is negative. In contrast, energy must be added from the surroundings to the system to initiate an *endothermic* reaction or process, in which the change of enthalpy is positive. The determination of the enthalpy change is the first step in understanding the reaction conditions. The direction of the reaction is then defined by **the second law of thermodynamics**.

The fundamentals of the second law were outlined in 1850 by the German physicist and mathematician Rudolf Clausius (1822–1888). The main idea behind the law is that heat flows from hot to cold bodies until both objects have the same temperature, and the one-directional heat transfer from warmer to colder objects cannot be reversed. Clausius also introduced the concept of **entropy**—a measure of a system disorder (randomness). Lord Kelvin (1824–1907) hypothesized about "the thermal death of Universe" as a final stage of evolution with a uniform temperature and maximum entropy, from which it would not be possible to extract any work. On a local scale, this state pertains to a thermodynamic equilibrium. A system is in stable equilibrium when no thermodynamic processes are possible.

The third law of thermodynamics states that the entropy of a pure, perfectly crystalline substance is close to zero at the absolute zero temperature. Hence, entropy is the additional state function defined by thermodynamic laws. The Austrian physicist Ludwig Boltzmann (1844–1906) showed that entropy can be measured in absolute numbers and defined in terms of the number of possible arrangements of particles constituting the system

$$S_{sys} = k_B \ln \Omega, \tag{6.7}$$

where k_B is Boltzmann's constant and Ω is the number of possible arrangements.

When an exothermic process takes place in a system, the heat transferred from the system to the surroundings will enhance the motion of molecules, thereby increasing the disorder and the entropy of the surroundings. The opposite is true for an endothermic reaction, when heat from the surroundings is absorbed into the system. For a constant pressure process, the heat change is equal to the enthalpy change of the system. For a given amount of heat, however, the change in entropy also depends on the temperature difference between the system and the surroundings. If the temperature of the surroundings is high, the amount of heat released by an exothermic reaction will have little effect on the entropy change of the surroundings. However, at a low temperature of the surroundings, the addition of the same amount of heat would cause a more drastic increase in the entropy. Hence, there is an inverse relationship between the change in entropy ΔS of the surroundings and the change in enthalpy of the system:

$$\Delta S_{\text{surr}} = -\Delta H_{\text{sys}}/T \tag{6.8}$$

The second law tells us that a spontaneous reaction must lead to an increase in the entropy of the universe

$$\Delta S_{\text{univ}} = \Delta S_{\text{surr}} + \Delta S_{\text{sys}} > 0 \tag{6.9}$$

$$\Delta S_{\text{univ}} = -\Delta H_{\text{sys}}/T + \Delta S_{\text{sys}} > 0 \tag{6.10}$$

Multiplying both sides of the preceding equation by T yields

$$T\Delta \text{S}_{\text{univ}} = -\Delta H_{\text{sys}} + T\Delta S_{\text{sys}} > 0 \tag{6.11}$$

and

$$-T\Delta \text{S}_{\text{univ}} = \Delta H_{\text{sys}} - T\Delta S_{\text{sys}} < 0 \tag{6.12}$$

Now we have a criterion for a spontaneous reaction that is expressed only in terms of the properties of the system. The new variable corresponds to the amount of free energy, which can be transformed into other types of energy

$$G = H - TS \equiv U + PV - TS \tag{6.13}$$

G must be less than zero for a spontaneous reaction to occur. Because G contains the units of energy, it is also a state function, which is given a special name—**Gibbs free energy**—in honor of the famous American scientist Josiah Willard Gibbs (1839–1903), who made important theoretical contributions in various fields of chemistry, physics, and mathematics. Similarly, **Helmholtz energy**, named in honor of the German physicist Hermann von Helmholtz (1821–1894), is also a state property expressing the amount of free energy available at a constant volume

$$A = U - TS \tag{6.14}$$

All the derived state functions are extensive variables, which means that their magnitude depends on the amount of substance. They can be transformed into intensive variables by dividing by number of

Table 6.1 Overview of State Functions and their SI Units

State Function	Extensive		Intensive	
	Abbreviation	SI Unit	Abbreviation	SI Unit
Energy	U	J [kg m s^{-2}]	$\overline{U}$	J mol^{-1}
Entropy	S	J K^{-1}	$\overline{S}$	J K^{-1} mol^{-1}
Enthalpy	H	J	$\overline{H}$	J mol^{-1}
Gibbs energy	G	J	$\overline{G}$	J mol^{-1}
Helmholtz energy	A	J	$\overline{A}$	J mol^{-1}
Volume	V	m^3	$\overline{V}$	J bar^{-1}

moles, and this is signified by writing a bar sign over the variable abbreviation (Table 6.1). Other state functions (temperature, pressure, density) are intensive variables independent of the amount of substance in a system.

6.2 FREE ENERGY OF REACTION

The free energy change of a reaction at constant temperature is given by

$$\Delta\overline{G} = \Delta\overline{H} - T\Delta\overline{S} \tag{6.15}$$

Except for entropy, for which a zero value is defined by the third law, other state properties involve internal energies or enthalpies with undefined zero values; therefore, they cannot be expressed as absolute values. This is, however, not a problem for thermodynamics, which deals only with relative changes of state properties accompanying a chemical reaction or a process.

It is, however, convenient to think of substances having absolute values of their state properties. By convention, an arbitrary absolute value of zero is assigned to properties of elements in their stable form at 298.15 K (25 °C) and 1 bar (10^5 Pa). With this arbitrary base level it is now possible to define state properties of a substance in terms of the property change accompanying the formation of that substance from elements at 298.15 K and 10^5 Pa.

To determine the Gibbs free energy change of a reaction at certain P and T, we need first to define the free energy of formation of reactants and products—$\Delta\overline{G}_\mathrm{f}^0$—in standard conditions, and to calculate the *standard free energy change of the reaction*—$\Delta\overline{G}_\mathrm{r}^0$. Considering the reaction between solid phases A, B, C, and D

$$a\mathrm{A} + b\mathrm{B} \rightarrow c\mathrm{C} + d\mathrm{D} \tag{6.16}$$

where a, b, c, and d are numbers of moles

$$\Delta\overline{G}_\mathrm{r}^0 = \left[c\Delta\overline{G}_\mathrm{f,C}^0 + d\Delta\overline{G}_\mathrm{f,D}^0\right] - \left[a\Delta\overline{G}_\mathrm{f,A}^0 + b\Delta\overline{G}_\mathrm{f,B}^0\right] \tag{6.17}$$

and for each solid phase

$$\Delta\overline{G}_\mathrm{f}^0 = \Delta\overline{H}_\mathrm{f}^0 - T\Delta\overline{S}_\mathrm{f}^0 \tag{6.18}$$

As mentioned, the standard free energy in the formation of a substance is the free energy change that occurs when 1 mol of the compound is synthesized from its elements in their standard states. However, when a chemical reaction starts, the standard conditions no longer exist because neither reactant nor product remains at 1 mol concentration. Hence $\Delta\overline{G}_\mathrm{r}$ must be calculated as

$$\Delta\overline{G}_\mathrm{r} = \Delta G_\mathrm{r}^0 + RT \ln Q_\mathrm{r} \tag{6.19}$$

where Q_r is a reaction quotient, which changes as the reaction progresses. For the progressing reaction at standard temperature, the term $\Delta\overline{G}_\mathrm{r}^0$ is fixed, but the term $RT \ln Q$ is not. However, when an equilibrium is reached, $\Delta\overline{G}_\mathrm{r}$ becomes zero and Q_r equals an equilibrium constant K of the reaction. Hence, we can write

$$\Delta\overline{G}_r = 0 = \Delta G_r^0 + RT \ln K \tag{6.20}$$

or

$$\Delta G_r^0 = -RT \ln K \tag{6.21}$$

The equilibrium constant K for Reaction (6.16) corresponds to

$$K = \frac{a_C^c \times a_D^d}{a_A^a \times a_B^b}, \tag{6.22}$$

where a_i^j is activity of the compound i raised to a power of j, which is the number of moles of that compound.

Values for standard entropy and enthalpy are tabulated for virtually all minerals and gases of interest to geologists (e.g., Robie et al., 1979; Belonoshko et al., 1992; Holland and Powell, 1990, 1998, 2011). From these data, the free energy change of formation of a solid can be obtained for any P and T with the help of an additional state property—**heat capacity** (C)—namely the amount of heat necessary to raise the temperature of a unit mass (usually 1 mol) of the substance by 1 K. The heat measured at a constant volume is called the *isochoric heat capacity*—$\overline{C}_V$, whereas that measured at a constant pressure is the *isobaric heat capacity*, or $\overline{C}_P$

$$\overline{C}_V = (dQ/dT)_V \tag{6.23}$$

$$\overline{C}_P = (dQ/dT)_P \tag{6.24}$$

Thermodynamic calculations prefer molar heats measured at constant pressure that are higher than those measured at constant volume. Both kinds of heat capacities are correlated by the equation

$$\overline{C}_P - \overline{C}_V = R \tag{6.25}$$

where R is the gas constant.

Various empirical fits of isobaric heat capacities of minerals against temperature are tabulated in thermodynamic databases. For instance, Robie and Hemingway (1995) used a five-term polynomial of Haas and Fisher (1976)

$$\Delta\overline{C}_P = a + bT + cT^{-2} + dT^{-0.5} + eT^2 \tag{6.26}$$

where a–e are empirical constants. Though the equation reproduces experimental data with a high precision, it yields inappropriately high C_P values at high temperatures owing to the quadratic term eT^2. To overcome this problem, Berman and Brown (1985) proposed the reduced expression

$$\Delta\overline{C}_P = a + bT^{-2} + cT^{-0.5} + dT^{-3} \tag{6.27}$$

which can also be extrapolated to higher temperatures. Holland (1981) and Holland and Powell (1998) recommended another reduced polynomial form

$$\Delta\overline{C}_P = a + bT + cT^{-2} + dT^{-0.5} \tag{6.28}$$

and Chatterjee (1991) used the seven-term polynomial

$$\Delta\overline{C}_P = \Delta a + \Delta bT + \Delta cT^{-2} + \Delta dT^2 + \Delta eT^{-1/2} + \Delta fT^{-1} + \Delta gT^{-3} \tag{6.29}$$

The isobaric heat capacity is related to enthalpy by the equation

$$\Delta\overline{H}_{\mathrm{f},T} = \Delta\overline{H}_{\mathrm{f}}^{0} + \int_{298}^{T} \Delta\overline{C}_{P}\mathrm{d}T \tag{6.30}$$

and the relationship of the isobaric heat capacity and entropy can be expressed as

$$\Delta\overline{S}_{\mathrm{f},T} = \Delta S_{298} + \int_{298}^{T} \left[\frac{\Delta\overline{C}_{P}}{T}\right]\mathrm{d}T \tag{6.31}$$

Because

$$\overline{G}_{T,P} = \overline{G}_{T}^{0} + \int_{1}^{P} \overline{V}\mathrm{d}P \tag{6.32}$$

the combination of Equations (6.15), (6.31), and (6.32) reads

$$\Delta\overline{G}_{\mathrm{f},T,P} = \Delta\overline{H}_{\mathrm{f}}^{0} + \int_{298}^{T} \Delta\overline{C}_{P}\mathrm{d}T - T\left(\Delta S_{298}^{0} + \int_{298}^{T} \left(\frac{\Delta\overline{C}_{P}}{T}\right)\mathrm{d}T\right) + \int_{1}^{P} \Delta\overline{V}_{\mathrm{f}}\mathrm{d}P \tag{6.33}$$

Substituting the reduced polynomial expression for $\Delta\overline{C}_{P}$ (Equation (6.28)) and integration gives the general equation of the Gibbs free energy change of reactions between solid phases, assuming their volumes do not vary significantly with changing pressure

$$\begin{aligned}\Delta G_{\mathrm{f},T,P} = {} & \Delta H_{\mathrm{f}}^{0} - T\Delta S^{0} + a(T-298) + \frac{b}{2}(T^{2}-298^{2}) - c\left(\frac{1}{T}-\frac{1}{298}\right) + 2d\left(T^{0.5}-298^{0.5}\right) \\ & - T\left[a\ln\left(\frac{T}{298}\right) + b(T-298) - \frac{c}{2}\left(\frac{1}{T^{2}}-\frac{1}{298^{2}}\right) - 2d\left(\frac{1}{T^{0.5}}-\frac{1}{298^{0.5}}\right)\right] + \int_{1}^{P} \Delta\overline{V}_{\mathrm{f}}\mathrm{d}P\end{aligned} \tag{6.34}$$

The last term of this equation has not been integrated. Because the compressibility of minerals is small, their volume changes are negligible, and if pressure remains constant, the term becomes zero. The equation, however, cannot be used for the chemical reactions involving a fluid phase, because in that case ΔV_{r} and ΔS_{r} are strongly dependent on P and T.

6.3 GAS FUGACITY

Fugacity—f [bar]—is an additional intensive state function introduced by Lewis (1908) to express the idealized partial pressure of a gas in a nonideal gaseous mixture. The concept of fugacity was accepted after publication of the textbook written by Lewis and Randall (1923). Eugster (1957) was the first to introduce the concept of gas fugacity in petrology. He experimented with the externally controlled oxygen fugacity. Eugster observed the influence of the oxidation potential on reactions giving rise to specific mineral assemblages, and designated the oxidation potential of oxygen as the partial oxygen pressure—$p(O_2)$—renamed later as the oxygen fugacity (Eugster and Wones, 1962).

Free-energy change of a gas at high pressures can be derived easily from the functional dependence between the Gibbs free energy and volume expressed by Equation (6.32). For an ideal gas, integration results in

$$\overline{G}_{T,P} = \overline{G}_T^0 + RT \ln P = \overline{G}_T^0 + RT \ln \frac{f}{f^0} \tag{6.35}$$

where $\overline{G}_T$ is the molar Gibbs free energy of pure gas at temperature T, f^0 is the fugacity of the pure gas species in standard conditions, and f is the fugacity of the gas species at given P and T. After defining f^* as

$$f^* = \frac{f}{f^0}$$

Equation (6.35) simplifies to

$$\overline{G}_{T,P} = \overline{G}_T^0 + RT \ln f^* \tag{6.36}$$

f^* is a dimensionless variable designated in literature as **activity**—a. The activity of a pure nonideal fluid in optional P–T conditions is equal to the ratio of the fugacity of the pure fluid to its fugacity in the standard state. If a gas under pressure of 1 bar is chosen as the standard state, the fugacity of pure nonideal gas in optional P–T conditions is equal to its activity.

Under high pressures, however, gases do not behave ideally, so they will not obey conditions of Equation (6.35). To compensate for the nonideal behavior, a correction function called the **fugacity coefficient**, ϕ, was introduced. For a pure gas

$$\phi = \frac{f}{P} \tag{6.37}$$

Because f and P (total pressure) are in bars, the fugacity coefficient is a dimensionless variable dependent on pressure and temperature.

Geofluids are in most cases mixtures of gases, where the concentration of a fluid species i is expressed as the molar fraction X_i ($\Sigma X_i = 1$). The molar Gibbs free energy of the ith species is similar to that in Equation (6.35):

$$\overline{G}_{T,P,i} = \overline{G}_{T,i}^0 + RT \ln \frac{f_i''}{f_i^0} \tag{6.38}$$

where f_i'' is the fugacity of the ith species in the mixture. Now the activity of the ith species in the mixture becomes important

$$a_i \equiv \frac{f_i''}{f_i} \tag{6.39}$$

Activity coefficient γ_i of the ith species in the gaseous mixture is

$$\gamma_i = a_i / X_i \tag{6.40}$$

Combination of Equations (6.39) and (6.40) yields

$$\gamma_i X_i = \frac{f_i''}{f_i} \tag{6.41}$$

and substituting the fugacity f_i by the fugacity coefficient ϕ (Equation (6.37)) one gets the basic relationship between the fugacity coefficient of gas species in the gaseous mixture (ϕ_i''), temperature T, pressure P, and molar concentration X_i of the gas

$$\frac{f_i''}{PX_i} = \phi_i \gamma_i \equiv \phi_i'' \quad (6.42)$$

or

$$\frac{f_i''}{p_i} \equiv \phi_i'' \quad (6.43)$$

where p_i stands for the partial pressure of the ith gas species, and P is the total pressure. The fugacity of a pure gas species, or that in a gaseous mixture, can be expressed in pressure units (bar, Pa). However, as these values sometimes approach infinitesimally small numbers, an expression in the form of a decadic logarithm ($\log f_i$) is preferred for practical reasons.

For a reaction between solid and gas phases, Equation (6.32) becomes

$$\Delta\overline{G}_{T,P} = \Delta\overline{G}_T^0 + \int_1^P \Delta\overline{V} \mathrm{d}P + nRT \ln f \quad (6.44)$$

where the $\Delta\overline{V}$ volume integral refers to the volume change of solid phases, and the term $RT \ln f$ accounts for the free-energy change due to volume expansion of gas. For a reaction between gases, Equation (6.44) becomes

$$\Delta\overline{G}_{T,P} = \Delta\overline{G}_T^0 + nRT \ln f \quad (6.45)$$

or

$$\Delta\overline{G}_{T,P} = \Delta\overline{G}_T^0 + RT \ln K \quad (6.46)$$

Equation (6.44) can be solved in two ways. If data on compressibility $\alpha(T)$ and expandability $\beta(P)$ of minerals are not available, the volume integral of solid phases can be approximated by

$$\int_1^P \Delta\overline{V} \mathrm{d}P = \Delta\overline{V}_{298}^0 (P-1) \quad (6.47)$$

where P stands for the pressure in bars, and the term $\Delta\overline{V}$ represents the volume difference between solid phases in standard conditions. If the compressibility and expandability data are available, the volume integral can be expressed as (Chatterjee, 1991)

$$\int_1^P \Delta\overline{V} \mathrm{d}P = \begin{bmatrix} \Delta\overline{V}_{298}^0 + \Delta\left(\alpha_0 V_{298}^0\right)(T-298) + \frac{1}{2}\Delta\left(\alpha_1 V_{298}^0\right)\left(T^2 - 298^2\right) - \\ (P-1)\left\{\frac{1}{2}\Delta\left(\beta_0 V_{298}^0\right) + \frac{1}{6}\Delta\left(\beta_1 V_{298}^0\right)(P+2)\right\} \end{bmatrix}(P-1) \quad (6.48)$$

where α_0 and α_1 are coefficients of the isobaric thermal expansion, β_0 and β_1 are coefficients of isothermal compressibility of a solid phase, V_{298}^0 is the volume of the solid phase in the standard state, and Δ is the change of the corresponding term for the whole reaction.

6.4 CALCULATING THE FLUID COMPOSITION FROM THERMODYNAMIC DATA

H_2O, H_2, CH_4, CO_2, CO, O_2, COS, SO_2, N_2, noble gases, and gaseous hydrocarbons are stable gas species in the Earth's crust and mantle. The composition of metamorphic and magmatic fluids in the crust is controlled by chemical reactions in the C–O–H system with graphite. CO_2 is the dominant reaction product at high, CH_4 at low, and H_2O at intermediate oxygen fugacities (Holloway and Reese, 1974; Ohmoto and Kerrick, 1977; Holloway, 1987). Higher concentrations (>0.1 mol.%) of sulfur-bearing gases, such as SO_2 or H_2S, are exceptional, and COS has not been detected up to now in measurable amounts in fluid inclusions.

Compared to the crust, the Earth's mantle is strongly oxidized and dehydrated. For this reason, CO_2 becomes the dominant carbon species accompanied by small amounts of N_2 (Pasteris, 1987; Andersen et al., 1995), CO (Bergman and Dubessy, 1984; Huraiová et al., 1991), H_2S (Murck et al., 1978), H_2O (Andersen et al., 1984), and noble gases (Porcelli et al., 1986; Villa et al., 1991; Rocholl et al., 1996; Matsumo et al., 1996, 1998).

Oxygen concentrations in fluid inclusions are extremely low and thus undetectable, except for the fluids trapped in minerals around and within the Oklo uranium deposit (Gabon), where Dubessy et al. (1988) identified molecular hydrogen and oxygen formed by radiolysis of water to H_2 and O_2 during spontaneous decay of radioactive elements.

In spite of the small quantities, the oxygen concentration is an important parameter of geofluids, because it controls their oxidation state. Absolute concentrations of oxygen can be calculated from the thermodynamic parameters and equilibrium constants of oxidation–reduction (redox) reactions discussed in the following chapters.

6.4.1 C–O SYSTEM

The oxidation state of the upper mantle is controlled by the reaction (e.g., Bergman and Dubessy, 1984)

$$CO + \tfrac{1}{2}O_2 \leftrightarrow CO_2 \tag{6.49}$$

If the CO_2/CO ratio is known, for instance, by Raman spectrometric analysis of fluid inclusions in mantle minerals, oxygen fugacity can be calculated from the equilibrium constant of Reaction (6.49)

$$K = \frac{f''_{CO_2}}{f''_{CO}\sqrt{f''_{O_2}}} \tag{6.50}$$

It is worth remembering that the fugacities of the reaction products are always placed in the nominator, whereas those of reactants are in the denominator. All fugacities are exponentiated by the number of moles of the gas species that is by unity in the case of CO_2 and CO, and by 0.5 in the case of O_2. Substituting with fugacity coefficients (Equation (6.42)) we get

$$K = \frac{\varphi''_{CO_2}PX_{CO_2}}{\varphi''_{CO}PX_{CO}\sqrt{f''_{O_2}}} \tag{6.51}$$

where X_{CO_2} and X_{CO} are CO_2 and CO molar fractions, respectively. Rearrangement of Equation (6.51) results in

$$f''_{O_2} = \left(\frac{\varphi''_{CO_2} X_{CO_2}}{K \varphi''_{CO} X_{CO}}\right)^2 \tag{6.52}$$

To calculate the oxygen fugacity in a gaseous mixture, one needs to know the relationship between the equilibrium constant of Reaction (6.49) and temperature (note that pressure is canceled out in this type of reaction). The T–K relationship for Reaction (6.49) was derived by Ohmoto and Kerrick (1977)

$$\log K = \frac{14,751}{T} - 4.535 \tag{6.53}$$

With experimentally determined fugacity coefficients of pure gas species (e.g., Belonoshko et al., 1992) and the known composition of the gaseous mixture f''_{O_2} can now be easily calculated for optional P and T (Problems 6.1 and 6.2).

The equilibrium constant of a reaction involving fluid species can also be derived from fundamental thermodynamic data. Standard molar Gibbs free energy change of Reaction (6.49) can be expressed using the change of standard enthalpy (ΔH^0_T) and that of standard entropy (ΔS^0_T). In equilibrium conditions and 1 bar pressure

$$\Delta G^0_{r,T} = \Delta H^0_{r,T} - T\Delta S^0_{r,T} \tag{6.54}$$

Changes to the standard entropy and the standard enthalpy are calculated from the mass balance of the reaction

$$\Delta H^0_{r,T} = cH^0_{T,CO_2} - aH^0_{T,CO} - bH^0_{T,O_2} \tag{6.55}$$

$$\Delta S^0_{r,T} = cS^0_{T,CO_2} - aS^0_{T,CO} - bS^0_{T,O_2} \tag{6.56}$$

where a, b, and c refer to numbers of moles of CO, O_2, and CO_2, respectively. Reorganization of Equation (6.46) and conversion of natural logarithm to decadic logarithm

$$\Delta \overline{G}_{r,T} = 0 = \Delta \overline{G}^0_{r,T} + 2.30259\, RT \log K \tag{6.57}$$

yields the basic expression for the calculation of the equilibrium constant of a reaction between gases

$$\log K = \frac{-\Delta G^0_{r,T}}{2.30259RT} \tag{6.58}$$

Substituting with Equation (6.54) results in

$$\log K = \frac{-\Delta H^0_{r,T} + T\Delta S^0_{r,T}}{2.30259RT} \tag{6.59}$$

and its rearrangement yields the final equation

$$\log K = \frac{1}{T}\left(\frac{-\Delta H^0_{r,T}}{2.30259R}\right) + \left(\frac{\Delta S^0_{r,T}}{2.30259R}\right) \tag{6.60}$$

PROBLEM 6.1 CALCULATING EQUILIBRIUM CONSTANT OF A REACTION

Calculate the equilibrium constant of the reaction $CO + ½O_2 \leftrightarrow CO_2$ at 1 bar and 1447 K using thermodynamic data for individual gas species and compare it with this derived by Ohmoto and Kerrick (Equation (6.53)).

Solution

Input data generated by the SUPERFLUID computer program (Belonoshko et al., 1992) are listed in Table 6.2.

Calculate changes of standard molar Gibbs energy of the reaction using Equations (6.54)–(6.56)

$$\Delta H_T^0 = -334{,}889.91 - (-73{,}548.45) - 0.5 \times 38{,}664.69 = -280{,}673.81\,\mathrm{J\,mol^{-1}}$$

$$\Delta S_T^0 = 289.9405 - 247.0301 - 0.5 \times 256.635 = -85.4071\,\mathrm{J K^{-1} mol^{-1}}$$

$$\Delta G_T^0 = -280{,}673.81 - 1447 \times (-85.4071) = -157{,}089.73\,\mathrm{J}$$

Substituting ΔG_T^0 into Equation (6.58) yields

$$\log K = 157{,}089.73/(2.30259 \times 8.3143 \times 1447) = 5.6707$$

$$K = 4.66849 \times 10^5$$

After substituting ΔH_T^0 and ΔS_T^0 into Equation (6.60) we get

$$\log K = \frac{14{,}661}{T} - 4.4612 \qquad (6.61)$$

The derived equation is similar to that proposed by Ohmoto and Kerrick (1977). Differences between Equations (6.53) and (6.61) are due to various calibrations of the standard state referred to 1 atm or 1 bar, respectively.

Table 6.2 Input Data for Calculation of Equilibrium Constants at 1447 K

Species	H_T^0 (J mol^{-1})	S_T^0 (J K^{-1} mol^{-1})	ϕ_i''
CO_2	−334,889.91	289.9405	3.995563
CO	−73,548.45	247.0301	3.480044
O_2	38,664.69	256.6350	2.792742
H_2O	−196,186.52	248.7957	1.346790
H_2	34,577.62	177.5855	1.835063
CH_4	−1216.38	276.5741	4.007620

PROBLEM 6.2 CALCULATING OXYGEN FUGACITY FROM FLUID COMPOSITION

Calculate the oxygen fugacity in a fluid inclusion trapped in quartz from the pincinite KP-31 at 1447 K and 5.5 kbar. The fluid inclusion composition measured by Raman spectroscopy is as follows: CO_2—95.43; CO—1.78; N_2—1.94; H_2—0.85 mol.%.

Solution

Substituting fugacity coefficients for CO_2 and CO from Table 6.2 combined with K from the preceding exercise and substituting the CO_2 and CO molar concentrations in Equation (6.52) yields $f''_{O_2} = 1.7149 \times 10^{-8} (\log f''_{O_2} = -7.8)$. The same calculation with slightly different fluid composition calculated using other Raman scattering cross sections yields the $\log f''_{O2}$ value equal to −7.6. Hence, the estimated average $\log f''_{O_2}$ value corresponds to -7.7 ± 0.1.

6.4.2 GRAPHITE–C–O–H SYSTEM

The procedure employed for the calculation of the fluid speciation in the system C–O can also be applied to the H_2O–CO_2–CH_4 fluids typical of the continental crust. The oxidation state of the mixed-volatile crustal fluid is controlled by the reaction

$$CH_4 + 2O_2 \leftrightarrow CO_2 + 2H_2O \tag{6.62}$$

with the equilibrium constant

$$K = \frac{f''_{CO_2} f''^2_{H_2O}}{f''_{CH_4} f''^2_{O_2}} \tag{6.63}$$

The temperature dependence of the equilibrium constant for the Reaction (6.63) was derived by Ohmoto and Kerrick (1977)

$$\log K = \frac{41{,}997}{T} + 0.719 \log T - 2.404 \tag{6.64}$$

The reaction constant can be used to calculate oxygen fugacity at temperatures greater than 350 °C. At lower temperatures, equilibrium between CO_2 and CH_4 cannot be achieved, and composition of the gaseous mixture has no thermodynamic meaning (Dubessy et al., 1989).

The composition of crustal fluids in metapelitic rocks is buffered by additional graphite-consuming and water-forming reactions (Holloway, 1987)

$$C + O_2 \leftrightarrow CO_2 \quad K_1 = f_{CO_2} / (aC \times f_{O_2}) \tag{6.65}$$

$$C + \tfrac{1}{2}O_2 \leftrightarrow CO \quad K_2 = f_{CO} / \left(aC \times f_{O_2}^{1/2}\right) \tag{6.66}$$

$$C + 2H_2 \leftrightarrow CH_4 \quad K_3 = f_{CH_4} / \left(aC \times f_{H_2}^2\right) \tag{6.67}$$

$$H_2 + \tfrac{1}{2}O_2 \leftrightarrow H_2O \quad K_4 = f_{H_2O} / \left(f_{H_2} \times f_{O_2}^{1/2}\right) \tag{6.68}$$

Equilibrium constants for the reactions can be expressed as

$$K_1 = f_{CO_2}/(a_C \times f_{O_2}) \tag{6.69}$$

$$K_2 = f_{CO}/\left(a_C \times f_{O_2}^{1/2}\right) \tag{6.70}$$

$$K_3 = f_{CH_4}/\left(a_C \times f_{H_2}^2\right) \tag{6.71}$$

$$K_4 = f_{H_2O}/\left(f_{H_2} \times f_{O_2}^{1/2}\right) \tag{6.72}$$

The sum of mole fractions of pure gas species must be equal to unity. The infinitesimal amount of oxygen is omitted from further calculations, hence

$$1 = X_{CO_2} + X_{CO} + X_{CH_4} + X_{H_2} + X_{H_2O} \tag{6.73}$$

Substituting fugacity coefficients defined by Equation (6.42) we get

$$P = (f_{CO_2}/\phi_{CO_2}) + (f_{CO}/\phi_{CO}) + (f_{CH_4}/\phi_{CH_4}) + (f_{H_2}/\phi_{H_2}) + (f_{H_2O}/\phi_{H_2O}) \tag{6.74}$$

and further rearrangement of Equations (6.65)–(6.68) with the graphite activity set to unity (fully ordered graphite) yields the quadratic function with two unknowns (f_{O_2}, f_{H_2})

$$0 = \left(K_3\phi_{H_2}^2/\phi_{CH_4}\right)X_{H_2}^2 + \left(1 + \left[K_4\phi_{H_2}f_{O_2}^{1/2}\right]/\phi_{H_2O}\right)X_{H_2} + \left(\left[K_1 f_{O_2}/\phi_{CO_2}\right] + \left[K_2 f_{O_2}^{1/2}/\phi_{CO}\right] - 1\right) \tag{6.75}$$

which can be solved for a real root at a given f_{O_2} to obtain the mole fraction and the fugacity of hydrogen in the gaseous mixture (Equation (6.42)). Substitution of the calculated hydrogen fugacity in Equations (6.70) and (6.71) yields f_{CH_4} and f_{H_2O}, which can be inserted into Equation (6.42) to calculate mole fractions of the respective components. Substituting f_{O_2} in Equations (6.65) and (6.66) yields CO_2 and CO fugacities, respectively.

Graphite is stable in given $P - T - f_{O_2}$ conditions, if the sum of the calculated mole fractions of pure species is equal to unity (Equation (6.73)). If this prerequisite is fulfilled, the composition of the fluid can be expressed in terms of atomic percentages according to the following scheme

$$\mathrm{C} = X_{CH_4} + X_{CO} + X_{CO_2} \tag{6.76}$$

$$\mathrm{O} = 2X_{CO_2} + X_{CO} + X_{H_2O} \tag{6.77}$$

$$\mathrm{H} = 4X_{CH_4} + 2X_{H_2} + X_{H_2O} \tag{6.78}$$

The obtained C, O, and H values can be normalized to unity and graphically displayed in a triangular diagram.

The unknown oxygen fugacity in Equation (6.75) must be fixed by another external buffer. For instance, Skippen and Marshall (1991) used equilibrium among biotite, magnetite, and K-feldspar to constrain f_{O_2} and f_{H_2}. Another approach was proposed by Connolly and Cesare (1993) who fixed the H/O ratio in crustal C–O–H fluids to 2, assuming that water is the main fluid species produced during the metamorphism of metapelites.

PROBLEM 6.3 CALCULATING WATER AND METHANE CONTENTS IN MIXED-VOLATILE FLUID

Calculate water and methane concentrations in equilibrium with the CO_2–CO–H_2 fluid coexisting with pincinite KP-31. Use the inclusion fluid composition measured by Raman spectroscopy and the oxygen fugacity calculated from CO_2–CO equilibrium at 1447 K and 5.5 kbar (Problem 6.2).

Solution

Calculate the equilibrium constant of the reaction $H_2 + ½O_2 \leftrightarrow H_2O$. Changes of the standard molar Gibbs energy of the reaction correspond to

$$\Delta H_T^0 = -196,186.52 - 34,577.62 - 0.5 \times 38,664.69 = -250,096.49\,\mathrm{J\,mol^{-1}}$$

$$\Delta S_T^0 = 248.7957 - 177.5855 - 0.5 \times 256.635 = -57.11\,\mathrm{J\,K^{-1}\,mol^{-1}}$$

$$\Delta G_T^0 = -250,096.49 - 1447 \times (-57.11) = -167,462.2\,\mathrm{J}$$

$$\log K = -(-157,089.73)/(2.30259 \times 8.3143 \times 1447) = 6.0451$$

$$K = 1.10952 \times 10^6$$

Equilibrium constant of the reaction $H_2 + ½O_2 \leftrightarrow H_2O$ can be written as

$$K = \frac{f''_{H_2O}}{f''_{H_2}\sqrt{f''_{O_2}}} = \frac{\phi''_{H_2O} P X_{H_2O}}{\phi''_{H_2} P X_{H_2}\sqrt{f''_{O_2}}}$$

After rearrangement we get

$$X_{H_2O} = \frac{K \phi''_{H_2} X_{H_2}\sqrt{f''_{O_2}}}{\phi''_{H_2O}}$$

Substitution of fugacity coefficients for pure gas species (Table 6.2) K and f_{O_2} values yields

$$X_{H_2O} = \frac{1,109,520 \times 1.819413 \times 0.0084 \times \sqrt{1.7149 \times 10^{-8}}}{1.336483} = 1.66$$

Calculate the equilibrium CH_4 content from Reaction (6.62). Calculate the equilibrium constant either from thermodynamic data in Table 6.2 ($\log K = 28.87989$) or from Equation (6.64) ($\log K = 28.89187$). Rearrangement of Equation (6.63) yields

$$X_{CH_4} = \frac{\phi''_{CO_2} X_{CO_2}\left(\phi''_{H_2O} P X_{H_2O}\right)^2}{K \phi''_{CH_4}\left(f''_{O_2}\right)^2} = 2.4 \times 10^{-6}$$

After normalization to 100, we get the final fluid composition (mol.%): H_2O—62.4; CO_2—35.89; H_2—0.32; N_2—0.73; CO—0.67; CH_4—0.0002.

6.4.3 C–O–H–S SYSTEM

Dubessy et al. (1989) proposed additional reactions for constraining the sulfur fugacity of crustal fluids in equilibrium with graphite. If H_2S is present in the gaseous mixture, the sulfur fugacity is controlled by the reaction

$$2H_2 + S_2 \leftrightarrow 2H_2S \tag{6.79}$$

whereas f_{H_2} is controlled by the reaction (Hollister and Burruss, 1976)

$$CH_4 + 2H_2O \leftrightarrow CO_2 + 4H_2 \tag{6.80}$$

Using the CO_2/CH_4 ratio determined by microthermometry, Raman spectroscopy, or mass spectroscopy, the hydrogen fugacity of geothermal fluids and volcanic emanations can be estimated from plots presented by Giggenbach (1980).

Composite C–O–H–S fluid speciation can be calculated from mole fractions of gas species

$$\begin{aligned} C &= X_{CH_4} + X_{CO_2} + X_{CO} \\ O &= 2X_{CO_2} + X_{H_2O} + X_{CO} + 2X_{O_2} \\ H &= 4X_{CH_4} + X_{H_2O} + 2X_{H_2} + 2X_{H_2S} \\ S &= X_{H_2S} + 2X_{S_2} \end{aligned} \tag{6.81}$$

The f_{O_2} and f_{CO_2} values can be substituted in Equation (6.65) with the equilibrium constant derived by Ohmoto and Kerrick (1977) to check if the calculated fluid composition is in equilibrium with graphite

$$\log K = \frac{20,586}{T} + 0.0421 + 0.028\left(\frac{P-1}{T}\right) \tag{6.82}$$

Graphite activity can be expressed as

$$a_C = f_{CO_2}/K \times f_{O_2} \tag{6.83}$$

Graphite is stable, if the activity is less than unity, and unstable at $a_C > 1$. At temperatures below 400 °C, graphite is not fully ordered, thus resulting in $a_C < 1$. The degree of graphite ordering determined by Raman spectroscopy can be used as an independent temperature constraint.

PROBLEM 6.4 CALCULATING TEMPERATURE OF GRAPHITE PRECIPITATION

Check whether graphite precipitates on cooling of the fluid inclusion with $X_{CO_2} = 0.9543$, $X_{H_2} = 0.0085$, $X_{CO} = 0.0178$, $X_{N_2} = 0.0194$, with the density of 0.846 g cm^{-3} sealed at 1447 K and 5.5 kbar at oxygen fugacity calculated in Problem 6.2, namely 1.7149×10^{-8}.

Solution

The problem can be solved by substituting the gas composition into one of the graphite-buffering reactions (Equations (6.65)–(6.67)) and by calculating the graphite activity along an isochore derived from EOS for gaseous mixtures (Holloway, 1977, 1981, 1987 and sample calculation in Problem 3.1). Using Reaction (6.65) and the corresponding equilibrium constant (Equation (6.82)), one

PROBLEM 6.4 CALCULATING TEMPERATURE OF GRAPHITE PRECIPITATION—CONT'D

can demonstrate that fully ordered graphite ($a_C = 1$) is in equilibrium with the observed anhydrous fluid at ~757 °C and 3770 bar, together with a small amount of methane (2.22 mol.%), whose concentration gradually increases on progressive cooling.

T (K)	P (bar)	f_{O_2}	a_C	X_{CH_4}
1447	5500	1.7149×10^{-8}	0.005	6.3×10^{-6}
1273	4790	2.9140×10^{-11}	0.030	9.7×10^{-5}
1173	4376	3.1684×10^{-13}	0.106	0.0007
1073	3956	1.4831×10^{-15}	0.475	0.0070
1029.5	3770	1.0388×10^{-16}	1.001	0.0222

The preceding calculations also permit formation P–T conditions to be calculated, providing the sulfur and oxygen fugacities are buffered by independent mineral–fluid reactions. Knowledge of the density and composition of the inclusion fluid permits the formation P–T conditions to be fixed along an isochore by the calculated f_{O_2} or f_{S_2} values. The C–O–H equilibria can be calculated also with help of GEOFLUID (Larsen, 1993), C–O–H (Huizenga, 2005), and GFluid (Zhang and Duan, 2010) computer programs.

The mass conservation method outlined by Ohmoto and Kerrick (1977) can be extended to calculate fugacities and speciation of the fluid that is not in equilibrium with graphite. This is particularly useful for mantle fluids. The mass conservation method was applied by Huraiová et al. (2005) to reconstruct the composition of fluid inclusions trapped in glassy orthopyroxene granodiorite (pincinite) ejected as xenoliths in alkali basalts of the Pannonian Basin.

The pressure-independent equilibrium constant of a redox reaction in graphite-unsaturated mixed-volatile fluid can be expressed in terms of the rearranged Equation (6.58) solved for the Gibbs free energy change

$$K = \exp\left(\frac{-\Delta G_T^0}{RT}\right) \tag{6.84}$$

where ΔG_T^0 is the standard Gibbs free energy change of a reaction at 1 bar and temperature T. Standard enthalpies, entropies, and fugacity coefficients (ϕ) of pure gas species can be derived from the equation of state for a mixed-volatile C–H–O–S–N–Ar fluid (Belonoshko and Saxena, 1992) included in the SUPERFLUID computer program (Belonoshko et al., 1992).

The fluid speciation of a carbon-unsaturated fluid is controlled by Reactions (6.49) and (6.62), as well as additional reactions

$$COS \leftrightarrow CO + 0.5S_2 \tag{6.85}$$

$$0.5S_2 + O_2 \leftrightarrow SO_2 \tag{6.86}$$

$$H_2S + 0.5O_2 \leftrightarrow H_2O + 0.5S_2 \tag{6.87}$$

Mole fractions of all species expressed in terms of f_{O_2}, f_{S_2}, X_{H_2O}, and X_{CO_2} can now be combined with $\Sigma X_i = 1$ and rearranged into a quadratic equation

$$AX_{H_2O}^2 + BX_{H_2O} + C = 0 \tag{6.88}$$

where

$$A = k_5 X_{CO_2}$$

$$B = k_6 + k_2 + 1$$

$$C = X_{CO_2} + k_1 X_{CO_2} + k_1 k_3 X_{CO_2} + k_4 - 1$$

and

$$k_1 = \frac{\phi_{CO_2}}{K_1 \phi_{CO} \sqrt{f_{O_2}}}$$

$$k_2 = \frac{\phi_{H_2O}}{K_2 \phi_{H_2} \sqrt{f_{O_2}}}$$

$$k_3 = \frac{\phi_{CO} \sqrt{f_{S_2}}}{K_3 \phi_{COS}}$$

$$k_4 = \frac{K_4 f_{O_2} \sqrt{f_{S_2}}}{P \phi_{SO_2}}$$

$$k_5 = \frac{\phi_{CO_2} \phi_{H_2O}^2 P^2}{K_5 \phi_{CH_4} f_{O_2}^2}$$

$$k_6 = \frac{\phi_{H_2O} \sqrt{f_{S_2}}}{K_6 \phi_{H_2S} \sqrt{f_{O_2}}}$$

ϕ_i is the fugacity coefficient of a pure fluid species, $K_{1\text{–}6}$ are equilibrium constants of the reactions listed above, and P in k_4 is the total pressure (bar).

The real root of Equation (6.88) corresponds to H_2O concentration in the mixed-volatile fluid

$$X_{H_2O} = \frac{-B + \sqrt{B^2 - 4AC}}{2A}$$

and concentrations of remaining fluid species can be obtained according to the following scheme

$$X_{CO} = k_1 X_{CO_2}$$

$$X_{H_2} = k_2 X_{H_2O}$$

$$X_{CH_4} = k_5 X_{H_2O}^2 X_{CO_2}$$

$$X_{H_2S} = k_6 X_{H_2O}$$

$$X_{SO_2} = k_4$$

$$X_{COS} = k_1 k_3 X_{CO_2}$$

It is assumed that the relative sulfur-to-carbon ratio in fluid inclusions does not change during the water leakage and the hydrogen diffusion from and into fluid inclusions. Hence the S/C ratio can serve as an external buffer of the fluid composition at the $P-T-f_{O_2}$ conditions of interest. The number of S and C atoms in the fluid inclusions can be measured directly by Raman spectroscopy

$$nS = X_{H_2S} + X_{COS} + X_{SO_2} \tag{6.89}$$

$$nC = X_{CO_2} + X_{CH_4} + X_{CO}$$

The complete set of mass conservation equations (6.85)–(6.89) can be solved simultaneously using the SOLVER tool in an EXCEL® spreadsheet. It should be noted that the calculation of the complete fluid speciation, including f_{S_2} and f_{O_2}, is possible only if water and sulfur concentrations are constrained by hydrogen and one of the sulfur-bearing compounds (SO_2 or H_2S) determined by the Raman spectrometric analysis of fluid inclusions. If these species are below detection limits, the f_{S_2} and f_{O_2} values must be constrained externally, by the coexisting sulfide and silicate mineral assemblages.

6.4.4 OXYGEN BUFFERS

Eugster (1957) introduced an experimental buffer in geology, because he was upset by the fact that traditional experiments always led to an oxidized magnetite–sanidine assemblage instead of Fe–biotite–annite. Eugster eliminated the problem by enclosing the capsule with reactants within mineral assemblage capable of maintaining the oxygen potential at a lower level needed for the annite synthesis. He called the reactions controlling the oxygen partial pressure the **equilibrium oxygen buffers**. In most of these reactions, pure mineral species reacted to liberate 1 mol of oxygen. The concept of oxidation state was later popularized by Wones (1981).

The practical significance of the oxygen fugacity and experimental buffers can be best explained on the example of iron—the fourth most abundant element on Earth. Iron can be present in three oxidation states: Fe^{3+}, Fe^{2+}, and Fe^0. Low concentrations of oxygen, for instance in the Earth's core or meteorites, permit the iron to be present in the metallic Fe^0 form. Higher f_{O_2} values in the silicate upper mantle stabilize iron as divalent Fe^{2+} cation according to the reaction (Frost, 1991)

$$Fe_2SiO_4(\text{fayalite}) \leftrightarrow 2Fe^0(\text{metallic iron}) + SiO_2(\text{quartz}) + O_2 \tag{6.90}$$

This reaction is called the quartz–iron–fayalite (QIF) oxygen buffer.

At a higher concentration of free oxygen, iron can be present in the form of both Fe^{3+} and Fe^{2+} cations, mainly in magnetite

$$2Fe_3O_4(\text{magnetite}) + 3SiO_2(\text{quartz}) \leftrightarrow 3Fe_2SiO_4(\text{fayalite}) + O_2 \tag{6.91}$$

The reaction is designated as the quartz–magnetite–fayalite (QMF) oxygen buffer.

Finally, at maximum oxygen concentrations, iron can be stable only as trivalent Fe^{3+} cation incorporated dominantly in hematite. The transformation of magnetite to hematite

$$6Fe_2O_3 \leftrightarrow 4Fe_3O_4 + O_2 \tag{6.92}$$

is designated as the magnetite–hematite (MH) oxygen buffer. The QMF and the QIF demarcate the upper and lower limits, respectively, for fayalite stability, whereas the MH defines the upper stability

Table 6.3 Selected Oxygen Buffers

Reaction	Abbreviation	Source
$2Fe_xO \leftrightarrow 2_xFe + O_2$	IW	Frost (1991)
$2x/(4x-3)\ Fe_3O_4 \leftrightarrow 6/(4x-3)\ Fe_xO + O_2$[a]	WM	Frost (1991)
$0.5Fe_3O_4 \leftrightarrow 1.5Fe + O_2$	IM	Frost (1991)
$2CoO \leftrightarrow 2Co + O_2$	CCO	Frost (1991)
$2NiO \leftrightarrow 2Ni + O_2$	NNO	Frost (1991)
$MgCO_3 + MgSiO_3 \leftrightarrow Mg_2SiO_4 + C + O_2$	EMOG	Katsura and Ito (1992)
$2Fe_2O_3 + 4TiO_2 \leftrightarrow 4FeTiO_3 + O_2$	RI	Carmichael and Nicholls (1967)

Abbreviations: I, iron; W, wüstite; M, magnetite; C, cobalt; CO, cobalt oxide; N, nickel; NO, nickel oxide; EMOG, enstatite–magnesite–olivine–graphite; RI, rutile–ilmenite.
[a] $x = 0.947$.

limit for magnetite in the Fe–Si–O system. An overview of additional oxygen buffers frequently used in petrology and mineralogy is given in Table 6.3.

The calculation of oxygen fugacity can be made using a general oxygen-buffering equation

$$bB = cC + dD + nO_2 \tag{6.93}$$

where the reactant—b moles of solid phase B—breaks down to form c moles of phase C, d moles of phase D, and n moles of oxygen.

The equilibrium constant of Reaction (6.93) can be written as

$$K = \frac{a_C^c a_D^d \left(f_{O_2}^*\right)^n}{a_B^b} \tag{6.94}$$

With activities of pure mineral species equal to unity, the reaction produces exactly 1 mol of oxygen ($n = 1$) and reduces to

$$K = f_{O_2}^* = f_{O_2} \tag{6.95}$$

because $f_{O_2}^* = f_{O_2}/f_{O_2}^0$, where $f_{O_2}^0$ is equal to 1 bar. Substituting Equation (6.47) in Equation (6.44) and converting natural logarithm to decadic logarithm we arrive at

$$\log f_{O_2} = \frac{-\Delta G_T^0 - \Delta V_{298,s}^0 (P-1)}{2.30259RT} \tag{6.96}$$

where P is total pressure (bar). Equation (6.96) permits the calculation of oxygen fugacity at optional P and T, using data from an internally consistent thermodynamic database (e.g., Robie et al., 1979; Holland and Powell, 1990, 1998, 2011; Berman, 1988; Fabrichnaya, 1995; Gottschalk, 1997; and others). An example given in Figure 6.1 illustrates the change of fluid composition buffered by the QMF assemblage and graphite along a crustal profile from 5 to 25 km. The diagram explains why CO_2 is the dominant fluid species in granulites, H_2O in amphibolites, and CH_4 in low-grade metapelites.

The amount of free oxygen available during geological processes is very small. For instance, the oxygen fugacity on the level of a middle oxidation grade controlled by the reaction between quartz,

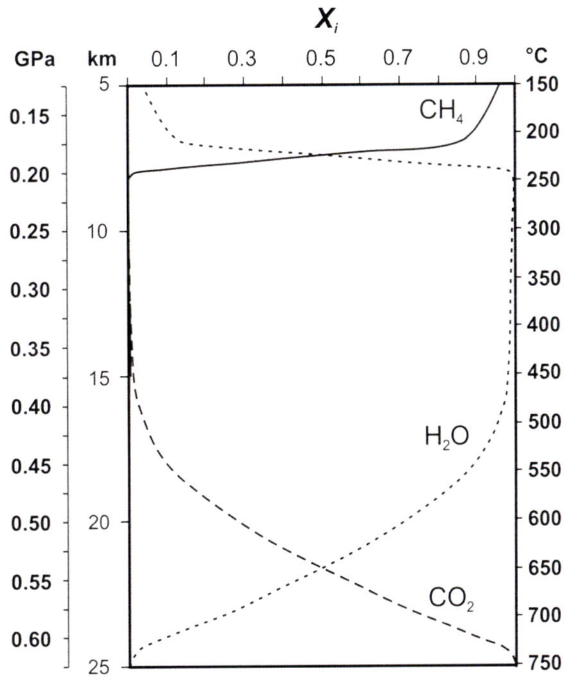

FIGURE 6.1

Idealized crustal profile (Hurai, 2010), showing the composition of a mixed-volatile fluid in equilibrium with graphite, buffered by α-quartz, magnetite, and fayalite (αQMF buffer), assuming a lithostatic load and the average rock density of 2.5 g cm^{-3}. Calculation was made using the thermodynamic database of Holland and Powell (1998) and the PERPLE_X computer program (Connolly, 1990, 1995).

fayalite, and magnetite (αQMF buffer) at the temperature of 500 °C corresponds only to 10^{-22} bar and is an equivalent of one oxygen molecule per 1 m^3 volume. Hence, the gas fugacity is always expressed in log units for practical reasons.

Generally, the gas fugacity increases with temperature. Iso-fugacity slopes of fugacity buffers are roughly similar, because the enthalpy change during oxidation is also similar. Owing to the dominant effect of temperature and small differences among individual buffers, f_{O_2} values are normalized to a selected buffering reaction. This permits comparison of f_{O_2} changes in a greater detail, while also eliminating the influence of temperature. The normalized $\log f_{O_2}$ values are designated as $\Delta\log f_{O_2}$ values.

Another method of calculating oxygen fugacity employs so-called reversal data on the enthalpy and entropy of a reaction ($\Delta H^0_{\langle T\rangle}, \Delta S^0_{\langle T\rangle}$), and volume changes of solid phases of the reaction in standard conditions ($\Delta V^0_{298,s}$). This approach is used when heat capacities of solid phases are unknown. The calculation is based on the simple empirical equation derived by Frost (1991)

$$\log f_{O_2} = -\left[\frac{\Delta H^0_{\langle T\rangle}}{2.3026R}\right]\frac{1}{T} + \frac{\Delta S^0_{\langle T\rangle}}{2.3026R} + \left[\frac{-\Delta V^0_{298,s}}{2.3026R}\right]\frac{(P-1)}{T} \tag{6.97}$$

Table 6.4 Constants of Equation (6.98) for Various Oxygen Fugacity Buffers (Frost, 1991; Katsura and Ito, 1992)

Buffer	*A* (K)	*B*	*C* ($K\ bar^{-1}$)	Temperature Range (°C)
αQIF	−29,435.7	7.391	0.044	150–573
βQIF	−29,520.8	7.492	0.050	573–1200
IW	−27,489.0	6.702	0.055	565–1200
WM	−32,807.0	13.012	0.083	565–1200
IM	−28,690.6	8.13	0.056	300–565
CoCoO	−24,332.6	7.295	0.052	600–1200
αQMF	−26,455.3	10.344	0.092	400–573
βQMF	−25,096.3	8.735	0.110	573–1200
NiNiO	−24,930.0	9.36	0.046	600–1200
MH	−25,497.5	14.330	0.019	300–573
MH	−26,452.6	15.455	0.019	573–682
MH	−25,700.6	14.558	0.019	682–1100
EMOG	−23,872.0	7.61	0.0064	800–1100

where expressions in parentheses are temperature- and pressure-independent constants. Hence, Equation (6.97) can be written in a simplified form:

$$\log f_{O_2} = A/T + B + C(P-1)/T \tag{6.98}$$

where T is the temperature (K), and P is the pressure (bar). The A, B, and C constants of Reaction (6.98) are summarized in Table 6.4.

PROBLEM 6.5 CALCULATING OXYGEN FUGACITY FROM A MINERAL BUFFER

Calculate the oxygen fugacity of the βQMF buffer at 1000 bar pressure, within the temperature range of 600–100 °C, using three different methods outlined above. The volumes of solid phases in their standard states—$\Delta V^0_{298,s}$ ($J\ bar^{-1}\ mol^{-1}$)—are as follows (Holland and Powell, 2011): magnetite 4.452, fayalite 4.631, quartz 2.269.

Solution

At a given temperature, Reaction (6.91) and Equations (6.44) and (6.48) supplemented by thermodynamic data (Table 6.5) can be used to calculate the change of Gibbs energy $-\Delta G^0_T$. If complete thermodynamic data for the calculation of volume changes of solid phases at given P and T are unknown, volumes in the standard state can be employed (Equation (6.96)). Calculation of the change of molar volumes of solid phases of the reaction at standard conditions—$\Delta V^0_{298,s}$—is governed by the mass balance of the reaction, where fayalite is the reaction product, and magnetite and quartz are reactants. Hence, the total volume change corresponds to

$$\Delta V^0_{298,s} = 3\times 4.631 - 2\times 4.452 - 3\times 2.269 = -1.818\,J\,bar^{-1}\,mol^{-1}$$

The third method is based on Equation (6.97) and data from Frost (1991). The results of all three calculation methods are summarized in Table 6.6.

Table 6.5 Input Data for Oxygen Fugacity Calculations Along βQMF Buffer at $P = 1000$ Bar (Thermodynamic Dataset from Holland and Powell (1998))

	Quartz			Fayalite		
T (°C)	H_T^0 (J mol^{-1})	S_T^0 (J K^{-1} mol^{-1})	$V_{T,S}$ (J bar^{-1})	H_T^0 (J mol^{-1})	S_T^0 (J K^{-1} mol^{-1})	$V_{T,S}$ (J bar^{-1})
600	−804,909	107.11	2.3679	−1,233,550	323.71	4.7318
700	−798,034	114.56	2.3691	−1,215,040	343.78	4.7513
800	−791,023	121.42	2.3703	−1,196,170	362.24	4.7709
900	−783,904	127.76	2.3715	−1,176,960	379.36	4.7907
1000	−776,696	133.66	2.3728	−1,157,440	395.33	4.8107
1100	−769,415	139.16	2.3740	−1,137,630	410.31	4.8307

	Magnetite			Oxygen		
T (°C)	H_T^0 (J mol^{-1})	S_T^0 (J K^{-1} mol^{-1})	$V_{T,S}$ (J bar^{-1})	H_T^0 (J mol^{-1})	S_T^0 (J K^{-1} mol^{-1})	T (°C)
600	−841,105	362.63	4.5894	79,484	238.92	600
700	−821,201	384.21	4.6152	82,923	242.65	700
800	−801,053	403.92	4.6413	86,415	246.07	800
900	−780,709	422.05	4.6676	89,951	249.22	900
1000	−760,207	438.82	4.6941	93,527	252.14	1000
1100	−739,575	454.42	4.7207	97,137	254.87	1100

Table 6.6 Results of Oxygen Fugacity Calculations Using Data from Table 6.5

T (°C)	ΔH_T^0 (J mol^{-1})	ΔS_T^0 (J K^{-1} mol^{-1})	$\Delta V_{T,S}$ (J bar^{-1})	ΔG_T^0	$\log f_{O_2}$	$\log f_{O_2}$	$\log f_{O_2}$
					With $\Delta V_{T,S}$	With $\Delta V_{298,S}$	(Frost, 1991)
					Equations (6.44) and (6.48)	Equation (6.96)	Equation (6.98)
600	475,771	163.48	−2.0869	333,052	−19.80	−19.82	−19.89
700	474,307	161.89	−2.0838	316,786	−16.89	−16.91	−16.94
800	473,080	160.71	−2.0807	300,638	−14.53	−14.55	−14.55
900	472,201	159.92	−2.0776	284,615	−12.58	−12.59	−12.57
1000	471,709	159.51	−2.0745	268,648	−10.94	−10.95	−10.89
1100	471,642	159.47	−2.0713	252,694	−9.53	−9.54	−9.46

The most precise results are obtained by the first method, but the second one also yields reasonable results if the complete thermodynamic data are not available. The third, simplest method is recommended for the cases when only normalization toward a specific buffer is required.

PROBLEM 6.6 CALCULATING OXIDATION STATE OF INCLUSION FLUID

Calculate the oxidation state of a CO_2–CO(–N_2–H_2) inclusion in pincinite KP-31 at 1447 K and 5.5 kbar ($\log f''_{O_2} = -7.77$, Problem 6.2) relative to the βQMF buffer (Frost, 1991).

Solution

The normalized value is

$$\Delta\log f_{O_2}(\beta QMF) = \log f_{O_2}(CO_2 - CO) - \log f_{O_2}(\beta QMF) = -7.77 - (-8.19) = 0.42$$

In other words, the inclusion fluid is enriched in oxygen by 0.42 log units compared to the βQMF buffer. The deviation between the calculated fugacity and the reference buffer should not be greater than 2.5 log units; otherwise, another buffering reaction must be selected.

6.4.5 SULFUR BUFFERS

The term *sulfidation state* defined by Barton (1970) is basically the same as the oxidation state defined by Wones (1981). As in the case of oxygen, the sulfur fugacity at any *P* and *T* may be compared to a standard mineral reaction (buffer). Einaudi et al. (2003) introduced an informal terminology of sulfidation states based on the sulfidation reactions among minerals in the system Cu–Fe–As–S and its subsystems, which involves porphyry copper deposits, porphyry-related veins, and epithermal vein and metasomatic deposits (Figure 6.2).

Boundaries between individual sulfidation states are defined by the following reactions:

$$2FeS\,(\text{pyrrhotite}) + S_2 \leftrightarrow 2FeS_2\,(\text{pyrite}) \tag{6.99}$$

$$5CuFeS_2\,(\text{chalcopyrite}) + S_2 \leftrightarrow Cu_5FeS_4\,(\text{bornite}) + 4FeS_2 \tag{6.100}$$

The last reaction approximately coincides with

$$0.67Cu_{12}As_4S_{13}\,(\text{tennantite}) + S_2 \leftrightarrow 2.67Cu_3AsS_4\,(\text{enargite}) \tag{6.101}$$

which represents the transition between the porphyry copper deposits and the porphyry-related base-metal veins. Furthermore, the reaction

$$0.47FeAsS\,(\text{arsenopyrite}) + 1.41CuFeS_2 + S_2 = 0.12Cu_{12}As_4S_{13} + 1.88FeS_2 \tag{6.102}$$

represents the lower limit of the intermediate sulfidation state of epithermal deposits defined by Reaction (6.99). Reactions

$$2FeAs_2 + 2FeS + S_2 \leftrightarrow 4FeAsS \tag{6.103}$$

and

$$2FeAs_2\,(\text{löllingite}) + S_2 \leftrightarrow 2FeAsS\,(\text{arsenopyrite}) + 2As\,(\text{native arsenic}) \tag{6.104}$$

represent the upper limits for low and very low sulfidation states, respectively. The sulfur condensation curve acts as a ceiling to f_{S_2} at any given temperature.

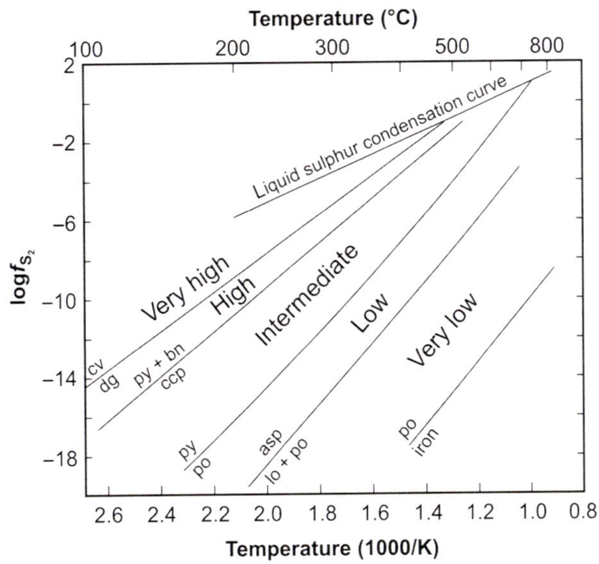

FIGURE 6.2

Relative sulfidation states of hydrothermal fluids (simplified after Einaudi et al., 2003) as defined by sulfur-buffering reactions (Barton and Skinner, 1979). Abbreviations (Whitney and Evans, 2010): cv, covellite; dg, digenite; bn, bornite; ccp, chalcopyrite; py, pyrite; po, pyrrhotite; asp, arsenopyrite; lo, löllingite.

It should be noted that the term *sulfidation state* is not strictly parallel to the high- and low-sulfidation classes of epithermal deposits, which are defined according to a number of other than thermodynamic and paragenetic criteria: fluid inclusions, character of alterations, and volcano-tectonic settings (White and Hedenquist, 1990; Hedenquist and Lowenstern, 1994; Sillitoe and Hedenquist, 2003). Table 6.7 summarizes possible mineral assemblages for various sulfidation states derived by the analysis of thermodynamic data.

Table 6.7 Possible Mineral Assemblages at Different Sulfidation States

	Sulfidation States		
Systems	**High-to-Very High**	**Intermediate**	**Low-to-Very Low**
Fe–S	py, native S	py, po	po
As–S	orp, rlg	native As	native As
Fe–As–S		asp	asp, lo
Cu–S	dg, cv	dg, cc	cc
Cu–Fe–S	ccp	ccp, bn	ccp, bn
Cu–As–S	eng	tnt	tnt
Fe–O	hem	mt	mt

Abbreviations (Whitney and Evans, 2010): asp, arsenopyrite; bn, bornite; cc, chalcocite; ccp, chalcopyrite; cv, covellite; dg, digenite; eng, enargite; hem, hematite; lo, löllingite; mt, magnetite; orp, orpiment; po, pyrrhotite; py, pyrite; rlg, realgar; tnt, tennantite.

6.5 CALCULATING PHASE DIAGRAMS AND MINERAL-FLUID EQUILIBRIA

6.5.1 COMPUTATIONAL METHODS

Thermodynamic calculations of multicomponent systems are too complex to be performed only with a pocket calculator or the EXCEL® spreadsheet. Fortunately, there are numerous computer programs that perform the tedious calculations automatically and flawlessly. Moreover, they also compute phase diagrams of multicomponent systems.

There are two basic approaches to computing complex mineral-fluid equilibria (Connolly, 1992). The oldest and most basic method is the **constrained Gibbs energy minimization**, by which the stable mineral assemblage at specified *P–T–X* coordinates is determined. Classical minimization strategies (e.g., De Capitani and Brown 1987; Eriksson and Hack, 1990) involve two major components: one component is to establish the equilibrium of phase assemblage, the second component determines whether the equilibrium is stable at given *P* and *T* (Figure 6.3). Minimization can be used to construct a phase diagram by computing stable assemblages on a grid defined by optional variables, whereas the phase boundaries are determined by the interpolation between the grid points. Only geometric elements that project as lines (univariant curves) and points (invariant points) are shown in the conventional *P–T* projection. The difficulty with minimization technique is that a phase diagram contains an infinite number of possible *P–T–X* coordinates, and with the increasing number of components the computation would involve an astronomical number of operations even with a very coarse grid. However, the minimization can be used to calculate any two-dimensional section through a multidimensional space. Thus, it is possible to compute how a multicomponent system behaves if only one compositional variable is changed. Moreover, this approach has the advantage of being capable of treating complex solutions and systems with any number of components. The major drawback is that the method assumes bulk equilibrium, and it is thus not applicable to systems where mineral zonation occurs or where the phase rule is violated (Connolly, 1992).

The second approach involves **combinatorial methods**, which make use of the fact that despite an infinite number of *P–T–X* coordinates in the phase diagram, there are only a finite number of phase equilibria. Hence, phase assemblages are enumerated, and their stability is tested. The necessity for combinatorial methods arises from difficulties involved in the evaluation of all the changes that occur on the minimum $\overline{G}$ surface as a function of *P* and *T* (Connolly, 1992).

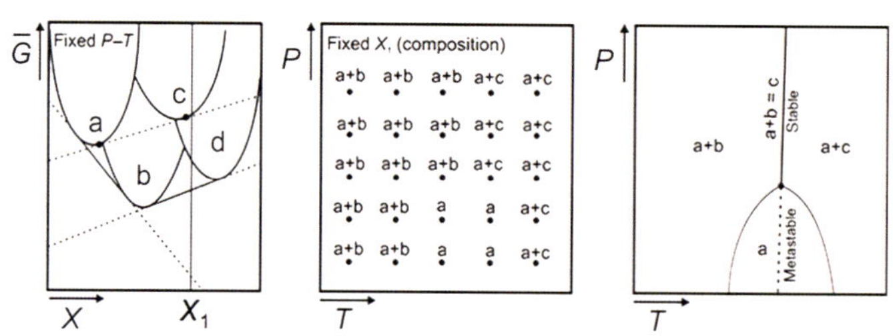

FIGURE 6.3

Minimization strategy and construction of a phase diagram of a system of four phases (Connolly, 1992). In the *P–T–X* coordinates on the left diagram, the a + c phase assemblage is metastable with respect to phases b and d. Reproduced with kind permission from J.A.D. Connolly.

Several computing strategies have been developed to overcome these problems. Only univariant or divariant phase assemblages are tested using GEO-CALC (Berman et al., 1987) and THERMOCALC (Powell and Holland, 1988; Powell et al., 1998) programs, respectively. Neither of these strategies, however, defines higher variance equilibria, and therefore they cannot be used to construct true phase diagrams. Moreover, these programs allow only calculations in the $P-T-X_{CO_2}$ space, namely with the H_2O–CO_2 fluid composition. Complete prediction of fluid-mineral phase composition, including fluid immiscibility, can also be calculated using the HCh modeling package (Shvarov 1999), which employs the Gibbs free energy minimization method. The program has recently been available at http://www.geol.msu.ru/deps/geochems/soft/index_e.html.

A simple linear algorithm for evaluating thermodynamic surfaces (Connolly and Kerrick, 1987) is used in the PERPLE_X program (Connolly, 1990) to determine all phase equilibria regardless of the variance for systems with an unlimited number of components. The PERPLE_X algorithm is numerically exact only for phases with fixed composition. For phases of variable compositions, such as solid solutions in minerals and melts, the program approximates the continuous $\overline{G}-X$ surface by a series of arbitrarily defined discrete compounds referred to as *pseudocompounds*. As a result, the $\overline{G}-X$ surface of the solution is approximated as a polyhedron, the vertices of which correspond to pseudocompounds (Figure 6.4). The program determines which of the pseudocompounds are stable and thereby approximates the stability field of the solution. The accuracy of the determination of the solution's composition depends on the pseudocompound spacing. The user controls the number and positions of the pseudocompounds, and thereby the accuracy of the approximation (Connolly, 1992).

The PERPLE_X algorithm can be used to construct any phase diagram using a noncombinatorial approach. **Compositional diagrams** show phase relations of a system as a function of its independent compositional variables with all other variables (P, T) held constant. The compositional variables are plotted against a potential variable (P, T, X_{CO_2}, X_O, fugacity, activity) in a **mixed-variable diagram**. Finally, **Schreinemaker's diagram** projects invariant and univariant phase relations of a multicomponent system onto a two-dimensional potential variable coordinate frame (Connolly, 1992).

There are no restrictions regarding solution models, fluid equation of states, fluid compositions, and thermodynamic mineral databases used in the PERPLE_X. Hence, the computer code is a versatile and flexible tool for solution of specific problems of fluid thermodynamics in all kinds of hydrothermal, metamorphic, and magmatic settings. Extensive tutorials, commented examples, program documentation, comments, and other supporting files relevant to understanding how the PERPLE_X works are

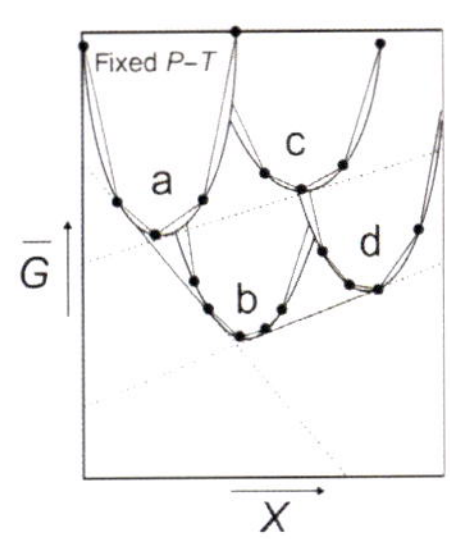

FIGURE 6.4

Noncombinatorial strategy used in the PERPLE_X program to evaluate thermodynamic surfaces and to calculate phase diagrams using pseudocompounds (Connolly, 1992). Reproduced with kind permission from J.A.D. Connolly.

available in http://www.perplex.ethz.ch/. Only basic information about program structure and primitive examples pertaining to the calculation of mixed-volatile fluid equilibria and construction of phase diagrams are provided in the next chapters. If not otherwise indicated, all information was extracted from tutorials posted on the website cited.

6.5.2 USING THE PERPLE_X

The PERPLE_X program package consists of several stand-alone FORTRAN programs, which run on Mac, Windows, Linux, and UNIX platforms. **BUILD** is a problem definition program, which creates input (*name*.dat) for **VERTEX**. This program calculates various types of phase diagrams using internally consistent thermodynamic datasets (e.g., *hp98ver.dat*—Holland and Powell, 1998), solid solution models of minerals (*solution_model.dat*) and equations of states for C–O–H–N–S fluids included in the COHSRK program. VERTEX generates output summaries that can be further processed by **WERAMI**. The program permits the user to extract information from a phase diagram section at a point (as a numerical output), through a region, or along a line or curve. A total of 38 parameters, including enthalpy, density, heat capacity, composition, fugacity, volume, mol.% or mass% of component in a solid solution and others, can be extracted and plotted in the form of isolines. The outputs can be then plotted with **PSTABLE**, **PSVDRAW**, or **PSSECT** as PostScript files, or using external PyWERAMI or MATLAB programs as editable .eps files. The PostScript files can be either viewed by Ghostview or imported in and processed with common graphical editors.

There are also several useful stand-alone programs in the PERPLE_X package: **FRENDLY** is a general thermodynamic calculator, which extracts data from thermodynamic datasets and calculates equilibrium coordinates for a reaction or thermodynamic properties for a phase or reaction relative to standard state at various T and P. The program also modifies thermodynamic data, which can be stored as a new entry in the selected thermodynamic data file. **CTRANSF** redefines components in thermodynamic datasets; for example, FeO can be converted to Fe or Fe_2O_3, H_2O to H_2, and so forth, to meet some special requirements. Other stand-alone programs included in the PERPLE_X program package permit direct calculations of mixed-volatile fluid equilibria. **COHSRK** is a program to call various subroutines to calculate fluid properties and speciation as a function of P, T, X_{CO_2}, f_{O_2}, f_{S_2}, X_O, X_C, a_C, or an O_2–S_2 buffer. **SPECIES** is a program for computing isobaric-isothermal C–O–H–S fluid speciation and properties as a function of X_O. **ISOCOH** computes P–T trajectories and isochores of graphite-saturated C–O–H fluids from specified trapping conditions down to a temperature at which equilibrium speciation is assumed to cease.

Figure 6.5 shows the protocol for calculating the molar volume and speciation of a mixed-volatile C–O–H–S fluid buffered by graphite and pyrite-pyrrhotite using the COHSRK program. The equation of state used is a hybrid EOS of Connolly and Cesare (1993) based on the HSMRK EOS of Kerrick and Jacobs (1981) and MRK EOS modified by de Santis et al. (1974) and Holloway (1977, 1981). The algorithm for solving the fluid speciation in terms of X_O is described in the paper of Connolly (1995). This type of computation can serve as a quick alternative to the manual mass conservation method invented by Ohmoto and Kerrick (1977).

The example shown in Figure 6.6 solves a similar problem, but now the SPECIES program uses atomic proportion of oxygen to hydrogen (X_O) as an independent variable and calculates the fluid composition over a range of X_O values. Plot information generated by this program is written into the *species.plt* file, which can be converted in PSVDRAW to a vector PostScript format (.ps) and further edited in graphical programs (Figure 6.7). The diagram demonstrates that in graphite-buffered fluids,

```
Select fluid equation of state:

   0 - X(CO2) Modified Redlich-Kwong (MRK/DeSantis/Holloway)
   1 - X(CO2) Kerrick & Jacobs 1981 (HSMRK)
   2 - X(CO2) Hybrid MRK/HSMRK
   3 - X(CO2) Saxena & Fei 1987 pseudo-virial expansion
   4 - Bottinga & Richet 1981 (CO2 RK)
   5 - X(CO2) Holland & Powell 1991, 1998 (CORK)
   6 - X(CO2) Hybrid Haar et al 1979/CORK (TRKMRK)
   7 - f(O2/CO2)-f(S2) Graphite buffered COHS MRK fluid
   8 - f(O2/CO2)-f(S2) Graphite buffered COHS hybrid-EoS fluid
   9 - Max X(H2O) GCOH fluid Cesare & Connolly 1993
  10 - X(O) GCOH-fluid hybrid-EoS Connolly & Cesare 1993
  11 - X(O) GCOH-fluid MRK Connolly & Cesare 1993
  12 - X(O)-f(S2) GCOHS-fluid hybrid-EoS Connolly & Cesare 1993
  13 - X(H2) H2-H2O hybrid-EoS
  14 - EoS Birch & Feeblebop (1993)
  15 - X(H2) low T H2-H2O hybrid-EoS
  16 - X(O) H-O HSMRK/MRK hybrid-EoS
  17 - X(O) H-O-S HSMRK/MRK hybrid-EoS
  18 - X(CO2) Delany/HSMRK/MRK hybrid-EoS, for P > 10 kb
  19 - X(O)-X(S) COHS hybrid-EoS Connolly & Cesare 1993
  20 - X(O)-X(C) COHS hybrid-EoS Connolly & Cesare 1993
  21 - X(CO2) Halbach & Chatterjee 1982, P > 10 kb, hybrid-Eos
  22 - X(CO2) DHCORK, hybrid-Eos
  23 - Toop-Samis Silicate Melt
  24 - f(O2/CO2)-N/C Graphite saturated COHN MRK fluid
  25 - H2O-CO2-NaCl Aranovich and Haefner 2004
12

Choose a buffer:

  1 - Pyrite + Pyrrhotite
  2 - Pyrrhotite
  3 - f(S2)

1

Compute f(H2) & f(O2) as the dependent fugacities
(do not unless you project through carbon) (Y/N)?

n

Reduce graphite activity (Y/N)?

n

Enter a zero for pressure to quit.


Enter p(bar), T(K),    X(O)   :
3000 673 0.2


Fluid equation of state: X(O)-f(S2) GCOHS-fluid hybrid-EoS Connolly & Cesare 199
3

          p(bar)       =    3000.0
          T(K)         =    673.00

          log[f(O2)] =   -29.641
          log[f(S2)] =   -7.2233
          a(gph/dia) =     1.0000

                      Speciation/Fugacities

             H2O            CO2            CO             CH4
     x    0.65696       0.14232E-02   0.16288E-05    0.32105
     f     568.46        14.331        0.37548E-01    6658.1

             H2             H2S            O2             SO2
     x    0.86648E-03   0.19699E-01   0.17275E-33    0.18310E-12
     f     18.164        117.25        0.22861E-29    0.12324E-08

             COS
     x    0.36323E-06
     f    0.45148E-02

     Sum of species fractions:     1.000000000


                      Atomic Proportions

        C             H              O              S             N
   0.88561E-01   0.72482       0.18121       0.54101E-02     0.0000

     Back-calculated X(O) =   0.200000000
     Back-calculated X(S) =   0.575716053E-01
     Back-calculated X(C) =   0.885614105E-01
     S/H =   0.74640E-02


     Molar Volume (cm3/mol) =     18.857
```

FIGURE 6.5

Example 1 dialog with the COHSRK program to calculate the speciation of fluid buffered by graphite and pyrite–pyrrhotite at 3000 bar, 673 K, and $X_O = 0.2$. *x* are mole fractions of fluid species, *f* is fugacities (bar). Hereafter, X_O is defined as $nO/(nO + nH)$, where *n* is the number of atoms.

```
Select fluid equation of state:

   0 - X(CO2) Modified Redlich-Kwong (MRK/DeSantis/Holloway)
   1 - X(CO2) Kerrick & Jacobs 1981 (HSMRK)
   2 - X(CO2) Hybrid MRK/HSMRK
   3 - X(CO2) Saxena & Fei 1987 pseudo-virial expansion
   4 - Bottinga & Richet 1981 (CO2 RK)
   5 - X(CO2) Holland & Powell 1991, 1998 (CORK)
   6 - X(CO2) Hybrid Haar et al 1979/CORK (TRKMRK)
   7 - f(O2/CO2)-f(S2) Graphite buffered COHS MRK fluid
   8 - f(O2/CO2)-f(S2) Graphite buffered COHS hybrid-EoS fluid
   9 - Max X(H2O) GCOH fluid Cesare & Connolly 1993
  10 - X(O) GCOH-fluid hybrid-EoS Connolly & Cesare 1993
  11 - X(O) GCOH-fluid MRK Connolly & Cesare 1993
  12 - X(O)-f(S2) GCOHS-fluid hybrid-EoS Connolly & Cesare 1993
  13 - X(H2) H2-H2O hybrid-EoS
  14 - X(CO2) Pitzer & Sterner 1994; Holland & Powell mixing 2003
  15 - X(H2) low T H2-H2O hybrid-EoS
  16 - X(O) H-O HSMRK/MRK hybrid-EoS
  17 - X(O) H-O-S HSMRK/MRK hybrid-EoS
  18 - X(CO2) Delany/HSMRK/MRK hybrid-EoS, for P > 10 kb
  19 - X(O)-X(S) COHS hybrid-EoS Connolly & Cesare 1993
  20 - X(O)-X(C) COHS hybrid-EoS Connolly & Cesare 1993
  21 - X(CO2) Halbach & Chatterjee 1982, P > 10 kb, hybrid-Eos
  22 - X(CO2) DHCORK, hybrid-Eos
  23 - Toop-Samis Silicate Melt
  24 - f(O2/CO2)-N/C Graphite saturated COHN MRK fluid
  25 - H2O-CO2-NaCl Aranovich et al. 2010
12

Choose a buffer:

  1 - Pyrite + Pyrrhotite
  2 - Pyrrhotite
  3 - f(S2)

1

Compute f(H2) & f(O2) as the dependent fugacities
(do not unless you project through carbon) (Y/N)?

n

Reduce graphite activity (Y/N)?

n
Plot (log) species fractions (y/n)?
(if you answer no, then fugacities are plotted).
y

Enter pressure (bar), temperature (K):
3000 673

Enter min, max, and increments for the independent variable (  X(O)     ):
0 1 0.1
Enter min-max (log) limits for the dependent variables, in the plot file,
values that exceed these limits will be set to the limits established here:
-5 0

values which follow are   X(O)     and the mole fractions of:
 H2O CO2 CO  CH4 H2  H2S SO2 COS
and log10(fO2).


the plot file includes the log of the  C-O-H-S-N atomic proportions and
the log of the species fractions.

0.100000E-07  0.394596E-07  0.183631E-16  0.367754E-12  0.973003      0.281153E-
02  0.241855E-01  0.182785E-26
0.434188E-13  -43.5855
0.100000      0.357052      0.957842E-03  0.209227E-05  0.616325      0.173575E-
02  0.239271E-01  0.109390E-12
0.344533E-06  -29.9031
0.200000      0.656962      0.142322E-02  0.162880E-05  0.321047      0.866477E-
03  0.196989E-01  0.183100E-12
0.363232E-06  -29.6409
0.300000      0.917513      0.767479E-03  0.478876E-06  0.729917E-01  0.230888E-
03  0.849599E-02  0.102546E-12
0.125400E-06  -29.5192
0.400000      0.854946      0.143729      0.801804E-05  0.549095E-03  0.227084E-
04  0.743654E-03  0.191485E-10
0.201003E-05  -27.3202
0.500000      0.664086      0.334301      0.214332E-04  0.685162E-03  0.363930E-
04  0.864696E-03  0.435060E-10
0.476188E-05  -27.1748
0.600000      0.497735      0.500783      0.355411E-04  0.606833E-03  0.427249E-
04  0.789711E-03  0.625289E-10
0.691871E-05  -27.0892
0.700000      0.351362      0.647527      0.487693E-04  0.400134E-03  0.402787E-
04  0.612878E-03  0.777094E-10
0.845964E-05  -27.0177
0.800000      0.221349      0.777954      0.607116E-04  0.190787E-03  0.309822E-
04  0.404150E-03  0.901870E-10
0.955710E-05  -26.9561
0.900000      0.104938      0.894719      0.713702E-04  0.486944E-04  0.169888E-
04  0.195613E-03  0.100693E-09
0.103539E-04  -26.9028
 1.00000      0.997900E-07  0.999908      0.808672E-04  0.482818E-16  0.180440E-
10  0.187381E-09  0.109704E-09
0.109460E-04  -26.8563

Change EoS, buffer, or graphite activity (y/n)?
```

FIGURE 6.6

Example 2 dialog with the SPECIES program to calculate speciation (mole fractions) of C–O–H–S fluid buffered by graphite and pyrite–pyrrhotite assemblage at 3000 bar, 673 K, and X_O from 0 to 1.

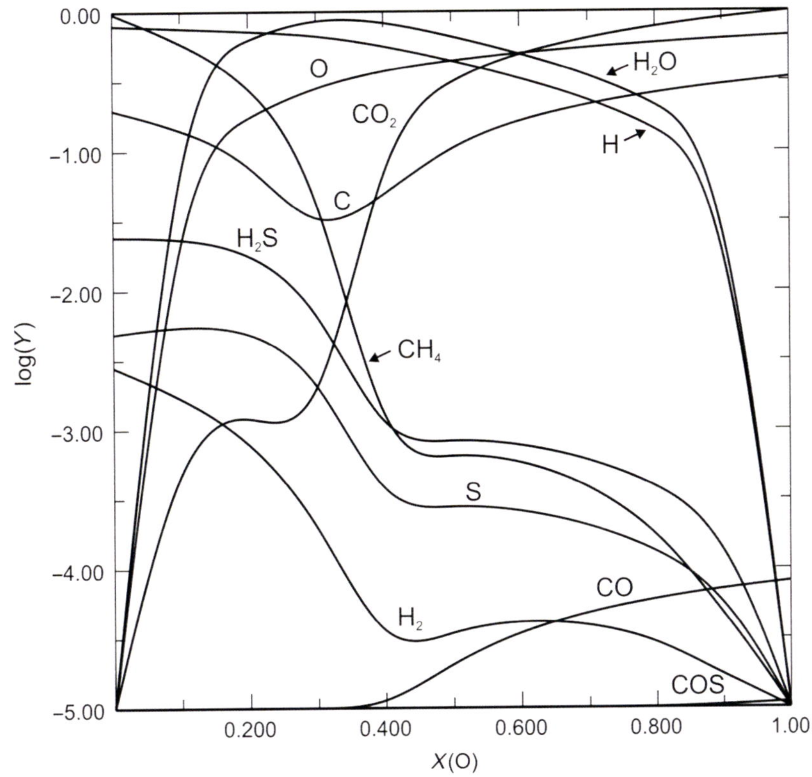

FIGURE 6.7

Plot output from the Example 2 dialog. It displays only values within the user-defined range from log(−5) to log(0), that is, 0.00001–1. Note that $X_{(O)}$ values are defined as the independent variable, and O as the dependent variable. This is because X_O is here defined as $nO/(nO+nH)$, where n is the number of atoms, whereas O represents the atomic proportion of oxygen relative to C, H, and S.

maximum H_2O content appears at $X_O = 1/3$, which corresponds to the stoichiometric proportion of O and H in H_2O, namely $1/(1+2)$. At lower X_O, CH_4 content rapidly increases at the expense of CO_2, whereas the opposite is true for the X_O values between 1/3 and 1. Hence, most geological problems can be approximated by the binary H_2O–CO_2 and H_2O–CH_4 systems. The X_{CO_2}, X_{CH_4}, and X_O values can be roughly correlated with the equations

$$\begin{aligned} X_O &\cong 2/3X_{CO_2} + 1/3 \\ X_{CO_2} &\cong (X_O - 1/3)/(1 - 1/3) \quad \text{for } X_O > 1/3 \end{aligned} \tag{6.105}$$

and

$$\begin{aligned} X_O &\cong (2/3 - 1)X_{CH_4} + 1/3 \\ X_{CH_4} &\cong (X_O - 1/3)/(2/3 - 1) \quad \text{for } X_O < 1/3 \end{aligned} \tag{6.106}$$

Example 3 shows the calculation of a Schreinemaker's phase diagram of the Fe–O–S system at 300 °C and 100 bar, with two independent variables, log f_{O_2} and log S_{O_2}. Figure 6.8 exemplifies the creation of

```
Microsoft(R) Windows DOS
(C)Copyright Microsoft Corp 1990-2001.

D:\PROGRAMY\PERPLE~3>build

Perple_X version 6.6.5.9, compiled 5/8/2011.

NO is the default (blank) answer to all Y/N prompts

Enter a name for this project (the name will be the prefix
for all output file names) [default = my_project]:
example3

The project definition file will be: example3.dat

Enter thermodynamic data file name [default = hp02ver.dat]:
sup92ver.dat

Enter the computational option file name [default = perplex_option.dat]:
See: www.perplex.ethz.ch/perplex_options.html

Reading computational options from: perplex_option.dat

Warning: value3 of the iteration keyword must be  >2 and <8
value3 will be assigned its default value [4].

The current data base components are:
 NA2O  MGO   AL2O3 SIO2  K2O   CAO   TIO2  MNO   FEO   O2    H2O   CO2
 S2    F     CL    CUO   PBO   ZNO   BAO   ZRO2  NIO   SNO   AGO   SRO
 HGO
Transform them (Y/N)?
y
Enter new component name, < 6 characters, left justified:
FE
Enter old component to be replaced with FE    :
FEO
Enter other components (< 11) in FE    1 per line, <enter> to finish:
O2

Enter stoichiometric coefficients of:
  FEO   O2
in FE    (in above order):
1 -0.5
 FE    =   1.00 FEO    -0.50 O2
Is this correct (Y/N)?
y

The current data base components are:
 NA2O  MGO   AL2O3 SIO2  K2O   CAO   TIO2  MNO   FE    O2    H2O   CO2
 S2    F     CL    CUO   PBO   ZNO   BAO   ZRO2  NIO   SNO   AGO   SRO
 HGO
Transform them (Y/N)?
n

Calculations with a saturated phase (Y/N)?
The phase is:   FLUID
Its compositional variable is: Y(CO2), X(O), etc.
n

Calculations with saturated components (Y/N)?
n

Use chemical potentials, activities or fugacities as independent
variables (Y/N)?
y

Specify a component whose chemical potential, activity or fugacity is
to be independent, enter a blank to finish:
 NA2O  MGO   AL2O3 SIO2  K2O   CAO   TIO2  MNO   FE    O2    H2O   CO2
 S2    F     CL    CUO   PBO   ZNO   BAO   ZRO2  NIO   SNO   AGO   SRO
 HGO
O2

Component O2    is to be characterized by:

     1 - chemical potential [default]
     2 - log10(fugacity)
     3 - log10(activity)
2

Phase O2         will be used to define O2     fugacity

The log10(O2     fugacity) variable is named: f_O2

Specify a component whose chemical potential, activity or fugacity is
to be independent, enter a blank to finish:
 NA2O  MGO   AL2O3 SIO2  K2O   CAO   TIO2  MNO   FE    H2O   CO2   S2
 F     CL    CUO   PBO   ZNO   BAO   ZRO2  NIO   SNO   AGO   SRO   HGO
S2

Component S2    is to be characterized by:

     1 - chemical potential [default]
     2 - log10(fugacity)
     3 - log10(activity)
2
```

(a)

FIGURE 6.8

Example 3 dialog with the BUILD program to create a problem definition file for the calculation of phase diagram of the Fe–O–S system at 300 °C and 100 bar with VERTEX.

Continued

```
Phase O2       will be used to define O2    fugacity

The log10(O2    fugacity) variable is named: f_O2


Specify a component whose chemical potential, activity or fugacity is
to be independent, enter a blank to finish:
 NA2O  MGO   AL2O3 SIO2  K2O   CAO   TIO2  MNO   FE    H2O   CO2   S2
 F     CL    CUO   PBO   ZNO   BAO   ZRO2  NIO   SNO   AGO   SRO   HGO
S2

Component S2    is to be characterized by:

     1 - chemical potential [default]
     2 - log10(fugacity)
     3 - log10(activity)
2

Phase S2       will be used to define S2    fugacity

The log10(S2    fugacity) variable is named: f_S2


Select thermodynamic components from the set:
 NA2O  MGO   AL2O3 SIO2  K2O   CAO   TIO2  MNO   FE    H2O   CO2   F
 CL    CUO   PBO   ZNO   BAO   ZRO2  NIO   SNO   AGO   SRO   HGO
Enter names, 1 per line, enter a blank line to finish:
FE


Specify computational mode:

     1 - Unconstrained minimization
     2 - Constrained minimization on a 2d grid [default]
     3 - Constrained minimization on a 1d grid
     4 - Output pseudocompound data
     5 - Phase fractionation calculations

Unconstrained optimization should be used for the calculation of composition,
mixed variable, and Schreinemakers diagrams. Gridded minimization can be used to

construct phase diagram sections.

2

The data base has P(bars)  and T(K)     as default independent potentials.
Make one dependent on the other, e.g., as along a geothermal gradient (y/n)?
n

Select x-axis variable:
     1 - P(bars)
     2 - T(K)
     3 - f_O2
     4 - f_S2
4

Enter minimum and maximum values, respectively, for: f_S2
-20 0

Select y-axis variable:
     2 - T(K)
     3 - f_O2
     4 - P(bars)

3

Enter minimum and maximum values, respectively, for: f_O2
-45 -25

Specify sectioning value for: T(K)
573

Specify sectioning value for: P(bars)
100

For gridded minimization, grid resolution is determined by the number of levels
(grid_levels) and the resolution at the lowest level in the X- and Y-directions
(x_nodes
and y_nodes) these parameters are currently set for the exploratory and autorefi
ne cycles
as follows:

stage          grid_levels  xnodes  ynodes    effective resolution
exploratory         1          20      20          20 x  20 nodes
auto-refine         4          60      60         473 x 473 nodes

To change these options edit or create the file perplex_option.dat
See: www.perplex.ethz.ch/perplex_options.html#grid_parameters
```

(b)

FIGURE 6.8—Cont'd

Continued

```
Specify component amounts by weight (Y/N)?
n

Enter molar  amounts of the components:
 FE
for the bulk composition of interest:
1

Generate print file output (Y/N)?
y

Exclude phases (Y/N)?
y
Do you want to be prompted for phases (Y/N)?
n
Enter names, 1 per line, enter a blank line to finish:
Frs-Ox

Do you want to treat solution phases (Y/N)?
n

Enter calculation title:
example3_
```

(c)

FIGURE 6.8—Cont'd

a problem definition file, which is further elaborated by VERTEX. The computational mode chosen was a two-dimensional gridded minimization, which allows construction of phase diagram sections. As for other diagrams calculated by gridded minimization, output from VERTEX can be read with PSSECT and exported as an editable .ps file. The raw file generated by PSSECT is shown in Figure 6.9.

At the outset, a new thermodynamic component named FE must be defined in BUILD as a combination of 1 mol of FEO and −0.5 mol of O_2. The new component can then be combined with two independent variables, $\log f_{O_2}$ and $\log f_{S_2}$, plotted along vertical and horizontal axes, respectively. The SUPCRT92 thermodynamic database of Helgeson et al. (1978) updated and modified by Johnson et al. (1992) was used and the Frs-Ox phase (ferrous oxide) was excluded from the calculation. The database (*sup92ver.dat*) has a benefit of containing, apart from oxides and silicates, thermodynamic data on sulfides, sulfates, carbonates, and halogenides, and is thus specially tailored to the solution of problems inherent to the deposition of minerals in hydrothermal settings.

6.5.3 ESTIMATING f_{O_2} AND f_{S_2} FROM HYDROTHERMAL MINERALS

The sulfide–oxide–sulfate assemblage of the Tisovec skarn, Slovakia, genetically associated with Miocene andesites and diorites, could serve as an illustrative example of the thermodynamic analysis performed with the aim of calculating the speciation of equilibrium gas, and to evaluate the possible effect of an increased SO_2 concentration on the sulfur isotope thermometry and the sulfur isotope composition of the coexisting fluid (Hurai and Huraiová, 2011).

The uppermost $\log f_{S_2}$ limit of ~−10.5 is defined by the composition of pyrrhotite (95 mol.% FeS) associated with pyrite, and by the composition of associated arsenopyrite, showing a minimum of 31 at.% As (Figure 6.10). The intersection of arsenopyrite and pyrrhotite isopleths also defines the temperature of ~325 °C. The limits of the oxygen and sulfur fugacities were also provided by the presence of chalcopyrite within the stability field of anhydrite, and the absence of hematite.

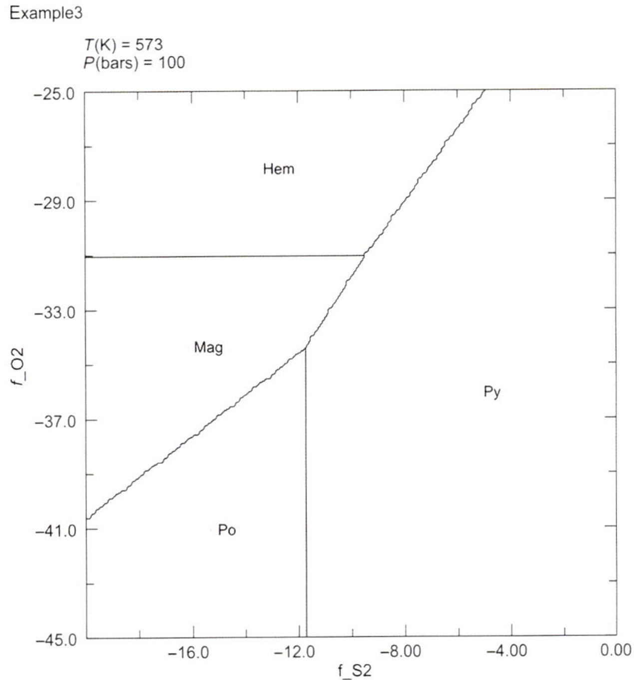

FIGURE 6.9

Schreinemaker's type of phase diagram of the Fe–O–S system at 300 °C and 100 bar created by BUILD, VERTEX, and PSSECT using the SUPCRT92 thermodynamic database (Helgeson et al., 1978; Johnson et al., 1992). f_{S_2} and f_{O_2} refer to log f_{S_2} and log f_{O_2}, respectively. Abbreviations: Mag, magnetite; Hem, hematite; Po, pyrrhotite; Py, pyrite denotes stability fields of the respective minerals.

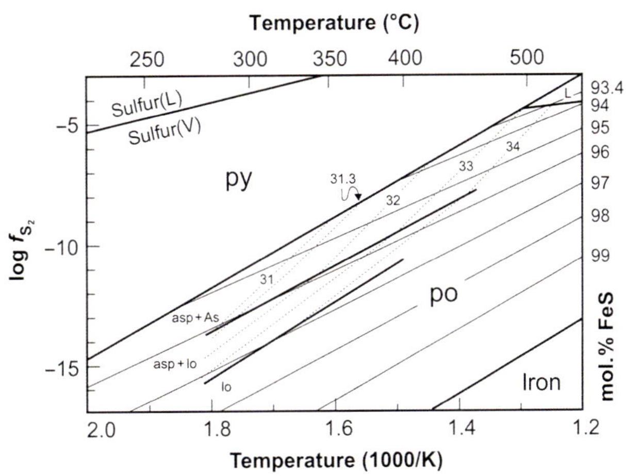

FIGURE 6.10

Portion of phase diagrams of Fe–S and Fe–As–S systems at atmospheric pressure compiled from Toulmin and Barton (1964), Kretschmar and Scott (1976), and Osadchii and Chareev (2006). The dashed lines are isopleths for atomic percent of As in arsenopyrite (asp), thin solid lines are isopleths of mol.% FeS in pyrrhotite (po), thick solid lines are phase boundaries. Other abbreviations (Whitney and Evans, 2010): py, pyrite; lo, löllingite; As, native arsenic; L, liquid; V, vapor.

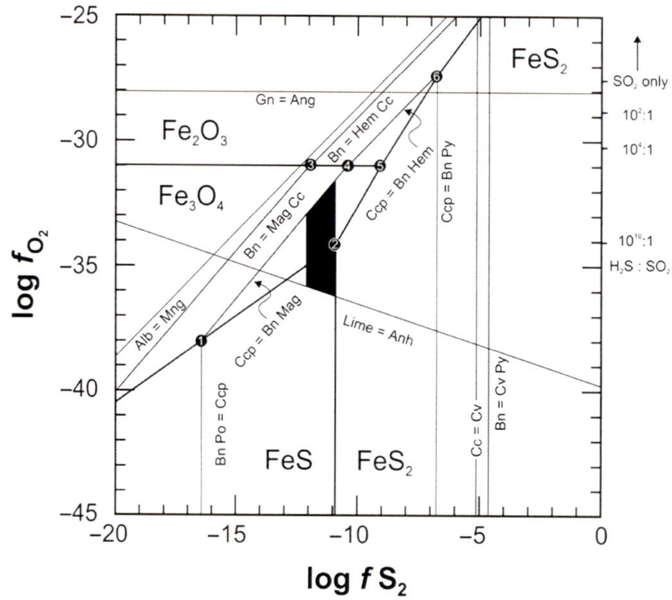

FIGURE 6.11

Schreinemaker's projection of the Fe–Cu–S–O system at 300 °C and 1 bar, with selected univariant curves from the Ca–S–O, Mn–S–O, and Pb–S–O systems calculated using PERPLE_X '06 program package and thermodynamic database included in the SUPCRT92 program (Helgeson et al., 1978; Johnson et al., 1992). For calculation of equilibrium gas composition displayed along right vertical axis as H_2S/SO_2 ratios, a hybrid EOS accommodated to the H–O–S system (Connolly and Cesare, 1993) was used (option 17 in BUILD). The following stability fields are contoured: alabandite (alb), anglesite (ang), anhydrite (anh), bornite (bn), covellite (cv), chalcocite (cc), chalcopyrite (ccp), galena (gn), hematite (hem), lime–CaO, magnetite (mag), manganosite (mng), pyrite (py), pyrrhotite (po), and invariant points: (1) mag-po-bn-ccp, (2) mag-po-py-ccp, (3) mag-hem-bn-cc, (4) mag-hem-bn-ccp, (5) mag-hem-py-ccp, and (6) hem-py-bn-ccp. Due to limited space, some reactions were written such that the high log f_{S_2} assemblage or phase is on the right of the = sign. The black area corresponds to the range of f_{O_2}–f_{S_2} parameters defined by the observed mineral assemblages (for further explanation see the text).

The anhydrite–magnetite–chalcopyrite–pyrrhotite–arsenopyrite–alabandite assemblage and the absence of bournonite bracketed f_{O_2} values to the range from -33 to -36 log units, and those of f_{S_2} to -12.5 ± 0.5 log units. Crystallization temperatures of 300–350 °C were constrained by the chlorite thermometer, sulfur isotope thermometry applied to the galena–sphalerite pair (assuming only H_2S present in equilibrium gas), and infrared fluid inclusion thermometry.

The Schreinemaker's phase diagram can be calculated using the dialog shown in Figure 6.8. At 300 °C, the actual oxidation and sulfidation states must have been slightly below the mag–po–py–ccp invariant point (2). In spite of the crystallization of primary hydrothermal anhydrite, the coexisting gas was still dominated by hydrogen sulfide, with the H_2S contents exceeding that of SO_2 by about 10 orders of magnitude. Hence, the negligible SO_2 concentration did not impose a measurable effect on the calculated $\delta^{34}S$ value of the coexisting H_2S gas, whose $\delta^{34}S$ values were slightly above typical mantle values (2.0–2.5‰ CDT). The carbonate-replacement skarn at Tisovec was classified as a medium-sulfidation type according to the assemblage of sulfidic minerals, high Fe content in sphalerite, and the calculated sulfur and oxygen fugacities.

6.5.4 CALCULATING MAGMATIC OXYTHERMOMETER

Any reaction liberating 1 mol of oxygen can be used as an oxygen buffer. Silicates and oxides containing ferric and ferrous iron are the most commonly used minerals for this purpose. Alternatively, reactions between Mg-silicates and carbonates, such as the EMOD/EMOG buffer $MgSiO_3 + MgCO_3 \leftrightarrow Mg_2SiO_4 + C + O_2$ (Eggler and Baker, 1982), can also be employed to constrain oxygen fugacity in the mantle. However, a single reaction cannot serve as an exact oxybarometer, because it defines only iso-fugacity lines along a divariant P–T plane. Hence, either another oxygen-buffering or temperature-dependent reaction must be applied to reduce the variance of the system, or the oxygen buffer is normalized to a constant pressure, when comparisons are to be made. For instance, the widely used ilmenite–magnetite oxythermometer (Buddington and Lindsley, 1964; Sauerzapf et al., 2008) is based on the temperature-dependent exchange $Fe^{2+} + Ti^{4+} \leftrightarrow 2Fe^{3+}$ between the magnetite–ulvöspinel and hematite–ilmenite solid solution series combined with the magnetite–hematite oxygen buffer—an equilibrium involving Ti-rich end-members

$$6FeTiO_3(\text{ilmenite}) + 2Fe_3O_4(\text{magnetite}) \leftrightarrow 6Fe_2TiO_4(\text{ulvospinel}) + O_2$$

In contrast, the rutile–ilmenite oxygen buffer (Carmichael and Nicholls, 1967; Zhao et al., 1999) is represented only by a single reaction

$$2Fe_2O_3(\text{in ilmenite}) + 4TiO_2(\text{rutile}) \leftrightarrow 4FeTiO_3(\text{in ilmenite}),$$

which must be combined with an independent thermometer.

Unique magmatic assemblage of orthopyroxene granodiorite was found in pincinite, ejected by a Pliocene phreato-magmatic basaltic eruption in the northern Pannonian Basin (Huraiová et al., 2005). The pincinite xenoliths are composed of quartz, plagioclase, and a large amount of interstitial silicate glass with orthopyroxene and ilmenite (Figure 6.12). The Fe-bearing minerals crystallizing from the intergranular silicate melt permit calculation of a magmatic oxythermometer. In addition, high-density gas is present in primary fluid inclusions in quartz, from which an independent value of oxygen fugacity can be calculated (Problem 6.2). Finally, the water content in the interstitial glass determined by laser Raman spectroscopy provides an additional control on the estimated temperature and pressure.

The mineral composition of the two investigated pincinites and representative analyses used for thermodynamic calculations are given in Tables 6.8 and 6.9. Generally, the absence of OH-bearing minerals, major and trace element abundances, and Sr–O isotope ratios in the pincinites are diagnostic of a high-temperature anorogenic parental magma originated by a dehydration melting of quartz-feldspathoid crust surrounding the alkali basalt reservoir. The anatexis led to generation of a melt-depleted granulitic crust, and the pincinites were interpreted as Pliocene equivalents of igneous charnockites and enderbites quenched at supersolidus temperatures. Hence, the rocks are unaffected by the subsolidus re-equilibration and metamorphic overprint, but the interstitial glass could lose water during isothermal decompression in ascending basaltic magma. In addition, composition of inclusion fluids could be affected by hydrogen diffusion. Identification of these processes and exact

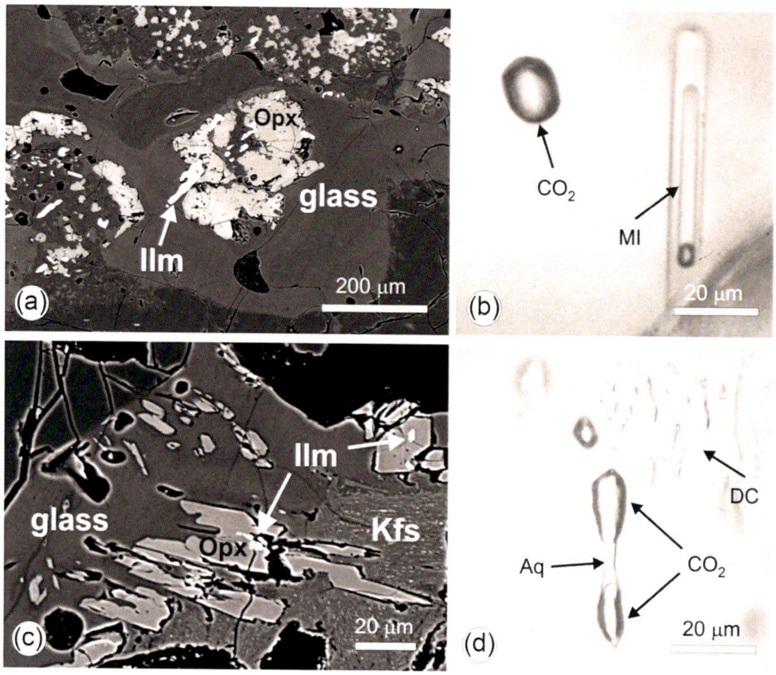

FIGURE 6.12

(a) Back-scattered-electron image (BSE) of orthopyroxene–ilmenite intergrowths in intergranular glass of peraluminous pincinite (KP-31). (b) Monophase CO_2-rich inclusion in quartz and tubular two-phase silicate melt inclusion in apatite from peraluminous pincinite. (c) BSE image of metaluminous pincinite (KP-30) shows orthopyroxene crystals with ilmenite inclusions, and newly formed potassium feldspar growing from interstitial melt. (d) CO_2-rich aqueous inclusion in quartz from the metaluminous pincinite. DC denotes the decrepitation crack emanating from the tip of the inclusion, indicative of water leakage. Aq and CO_2 denote aqueous- and CO_2-phases within the fluid inclusion.

determination of temperature and pressure of the pincinite formation is essential for estimating the depth of the basaltic magma reservoir in the Pannonian Basin.

Formation temperature and oxygen fugacity can be determined using electron microprobe and ^{57}Fe Mössbauer spectroscopy data on the orthopyroxene–ilmenite–silicate melt assemblage (Figure 6.12a and c) in the intergranular glass of both metaluminous and peraluminous pincinite types. In addition, pressure could be derived from isochores of primary fluid inclusions that occur in quartz (Figure 6.12b and d). A complete fluid speciation in equilibrium with the orthopyroxene–ilmenite–silicate melt assemblage can be calculated and compared with that observed in fluid inclusions. The following text is concerned with the formulation and calculation of the oxythermometer buffered by the reaction between orthopyroxene, ilmenite, and silicate melt.

Oxygen fugacity of the orthopyroxene–ilmenite–silicate melt assemblage is controlled by the following reaction:

$$2Fe_2O_3(\text{ilmenite}) + 4SiO_2(\text{melt}) \leftrightarrow 2Fe2Si_2O_6(\text{orthopyroxene}) + O_2 \tag{6.107}$$

Table 6.8 Representative Compositions of Minerals and Silicate Glass (Mass%) and Crystallochemical Formulae of Minerals (Mole Fractions of End-Members) in Pincinites Used for Oxythermometry

	Peraluminous Pincinite KP-31			Metaluminous Pincinite KP-30		
Phase	**Glass**	**Opx**	**Ilm**	**Glass**	**Opx**	**Ilm**
Anal. No.	#21	#20	#19	#22	#20	#27
SiO_2	68.34	49.72	0.06	70.29	50.10	0.03
TiO_2	0.82	0.17	48.20	0.84	0.83	45.11
Al_2O_3	13.27	0.77	0.25	13.43	7.72	0.87
Cr_2O_3	0.00	0.00	0.01	0.01	0.00	0.00
Fe_2O_3	0.34[a]		9.09	0.81[a]		17.80
FeO	4.79	32.66	37.20	2.05	15.82	27.93
MnO	0.16	1.15	0.57	0.04	0.43	0.40
MgO	0.74	14.17	3.15	0.26	25.44	6.85
CaO	1.91	1.45	0.03	0.73	0.60	0.05
Na_2O	2.36	0.07		0.79	0.02	
K_2O	3.08			6.18		
Cl	0.08			0.06		
F	0.00			0.00		
Total	95.89	100.16	98.56	95.49	100.96	99.04
Hematite			0.087			0.165
Geikielite			0.118			0.251
Pyrophanite			0.012			0.008
Ilmenite			0.783			0.576
Enstatite		0.436			0.741	
Ferrosilite		0.564			0.259	

[a] *Fe_2O_3 content inferred from the maximum Fe^{3+}/Fe^{2+} ratio determined by ^{57}Fe Mössbauer spectroscopy.*

A basic formulation of oxybarometer can be obtained by rearranging the general equation for a Gibbs free energy change (Equation (6.44))

$$\log f_{O_2} = \frac{-\Delta G_T^0 \int_1^P \Delta V_{T,S}(P)\,dP}{2.3026RT} - \log \frac{a^2_{Fe_2Si_2O_6}}{a^2_{Fe_2O_3} a^4_{SiO_2}} \tag{6.108}$$

where R is the gas constant, T stands for temperature (K), P is pressure (bar), $\Delta V_{T,S}$ is the volume difference between solid products and reactants (J bar^{-1}), and G_T^0 is the Gibbs free energy (J) of the reaction at T and 1 bar.

The ferric iron content of ilmenite was derived from electron microprobe data assuming ideal stoichiometry (Stormer, 1983; Kolker and Lindsley, 1989). The activity of the hematite component in ilmenite ($a_{Fe_2O_3}$) was obtained from a nonideal mixing model for the rhombohedral oxide solid solution series (Ghiorso, 1990; Ghiorso and Sack, 1991). The activities of ferrosilite

Table 6.9 Ilmenite and Orthopyroxene End-Members in Peraluminous and Metaluminous Pincinites

Pincinite	Peraluminous KP-31	Metaluminous KP-30
Assemblage	Qtz–Plg–Opx–Ilm–Glass (Mol.%)	Qtz–Plg–Kfs–Opx–Ilm–Glass (Mol.%)
Ilmenite end-members		
Ilmenite $FeTiO_3$	76.0–79.4	57.6–69.4
Hematite Fe_2O_3	8.2–10.0	16.5–19.5
Geikielite $MgTiO_3$	8.8–13.7	11.6–25.1
Pyrophanite $MnTiO_3$	1.1–1.7	0.8–1.3
Corundum Al_2O_3	<1.2	<2.4
Orthopyroxene end-members		
Ferrosilite $Fe_2Si_2O_6$	55.2–71.4	37.8–43.4
Enstatite $Mg_2Si_2O_6$	25.9–41.5	54.0–60.8
Wollastonite $Ca_2Si_2O_6$	1.1–3.6	0.9–2.7

($a_{Fe_2Si_2O_6}$) in the orthopyroxene and that of SiO_2 in the silicate melt were calculated from mixing models of Sack and Ghiorso (1989) and Ghiorso and Sack (1995), respectively. A relevant web application (Melts Supplemental Calculator) can be found at http://melts.ofm-research.org/CalcForms/index.html. Another set of independent oxygen fugacities was calculated from the thermodynamic model for silicate melts (Ghiorso and Sack, 1995), using electron microprobe analyses and Fe^{3+}/Fe^{2+} ratios in the silicate glass determined by ^{57}Fe Mössbauer spectroscopy. The thermodynamic database of Berman (1988) was used in both calculations to maintain the consistency of mixing models.

The two oxybarometer divariant planes intersect in a univariant line, steeply oblique to the temperature axis (Figure 6.13), thus defining the temperature within ±5 °C and $\pm 0.2 \log f_{O_2}$ units for the metaluminous pincinite, and ±30 °C and $\pm 0.1 \log f_{O_2}$ units for the peraluminous pincinite, within the pressure range of 1–7 kbar, corresponding to the presumed thickness of continental crust beneath the Pannonian Basin (approximately 30 km according to geophysical observations). To infer pressure and to calculate the depth of formation, the oxythermometer must be intersected with isochores of fluid inclusions trapped in quartz. This also more precisely defines the $\log f_{O_2}$–P–T parameters, which correspond to −10.35, ~5.5 kbar, and ~1160 °C for the peraluminous pincinite, and to −14.78, ~3 kbar, and ~755 °C for the metaluminous pincinite.

Preliminary interpretation of our data suggests that the two investigated xenoliths were yanked by traversing basaltic magma from two magmatic reservoirs emplaced at different depths. But is this interpretation correct? We must exercise caution because there are at least two controversies involved in our data: (1) the equilibrium temperature of the orthopyroxene–ilmenite–melt assemblage in the peraluminous pincinite is too high and substantially exceeds even the dry granite liquidus (Holtz and Johannes, 1994; Holtz et al., 2001); (2) the theoretical composition of the equilibrium fluid significantly differs from that observed within the inclusions. This is demonstrated by the oxygen fugacity calculated from Raman spectroscopic data on fluid inclusions in the peraluminous pincinite

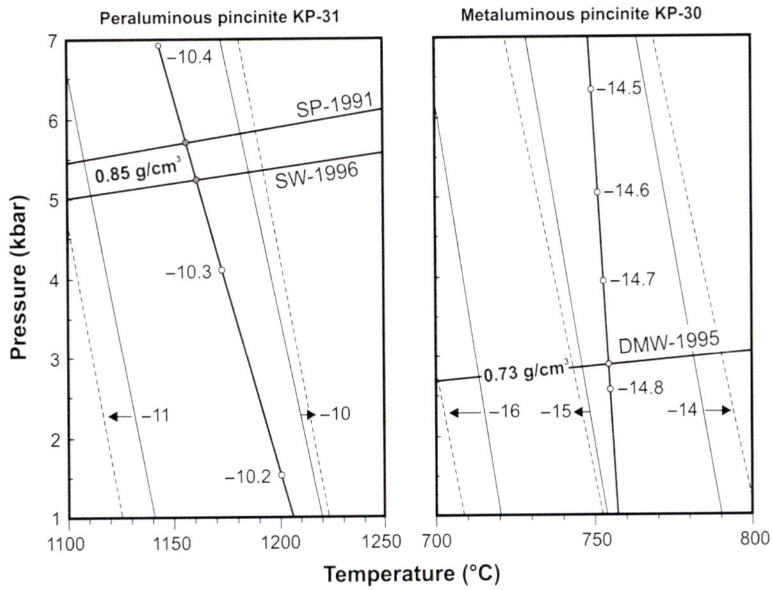

FIGURE 6.13

The location of P–T–f_{O_2} equilibrium conditions for minerals and melts listed in Tables 6.8 and 6.9. Steep-oblique thin lines are plane projections of oxygen fugacity isolines defined by the orthopyroxene–ilmenite–melt assemblage (dashed lines) and by the Fe^{3+} content in the silicate glass (solid lines). Both oxybarometers intersect along a univariant line (which is a thick, solid, subvertical line with circles that denotes oxygen fugacities in log f_{O_2} units). Subhorizontal solid lines are fluid inclusion isochores calculated from various equations of state: SP-1991 (Sterner and Pitzer, 1994), SW-1996 (Span and Wagner, 1996) for CO_2-rich fluid inclusions in quartz in peraluminous pincinite, and DMW-1995 (Duan et al., 1995) for H_2O–CO_2 inclusions in metaluminous pincinite.

(Problem 6.6), which is ~2.7 orders of magnitude higher than that derived from the orthopyroxene–ilmenite–melt equilibrium. Moreover, the calculated water content in equilibrium with hydrogen present in the fluid inclusions (Problem 6.3) indicates that as much as 62 mol.% H_2O must have been removed from the inclusions by diffusion or leakage. Absence of re-equilibration cracks around the inclusions (Figure 6.12b), however, does not support the hypothetical loss of the inclusion content via cracks. A further complication arises from the fact that for any possible water concentration the fluid composition in equilibrium with the orthopyroxene–ilmenite–melt assemblage must be significantly enriched in reduced gas species (CO, CH_4, and H_2) compared to the gas composition observed in fluid inclusions. Last but not least, methane is missing in the inclusions, though it should theoretically be formed on cooling by the reaction between hydrogen and carbonic species (Problem 6.4). Hence, one must admit that either our calculations are wrong, or hydrogen diffused inward to the CO_2-rich inclusions due to a negative oxygen fugacity gradient established in the time interval between the crystallization of early quartz and that of the late interstitial orthopyroxene–ilmenite assemblage. Obviously, Roedder's rules have been violated, and fluid inclusion densities can be used to constrain pressures with caution. Independent P–T constraints are thus unconditionally desirable, and these will be provided by the pseudosection modeling discussed in detail in Section 6.5.6.

6.5.5 CALCULATING METAMORPHIC MINERAL–FLUID EQUILIBRIA

This section deals with calculation of a metamorphic H_2O–CO_2 buffer in ruby-bearing metacarbonates with graphite. The example was selected because the mineral assemblage is composed of essentially pure phases, and thus there is no need to employ solution models. Second, even if graphite is present in carbonate rocks, the graphite-buffered fluid remains highly oxidized and so can be modeled using the simple H_2O–CO_2 fluid. Last but not least, an independent control on the fluid composition and density was provided by fluid inclusions.

The hand specimen studied (Figure 6.14a) is a Cambrian marble from the Luc-Yen mining district in Vietnam, which is one of the most important sources of high-quality gem rubies and sapphires in southeast Asia. The mining district is situated along the northeastern flank of the Day Nui Con Voi Range in the Red River Shear Zone. The origin of the ruby and the associated assemblage is linked with the activity of the Red River shear zone between 35 and 40 Ma, coincidental with the deformation of the Asian plate after the collision with India (the Himalayan orogen). Gem ruby deposits are believed to have originated during a retrograde metamorphic path at temperatures of 620–670 °C and a 2.6–3.3 kbar pressure in the presence of evaporites, which played a key role in their formation (Garnier et al., 2002; Giuliani et al., 2003). Microscopic and electron microprobe investigation of thin sections revealed hydroxylclinohumite, graphite, pyrrhotite, chalcopyrite, pyrite, muscovite,

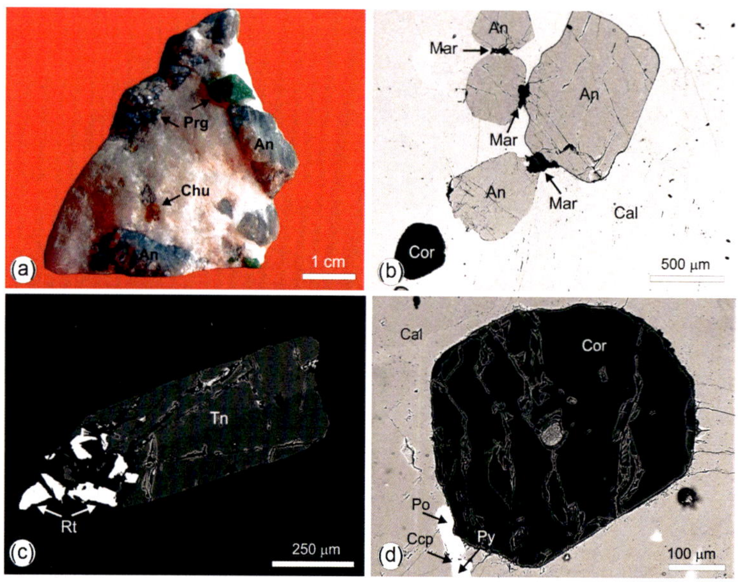

FIGURE 6.14

Mineral assemblage of clinohumite- and ruby-bearing marbles from the Luc-Yen district, Vietnam. (a) Macroscopic view of a hand specimen showing green pargasite (Prg), yellow clinohumite (Chu), and gray anorthite (An) crystals in calcite matrix. (b) Back-scattered electrons image (BSE) documenting corundum (Cor), anorthite (An), and margarite (Mar) in calcite (Cal) matrix. (c) BSE image showing the breakdown of titanite (Tn) to rutile (Rt). (d) BSE image of corundum crystal (Cor) with attached pyrite (Py), pyrrhotite (Po), and chalcopyrite (Ccp).

phlogopite, margarite, clinochlore, titanite, rutile, dolomite, ankerite, spinel, thorianite, fluorapatite, forsterite, pargasite, zircon, ruby, and sapphire (Hurai et al., 2014). Formation *P–T* conditions will be constrained using the ruby–margarite–anorthite (Figure 6.14b), titanite–rutile (Figure 6.14c), and pyrite–pyrrhotite assemblages (Figure 6.14d).

As the first step, mixed-variable diagrams, namely isobaric $T–X_O$ sections of the CaO–SiO_2–Al_2O_3–TiO_2 system saturated with a G–C–O–H fluid are to be calculated to identify the stability field of the observed mineral assemblage. The problem definition file (Figure 6.15) was created using the following specifications: (1) thermodynamic data from *hp02ver.dat*; (2) saturated FLUID phase; (3) saturated CAO and O_2 components; (4) EOS for graphite-saturated fluid (option 10); (5) unconstrained minimization—section and Schreinemaker type diagrams; and (6) exclude unnecessary phases as indicated in the caption to Figure 6.15.

Two isobaric mixed-variable diagrams obtained with VERTEX and converted to a Postscript format by the PSVDRAW are shown in Figure 6.16. Upper and lower temperature limits (450–520 °C) are constrained by the composition of pyrrhotite (92.9–93.5 mol.% FeS) coexisting with pyrite. At 1000 bar, the corundum–margarite–anorthite–rutile–titanite assemblage is stable only in H_2O-rich fluid (0–70 mol.% H_2O) at a narrow range of temperatures between 450 and 467 °C. With increasing pressure, the stability field widens and the molar proportion of CO_2 in the fluid increases up to a maximum of 78 mol.% at 450 °C. The stability field is defined by two fluid-buffering reactions: (1) `ru+q+cc=sph+CO2` and (2) `Ma=cor+an+H2O`, and by the absence of diaspore. The lowest X_O limit, approximately 0.33, is constrained by the absence of wollastonite due to reaction of quartz with $CaCO_3$: `q+cc=wo+CO2`. The oxidizing H_2O–CO_2 fluid inclusion composition in ruby was also confirmed by Raman spectroscopy (Figure 6.17), which did not reveal any other component than CO_2. Water in the fluid inclusions was corroborated only indirectly, by the occurrence of ubiquitous diaspore, proving the retrograde reaction: `cor+H2O=dsp`. This is, however, bad news for our interpretations, because the retrograde reaction indicates the violation of Roedder's rules, thus invalidating the inclusion fluid composition and isochores as representatives of the original trapping conditions.

Owing to the oxidizing character of the equilibrium fluid, we can now simplify our further calculations by increasing the lower limit of X_O in the GCOHF solution model to the value of 0.35, which eliminates unrealistic assemblages and makes the phase diagrams more synoptic.

An isobaric $T–X_O$ diagram is not very suitable for plotting the fluid inclusion isochores, particularly if their H_2O content is unknown. However, Schreinemaker's *P–T* sections can also be calculated using the PERPLE_X program package. For this purpose, H_2 and O_2 must be created as end-member thermodynamic components, and further treated using a fluid solution model of the G–C–O–H system. In such a diagram, X_O contents can be deciphered from projections of curves of a higher variance.

In the first step, the thermodynamic database of Holland and Powell (1998) (*hp02ver.dat*) must be redefined using the CTRANSF program (Figure 6.18). Then the problem definition file has to be created with BUILD using the following specifications: (1) thermodynamic data from modified *hp02ver.dat*, now represented by *ctransf.dat*; (2) no FLUID phase saturation; (3) saturation with C only; (4) solution model GCOHF; (5) select O_2 and H_2 as thermodynamic components; (6) EOS for GCOH fluid (option 10); (7) compute f_{O_2} and f_{H_2} as dependent fugacities; (8) unconstrained minimization—section and Schreinemaker type diagrams; (9) exclude unnecessary phases, except phases of interest and fluid end-members H_2 and O_2, and be sure to exclude H_2O, CO_2, CO, and CH_4. The abridged dialog with BUILD is reproduced in Figure 6.19.

```
The project definition file will be: example4.dat

Enter thermodynamic data file name [default = hp02ver.dat]:
hp02ver.dat

Enter the computational option file name [default = perplex_option.dat]:
See: www.perplex.ethz.ch/perplex_options.html
perplex_options.dat

Warning: the Perple_X option file: perplex_options.dat

was not found, default option values will be used.


The current data base components are:
 NA2O  MGO   AL2O3 SIO2  K2O   CAO   TIO2  MNO   FEO   NIO   ZRO2  CL2
 O2    H2O   CO2
Transform them (Y/N)?
n

Calculations with a saturated phase (Y/N)?
The phase is:    FLUID
Its compositional variable is: Y(CO2), X(O), etc.
y

Select the independent saturated phase components:
 H2O   CO2
Enter names, 1 per line, enter a blank line to finish:

For C-O-H fluids it is only necessary to select volatile species present in
the solids of interest. If the species listed here are H2O and CO2, then to
constrain O2 chemical potential to be consistent with C-O-H fluid speciation
treat O2 as a saturated component. Refer to the Perple_X Tutorial for details.

H2O
CO2


Calculations with saturated components (Y/N)?
y

**warning ver015** if you select > 1 saturated component, then the order you
enter the components determines the saturation heirarchy and may effect your
results (see Connolly 1990).


Select < 6 saturated components from the set:
 NA2O  MGO   AL2O3 SIO2  K2O   CAO   TIO2  MNO   FEO   NIO   ZRO2  CL2
 O2
Enter names, 1 per line, enter a blank line to finish:
CAO
O2


Use chemical potentials, activities or fugacities as independent
variables (Y/N)?
n

Select thermodynamic components from the set:
 NA2O  MGO   AL2O3 SIO2  K2O   TIO2  MNO   FEO   NIO   ZRO2  CL2
Enter names, 1 per line, enter a blank line to finish:
AL2O3
SIO2
TIO2


Select fluid equation of state:

    0 - X(CO2) Modified Redlich-Kwong (MRK/DeSantis/Holloway)
    1 - X(CO2) Kerrick & Jacobs 1981 (HSMRK)
    2 - X(CO2) Hybrid MRK/HSMRK
    3 - X(CO2) Saxena & Fei 1987 pseudo-virial expansion
    4 - Bottinga & Richet 1981 (CO2 RK)
    5 - X(CO2) Holland & Powell 1991, 1998 (CORK)
    6 - X(CO2) Hybrid Haar et al 1979/CORK (TRKMRK)
    7 - f(O2/CO2)-f(S2) Graphite buffered COHS MRK fluid
    8 - f(O2/CO2)-f(S2) Graphite buffered COHS hybrid-EoS fluid
    9 - Max X(H2O) GCOH fluid Cesare & Connolly 1993
   10 - X(O) GCOH-fluid hybrid-EoS Connolly & Cesare 1993
   11 - X(O) GCOH-fluid MRK Connolly & Cesare 1993
   12 - X(O)-f(S2) GCOHS-fluid hybrid-EoS Connolly & Cesare 1993
   13 - X(H2) H2-H2O hybrid-EoS
   14 - X(CO2) Pitzer & Sterner 1994; Holland & Powell mixing 2003
   15 - X(H2) low T H2-H2O hybrid-EoS
   16 - X(O) H-O HSMRK/MRK hybrid-EoS
   17 - X(O) H-O-S HSMRK/MRK hybrid-EoS
   18 - X(CO2) Delany/HSMRK/MRK hybrid-EoS, for P > 10 kb
   19 - X(O)-X(S) COHS hybrid-EoS Connolly & Cesare 1993
   20 - X(O)-X(C) COHS hybrid-EoS Connolly & Cesare 1993
   21 - X(CO2) Halbach & Chatterjee 1982, P > 10 kb, hybrid-Eos
   22 - X(CO2) DHCORK, hybrid-Eos
   23 - Toop-Samis Silicate Melt
   24 - f(O2/CO2)-N/C Graphite saturated COHN MRK fluid
   25 - H2O-CO2-NaCl Aranovich et al. 2010
10
```

(a)

FIGURE 6.15

Example 4 dialog with BUILD to create problem definition file for the calculation of isobaric T–X_O sections of the CaO–Al_2O_3–SiO_2–TiO_2 system saturated with graphite-buffered fluid phase. Excluded phases are as follows: lrn, gr, and, ky, sill, tpz, spu, zo, cz, law, geh, rnk, ty, cats, pswo, prl, kao, pre, trd, crst, coe, stv, me, wrk, lmt, heu, stlb, lime, arag, silL, aanL, h2oL, qL, qGL, coGL, woGL, h2oGL, mthm, qfm, sil8L, q8L.

Continued

```
Compute f(H2) & f(O2) as the dependent fugacities
(do not unless you project through carbon) (Y/N)?

n

Reduce graphite activity (Y/N)?

n

Specify computational mode:

      1 - Unconstrained minimization
      2 - Constrained minimization on a 2d grid [default]
      3 - Constrained minimization on a 1d grid
      4 - Output pseudocompound data
      5 - Phase fractionation calculations

Unconstrained optimization should be used for the calculation of composition,
mixed variable, and Schreinemakers diagrams. Gridded minimization can be used to

construct phase diagram sections.

1

The data base has P(bar)   and T(K)     as default independent potentials.
Make one dependent on the other, e.g., as along a geothermal gradient (y/n)?
n


Specify number of independent potential variables:
      0 - Composition diagram [default]
      1 - Mixed-variable diagram
      2 - Sections and Schreinemakers-type diagrams
2

Select x-axis variable:
      1 - P(bar)
      2 - T(K)
      3 - X(O)
3

Enter minimum and maximum values, respectively, for: X(O)
0.001 0.999

Select y-axis variable:
      2 - T(K)
      3 - P(bar)
2

Enter minimum and maximum values, respectively, for: T(K)
573 873

Specify sectioning value for: P(bar)
1000

Generate print file output (Y/N)?
y

Summary of valid make definitions:

             AL2O3  SIO2   O2
mthm         0.00   0.00   1.00
qfm          0.00   0.00   1.00
sil8L        1.60   1.60   0.00
q8L          0.00   4.00   0.00



Exclude phases (Y/N)?
y
Do you want to be prompted for phases (Y/N)?
y
Exclude lrn        (Y/N)?
..
...abridged...
Exclude sil8L      (Y/N)?
y
Exclude q8L        (Y/N)?
y

Do you want to treat solution phases (Y/N)?
n
```

(b)

```
Enter calculation title:
example4
```

FIGURE 6.15—Cont'd

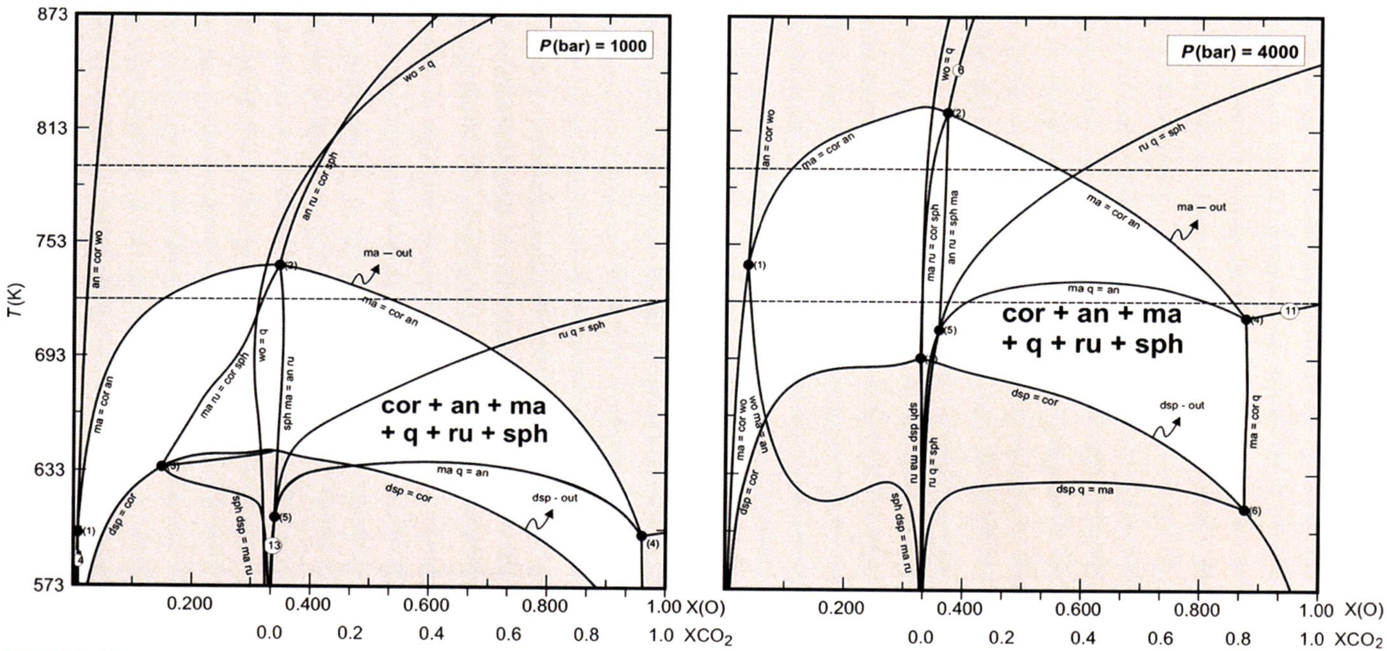

FIGURE 6.16

Stability field of the corundum–anorthite–margarite–rutile–titanite assemblage in ruby-bearing marbles displayed as T–X_O sections of the (mixed-variable diagram) at constant pressure (Example 4, Figure 6.15). The field is further constrained by the absence of diaspore. The horizontal dashed lines are temperature limits obtained from the composition of pyrrhotite (92.9–93.5 mol.% FeS), coexisting with pyrite. At 1000 bar, the assemblage is stable only in H_2O-rich fluid (0–70 mol.% H_2O) at a narrow range of temperature between 450 and 467 °C. At 4000 bar, the stability field is wider and the molar proportion of CO_2 in the fluid increases up to a maximum of 78 mol.% at 450 °C. As the system is saturated with CaO, some reactions lack calcite. For instance, `ru+q = sph` should be read as `ru+q+cc = sph+CO2`. Reactions are written such that the high-temperature assemblage is on the right of the = sign. The diagrams are outputs from PSVDRAW redrafted in CorelDraw.

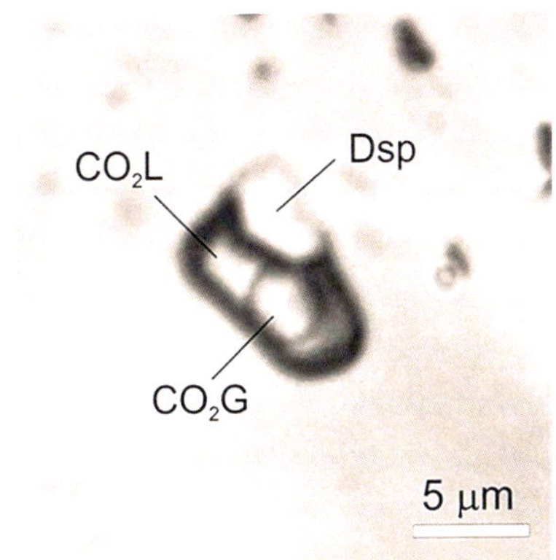

FIGURE 6.17

Primary fluid inclusion in ruby from marbles of the Luc-Yen district, Vietnam, containing CO_2 liquid and vapor phases (CO_2L, CO_2G), and diaspore daughter crystal (Dsp), which originated by reaction of the corundum host with the inclusion water.

Simplified output from VERTEX and PSVDRAW overlapped by the isochoric envelope of primary fluid inclusions in ruby is shown in Figure 6.20. Now we can summarize our observations and draw conclusions. Within the temperature range defined by the composition of pyrrhotite associated with pyrite, the margarite–corundum–anorthite–rutile–titanite assemblage is stable only within a narrow compositional range of the equilibrium fluid, namely $X_{CO_2} = 0.43 - 0.53$, and pressures between ~1800 and 4000 bar. The isochoric fluid envelope projects below the *P–T* field defined by mineral assemblages due to lowering the inclusion fluid density by retrograde reaction of H_2O with the ruby host to diaspore were also observed in earlier works (Giuliani et al., 2003). Hence, fluid inclusions in corundum formed in crustal conditions are most likely nonrepresentative of the original trapping conditions. The inferred temperatures (450–520 °C) are somewhat lower than those specified in other published papers (620–670 °C), whereas the range of pressures constrained by the thermodynamic modeling (1.8–4 kbar) is consistent with the published data (2.6–3.3 kbar).

6.5.6 THERMOBAROMETRY IN PSEUDOSECTIONS

In preceding sections we learned procedures for phase equilibria calculations for simple and complex systems in various projections using pure mineral phases and variable fluid composition. This section explains how the phase equilibria can be computed for a specific bulk composition of a system (pseudosection) without and with solution models for mineral phases. Furthermore, we demonstrate how subtle changes in the bulk composition as well as in total water content influence the computed phase boundaries.

The pincinite samples treated in previous chapters have been selected for the thermodynamic analysis because of (1) suspicious *P–T* data obtained from the orthopyroxene–ilmenite–melt oxythermometer; (2) relatively simple mineral assemblage originating at high temperature; (3) rapid quenching prior to undergoing a low-temperature recrystallization and/or interaction with fluids; (4) the possibility of direct determination of water content in interstitial glass; (5) independent

```
D:\PROGRAMY\PERPLE~3>ctransf

Perple_X version 6.6.5.9, compiled 5/8/2011.

NO is the default answer to all Y/N prompts

Enter thermodynamic data file name [default = hp02ver.dat]:
hp02ver.dat

Output will be written to file: ctransf.dat

The current data base components are:
 NA2O  MGO   AL2O3 SIO2  K2O   CAO   TIO2  MNO   FEO   NIO   ZRO2  CL2
 O2    H2O   CO2
Transform them (Y/N)?
y
Enter new component name, < 6 characters, left justified:
C
Enter old component to be replaced with C     :
O2
Enter other components (< 11) in C      1 per line, <enter> to finish:
CO2

Enter stoichiometric coefficients of:
  O2    CO2
in C     (in above order):
-1 1
 C     =  -1.00 O2     1.00 CO2
Is this correct (Y/N)?
y

The current data base components are:
 NA2O  MGO   AL2O3 SIO2  K2O   CAO   TIO2  MNO   FEO   NIO   ZRO2  CL2
 C     H2O   CO2
Transform them (Y/N)?
y
Enter new component name, < 6 characters, left justified:
O2
Enter old component to be replaced with O2    :
CO2

CO2   is a possible saturated phase component. Is the new component O2
also a possible saturated phase component (Y/N)?
y
Enter other components (< 11) in O2     1 per line, <enter> to finish:
C

Enter stoichiometric coefficients of:
  CO2   C
in O2    (in above order):
1 -1
 O2    =   1.00 CO2   -1.00 C
Is this correct (Y/N)?
y

The current data base components are:
 NA2O  MGO   AL2O3 SIO2  K2O   CAO   TIO2  MNO   FEO   NIO   ZRO2  CL2
 C     H2O   O2
Transform them (Y/N)?
y
Enter new component name, < 6 characters, left justified:
H2
Enter old component to be replaced with H2    :
H2O

H2O   is a possible saturated phase component. Is the new component H2
also a possible saturated phase component (Y/N)?
y
Enter other components (< 11) in H2     1 per line, <enter> to finish:
O2

Enter stoichiometric coefficients of:
  H2O   O2
in H2    (in above order):
1 -0.5
 H2    =   1.00 H2O   -0.50 O2
Is this correct (Y/N)?
y

The current data base components are:
 NA2O  MGO   AL2O3 SIO2  K2O   CAO   TIO2  MNO   FEO   NIO   ZRO2  CL2
 C     H2    O2
Transform them (Y/N)?
n
```

FIGURE 6.18

Dialog with CTRANSF to create modified thermodynamic database *ctransf.dat* containing new O_2, C, and H_2 components.

```
Enter thermodynamic data file name [default = hp02ver.dat]:
ctransf.dat

Enter the computational option file name [default = perplex_option.dat]:
See: www.perplex.ethz.ch/perplex_options.html


Reading computational options from: perplex_option.dat


Warning: value3 of the iteration keyword must be  >2 and <8
value3 will be assigned its default value [4].


The current data base components are:
 NA2O  MGO   AL2O3 SIO2  K2O   CAO   TIO2  MNO   FEO   NIO   ZRO2  CL2
 C     H2    O2
Transform them (Y/N)?
n

Calculations with a saturated phase (Y/N)?
The phase is:   FLUID
Its compositional variable is: Y(CO2), X(O), etc.
n

Calculations with saturated components (Y/N)?
y

**warning ver015** if you select > 1 saturated component, then the order you
enter the components determines the saturation heirarchy and may effect your
results (see Connolly 1990).


Select < 6 saturated components from the set:
 NA2O  MGO   AL2O3 SIO2  K2O   CAO   TIO2  MNO   FEO   NIO   ZRO2  CL2
 C     H2    O2
Enter names, 1 per line, enter a blank line to finish:
C


Use chemical potentials, activities or fugacities as independent
variables (Y/N)?
n

Select thermodynamic components from the set:
 NA2O  MGO   AL2O3 SIO2  K2O   CAO   TIO2  MNO   FEO   NIO   ZRO2  CL2
 H2    O2
Enter names, 1 per line, enter a blank line to finish:
AL2O3
SIO2
CAO
TIO2
H2
O2


**warning ver016** you are going to treat a saturated (fluid) phase component
as a thermodynamic component, this may not be what you want to do.


**warning ver016** you are going to treat a saturated (fluid) phase component
as a thermodynamic component, this may not be what you want to do.


Select fluid equation of state:

    0 - X(CO2) Modified Redlich-Kwong (MRK/DeSantis/Holloway)
    1 - X(CO2) Kerrick & Jacobs 1981 (HSMRK)
    2 - X(CO2) Hybrid MRK/HSMRK
    3 - X(CO2) Saxena & Fei 1987 pseudo-virial expansion
    4 - Bottinga & Richet 1981 (CO2 RK)
    5 - X(CO2) Holland & Powell 1991, 1998 (CORK)
    6 - X(CO2) Hybrid Haar et al 1979/CORK (TRKMRK)
    7 - f(O2/CO2)-f(S2) Graphite buffered COHS MRK fluid
    8 - f(O2/CO2)-f(S2) Graphite buffered COHS hybrid-EoS fluid
    9 - Max X(H2O) GCOH fluid Cesare & Connolly 1993
   10 - X(O) GCOH-fluid hybrid-EoS Connolly & Cesare 1993
   11 - X(O) GCOH-fluid MRK Connolly & Cesare 1993
   12 - X(O)-f(S2) GCOHS-fluid hybrid-EoS Connolly & Cesare 1993
   13 - X(H2) H2-H2O hybrid-EoS
   14 - X(CO2) Pitzer & Sterner 1994; Holland & Powell mixing 2003
   15 - X(H2) low T H2-H2O hybrid-EoS
   16 - X(O) H-O HSMRK/MRK hybrid-EoS
   17 - X(O) H-O-S HSMRK/MRK hybrid-EoS
   18 - X(CO2) Delany/HSMRK/MRK hybrid-EoS, for P > 10 kb
   19 - X(O)-X(S) COHS hybrid-EoS Connolly & Cesare 1993
   20 - X(O)-X(C) COHS hybrid-EoS Connolly & Cesare 1993
   21 - X(CO2) Halbach & Chatterjee 1982, P > 10 kb, hybrid-Eos
   22 - X(CO2) DHCORK, hybrid-Eos
   23 - Toop-Samis Silicate Melt
   24 - f(O2/CO2)-N/C Graphite saturated COHN MRK fluid
   25 - H2O-CO2-NaCl Aranovich et al. 2010
 10

Compute f(H2) & f(O2) as the dependent fugacities
(do not unless you project through carbon) (Y/N)?
y
```

(a)

FIGURE 6.19

Example 4 dialog with BUILD to create the problem definition file for calculation of the Schreinemaker *P–T* projection of the graphite-saturated $CaO–TiO_2–Al_2O_3–SiO_2–O_2–H_2$ system.

Continued

```
Reduce graphite activity (Y/N)?

n

Specify computational mode:

     1 - Unconstrained minimization
     2 - Constrained minimization on a 2d grid [default]
     3 - Constrained minimization on a 1d grid
     4 - Output pseudocompound data
     5 - Phase fractionation calculations

Unconstrained optimization should be used for the calculation of composition,
mixed variable, and Schreinemakers diagrams. Gridded minimization can be used to

construct phase diagram sections.

1

The data base has P(bar)   and T(K)     as default independent potentials.
Make one dependent on the other, e.g., as along a geothermal gradient (y/n)?
n


Specify number of independent potential variables:
     0 - Composition diagram [default]
     1 - Mixed-variable diagram
     2 - Sections and Schreinemakers-type diagrams
2

Select x-axis variable:
     1 - P(bar)
     2 - T(K)
2

Enter minimum and maximum values, respectively, for: T(K)
573 873

Enter minimum and maximum values, respectively, for: P(bar)
1000 4000

Generate print file output (Y/N)?
y

Summary of valid make definitions:

          AL2O3 SIO2  O2
mthm      0.00  0.00  1.00
qfm       0.00  0.00  1.00
sil8L     1.60  1.60  0.00
q8L       0.00  4.00  0.00



Exclude phases (Y/N)?
y
Do you want to be prompted for phases (Y/N)?
y
Exclude lrn      (Y/N)?
```

Exclude unnecessary phases, except the phases of interest and fluid endmembers (O2, H2). Be sure to exclude H2O, CO2, CO, CH4

```
Exclude q8L      (Y/N)?
y

Do you want to treat solution phases (Y/N)?
y
Enter solution model file name [default = solution_model.dat]:


**warning ver025** 0 endmembers for Bio(TCC)   The solution will not be consider
ed.
**warning ver025** 0 endmembers for Opx(HP)    The solution will not be consider
ed.
```

... Abridged...

```
Select phases from the following list, enter 1 per line,
enter a blank line to finish

  melt(HP)    pMELTS(G)   MELTS(GS)   F           F(salt)     GCOHF

GCOHF


Enter calculation title:
example4_
```

(b)

FIGURE 6.19—Cont'd

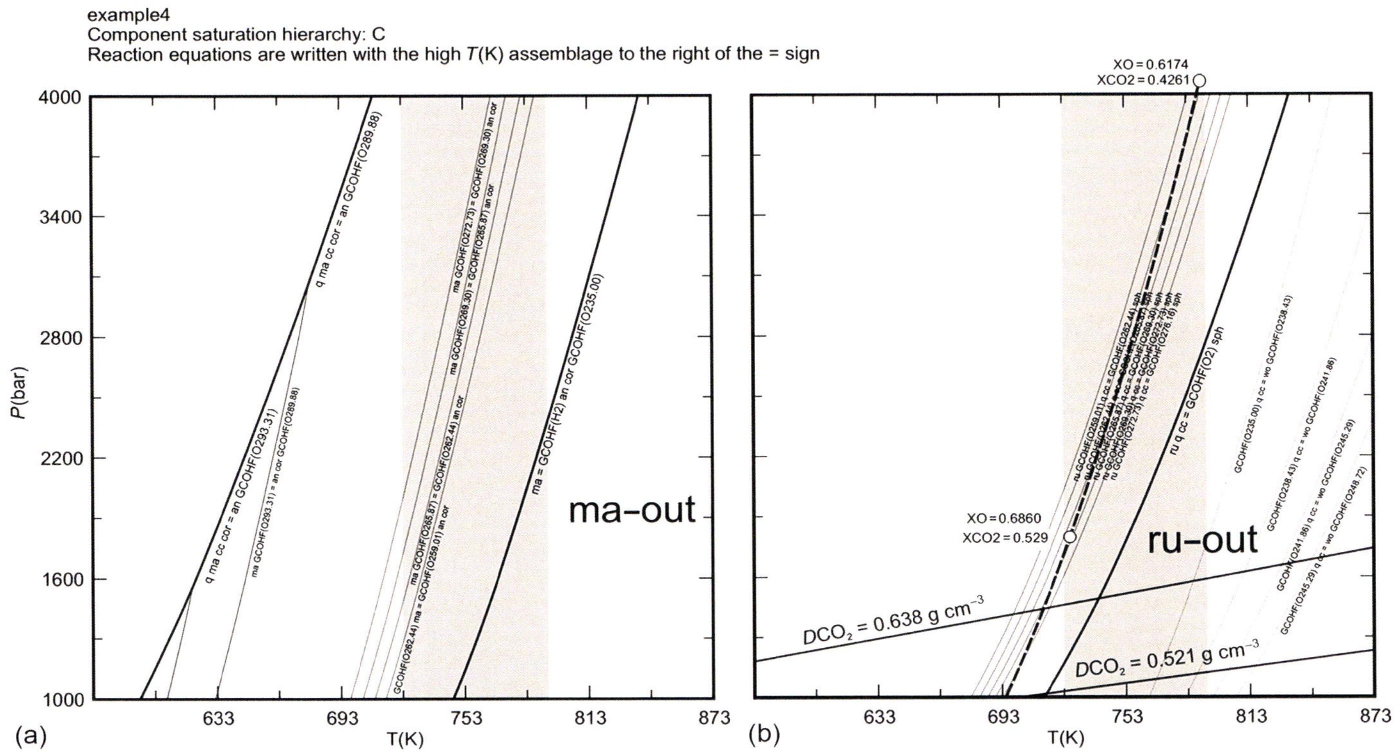

FIGURE 6.20

Schreinemaker's P–T projection of the graphite-saturated CaO–TiO_2–Al_2O_3–SiO_2–O_2–H_2 system relevant to the phase assemblage in ruby-bearing metacarbonates of the Luc-Yen district, Vietnam. Selected projections are shown of divariant planes delineating stability limits of (a) margarite, (b) rutile and wollastonite at different X_O values. The thick curves represent univariant reactions. Assuming that fluid composition was held constant, the stability of the margarite–corundum–anortite–rutile–sphene assemblage is possible only near the thick-dashed curve in (b), which represents an intersection of two divariant reactions (1) `ru+q+cc+GCOHF1 = sphene+GCOHF2` and (2) `ma+GCOHF1 = an+cor+GCOHF2`. Selected projection curves are also shown for the reaction: `q+cc+GCOHF1 = wo+GCOHF1` illustrating that wollastonite must be unstable due to excess water in the fluid at the range of X_O values defined by reactions 1 and 2. This is in accordance with the absence of wollastonite in the samples studied. The shaded rectangle delineates temperatures derived from the composition of pyrrhotite associated with pyrite. Two oblique subhorizontal solid lines represent P–T limits defined by the isochoric envelope constructed from homogenization temperatures of CO_2-rich inclusions in ruby and calculated using EOS for pure CO_2 (Span and Wagner, 1996). See the text for more details.

constraint on possible *P–T* conditions by fluid inclusion isochores; and (6) independent constraint on oxygen fugacities provided by the fluid inclusion composition.

The problem definition file created with BUILD has the following specifications: (1) thermodynamic data from *hp02ver.dat*; (2) no saturated FLUID phase; (3) saturation with SIO_2 (constrained saturation) and O_2 (unconstrained saturation); (4) EOS for GCOH fluid (option 8) with low graphite activity (0.01); (5) αQMF buffer offset by $-2\log f_{O_2}$ units; (6) compute f_{O_2} and f_{H_2} as **independent** fugacities; (7) minimization constrained by the composition of intergranular glass with 2 mass% H_2O (Table 6.8) normalized to 100%; and (8) solution models used: Opx(HP), melt(HP), Pl(h), San, and Ilm(WPH). The abridged dialog with BUILD is shown in Figure 6.21.

To activate the Ilm(WPH) solution model, which enables the calculation of hematite, geikielite, and ilmenite end-member abundance, two new thermodynamic entities (`oilm=1 ilm` and `dilm=1 ilm`) must be created before running VERTEX in the make-definition segment of the *hp02ver.dat* with the excess functions noted in the heading to the Ilm(WPH) model in the *solution_model.dat*. The melt(HP) solution model has a total of eight end-members and their combination sometimes results in an astronomical number of pseudocompounds (>10,000) and extremely long computation times (several hours). To avoid this situation, it is recommended to do the following: (1) accommodate subdivision ranges in the melt(HP) solution model; (2) reduce grid levels from four to three or two; and (3) reduce the number of nodes in the auto-refine level (e.g., from 60 to 40 or less). The last two changes have to be done by modifying the *perplex_option.dat* file in any text-processing program. After the initial exploratory computation is performed and the relevant phase fields are identified, the final computation should be done within the reduced *P–T* frame with full resolution at four grid levels and with the increased number of nodes in the auto-refinement stage.

A Postscript file created by the PSSECT from the output from the VERTEX and redrafted in a graphical editor is shown in Figure 6.22. The constrained minimization reveals that the stability field of the Opx–Ilm–Kfs–melt assemblage of the metaluminous pincinite closes at 3600 bar and 865 °C, thus defining the uppermost *P–T* limit. The lowermost *P–T* limit (855 °C and 3260 bar) is located at the intersection of the stability field with the isochore of the densest CO_2–H_2O fluid inclusion in quartz. Hence, the combined fluid inclusion and thermodynamic constraints demarcate the possible range of equilibrium *P–T* conditions for the Opx–Ilm–Kfs–melt assemblage to the values of 855–865 °C and 3.3–3.6 kbar.

A phase diagram for the peraluminous pincinite calculated in the same manner is shown in Figure 6.23. The diagram was calculated with the specifications previously noticed, but the offset from the αQMF buffer was set to zero. The electron microprobe analysis of the intergranular glass from Table 6.8 with 2 mass% of H_2O, all normalized to 100, was used as the bulk compositional constraint. The main reason we use the composition of the intergranular glass instead of that of whole rock stems from the fact that we are not able to decide whether the investigated pincinites contain restitic quartz and/or plagioclase, and if they do or do not represent a flotation cumulate. In both cases, the bulk composition is nonrepresentative of the parental magma.

With the absence of potassium feldspar, the possible range of equilibrium *P–T* conditions for the Opx–Ilm–melt assemblage of the peraluminous pincinite is much broader than that in the metaluminous pincinite. Maximum temperatures are limited by the Opx-out boundary, for example, ~960 °C at 7 kbar, whereas the isochore for the densest CO_2 inclusion in quartz delineates the minimum *T* and *P* (855 °C, 5.6 kbar) at the intersection with the Kfs-out boundary.

```
Calculations with a saturated phase (Y/N)?
The phase is:   FLUID
Its compositional variable is: Y(CO2), X(O), etc.
n

Calculations with saturated components (Y/N)?
y

**warning ver015** if you select > 1 saturated component, then the order you
enter the components determines the saturation heirarchy and may effect your
results (see Connolly 1990).


Select < 6 saturated components from the set:
 NA2O  MGO   AL2O3 SIO2  K2O   CAO   TIO2  MNO   FEO   NIO   ZRO2  CL2
 O2    H2O   CO2
Enter names, 1 per line, enter a blank line to finish:
SIO2
O2


Use chemical potentials, activities or fugacities as independent
variables (Y/N)?
n

Select thermodynamic components from the set:
 NA2O  MGO   AL2O3 K2O   CAO   TIO2  MNO   FEO   NIO   ZRO2  CL2   H2O
 CO2
Enter names, 1 per line, enter a blank line to finish:
AL2O3
CAO
NA2O
K2O
FEO
MGO
TIO2
H2O


**warning ver016** you are going to treat a saturated (fluid) phase component
as a thermodynamic component, this may not be what you want to do.


Select fluid equation of state:

    0 - X(CO2) Modified Redlich-Kwong (MRK/DeSantis/Holloway)
    1 - X(CO2) Kerrick & Jacobs 1981 (HSMRK)
    2 - X(CO2) Hybrid MRK/HSMRK
    3 - X(CO2) Saxena & Fei 1987 pseudo-virial expansion
    4 - Bottinga & Richet 1981 (CO2 RK)
    5 - X(CO2) Holland & Powell 1991, 1998 (CORK)
    6 - X(CO2) Hybrid Haar et al 1979/CORK (TRKMRK)
    7 - f(O2/CO2)-f(S2) Graphite buffered COHS MRK fluid
    8 - f(O2/CO2)-f(S2) Graphite buffered COHS hybrid-EoS fluid
    9 - Max X(H2O) GCOH fluid Cesare & Connolly 1993
   10 - X(O) GCOH-fluid hybrid-EoS Connolly & Cesare 1993
   11 - X(O) GCOH-fluid MRK Connolly & Cesare 1993
   12 - X(O)-f(S2) GCOHS-fluid hybrid-EoS Connolly & Cesare 1993
   13 - X(H2) H2-H2O hybrid-EoS
   14 - X(CO2) Pitzer & Sterner 1994; Holland & Powell mixing 2003
   15 - X(H2) low T H2-H2O hybrid-EoS
   16 - X(O) H-O HSMRK/MRK hybrid-EoS
   17 - X(O) H-O-S HSMRK/MRK hybrid-EoS
   18 - X(CO2) Delany/HSMRK/MRK hybrid-EoS, for P > 10 kb
   19 - X(O)-X(S) COHS hybrid-EoS Connolly & Cesare 1993
   20 - X(O)-X(C) COHS hybrid-EoS Connolly & Cesare 1993
   21 - X(CO2) Halbach & Chatterjee 1982, P > 10 kb, hybrid-Eos
   22 - X(CO2) DHCORK, hybrid-Eos
   23 - Toop-Samis Silicate Melt
   24 - f(O2/CO2)-N/C Graphite saturated COHN MRK fluid
   25 - H2O-CO2-NaCl Aranovich et al. 2010
8

Modify default buffer (max H2O) (Y/N)?
y

Select buffer:

 1 - aQFM, 298-1200K
 2 - Maximum H2O content, 523-1273K, .5-30kbar
 3 - user specified f(O2)
 4 - aQ-Ru-Cc-Tn-Gph
 5 - ln(f(O2))= a + (b + c*p)/t + d/t**2 + e/t**3

1

Modify calculated fO2 by a constant (Y/N)?

y

Enter constant in units of log10(fO2):

-2

Compute f(H2) & f(O2) as the dependent fugacities
(do not unless you project through carbon) (Y/N)?

n

Reduce graphite activity (Y/N)?

y
```

FIGURE 6.21

Abridged problem definition file created by BUILD (Example 5) to calculate a *P–T* projection of the NCKFMASTH system constrained by the composition of interstitial glass of the metaluminous pincinite.

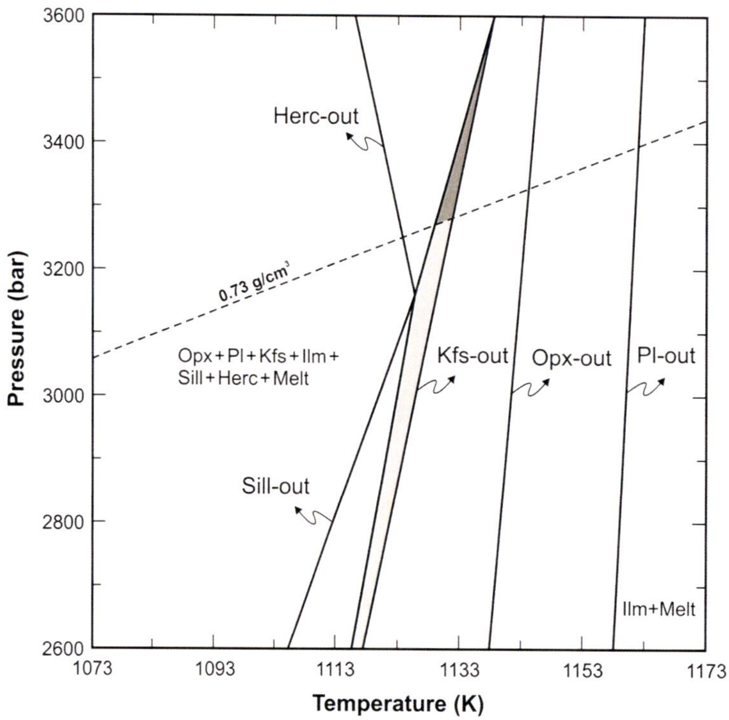

FIGURE 6.22

P–T section of a quartz-saturated NCKFMASTH system constrained by the composition of interstitial glass in the metaluminous pincinite (mass%: Na_2O—0.81; CaO—0.75; K_2O—6.37; FeO—2.58; MgO—0.27; Al_2O_3—13.84; SiO_2—72.44; TiO_2—0.87; H_2O—2.0) calculated for log $f_{O_2} = \alpha$QMF—2 log units and carbon activity equal to 0.01. The thermodynamic database of Holland and Powell (1998) solution models for melt—melt(HP); orthopyroxene—Opx(HP); sanidine—San; plagioclase—Pl(h); and ilmenite—Ilm(WPH); and the EOS for GCOH fluid (Connolly and Cesare, 1993) were used (option 8 in BUILD). The contours of stable assemblages in the plot output generated by the VERTEX and the PSSECT were redrafted and smoothed in the CorelDraw. Other abbreviations: herc, hercynite; sill, sillimanite. Isochore for the densest CO_2–H_2O–NaCl inclusion in quartz (0.73 g cm^{-3}) calculated using the EOS by Duan et al. (1995).

Note that with the slightly changing composition of the intergranular melt the stability field of the Opx–Ilm–Kfs–melt assemblage extends to higher pressures and temperatures. At fixed bulk composition but different H_2O contents, the phase boundaries also change significantly. For instance, the Kfs-out boundary shifts to lower temperatures at a slightly increased H_2O content, and the Opx–Ilm–Kfs–melt stability field diminishes.

The phase diagram for peraluminous pincinite can be used to demonstrate possibilities offered by the solution models. In this case, various phase parameters, for instance, molar fractions of end-members, mass percentages of oxides, water content in silicate melt, can be plotted as isolines in the *P–T* space. In an optimum case, the end-member composition in coexisting minerals could serve

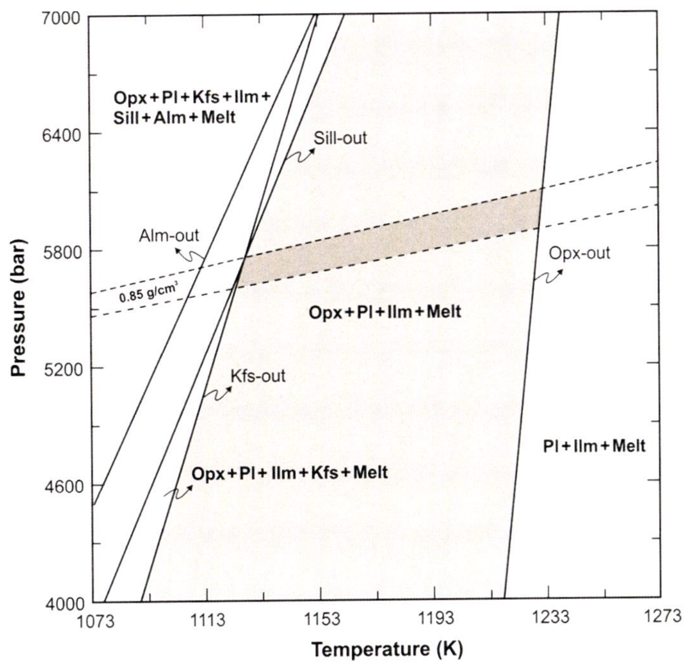

FIGURE 6.23

P–T section of a quartz-saturated NCKFMASTH system constrained by the composition of interstitial glass in the peraluminous pincinite (mass%: Na_2O—2.42; CaO—1.96; K_2O—3.16; FeO—5.11; MgO—0.76; Al_2O_3—13.61; SiO_2—70.09; TiO_2—0.84; H_2O—2.0), calculated for $\log f_{O_2} = \alpha$QMF and carbon activity equal to 0.01. The thermodynamic database of Holland and Powell (1998) includes solution models for melt—melt(HP); orthopyroxene—Opx(HP); sanidine—San; plagioclase—Pl(h); and ilmenite—Ilm(WPH); and the EOS for GCOH fluid (Connolly and Cesare, 1993), which were used (option 8 in BUILD). Contours of stable assemblages in the plot output generated by VERTEX and PSSECT were redrafted and smoothed in CorelDraw. Other abbreviations include: alm, almandine; sill, sillimanite. Two isochores for the densest CO_2 inclusion in quartz (0.85 g cm^{-3}) were calculated using the EOS by Sterner and Pitzer (1994) and Span and Wagner (1996).

as a thermobarometer defining more precisely the equilibrium *P–T* conditions within the corresponding stability field, or along phase boundary.

Phase properties can be extracted from the VERTEX output file using WERAMI. The dialog with the program is intuitive and need not be reproduced here. However, a sample calculation of molar proportion of hematite end-member in ilmenite is discussed in Problem 6.7. The output from WERAMI is stored as a *.tab* file, which can be converted to an editable Postscript file by PSTABLE. The contour interval set by PSSTABLE, however, cannot be changed. Moreover, the output is in a black-and-white format, which is sometimes less synoptic. The PyWERAMI program developed by Ondrej Lexa and maintained at http://petrol.natur.cuni.cz/~ondro/pywerami:home is an easy-to-use alternative for plotting contours with user-defined intervals and increments. Several examples of the

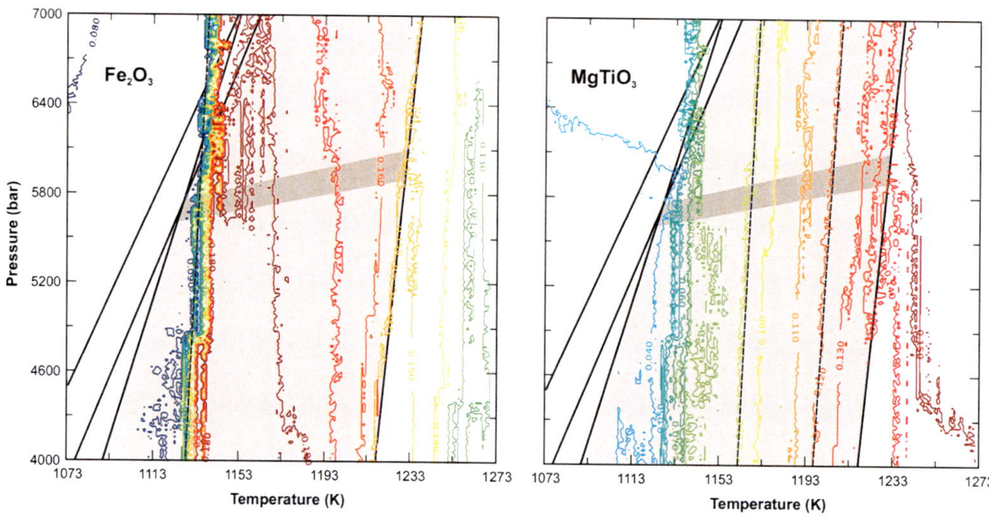

FIGURE 6.24

Isopleths of molar proportions of hematite and geikielite end-members in ilmenite from peraluminous pincinite calculated using WERAMI and PyWERAMI. The shaded field corresponds to the stable Opx–Ilm–Pl–melt assemblage. The darker shaded subfield is that delineated by isochores for the densest CO_2 inclusion. The subvertical dashed lines in the left diagram mark the highest hematite concentration determined in ilmenite from peraluminous pincinite. Two subvertical dashed lines in the right diagram demarcate the range of geikielite molar concentrations in the same ilmenite. Figure 6.23 identifies phases in other fields.

PyWERAMI outputs overlapped onto the phase diagram of peraluminous pincinite are shown in Figures 6.24 and 6.25.

The calculated isopleths, however, did not help much in the refinement of the possible equilibrium *P–T* conditions of the Opx–Ilm–melt assemblage because the phase compositions calculated using the solution models did not match those determined by electron microprobe. For instance, 8–10 mol.% of hematite and 9–14 mol.% of geikielite components were determined in ilmenite from the peraluminous pincinite by electron microprobe, but these intervals do not overlap in the contour plots in Figure 6.24. The molar proportion of enstatite (26–44 mol.%) determined by electron microprobe in orthopyroxene stretches over the whole stability field of the Opx–Ilm–melt assemblage. In addition, the composition of orthopyroxene in metaluminous pincinite with as much as 54–61 mol.% of the enstatite component cannot be reproduced by any thermodynamic model and bulk composition, oxygen fugacity, or water content. The phase composition in both pincinite types seems to be controversial, because the Mg-rich ilmenite and orthopyroxene crystallized from the metaluminous pincinite in the presence of potassium feldspar, and so at a lower temperature. In contrast, Fe-rich ilmenite and orthopyroxene crystallized in the peraluminous pincinite at a higher temperature.

Now let us summarize the results of the thermodynamic modeling. It was unequivocally confirmed that the Opx–Ilm–melt oxythermometer yielded unrealistic results inconsistent with the liquidi and H_2O solubility curves in the granite–H_2O system (e.g., Holtz and Johannes, 1994). For instance, extreme pressures of several GPa would be needed to dissolve 2 mass% H_2O in the peraluminous pincinite melt at a temperature of ~1160 °C calculated from the oxythermometer (Figure 6.26).

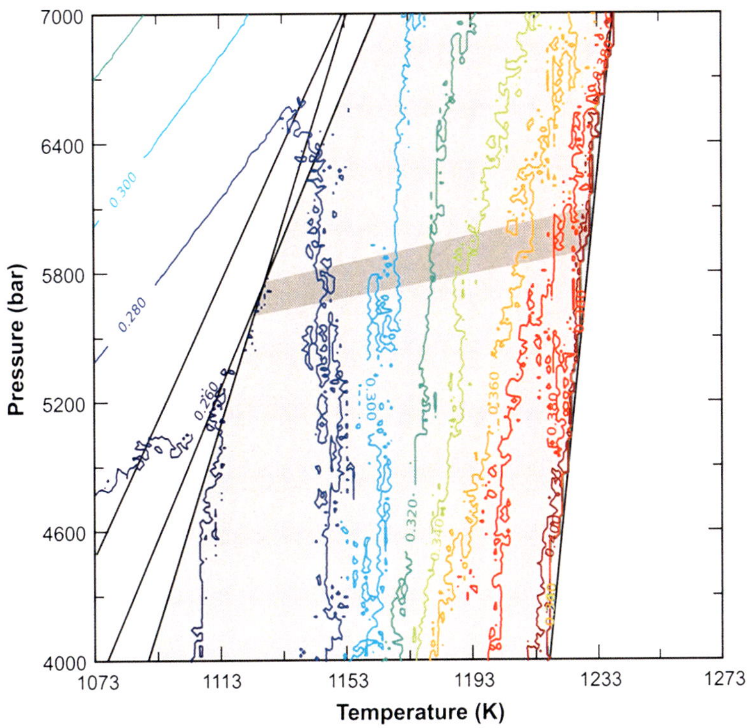

FIGURE 6.25

Isopleths of the molar proportions of enstatite in orthopyroxene from peraluminous pincinite calculated using WERAMI and visualized by PyWERAMI. The shaded region is the stability field of the Opx–Ilm–Pl–melt assemblage. The dark-shaded subfield is that delineated by isochores of the densest CO_2 inclusion. The identification of phases in other fields is depicted in Figure 6.23.

Similarly, the equilibrium temperature of ~750 °C indicated by the Opx–Ilm–melt oxythermometer in the metaluminous pincinite is too low to account for the water-oversaturation of the equilibrium fluid documented by the presence of aqueous liquid in fluid inclusions, because at this temperature and a pressure of 2.8 kbar constrained by the fluid inclusion isochores, at least 4 mass% of water must have been dissolved in the melt. In that case, 50 mass% of water would have to be lost by the devolatilization of the interstitial glass from the metaluminous pincinite. It is unreasonable to suppose that the interstitial melt would not be outgassed in an equal extent from the peraluminous pincinite. Hence, the granite liquidus with 4 mass% of water defines the uppermost *P–T* limit for both pincinite types. With 8 kbar of pressure imposed by the lithostatic load of a 28–30 km thick crust in the Pannonian Basin, the temperature of the pincinite equilibration should not exceed 820 °C.

Obviously, in spite of the discrepancies between calculated and measured phase compositions, the thermodynamic modeling much better constrained the possible formation *P–T* conditions than did the Opx–Ilm–melt oxythermometer. This is probably because divariant planes of two involved oxybarometers intersect at a very small angle, resulting in a large uncertainty of the oxythermometer. Hence, it is better to formulate any oxythermometer as an intersection of an oxygen-buffering reaction with

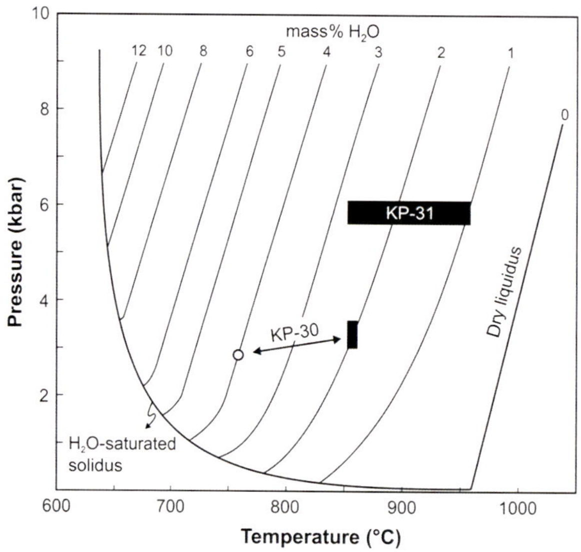

FIGURE 6.26

Formation *P–T* conditions of peraluminous (KP-31) and metaluminous (KP-30) pincinite xenoliths plotted on liquidus curves for a given amount of dissolved H_2O (solid lines) and H_2O-solubility curves in the Qtz–Ab–Or–H_2O system for minimum and eutectic compositions (Holtz and Johannes, 1994). The open circle denotes equilibrium *P–T* conditions of the metaluminous pincinite calculated using the Opx–Ilm–melt oxythermometer. The solid rectangles are *P–T* conditions constrained by the pseudosection modeling, isochores of fluid inclusions, and water content in the interstitial glass.

an independent thermometer, as was done in the case of the commonly used ilmenite–magnetite oxythermometer.

Pseudosection calculation showed that formation temperatures of the two pincinite types are roughly similar, probably ranging between 850 and 900 °C. This is a more reasonable interpretation considering the qualitatively similar mineral assemblage. The estimated temperatures are also consistent with the water content of about 2 mass% determined in the interstitial glass by Raman spectroscopy. Regarding the possible effects of hydrogen diffusion and water leakage on the position of fluid inclusion isochores, and also bearing in mind the possible extension of the Opx–Kfs–Ilm–melt stability field up to 5.6 kbar (Figure 6.25), it seems reasonable to suppose that both pincinite types might have formed at the same depth, and their different phase, fluid composition, and oxidation state reflect the temperature, oxygen fugacity, and water activity gradients between the basaltic magma reservoir and the surrounding crustal rocks. The depth of the magmatic reservoir is now defined more precisely by the lithostatic load between 5.6 and 6 kbar constrained by the thermodynamic modeling. Assuming the average density of crustal rocks corresponds to about 2.75 g cm^{-3}, the lithostatic load equals a depth of 21–22 km (see Problem 4.11 for details of the calculation of burial depth).

Calculations with different model parameters showed that variable oxygen fugacities (e.g., 2logf_{O_2} units below or above the αQMF buffer) do not affect positions of the calculated phase boundaries.

In contrast, minute differences in the bulk water content significantly shift the phase boundaries even at fixed bulk composition. The same effect is also imposed by changes in the bulk rock composition itself. Obviously, determination of these two parameters is critical in any pseudosection modeling. Though the analysis of the bulk rock composition is trivial, estimation of the bulk water content pertaining to the formation *P–T* conditions is impossible. For this reason, pseudosections of metamorphic rocks are usually calculated assuming saturation with water or excess COH fluid buffered by graphite. This approach may lead to significant errors, particularly in high-grade metamorphic rocks, which underwent dehydration melting, extraction of the partial melt from the system, and retrograde reactions accompanied by input of hydrothermal fluids.

The problem of unknown water content can sometimes be solved analyzing mineral assemblages along $T - X_{H_2O}$ sections, if the pressure is constrained as the dependent variable by isochores of fluid inclusions. This can easily be done with BUILD, where *P* and *T* can be defined along a geothermal gradient or isochore. In this case, the user is asked to prompt parameters of a linear equation (the slope and the offset from zero), which can be easily calculated by data regression in EXCEL® spreadsheet. This approach was used by Hurai et al. (2010) to estimate the minimum water content of the ultra-high-pressure metapelite from the Pohorje Mountains of Slovenia along the retrograde *P–T* path. Here, the following was used for the pseudosection calculation: the liberation of free fluid phase proven by aqueous inclusions in retrograde garnet, crystallization of staurolite, and the absence of chlorite constrained the bulk water content along the retrograde trajectory to a range between 1.9 and 2.4 mass%.

Another key assumption in pseudosection modeling is that the system is in overall equilibrium. If this were true, the rock should contain only $c+2$ phases and all minerals should be homogeneous. Obviously, this is rarely true in real rocks, where minerals often exhibit compositional zoning and relics of preserved prograde and retrograde equilibria. Moreover, heterogeneity at a greater scale makes it difficult to select a representative bulk composition. In spite of this, pseudosection modeling yields useful constraints on the history and *P–T* evolution of metamorphic and magmatic rocks, if supplemented by conventional thermobarometers, fluid inclusion data, and other independent constraints.

PROBLEM 6.7 CALCULATING ISOPLETHS OF END-MEMBERS IN SOLID SOLUTIONS

Calculate isopleths for molar fractions of hematite end-member in ilmenite and that of enstatite in orthopyroxene using the WERAMI program.

Solution

Select the option "mol" for the phase composition in the *perplex_option.dat* and run VERTEX with the problem definition file of Example 5 (Figure 6.21) created by BUILD. In the output file, molar proportions of FeO, MgO, and other oxides will be calculated. Run WERAMI and select the option 8 (Composition mol or wt of a solution). You will be asked to enter the name of a solution model, the molar proportion of a component, and its weighing factor.

In the case of ilmenite, the molar proportion of FEO also includes Fe^{3+}. Hence,

$$Fe_{tot} = Fe^{3+} + Fe^{2+} \qquad (6.109)$$

Continued

PROBLEM 6.7 CALCULATING ISOPLETHS OF END-MEMBERS IN SOLID SOLUTIONS—CONT'D

The sum of molar proportions of geikielite ($MgTiO_3$) and ilmenite ($Fe^{2+}TiO_3$) is equal to the total amount of available TiO_2

$$\mathrm{Ti} = \mathrm{Mg} + \mathrm{Fe}^{2+} \tag{6.110}$$

Solving Equation (6.110) for Fe^{2+} and substituting in Equation (6.109) yields

$$\mathrm{Fe}^{3+} = \mathrm{Fe}_{\mathrm{tot}} + \mathrm{Mg} - \mathrm{Ti}$$

and for molar fraction of the hematite end-member

$$\mathrm{hem} = 0.5\mathrm{Fe}^{3+} = 0.5\mathrm{Fe}_{\mathrm{tot}} + 0.5\mathrm{Mg} - 0.5\mathrm{Ti} \tag{6.111}$$

Now we can easily trick the WERAMI by setting three components (FEO, MGO, TiO_2) with weighing factors 0.5, 0.5, and −0.5, respectively, into numerator and zero components in the denominator to obtain the molar fraction of hematite component in ilmenite.

The procedure is simpler for orthopyroxene, which consists of only two end-members with cations of the same valence. Hence, we can calculate directly the enstatite molar fraction by setting MGO into the numerator, and MGO and FEO into the denominator, all with the weighing factor equal to unity for all components. The same procedure can be used to calculate the albite content in plagioclase using the Pl(h) solution model.

CHAPTER

7 RAMAN AND INFRARED SPECTROSCOPIC ANALYSIS

CHAPTER OUTLINE

Raman spectroscopy records the interaction of electromagnetic radiation with molecular vibrations as a shift in the vibration from an incident source. In contrast, infrared spectroscopy records the wave number at which molecular rotations give rise to absorption. In most cases, the infrared and Raman spectra are complementary because the vibrations and rotations invisible in the infrared spectrum are visible in the Raman spectrum. With the development of desktop devices and sensitive detectors, the Raman and infrared spectroscopic microanalysis have become routine nondestructive methods of geofluid research.

The inelastic scattering of light was predicted by the Austrian theoretical physicist Adolf Smekal (1895–1959) and published in 1923 (Smekal, 1923). The Raman effect, sometimes called the Smekal–Raman effect, was named after one of its discoverers, the Indian physicist Chandrasekhara Venkata Raman (1888–1970), who observed that the visible wavelength of a small fraction of light scattered by liquids differs from that of the incident beam (Raman and Krishnan, 1928). Russian scientists Grigory S. Landsberg (1890–1957) and Leonid I. Mandelstam (1879–1944) independently discovered the combinatorial scattering of light in crystals at the same time (Landsberg and Mandelstam, 1928). Nevertheless, Raman was awarded the 1930 Nobel Prize for the discovery of the inelastic scattering.

Geofluids. http://dx.doi.org/10.1016/B978-0-12-803241-1.00007-1

The fundamental theory of inelastic scattering was outlined by the Czech physicist George Placzek (1905–1955) and published in 1934 (Placzek, 1934).

The first attempts to analyze fluid inclusions using a Raman spectrometer were made by Delhaye and Dhamelincourt (1975), Rosasco and Roedder (1975), and Dhamelincourt et al. (1979). A review of the principles of Raman spectroscopy and its applications to geosciences in general was provided by Nasdala et al. (2004) and Dubessy et al. (2012). Burke (1995, 2001), Burruss (2003), and Frezzotti et al. (2012b) reviewed applications of Raman spectroscopy in geofluid research. Because the principles of both spectroscopic methods are very similar, they are often treated together (Colthup et al., 1990; Larkin, 2011; and others).

7.1 FUNDAMENTALS

Electromagnetic radiation is characterized by the length of one wave (wavelength, λ), the number of vibrations per time unit (frequency, ν), and the number of waves per length unit (wave number, $\bar{\nu}$). All three parameters are related by the following equation

$$\bar{\nu} = \frac{\nu}{c/n} = \frac{1}{\lambda} \tag{7.1}$$

where c is the velocity of light in a vacuum, and c/n is the velocity of light in a medium with a refractive index n. The frequency ν is independent of the medium and is expressed in cycles per second (Hertz, Hz). The wave number is sometimes referred to as the frequency in cm^{-1} meant as the frequency in Hz, divided by the velocity of light in cm/s. Ultraviolet, visible, and infrared regions of the electromagnetic spectrum assume the values of all three parameters as shown in Table 7.1.

The use of absolute wave numbers is impractical in Raman spectroscopy because the absolute wave number of each Raman band depends on the wave number of the incident light. Hence, it has become usual to express the Raman shifts using the relative wave number, that is, the wave number difference between the incident and the scattered light.

If a molecule interacts with an electric field, a transfer of energy from the field to the molecule can occur only when Bohr's frequency condition is satisfied (Ferraro et al., 2003)

$$E = h\nu = hc\bar{\nu} \tag{7.2}$$

Table 7.1 Frequency, Wavelengths, and Wave Numbers of Electromagnetic Radiation (Colthup et al., 1990)

Region	λ (cm)	$\bar{\nu}$ (cm^{-1})	ν (Hz)
Ultraviolet			
Far	$1\times10^{-6}-2\times10^{-5}$	1,000,000–50,000	$3\times10^{16}-1.5\times10^{15}$
Near	$2\times10^{-5}-3.8\times10^{-5}$	50,000–26,300	$1.5\times10^{15}-7.9\times10^{14}$
Visible	$3.8\times10^{-5}-7.8\times10^{-5}$	26,300–12,800	$7.9\times10^{14}-3.8\times10^{14}$
Infrared			
Near	$7.5\times10^{-5}-2.5\times10^{-4}$	12,800 – 400	$3.8\times10^{14}-1.2\times10^{14}$
Middle	$2.5\times10^{-4}-5\times10^{-3}$	4000–200	$1.2\times10^{14}-6\times10^{12}$
Far	$5\times10^{-3}-1\times10^{-1}$	200–10	$6\times10^{12}-3\times10^{11}$

where h is the Planck's constant and c is the velocity of light in a vacuum. This photon energy may be absorbed or emitted by a molecule, in which case the molecular energy will be changed by an amount of ΔE. If the molecule gains energy, ΔE is positive, and a photon is absorbed. If the molecule loses energy, ΔE is negative, and a photon is emitted.

Upon irradiation with a beam of light, a molecule can be transferred to a higher energy excited level through the absorption of a light photon and the excitation of a vibrational phonon. A necessary prerequisite for the absorption is that the energy of the incoming photon is equivalent to the energy difference between two allowed vibrational states of the molecule. In this case, the molecule absorbs ΔE from the light in the middle- to far-infrared range at each vibrational transition. Such absorption is thus detected using the **infrared absorption** spectroscopic technique (Figure 7.1).

The origin of Raman spectra is markedly different from that of infrared spectra (Nasdala et al., 2004). The phonon energy of visible, ultraviolet, and near-infrared light is much higher than the energy differences between the vibrational states of molecules and cannot be thus absorbed through the phonon excitation. The incident light will excite the system to a virtual high-energy state, from which the system recovers immediately. The diffusely scattered light observed in the direction perpendicular to the incident beam has the same photon energy and thus the same frequency as that of the incident light, thus producing a **Rayleigh scattering**. It may also happen very rarely ($\sim 10^{-5}$ of the incident beam) that the system reach a higher or lower energy state compared to the initial value. If the system gains energy by a phonon excitation, the scattered photon loses the same energy portion $[E = h(\nu_0 - \nu_1)]$, thus giving rise to a **Stokes-type scattering**. It is also possible that an already vibrating system has excited, and it may instead lose the vibrational energy through the interaction with the light. Here, the scattered photon has an increased energy $[E = h(\nu_0 + \nu_1)]$, thus corresponding to the **anti-Stokes-type scattering**. The photon energy difference (energy shift = Raman shift) corresponds to the energy difference between vibrational levels of the molecule. The intensity ratio of the Stokes and the anti-Stokes Raman light depends on the temperature. It is customary to record only Stokes lines, which are always higher in intensity than those of anti-Stokes lines.

According to the classical theory, the Raman scattering can be explained in terms of the electric field strength E of the electromagnetic wave (laser beam), which fluctuates with time τ (Ferraro et al., 2003)

$$E = E_0 \cos 2\pi\nu\tau \tag{7.3}$$

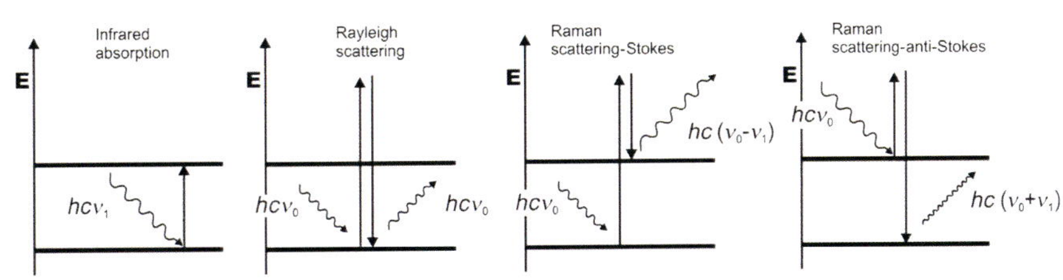

FIGURE 7.1

Elucidation of the interaction between electromagnetic irradiation and molecule using a simplified level diagram (Nasdala et al., 2004). Reprinted with kind permission from the European Mineralogical Union.

where E_0 is the vibrational amplitude and ν is the laser frequency. If a diatomic molecule is irradiated, an electric dipole moment P_d is induced

$$P_d = \alpha E \tag{7.4}$$

where α is a proportional constant called polarizability. If the molecule is vibrating with a frequency ν_1, the nuclear displacement q corresponds to

$$q = q_0 \cos 2\pi\nu_1\tau \tag{7.5}$$

where q_0 is the vibrational amplitude. For a small amplitude of vibration, α and q are correlated by a linear function

$$\alpha = \alpha_0 + \left(\frac{\partial\alpha}{\partial q}\right)_0 q_0 + \cdots \tag{7.6}$$

where α_0 is the polarizability at the equilibrium position, and $(\partial\alpha/\partial q)_0$ is the rate of change of α with respect to change in q at the equilibrium position.

A combination of the previous equations yields

$$P_d = \alpha_0 E_0 \cos 2\pi\nu_0\tau + \tfrac{1}{2}(\partial\alpha/\partial q)_0 E_0 [\cos\{2\pi(\nu_0 + \nu_1)\tau\} + \cos\{2\pi(\nu_0 - \nu_1)t\}] \tag{7.7}$$

where the first term represents an oscillating dipole radiating light of the frequency ν_0 of the Rayleigh scattering, the second term corresponds to the frequency $\nu_0+\nu_1$ of the anti-Stokes lines, and the third term relates to the frequency $\nu_0-\nu_1$ of the Stokes lines.

Each band in the Raman spectrum corresponds to the interaction of the incident light with a certain vibration of the nuclei. These are affected by sizes, valences, and masses of atoms in the sample, bond forces between them, and the symmetry of their arrangement in the crystal structure or molecule. Hence, Raman spectra are sample-specific and can be thus used for the identification and structural characterization of unknown samples.

Bands in a Raman spectrum can be subdivided into (Nasdala et al., 2004):

- Fundamental molecular vibrations and rotations.
- Overtones of a given frequency (twice the frequency of the fundamental tone).
- Combination tones (the sum or difference between two simpler frequencies).
- Overtone or combination tones interfering with fundamental tones (Fermi resonance effect).

The symbol *A* denotes single vibrations that are symmetrical whereas *B* denotes vibrations that are antisymmetrical with respect to a rotation about the principal axis of the atomic group. *E* and *F* denote doubly and triply degenerate vibrations, respectively. The degenerate mode is that of two or three different vibrations having the same frequency. Subscripts g (*gerade*) and u (*ungerade*) specify vibrations that are symmetric and antisymmetric with respect to the symmetry center. The g-type vibrations are Raman active, and u-types are infrared active. Finally, superscripts $'$ and $''$ specify vibrations symmetrical and antisymmetrical with respect to a mirror plane. Basic types of vibrations are described using Greek symbols: ν_s—symmetric stretching, ν_{as}—antisymmetric stretching, δ—bending, ρ—rocking, π, τ—twisting, ω—wagging.

A common problem of Raman spectroscopy is the emission of photoluminescence excited by a laser beam. The intensity of the photoluminescence may surpass the intensity of the Raman scattered light by several orders of magnitude, thus making it impossible to obtain a Raman spectrum. It is, however, often possible to find a suitable excitation wavelength with which the Raman spectrum lies in the spectral range that is not affected by the luminescence signal. Luminescence is often characterized by broad-band emissions, which are possible to correct. However, in the case of narrow luminescence emissions, it may be difficult to distinguish between the Raman and luminescence bands. Reliable distinction is possible by obtaining Raman spectra with different excitation wavelengths, in which bands with equal Raman shifts will appear in all of the spectra. By contrast, the luminescence emissions must show different apparent Raman shifts in different excitation wavelengths. Infrared absorption is not subject to interference by the fluorescence emissions.

Spatial resolution of the Raman spectroscopy is about twice the excitation wavelength, thus allowing for the detection of objects as small as 1–1.5 μm using blue or green lasers. Although the IR absorption occurs within the entire optical path of the spectrometer, the Raman scattering can be measured in a restricted volume using the confocal laser beam. The spatial resolution is improved by cutting the truncated cone at its broader base, i.e., only the light scattered from areas in or slightly below and above the focal plane are allowed to pass into the detector, whereas the rest is rejected by a narrow confocal diaphragm, a so-called confocal hole. Modern confocal systems can operate at a lateral resolution of ~1 μm and depth resolution of about ±2 μm, thus resulting in an effective volume resolution of ~5 μm^3 for transparent samples. However, the maximum lateral and depth resolution is achieved when the laser beam is focused on the sample surface (Figure 7.2). Focusing the laser beam inward the sample leads to a significant decrease in the effective volume resolution. The expansion of the effective volume occurs when fluid inclusions at various depths below the sample surface are measured. The confocal beam produces a signal from a sample volume with an approximately inverted conical shape, 2 μm wide and 5–10 μm deep (Nasdala et al., 2010; Thomas et al., 2006a).

The following compounds can be detected by Raman spectrometry in the gaseous phase of fluid inclusions: CO_2, CH_4, higher gaseous hydrocarbons, CO, N_2, H_2, O_2, NH_3, H_2S, SO_2. Most solid sulfates,

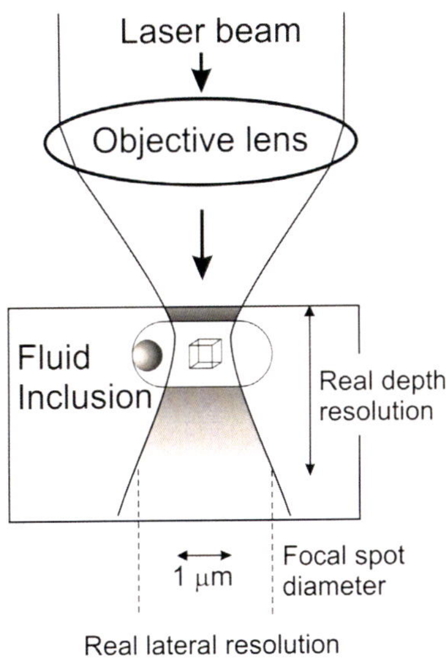

FIGURE 7.2

Simplified sketch of the laser beam path of a confocal Raman spectrometer, showing effective volume resolution of the beam focused on a fluid inclusion. The analyzed volume is shown shaded. Note that the real lateral resolution exceeds the focal spot size. Modified from Nasdala et al., 2010.

carbonates, borates, silicates, and bitumens occurring as daughter phases in fluid inclusions are also Raman-active. Gas and salt hydrates formed in frozen fluid inclusions (Baumgartner and Bakker, 2009a, 2010; Dubessy et al., 1982), and anions, such as SO_4^{2-}, HS^-, HSO_4^-, and NH_4^- that are dissolved in an aqueous phase of fluid inclusions are also analyzed. In contrast, Raman inactive are monoatomic ions, such as Na^+, K^+, Ca^{2+}, Cl^-, as well as chlorides of alkali metals, for instance, halite and sylvite. However, these electrolytes affect the hydrogen bonding in liquid water, and this allows the estimation of the total salinity of the aqueous phase (Baumgartner and Bakker, 2009b; Dubessy et al., 2002b). Infrared spectroscopy is most frequently used for distinguishing between molecular water and hydroxyl groups in silicate melt inclusions, and for the identification of liquid hydrocarbons and bitumens.

7.2 GASES AND THEIR MIXTURES

The number of potentially occurring vibrations (Z) depends on the degree of freedom of the system. For a single nonlinear gas molecule composed of n atoms, there are $3n-6$ degrees of vibrational freedom (three for each atom, minus three rotational and three translational principal movements in the three dimensions of space). Linear molecules, such as H_2, N_2, or O_2, have $3n-5$ degrees of freedom (Ferraro et al., 2003; Nasdala et al., 2004).

Carbon dioxide (O═C═O) provides an instructive example of a more complex gas molecule with $3(3)-5=4$ degrees of freedom and, therefore, four vibrations, which include an in-phase and out-of-phase stretching and two mutually perpendicular bending vibrations (Figure 7.3). The g-type symmetric stretch at 1339.5 cm^{-1} is Raman-active, whereas the u-type ν_2 (667.3 cm^{-1}) and ν_3 (2349.1 cm^{-1}) modes are infrared active. However, instead of a single symmetric stretching mode ν_1, two strong Raman modes at 1286 and 1389 cm^{-1}, so-called first Fermi dyad (doublet), are observed. This doublet is due to the Fermi resonance—an anharmonic mixing of the overtone of the bending mode $2\nu_2$ with ν_1.

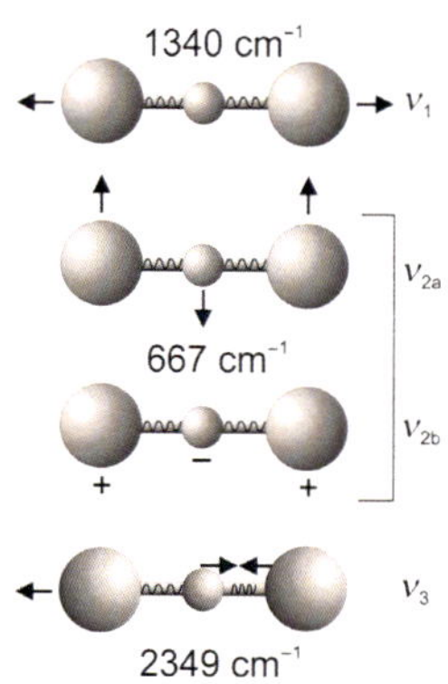

FIGURE 7.3

Internal vibrations of CO_2 molecule (Ferraro et al., 2003): ν_1—symmetric (in-phase) stretching; ν_2—symmetric bending; ν_3—antisymmetric (out-of-phase) stretching. The + and − signs denote vibrations going upward and downward, respectively, in direction perpendicular to the paper plane. Reprinted from Ferraro, J.R., Nakamoto, K., Brown, C.W., 2003. Introductory Raman Spectroscopy, First Edition. Copyright 2003, with permission from Elsevier.

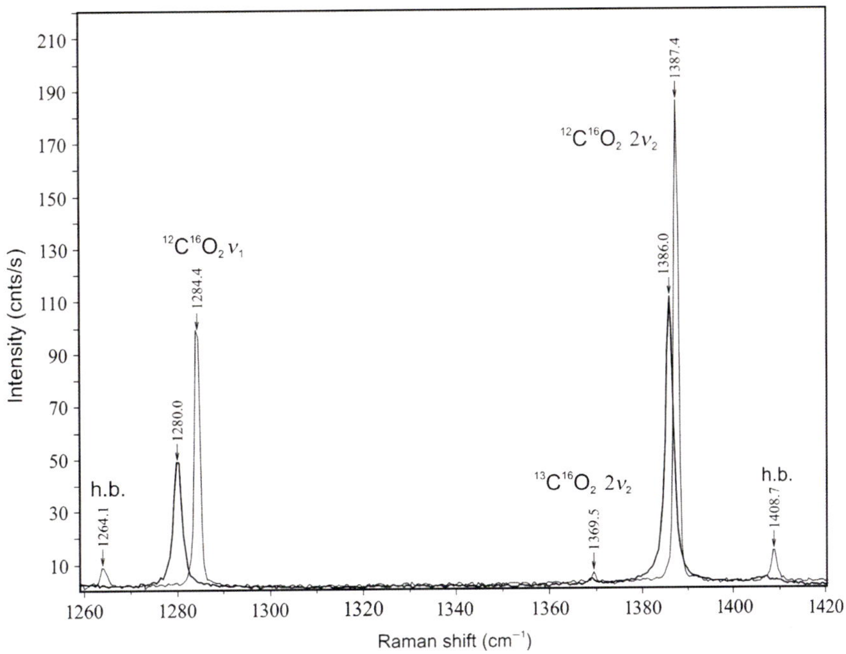

FIGURE 7.4

Raman spectra of CO_2 in superdense CO_2 (1.197 g/cm^3, 36.8 cm^3/mol, solid line) and that in a medium-density CO_2–CH_4 mixture (130 cm^3/mol, stippled line) excited by an Ar^+ 514 nm laser. Shown are suppressed hot bands in the superdense CO_2 and the Fermi dyad shifts proportional to the fluid density. h.b. designates hot bands.

Small satellite bands at 1264 and 1409 cm^{-1} represent modes of the second Fermi dyad (Rudolph et al., 2006). These bands were already observed in fluid inclusions containing CO_2 in very early papers dealing with Raman spectrometry of fluid inclusions (Dhamelincourt et al., 1979; Rosasco et al., 1975). Two additional low-intensity bands can be discerned at outer sides of both main bands at ~1260 and ~1405 cm^{-1} (Figure 7.4). These frequencies are usually assigned to hot bands, originating due to thermal energy of the CO_2 molecule. The hot bands tend to diminish, peak positions shift to lower wave numbers, and the distance between Fermi dyad also increases with the increasing fluid density. This phenomenon is used for the estimation of inclusion fluid density and pressure (e.g., Rosso and Bodnar, 1995; Kawakami et al., 2003; Wang et al., 2011).

High-resolution scans (Figure 7.5) reveal further minor bands, which distort the symmetry of the CO_2 molecule. Local maximum with low intensity at 1369 cm^{-1} belongs to the $2\nu_2$ mode of $^{13}C^{16}O_2$ (Rosso and Bodnar, 1995). The $^{13}C^{16}O_2$ and the most abundant $^{12}C^{16}O_2$ molecules are called *isotopologues*—molecular entities that differ only in their isotopic composition. The two ν_1 and $2\nu_2$ modes of the $^{12}C^{16}O^{18}O$ molecule can also be discerned at 1258 and 1364 cm^{-1}, respectively. However, discrepancy exists regarding the exact position of the ν_1 mode of $^{13}C^{16}O_2$. Morizet et al. (2009) observed that the distance between the ν_1 and $2\nu_2$ modes of $^{12}CO_2$ and $^{13}CO_2$ is exactly 100 cm^{-1}. Indeed, all high-density CO_2 inclusions exhibit two low-intensity peaks at this distance, of which the lower intensity band at around 1270 cm^{-1} should correspond to the ν_1 mode of $^{13}CO_2$. In contrast, Irmer and Graupner

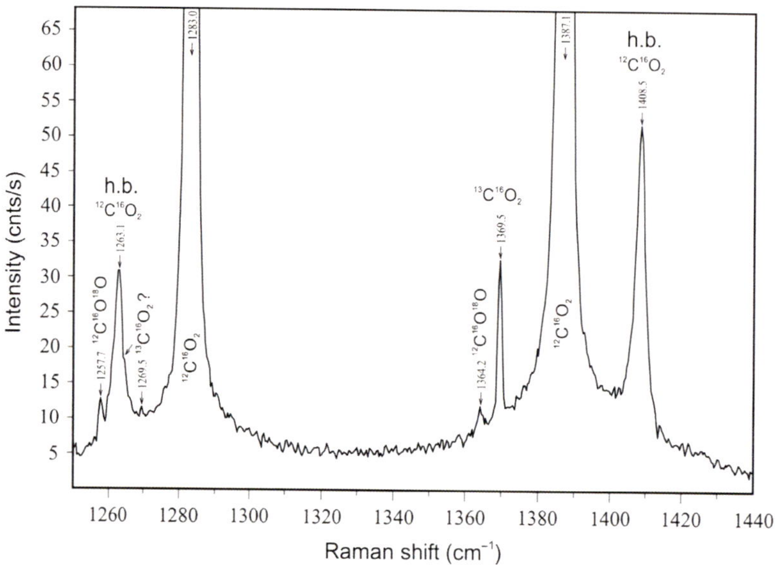

FIGURE 7.5

Detailed scan (2 × 20 s, 1800 gr/mm grating, 532 nm excitation line, 240 mW laser power) of CO_2 vibrations in high-density CO_2–CH_4–N_2 inclusion reveals several CO_2 isotopologues. Raman band assignment is according to Irmer and Graupner (2002).

(2002) determined the separation distance of the Fermi dyad to be ~103 cm^{-1}. Thus, the ν_1 $^{13}CO_2$ band is overlapped by the hot band of the $^{12}CO_2$. In this case, however, nature of the small but distinct band that occurs virtually in all high-density CO_2-rich fluid mixtures at ~1270 cm^{-1} remains unclear.

The Raman spectrum of methane is quite complicated. In the low-frequency region, a clear band at 1532 cm^{-1} is accompanied by two peaks that lie symmetrically on either side of this band (Atamas et al., 2004). Two vibrational modes in the high-frequency region are the symmetric stretching mode and the twisting or antisymmetric stretching mode, usually denoted as ν_1 and ν_2, respectively (Lin et al., 2007b; Mernagh and Bastrakov, 2013). Raman spectra of methane-rich inclusions show usually only one major band at ~2917 cm^{-1}, corresponding to ν_1, whose position is pressure-dependent and shifts systematically to lower wave numbers as the internal pressure in and density of fluid inclusions increase. In addition, two other low-intensity bands occur at ~3020–3030 and 3070–3090 cm^{-1} (Figure 7.6). These bands have been previously interpreted as those belonging to higher hydrocarbons; however, recent investigations of pure methane gas in pressure cells clearly showed that they represent the ν_3 and $2\nu_2$ vibrations (Brunsgaard Hansen et al., 2002). Hence, the ν_2 vibration should theoretically occur at ~1635 cm^{-1}. The ν_3 vibration band tends to be weaker and broader with the increasing methane density. In CH_4–H_2S mixtures, Mernagh and Bastrakov (2013) observed two distinct bands at 2578–2580 and 2609 cm^{-1}, and assigned them to the $2\nu_4$ and ν_1 vibrations of gaseous CH_4 and H_2S, respectively. The presumable $2\nu_4$ mode of CH_4 should thus be practically identical with that of H_2S liquid.

Raman spectroscopic evidence for abiogenic higher hydrocarbons in fluid inclusions is rare (Arai et al., 2012; Dele-Dubois et al., 1980). Most of them refer to the Fischer–Tropsch synthesis during

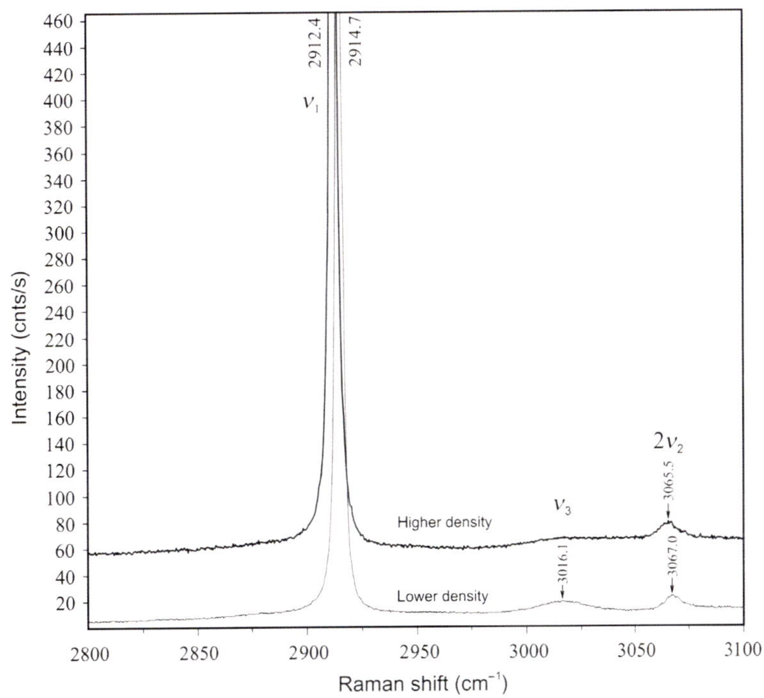

FIGURE 7.6

Raman spectrum of methane in low-density CO_2–CH_4 inclusion and that in a higher density CO_2–CH_4–N_2 inclusion. Illustrated are shifts of ν_1, $2\nu_2$, and ν_3 vibration modes in the C–H stretching region, and the suppression of the ν_3 vibration mode in the higher density inclusion.

evolution of peralkaline igneous intrusions (Konnerup-Madsen, 2001; Krumrei et al., 2007; Potter and Konnerup-Madsen, 2003; Potter et al., 2004). Raman spectra of the C2–C4 *n*-alkanes (Atamas et al., 2004) are fairly similar in the high-frequency region. Ethane shows two high-intensity bands at 2895 and 2955 cm^{-1}, whereas two symmetrical bands with maxima at 1000 and 1440 cm^{-1} are observed in the low-frequency region. Propane shows a three-component band in the high-frequency region: the feature at 2883 cm^{-1} is the most intense, whereas asymmetric stretching modes give rise to the bands at 2937 and 2973 cm^{-1}. Raman spectrum of butane is even more complex in the high-frequency region, consisting of at least three sub-bands at 2885, 2927, and 2960 cm^{-1}. Moreover, the bands at 2885 and 2960 display shoulders at 2865 and 2987 cm^{-1}, respectively. The increasing number of bands of the gaseous *n*-alkanes reflects the increasing complexity of the hydrocarbon molecular structure and disharmonic interactions between CH_3 and CH_2 stretching vibrations.

Raman spectrum of nitrogen should exhibit only one single ν_1 vibrational band. However, high-density nitrogen-rich inclusions always show a smaller satellite band, whose position and intensity change proportionally with the main vibrational band assigned to the $^{14}N_2$ molecule (Figure 7.7). A simple calculation can be performed to reconcile, whether or not the satellite band corresponds to a nitrogen isotopologue, either $^{15}N_2$ or ^{14}N–^{15}N (Problem 7.1).

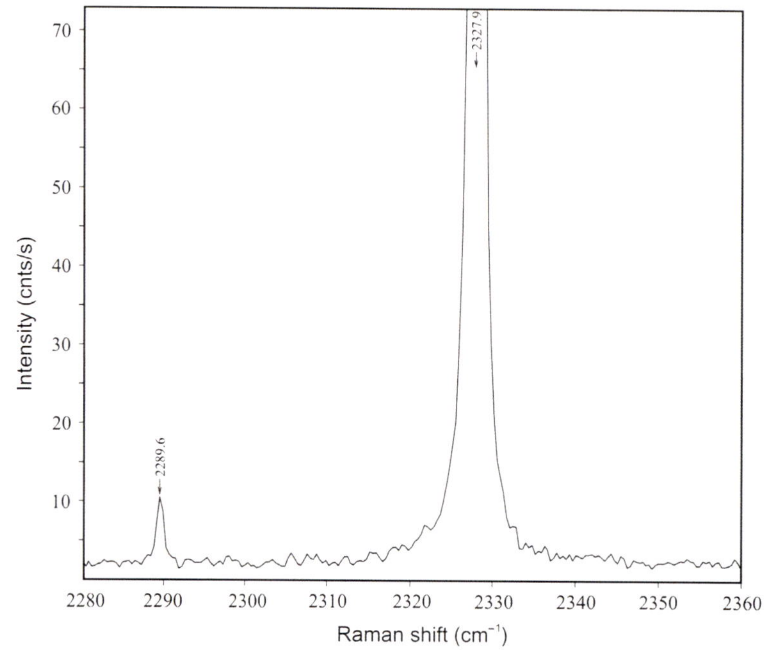

FIGURE 7.7

Raman spectrum of nitrogen in a high-density N_2–CO_2 inclusion ($XN_2 = 0.868$, $d = 0.942$ g/cm^3) excited with a 532 nm Nd-YAG laser.

PROBLEM 7.1 CALCULATING POSITIONS OF VIBRATION MODES OF SIMPLE GAS MOLECULE

Positions of vibrational modes in simple symmetric, diatomic molecules can be inferred from the relationship of Dunham (1932)

$$\frac{\varpi_1}{\varpi_2} = \sqrt{\frac{\mu_2}{\mu_1}} \tag{7.8}$$

where ϖ_i is the normal mode wave number and μ_I is the reduced mass calculated as

$$\mu = \frac{m_1 m_2}{m_1 + m_2} \tag{7.9}$$

m is the isotopic mass and 1 or 2 refers to different isotopologues of the same molecule. Calculate ν_1 band position of the ^{14}N–^{15}N isotopologue in the Raman spectrum in Figure 7.7.

Solution

The reduced masses of $^{14}N_2$ and ^{14}N–^{15}N correspond to 7 and 7.24, respectively (Equation (7.9)). Substituting the reduced masses and the position of the major vibrational band assigned to the $^{14}N_2$ isotopologue into Equation (7.8) we get the position of the ν_1 band of the ^{14}N–^{15}N isotopologue

Continued

$$\varpi_1 = 2327.9\sqrt{\frac{7}{7.24}} = 2289\ \text{cm}^{-1}$$

The calculated ν_1 vibrational band position of the $^{14}N{-}^{15}N$ isotopologue exactly matches the unknown band observed in the Raman spectrum in Figure 7.7. It is worth of remembering that according to the Dunham's equation, vibrational bands of heavier isotopologues always occur at lower wave numbers than that of the main ν_1 vibration. The $^{15}N_2$ molecule should vibrate at ~2250 cm^{-1} (Bendtsen, 2001), but such a signal has never been observed in natural N_2-rich inclusions.

Hydrogen sulfide (H_2S) and carbon monoxide (CO) show single vibrational modes accompanied by minor satellite bands corresponding to their most abundant isotopologues. However, owing to a small concentration of these gases in natural fluids, the isotopologue bands are only rarely discernible by the Raman spectrometry.

Hydrogen has been identified in fluid inclusions using the Raman spectroscopy by Konnerup-Madsen et al. (1985), Bottrell et al. (1988a), Dubessy et al. (1988), Peretti et al. (1992), Derome et al. (2003), Huraiová et al. (2005), Bakker and Elburg (2006), and Naumov et al. (2008). Raman spectrum of hydrogen is more complex than that of other gases. A total of seven bands have been discerned (Dubessy et al., 1988), of which the strongest band at 4156 cm^{-1} is diagnostic of low hydrogen concentration in gaseous mixtures (Figure 7.8). Hydrogen contents in fluid inclusions must be measured repeatedly with different exposure times to exclude the possible generation of artificial hydrogen triggered by the incident laser beam in a multicomponent fluid mixture.

Determination of the composition of a gaseous mixture requires determination of peak areas and knowledge of the so-called Raman-scattering cross sections (RSCSs) that are measures of the Raman activity (Schrötter and Klöckner, 1979). For practical purposes, empirical "relative" RSCSs (Σ) are defined as the ratio of the RSCS to that of N_2 at 2331 cm^{-1} (Burke, 2001; Burruss, 2003). The Σ values are relative normalized RSCSs used only for comparison of results measured at different wavelengths. The calculation of a gas composition from peak areas employs relative RSCSs (σ) recalculated to the λ value of the incident laser beam - RSCSs (σ).

The λ value of a laser is correlated with RSCSs through the equation

$$\Sigma_i = \sigma_i \left[\frac{(\nu_0 - \nu_1)^{-4}}{(\nu_0 - 2331)^{-4}}\right] \times \left[1 - \exp\left(\frac{-hc\nu_i}{k_B T}\right)\right] \tag{7.10}$$

where Σ_i and σ_i are different scattering values for a Raman shift ν_i (cm^{-1}), ν_0 is the laser wavelength used (in cm^{-1}), h is the Planck's constant, c is the velocity of light, k_B is the Boltzmann's constant, and T is temperature (Kelvin). This equation can be used to calculate σ values for optional laser wavelengths (Table 7.2).

A simple formula based on the polarizability theory of Raman scattering (Placzek, 1934) can then be employed to calculate molar fractions of species present in fluid inclusions (Dubessy et al., 1989). For a polycomponent fluid mixture consisting of A, B,. . .,X species

$$X_A = \frac{I^A/\sigma^A\xi^A}{I^A/\sigma^A\xi^A + I^B/\sigma^B\xi^B + \cdots + I^X/\sigma^X\xi^X}, \tag{7.11}$$

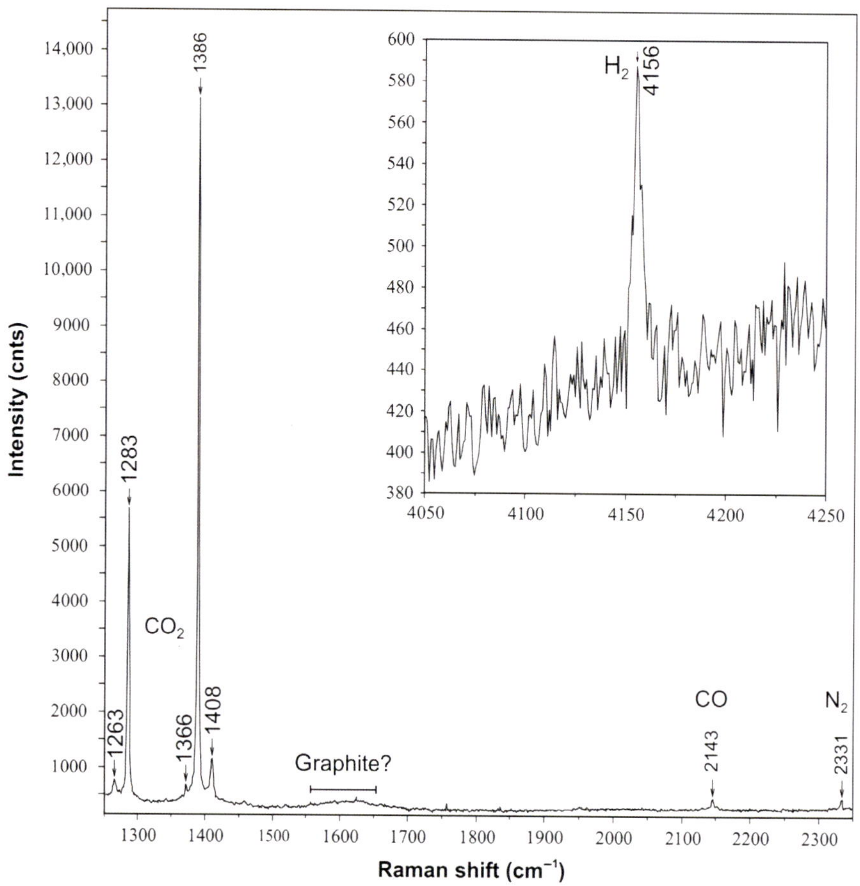

FIGURE 7.8

Raman spectrum of a CO_2-rich inclusion in quartz from pincinite xenolith in Pliocene alkali basalt (sample KP-31, Pinciná maar, Western Carpathians, Slovakia) excited with a 266 nm Nd-YAG laser, showing dominant bands of the CO_2 Fermi dyad accompanied by minor bands of CO, N_2, and H_2. Molar concentrations of the minor gas species calculated using absolute band areas and Raman scattering cross sections correspond to $X_{H_2} = 0.0085$, $X_{CO} = 0.0178$, $X_{N_2} = 0.0194$.

where X is the mole fraction of gas component, I is the peak area of Raman band, σ is the cross section for the Raman band of the corresponding component, and ξ is the instrument efficiency factor at the frequency of the corresponding band defined usually by comparing the spectra of standard gas mixtures or natural fluid inclusions with known compositions. Problem 7.2 illustrates the determination of the gas composition of a CO_2–CH_4–N_2–H_2S inclusion in quartz from the Raman spectrum in Figure 7.9.

Table 7.2 Compilation of Relative RSCSs of Selected Gas Species Relative to N_2 (RSCS $N_2 = 1$)

Species	Raman Shift $\Delta\nu_i$ (cm^{-1})	Relative RSCS Σ_i	Laser λ (nm)-Dependent RSCS (σ_i)					Reference
			266	488	514	532	633	
SO_2	1151	4.03	4.60	5.18	5.26	5.32	5.64	Burke (2001)
$^{12}CO_2\ \nu_1$	1285	0.80	0.90	1.00	1.01	1.02	1.08	Burke (2001)
		0.63	0.71	0.79	0.80	0.81	0.85	Kiefer et al. (2008)
$^{13}CO_2\ 2\nu_2$	1370	1.23	1.37	1.51	1.53	1.54	1.62	Burke (2001)
$^{12}CO_2\ 2\nu_2$	1388	1.23	1.37	1.51	1.52	1.54	1.61	Burke (2001)
		0.88	0.98	1.08	1.09	1.10	1.15	Kiefer et al. (2008)
O_2	1555	1.03	1.12	1.22	1.23	1.24	1.29	Burke (2001)
		0.83	0.91	0.98	0.99	1.00	1.04	Kiefer et al. (2008)
CO	2143	0.90	0.92	0.94	0.94	0.94	0.95	Burke (2001)
		0.86	0.88	0.90	0.90	0.90	0.91	Kiefer et al. (2008)
H_2S	2611	6.80	6.59	6.39	6.37	6.35	6.25	Burke (2001)
iso-C_4H_{10}	2880	9.62	9.03	8.51	8.44	8.40	8.14	Kiefer et al. (2008)
	794	1.96	2.32	2.71	2.76	2.80	3.02	Kiefer et al. (2008)
HCl	2886	3.20	3.00	2.83	2.80	2.79	2.70	El-Diasty (2011)
n-C_4H_{10}	2890	17.91	16.80	15.80	15.68	15.60	15.12	Kiefer et al. (2008)
	827	1.34	1.58	1.84	1.88	1.90	2.05	Kiefer et al. (2008)
C_3H_8	2908	22.61	21.17	19.87	19.71	19.60	18.98	Kiefer et al. (2008)
	2890	20.40	19.14	18.00	17.86	17.77	17.22	Burke (2001)
	870	1.14	1.34	1.55	1.58	1.60	1.72	Kiefer et al. (2008)
$CH_4\ \nu_3$	3020	4.82	4.45	4.13	4.09	4.06	3.91	El-Diasty (2011)
$CH_4\ \nu_1$	2917	8.63	8.07	7.57	7.51	7.47	7.22	Burke (2001)
		9.94	9.30	8.72	8.65	8.60	5.32	Kiefer et al. (2008)
		8.56	8.01	7.51	7.45	7.40	7.16	El-Diasty (2011)
	1535	0.08	0.09	0.09	0.10	0.10	0.10	Kiefer et al. (2008)
C_2H_6	2954	14.95	13.92	13.00	12.89	12.81	12.37	Burke (2001)
	2914	17.33	16.21	15.21	15.09	15.00	14.52	Kiefer et al. (2008)
	993	0.88	1.02	1.17	1.19	1.20	1.29	Kiefer et al. (2008)
NH_3	3334	6.40	5.70	5.10	5.03	4.98	4.70	El-Diasty (2011)
		6.32	5.63	5.03	4.96	4.91	4.63	Burke (2001)
H_2O	3657	3.29	2.82	2.43	2.38	2.35	2.17	Burke (2001)
H_2	4156	3.54	2.86	2.32	2.26	2.21	1.98	Burke (2001)
		6.24	5.05	4.09	3.98	3.90	3.49	Kiefer et al. (2008)
		4.00	3.23	2.62	2.55	2.50	2.24	El-Diasty (2011)

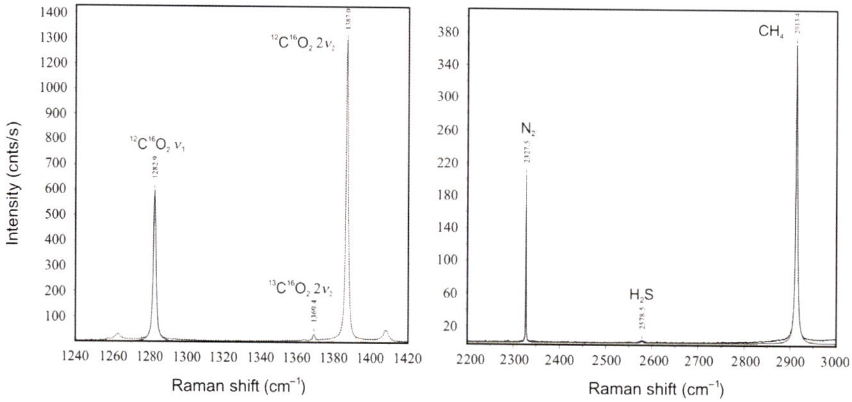

FIGURE 7.9

Raman spectrum of a mixed-volatile gas inclusion in quartz from calcite veins in sandstones of the Tertiary accretion wedge of Western Carpathians (Zborov – 1 drillhole, 3409 m) excited at 532 nm. CO_2 solid in the inclusion melted at −61.4 °C, and total homogenization to liquid occurred at −12.7 °C. Thin curves are Gauss–Lorentz functions fitted to obtained spectroscopic data and used for the gas composition calculation.

PROBLEM 7.2 CALCULATING FLUID COMPOSITION

Calculate relative molar concentrations (molar fractions) of gas species recorded by the Raman spectrum in Figure 7.9, using RSCSs for the 532 nm excitation wavelength (Table 7.2). Relative band areas measured using an integral operation routine in the LabSpec5 computer program from Horiba Inc. are as follows: $\nu_1(CO_2)$—3390, $\nu_2(CO_2)$ (including also minor $^{13}CO_2$ band at 1369 cm^{-1}, and excluding hot bands)—6668 + 362, N_2—802, H_2S—120, CH_4—6633 a.u.

Solution

Spectral efficiencies ξ are set to unity. The CO_2 concentration must be calculated using the sum of the ν_1 and $2\nu_2$ peak areas as well as the sum of corresponding σ's (Burke, 2001; Nasdala et al., 2004). In other words, the Fermi dyad must be treated as one Raman band. With RSCSs from Burke (2001) and Equation (7.11) the following reduced peak areas are obtained:

$$CO_2 = (3\,390 + 6\,668 + 362)/(1.02 + 1.54) = 4070$$

$$N_2 = 802/1 = 802$$

$$H_2S = 120/6.35 = 19$$

$$CH_4 = 6633/7.47 = 888$$

$$\text{Total of the reduced peak areas: } 5779$$

Assuming that the low-density σ values are applicable also to higher density fluids, the inclusion gas composition (mol.%) is as follows: CO_2—70.4, N_2—13.9, H_2S—0.3, CH_4—15.4.

Inconsistencies exist regarding the calculation of the CO_2 molar fraction. Seitz et al. (1996) proposed using only the $2\nu_2$ band to calculate the fluid composition of the CO_2–CH_4 mixtures, instead of the sum of two CO_2 bands. Burke (2001) insisted on using both bands, but he set the spectral efficiency factors 0.5 and 1 for CO_2 and other gases, respectively, to calculate the gas composition in the CO_2–CH_4–N_2 mixture. The 1388/1286 cm^{-1} peak height/area ratio depends on the density, internal pressure, and temperature of the inclusion gas, and it is different in coexisting CO_2-gas and liquid phases, e.g., ~2.4 in the high-density CO_2 liquid, and ~1.5 in the low-density CO_2 gas (van den Kerkhof, personal comm.). Hence, due to the interference of the CO_2-vapor and CO_2-liquid lines, measurements should be performed in homogenized inclusions, preferentially at 40 °C.

Note also that σ values for CO_2 and CH_4 differ significantly in the dataset compilation by Burke (2001) and other compilations (El-Diasty, 2011; Kiefer et al., 2008). This contributes to the uncertainty of Raman spectroscopic data. Calibration of each Raman probe using spectral efficiency factor from synthetic or natural fluid inclusions of known compositions is an absolute necessity to reduce the uncertainty in the calculation of gas mixture compositions.

As an alternative to microthermometric measurements, densities of CO_2 in fluid inclusions can be determined using the separation between bands in the Fermi dyad (Rosso and Bodnar, 1995; Yamamoto et al., 2007). Kawakami et al. (2003) investigated the ability of Raman spectroscopy for the determination of the CO_2 density in natural fluid inclusions. Relationship between distance (Δ, cm^{-1}) between two Fermi's resonance bands and the density of CO_2 (d, g/cm^3) can be expressed as

$$d = -0.03238697\Delta^3 + 10.08428\Delta^2 - 1046.189\Delta + 36163.67 \tag{7.12}$$

The equation is valid for fluid densities from 0.1 to 1.21 g/cm^3.

Another polynomial fit valid for the CO_2 densities between 0.1 and 1.24 g/cm^3 was provided by Yamamoto and Kagi (2006):

$$\begin{aligned} d = & -0.00111808(\Delta - 100)^8 + 0.04498451(\Delta - 100)^7 \\ & - 0.7727143(\Delta - 100)^6 + 7.4128146(\Delta - 100)^5 - 43.468301(\Delta - 100)^4 \\ & + 159.54433(\Delta - 100)^3 - 357.7651(\Delta - 100)^2 + 448.2404(\Delta - 100) - 240.461 \end{aligned} \tag{7.13}$$

The method was tested using fused capillary capsules with known CO_2 density (Wang et al., 2011). The least-square fit to the experimental data is as follows:

$$\begin{aligned} d = & -0.04258891551\Delta^3 + 13.25586152\Delta^2 - 1374.824414\Delta \\ & + 47513.64243 \ (\sigma = 0.0253 \text{ g/cm}^3) \end{aligned} \tag{7.14}$$

The equation can also be applied to CO_2–H_2O mixtures, and to CO_2–CH_4 mixtures with less than 10 mol.% CH_4.

Bakker (2011) showed that the homogeneous CO_2 measured at variable temperatures indicated a variation of the measured Fermi dyad split with internal pressure, resulting in an uncertainty of about 0.3 cm^{-1}. Hence, the use of a simple polynomial equation to calculate the density of CO_2 directly from the peak difference yields incorrect results. From the most optimal uncertainty of 0.3 cm^{-1}, the recalculated uncertainty in the CO_2 density is about ±0.2 g/cm^3. Hence, the peak position must be corrected at least to the temperature of measurement, and a nonlinearity of the Raman spectrometer must also be considered.

This approach was employed by Lin et al. (2007a), who precisely determined the position of symmetric stretching ν_1 band of methane at 2915 cm^{-1} using simultaneous calibration against two emissions from a neon lamp. Afterward, they used the corrected peak position and the temperature of measurement to calculate the internal fluid inclusion pressure. Fluid density and isochore can then be inferred from the equation of state for pure methane (e.g., Setzmann and Wagner, 1991). Finally, true trapping pressure of the fluid inclusion can be calculated from the isochore using independently determined crystallization temperature. A similar approach can also be applied to gaseous methane-dominated CH_4–H_2O inclusions with an external calibration provided by emission lines of He–Ne laser (Lu et al., 2007).

PROBLEM 7.3 CALCULATING FLUID DENSITY

Calculate the density of superdense CO_2 inclusion (1.197 g/cm^3) using the Raman spectrum in Figure 7.4.

Solution

The separation of Fermi distance corresponds to 106 cm^{-1}. The calculated densities are as follows: 1.207 g/cm^3 (Kawakami et al., 2003), 1.222 g/cm^3 (Yamamoto and Kagi, 2006), and 1.034 g/cm^3 (Wang et al., 2011). Hence, the equation of Kawakami et al. is the best choice to reproduce the density of the superdense CO_2 calculated from microthermometric data.

7.3 AQUEOUS FLUIDS

The observation of specific phase transitions (salt hydrate dissolution, ice melting, clathrate dissociation, and salt disappearance) as a function of temperature is a commonly used method to determine the salinity of aqueous fluid inclusions (Section 4.6.3). In some cases, however, this method cannot be employed because optical observations are impossible or unclear, the number of phases is not sufficient to thermodynamically constrain the system (e.g., single-phase metastable inclusions), and the vapor bubble disappears upon cooling due to the change of the molar volume when ice nucleates. Moreover, fluid inclusions in some soft and cleavable minerals easily stretch, leak, or decrepitate on freezing or heating. The microthermometric method is recently restricted to aqueous chloride solutions of Na, K, Ca, and Mg, although Borisenko (1977, 1982) showed that concentrations of other carbonates, hydrocarbonates, sulfates, and bromides can also be determined with reasonable precision. Peretyazhko et al. (2000) demonstrated that the cryometric method can be used to determine the concentration of boric acid by measuring the temperatures of ice and sassolite dissolution.

As an alternative to the microthermometry, a Raman spectrometric technique is also employed to determine the salinity of fluid inclusions. A concentration of dissolved chlorides can be inferred from the shape of the O–H stretching band, which changes with the addition of dissolved chlorides (Bakker, 2004; Caumon et al., 2013; Dubessy et al., 2002b; Mernagh and Wilde, 1989; Rull, 2002; Sun et al., 2010). A similar approach can be applied to the determinations of dissolved carbonates (Sun and Qin, 2011), boric acid (Sirbescu et al., 2013; Thomas, 2002a), and sulfates (Thomas and Davidson, 2012c) in fluid and melt inclusions by measuring the relative intensity ratios of two Raman sub-bands at about 3230 and 3430 cm^{-1} in the O–H stretching region and those of the H_3BO_3, CO_3^{2-}, HCO_3^{-}, and SO_4^{2-} species located at 875, 1380, 1017, and 983 cm^{-1}, respectively. For the determination of carbonates in fluid inclusions in quartz, the weaker Raman band at 1380 cm^{-1} is preferred, because the main ν_1 CO_3^{2-} band at 1066 cm^{-1} is strongly affected by two quartz bands at 1066 and 1080 cm^{-1} (Thomas and Davidson, 2012c).

The anisotropy of a host mineral significantly affects the concentrations of dissolved species measured by Raman spectroscopy (Baumgartner and Bakker, 2009b; Caumon et al., 2013;

Dubessy et al., 2002b; Sun et al., 2010). Depending on the crystallographic orientation, the salinity of inclusions in quartz displayed variations higher than 50%. The effect of host mineral anisotropy can be minimized by using quartz wafers cut perpendicularly to the *c*-axis, or by placing the crystal optic plane at the extinction angle, that is, parallel or perpendicular to the polarization plane of the laser, or by using an unpolarized laser radiation. In either case, an external calibration must be made with standard solutions in cuvettes covered by quartz plates in the appropriate crystallographic orientation (Sirbescu et al., 2013), or by using fused-silica capillary capsules (Caumon et al., 2013).

Additional dissolved salts also affect the Raman spectrum of the measured species in multicomponent aqueous solutions. Sirbescu et al. (2013) have shown that NaCl, LiCl, and LiF do not interfere with the H_3BO_3 analysis. In contrast, Li_2CO_3 significantly reduced the $\nu_s B^{[3]}$–O band at 878 cm^{-1}, probably due to the pH change and the concomitant modification of the boron speciation. It is likely that other alkali carbonates and bicarbonates would exhibit a similar effect.

As an example, we show Raman spectra of a sulfate- and carbonate-rich fluid inclusion ($T_h \sim 350$ °C) in quartz-specularite pockets in an aplite vein in granodiorite (Oppach/Lusatia, Germany), which contains 14% SO_4^{2-} and 4.3% CO_3^{2-} related to alkalis (Figure 7.10). The Raman spectrum shows a distinct sulfate band at 983 cm^{-1} and the carbonate band at 1382 cm^{-1}. The ratio of the integral intensities of sulfate and carbonate bands versus the integral intensity of sub-bands at 3264 and 3450 cm^{-1} in the high frequency OH–H_2O region in the same inclusion is employed for the determination of salinity.

The determination of boric acid content is carried out in the same way. Figure 7.11 documents a fluid inclusion ($T_h \sim 450$ °C) in quartz from the Pechtelsgrün tungsten deposit (Vogland, Germany) with distinct H_3BO_3, carbonate and water bands. The aqueous solution of the inclusion contains 0.63% H_3BO_3 and 10.2% CO_3^{2-}.

Raman spectroscopy can also be employed to quantifying the gas content dissolved in aqueous inclusions and to calculating their densities and isochores (Dubessy et al., 2001, 2002a; Guillaume et al., 2003; Pironon et al., 2003). Dubessy et al. (2001) calibrated CH_4/H_2O ratios using the areas of symmetric stretching ν_1 band of CH_4 dissolved in the aqueous solution (2908 cm^{-1}), and those of the bending (~1640 cm^{-1}) and stretching modes of H_2O (3000–3700 cm^{-1}) in synthetic fluid inclusions. In the following step, they iteratively calculated the fluid pressure at the homogenization temperature T_h using the thermodynamic models of Duan et al. (1992a, b). Finally, they constructed isochores from the starting point represented by the homogenization temperature and pressure (two-phase envelope) parallel to the isochore of the methane-free fluid of the H_2O–NaCl system with the same T_h and salinity as the methane-bearing inclusions according to empirical equations derived by Zhang and Frantz (1987). Dubessy et al. (2001) also provided a protocol for the determination of bulk composition of H_2O–CH_4–NaCl inclusions, consisting of liquid and vapor phases. This protocol is particularly useful in the cases, when the total homogenization cannot be determined because of the fluid inclusion stretching or decrepitation.

Lu et al. (2008) inferred a linear calibration curve for the Raman signal of gaseous methane dissolved in pure water as a function of molality up to 0.22 mol CH_4/kg H_2O. Caumon et al. (2014) measured concentrations of CH_4 in H_2O–NaCl aqueous solutions in a wide range of temperatures (60–180 °C) and pressures (30–1000 bar) using area ratios of CH_4 and H_2O Raman bands, as exemplified in Figure 7.12. They found out that the calibration curve is independent of the salinity and can thus be readily applied also to fluid inclusions with the unknown or poorly constrained NaCl content. The following second-order polynome was fit to the experimental data

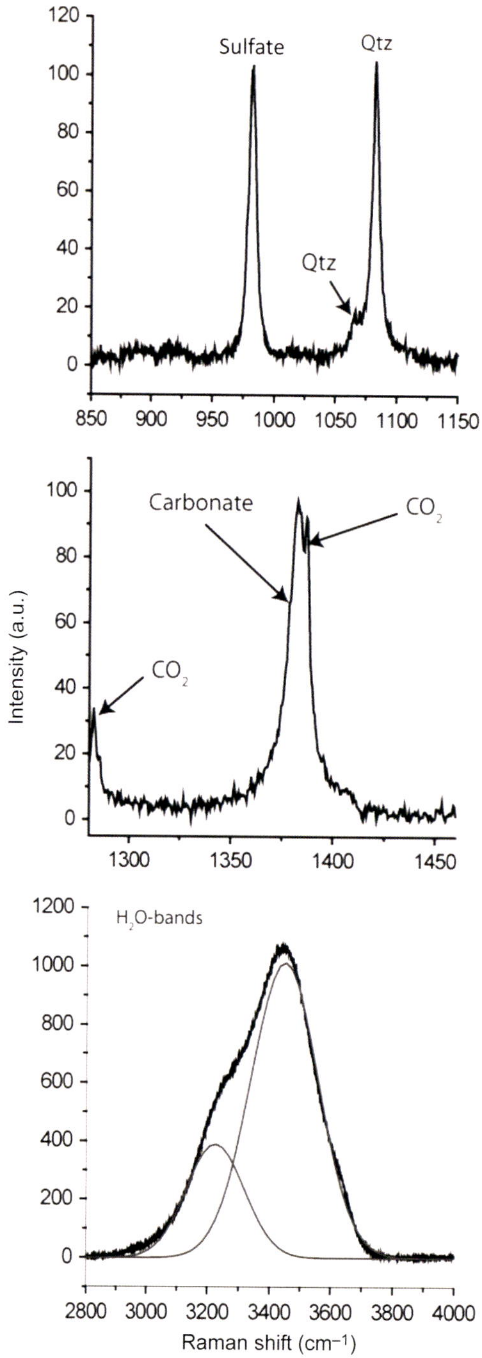

FIGURE 7.10

Raman spectrum of sulfate- and carbonate-rich fluid inclusion ($T_h \sim 350$ °C) in quartz crystals from quartz–specularite pockets in an aplite vein in granodiorite (Oppach/Lusatia, Germany). The OH-stretching band is deconvoluted using two Gauss–Lorentzian components.

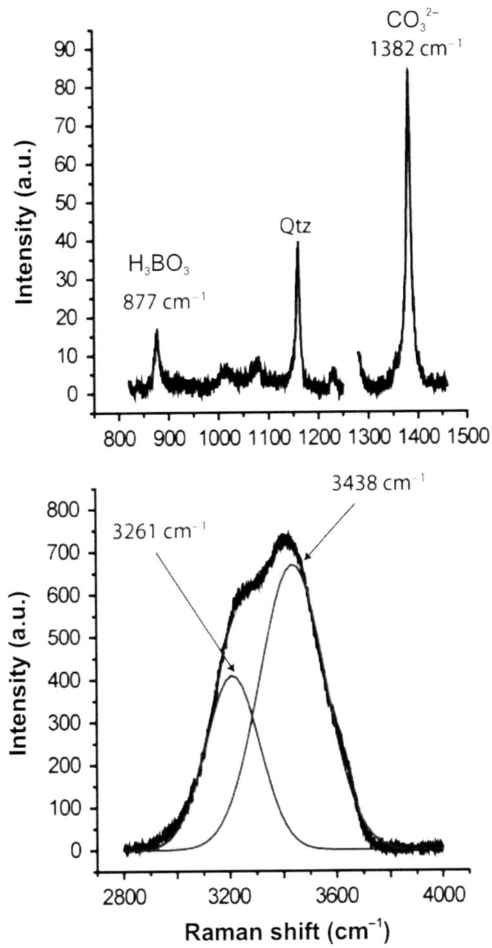

FIGURE 7.11

Raman spectrum of fluid inclusion ($T_h \sim 450°C$) in quartz from the Pechtelsgrün tungsten deposit (Vogland, Germany) with distinct H_3BO_3, carbonate, and water bands. The OH-stretching band is deconvoluted using two Gauss–Lorentzian components at 3261 and 3438 cm^{-1}.

$$mCH_4 = (0.008 \pm 0.001) + (50.2 \pm 0.3) \times A_{CH_4}/A_{H_2O} - (660 \pm 20) \times (A_{CH_4}/A_{H_2O})^2 \quad (7.15)$$

where mCH_4 is the solubility of CH_4 in the aqueous solution (mol/kg of water), and A_{CH_4}/A_{H_2O} is the methane-to-water peak area ratio in the liquid phase. Once the methane content is known, the isochore can be more precisely localized using an appropriate EOS. Molar fraction of water in methane gas Y_{CH_4} can also be calculated

$$Y_{H_2O} = (5.19 \pm 0.02) \times A_{CH_4}/(A_{CH_4} + A_{H_2O}) \quad (7.16)$$

where $A_{CH_4}/(A_{CH_4} + A_{H_2O})$ is the methane-to-(water+methane) peak area ratio in the gas phase.

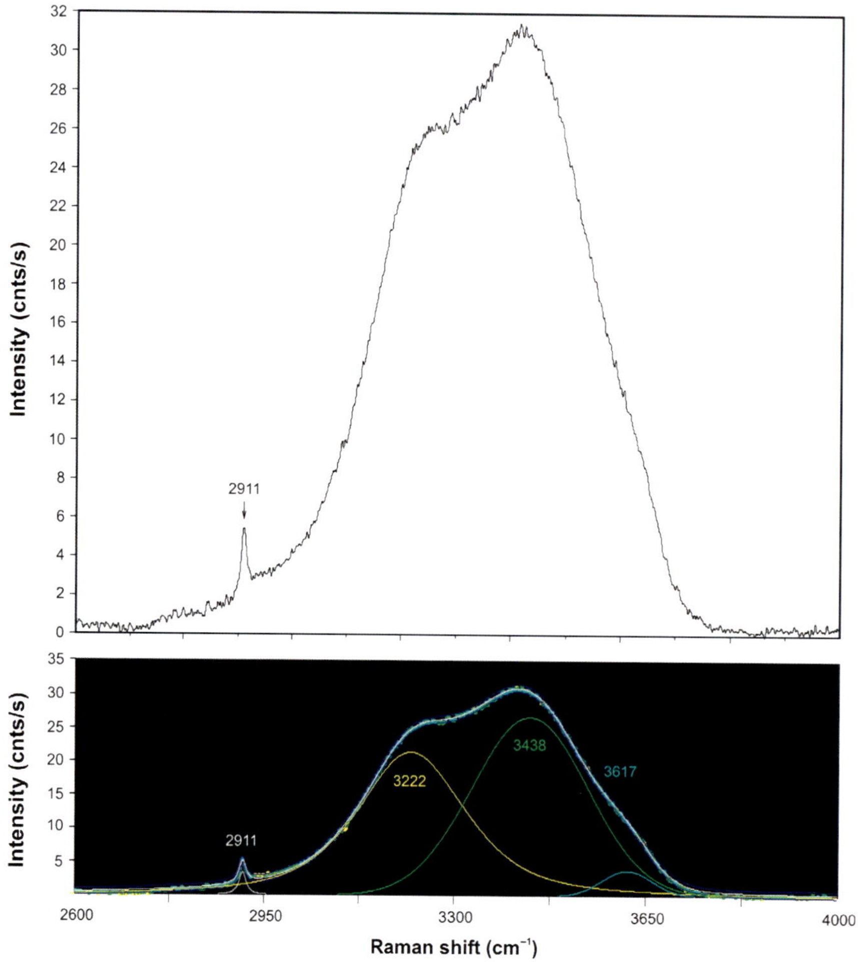

FIGURE 7.12

Raman spectrum of an aqueous inclusion in barite from a calcite veinlet in flysch sandstones of Outer Carpathians (Poland), showing a sharp peak of CH_4 (2911 cm^{-1}) dissolved in aqueous liquid and a broad asymmetric band of water between 2800 and 3800 cm^{-1}. CH_4 content calculated according to Equation (7.15) from band area ratios (82 a.u. for the methane peak, 14,609 a.u. for three Gauss–Lorentzian components at 3222, 3438, and 3617 cm^{-1} of the OH-stretching region) corresponds to 0.269 ± 0.002 moles of CH_4/kg H_2O. Deconvolution and peak area calculations were made using the Peakfit program from Seasolve Inc.

7.4 GRAPHITE AND CARBONACEOUS SUBSTANCES

Raman spectrometry has been successfully used to characterize carbonaceous materials occurring in sedimentary rocks (Jehlička et al., 1997; Liu et al., 2013a). Graphite and bitumens are closely genetically associated with gaseous hydrocarbons and aqueous solutions, and frequently also occur as daughter phases in fluid inclusions (Figure 7.13).

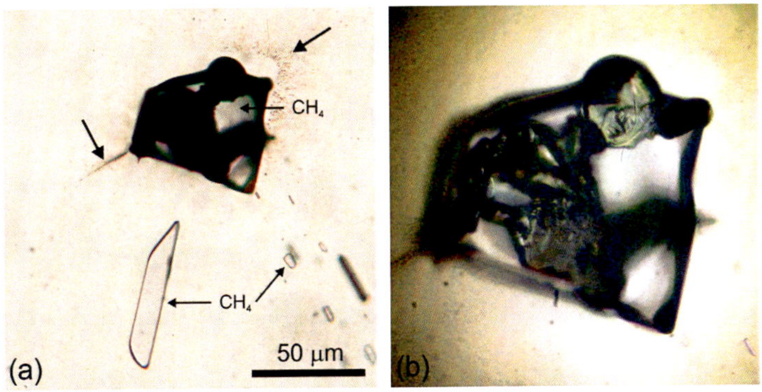

FIGURE 7.13

(a) Composite methane–bitumen inclusion in quartz from calcite veinlets in flysch sediments (Vyšné Ružbachy, central Carpathian Palaeogene fore-arc basin, Slovakia). Thin arrows designate liquid methane (CH_4), thicker arrows mark decrepitation cracks oriented parallel to the *c*-axis of the host quartz and decorated with minute satellite inclusions. (b) An enlarged view of the methane–bitumen inclusion in reflected light reveals bitumen particles with different reflectances.

The first study dedicated to the Raman spectroscopy of graphitic carbon was published by Tuinstra and Koenig (1970). The shape of the graphite Raman spectrum depends on the degree of crystallinity (Figure 7.14). Well-ordered graphite usually has only one maximum at ~1575–1582 cm^{-1} (G band) due to the symmetric plane-parallel vibration of aromatic chains. The band at ~1355 cm^{-1} called defect or disorder band (D band) is typical of a less-ordered graphite, and its intensity increases with the number of vacancies in distorted aromatic ring chains (Beyssac and Lazzeri, 2012). The D peak can be activated even by a small amount of impurities and structural defects. Its frequency is shifted when the excitation laser energy is increased, e.g., from ~1350 cm^{-1} at 514 nm to 1308 cm^{-1} at 785 nm. The D band is, in fact, composed of two overlapping components, which can be discerned using the band component analysis (Figure 7.15).

The second-order spectrum of perfectly ordered graphite consists of two neighboring bands at 2690 and 2730 cm^{-1}. The two bands gradually converge with the decreasing degree of ordering. Amorphous graphite shows only one peak at ~2700 cm^{-1}. In contrast, the two-dimensional peak splits in the high-temperature (> ~500 °C) graphitic carbon, indicating a possible three-dimensional ordering.

The Raman spectrometry of carbonaceous matter is used as a quantitative geothermometer and a peak-temperature indicator for sediments and metamorphites (Beyssac et al., 2002a,b). Within the temperature range of 330–650 °C, graphitization results in the irreversible polymerization and reorganization of the aromatic skeleton, for which Raman spectroscopy provides sensitive quantitative information. Raman spectra of carbonaceous materials from sediments that underwent advanced diagenesis (~200 °C) to low-grade (~320 °C) metamorphism are more complex, showing several defect bands at ~1200 cm^{-1} (D_4), ~1350 cm^{-1} (D_{12}), ~1500 cm^{-1} (D_3), and ~1620 cm^{-1} (D_2) (Lahfid et al., 2010). Evolution of these bands relative to the graphite G band is proportional with the increasing metamorphic grade.

Graphite can be employed as a geothermometer also in granites and granite-related systems, where μm-sized graphite crystals are often observed (Seifert et al., 2009, 2010, 2011), for instance, in rapakivi granites of Finland, Precambrian granites of the Bornholm island, Denmark, in pegmatitic granite

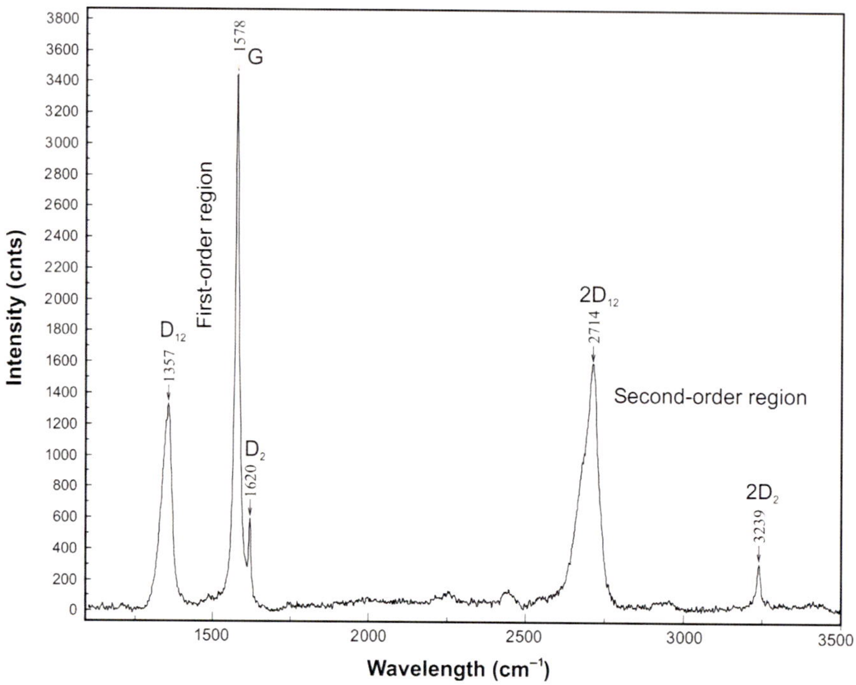

FIGURE 7.14

Background-corrected Raman spectrum (300 mW, 532 nm Nd-YAG laser) of the graphite inclusion in zircon from the diamondiferous gneiss of the Tromsø Nappe, Norway (courtesy of M. Janák), showing the mixture of disordered graphite with defect D bands and ordered graphite diagnostic of the G band at 1578 cm^{-1}. Note the asymmetry of the D_{12} and $2D_{12}$ bands.

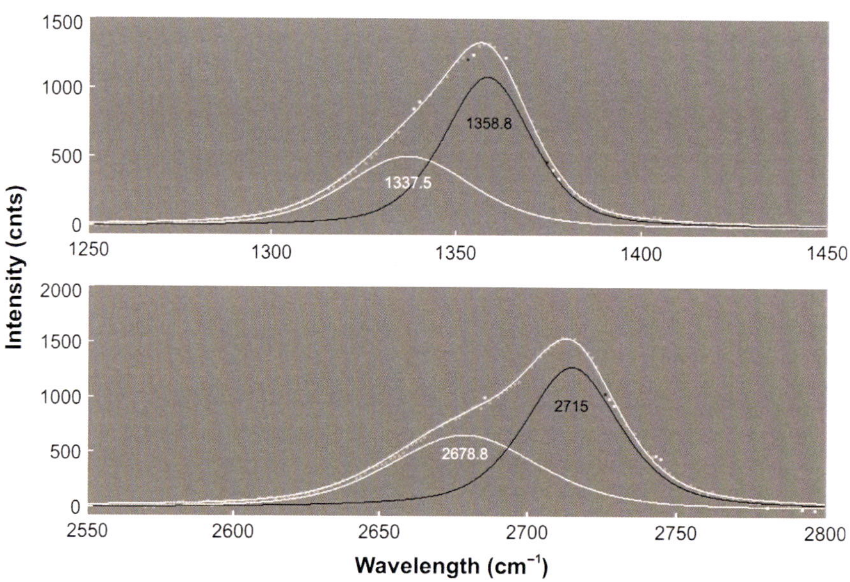

FIGURE 7.15

Deconvolution of D_{12} bands from the previous picture using a mixed Gauss–Lorentzian function (PeakFit© program from SeaSolve Software, Inc., version 4.12) shows that both D_{12} bands are in fact doublets consisting of two overlapping bands.

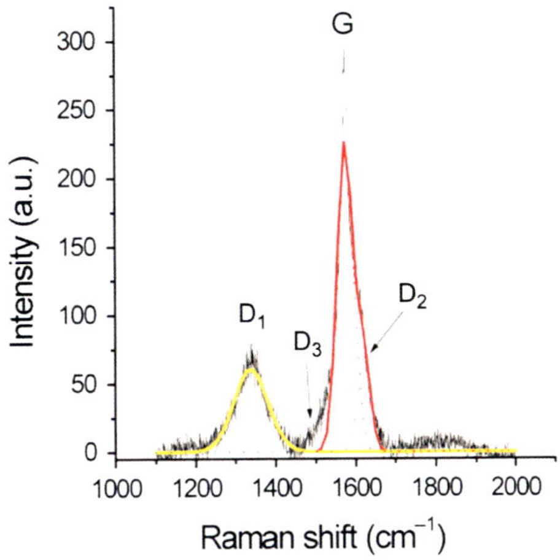

FIGURE 7.16

Raman spectrum of magmatic graphite from the Greifenstein granite. The spectrum was recorded with a Jobin-Yvon LabRam HR800 spectrometer, 1800 gr/mm grating, using an Olympus optical microscope and a long-working-distance LMPlanFI 100 ×/0.80 objective, the 488 nm excitation of a Coherent Ar^{+} laser (Innova 70C) with a power of 300 mW (about 45 mW on sample), 100 μm confocal hole. The laser energy in the measuring spot was reduced by $D_{0.3}$ to D_4 density filters.

and rose quartz from the Borborema pegmatite field, and also in the Variscan Greifenstein granite of Erzgebirge/Germany (Thomas and Davidson, 2012c). Graphite here is a very important fluid-deposited phase, which bears information on the oxidation state of the magma.

Careful study showed that the graphite geothermometer of Beyssac et al. (2002a), after some modification, can also be successfully applied to magmatic systems and associated hydrothermalites. Raman spectrum of the magmatic graphite from the Greifenstein granite (Figure 7.16) demonstrates that the graphite has a near triperiodic sp^3 structure formed at the temperature of about 700 °C.

The Raman spectra can be scanned from petrographic thin sections, perpendicular to the graphite (002) stacking plane, if possible. The laser beam is to be focused beneath the graphite surface to reduce the influence of the preparation damage on the graphite structure (Pasteris, 1989). In the case of "reaction" of the laser with the graphite, the laser power should be reduced using D-density filters. For independent temperature estimation, fluid and melt inclusion data as well as the Ti-in-quartz geothermometer (Wark and Watson, 2006) can be used as independent temperature constraints in magmatic systems (Seifert et al., 2009, 2010, 2011).

Figure 7.17 shows that the $R2$ parameter is linearly correlated with temperature ($R^2 = 0.95$) according to the equation

$$10{,}000/(t + 273.15) = 9.82794 + 9.5697R2 \tag{7.17}$$

where t is temperature in °C. Using this equation, formation temperatures of magmatic and hydrothermal graphite can be estimated up to 780 °C. However, the equation cannot be extrapolated to higher temperatures, owing to a larger scatter of the $R2$ values.

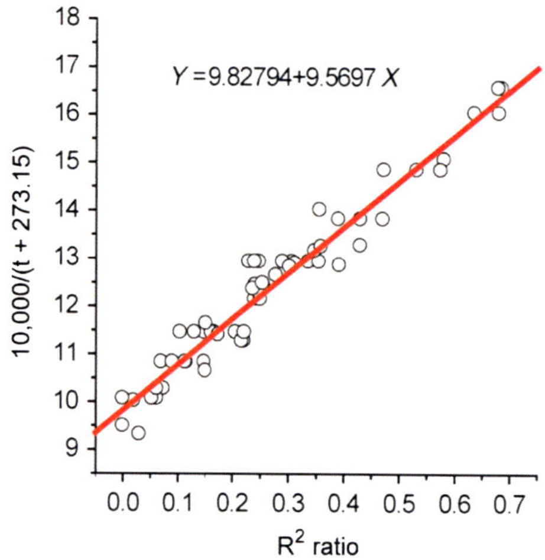

FIGURE 7.17

Correlation between the $R2$ parameter $(D_1/(G+D_1+D_2)_{area}$ and formation temperatures of magmatic graphite in granitic systems.

PROBLEM 7.4 GRAPHITE THERMOMETRY

Calculate the temperature from the graphite thermometer (Beyssac et al., 2002b) using the data shown in Figure 7.14.

Solution

Deconvoluted Raman spectrum using a mixed Gauss–Lorentzian function yielded the following relative peak areas: D_1 band (1337.5 cm^{-1})—24,093, D_{12} band (1358.8 cm^{-1})—37,408, G band (1577.5 cm^{-1})—75,004, D_2 band (1619 cm^{-1})—5293.

$$R2 = (37{,}408 + 24{,}093)/(24{,}093 + 37{,}408 + 75{,}004 + 5293) = 0.43$$

$$T(°\mathrm{C}) = -445 \times R2 + 641 = 448\ °\mathrm{C} \tag{7.18}$$

The thermometer is valid for temperatures between 330 and 650 °C, and the uncertainty corresponds to ±50 °C. At higher temperatures, the $R2$ parameter remains fixed at ~0.05.

Note that the fitting parameterization used for the medium-grade graphite thermometry (Beyssac et al., 2002b) cannot simply be extended to lower temperatures (Rahl et al., 2005). For lower metamorphic to diagenetic grades between 200 and 320 °C, Lahfid et al. (2010) found out that two band area ratios, $R_{A1} = (D_1+D_4)/(D_1+D_2+D_3+D_4+G)$ and $R_{A2} = (D_1+D_4)/(D_2+D_3+G)$ increase linearly with increasing temperatures according to equations

$$R_{A1} = 0.0008T + 0.3758 \tag{7.19}$$

$$R_{A2} = 0.0045T + 0.27$$

where T is temperature in °C, D_1 to 4 denote the graphite defect bands located at ~1350, 1620, 1500, and 1200 cm^{-1}, respectively, and D_1 corresponds to D_1+D_{12}.

The distance and heights of D_{12} and G bands can be correlated with the maturation of sedimentary graphitic substance—vitrinite, which is a woody plant debris modified by heating during burial or hydrothermal overprint (Liu et al., 2013a). Hence, the Raman reflectance parameter ($RmcR_0\%$) is the equivalent of the mean vitrinite reflectance commonly used to estimate the sample maturation in coal and sedimentary petrology (e.g., Barker, 1983; Burnham and Sweeney, 1989). The corresponding empirical equations are as follows (Liu et al., 2013a)

$$RmcR_0\% = 0.0537\ d(\mathrm{G}-\mathrm{D}) - 11.21 \tag{7.20}$$

where $d(\mathrm{G}-\mathrm{D})$ is the distance (in cm^{-1}) between G and D_{12} bands shown in Figure 7.15. The equation is applicable to graphites in less mature sedimentary rocks.

From over-mature to a super-anthracite stage, the distance between G and D_{12} bands remains constant or even decreases, because the molecular vibration mode is modified in the aromatic ring structure of solid organics. In this case, the peak/height ratio parameter $h(D_h/G_h)$ is used to calculate the Raman reflectance according to equation

$$RmcR_0\% = 1.1659\ h(D_\mathrm{h}/G_\mathrm{h}) + 2.7588 \tag{7.21}$$

The method can be applied to calculate the maturation of all kinds of solid bitumens and organic materials, including kerogen, pyrobitumen, micronite, carbonized fossils of animals and plants, and solid bitumens in fluid inclusions. Care must be exercised, however, when measuring Raman spectra on polished surfaces, where structural disorder of the graphite may occur due to mechanical thinning and polishing (Nasdala et al., 2004; Wang et al., 1989).

Owing to universality and simplicity, the vitrinite reflectance determination is the critical parameter in reconstructions of thermal history of sedimentary basins. It is generally accepted that an intense oil generation from a marine sapropel (oil window) starts at a vitrinite reflectance of 0.5–0.65% and oil concentrations are at their maximum at $R_\mathrm{m} = 0.85$–1.1% (Tissot and Welte, 1984). Apart from the temperature, the model of Burnham and Sweeney (1989) also involves the reaction kinetics of the vitrinite modification, thus incorporating the additional parameter—time. Barker (1983) showed that temperature (T in °C) and mean vitrinite reflectance (R_m in %) of sedimentary organic matter sampled from liquid-dominated geothermal systems plotted along an exponential regression curve approximated by the equation

$$R_\mathrm{m} = 0.435 \times \mathrm{e}^{0.00683 \times T} \tag{7.22}$$

This indicated that the thermal stability of organic matter was reached after 1000 to 1,000,000 years. The reaction duration has little or no influence on the metamorphism of organic matter in liquid-dominated geothermal systems after about 10,000 years.

7.5 PETROLEUM AND BITUMEN

The laser Raman spectroscopy becomes a standard analytical technique for the characterization of gaseous hydrocarbons (Guilhaumou et al., 1988). However, as a tool in the petroleum research, its role is hindered by the high fluorescence and laser-induced photodegradation. The intense fluorescence is mostly caused by aromatic compounds (Bourdet et al., 2011). Ultraviolet (Loppnow et al., 2004; Shoute et al., 2002) or infrared (Edwards et al., 2007) excitations are employed to reduce the intense fluorescence background. Alternatively, extremely low laser power and exposure times can be used to record the Raman spectra excited in the visible region of the electromagnetic spectrum. For instance, Orange et al. (1996) irradiated petroleum inclusions for 12–24 h using the 633 nm excitation line of the He–Ne laser with 1–5 mW laser power on sample. The 514 nm excitation line of Ar^+ lasers sometimes also yields fluorescence-free spectra (Zhang et al., 2007), but low laser power (D_1 filter) and shorter acquisition times should be applied to avoid the petroleum degradation.

Assignments of Raman and infrared bands of important organic compounds are summarized in the handbook by Lin-Vien et al. (1991). An overview of characteristic Raman vibrational modes of selected hydrocarbon compounds is given in Table 7.3.

Saturated compounds do not exhibit first-order Raman bands above 1470 cm^{-1}. This is expected because they do not contain C=C bonds and other double bonds vibrating between 1500 and

Table 7.3 Selected Raman Vibration Absorption Bands of Organic Compounds

Wave Number Range (cm^{-1})		Functional Group	Mode	Compound	Source[a]
166	610	C–Br	Stretching	Bromobenzene	6
184	550	C–C–C	Out-of-plane bending		1
520	552	C–C–C	Ring deformation		2
608	996	C–H	Out-of-plane bending	Polycyclic aromatic hydrocarbons	1
620	625	C–C	Bending	Aromatic hydrocarbons	2
620	715	S–C	Stretching		6
700	1000	C–C	Ring breathing/ stretching	Cycloalkanes	6
710	760	CH_2	Rocking	Saturated hydrocarbons	3
721	742	C–C	Stretching		2
837	905	C–C		*n*-Alkanes	6
800	1065	S=C	Stretching		6
910	930	CH_3	Wagging	Double-branched hydrocarbons	3
950	970	CH_3	Wagging	Single-branched hydrocarbons	3
990	1060		Ring deformation	Benzene and derivatives	5, 6
1019	1247	C–H	In-plane-bending/ rocking	Mono- and polycyclic aromatics	1
1020	1200	C–C	Stretching/Bending	Monocyclic aromatic hydrocarbons	3
1110	1155	$-SO_2-$	Symmetric stretching		5
1240	1472	CH_2	Bending		2
1260	1330	$-SO_2-$	Asymmetric stretching		5

Continued

Table 7.3 Selected Raman Vibration Absorption Bands of Organic Compounds—Cont'd

Wave Number Range (cm^{-1})		Functional Group	Mode	Compound	Source[a]
1270	1633	C–C	Stretching	Polycyclic aromatic hydrocarbons	1
1284	1295	CH_2	Scissors		2
1300	1340	CH_2	Wagging	Saturated hydrocarbons	3
1330	1450	C–H	Bending	Saturated hydrocarbons	3
1442	1450	N–H	Bending		2
1446	1473	CH_2		Saturated hydrocarbons	5
1600	1605	C=C	Stretching	Monocyclic aromatic hydrocarbons	2, 3
1630	1680	C=C	Stretching	Alkene, amber	2
2100	2160	C≡C		Alkylalkynes	5
2190	2300	C≡C		Di-alkylalkynes	5
2550	2600	S–H	Stretching	Thiols	4, 5, 6
2840	2865	CH_2	Symmetric stretching	Lipids, saturated hydrocarbons	3, 5
2850	2930	C–H	Stretching		2
2850	2700	C–H		Aliphatic aldehydes	5
2862	2885	CH_3	Symmetric stretching	Saturated hydrocarbons, lipids	3
2883	2884	CH_3	Symmetric stretching	Saturated hydrocarbons	5
2912	2949	CH_2	Asymmetric stretching	Saturated hydrocarbons	5
2927	3085	C–H	Stretching	Polycyclic aromatic hydrocarbons	1
2965	2969	CH_3	Asymmetric stretching	Saturated hydrocarbons	5
2980	2990	$=CH_2$		$C=CH_2$ derivatives	5
3020	3070	C–H		Benzene derivatives	5
3057	3060	C–H		Alkylbenzene, benzene	5
3075	3080	C=CH	Symmetric stretching	Exocyclic methylene group	2
3250	3550	NH_2		Primary free ammines, 2 bands	5
3250	3400	NH_2		Primary associated ammines, 2 bands	5
3270	3350	≡C–H		Alkyl-alkynes	5
3300	3355	N–H		Secondary ammines	5
3400	3490	N–H		Secondary amides	5

[a]*References: (1) Colangelli et al. (1992). (2) Edwards et al. (2007). (3) Loppnow et al. (2004). (4) Zhang et al. (2007). (5) Milata et al. (2008). (6) Orange et al. (1996).*

1750 cm^{-1}. The strongest Raman band of saturated hydrocarbons is the C–H deformation mode at 1430 cm^{-1} and its overtone at 2860 cm^{-1} (Loppnow et al., 2004). Petroleum with a high content of saturated hydrocarbons exhibits the strongest bands between 2700 and 2970 cm^{-1}, corresponding to C–H stretching and bending modes (Zhang et al., 2007).

Raman spectroscopy better evidences C–C vibrations than the C–H ones in aromatic hydrocarbons. Monocyclic aromatic hydrocarbons show the strongest bands at 1020–1200 cm^{-1} and one or more bands at around 1600 cm^{-1}. Bicyclic naphthalene and tricyclic anthracene exhibit similar spectra, showing a stronger C=C ring stretch deformation at 1380 and 1410 cm^{-1}, respectively, and a weaker C=C ring stretch at 1630 cm^{-1}. For other polycyclic aromatic hydrocarbons (PAHs) with multiple aromatic rings the Raman spectra are similar, consisting of 2–3 strong bands, one at ~1400 cm^{-1}, and one or two others at ~1600 cm^{-1}, which resemble the disorder and order bands, respectively, of pure graphite. Hence, it is hypothesized that defects in graphitic materials should be similar in size to those in PAHs (Castiglioni et al., 2001). Note, however, that small differences in the excitation wavelength have a large effect on relative intensities of the 1400 and 1600 cm^{-1} bands in aromatic hydrocarbons.

Single hydrogen atom in benzene rings is indicated by the bands between 835 and 910 cm^{-1}. When two adjacent peripheral atoms are present, bands occur between 800 and 855 cm^{-1}. Three adjacent hydrogen atoms at the edges of molecules are indicated by bands between 765 and 800 cm^{-1}. Finally, strong features between 740 and 770 cm^{-1} indicate the presence of four adjacent peripheral hydrogen atoms (Colangelli et al., 1992; Loppnow et al., 2004). The typical phenyl (C_6H_5) band occurs at 988 cm^{-1} in some petroleum inclusions, whereas weak bands between 2970 and 3200 cm^{-1} are attributed to methyl (–CH_3) and methylene (>CH_2) groups (Zhang et al., 2007).

Complex infrared absorption spectra can be interpreted using the following rules (Coates, 2000), which can also be partly applied to interpretations of Raman vibrational spectra:

1. Well-defined bands in the region **2700–3200 cm^{-1}** are normally assigned to the **C–H stretching**. The bands above 3000 cm^{-1} are diagnostic of unsaturated aromatic compounds (C=C). Isolated bands at 3010 and/or 3040 cm^{-1} correspond to simple olefinic unsaturation. If the main band occurs below 3000 cm^{-1}, the compound is probably aliphatic. Main absorptions at 2860 and 2935 cm^{-1} accompanied by those at 1470 and 720 cm^{-1} are diagnostic of a long linear aliphatic chain.
2. Characteristic broad-band profiles between **3250 and 3650 cm^{-1}** are diagnostic of **hydroxy** or **amino** groups.
3. The region between **1650 and 1850 cm^{-1}** is diagnostic of **carbonyl** compounds (C=O), but certain C=N (imino-) and N=N (azo-) groups may also interfere.
4. Weak-to-moderate bands in the region of **1620–1670 cm^{-1}** are diagnostic of **unsaturated** hydrocarbons. A narrow band centered around 1650 cm^{-1} reveals the olefinic unsaturation (C=C). A lowering and intensification of this band is characteristic of the conjugation with another double bond (C=C or C=O) or an aromatic ring. The test for unsaturation usually involves a check for C–H peaks above 3000 cm^{-1} (step 1). Single or multiple bands between 800 and 1000 cm^{-1} (C–H out-of-plane bending region) combined with the 1650 and >3000 cm^{-1} bands can also be used as the sign of the unsaturation.
5. Well-defined bands in the region between **1495 and 1615 cm^{-1}** are diagnostic of **aromatic rings**. Typically two sets of bands occur around 1500 cm^{-1} and the other one at around 1600 cm^{-1}. The aromatic rings are supported by a series of weak-to-moderate bands in the region of 3000–3150 cm^{-1} assigned to the aromatic C–H stretching. For simple aromatic compounds,

a series of weak bands may be observed between 1700 and 2000 cm^{-1}. The C–H bending vibrations can also be used to support the presence of an aromatic structure. Medium-to-strong bands between 670 and 850 cm^{-1} are assigned to the C–H out-of-plane bending of an aromatic ring. A single strong band in this region may support simple *ortho*- (~750 cm^{-1}) or *para*- (~830 cm^{-1}) substitutions.

6. The region between **1990 and 2300 cm^{-1}** is diagnostic of **multiple bonded nitrogen** compounds, such as cyanides, cyanates, and thiocyanates.
7. The region between **430 and 710 cm^{-1}** is diagnostic of C–S and S–S stretching in **thio compounds** (thiols, disulfides, polysulfides).
8. The region between **850 and 1050 cm^{-1}** is diagnostic of the P–O–C stretching in aliphatic (990–1050 cm^{-1}) and aromatic (850–995 + 1190–1240 cm^{-1}) **phosphates**.

Raman spectra of bitumens usually show two main peaks centered at ~1400 and 1620 cm^{-1}, corresponding to the C–C stretching in aromatic hydrocarbons. Additional weak bands between 2970 and 3200 cm^{-1} may be attributed to methyl- and methylene groups (Zhang et al., 2007).

Raman spectra of bitumens from sedimentary rocks and fluid inclusions (Figure 7.18) show well-defined C–C stretching modes of PAHs at 1336–1345 and 1582–1596 cm^{-1}. A minor band centered at about 1050 cm^{-1} in the bitumen from calcareous breccia could be due to a small amount of monocyclic aromatic hydrocarbons. The sharp bands at 431, 583, and 953 cm^{-1} are reminiscent of those in apatite. However, the major band at 953 cm^{-1} is significantly downshifted from the main PO_4^{3-} vibration typical of a crystalline apatite (959–965 cm^{-1}). Hence, we infer that the band at 953 cm^{-1} could correspond to the P–O–C stretching of aromatic phosphate groups.

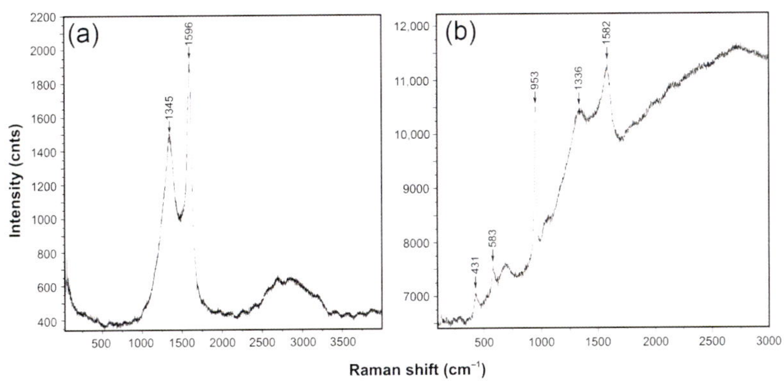

FIGURE 7.18

Raman spectra of natural bitumens. (a) Black pyrobitumen in the liquid methane inclusion in quartz from the fissure in Paleogene sandstones documented in Figure 7.13. The spectrum contains only bands diagnostic of aromatic C–C bonds. (b) Bitumen from the calcareous breccia of the west Atlantic continental margin. Apart from the aromatic C–C bonds (1336 and 1582 cm^{-1}), the sharp peak at 953 cm^{-1} diagnostic of aromatic phosphate groups also occurs. Both spectra were obtained with a 532 nm laser excitation, 1800 gr/mm grating, and 2 × 10 s acquisition time. Note the relatively strong fluorescence of the bitumen from the west Atlantic calcareous breccia.

7.6 SILICATE MELTS

A homogeneous molten magmatic rock is called melt. The melt quenches to glass at a rapid cooling or crystallizes to well-defined mineral phases on very slow cooling. Most natural melts are aluminosilicate melts dominated by SiO_2 and Al_2O_3, but carbonatite and sulfide melts also occur associated with alkalic and mafic silicate rocks, respectively. Due to a low viscosity at high temperature, volatile melts possess attributes of fluid, although they are solid at room temperature.

Cations in silicate melts are subdivided into network formers and network modifiers (Bottinga and Weill, 1972). On short distances of several atomic bonds, silicate melts exhibit structural ordering typical of minerals, when network-forming cations (Si, Al, Fe^{3+}) occupy central positions in tetrahedral entities linked together by bridging oxygen atoms. In contrast, network-modifying cations (alkali metals, alkalic earth elements, Fe^{2+}, Fe^{3+}, Mn, P^{5+}, Ti^{4+}, REE) occupy central positions in nontetrahedral polyhedrons with nonbridging oxygens on their corners (Mysen, 1991). The degree of melt polymerization is described by an empirical parameter NBO/T—the ratio between nonbridging oxygens and tetrahedral-coordinated cations (Mysen, 1991; Mysen and Virgo, 1989). Anhydrous rhyolite glass practically does not contain any nonbridging oxygens, thus being almost fully polymerized with the corresponding NBO/T value below 0.1. Andesitic and basaltic glasses are less polymerized, having typical NBO/T ratios around 0.2 and 0.4–1.2, respectively (Mysen, 1988).

The NBO/T ratio reflects the distribution of Q^n species in silicate melts. The Q^n notation (Rossano and Mysen, 2012) indicates tetrahedra where the number of bridging oxygens corresponds to the superscript n $(0-4)$. The Q^4 is ideal, fully polarized, three-dimensional silicate structure, where all four oxygens are bonded to Si. Q^3 species have three bridging and one nonbridging oxygen, being thus equivalent to a two-dimensional structure of sheet silicates (T_2O_5, where T designates the tetrahedral-coordinated cation). Q^2 species have two bridging and two nonbridging oxygens equivalent to the chain structure of metasilicates-pyroxenes (TO_3). Q^1 species have one bridging and three nonbridging oxygens as an equivalent to pyrosilicates (T_2O_7), where two tetrahedra are linked by one shared oxygen. Finally, all oxygens are nonbridging in Q^0 species, having the least polymerized structure equivalent to those of nesosilicates (TO_4) with isolated tetrahedra. The abundance of individual Q species in natural melts varies with their composition. Diagnostic Raman bands of individual Q species, as well as other functional groups present in natural and synthetic silicate glasses are summarized in Table 7.4.

Raman spectra of various silicate glasses are shown in Figures 7.19–7.21. The most prominent features in the low-frequency region below 1500 cm^{-1} are SiO_4 breathing modes in Q^4 units or T–O–T symmetric bending bands between 460–500 cm^{-1}, and the asymmetric stretch of $Si–O^-$ bonds in Q^2–Q^4 units between 900 and 1100 cm^{-1}. The weak band at 1628 cm^{-1} corresponds to the H–O–H bending vibration of dissolved molecular water while the broad asymmetric band between 3000 and 3600 cm^{-1} corresponds to the stretching of molecular water and hydroxyl groups.

7.7 MAGMATIC VOLATILES AND FLUXING SPECIES

High spatial resolution and accuracy of Raman spectroscopy enables the quantification of water contents and speciation in natural and experimental glasses and also the water tracing in nominally anhydrous minerals, such as quartz and garnet. In some cases, the estimation of bulk water content of water-bearing minerals is also possible.

Table 7.4 Overview of Raman Bands Determined in Synthetic and Natural Silicate Glasses

Raman Band (cm^{-1})	Interpretation	Source[a]
440–495	Oxygen breathing in SiO_4 units in Q^4	1
<500	Metal-oxide vibrations	1
520–550	Fluoroberyllate ions ($Be_2F_7^{2-}$, BeF_4^{2-})	2
550–590	Si–O–Si bending in Q^3	1
565–595	Si–O–Si bending in Q^2	1
605	Oxygen breathing in SiO_4 units in Q^4	1
700	Si–O–Si bending in Q^0/Q^1	1
776	MBF_4 (M = Li, Na, K, Rb, Cs)	5
780	Ti–O vibrations of fourfold coordinated Ti^{4+}	1
800	O vibrations perpendicular to the Si–Si bond in Q^4	1
850	Antisymmetric stretch $Si–O^-$ (nonbridging O) in Q^0/Q^1	1
871	$HBF_2(OH)_2$	5
880	Ti–O vibrations of fivefold coordinated Ti^{4+}	1
900	Antisymmetric stretch $Si–O^-$ (nonbridging O) in Q^1	1
950	P–O stretch in PO_4 units in Q^0	1
960	Antisymmetric stretch $Si–O^-$ (nonbridging O) in Q^2	1
970	Si–OH stretch	1
1000	P–O stretch in P_2O_7 units in Q^1/S–O stretch in SO_4 units	1
1020	P–O stretch in PO_3 units in Q^1	1
1050	Antisymmetric stretch $Si–O^-$ (nonbridging O) in Q^2	1
1060	Antisymmetric Si–O stretch in SiO_4 units in Q^4	1
1070	C–O symmetric stretch in CO_3 groups	1
1100	Antisymmetric stretch $Si–O^-$ (nonbridging O) in Q^3	1
1150	Bridging P–O–Si bonding in Q^2	1
1200	Antisymmetric Si–O stretch in SiO_4 units in Q^4	1
1280	Molecular CO_2	1
1380	Double-bonded P in Q^3/molecular CO_2 in melt	1
1650	H–O–H bending molecular H_2O	1
2070	Nitrosyl NO groups	1
2320	Nitrosyl NO groups/molecular N_2	1
2550	H–S stretch in the HS^- complex	1
2800–3800	O–H + H_2O stretch (wide band)	1
2900	C–H stretch in CH_3/CH_4	1
2950	Ammide NH_2^+ groups	1
3165	H_3BO_3	4
3247	H_3BO_3	4
3250	Ammine NH_2^- groups	1
3300	Ammonium NH_3 groups	1
3350	(C–H)-bearing groups with triple bonded C/ammine NH_2^- groups	1
3594–3620	O–H stretch in NaOH	3, 6
4130	Molecular H_2	1

[a] *References: (1) Rossano and Mysen (2012) and references therein. (2) Thomas (unpublished). (3) Thomas and Davidson (2012a). (4) Thomas et al. (2012a). (5) Thomas et al. (2005). (6) Thomas et al. (2011a).*

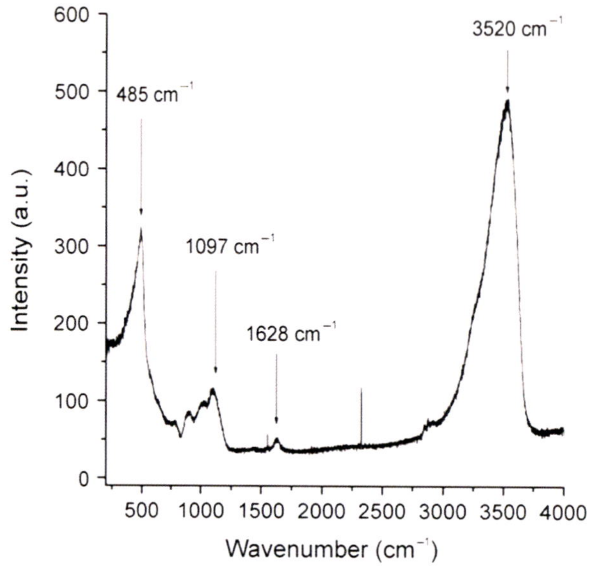

FIGURE 7.19

Raman spectrum of albite glass (AB84 with 11.95 mass% H_2O; Thomas, 2000). The broad band at 485 cm^{-1} results from the T–O–T symmetric bending. Bands centered at about 1090 cm^{-1} represent T–O and T–O–T bending and stretching modes while the weak band at 1628 cm^{-1} corresponds to H–O–H bending vibrations of dissolved molecular water. The broad asymmetric band at 3520 cm^{-1} corresponds to stretching of molecular water and hydroxyl groups. The unindexed sharp peak at ~2300 cm^{-1} corresponds to the atmospheric N_2.

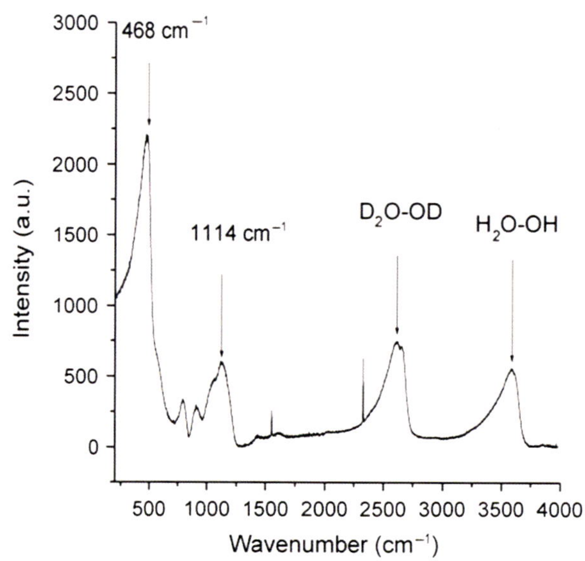

FIGURE 7.20

Raman spectrum of a synthetic H_2O- and D_2O-bearing (about 5 mass% each) rhyolitic glass (Leschik et al., 2004) with characteristic broad asymmetric H_2O–OH and D_2O–OD stretching modes at 3100–3750 cm^{-1} and 2250–2900 cm^{-1}, respectively (Thomas and Davidson, 2006, 2007). The unindexed sharp peak at ~2300 cm^{-1} corresponds to the atmospheric N_2.

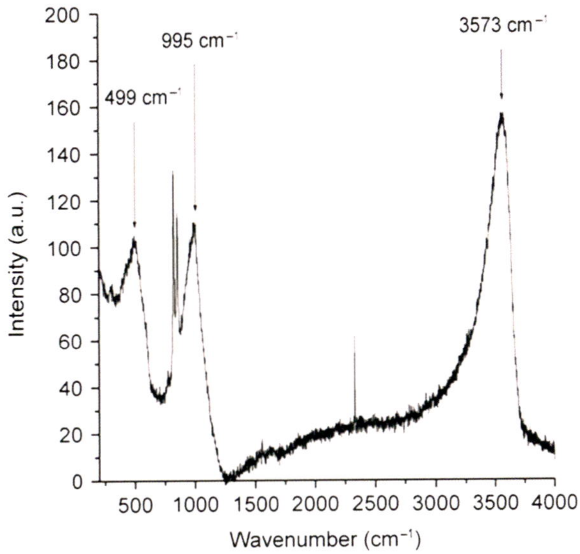

FIGURE 7.21

Raman spectrum of basaltic glass in a melt inclusion trapped in olivine from the Chikurachki Volcano, Kuril arc, Russia, containing 6.1 mass% H_2O_T. Unindexed sharp bands at 824 and 856 cm^{-1} are those of forsterite olivine host. The band at 3573 cm^{-1} represents molecular water and hydroxyl groups dissolved in the inclusion glass (Thomas and Davidson, 2006). The unindexed sharp peak at ~2300 cm^{-1} corresponds to the atmospheric N_2.

Raman spectroscopy had not been applied before 1999 to the quantitative determination of water content in silicate melts or glasses, although it had been used for the identification of daughter minerals in inclusions, as well as in structural studies of volatile glass (Ihinger et al., 1994). Though the micro-Raman spectrometry was proposed for quantifying the OH-content in silica glass down to the 10 ppm level (Galeener and Mikkelsen, 1981), this technique was not suitable for the study of melt inclusions due to the low spatial resolution. However, with the introduction of a new generation of sensitive confocal Raman spectrometers, the determination of water content in natural glasses and melt inclusions became reality and the technique has developed into a routine and versatile analytical method (Behrens et al., 2006; Chabiron et al., 2004; Di Muro et al., 2005; Le Losq et al., 2012; Mercier et al., 2010; Severs et al., 2007; Thomas, 2000, 2002b; Thomas and Davidson, 2006; Thomas et al., 2005, 2006a; Zajacz et al., 2005). Recently, this micro-Raman technique can be used for the accurate, rapid, and high-precise determination of the total water content of glasses over a wide concentration range, from 0.1 to 35 mass% H_2O. Thomas et al. (2008b) showed that under optimal conditions, water concentrations of less than 10 ppm can be measured using reference glasses with 37 ± 8 ppm H_2O determined by the proton–proton scattering technique. With some limitations (Behrens et al., 2006), the quantitative determination of water speciation, that is, the ratio of molecular water (H_2O_m) to hydroxyl group, OH (Chabiron et al., 2004; Di Muro et al., 2006), as well as that of D_2O_m/OD ratios (Thomas et al., 2006a) are also possible. The modern generation of Raman spectrometers equipped with a high-precision motorized *XY* mapping stage permits the collection of H_2O-concentration profiles, which enable the visualization of water diffusion in glasses, and H_2O concentration distribution in melt inclusions. As an example, a pseudoholographic Raman map is illustrated in Figure 7.22, in which each

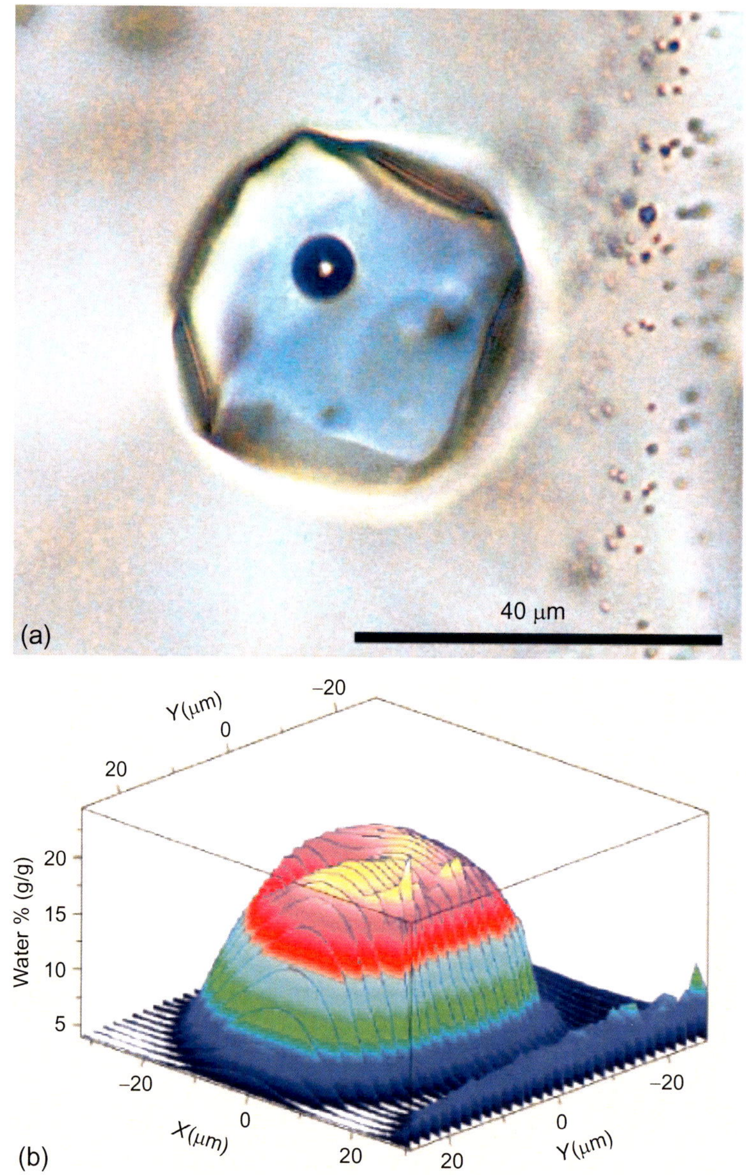

FIGURE 7.22

(a) Water-rich (25 mass%) melt inclusion in the pegmatite quartz from Ehrenfriedersdorf, Germany, rehomogenized at 600 °C and 1 kbar in a cold-seal pressure vessel pressurized with CO_2. (b) Three-dimensional diagram of the water distribution measured by Raman spectroscopy. The distribution image was taken with a LabRam HR800 spectrometer using a 488 nm excitation laser line and a 50 μm pinhole. The image is composed of a total of 750 single points, 2 μm in diameter, each containing a complete Raman spectrum in the 2800–4000 cm^{-1} frequency range. For the quantification of water content, the integral intensity of the broad asymmetric H_2O–OH band between 3100 and 3750 cm^{-1} and the calibration procedure of Thomas et al. (2006a) was used. The warmest colors correspond to the highest water concentrations, reaching a maximum of 25 mass% (in yellow). The blue ridge at the right side is caused by a trail of secondary fluid inclusions positioned perpendicularly to the investigated quartz chip.

one µm-wide pixel contains a complete quantitative spectrum of the dissolved water recalculated in terms of the H_2O concentration expressed in mass%.

The advantage of the Raman spectroscopic method in the study of water contents in melt inclusions over the traditional infrared spectroscopy stems from the simplicity of sample preparation, the low matrix effect, an ability to measure large concentration ranges and to analyze selectively discrete volumes within the sample (Le Losq et al., 2012). Melt inclusions ranging from granitic to basaltic glass compositions can be measured without prior knowledge on the composition and density of the analyzed glass (Figures 7.19–7.21).

The confocal micro-Raman spectroscopy permits the H_2O content to also be determined in unexposed melt inclusions (Thomas et al., 2006a). Fourier-transform infrared spectroscopy can be principally used for the same purpose (Befus et al., 2012). The method is based on the negatively correlated dependence of the integral intensity of the water–OH stretching band (2800–3800 cm^{-1}) with the distance from the polished mineral surface (Figure 7.23). The investigation of unexposed inclusions permits the study of H_2O-rich, unstable inclusion glasses, which would be spontaneously and irreversibly decomposed during the sample grinding and polishing.

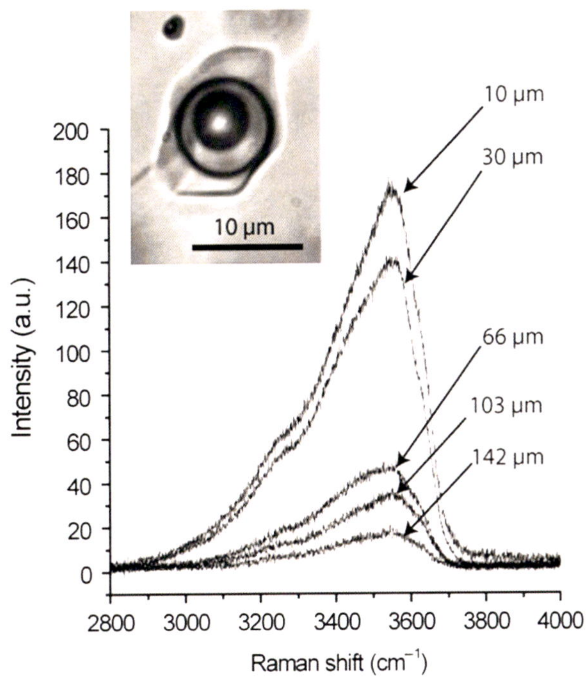

FIGURE 7.23

Raman spectra in the high-frequency range (2800–4000 cm^{-1}) of water-saturated silicate glass in five water–silicate melt inclusions located in different depths of doubly polished quartz wafer. Insert illustrates one of these inclusions composed of a silicate glass (most external phase), aqueous solution, and a vapor bubble as seen at room temperature.

The technique can be used to determine water content and speciation in several melt inclusions located in different depths within the investigated mineral. The inclusion melts with various water contents would create different slopes of linear trends fitted to the depth versus integral intensity correlation diagram (Table 7.5, Figure 7.24). In a limiting case, the water content of a single inclusion can be determined provided the inclusion is not exactly at a midpoint of the doubly polished wafer. The Raman spectrum is then recorded twice, from different sides of the polished wafer. The larger is the difference between the distances of the inclusion to either surface; the more accurate is the water content determination. The position of single inclusion can also be changed by stepwise polishing of the wafer from one side.

The effect of the inclusion host lattice orientation must be also considered (Sirbescu et al., 2013). Hence, melt inclusions of only one single crystal grain can be directly used for the construction of correlation diagrams. Rotation of the sample or data from variably oriented samples yield erroneous results.

Two methods have been used to determine the H_2O contents of glasses and unexposed melt inclusions: (1) the internal normalization of the H_2O_T band (the H_2O+OH stretching band) to that related to the vibration of aluminosilicate network at 490 cm^{-1} (the T–O–T stretching band, Thomas, 2000; Chabiron et al., 2004; Di Muro et al., 2006), or those between 850 and 1250 cm^{-1} (T–O stretching band, Zajacz et al., 2005; Behrens et al., 2006; Severs et al., 2007), and (2) the external calibration

Table 7.5 Determination of the Total Water Content in Silicate Melt Inclusions in Quartz from Granite and Pegmatite (Thomas et al., 2006a)

Depth (μm)	Integral Intensity (a.u.)	Integral Intensity (%)	H_2O_T Measured (Mass%)
Quartz from topaz-albite granite, Zinnwald, Germany			
72	21,244	25.58	2.32
60	27,392	32.98	2.99
31	46,458	55.94	5.06
23	50,749	61.11	5.53
10	69,346	83.50	7.56
3	78,104	94.05	8.52
0	**83,044**	**100.00**	**9.06**
Quartz from pegmatite, Ehrenfriedersdorf, Germany			
72	95,896	41.69	10.46
61	108,442	47.14	11.82
51	125,431	54.53	13.68
45	129,538	56.31	14.12
32	157,639	68.53	17.19
26	168,397	73.20	18.36
18	184,454	80.18	20.11
0	**230,043**	**100.00**	**25.08**

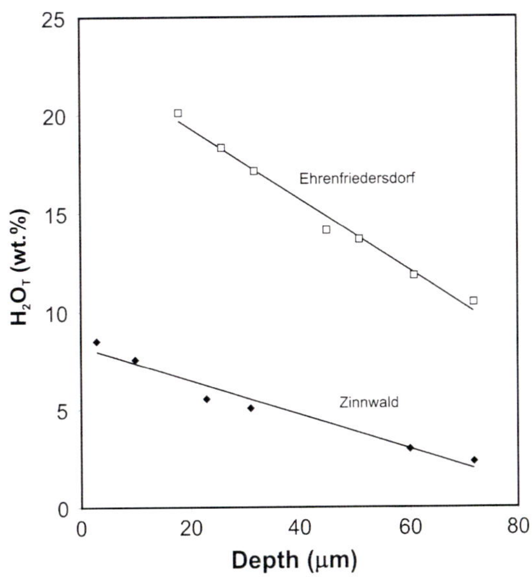

FIGURE 7.24

Correlation of the calculated apparent water concentration H_2O_T (mass%) versus inclusion depths for the data listed in Table 7.5. In both cases, the water content can be obtained by the extrapolation to a depth of zero µm. Note that different water concentrations result in the variable trendline slopes.

procedure using well-characterized glass standards—the comparator method (Thomas, 2000; Di Muro et al., 2006). Initially, the internal calibration procedure was employed because of uncertainty about the long-term stability of external standards.

Thomas et al. (2006a) demonstrated that the construction of a calibration curve for each experimental setup is not necessary. The fact that all values of the used standards are linearly correlated through zero point is a clear indication that every single measurement point can be taken for the calibration. The best results are obtained if water contents and compositions of the standard and the sample are similar. Using the comparator method, however, the water content in the sample is determined by a simple comparison with a known reference standard. However, if necessary, a calibration curve or bracketing by using two or more glass standards can also be done.

The water content in the glass can be very simply calculated because the integral intensity of the H_2O–OH stretching band between 2800 and 4000 cm^{-1} increases linearly with the total water content independently on the glass composition (Table 7.6, Figure 7.25). It is basically possible to also recalculate other Raman measurements, if a well-characterized standard glass (or melt inclusion standard) was used. As an example, for a large melt inclusion in olivine from the Chikurachki Volcano, Kurily arc, Russia (Gurenko et al., 2005), a total water concentration H_2O_T=6.1 mass% was previously obtained by the external calibration method, while Thomas (2002b) obtained in the same inclusion the value of 6.5 mass% using the ratio or internal calibration method.

Table 7.6 Summary of Raman Spectrometric Analyses of H_2O_T in Synthetic Glasses and Natural Silicate Melt Inclusions

Glass	H_2O_T[a] (Mass%)	Integral Intensity (3100–3750 cm^{-1})	SD (1σ)[b]	Source
KG (pp-scattering)	0.004	652	5	Thomas et al. (2008b)
KG (pp-scattering)	0.004	651	4	Thomas et al. (2008b)
KOG (FT-IR)	0.033	914	17	Thomas et al. (2008b)
PCD	0.170	2173		Courtesy of H. Westrich
Albite glass	**0.60**	6559	400	Courtesy of H. Behrens
1N	0.92	9092		Courtesy of H. Westrich
ANSI 96-11C	*1.10*	12,329	497	Webster et al. (1999)
AOQ	**1.86**	16,985	2457	Courtesy of F. Holtz
2N	2.02	19,229		Courtesy of H. Westrich
Albite glass	**2.80**	23,158		Courtesy of H. Behrens
Albite glass	**2.90**	28,110	2432	Courtesy of H. Behrens
P2 Granite	**4.00**	35,434		Courtesy of F. Holtz
Etna MI[c]	4.40	40,333	2795	Thomas et al. (2006a)
CHK, MI in Plg	4.61	42,896	8922	Courtesy of A. Gurenko
CHK, MI in Ol	4.86	45,229	14,111	Courtesy of A. Gurenko
M6N	**5.11**	49,026	3510	Courtesy of H. Westrich
Albite glass	**5.51**	54,168		Courtesy of R.J. Bodnar
ANSI 97-5E	*6.00*	56,994	1000	Webster et al. (1999)
CHK 13/1-2, MI in Ol	6.20	57,523	2761	Courtesy of Gurenko
Granite	**6.53**	63,905		Courtesy of H. Behrens
Albite glass	**8.48**	76,273	2223	Courtesy of R.J. Bodnar
ZW MI[c]	8.90	83,044	1816	Thomas et al. (2006a)
Granite	**8.96**	81,300	1616	Courtesy of H. Behrens
Qu8	10.0	92,490		
AB83	**11.71**	107,388	4638	Courtesy of H. Behrens
AB84	**11.95**	116,314	5876	Courtesy of H. Behrens
Qu8 MI[c]	25.3	230,043	13,279	Thomas et al. (2006a)
Qu8 type B MI	31.8	292,309	14,620	
ZW type B MI	41.8	388,406		Thomas et al. (2005)

[a]*H_2O contents in* **bold** *are determined by the Karl–Fischer titration, in* italics *by SIMS (1σ ranges from 0.2 to 0.4) in basalt glass.*
[b]*SD—standard deviation is obtained from five measurements; without SD if only one measurement was performed.*
[c]*MI—melt inclusions from the Etna volcano, the Zinnwald granite (ZW), the Ehrenfriedersdorf pegmatite (Qu8), and the Chikurachki volcano (CHK).*
Raman integral intensities were measured at the following conditions: 800 mm spectrometer with 1800 groves/mm grating, 488 nm excitation laser line, 450 mW laser power (36 mW on sample), confocal pinhole 150 μm, acquisition time 5 × 50 s, linear background correction.

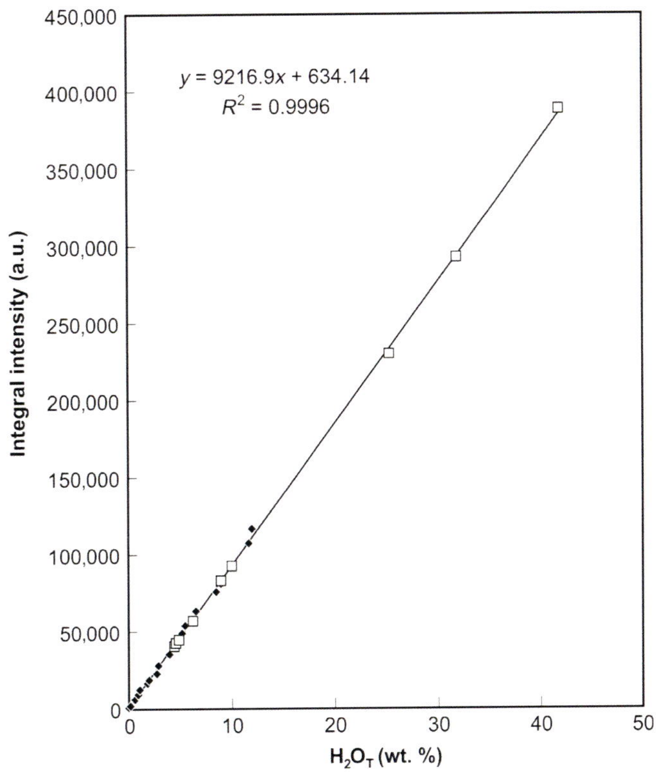

FIGURE 7.25

Correlation between total H_2O concentrations (mass%) in different synthetic glasses and natural silicate melt inclusions and their measured integral Raman intensity of the H_2O–OH stretching band between 3100 and 3750 cm^{-1} (Table 7.6). Solid symbols correspond to reference glasses in which water concentrations were determined by Karl–Fischer titration, SIMS, or proton-proton-scattering, whereas open symbols correspond to natural melt inclusions in olivine and plagioclase from Chikurachki Volcano (Kurily arc, Russia) and Etna (Sicilia, Italy), and pegmatite quartz from Ehrenfriedersdorf and Zinnwald (Erzgebirge, Germany).

Because we use only the H_2O–OH stretching band in the frequency region between 2800 and 4000 cm^{-1}, the possible compositional influence of the host glass is strongly suppressed (Di Muro et al., 2006). However, if the samples are embedded in epoxy, the integral limits may have to be changed, because the epoxy has a strong band at 3070 cm^{-1}, and a similar band can also be introduced by neon discharge tubes. In this case, the integration limits between 3100 and 3750 cm^{-1} should be used. However, in the case of extremely water-rich melt inclusion glasses ($\geq$30 mass% H_2O), it is recommended to use the whole asymmetric H_2O–OH band between 2800 and 4000 cm^{-1}, otherwise the calculated water content may be underestimated by >15% (Figure 7.26).

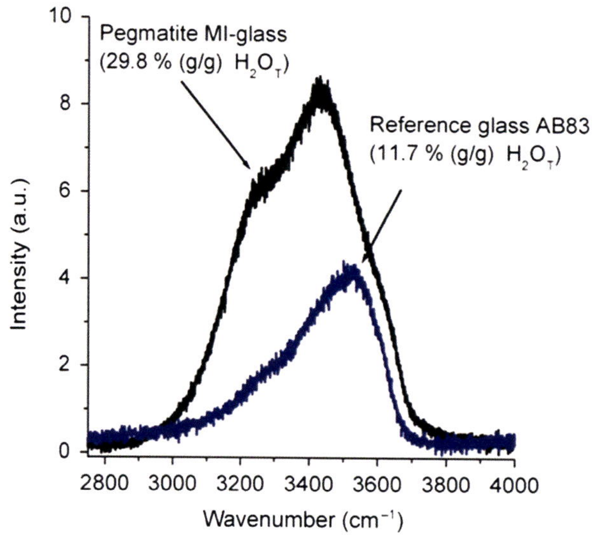

FIGURE 7.26

Comparison of the asymmetric H_2O–OH stretching band in the wave number range 2800 and 4000 cm^{-1} for a melt inclusion in pegmatite quartz from Ehrenfriedersdorf with 29.8 mass% H_2O, and the reference albite glass AB83 with the total water concentration of 11.71 mass% H_2O_T shows that in the case of a very high water concentration, the integral limits have a relatively large influence on the determined total water concentration. For the standard limits between 3100 and 3750 cm^{-1}, the total water concentration is underestimated by >15 mass%.

PROBLEM 7.5 CALCULATING WATER CONTENT IN SINGLE MELT INCLUSION

Calculate the water content in an unexposed glass inclusion in the pegmatite quartz from Strzegom/Striegau Poland (Thomas and Davidson, 2007) using the comparator method. The melt inclusion was located 160 and 7 μm deep, respectively, as measured from both surfaces of polished quartz wafer. The following integral intensities between 3100 and 3750 cm^{-1} have been derived from Raman spectra measured from both sides of the quartz wafer:

$$7\ \mu m{:}\ I = 2266\ \text{a.u.},\quad 160\ \mu m{:}\ I = 146.8\ \text{a.u.}$$

Solution

A two-point linear equation correlating depth and the decadic logarithm of integral intensity can be constructed

$$\log I = 3.4096 - 0.0078d$$

where I is the integral intensity, d is the depth of the inclusion from the surface of a polished slab in μm. Extrapolation to the depth of 0 μm, i.e., setting d to zero, yields an integral intensity of 2564.5. At the same setup conditions, a standard glass with 6.53 mass% of water has an integral intensity of 2950 a.u. After simple algebraic recalculation, we get the water content in the sample

Continued

$$\text{mass\% } H_2O = 2564.5 \times 6.53/2950 = 5.68$$

Note that the nonlogarithmic integral intensity I yields a somewhat different result. From the previous data it follows

$$I = 2363 - 13.851d$$

Hence, the water content for the zero distance from the surface reads

$$\text{mass\% } H_2O = 2363 \times 6.53/2950 = 5.23$$

Note that data should be fitted with a second-order polynomial equation, if the inclusion depth is greater than 175 μm. In this case, either more than two inclusions should be measured or the position of a single inclusion should be modified by a stepwise polishing.

Thomas (2002a,b) demonstrated that H_2O diffusion in silicate melt inclusions can also be traced with the confocal micro-Raman spectroscopy. Figure 7.27 shows the distribution of water content in a melt inclusion in quartz, 20 μm in diameter, homogenized for 20 h at 700 °C and one kbar in a cold-sealed pressure vessel. After quenching, the inclusion was exposed by polishing and the water concentration was measured stepwise from the center to the inclusion wall. One can see that the silicate glass is depleted in water by about 50% at a distance of 2.5 μm from the wall. This means that the water concentration of small melt inclusions can be significantly underestimated.

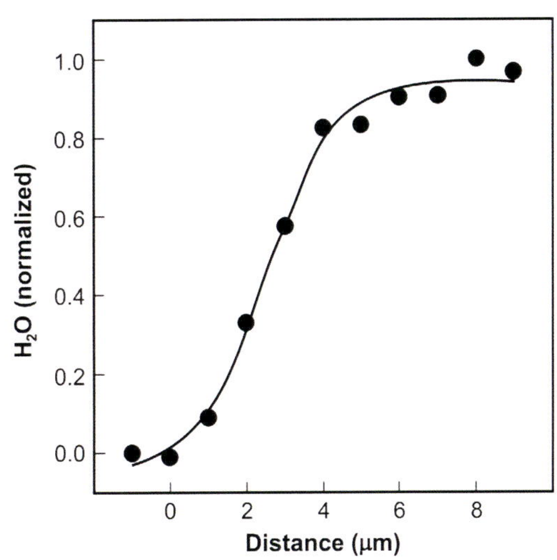

FIGURE 7.27

Gradient of water content (normalized) in a silicate glass inclusion with a diameter of 20 μm from the center of the inclusion (10 μm) to the wall (0 μm). The inclusion is similar to that shown in Figure 7.22, but homogenized in 20 h using the cold-seal pressure vessel technique at 700 °C, 1 kbar. Heating longer than 20 h increases the rate and extent of H_2O loss (Severs et al., 2007).

As the water concentration in magma decreases with the decreasing pressure during magma ascent, silicate melt inclusions in phenocryst minerals (olivine, quartz) re-equilibrate with the host magma by hydrogen diffusion. A correlation is expected between the water concentration and the melt inclusion radii because small inclusions re-equilibrate more rapidly than the larger ones. Careful measurements of water contents in variably sized melt inclusions permit initial water content in the magma as well as the magma ascent rates to be determined by Raman spectroscopy (Chen et al., 2013).

Thomas and Davidson (2012b) compiled data on water concentrations in melt inclusions in quartz from granites (4600 measurements) and pegmatites (2100 measurements). Generally, the water concentration in granite melt inclusion is less than 10 mass%, with two maxima located at 5 and 8 mass%. Some inclusions contain up to 30 mass% H_2O_T. Thomas and Davidson (2013) have interpreted these H_2O-rich melt inclusions as intergranular melts generated outside the crystals. Such low-viscosity and low-density melts can move rapidly through the crystal mush of the crystallizing granite magma. For pegmatites, a considerably larger extent of the water contents was determined, ranging from 2 to more than 50 mass%, with two distinct maxima at 10.5 ± 8.1 and 26.2 ± 14.3 mass%. The low water concentrations are typical for the silicate melt–H_2O–CO_2 system.

The comparator method considerably simplifies the determination of water in melt inclusions. One can also determine the water concentrations of gels in unexposed inclusions (Thomas and Davidson, 2012a) and extremely water-rich glasses (Thomas et al., 2000, 2005). Another advantage is that the calibration curve is not necessary, and hence, the total water content of an unknown glass can be directly calculated by the geometric comparison of the Raman spectrum with that of a reference glass with known water content.

Besides the quantification of water content, the ratio of molecular water to hydroxyl groups can also be determined using micro-Raman spectroscopy. The H_2O–OH stretching band can be deconvoluted into a total of five components (Di Muro et al., 2006): I—3240–3395 cm^{-1}; II—3420–3490 cm^{-1}; III—3520–3580 cm^{-1}; IV—3610–3660 cm^{-1}; and V—3710–3750 cm^{-1}, from which the H_2O_m/OH ratio can be calculated as the ratio between the sum of the integral intensities of the (I + II) and the (III + IV + V) bands.

PROBLEM 7.6 CALCULATING WATER CONTENT IN A GROUP OF MELT INCLUSIONS

Calculate the water content in a group of glass inclusions in quartz from the topaz–albite granite of Zinnwald, Germany (Thomas et al. 2006a), using the reference glass AB83 with 11.71 mass% and the integral intensity of the H_2O–OH band of 107,388 a.u. (Table 7.6). The depths and Raman integral intensities of the investigated melt inclusions are listed in Table 7.7.

Solution

(1) Construct the diagram correlating depths (d) with decadic logarithms of integral intensities of the H_2O–OH stretching band (I). (2) Derive smoothed log$I$ values for each inclusion from a linear equation fitted to the data ($\log I = 4.913282 - 0.008084d$). (3) Calculate the integral intensity for each inclusion corrected to depth according to the following equation: $I_{cor} = I_0 - I_{calc} + I$. (4) Calculate the water content of each inclusion by comparison with the AB83 standard. (5) Calculate the average water content and standard deviation for the whole group of investigated inclusions. The calculation procedure and result are displayed in Table 7.7.

Table 7.7 Depths and Raman Integral Intensities of Investigated Melt Inclusions

Depth (μm)	Integral Intensity (I)	log (I)	log (I_{calc})	I_{calc}	I_0–I_{calc}	I_{cor}	H_2O_T (Mass%)
72	21,244	4.327236	4.331234	21,440	60,459	81,703	8.91
60	27,392	4.437624	4.428242	26,807	55,093	82,485	8.99
31	46,458	4.667061	4.662678	45,992	35,908	82,366	8.98
23	50,749	4.705427	4.727350	53,376	28,523	79,272	8.64
10	69,346	4.841021	4.832442	67,990	13,910	83,256	9.08
3	78,104	4.892673	4.889030	77,452	4448	82,551	9.00
0			4.913282	81,900	0		
							8.93 ± 0.15
AB83	107,388						11.71

Silicate melts can dissolve volatiles in molecular forms (e.g., CO_2, H_2, N_2, SO_2, H_2S, CH_4), complexes (e.g., HS^-, OH^-), and/or anionic groups (e.g., PO_4^{3-}, BO_3^{3-}, CO_3^{2-}, HCO_3^-, SO_4^{2-}, NH_2^-, $Be_2F_7^{3-}$, BeF_4^{2-}). Besides water, pegmatite melts contain large concentrations of fluxing compounds, which drastically reduce the melt viscosity and promote the growth of large crystals by enhancing element diffusion rates and suppressing the formation of crystallization nuclei (Thomas et al., 2012b). Concentrations of these species in pegmatite-forming melts and highly fractionated granites may reach up to several tens of mass%, for instances up to 79.6 mass% H_3BO_3 (44.8 mass% B_2O_3), 10 mass% BeO (Thomas and Davidson (2010), 25 mass% Na_2SO_4 (Thomas and Davidson, 2012c), and 6.4 mass% F (Thomas et al., 2005).

The melts extremely enriched by fluxing (network-modifying) molecules usually originate by the melt–melt or melt–fluid immiscibility. In extreme cases, the generation of polyphase melt immiscibility is possible (Figure 7.28). Numerous examples of the fluid–melt(–melt) immiscibility in pegmatites, highly fractionated granites, and Cu–Mo porphyry deposits have been documented by Thomas et al. (2000, 2005, 2012a, 2012b), Davidson and Kamenetsky (2001), Kamenetsky et al. (2004), Borisova et al. (2012), and Thomas and Davidson (2013).

Aqueous fluids exsolved from highly fractionated hydrous melts preferentially concentrate volatile and fluxing species. As a consequence, exotic daughter phases crystallize within fluid and melt inclusions on cooling. Table 7.8 yields the overview of some exotic daughter phases identified in melt and fluid inclusions from evolved magmatic rocks and their derivatives, especially granites, pegmatites, and carbonatites. Some of these minerals are water-soluble, and therefore they can exist in nature only in hermetically sealed inclusions.

Except for halogenides, sulfides, and some oxides, daughter minerals in fluid inclusions usually exhibit high scattering power and they can thus be easily recognized by Raman spectroscopy, even if their spectrum is obscured by that of the host mineral. If a complete spectrum is not available, the anionic group can be roughly determined using the main symmetric or asymmetric stretching vibration. Single anions, such as chloride, sulfide, fluoride, and oxide exhibit the main vibration at wavelengths below 700 cm^{-1}, whereas complex anions vibrate at higher wave numbers (Table 7.9).

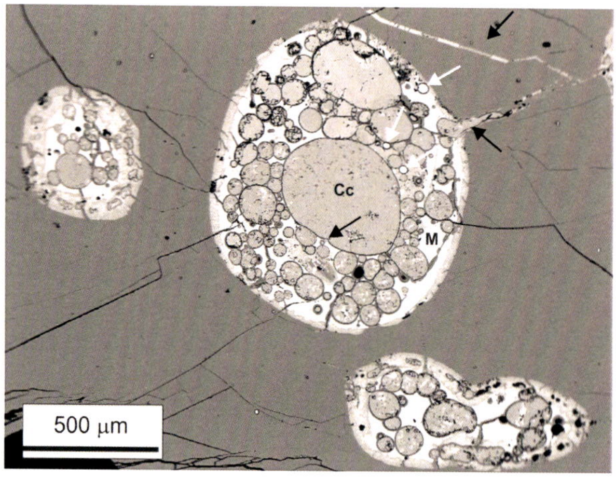

FIGURE 7.28

Three-phase melt immiscibility observed in melt inclusions in plagioclase megacryst from basaltic diatreme in the Hajnáčka village, Slovakia. The melt inclusion assemblage consists of P-rich calcic carbonatite (Cc), Ca-rich, Si-undersaturated (larnite- and wollastonite-normative) silicate glass matrix (M), and Ca-phosphate melt globules indicated by white arrows. Evidence for the carbonatite liquid is provided by bulging out from a crack into inclusion and deformation of the carbonate globules by newly formed Ca-plagioclase crystals, indicated by two solid arrows around the upper-right perimeter of the central inclusion.

Table 7.8 Diagnostic Raman Bands of Daughter Minerals Identified in Fluid and Melt Inclusions from Magmatic Rocks and Magmatogenic Deposits

Mineral	Formula	Main Diagnostic Bands (cm^{-1})	Source[a]
Alkalic tetrafluoroborate	(Li, Na, K, Rb, Cs) BF_4	772	3
Amblygonite	$LiAl(PO_4)(F, OH)$	601, 644, 1011	3
Ancylite-Ce	$Sr(Ce, Ca, La)(CO_3)_2(OH)_2 \cdot H_2O$	1083	9
Anhydrite	$CaSO_4$	1015–1018	15
Aphthitalite (glaserite)	$(K, Na)_3Na(SO_4)_2$	981–996	10, 15
Arcanite	K_2SO_4	983	15
Arsenolamprite	As	253	2
Arsenolite	As_2O_3	370, 781	4, 8
Avogadrite	$(K, Cs)BF_4$	774	4, 8
Barite	$BaSO_4$	461, 988, 1144	15
Bastnäsite	$(Ce, La)CO_3F$	1085, 1096	11
Berlinite	$AlPO_4$	1110	3
Beryllonite	$NaBePO_4$	1012, 1056	6
Billietite	$Ba(UO_2)_6O_4(OH)_6 \cdot 8H_2O$	830	9
Blodite	$Na_2Mg(SO_4)_2 \cdot 4H_2O$	992, 1188	15
Bromellite	BeO	678, 684, 772	2
Ca-hexaborate	$CaB_6O_{10} \cdot 4H_2O$	763, 855, 861, 895	3

Continued

Table 7.8 Diagnostic Raman Bands of Daughter Minerals Identified in Fluid and Melt Inclusions from Magmatic Rocks and Magmatogenic Deposits—Cont'd

Mineral	Formula	Main Diagnostic Bands (cm^{-1})	Source[a]
Cassiterite	SnO_2	633	3
Cebaite	$Ba_3Ce_2(CO_3)_5F_2$	267, 1091	11
Celestite	$SrSO_4$	656, 1000	15
Chiolite	$Na_5Al_3F_{14}$	531	1
Cryolite	Na_3AlF_6	552	3
Cryolithionite	$Na_3Li_3Al_2F_{12}$	563	1
Dawsonite	$NaAl(CO_3)(OH)_2$	192, 590, 1091, 3283	1
Diomignite	$Li_2B_4O_7 \cdot 5H_2O$	446, 543, 772,1028	2
Elpasolite	K_2NaAlF_6	326, 559	3
Ferricopiapite	$Fe_5(SO_4)_6(OH)_2 \cdot 20H_2O$	996	9
Fluorapatite	$Ca_5(PO_4)_3F$	962	3
Gaylussite	$Na_2Ca(CO_3)_2 \cdot 5H_2O$	1071	9
Glauberite	$Na_2Ca(SO_4)_2$	1002	15
Grandidierite	$(Mg, Fe)Al_3(BO_3)(SiO_4)O_2$	161, 343, 424, 577, 864, 949	4, 5
Graphite	C	1350, 1575	13
Gypsum	$CaSO_4 \cdot 2H_2O$	493, 1007, 3405	15
Hambergite	$Be_2BO_3(OH, F)$	148, 158, 988, 3419, 3484, 3600	2
Herderite	$CaBePO_4(F, OH)$	584, 594, 983, 1005, 3564, 3573	1
Hieratite	K_2SiF_6	655	1
Hübnerite	$MnWO_4$	881	3
Johannite	$Cu(UO_2)_2(SO_4)_2(OH)_2 \cdot 8H_2O$	834, 1043, 1090	9
Lacroixite	$NaAl(PO_4)F$	609, 623, 1001	3
Li-metaborate	$LiBO_2$	713, 1459	3
Nahcolite	$NaHCO_3$	1046	1
Natrite	Na_2CO_3	1078	14
Pectolite	$NaCa_2Si_3O_8(OH)$	652	15
Polyhalite	$K_2Ca_2Mg(SO_4)_4 \cdot 2H_2O$	987, 1013	15
Ramanite-(Cs)	$CsB_5O_8 \cdot 4H_2O$	548, 3396	5
Ramanite-(Rb)	$RbB_5O_8 \cdot 4H_2O$	554, 3374	5
Rb-tetrafluoroborate	$RbBF_4$	772	1
Rb-Zn tetrachloride	Rb_2ZnCl_4	291	1
Saléeite	$Mg(UO_2)_2(PO_4)_2 \cdot 10H_2O$	836, 989	9
Santite	$KB_5O_8 \cdot 4H_2O$	107, 556, 765, 918	5, 12
Sassolite	H_3BO_3	500, 880, 3246	2
Sborgite	$NaB_5O_8 \cdot 5H_2O$	121, 529, 856	5, 12
Sodium oxide monohydrate	$NaOH \cdot H_2O$	2950, 3633, 3662	14
Sphalerite	ZnS	349	3
Sulfur	S	157, 219–222, 473	13
Tantalite-(Mn)	$(Mn, Fe)Ta_2O_6$	887	14
Thenardite	Na_2SO_4	993	9, 15
Villiaumite	NaF	369	1
W-ixiolite	$(Fe, Mn, Nb)(Nb, W, Ta)O_4$	395, 526, 868	3
Zabuyelite	Li_2CO_3	95, 1091, 1456	7, 14

[a]*References: (1) Thomas (unpublished). (2) Thomas and Davidson (2010). (3) Rickers et al. (2006). (4) Downs (2006). (5) Thomas et al. (2008a). (6) Thomas et al. (2011b). (7) Anderson et al. (2001). (8) Thomas and Davidson (2012c). (9) Borisenko et al. (2011). (10) Hansteen and Burke (1994). (11) Fan et al. (2004). (12) Li et al. (1995). (13) Thomas et al. (2009). (14) Thomas et al. (2011a). (15) Naumov et al. (2008).*

Table 7.9 Limits of Main Stretching Vibrations (internal modes) of Some Anion Groups

Range (cm^{-1})	Group
<410	Chloride
<470	Sulfide
<580	Fluoride
290–720	Oxide
450–960	Borate
450–1100	Silicate
700–870	Uranyl
790–820	Arsenate
790–930	Wolframate
910–1120	Phosphate
970–1050	Sulfate
1030–1060	Hydrocarbonate
1050–1120	Carbonate
3250–3720	Hydroxyl

7.8 RAMAN MAPPING AND MODELING

Raman mapping can be performed using a microscope system equipped with a software-controlled motorized *XY* stage, which can move the sample point by point equidistantly relative to the fixed microscope lens. For the acquisition of one-dimensional data array (profile) or two-dimensional data array (map), an oversampling is often used when the incremental value is lesser than the focal spot diameter of the laser beam. This enables the detection of very small objects in a submicrometer range that usually escape attention in an optical microscope. The Raman mapping is thus a very useful tool for the identification of daughter minerals in multiphase fluid and melt inclusions.

The Raman mapping requires very long analytical times if the mapped areas are larger and/or the oversampling is applied. For instance, the mapping of an area 50 × 50 μm in size with a 1 μm increment, a 5 s acquisition time, and two accumulations in each point requires at least 7 h of the analytical time. The mapping duration can be significantly reduced if the computer screen is switched off, and both the spectral range and resolution are selected such that the whole spectrum is acquired in a single window. The optimal number of spectra in the multidimensional data array should be between 1000 and 10,000 (Nasdala et al., 2004). Long acquisition times (hours) can be significantly reduced (to minutes) by special techniques, such as the line scanning or fast macro-mapping modes for the screening large areas.

Each analyzed point scanned by the conventional point-by-point mapping contains a complete set of spectral information, which can be further treated using statistical methods and numerical calculations. After basic manipulation (smoothing, background correction, deconvolution), the final treatment is made using the direct classical least squares (DCLS) fitting procedure. The DCLS fitting can be done automatically by using a factor analysis algorithm of the clustering procedure, which identifies reference component spectra. The algorithm searches the entire dataset and locates clusters of similar spectra. Number of clusters (= phases present) together with the threshold value (% of the highest intensity in the dataset, which should be eliminated from the clustering procedure) are to be defined prior the analysis. Manual selection of the reference component spectrum either from the analyzed dataset or as

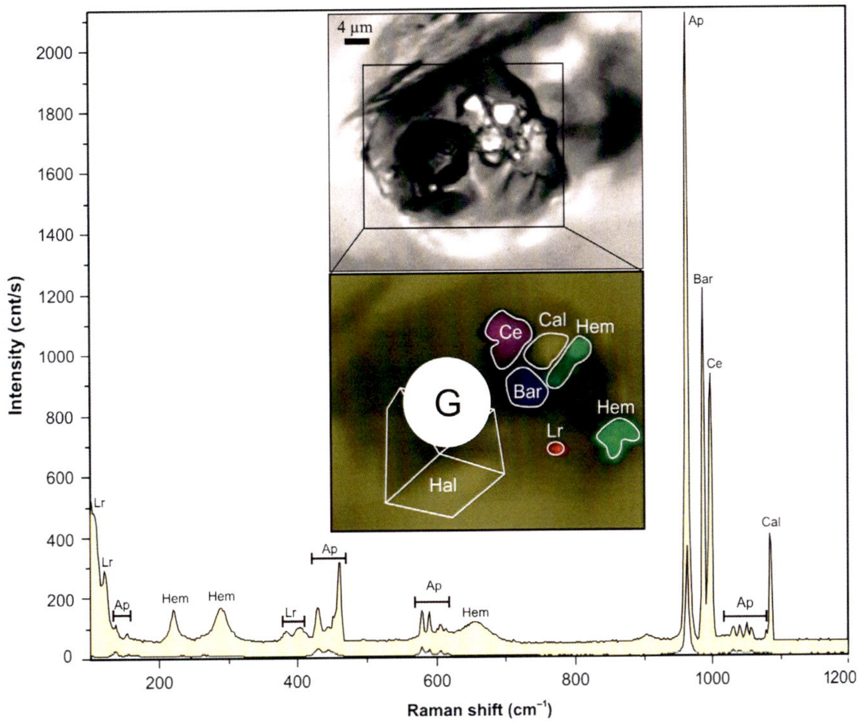

FIGURE 7.29

Uncorrected multidimensional data matrix recorded by a Labram HR800 Raman spectrometer from a multiphase aqueous fluid inclusion in apatite of the Evate carbonate–phosphate deposit, Mozambique. Raman mapping based on the factor analysis algorithm (DCLS) included in the Labspec5 computer program from Horiba™ revealed celestite (Ce), calcite (Cal), barite (Bar), hematite (Hem), and laurionite (Lr) daughter crystals precipitating from a chloride–sulfate aqueous solution trapped in an apatite (Ap) host. G denotes the vapor bubble. Note numerous Raman-inactive solid phases, including a large halite crystal (Hal). The laurionite crystal is invisible in the focal plane of the inserted photomicrograph, and it was revealed because of the cluster analysis of the Raman map.

an external spectrum is also possible. The normalized DCLS modeling is suitable for a general analysis of spectral array/data matrix and the characterization of phase distribution (Figure 7.29).

After deconvolution and fitting, statistical parameters such as full-width-at-half-maximum (FWHM) intensity and position are available for each point of the multidimensional data array, and these can be visualized as maps using a total of three different channels. In addition, the peak intensity ratio from blue and green channels can also be displayed simultaneously as a function of the relative abundance of two components in a solid solution or a mixed crystal aggregate.

Here we present an example of the statistical treatment of a multidimensional data array scanned from two carbonatite globules embedded in Ca-silicate glass (see Figure 7.28 for a view of the phases in composite melt pockets sealed within a plagioclase megacryst in alkali basalt). The region of carbonate ν_1 vibration mode was selected and deconvoluted using two Gauss–Lorentzian functions (Figure 7.30a). Two distinct clusters have been identified by the fitting procedure: one cluster at around

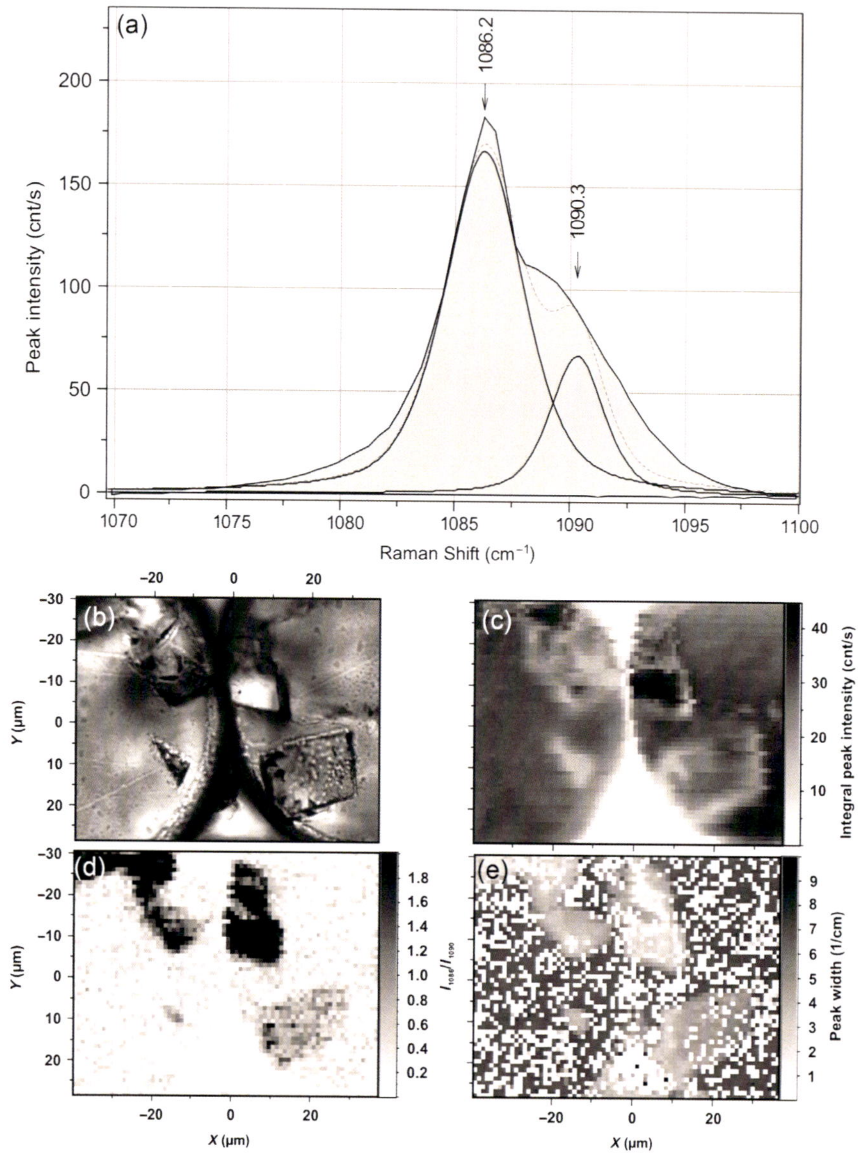

FIGURE 7.30

Raman mapping of carbonatite globules embedded in Ca-silicate glass matrix of a melt pocket sealed within plagioclase megacryst in alkali basalt (Figure 7.28). (a) Documents deconvolution of the data array to two distinct clusters corresponding to calcite crystals (1086 cm^{-1}) crystallizing along margins of quenched carbonatite glass globules (1090 cm^{-1}). (b) Shows the transmitted light image of the scene provided by a built-in CCD camera of the Labram 800HR system. (c) Shows the distribution of maximal intensities. (d) Shows the ratios between intensities of the 1086 and 1090 cm^{-1} bands. (e) Displays the distribution of bandwidths, documenting lower FWHM values for calcite crystals crystallizing along margins of quenched carbonatite melt globules. Note that scales for absolute I_{1086}/I_{1090} and FWHM values are present on right sides of (d) and (e). All maps have been prepared using the Labspec5 computer program from Horiba™.

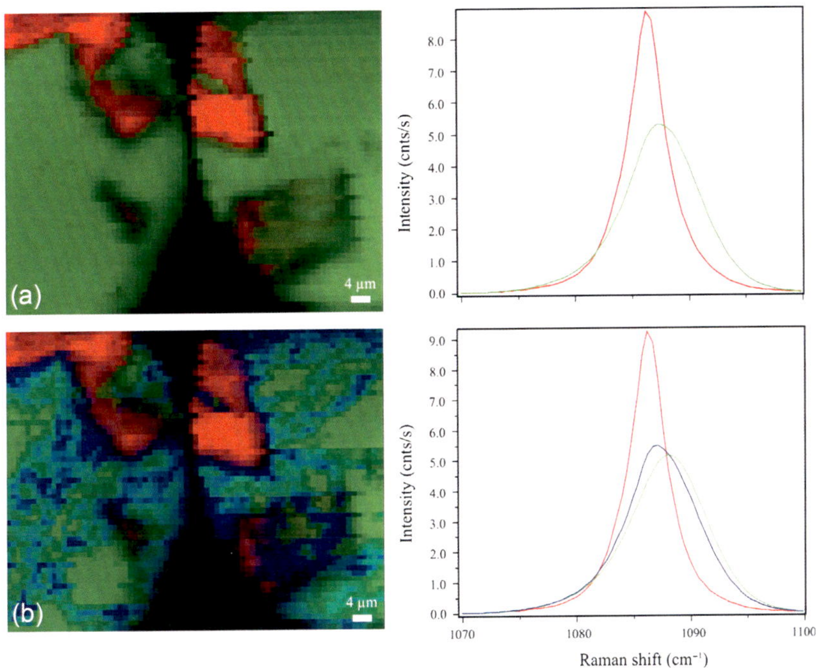

FIGURE 7.31

False-color maps of the dataset from Figure 7.30 produced by the factor analysis algorithm included in the Labspec5 software from Horiba™. The factor analysis based on two reference spectra (a) shows the difference between calcite (red pixels) and the associated quenched carbonate glass (green pixels), whereas the factor analysis based on three reference spectra visualizes phase boundaries, scratched surface, and possibly also strain within the quenched carbonatite glass (blue pixels).

1086 cm^{-1} with higher peak intensity and narrower bandwidth corresponds to $CaCO_3$ crystals with locally increased SrO concentration (~1 mass%), whereas the cluster at around 1090 cm^{-1} with the considerably lower intensities and higher FWHMs refers to quenched carbonatite melt with an increased MgO (~2 mass%) and P_2O_5 (~0.2 mass%) concentrations. The scanned area as shown by the built-in CCD camera of the Raman system is documented in Figure 7.30b, whereas the maps of integral band intensities, intensity ratios of the 1086 and 1090 bands, and FWHMs are displayed in Figure 7.30c, d, and e, respectively.

The previous dataset can also be treated using the DCLS fitting and the selection of reference spectra from the spectral array. Figure 7.31a shows the map based on two clusters, which clearly identifies calcite crystals within the quenched carbonate glass. The factor analysis based on three clusters (Figure 7.31b) visualizes phase boundaries, scratches on the nonideal sample surface, and possibly also some strain within the quenched carbonate glass globules.

CHAPTER

MISCELLANEOUS SPECTROMETRIC AND CHROMATOGRAPHIC METHODS

8

CHAPTER OUTLINE

8.1 CRUSH-LEACH ANALYSIS OF AQUEOUS INCLUSIONS

Crush-leach analysis is a destructive method that enables the direct quantitative determination of anions and cations in the aqueous phase of fluid inclusions. Sorby (1858) was the first scientist who crushed quartz crystals, leached them in pure water, and analyzed the soluble components liberated from fluid inclusions. Zirkel (1873) introduced an easy and sensitive check of sodium by reaction with silver nitrate and provided the first qualitative data on chlorine within fluid inclusions by analyzing water leachate from crushed quartz. Scherfel (1878) proved NaCl within fluid inclusions using the spectral line for sodium and the positive test for chlorine by reaction with silver nitrate.

The crush-leach method was revitalized in the twentieth century with the development of sensitive spectrometric (Raman spectroscopy, ion chromatography—IC) and microanalytical methods. Initially, the crush-leach technique required as much as 1 kg of a sample to yield a sufficient concentration of dissolved salt for conventional wet chemical analysis (Roedder, 1958). With improvements of analytical techniques (atomic absorption spectroscopy) and the lowering of the detection limits, the sample size has been significantly reduced. Thirty years later, Bottrell et al. (1988b) used only ~2 g samples for multiple crushes in steel or copper tubes. Moreover, they greatly improved the crush-leach technique by adding small amounts of La^{3+} into the leaching solution in the form of $LaCl_3$ to prevent the absorption of divalent cations on fresh quartz surfaces produced by the crushing. Monovalent ions (e.g., Li, Na, K, Cl, F, Br) do not absorb on surfaces of crushed minerals in water leach solutions.

Recently, major and minor ions have been analyzed by atomic absorption spectrometry (AAS), inductively coupled plasma mass spectrometry (ICP-MS), and inductively coupled plasma atomic

Geofluids. http://dx.doi.org/10.1016/B978-0-12-803241-1.00008-3

emission spectroscopy (ICP-AES) (Banks and Yardley, 1992). Anions are determined by IC. Total reflection X-ray fluorescence spectroscopy (TXRF) is still only rarely used in the analysis of geofluids, although it combines the advantages of low and very low detection limits (ppb to sub-ppb range), it requires only a small amount of sample (μL), and it is a relatively fast and inexpensive analytical procedure (Landenburger et al., 2012). Ratios of Rb and Sr isotopes can also be determined using the isotope dilution technique from bulk crush-leach solutions (Pettke and Diamond, 1995).

The bulk crush-leach analysis suffers from several complications (Banks and Yardley, 1992): (1) Solid impurities (inclusions) can swamp the signal from fluid inclusions, (2) leaching larger sample volumes yields signals from multiple, possibly genetically unrelated fluid inclusion generations. These problems can be partially circumvented by the analysis of individual fluid inclusions using microbeam and laser ablation techniques, or by the selection of clear crystals with only one type of fluid inclusions.

Furthermore, the crush-leach analysis yields only proportional data because the volume of the extracted fluid and analyzed fluid inclusions is unknown. This can be avoided by the recalculation of the proportional analytical data to absolute concentrations (in mass% NaCl equation) derived from microthermometry.

The charge balance error (CBE) is calculated to test whether the crush-leach analyses are correct or not (Fritz, 1994)

$$\mathrm{CBE}\,(\%) = 100\left(\frac{\sum zm_{\mathrm{C}} - \sum zm_{\mathrm{A}}}{\sum zm_{\mathrm{C}} + \sum zm_{\mathrm{A}}}\right) \tag{8.1}$$

where m_{C} is the molality of cation, m_{A} is the molality of anion, and z is the charge.

If CBE $<5\%$, the results are appropriate for further interpretation. An excess of negative charges (anions) most frequently indicates an adsorption of Ca^{2+} and other divalent cations. Elemental concentrations are usually normalized to 10 mass% NaCl equivalents (60,663 ppm Cl) and then adjusted to the measured Cl concentration (Cl in ppm) by multiplying with the correction factor CF (Banks et al., 2002)

$$\mathrm{CF} = (\text{total salinity in ppm} \times M_{\mathrm{Cl}}/M_{\mathrm{NaCl}})/\mathrm{Cl\ in\ ppm} \tag{8.2}$$

where M stands for atomic/molecular masses of Cl (35.453) and NaCl (58.443), respectively.

Analyses of solute compositions in fluid inclusions corroborated that the dominant cations in crustal fluids are Na, K, and Ca followed by Fe and Mg. The dominant anion is almost exclusively Cl^-, with subordinate amounts of SO_4^{2-}, HCO_3^-, and NO_3^- (Wilkinson, 2001), but hydrothermal fluids exsolved from granitic melts may be also dominated by HCO_3^- and H_3BO_3 (Thomas and Davidson, 2010).

8.2 BULK GAS ANALYSIS

Raman spectroscopy offers a simple and versatile nondestructive method for the determination of relative concentrations of gases in fluid inclusions. Its alternative is the destructive bulk analysis of fluid inclusion gases by mass spectrometry or gas chromatography coupled with thermal or mechanical crushing of the samples. Karpinskiy (1880) was the first to attest CO_2 under high pressure in fluid inclusions opened by grinding the sample in mercury and adsorbing the liberated CO_2 gas in the $Ba(OH)_2$ solution. Hawes (1881) liberated volatiles from fluid inclusions by vacuum heating. Hence, both researchers set fundamentals of the "crushing stage" technique frequently employed in the twentieth century prior to the introduction of nondestructive microbeam methods of gas analysis. The analysis of gas liberated from fluid inclusions by thermal or mechanical crushing gained considerable attention in the late 1980s and early 1990s (Kesler et al., 1986; Norman and Sawkins, 1987; Bray et al., 1991; Burlinson, 1991; Graney et al., 1991; Landis and Hofstra, 1991; Moon, 1991; Norman et al., 1985, 1991; Shepherd et al., 1991; Takenouchi, 1991; Walder et al., 1991; Salvi and Williams-Jones, 1997).

Fluid inclusion volatiles are released under vacuum by crushing or heating. The released volatiles are analyzed in a dual quadrupole mass spectrometer or a gas chromatograph. The gases routinely reported in the geological literature include He, Ar, H_2, O_2, N_2, H_2S, H_2O, CO_2, SO_2, CH_4, C_2H_4, C_2H_6, C_3H_6, C_3H_8, C_4H_8, C_4H_{10}, and benzene. The bulk gas analysis suffers from complications inherent to other bulk methods: contamination with volatiles bound to structural defects, loss of extracted volatiles due to absorption, and a possibility to obtain only an average bulk composition from various fluid inclusion generations present in the analyzed mineral. On the other hand, there are also some advantages: (1) analysis of gas species undetectable by microthermometry and Raman spectrometry (Ar), (2) possibility to analyze gases included in opaque minerals, (3) superior detection limits uncomparable with the Raman spectrometry.

The bulk gas composition can be employed to discriminate various fluid sources, to identify boiling and mixing processes, to constrain oxidation states and chemical indicators for modeling fluid-rock equilibria (Blamey, 2012), and to provide independent geothermometers (Darling, 1998).

8.3 LASER ABLATION ANALYSIS

The laser-ablation inductively coupled-plasma mass spectrometry (LA-ICPMS) has recently become widely used as a powerful and efficient multielement microanalytical technique in fluid and melt inclusion studies. One of its key strengths is the ability to analyze a wide concentration range from major (tens of mass percent) to trace (ng/g—ppb) levels in microscopic inclusions, several tens of μm in diameter. Wilkinson et al. (1994), Irwin and Roedder (1995), and Shepherd and Chenery (1995) were the first to study the elemental composition of individual fluid inclusions by the laser ablation. Melt inclusions have been studied for the first time using the laser ablation technique by Kamenetsky et al. (1997) and Taylor et al. (1997). Ultrashort pulses from a femtosecond laser have recently enabled also the liberation of unaltered oil from individual inclusions, and on-line characterization of C4-19 alkanes and aromatic hydrocarbons using a coupled-gas chromatograph-mass spectrometer (Volk et al., 2010). Laser-induced breakdown spectroscopy (LIBS) combined with microthermometry and Raman spectroscopy enabled characterization of individual aqueous inclusions in quartz from Alpine fissures (Fabre et al., 2002).

All elements of the periodic table except for noble gases, F, O, N, H, and polyatomic species can be determined by LA-ICPMS. A major advantage of fluid inclusion analyses by LA-ICPMS over bulk methods is that single fluid inclusions in specific fluid inclusion assemblages can be easily targeted, thus permitting mixing trends within one fluid generation to be recognized.

The fluid inclusion analysis by LA-ICPMS involves several steps. The content of the inclusion is liberated by drilling it out of the host mineral. The ablated aerosol is transported from the ablation chamber in helium carrier gas, then mixed with argon and introduced in the mass spectrometer. The aerosol particles are atomized and ionized in the inductively coupled plasma, the ions are then extracted from the plasma and filtered according to their mass-charge ratio, and finally registered by a detector system of a mass spectrometer (Heinrich et al., 2003; Pettke, 2008; Pettke et al., 2012).

An argon-fluoride excimer laser system ($\lambda = 193$ nm) with imaging optics for controlled UV ablation and simultaneous viewing is the most commonly used for the representative sampling and quantitative multielement analysis of fluid, melt and mineral inclusions (Günther et al., 1997). Quadrupole and single-focusing magnetic-sector mass spectrometers are typically employed for the analysis of trace elements. In comparison to the quadrupole mass filters, double-focusing magnetic-sector mass spectrometers offer an advantage of variable resolution by using several sets of slit positions. This is important for reducing polyatomic mass interferences. Moreover, flat-topped peak shapes allow better precision, and this mass spectrometer is specially designed for measurements of isotope mass ratios. The quadrupole mass spectrometer allows for fast mass jumping between elements in the range from Li to U (a real advantage for the analysis of fast transient signals produced from fluid inclusions) and returns limits of detection as low as 0.01 μg/g for the best-detected elements. Between 5 and 20 elements are typically analyzed for average-sized liquid inclusions, but this number is reduced for smaller inclusions or vapor inclusions, because the resulting signals are shorter. For 20 elements at a typical dwell time of 10 ms and the quadrupole settling time of 3 ms, a total analytical duration of 260 ms results for a single sweep with all elements measured once. A signal duration of about 5 s would result in 20 measurements per element, and longer signals correspondingly return more data. If it is desirable to optimize recordings for specific elements, heavy metals in fluid inclusions for instance, the dwell time can be increased, resulting in lower detection limits. However, this will increase the sweep duration; hence, a compromise between fast-scanning protocols at 10 ms dwell time and improved detection limits must be defined for each analytical session (Pettke et al., 2012).

Bromine and chlorine are common geofluid constituents that can be used for tracing the origin of fluids in sedimentary and hydrothermal/magmatic systems, because they have relatively low abundances in the majority of silicate minerals. However, their analysis by LA-ICPMS has been hampered by relatively low ionization efficiencies that lead to lower sensitivity, and thus increased detection limits. The energy required to positively ionize fluorine (17.42 eV) makes it impossible for the quantification via argon plasma with the ionization energy of 15.76 eV. Br, Cl, and I are not routinely measured by LA-ICPMS, not only because of their high ionization energies (11.81, 12.97, and 10.45 eV, respectively), but also because of the lack of suitable standards (Hammerli et al., 2013). Despite these obstacles, Seo et al. (2011), Leisen et al. (2012a), and Hammerli et al. (2013) have demonstrated the feasibility of Cl and Br analyses by the LA-ICPMS technique.

The amount of each ablated analyte depends primarily on the size of the fluid inclusion and on the ion concentration, thus the signal intensity for a given element cannot be interpreted directly in terms of an absolute concentration. Quantitative analysis than requires the combined use of an internal and an external standard. For the internal standard, concentration of one element must be known from an independent observation. This information can be gained from microthermometric measurements, for instance. For low-to-moderate salinities, the calculation method based on the Pitzer's model (Pitzer, 1973) for aqueous ionic fluids can be employed, in which the ice melting temperature is directly linked to the water activity and therefore to the composition of the aqueous solution (Leisen et al., 2012a). The model, however, has not yet been parameterized for high-salinity fluids containing halite and sylvite as daughter phases. In this case, the Na and Cl concentrations are inferred from temperatures of melting of last two solids (e.g., ice + halite, halite + sylvite) by an approximation to the $NaCl–H_2O$ (Bodnar, 1993; Lecumberri-Sanchez et al., 2012), $NaCl–KCl–H_2O$ (Hall et al., 1988), $NaCl–CaCl_2–H_2O$

(Vanko et al., 1988; Oakes et al., 1990), $NaCl–MgCl_2–H_2O$ (Dubois and Marignac, 1997), NaCl–LiCl–H_2O (Dubois et al., 2010), or other specific sulphate and hydrocarbonate, NaCl-containing systems (Borisenko, 1977, 1982).

Charge-balance and mass-balance methods are employed to calculate absolute concentrations of other ions, providing the total concentration expressed relative to NaCl is known from the microthermometric measurements. Chlorine cannot be selected as an internal standard because the background signal is high and the signal is weak due to its high ionization energy. In the charge balance method, the ice melting temperature interpreted in terms of equivalent mass% NaCl is used to derive the Cl concentration, considering that chloride is the only or the major anion. Electric charge balance equation is written as (Allan et al., 2005)

$$m_{Cl^-} = m_{Na^+} + \sum_i (v_i \times m_i) \tag{8.3}$$

where m_i is the concentration of cation i (mol/kg H_2O), m_{Na^+} is the concentration of Na, m_{Cl^-} is the concentration of Cl, and v_i is the charge absolute value. Using calibration curves with respect to Na, the m_i/m_{Na^+} ratio can be derived. Then, Equation (8.3) is used to calculate the Na concentration

$$m_{Na^+} = m_{Cl^-}\left(1 + \frac{\sum_i (v_i \times m_i)}{m_{Na^+}}\right)^{-1} \tag{8.4}$$

The other element concentrations are deduced from their respective molal ratios to Na.

The mass-balance method is based on the observation of temperature contours of the vapor-saturated NaCl liquidi at the ice melting temperature in the H_2O–NaCl–KCl and $H_2O–NaCl–CaCl_2$ systems. Because the curves run approximately at right angles away from the H_2O–NaCl pseudobinary join, Heinrich et al. (1992) derived a general equation for other mono- and di-chloride salts showing similar behavior

$$m_{NaCl}^{eq.} \approx m_{NaCl}^{true} + 0.5\sum_i m_i \tag{8.5}$$

where superscript "eq." denotes the mass% NaCl equivalent that is deduced from the temperature of ice melting, interpreted in terms of the H_2O–NaCl system, superscript "true" denotes the actual NaCl content of the fluid inclusion, and m_i is the concentration of all other salts in the fluid.

For a given element, a series of silica capillaries containing aqueous liquid standards with various concentrations of Na and other elements must be prepared and ablated to construct a calibration curve. The difference of matrix between standard solutions sealed in silica glass and natural fluid inclusions in quartz has no effect on the intensity ratios (Günther et al., 1997). A calibration curve correlating Na/E (intensity) and Na/E (molar ratio) is then established, which allows the direct determination of the Na/E ratio in natural fluid inclusions from the Na/E intensity ratio measured by the LA-ICPMS. Because the Na concentration is known, the concentration of another element of interest can be calculated from the calibration curve using the following equation (Leisen et al., 2012a)

$$c_E = c_{Na} \times \left(\alpha \frac{I_{Na}}{I_E} + \beta\right)^{-1} \tag{8.6}$$

where c_{Na} and c_E are respective concentrations of Na and element E in fluid inclusions, I is the LA-ICPMS signal intensity in the fluid inclusions, and α and β are, respectively, the slope and the y-intercept of the calibration curve.

The uncertainty of the measured concentration is calculated from the following equation

$$\Delta c_i = c_i \times \left(\frac{\Delta c_{Na}}{c_{Na}} + \frac{\Delta I_{Na}}{I_{Na}} + \frac{\Delta I_i}{I_i} \right) \tag{8.7}$$

where c_{Na} and c_i are the concentrations of Na and the element i in the fluid inclusion, I are the LA-ICPMS signal intensities for ^{23}Na, and an element i, $\Delta c_{Na}/c_{Na}$ stands for the uncertainty on Na concentration and $\Delta I/I$ is the relative standard deviation. Because LA-ICPMS signals from fluid inclusions are transient and they do not exhibit a plateau during ablation, standard deviations of the fluid inclusion signal intensities cannot be estimated in the same way as for the ablation of solids. Instead, they are estimated from the variation of mass signal intensities on capillary standards according to the method described by Leisen et al. (2012b).

The LA-ICPMS has become a powerful *in situ* microanalytical technique also for the determination of trace elements and isotope ratios in melt inclusions. The advantage of the LA-ICPMS technique is in the simple sample preparation, large sample throughput, matrix- and concentration insensitive external calibration, and a possibility to analyze unexposed melt inclusions undamaged by the polishing. Mass spectrometers with improved sensitivities enabled determinations of fluid/melt partition coefficients (Zajacz et al., 2008) and lead isotope ratios in melt inclusions (Zhang et al., 2014). Certified NIST silicate glasses can be used as external standards (Rocholl et al., 1997).

Two research strategies have been recently followed to analyze either rehomogenized and quenched melt inclusions or to ablate the entire inclusion along with the host. The second method offers the opportunity to determine the composition of crystallized melt inclusions without their homogenization, by the deconvolution of the mixed ablation signal (Liu et al., 2013b). However, the quantification of LA-ICPMS signals requires also the concentration of one element or an elemental ratio at the time of the melt inclusion entrapment to be known as an internal standard, to isolate the melt inclusion composition from the mixed ablation signal. Moreover, all major and minor elements must be recorded in the same LA-ICPMS signal to allow the normalization of the mixed signal to a fixed total of oxides (Halter et al., 2002; Pettke et al., 2004). Hence, the quality of obtained data is determined by the reliability with which the internal standard element concentration can be determined.

8.4 SELECTED APPLICATIONS

8.4.1 CATION-EXCHANGE THERMOMETRY

Assessment of the formation temperature of aqueous fluids is usually based on chemically reactive, nonconservative species, such as Na, K, Mg, Ca, and chemical equilibria between water and minerals. For instance, the principle of the Na/K geothermometer for hydrothermal systems is inferred from the temperature-dependent variations in Na and K contents controlled by the exchange of these elements between coexisting feldspars (Poty et al., 1974; Lagache and Weisbrod, 1977; Nicholson, 1993)

$$K^+ + \text{albite} \leftrightarrow K - \text{feldspar} + Na^+ \tag{8.8}$$

and the equilibrium constant K given in terms of thermodynamic activities

$$K = \frac{a_{K-\text{spar}} \times a_{Na^+}}{a_{\text{albite}} \times a_{K^+}} \tag{8.9}$$

If the K- and Na-feldspars are represented by pure endmember phases, Equation (8.9) reduces to

$$K \cong a_{Na^+}/a_{K^+} \tag{8.10}$$

where a_i are activities of the solute species i. The temperature dependence of the equilibrium constant can be approximated by the van't Hoff equation

$$\ln K = -\frac{\Delta H^0}{RT} + \frac{\Delta S^0}{R} \tag{8.11}$$

or

$$\log K = -\frac{\Delta H^0}{2.303RT} + C \tag{8.12}$$

where ΔH^0 and ΔS^0 are the standard enthalpy and entropy of the reaction, respectively, at temperature T (K), and R is the gas constant. The equation assumes a linear relationship between ln K and $1/T$, if ΔH^0 and ΔS^0 are maintained stable over a temperature interval, normally between 0 and 350 °C, typical for most hydrothermal solutions. The Na/K activity ratio can be approximated to that of respective Na and K concentrations in the solution (Lagache and Weisbrod, 1977), and solving Equation (8.12) for temperature (t in °C) results in the classical formulation of the Na/K thermometer

$$t = \frac{a}{b \log \frac{Na}{K} + c} - 273.15 \tag{8.13}$$

where a and b are coefficients inferred from the linear regression of log K versus $1/T$ values (Equation 8.12).

Temperatures calculated from various identical cation-exchange geothermometers are usually different due to errors in calibrations, coefficients, and chemical analyses, and hence, the thermometers are gradually improved. For instance, a total of 15 various calibrations of the Na/K thermometers is recently known, and each is tailored to specific temperature ranges, fluid chemistries, and concentrations (Table 8.1). The Na/K thermometer can be used for low-Ca fluids and for temperatures exceeding 180 °C, because the equilibrium between feldspars and aqueous fluids is not attained at lower temperatures.

For Ca-rich fluids, the classical Na–K–Ca geothermometer (Fournier and Truesdell, 1973) can also be employed for the interpretation of fluid inclusions in minerals

$$t = \left[\frac{1647}{\log\left(\frac{Na}{K}\right) + \beta \log\left(\frac{\sqrt{Ca}}{Na} + 2.06\right) + 2.24}\right] - 273.15 \tag{8.14}$$

where $\beta = 1/3$ stands for $\log\left[\sqrt{Ca/Na} + 2.06\right] < 0$, and $\beta = 4/3$ for $\log\left[\sqrt{Ca/Na} + 2.06\right] > 0$. If the calculated $T > 100$ °C, $\beta = 1/3$ should be used. For fluids with higher Mg concentrations, the Na–K–Ca geothermometer yields erroneous results, and the correction for Mg is necessary (Fournier and Potter, 1979). Na/Li, K/Mg, and K/Ca thermometers are also available (Table 8.1). The complete set of cation-exchange geothermometers is included in the SolGeo computer program (Verma et al., 2008).

Table 8.1 Cation-Exchange Geothermometers

Geothermometer[a]	*a*	*b*	*c*	Range (°C)	Source
Na/K	777	1	0.7		Fournier and Truesdell (1973)
Na/K	855.6	1	0.8573	100–275	Truesdell (1976)
Na/K	1217	1	1.483		Fournier (1979)
Na/K	833	1	0.780		Tonani (1980)
Na/K	933	1	0.993	25–250	Arnórsson (1983)
Na/K	1217	1	1.483	250–350	Arnórsson (1983)
Na/K	1178	1	1.47		Nieva and Nieva (1987)
Na/K	1289	1	1.615		Verma and Santoyo (1997)
Na/K	2.0446	1	126.01	91–516	Bayram (2001)
Na/K	883	1	0.894	30–350	Díaz-Gonzáles and Santoyo (2008)
Na/K	883	1	0.908	30–350	Díaz-Gonzáles and Santoyo (2008)
Na/Li	1000	1	0.38	200–325	Michard and Fouillac (1981)
Na/Li	2002	1	−1.322		Sanjuan et al. (2010)
Na/√Ca	1096.7	1	2.37		Tonani (1980)
K/Na	1390	−1	1.75		Giggenbach (1988)
K/√Ca	1930	1	2.92		Tonani (1980)
K/Mg	4410	−1	14		Giggenbach (1988)

[a]*Concentrations in mg/kg (ppm).*

8.4.2 ORIGIN OF FLUIDS CONSTRAINED BY SOLUTE CHEMISTRY

An assessment of the fluid provenance is the second major application field of the bulk solute chemistry, which, contrasting with the cation-exchange thermometry, also employs concentrations of chemically inert, nonreactive constituents, such as halogens, Rb, and Cs. Simple discrimination diagrams can be used to reveal the seawater component in hydrothermal fluids, as well as cationic exchange mechanisms taking place during fluid-rock interactions. An average salinity of seawater unmodified by the evaporation corresponds to a 3.2 mass% NaCl equivalents, but it may be significantly increased by evaporation in semiclosed basins. Typical minerals precipitate from the residual brine at various stages of seawater evaporation (Figure 8.1). Bitter brines remaining after the seawater evaporation are stored in sediments as formation (connate) waters, or they infiltrate the subjacent crust along faults and rifts. The brine composition is further modified during diagenesis and anchimetamorphism by specific chemical reactions. In such a way, the chemical composition of the residual pore fluid becomes enriched or depleted in Na, Ca, and Mg relative to the unmodified seawater.

An excess or deficit of specific elements in fluid inclusion leachates can be calculated to reveal the nature of fluid–rock interactions (Davisson and Criss, 1996; Heijlen et al., 2001)

$$\mathrm{Ca_{ex}} = \left[\mathrm{Ca_m} - (\mathrm{Ca/Cl})_{sw} \times \mathrm{Cl_m}\right] \times 2/40.08 \tag{8.15}$$

$$\mathrm{K_{ex}} = \left[\mathrm{K_m} - (\mathrm{K/Cl})_{sw} \times \mathrm{Cl_m}\right]/39.1$$

$$\mathrm{Na_{def}} = \left[(\mathrm{Na/Cl})_{sw} \times \mathrm{Cl_m} - \mathrm{Na_m}\right]/22.9899$$

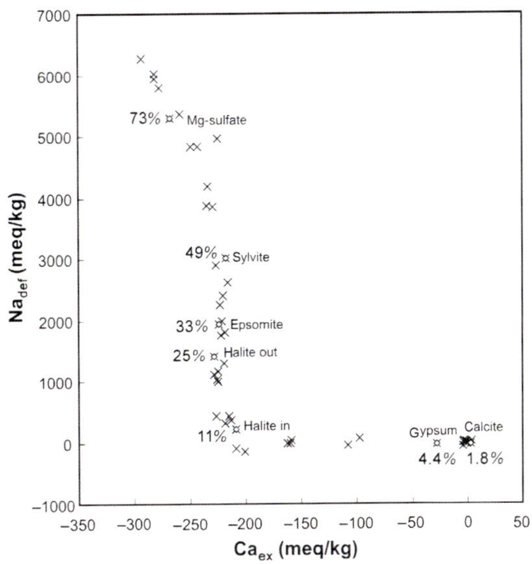

FIGURE 8.1

Seawater evaporation path expressed in terms of Ca-excess and Na-deficit values in the residual bitter brine (McCaffrey et al., 1987) relative to initial seawater composition (Horita et al., 1991). Critical minerals that precipitate at various stages of the evaporation are also shown. The degree of evaporation is expressed as a percent of seawater evaporated.

where subscripts "ex" and "def" refer to the excess and deficit of the corresponding element relative to seawater (sw). The elemental concentrations are expressed in mg/L (ppm). The subscript "m" stands for the measured concentration of an element. The final excess or deficit values are expressed in milliequivalents per kilogram (meq/kg) because they are divided by the atomic mass of the corresponding cation.

Formation waters in diagenetically altered sediments are enriched in Ca relative to seawater and the seawater evaporation trend created a trendline called the basinal fluid line (BFL) (Davisson and Criss, 1996)

$$Ca_{ex} = 0.967(Na_{def}) + 140.3 \tag{8.16}$$

Other diagenetic processes, such as the dissolution of calcite matrix, potassium feldspars, and evaporitic minerals, dedolomitization, and albitization of plagioclase, are manifested by the depletion/enrichment in specific major element(s), and can be thus revealed easily by crush-leach analyses (Figure 8.2).

Another method of distinguishing between different hydrothermal fluid types relies on the halogen content of fluid inclusions. The Cl/Br and Na/Br ratios are widely used as geochemical tracers of the origin of salinity in crustal fluids (Banks et al., 2000; McCaig et al., 2000; Lüders et al., 2005). Because Cl and Br are rarely abundant in minerals (except for Cl in evaporites), these elements are generally conservative in solution and do not fractionate to a larger extent during a fluid–rock interaction (McCaffrey et al., 1987; Fontes and Matray, 1993; Worden, 1996). Hence, the Cl/Br and Na/Br values, as well as absolute halogen concentrations, are indicative of the origin of salinity, reflecting a seawater evaporation, evaporite dissolution, or exsolution from magma (Yardley et al., 2000). In addition,

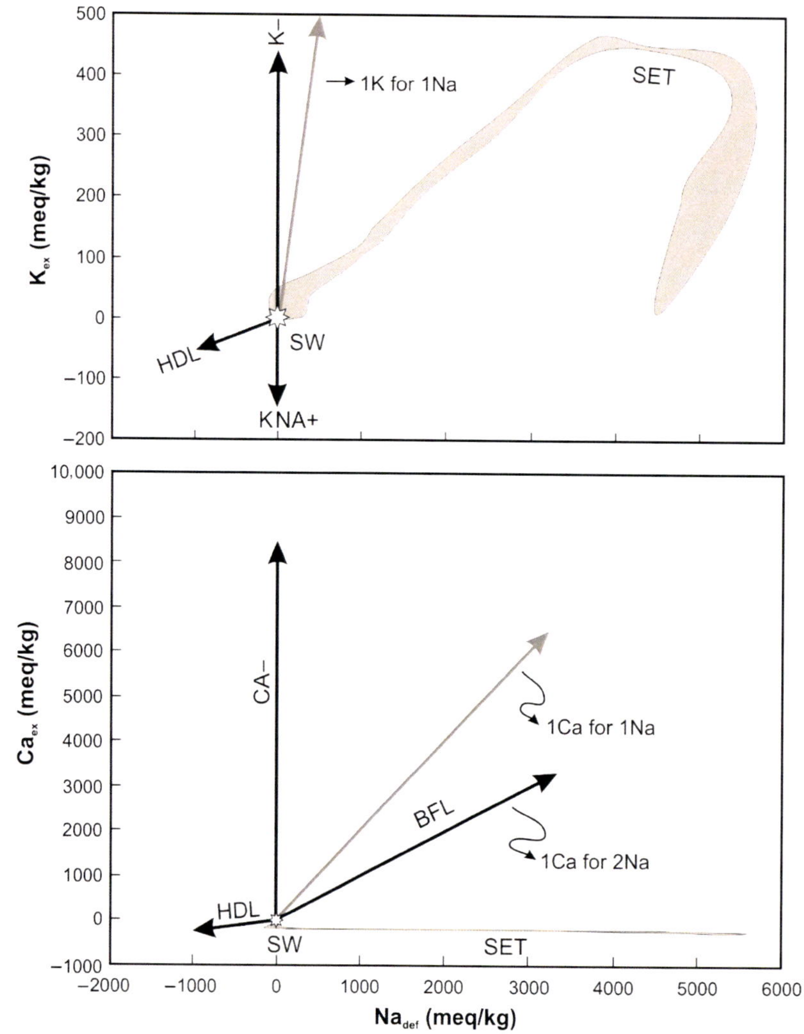

FIGURE 8.2

Correlations of excess and deficit concentrations of Na, K, and Ca in pore fluids of sedimentary basins relative to recent seawater (SW). Arrows designate trends of various exchange reactions: BFL is the basinal fluid line reflecting the cationic exchange of 1Ca for 1Na during albitization of plagioclase according to the reactions: $CaAl_2Si_2O_8 + SiO_2 + Na^+ + 4H^+ \rightarrow NaAlSi_3O_8 + Ca^{2+} + Al^{3+} + 2H_2O$, or $CaAl_2Si_2O_8 + Na^+ + H^+ + 2SiO_2 + 0.5H_2O \rightarrow NaAlSi_3O_8 + 0.5Al_2Si_2O_5(OH)_4 + Ca^{2+}$. The 2Ca for 1Na line represents the trend of the plagioclase albitization according to the reaction $CaAl_2Si_2O_8 + 4SiO_2 + 2Na^+ \rightarrow 2NaAlSi_3O_8 + Ca^{2+}$. The 1K for 1Na line designates the trend produced by the albitization of potassium feldspar. HDL is the halite dissolution line. CA corresponds to the calcite or anhydrite dissolution trend, or that of dedolomitization. K is the K-feldspar dissolution trend. KNA+ is the trend diagnostic of the concomitant illitization and K-feldspar precipitation. SET is the seawater evaporation trend calculated from data in McCaffrey et al. (1987). Upper diagram reproduced from Heijlen, W., Muchez, P., Banks, D.A., 2001. Origin and evolution of high-salinity, Zn-Pb mineralizing fluids in the Variscides of Belgium. Miner. Deposita 36, 2001, 165–176, Fig. 8b, with kind permission from Springer Science and Business Media. Lower diagram reprinted from Davisson, M.L., Criss, R.E., 1996. Na-Ca-Cl relations in basinal fluids, Geochim. et Cosmochim. Acta 60, 847–859, Copyright 1996, with permission from Elsevier.

because evaporites and seawater are by far the largest halogen reservoirs, the halogen systematics inferred from crush-leach data is sensitive to the involvement of these reservoirs in ore-forming solutions (Wilkinson, 2001).

Simple Cl/Br versus Na/Br plots have been frequently used as the evidence for the evaporated seawater or dissolved evaporite components in hydrothermal fluids. Because Br is an incompatible element, its concentration in the residual seawater gradually increases with the progressive seawater evaporation, thus leading to the lowering of the Cl/Br and Na/Br ratios. In contrast, halite dissolution results in Br-depletion compared to seawater owing to negligible Br substitution in halite and other typical evaporite minerals (calcite, sylvite, sulphates). It should be noted, however, that the Cl/Br versus Na/Br plot alone can be misleading owing to the fact that all crustal fluids are dominated by NaCl, and they thus plot close to the 1:1 line, though having nothing to do either with the evaporated seawater or the halite fractionation (Figure 8.3).

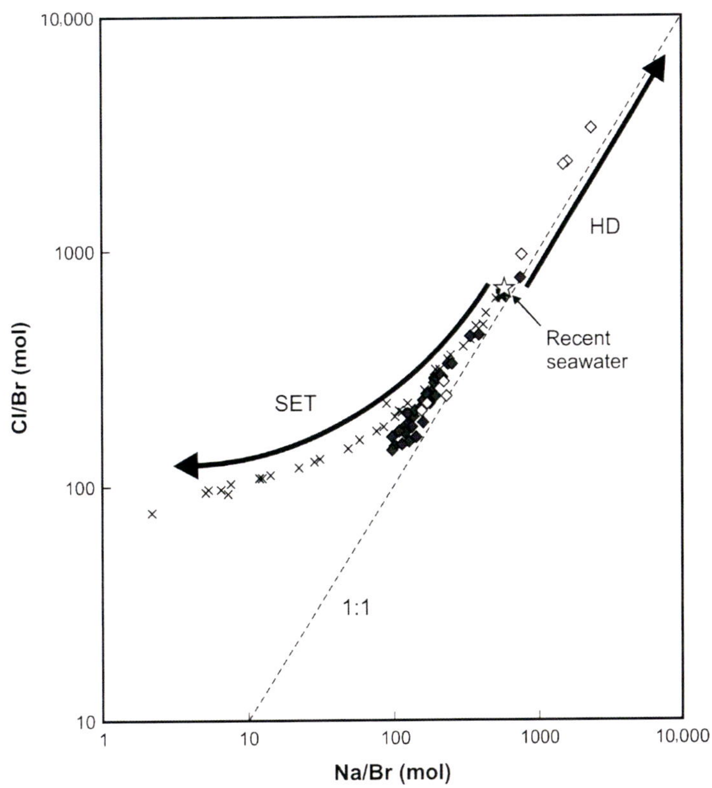

FIGURE 8.3

Molar ratios of Na, Br, and Cl during seawater evaporation (McCaffrey et al., 1987), with projection points of crush-leach analyses from siderite ± barite polymetallic deposits of the Gemeric Unit, Western Carpathians (Hurai et al., 2002a, 2008a, 2011). Solid diamonds are siderite, open diamonds are barite. HD is the halite dissolution trend; SET is the seawater evaporation trend. Composition of the present-day seawater is taken from Horita et al. (1991).

PROBLEM 8.1 ELEMENT BALANCE DURING SEAWATER EVAPORATION

Composition of modern seawater (Horita et al., 1991) is as follows: Na—485; K—10.6; Ca—10.6; Cl—565 mmol/kg H_2O. Calculate excess/deficit values of these elements during gypsum precipitation (13.2% evaporation), when Na, K, Ca, and Cl contents correspond to 5020, 152, 3.5, and 5750 mmol/kg H_2O, respectively.

Solution

Because we have compositions expressed in mmol/kg H_2O, the corresponding excess values can be calculated using the following equations

$$M_{\text{ex}} = 2 \times \left[M_{\text{m}} - (M/\text{Cl})_{\text{sw}} \times \text{Cl}_{\text{m}}\right] \quad \text{for divalent cations} \tag{8.17}$$

and

$$M_{\text{ex}} = M_{\text{m}} - (M/\text{Cl})_{\text{sw}} \times \text{Cl}_{\text{m}} \quad \text{for monovalent cations} \tag{8.18}$$

Similarly, deficit values (for Na) are calculated according to

$$M_{\text{def}} = (M/\text{Cl})_{\text{sw}} \times \text{Cl}_{\text{m}} - M_{\text{m}} \tag{8.19}$$

where subscripts "sw" and "m" refer to elemental concentrations in seawater and the measured concentrations in the evaporated seawater, respectively, and M is the measured element. The calculated Na_{def}, K_{ex}, and Ca_{ex} values correspond to −84, 44, and −209 meq/kg, respectively.

SUGGESTED FURTHER READING

Blamey, N.J.F., 2012. Composition and evolution of crustal, geothermal and hydrothermal fluids interpreted using quantitative fluid inclusion gas analysis. J. Geochem. Explor. 116–117, 17–27.

Heinrich, C.A., Pettke, T., Halter, W.E., Aigner-Torres, M., Audétat, A., Günther, D., Hattendorf, B., Bleiner, D., Guillong, M., Horn, I., 2003. Quantitative multi-element analysis of minerals, fluid and melt inclusions by laser-ablation inductively-coupled-plasma mass-spectrometry. Geochim. Cosmochim. Acta 67, 3473–3496.

Pettke, T., Halter, W.E., Webster, J.D., Aigner-Torres, M., Heinrich, C.A., 2004. Accurate quantification of melt inclusion chemistry by LA-ICPMS: a comparison with EMP and SIMS and advantages and possible limitations of these methods. Lithos 78, 333–361.

Pettke, T., Oberli, F., Audétat, A., Guillong, M., Simon, A.C., Hanley, J.J., Klemm, L.M., 2012. Recent developments in element concentration and isotope ratio analysis of individual fluid inclusions by laser ablation single and multiple collector ICP-MS. Ore Geol. Rev. 44, 10–38.

CHAPTER

STABLE ISOTOPE GEOCHEMISTRY OF GEOFLUIDS

9

CHAPTER OUTLINE

The history of mass spectrometry began with the study of gas discharges in the mid-nineteenth century, which led to the discovery of positive ions and electrons. Improved capabilities in the separation of the positive ions enabled the discovery of neon isotopes in 1913 by Sir J. J. Thompson (1856–1940), who is also credited as the inventor of the mass spectrometer (Griffiths, 2008). Light stable isotope geochemistry as it is known now began with the discovery of deuterium (Urey et al., 1932). Soon after the discovery of stable isotopes it was noticed that significant fractionation occurs in nature between isotopes

Geofluids. http://dx.doi.org/10.1016/B978-0-12-803241-1.00009-5

of light elements (Maillard, 1936). Real progress of stable isotope geochemistry started when Waldmann (1943) and, independently, Bigeleisen and Mayer (1947) published the first equations for the isotopic exchange calculations. Urey (1947) theoretically described the fractionation between various isotopic substances, and McKinney et al. (1950) developed a mass spectrometer for routine measurements of small differences in the isotopic composition of naturally occurring substances. Later in the 1940s and 1950s, Urey and coworkers significantly improved the precision of the isotope ratio mass spectrometer constructed by Nier (1947) and proposed a reproducible standard extraction method for calcite (McCrea, 1950). Standards and protocols developed by Urey's group are, for the most part, followed today. Their paleotemperature scale for the determination of temperatures of ancient oceans from oxygen isotope compositions of fossil shells (Urey et al., 1948; Epstein et al., 1951) has been heralded as one of the outstanding scientific achievements of the twentieth century (Sharp, 2007).

9.1 TERMINOLOGY AND PRINCIPLES

Isotopes are atoms with an equivalent number of protons, but a different number of neutrons. Isotopes are formally described as ${}^{A}_{Z}E$, where E is the element abbreviation, A is the atomic mass (sum of protons Z and neutrons N), and Z stands for the atomic number equivalent to that of protons. Atomic mass is a weighted average of all natural isotopes. For instance, the atomic mass of sulfur is 32.064, which is a weighted average of 95.02% ^{32}S, 0.75% ^{33}S, 4.21% ^{34}S, and 0.02% ^{36}S. The atomic mass is expressed in atomic mass units (a.m.u), corresponding to 1/12 of the carbon mass, which is 1.660531×10^{-27} kg.

Isotope ratios of light elements are modified by isotope fractionation. Fractionation occurs during chemical exchange reactions and physical processes in the Earth's crust, biosphere, and atmosphere. Isotope fractionation is caused by large mass differences between isotopes of light elements, which result in slightly different physical properties of molecules with the same chemical but different isotope compositions called *isotopologues* (e.g., ${}^{1}_{1}H_2O$, ${}^{2}_{1}H_2O$, ${}^{3}_{1}H_2O$).

Isotopes of elements classically treated as light stable isotopes exhibit the following properties (Hoefs, 1997; Sharp, 2007):

1. They have low atomic mass.
2. The mass difference among isotopes is large.
3. They create covalent bonds.
4. They either exist in several oxidation states (e.g., C, N, S), create a number of compounds (O), or are important constituents of minerals and fluids (H).

Except for lithium, abundances of isotopes are governed by the Oddo–Harkins rule (1914–1917) stating that elements with an even atomic number are more abundant than those with an odd atomic number. Hence, the isotopes with even Z and N are the most abundant, whereas those with an uneven Z and N are the least abundant.

Isotope ratios may be expressed as absolute numbers (R), with the less abundant isotope in the numerator and the more abundant isotope in the denominator (e.g., $^{34}S/^{32}S$, D/H). The absolute numbers are, however, less illustrative because the absolute change of the isotope ratio in most natural samples occurs at best on the second decimal place. For this reason, isotope ratios of light elements are expressed as one part per thousand deviations (‰) from a standard. This expression is called δ-notation. Assuming that R_S is the absolute elemental isotope ratio in the standard and R_A is that in the investigated sample, the δ-notation can be written as

Table 9.1 Stable Isotope Standards (Gionfantini et al., 1995; Hoefs, 1997; Ding et al., 2001)

Element	Notation	Ratio	Standard	Compound	Isotope Ratio
Hydrogen	δD	D/H ($^2H/^1H$)	V-SMOW	H_2 gas	$155.76 \pm 0.05 \times 10^{-6}$
Lithium	δ^6Li	$^6Li/^7Li$	L-SVEC	Li-tetraborate	0.0832
Boron	$\delta^{11}B$	$^{11}B/^{10}B$	NBS 951	$Cs_2BO_2^+$	4.044
Carbon	$\delta^{13}C$	$^{13}C/^{12}C$	V-PDB	CO_2 gas	$11{,}237.2 \times 10^{-6}$
Nitrogen	$\delta^{15}N$	$^{15}N/^{14}N$	Air	Air	$3612 \pm 7 \times 10^{-6}$
Oxygen	$\delta^{18}O$	$^{18}O/^{16}O$	Air	Air	$2077 \pm 2 \times 10^{-6}$
Oxygen	$\delta^{18}O$	$^{18}O/^{16}O$	V-SMOW	CO_2 gas	$2005.2 \pm 0.45 \times 10^{-6}$
Oxygen	$\delta^{18}O$	$^{18}O/^{16}O$	V-PDB	CO_2 gas	2079×10^{-6}
Sulfur	$\delta^{34}S$	$^{34}S/^{32}S$	V-CDT	SO_2 gas	0.044163

$$\delta_A = 1000\left(\frac{R_A - R_S}{R_S}\right) \quad \text{or} \quad \delta_A = 1000\left(\frac{R_A}{R_S} - 1\right) [\text{‰}] \tag{9.1}$$

Commonly used standards and their isotope ratios are listed in Table 9.1. The negative δ_A value indicates that the investigated sample is depleted in the heavier isotope, meaning that it is isotopically lighter compared to the standard, or *vice versa*. The same is valid for the comparison of δ values between two compounds or minerals.

Chemical bonds created by lighter isotopes are weaker than those created by heavier isotopes. As a consequence, the lighter molecules will react faster than heavier ones, thereby resulting in the enrichment of reactants of chemical reactions in heavy isotopes. Usually the heavier isotope is concentrated in the ions, atoms, and molecules with a higher oxidation state and/or in a more condensed state (Urey, 1947). This process is called isotope fractionation. Most changes in isotope ratios in nature result from the equilibrium fractionation, which is caused by the mass-dependent translation, rotation, and vibration of molecules and atoms. Equilibrium isotope fractionation is dominantly affected by the vibrational energy, which is dependent on temperature. Hence, isotope fractionation is a linear inversion function of temperature and can thus be effectively used in stable isotope thermometry. In contrast, kinetic (nonequilibrium) fractionation is connected with fast, incomplete, or vector physical processes and reactions, such as evaporation, condensation, diffusion, and metabolism in living organisms. Owing to weaker chemical bonds, light isotopes are preferably concentrated in products of the interrupted, nonequilibrium chemical reactions. Particularly large isotope fractionation is connected with metabolic processes (photosynthesis, bacterial reduction/oxidation). For instance, plants cannot convert all atmospheric CO_2 to organic carbon. Thus, the resulting depletion in ^{13}C of the photosynthetic products is large. Similarly, the light ^{32}S isotope is dramatically concentrated in H_2S—the product of bacterial reduction of the sulfate dissolved in the oceanic water.

Isotope fractionation is expressed in terms of the fractionation factor α, correlating the isotope ratios R in compounds A and B (Hoefs, 1997; Sharp, 2007)

$$\alpha_{A-B} = \frac{R_A}{R_B} \tag{9.2}$$

Table 9.2 Comparison of Stable Isotope Fractionation Factor Obtained using Various Calculation Methods

δ_A	δ_B	α_{A-B}	$10^3 \ln \alpha_{A-B}$	$1000(\alpha_{A-B}-1)$	Δ_{A-B}
1	0	1.00100	0.9995	1	1
10	0	1.01000	9.9503	10	10
20	0	1.02000	19.803	20	20
10	5	1.00498	4.9628	4.98	5
20	15	1.00493	4.9140	4.93	5
30	10	1.01980	19.609	19.8	20
30	20	1.00980	9.7562	9.8	10

Considering $1000 \ln(1.00X) \cong X$, the fractionation factor α and the isotope composition of compounds A and B expressed using δ-notation can be approximated by the following equation

$$\delta_A - \delta_B = \Delta_{A-B} \cong 1000 \ln \alpha_{A-B} \cong 1000(\alpha_{A-B} - 1) \tag{9.3}$$

The approximation $1000 \ln \alpha_{A-B}$ is valid for the δ_A and/or δ_B values less than 10, whereas the approximation $1000(\alpha_{A-B} - 1)$ is accurate if one of the δ values is less than 10 (Table 9.2). In other cases, an extended equation, resulting from the combination and rearrangement of Equations (9.1) and (9.2), must be used

$$\alpha_{A-B} = \frac{1 + \dfrac{\delta_A}{1000}}{1 + \dfrac{\delta_B}{1000}} = \frac{1000 + \delta_A}{1000 + \delta_B} = \frac{R_A}{R_B} \tag{9.4}$$

Equilibrium fractionation factors are determined experimentally, empirically, or by theoretical calculation. Methods based on statistic and quantum mechanics were developed by Urey (1947) and Bigeleisen and Mayer (1947). Golyshev et al. (1981) introduced a lattice dynamic model for the calculation of equilibrium isotope fractionation factors in carbonate minerals. The increment computation was introduced by Schütze (1980) and later popularized by Zheng (1991, 1993a–c, 1999). The great advantage of the increment method is the capability of calculating oxygen isotope fractionation factors for crystalline minerals based on their structure, composition, cation valence, and ionic radii (Zheng, 2011). Chacko et al. (2001) and Horita and Clayton (2007) criticized the increment method due to inconsistencies in the low-temperature (<100 °C) oxygen isotope fractionation in carbonates. The most extensive compilations of experimental stable isotope fractionation factors were published by Friedman and O'Neil (1977), Chacko et al. (2001), and Beaudoin and Therrien (2004, 2009). Appendix IV summarizes selected fractionation factors relevant to mineral–fluid systems.

9.2 STABLE ISOTOPE COMPOSITION OF GEOFLUID RESERVOIRS

Variations in stable isotope composition of Earth result from the fluid exchange between the main geofluid reservoirs: hydrosphere, atmosphere, and deep-seated fluids from the mantle. The isotopic variations are caused by mixing, equilibrium and kinetic exchange, physical processes, and chemical reactions. All these factors are thoroughly discussed in any textbook dedicated to stable isotopes, but those written by Hoefs (1997) and Sharp (2007) provide the most excellent summaries and are thus recommended for further reading.

The pristine isotope composition of the mantle is difficult to determine directly, as it has been permanently modified by the generation and partial melting of oceanic and continental crust, assimilation of the crustal material and marine sediments along subduction zones, interaction with fluids and melts generated during the subduction, and by the permanent devolatilization taking place irrespective of the tectonic and magmatic processes (Kyser, 1986). Despite this, the mantle is relatively homogeneous, with the $\delta^{18}O$ values between 5‰ and 6‰, as inferred from mantle xenoliths, basalts of mid-oceanic ridges (MORBs), and oceanic islands (OIBs) unaffected by the assimilation of crustal material. The oxygen isotope ratios in the mantle are similar to those of enstatite chondrites and lunar samples. Because the Moon lies exactly on the terrestrial $\delta^{17}O$–$\delta^{18}O$ fractionation line, both bodies must have originated from the same material. Hence, there is a wide consensus that the range between 5‰ and 6‰ is the most representative of the oxygen isotope composition of the pristine mantle.

The determination of the hydrogen isotope composition of the mantle is a challenging task because any process bringing the mantle material to the surface induces a large fractionation of hydrogen isotopes. For instance, glass from submarine basalts yielded δD values between −120‰ and +13‰ V-SMOW (Sheppard and Epstein, 1970; Boettcher and O'Neil, 1980; Kyser, 1986). Kyser and O'Neil (1984) extrapolated measured ranges in δD values and water contents in the basaltic glass back to unmodified values unaffected by the devolatilization and mixing with seawater, and obtained a δD value of about −80‰ V-SMOW. Phlogopites from kimberlites exhibit δD values from −58‰ to −79‰ V-SMOW (Boettcher and O'Neil, 1980) that have been recently accepted as the representative values of the pristine mantle. *Primary magmatic water* is a fluid equilibrated with the pristine mantle at high temperatures at a low fluid–rock ratio, thus acquiring mantlelike isotopic signatures. Ore geologists are permanently in quest of the primary magmatic, mantle-derived water component in ore-forming fluids, though the pristine water unaffected by the interaction with the hydrosphere could not be preserved over geological time.

The pristine sulfur isotope composition of the mantle, near 0‰ V-CDT, is well constrained by that of iron meteorites, chondrites, and MORB basalts. An enrichment in ^{34}S results from the oxidation of sulfide to sulfate along mid-oceanic ridges and also from the contamination of the basaltic magma with sedimentary sulfates from the subducted material.

The $\delta^{13}C$ values between −5‰ and −6‰ V-PDB are accepted as representative of the pristine mantle. This is corroborated by the carbon isotope composition of peridotitic diamonds, showing a limited range of $\delta^{13}C$ values, mostly between −5‰ and −7‰ V-PDB, with outliers going far from this limit. The lower and higher outliers are considered as indicators of the mantle contamination by the subducted material. Volcanic emanations noninfluenced by the interaction with the continental crust have $\delta^{13}C$ values between −3‰ and −4‰. The enrichment is probably caused by the fractionation between CO_3^{2-} dissolved in the silicate melt and the molecular gaseous CO_2.

Hydrosphere is another large stable isotope fluid reservoir. The *meteoric water from atmospheric precipitations* is the principal fluid component in shallow hydrothermal systems emplaced in the continental crust, whereas the *marine water* plays the same role in submarine hydrothermal systems along seafloor spreading centers.

The present-day $\delta^{18}O$ and δD composition of ocean water is extremely uniform (0‰ V-SMOW), but it may change locally in semiclosed basins by dilution with isotopically lighter meteoric water in humid climatic zones or by evaporation in arid zones. Atmospheric circulation of water due to evaporation, freezing, and condensation produces the largest isotope variations because it takes place at low temperature under the assistance of a kinetic fractionation. During these processes, the initially homogeneous isotope composition of oceanic water changes to extremely negative values. For instance, atmospheric moisture at the South Pole exhibits $\delta^{18}O$ values down to −50‰ (Ciais et al., 1995). In contrast, rainfalls on isolated oceanic islands and coastal low-elevation areas in intertropical regions show δD and $\delta^{18}O$ values close to those of oceanic water (from −2‰ to −3‰). Globally, the δD and $\delta^{18}O$ values of the atmospheric precipitations are correlated by the global meteoric water line (GMWL) (Craig, 1961), with the best linear fit expressed by the equation $\delta D = 10 + \delta^{18}O$. For some regions, however, significant deviation from the GMWL occurs (Araguás-Araguás et al., 2000).

Biological processes significantly contribute to the redistribution of stable isotopes due to extreme kinetic fractionation associated with the fixation of carbon in organic tissues synthesized from atmospheric CO_2 and liquid water. The atmospheric CO_2 with the $\delta^{13}C$ value of −8‰ V-PDB represents a mixture of the ^{12}C-rich CO_2 produced by photorespiration of rain-forest plants and the ^{13}C-rich CO_2 degassed from the mantle by volcanic emanations. Living organisms use the atmospheric CO_2 in two ways: for the production of organic tissues and for the creation of protective exoskeletons by molluscs, brachiopods, worms, and foraminifers. The reduced and oxidized carbon reservoirs in sedimentary basins represent two carbon sources with contrasting isotope compositions.

Most atmospheric CO_2 is dissolved in the ocean as HCO_3^-, which dominates over other carbon-bearing species and is thus preferably consumed by marine organisms for building up their carbonate exoskeletons. Because HCO_3^- is dissolved in seawater is enriched in ^{13}C by +8‰ compared to the atmospheric CO_2 gas, carbonate shells exhibit $\delta^{13}C$ values equivalent to the PeeDee *Belemnitella* standard, which means zero ‰ V-PDB. However, various organisms produce variable carbon isotope compositions depending on the inhabited ecosystem. For instance, exoskeletons of planktonic foraminifers are enriched by ~1‰ compared to benthic foraminifers due to intense photosynthesis in the uppermost ocean layer, which leads to the depletion in the light carbon isotope stored in soft organic remains deposited at the bottom of the water column. Hence, typical marine carbonates unaffected by diagenetic processes, weathering, or hydrothermal alteration exhibit $\delta^{13}C$ values ranging between +1‰ and −3‰ V-PDB. Oxygen isotope composition of the calcareous shells and exoskeletons is largely determined by an equilibrium isotope fractionation between calcite and seawater. Because the fractionation factor α_{cc-H_2O} corresponds to 1.0286 at 25 °C, typical marine carbonates acquire the $\delta^{18}O$ values around +28.6 ± 1‰ V-SMOW (= ~0% V-PDB).

The fixation of carbon in organic tissues by photosynthesis occurs in several ways, but all are based on the production of the adenosine triphosphate (ATP) considered to be a source of energy for biological processes. The C3 (Calvin–Benson) pathway includes the carbon fixation from the atmospheric CO_2 in several steps under the catalytic assistance of the $NADPH_2$ and RuBisCo enzymes (Bassham et al., 1950). The C3 plants have emerged during Paleozoic times and still represent ~95% of the recent plant biomass, growing in moderate and humid climatic conditions.

However, with increasing temperatures, the RuBisCo enzyme tends to fix oxygen instead of CO_2. This results in the oxidation of photosynthetic carbon and its loss as isotopically light CO_2 gas by photorespiration. To circumvent this problem, some plants evolved an additional set of reactions that allows them to survive in arid and hot areas. During the C4 (Hatch–Slack) pathway, the RuBisCo is isolated from atmospheric oxygen in specially structured leaves and saturated with CO_2 released by decarboxylation of the malate or oxaloacetate (Slack and Hatch, 1967). The pathway is more energy-demanding than the previous cycle, because each CO_2 molecule has to be fixed twice, but this energy debt is counterbalanced by avoiding losing the photosynthetic carbon by the photorespiration.

Succulent plants from arid areas combine both pathways and fix the atmospheric carbon using a Crassulacean acid metabolism (CAM). They either fix atmospheric carbon in the manner of C3 plants by use of ribulose biphosphate carboxylase or in a time-separated C4-like sequence in which phosphoenolpyruvate is carboxylated, then reduced in the dark, forming malate, which accumulates in vacuoles. In the following light periods, the malate is decarboxylated, and the CO_2 thus formed is fixed by ribulose biphosphate carboxylase (O'Leary, 1981).

The dissolution of CO_2 in the cell sap and its diffusion in chloroplast, where carboxylation occurs, have little effect on the isotope fractionation. However, ribulose biphosphate carboxylase, the key carboxylating enzyme in all plants, strongly discriminates against $^{13}CO_2$, thus resulting in the large fractionation during the carboxylation, attaining 29‰ (Roeske and O'Leary, 1984). Owing to the different carboxylation mechanism, the C3 plants are isotopically lighter ($\delta^{13}C$ between −25‰ and −29‰) than C4 plants (from −12‰ to −16‰) (O'Leary, 1988).

Marine phytoplankton produces smaller variations in the carbon isotopes owing to smaller temperature fluctuations of global ocean. Because the phytoplankton uses isotopically heavier HCO_3^- dissolved in seawater, it is also enriched in ^{13}C ($\delta^{13}C$ between −15‰ and −20‰) compared to terrestrial plants.

About 95% of organic carbon deposited in sediments comes from the marine phytoplankton. Biodegradation of the organic remains by anaerobic bacteria breaks down the long hydrocarbon polymers and creates a chemically inert, insoluble mixture of aliphatic and aromatic hydrocarbons called *kerogen*. The formation of kerogen is accompanied by a ~10–15‰ depletion in ^{13}C. The resulting $\delta^{13}C$ values between −25‰ and −35‰ V-PDB in marine sediments strongly contrast with carbonate exoskeletons with $\delta^{13}C$ values between +1‰ and −3‰ V-PDB. Thermal degradation of kerogen leaves behind a low-ordered graphitic substance with $\delta^{13}C$ values between −20‰ and −32‰ V-PDB, which gradually converts with an increasing metamorphic grade to the well-ordered graphite with ~ −20‰ V-PDB at the boundary between greenschist and amphibolite facies. An abrupt enrichment in ^{13}C along this boundary, up to −10‰ V-PDB, is attributed to the crystallization of inorganic graphite deposited from CO_2–CH_4 fluids.

The continental and oceanic crust is the scene of a large-scale exchange of isotopes between fluid and rock reservoirs. As a result of the fluid–rock interactions and mixing, additional transient geofluid types are formed (Taylor, 1974):

Geothermal water is the meteoric water modified by cation- and isotope-exchange reactions with silicate rocks. It occurs in geothermal fields genetically associated with volcanism. The exchange of isotopes results in the ^{18}O enrichment, whereas δD values remain either preserved or only moderately enriched compared to those in contemporary meteoric waters in continental settings or seawater in submarine hydrothermal systems.

Formation (connate) water is the water stored in marine basins. It is composed of a fossil seawater modified by dilution with meteoric water and/or by diagenetic reactions. These reactions along with adsorption of light isotopes on surfaces of clay particles cause significant enrichment in heavy isotopes of oxygen and hydrogen compared to contemporary seawater or seawater–rainwater mixture.

Metamorphic water is produced by metamorphic reactions and fluid–rock interactions in orogens and subducted oceanic crust. Owing to a wide range of *P–T* conditions and variable proportions of different fluid components (meteoric, magmatic, marine), the isotope composition of metamorphic water fluctuates within wider intervals, between 5‰ and 25‰ for oxygen, and between −20‰ and −65‰ for hydrogen.

9.3 STABLE ISOTOPES IN FLUID INCLUSIONS

First direct analysis of stable isotopes in fluid inclusions was made after substantial improvement of precision of mass spectrometer (Rye and O'Neil, 1968). Carbon isotopes ($^{13}C/^{12}C$) in carbon dioxide, D/H, and $^{18}O/^{16}O$ ratios in inclusion water can recently be measured routinely in quartz and carbonates (Kazahaya and Matsuo, 1984; Ohba and Matsuo, 1988; Lecuyer and O'Neil, 1994; Simon, 2001; Orvošová et al., 2004). One of the most frequently used methodologies relies on the combination of direct and indirect methods, which means that the measurement of D/H ratios in the extracted inclusion fluid is complemented by the calculation of the oxygen isotope composition of the fluid from the $\delta^{18}O$ value of host quartz or calcite, using appropriate fractionation factors. However, the indirect approach is not ideal, as it is dependent on the precise determination of formation temperature. Moreover, isotopic disequilibria may occur between quartz and the fluid phase at temperatures below 250 °C (Matsuhisa et al., 1978; Zhang et al., 1989).

Despite possible kinetic effects, analysis of fluid inclusions combined with that of host carbonates becomes an important tool in paleoclimatic reconstructions based on the study of speleothems (Lecuyer and O'Neil, 1994; Matthews et al., 2000; Dennis et al., 2001; Sereffidin et al., 2002; McGarry et al., 2004). Comparison of the directly measured and calculated $\delta^{18}O$ values showed that unlike the temperature-corrected quartz data, fluid inclusions better reproduce the isotope compositions of paleofluids in epithermal hydrothermal systems (Naden et al., 2003).

Mass spectrometric analysis of fluid inclusions is difficult due to a very small volume of the fluid liberated by vacuum milling or thermal decrepitation. This complication recently seemed to be resolved by the invention of extremely sensitive continuous-flow extraction and analysis coupled with a gas chromatography column (Vohnhof et al., 2006; Dublyansky and Spötl, 2009), and laser absorption cavity ring-down spectroscopy (Arienzo et al., 2013; Affolter et al., 2014). However, other complications similar to those pertinent to the crush–leach analysis remain unresolved: namely, the contamination with volatiles bound to structural defects and fractures and a possibility to obtain only a bulk composition involving various fluid inclusion generations present in the analyzed mineral. Kinetic isotopic fractionation may also occur during the sample preparation due to either incomplete liberation of volatiles from fluid inclusions by thermal decrepitation, or the isotopic exchange with the host mineral. The $\delta^{18}O$ values can be influenced also by the diffusion of water molecules through quartz (Ohba et al., 1995). The δD values of water mechanically or thermally extracted from fluid inclusions in quartz can be significantly underestimated owing to an isotopic exchange between structurally bound and fluid inclusion water (Simon, 2001).

9.4 STABLE ISOTOPE THERMOMETRY

Geothermometry is one of the most useful applications of stable isotopes in geology. In principle, the equilibrium isotope fractionation factors between either two coexisting substances can be used to calculate the temperature at which they equilibrated. In practice, however, there is no certainty about the attainment of the isotope equilibrium between two compounds. Specific pressure and temperature conditions, as well as the presence of other reactants and catalysts, are usually needed to promote a reaction and the possible attainment of the chemical equilibrium. Sharp (2007) analyzed in detail and summarized the criteria, which must be fulfilled for the successful application of stable isotope geothermometers. In spite of many uncertainties, the temperature dependence of the stable isotope fractionation between two minerals, aqueous species, or gases has been widely used for the stable isotope geothermometry applied to minerals (Ohmoto and Rye, 1979; Valley, 2001; Seal, 2006; Marini et al., 2011).

Similar to conventional geothermometers based on the partitioning of trace or major elements, the isotope thermometry follows the principle of the temperature dependence of the equilibrium constant (Urey, 1947; Hoefs, 1997; Chacko et al., 2001). For any two minerals in the isotopic equilibrium, the exchange reaction relates the isotopic end-member compositions. For instance, for the exchange reaction of oxygen isotopes (Valley, 2001)

$$\mathrm{A^{18}O + B^{16}O = A^{16}O + B^{18}O}$$

the equilibrium constant of the reaction is proportional to the fractionation factor α

$$K = \left[\frac{(^{18}\mathrm{O}/^{16}\mathrm{O})\mathrm{A}}{(^{18}\mathrm{O}/^{16}\mathrm{O})\mathrm{B}}\right] = \alpha^{18}\mathrm{O}_{(\mathrm{A-B})}$$

$$1000\ \ln\left(-\Delta G^0/RT\right) = 1000\ \ln\ K = 1000\ \ln\alpha_{(\mathrm{A-B})} \cong \Delta^{18}\mathrm{O}_{(\mathrm{A-B})}$$

Thus, the fractionation is mostly dependent on temperature. Because isotopes differ only in their nucleus, the volume change of the exchange reaction is very small, and from the relation $\Delta G = -S\Delta T + V\Delta P$ it follows that there is little or no measurable effect of pressure on the isotope fractionation (Chacko et al., 2001).

The relationship between the isotope fractionation factor α and the temperature T [in K] is expressed by the polynomial equation

$$1000\ \ln\alpha_{\mathrm{A-B}} = a + b\frac{10^3}{T} + c\frac{10^6}{T^2} + d\frac{10^9}{T^3} + eT \qquad (9.5)$$

where A refers to a mineral and B corresponds to another mineral or fluid. Several terms of Equation (9.5) reduce to zero in specific mineral–fluid systems (Appendix IV). Simple polynomials of the first or second order permit the temperature to be inferred directly from isotope compositions of two coexisting minerals.

A plethora of geothermobarometers have been developed based on the direct equilibrium exchange of oxygen isotopes between calcite and silicates or oxides: albite, muscovite, phlogopite, anorthite, apatite, diopside, garnet, gehlenite, forsterite, rutile, quartz, and magnetite. Equilibrium isotope fractionation factors are fitted by linear equations, which, however, cannot be extrapolated below 600 °C (Chacko et al., 2001). Accuracy of the stable isotope thermometers increases with the increasing difference in the isotope ratio between two coexisting minerals. Silicates and particularly quartz are usually the most enriched in $^{18}\mathrm{O}$ due to strong bonds with silica in tetrahedral groups. In contrast, oxides and hydroxides are the most depleted due to weak covalent bonds. Hence, the combination of silicates and oxides yields the most precise temperature estimates, particularly for the quartz–magnetite, quartz–rutile, and quartz–hematite pairs (Figure 9.1). Corundum is the most extremely depleted in $^{18}\mathrm{O}$ and combined with any genetically associated refractory mineral (rutile, titanite, anorthite, quartz), it might serve as a geothermometer in calc–silicate metamorphic or skarn assemblages.

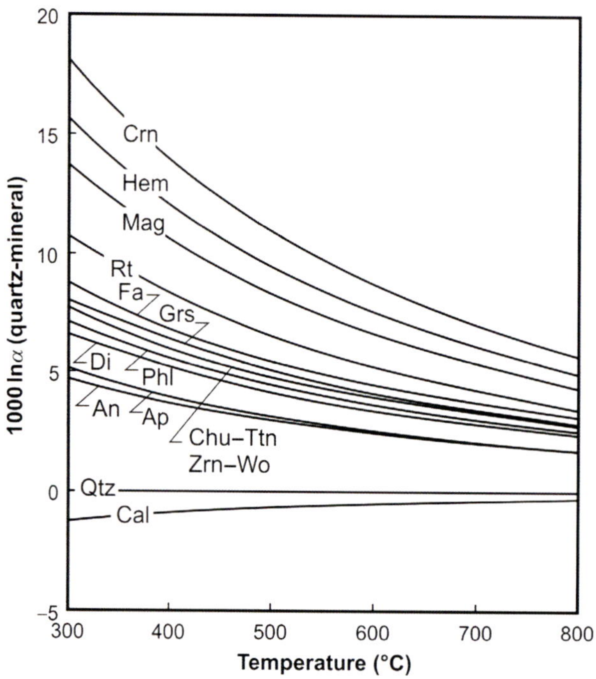

FIGURE 9.1

Oxygen isotope fractionation between various minerals occurring in calc-silicate rocks (Zheng, 1993a,b, 1996). Abbreviations: Crn, corundum; Hem, hematite; Mag, magnetite; Rt, rutile; Fa, fayalite; Grs, grossular; Di, diopside; Phl, phlogopite; An, anorthite; Ap, chlorapatite; Chu, clinohumite; Ttn, titanite; Zrn, zircon; Wo, wollastonite; Qtz, quartz; Cal, calcite.

Garnet–quartz, kyanite–quartz, or rutile–quartz pairs yield geologically reasonable temperatures in eclogitic mineral assemblages (Figure 9.2). However, keep in mind that oxygen isotope ratios in slowly cooling metamorphic or magmatic minerals, particularly in quartz, are affected by the diffusion-controlled retrograde isotope exchange, thus recording a closure temperature for the oxygen diffusion rather than the true formation temperature (Sharp and Kirschner, 1994; Zheng et al., 2002). Moreover, the oxygen isotope ratios in garnet are often overprinted by retrograde fluids (Xiao et al., 2000).

Quartz, calcite, feldspar, and plagioclase are also very sensitive to a secondary hydrothermal alteration that significantly modifies the pristine isotope composition. For instance, Milovský et al. (2012) found out that the oxygen isotope ratios in newly formed quartz, albite, phlogopite, and apatite from the thrusting plane between a calcareous nappe and its silicate metamorphic basement were in line with experimental and theoretical oxygen isotope fractionation factors, but two thirds of the Qtz–Ab, Qtz–Phl, and Qtz–Ap pairs did not yield temperatures consistent with those inferred from primary fluid inclusions. Oxygen isotope ratios of K-feldspars fluctuated randomly, thus indicating an isotopic disequilibrium, an inherited isotopic signature, or an overprint by percolating fluids. Newly formed calcite did not fit into any of the predicted fractionation schemes, showing constant $\delta^{18}O$ values buffered by the overlying sedimentary carbonates.

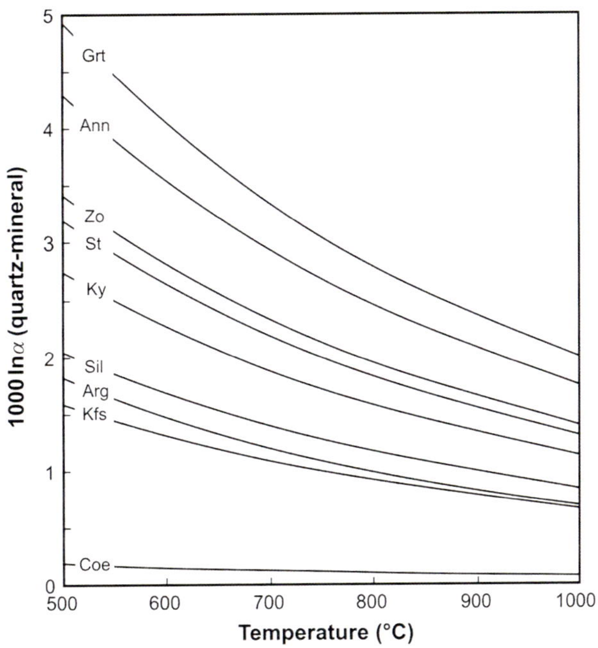

FIGURE 9.2

Oxygen isotope fractionation between various minerals occurring in high- to ultra-high-pressure silicate rocks (Zheng, 1993a–c, 1999). Abbreviations: Grt, garnet; Ann, annite; Zo, zoisite; St, staurolite; Ky, kyanite; Sil, sillimanite; Arg, aragonite; Kfs, K-feldspar; Coe, coesite.

Chacko et al. (2001) summarized available data on D/H fractionation factors and concluded that major unresolved discrepancies exist even for identical mineral–H_2O systems. The inconsistencies pertain not only to the magnitude of fractionation factors, but also to shapes of the fractionation curves. Hence, unlike the situation for oxygen, the lack of consensus on the hydrogen isotope fractionation factors limits their usability in the stable isotope thermometry and in reconstructions of the isotope composition of equilibrium water.

Sulfur isotopes in coexisting minerals can also be employed as geothermometers. As the heavier isotopes are concentrated during equilibrium chemical reactions in the ions with the higher oxidation states, ^{34}S will be enriched in sulfate relative to sulfite, in sulfite relative to elemental sulfur, and in sulfur relative to sulfide (Seal, 2006; Marini et al., 2011). Sulfides and sulfates often crystalize from hydrothermal solutions simultaneously at relatively low temperatures, thus recording a large fractionation between S_2^- and SO_3^{2-} dissolved in aqueous fluids (Figure 9.3). Moreover, sulfides exhibit sufficiently contrasting sulfur fractionation owing to large differences in atomic masses of the cations bound to the sulfur anion. Although the most precise formation temperatures are obtained using a sulfate–sulfide pair, for example, barite and galena, the combination of galena or stibnite with another sulfide (pyrite, pyrrhotite, or sphalerite) also yields satisfactory results. The galena–sphalerite pair is the most promising candidate for the precise isotope thermometry, because both minerals precipitate simultaneously from hydrothermal solutions within a narrow range of temperatures. In contrast, the

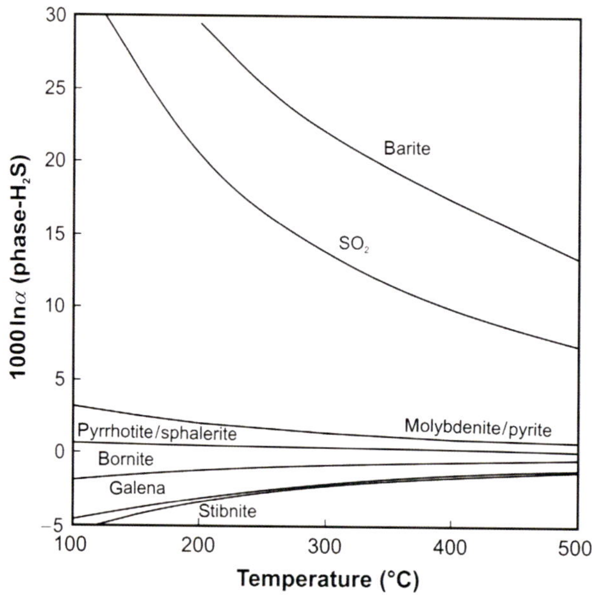

FIGURE 9.3

Experimentally determined sulfur isotope fractionations between selected sulfides, sulfates, H_2S, and SO_2 (Ohmoto and Rye, 1979; Li and Liu, 2006).

combination of pyrite with another sulfide only rarely yields satisfactory results owing to a large temperature range of the pyrite stability in hydrothermal conditions.

Fractionation between the oxidized (SO_2) and reduced sulfur (H_2S) remains large up to magmatic temperatures (Figure 9.3), and the sulfur dioxide can significantly affect the bulk sulfur isotope composition of the source even at small concentrations. Because the H_2S/SO_2 ratio is controlled by the oxygen and sulfur fugacities, as well as by the pH of hydrothermal solutions, a moderate change of each parameter may lead to the significant change in the $\delta^{34}S$ value of the precipitating sulfide. For instance, an increase in the log f_{O2} value by one logarithmic unit can cause a decrease in the $\delta^{34}S$ values of sulfur-bearing minerals by as much as 20‰ at 250 °C (Ohmoto, 1972). Hence, independent f_{S2} and f_{O2} constraints are needed to ascertain whether the sulfide assemblage crystalized in equilibrium, and whether the sulfur isotope composition of the coexisting minerals can be used to infer the bulk $\delta^{34}S$ value and temperature of the fluid source.

An example of such an approach is given in Chapter 6 (Section 6.5.3), where temperature and f_{S2} were constrained by the pyrrhotite–pyrite equilibrium, chlorite and arsenopyrite composition, and fluid inclusions in sphalerite from a carbonate-replacement exoskarn genetically related to a Miocene diorite intrusion (Hurai and Huraiová, 2011). The oxygen fugacity and the H_2S/SO_2 ratio in the fluid were inferred from independently constrained T–f_{S2} and thermodynamic modeling. The thermodynamic calculation showed that despite the crystallization of primary hydrothermal magnetite and anhydrite, the coexisting sulfur-bearing gas was still dominated by hydrogen sulfide, and the H_2S content exceeded that of SO_2 by about ten orders of magnitude. Hence, the $\delta^{34}S$ values of the coexisting sphalerite and

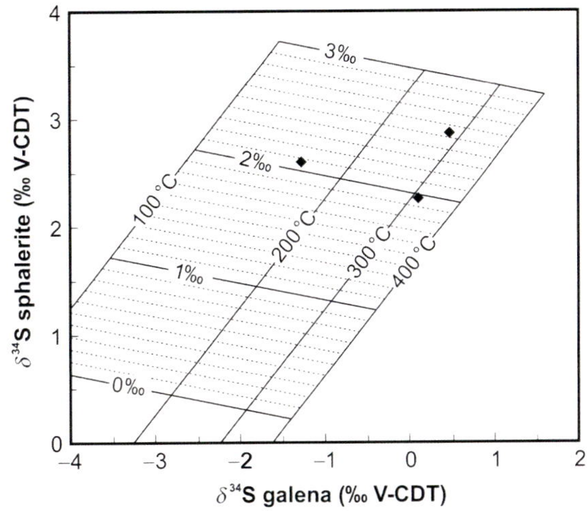

FIGURE 9.4

Sulfur isotope ratios in the coexisting galena and sphalerite projected on temperatures and δ^{34}S values of the equilibrium H_2S gas (Ohmoto and Rye, 1979). Solid diamonds are projection points of three galena–sphalerite pairs from the carbonate-replacement exoskarn near Tisovec (Slovakia), defining the δ^{34}S value of the parental fluid and the possible formation temperatures.

galena recalculated in terms of the mineral–H_2S system could be used to infer the temperature and the δ^{34}S value of the sulfur-bearing gas in equilibrium with the parental hydrothermal fluid (Figure 9.4).

Possibilities to employ carbon isotopes as geothermometers are limited to the coupled graphite–carbonate precipitation from oxidized fluids (Luque et al., 2012) because differences between carbon (and oxygen) isotopic compositions of coexisting carbonates are negligible even at low temperatures (Figure 9.5). In contrast, the carbon isotope fractionation between the fluid-deposited calcite and graphite is relatively large, and it remains at a level of ~2–3‰ up to mantle conditions (Scheele and Hoefs, 1992; Chacko et al., 2001). As originally suggested by Valley and O'Neil (1981), the graphite–calcite system serves as an effective geothermometer in polymetamorphosed carbonate rocks, where it records peak temperatures because the graphite is resistant to resetting during retrograde events (Dunn and Valley, 2007). Theoretical fractionation of carbon isotopes in the calcite–graphite system (Bottinga, 1968, 1969) was inconsistent with field observations and temperatures obtained from independent thermometers. Chacko et al. (1991) resolved problems with sluggishness of the direct calcite–graphite isotopic exchange reaction and proposed the empirical fit to experimental data valid for temperatures between 50 and 900 °C

$$\Delta_{\text{cc-gph}} = 4.007X - 0.301X^2 + 0.0139X^3 \tag{9.6}$$

where $\Delta_{\text{cc-gph}}$ is the difference in δ^{13}C values between the coexisting calcite and graphite, $X = 10^6/T^2$, and T is temperature in Kelvins. Comparison of carbon isotope fractionations between calcite and graphite in natural samples with independent petrological thermometers provided the basis for the following empirical calibration applicable over the range of 400–800 °C (Dunn and Valley, 2007)

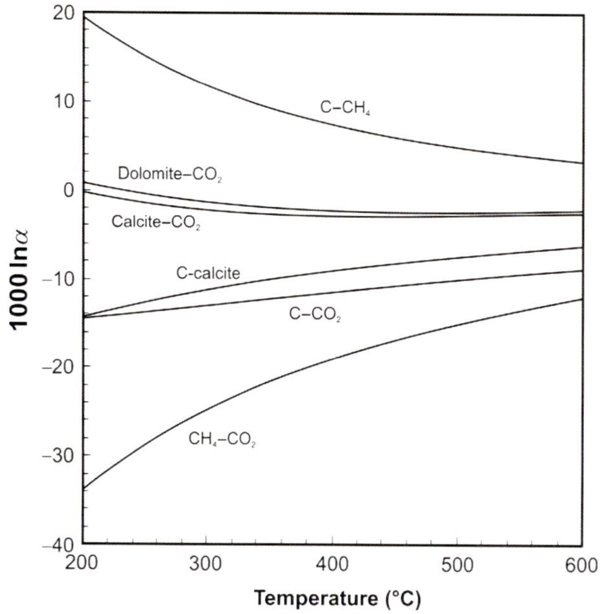

FIGURE 9.5

Experimentally determined carbon isotope fractionations between calcite, dolomite, graphite, CH_4, and CO_2 (Bottinga, 1968, 1969; Sheppard and Schwarcz, 1970).

$$\delta^{13}C_{cc-gph} = 5.81 \times 10^6/T^2 - 2.61 \quad (9.7)$$

The determination of formation temperatures is not primarily targeted by the stable isotope research on carbonates, because their crystallization temperatures can be more precisely constrained by the cation-exchange thermometry (Anovitz and Essene, 1987) or by the K/Na ratio in fluid inclusions (Chapter 8, Section 8.4.1). Once the temperature is known, the task reversal to the isotope thermometry can be solved, namely the isotope compositions of carbon-bearing species and water in the equilibrium parental fluid can be calculated from $^{18}O/^{16}O$ and $^{13}C/^{12}C$ ratios in the carbonate (Figure 9.6).

The identical approach can also be employed for other mineral–fluid systems. Similar to sulfur isotopes, however, the method is hampered by the missing information on the composition of the carbon-bearing species, which may occur in various molecular forms in oxidizing conditions: whereas CO_2 is the only carbon-bearing species dissolved in water as H_2CO_3 at high temperatures, HCO_3^- becomes dominant at temperatures below ~160 °C (Zheng, 1990a). Moreover, because the carbonate solubility increases with the decreasing temperature (Rimstidt, 1997), its precipitation from cooling hydrothermal fluids is triggered by the devolatilization of the carbon-bearing species. The devolatilization-induced changes in f_{O2} and pH values in turn strongly affect the carbon isotope ratios in precipitating carbonates (Ohmoto, 1972). The larger the concentration of carbon-bearing species is in hydrothermal fluids, the proportionally larger is also the variation of the $\delta^{13}C$ value in the deposited carbonate. Hence, the carbon isotope ratio in the hydrothermal carbonate is a function of not only the temperature, but the precipitation mechanism as well. Along with carbon, the oxygen isotope composition of the

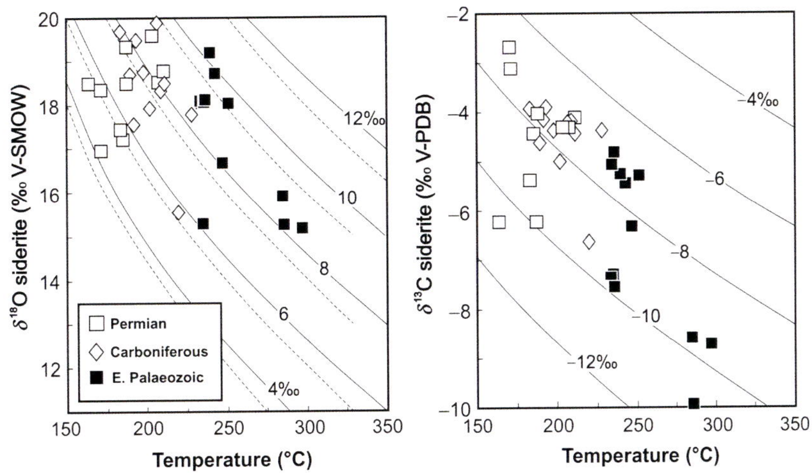

FIGURE 9.6

Oxygen and carbon isotope compositions of siderite-forming fluids in various lithologies of the Variscan basement of the Gemeric tectonic unit of Western Carpathians calculated from the K/Na thermometer (Verma and Santoyo, 1997) and the C–O isotope composition of siderite assuming an equilibrium oxygen isotope fractionation in the siderite–H_2O system (solid curves: Zheng, 1999, dashed curves: Friedman and O'Neil, 1977), and carbon isotope fractionation in the siderite–CO_2 system (Golyshev et al., 1981).

carbonate also changes, because oxygen isotopes are redistributed between the liberated carbon-bearing species (HCO_3^-, H_2CO_3), water, and the deposited carbonate.

The possibility of simultaneous determination of oxygen and carbon isotope ratios in hydrothermal carbonates brings an additional dimension in variation diagrams, where crystallization trends in $\delta^{13}C$ and $\delta^{18}O$ coordinates can be recognized and further quantified in terms of models reproducing the isotope fractionation in open or closed carbonate–H_2O–H_2CO_3 or carbonate–H_2O–HCO_3^- systems (Sections 9.6 and 9.10). As will be shown later, the three-component modeling significantly reduces temperatures and isotope compositions of equilibrium fluids bracketed by the fractionation in binary carbonate–H_2O and carbonate–CO_2/HCO_3^- subsystems.

PROBLEM 9.1 CALCULATING TEMPERATURE FROM STABLE ISOTOPE DATA

Calculate the temperature of crystallization of the quartz–magnetite assemblage in a metamorphic rock. The $\delta^{18}O$ value in quartz corresponds to 14.8‰, whereas that in magnetite is 5‰. Calculate the isotope composition of the equilibrium water.

Solution

Combination of the equation for T-dependent equilibrium fractionation of oxygen between quartz (Q) and water (Bottinga and Javoy, 1973)

$$1000 \ln \alpha_{Q-H_2O} = -3.7 + 4.1 \times 10^6/T^2 \qquad (9.8)$$

Continued

PROBLEM 9.1 CALCULATING TEMPERATURE FROM STABLE ISOTOPE DATA—CONT'D

and that of the magnetite–water system (Bottinga and Javoy, 1973)

$$1000 \ln \alpha_{Mt-H_2O} = -3.7 - 1.47 \times 10^6/T^2 \tag{9.9}$$

yields the equation for the equilibrium isotope fractionation between quartz and magnetite (Mt)

$$\begin{aligned} 1000 \ln \alpha_{Q-Mt} &= 1000(\ln \alpha_{Q-H_2O} - \ln \alpha_{Mt-H_2O}) \\ &= -3.7 + 4.1 \times 10^6/T^2 - (-3.7 - 1.47 \times 10^6/T^2) \\ &= 5{,}570{,}000/T^2 \end{aligned}$$

where T stands for temperature (K). Using the simplified approximation (Equation (9.3)) and substituting stable isotope ratios we get the equation with one unknown

$$14.8 - 5 = 5{,}570{,}000/T^2$$

which can be solved for T (480 °C). Substitution of the calculated temperature and the experimental $\delta^{18}O$ value for quartz or magnetite into Equation (9.8) or (9.9) yields the $\delta^{18}O$ value of the equilibrium water (+11.3‰).

PROBLEM 9.2 STABLE ISOTOPE THERMOMETRY FROM QUADRATIC EQUATION

Calculate the temperature from a second-order polynomial form of Equation (9.5), using $\delta^{18}O$ values in quartz and coexisting magnetite from the previous exercise. The oxygen isotope fractionation between quartz and magnetite is now defined by another equation inferred using the increment method (Zheng, 1991)

$$1000 \ln \alpha_{Q-Mt} = -4.35 + 8.22 \times 10^3/T + 1.22 \times 10^6/T^2$$

Solution

Employ the approximation in Equation (9.3), substitute the $\delta^{18}O$ values and rearrange the equation to a quadratic function

$$0 = ax^2 + bx + c$$

$$14.8 - 5 = -4.35 + 8220/T + 1{,}220{,}000/T^2$$

$$0 = -14.15T^2 + 8220T + 1{,}220{,}000$$

Calculate the discriminant D

$$D = b^2 - 4ac = 8220^2 - 4 \times (-14.15) \times 1{,}220{,}000 = 136{,}620{,}400$$

and the real root of the quadratic function

$$x_{1,2} = \frac{-b \pm \sqrt{D}}{2a}; \; x_1 = -\left(8220 + \sqrt{136{,}620{,}400}\right)/(2 \times -14.15)$$

$$= 703\text{K} = 430^\circ\text{C}$$

9.5 ISOTOPE FRACTIONATION IN OPEN AND CLOSED SYSTEMS

9.5.1 RAYLEIGH FRACTIONATION

Geological literature abounds with modifications of the equation derived by Lord Rayleigh (John William Strutt, 1842–1919) to describe the fractional distillation of liquid air (Rayleigh, 1896, 1902). The Rayleigh equation relates the partitioning of elements between two reservoirs as one reservoir decreases in size. The modified Rayleigh equation also describes the isotope fractionation process in an open system with molecules of two or more isotopic species provided that the material is continuously removed from the system, and the fractionation factor α accompanying the removal process remains fixed. Under these conditions, the evolution of the isotopic composition of the residual material is described as (Kendall and Caldwell, 1998)

$$R_A/R_0 = (X/X_0)^{\alpha-1} \tag{9.10}$$

where R_A is the isotope ratio (e.g., $^{18}O/^{16}O$) in the residuum, R_0 is the initial isotope ratio of reservoir, X_0 is the initial concentration of the removed element (e.g., oxygen), and X is the total concentration of the examined element at each step of the Rayleigh distillation. The equation can be written as

$$R_A = R_0 F^{(\alpha-1)} \tag{9.11}$$

where $F = X/X_0$ stands for the fraction of the element remaining (1—start, 0—end).

The isotope composition of extract also evolves with the advancing fractionation (Allégre, 2008)

$$R_B = \alpha R_0 F^{(\alpha-1)} \tag{9.12}$$

The combination of Equations (9.10), (9.2), and (9.4) yields the δ-notation for the isotope composition of a residuum

$$(\delta_A + 1000)/(\delta_0 + 1000) = F^{(\alpha-1)} \tag{9.13}$$

and after rearrangement

$$\delta_A = (\delta_0 + 1000)F^{(\alpha-1)} - 1000 \tag{9.14}$$

where α is the isotope fractionation factor between the extract B and the reservoir (δ_0).

Similarly, for the extract B at $0 < F < 1$ we get

$$\delta_B = (\delta_0 + 1000)\alpha F^{(\alpha-1)} - 1000 \tag{9.15}$$

In most cases $1 > F > 0.1$ and $0.9X < \alpha < 1.0X$, and the equations can be reduced to common forms frequently cited in literature

$$\delta_A = \delta_0 + 1000\left(F^{(\alpha-1)} - 1\right) \tag{9.16}$$

$$\delta_B = \delta_0 + 1000\left(\alpha F^{(\alpha-1)} - 1\right) \tag{9.17}$$

Equation (9.16) can be further simplified by the application of natural logarithm

$$(\delta_A - \delta_0)/1000 + 1 = F^{(\alpha-1)}$$

$$\ln[(\delta_A - \delta_0)/1000 + 1] = (\alpha - 1)\ln F$$

Using the Taylor series approximation [$\ln(1+X) \approx X$ for $X < 1$] one has

$$(\delta_A - \delta_0)/1000 = (\alpha - 1) \ln F$$

and after rearrangement

$$\delta_A = \delta_0 + 1000(\alpha - 1) \ln F \tag{9.18}$$

Similarly

$$\delta_B = \delta_0 + 1000 \ln \alpha + 1000(\alpha - 1) \ln F$$

and further simplification (Zheng, 1990b) results in

$$\delta_A = \delta_0 + 1000 \ln \alpha \ln F \tag{9.19}$$

and

$$\delta_B = \delta_0 + 1000 \ln \alpha(\ln F + 1) \tag{9.20}$$

The degree of fractionation F can now be calculated from Equations (9.18) or (9.19) as

$$F = \exp\left[\frac{\delta_A - \delta_0}{1000(\alpha - 1)}\right] \quad \text{or} \quad F = \exp\left[\frac{\delta_A - \delta_0}{1000 \ln \alpha}\right] \tag{9.21}$$

Since $1000(\alpha_{B-0} - 1) \cong 1000 \ln \alpha_{B-0} \cong \delta_{B\,ini} - \delta_0$ (Equation (9.3)), Equations (9.18) and (9.21) can be employed to calculate isotope compositions of the reservoir (δ_0), the residuum (δ_A), and the initial extract ($\delta_{B\,ini}$) at $0 < F < 1$

$$F = \exp\left[\frac{\delta_A - \delta_0}{\delta_{B\,ini} - \delta_0}\right] \tag{9.22}$$

$$\delta_A = \ln F(\delta_{B\,ini} - \delta_0) + \delta_0 \tag{9.23}$$

$$\delta_{B\,ini} = \frac{\delta_A + \delta_0(\ln F - 1)}{\ln F} \tag{9.24}$$

A comparison of δ_A, δ_B, and F values obtained from the equations previously derived is given in Table 9.3. Note that errors resulting from the algebraic approximations are within uncertainty limits of stable isotope ratio determinations for $F > 0.1$. However, substantial deviations occur and simplified Rayleigh equations are no longer applicable for $F < 0.1$.

In a strict sense, the term *Rayleigh fractionation* should refer only to open systems, where the removed isotopic species are in the thermodynamic and isotopic equilibrium with those remaining in the system, the reactant reservoir is infinite and well mixed, and it did not re-react with the extract. However, the term is commonly used also to describe finite, closed reservoirs and kinetic fractionations as well, because the situations may be computationally identical (Kendall and Caldwell, 1998). The Rayleigh equation can be employed for the investigation of a plethora of geological processes, where fluid and mineral phases are separated by fractionation, crystallization, evaporation, or condensation.

Table 9.3 Comparison of Isotopic Compositions of the Residuum A and the Extract B Calculated for the Initial Isotope Composition of the Reservoir $\delta_0 = 5‰$, $\alpha_{B\text{-reservoir}} = 1.01$ ($\delta_{B\,ini} = 15.05‰$) Assuming the Rayleigh Fractionation

	Equation								
	(9.14)	(9.16)	(9.18)	(9.19)	(9.23)	(9.15)	(9.17)	(9.18)	(9.20)
F	δ_{A1}	δ_{A2}	δ_{A3}	δ_{A4}	δ_{A5}	δ_{B1}	δ_{B2}	δ_{B3}	δ_{B4}
1.0	5.00	5.00	5.00	5.00	5.00	15.05	15.00	14.95	14.95
0.9	3.94	3.95	3.95	3.95	3.95	13.98	13.94	13.90	13.90
0.8	2.76	2.77	2.77	2.78	2.77	12.79	12.75	12.72	12.73
0.7	1.42	1.44	1.43	1.45	1.43	11.44	11.40	11.38	11.40
0.6	−0.12	−0.10	−0.11	−0.08	−0.11	9.88	9.85	9.84	9.87
0.5	−1.94	−1.91	−1.93	−1.90	−1.93	8.04	8.02	8.02	8.05
0.4	−4.17	−4.12	−4.16	−4.12	−4.16	5.79	5.79	5.79	5.83
0.3	−7.03	−6.97	−7.04	−6.98	−7.04	2.90	2.91	2.91	2.97
0.2	−11.05	−10.97	−11.09	−11.01	−11.09	−1.16	−1.13	−1.14	−1.06
0.1	−17.88	−17.76	−18.03	−17.91	−18.03	−8.06	−7.99	−8.08	−7.96
0.01	−40.23	−40.01	−41.05	−40.82	−41.05	−30.63	−30.46	−31.10	−30.87

9.5.2 BATCH FRACTIONATION

Considering a closed system, products of the phase separation remain in the reservoir until they are extracted as a "batch" after the process was completed. In this case, the total number of atoms of each isotope is fixed, and the fractionation must obey the law of mass conservation. Isotope ratios in the residuum (R_A) and extract (R_B) are then related by the general mass–balance equation (Taylor, 1977)

$$R_A F + R_B(1-F) = R_0 \quad (9.25)$$

Combination with Equation (9.2) yields

$$R_A = R_0/[F + \alpha(1-F)] \quad (9.26)$$

where α is the isotope fractionation factor between extract B and reservoir (δ_0). An extended δ-expression for the residuum A defined by Equations (9.24) and (9.4) reads

$$\delta_A = (\delta_0 + 1000)/[F + \alpha(1-F)] - 1000 \quad (9.27)$$

can be now reduced for $0.9X < \alpha < 1.0X$

$$\delta_A = [F + \alpha(1-F)] + 1000[F + \alpha(1-F)] = \delta_0 + 1000$$

Because $\delta_A[F + \alpha(1-F)] \cong \delta_A$, the previous equation is simplified to

$$\delta_A = \delta_0 + 1000 - 1000[F + \alpha(1-F)]$$

and after re-arrangement

$$\delta_A = \delta_0 - 1000\,(\alpha - 1) + 1000F(\alpha - 1) \quad (9.28)$$

Since $1000(X-1) \cong 1000 \ln X$

$$\delta_A = \delta_0 - (1-F)1000 \ln\alpha \tag{9.29}$$

Similarly, a simplified equation for the extract can be derived:

$$\delta_B = \delta_0 + 1000F(\alpha - 1) \tag{9.30}$$

The degree of fractionation F can be calculated as

$$F = \frac{\delta_A - \delta_0}{1000 \ln\alpha} + 1 \cong \frac{\delta_A - \delta_0}{1000(\alpha-1)} + 1 \cong \frac{\delta_B - \delta_0}{1000(\alpha-1)} \tag{9.31}$$

The combination with Equation (9.3), this yields

$$F = \frac{\delta_A - \delta_0}{\delta_B - \delta_A} + 1 \cong \frac{\delta_B - \delta_0}{\delta_B - \delta_A} \tag{9.32}$$

which can be solved for the isotope composition of extract at $F < 1$

$$\delta_B = \frac{F\delta_A - \delta_0}{F - 1} \tag{9.33}$$

The comparison of δ_A, δ_B, and F values calculated using various equations is given in Table 9.4.

Considerable differences between the two models do not exist for $F > 0.6$, but when the devolatilization nears completion, substantial deviations between the batch and Rayleigh models occur (Figure 9.7). However, the F values for oxygen below 0.6, called the calc–silicate limit (Valley, 1986), are very unlikely in nature, owing to the stoichiometry of decarbonatization reactions, when oxygen is bound in reaction products (silicates) and only a portion of it can be released as a volatile

Table 9.4 Comparison of Isotopic Compositions of the Residuum A and the Extract B Calculated for the Initial Isotope Composition of the Reservoir $\delta_0 = +5‰$, $\alpha_{B\text{-reservoir}} = 1.01$ ($\delta_{B0} = +15.05‰$) Assuming the Batch Fractionation

	Equation				
	(9.27)	(9.28)	(9.29)	(9.30)	(9.33)
F	δ_{A1}	δ_{A2}	δ_{A3}	δ_{B1}	δ_{B2}
1.0	5.00	5.00	5.00	15.00	–
0.9	4.00	4.00	4.00	14.00	13.96
0.8	2.99	3.00	3.01	13.00	12.96
0.7	1.99	2.00	2.01	12.00	11.97
0.6	1.00	1.00	1.02	11.00	10.97
0.5	0.00	0.00	0.02	10.00	9.98
0.4	−0.99	−1.00	−0.97	9.00	8.98
0.3	−1.99	−2.00	−1.97	8.00	7.99
0.2	−2.98	−3.00	−2.96	7.00	6.99
0.1	−3.95	−4.00	−3.96	6.00	6.00
0.0	−4.95	−5.00	−4.95	5.00	5.00

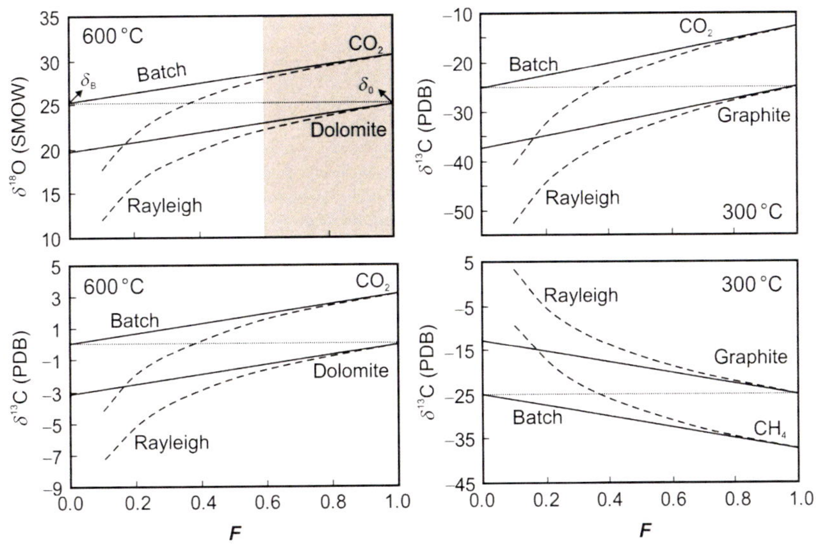

FIGURE 9.7

Isothermal Rayleigh and batch devolatilization during metamorphism of carbonate and graphitic metapelite. Calculated according to experimentally determined fractionation factors for oxygen isotopes in the system dolomite–CO_2 (O'Neil and Epstein, 1966), and those for carbon isotopes in the systems dolomite–CO_2, graphite–CO_2, and graphite–CH_4 (Ohmoto and Goldhaber, 1997). Note reversal trends during the devolatilization of the graphitic metapelite at various oxygen fugacities, giving rise to either isotopically heavy carbon dioxide or isotopically light methane.

oxidized component. In contrast, the Rayleigh-controlled process becomes important for C, H, and S, which are bound only in volatile components, thus experiencing F values down to zero (Baumgartner and Valley, 2001).

9.6 MAGMATIC DEVOLATILIZATION–CRYSTALLIZATION

9.6.1 CARBONATITE–CO_2 SYSTEM

If isotope compositions or total concentrations of two elements in the residuum are defined, only the temperature of fractionation and the isotope composition of the source remain unknown. Hence, an independently constrained temperature permits the stable isotope composition of the source prior to degassing to be deciphered. Similarly, once the isotope compositions of two elements in the source and those in the degassed material are known, the temperature of devolatilization and the degree of fractionation can be defined. In practice, this approach can be applied to calculate the temperature and degree of fractionation of degassed silicate or carbonatite melts. However, owing to a limited isotopic fractionation at magmatic temperatures, precision of the stable isotope geothermometer is rather poor. The situation is further complicated by the fact that devolatilization of natural systems always occurs between two extremes represented by open and closed systems (Figure 9.8). In carbonatite

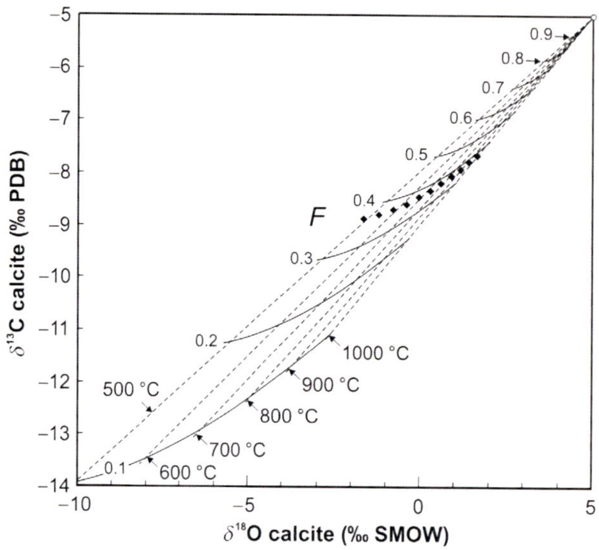

FIGURE 9.8

Composition of residual calcic carbonatite melt formed by the simple Rayleigh and batch devolatilization of mantle carbonatite ($\delta^{18}O = +5‰$, $\delta^{13}C = -5‰$). The model is calculated using fractionation factors for oxygen (Zheng, 1999) and carbon isotopes (Scheele and Hoefs, 1992) in the CO_2–calcite system, assuming no gravitational separation of calcite phenocrysts. Solid diamonds represent the lowermost limit for the batch devolatilization at $F=0$.

magmas, additional complications can arise from possible isotope fractionation between melt and cumulates, crystallization of aragonite at higher pressures, postmagmatic alterations, and presence of additional gas components, particularly water, in the exsolved volatile phase. Some of these obstacles can be partly circumvented by the application of three-component devolatilization/precipitation models discussed hereafter.

At an increased pressure, carbonatite melt dissolves a large amount of CO_2. When the magmatic reservoir is closed and pressure fluctuations do not occur, the dissolved CO_2 may be modified only by the settling of calcite phenocrysts deposited at the bottom of the magmatic reservoir (Pineau et al., 1973). In this case, isotopically lighter calcite is removed from the source, whereas CO_2 in the residuum becomes depleted in lighter isotopes. The evolution of stable isotope composition of the calcite phenocrysts fractionated from a carbonatite reservoir with a mantle isotope signature is shown in Figure 9.9. Note that the earliest phenocrysts are the most depleted compared to the unfractionated reservoir composition. At $F \cong 0.37$, the isotope composition of the fractionated phenocrysts overlaps that of the unfractionated reservoir, and with the advancing fractionation, the deposited calcite phenocrysts are enriched above the initial reservoir value. This situation, however, most likely does not occur in nature, because the Rayleigh fractionation of phenocrysts is no longer possible at 63% crystallization $[100(1-F)]$ of the magmatic reservoir.

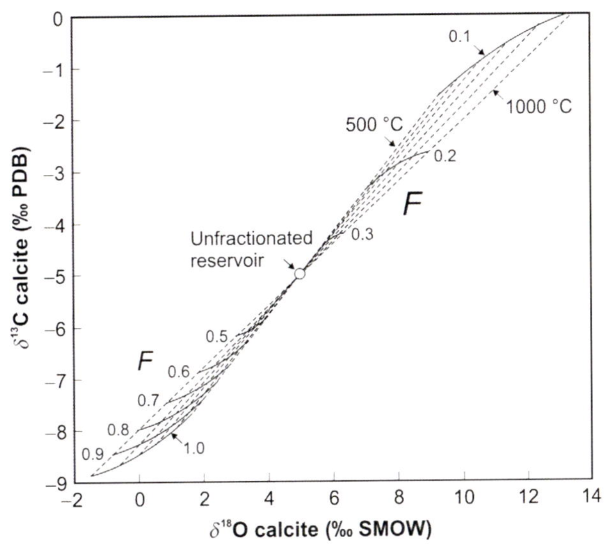

FIGURE 9.9

C–O isotope composition of calcite phenocrysts fractionated from a magmatic reservoir of mantle-derived carbonatite ($\delta^{18}O=+5‰$, $\delta^{13}C=-5‰$). The calculation was made using the Rayleigh model and fractionation factors for oxygen (Zheng, 1999) and carbon isotopes (Scheele and Hoefs, 1992) in the calcite–CO_2 system.

PROBLEM 9.3 CALCULATING TEMPERATURE AND DEVOLATILIZATION DEGREE OF CARBONATITE

Isotope composition of aragonite phenocrysts ($\delta^{18}O=23.6‰$ SMOW, $\delta^{13}C=-4.5‰$ PDB) in a carbonatite xenolith from a basaltic diatreme reflects pre-eruption signature of the parental melt (Hurai et al., 2013). Use the simple Rayleigh model to calculate the temperature and degree of CO_2-devolatilization in crustal magmatic reservoir, assuming that the carbonatite had been derived by the partial melting of marine carbonates with $\delta^{18}O=+30‰$ and $\delta^{13}C=0‰$ introduced in the mantle in subducted oceanic crust.

Solution

Because only CO_2 is degassed, the molecular proportions of oxygen and carbon during devolatilization remain constant. Hence, we can employ Equation (9.21) for the oxygen isotope fractionation in the system CO_2–calcite and that of carbon in the same system together with the approximation in Equation (9.3) to build up the quadratic function with one temperature variable

$$\frac{23.6-30}{1000\ln\alpha^{18/16O}_{CO_2-\text{calcite}}}=\frac{-4.5-0}{1000\ln\alpha^{13/12C}_{CO_2-\text{calcite}}} \tag{9.34}$$

where $1000\ln\alpha^{18/16O}=-3.45+10.01\times10^3/T-1.71\times10^6/T^2$ (Zheng, 1991), and $1000\ln\alpha^{13/12C}=-2.72+9.58\times10^3/T-3.46\times10^6/T^2$ (Scheele and Hoefs, 1992). The quadratic equation obtained by the rearrangement of Equation (9.34) reads

Continued

PROBLEM 9.3 CALCULATING TEMPERATURE AND DEVOLATILIZATION DEGREE OF CARBONATITE—CONT'D

$$0 = 1.883 \times T^2 - 16267 \times T + 14{,}449{,}000$$

Calculate the discriminant D

$$D = b^2 - 4ac = 155{,}785{,}421$$

and real root x of the quadratic function

$$x_{1,2} = \frac{-b \pm \sqrt{D}}{2a};\ x_1 = -\left(-16{,}267 - \sqrt{155{,}785{,}421}\right)\Big/(2 \times 1.883) = 1005\,\text{K} = 732\,°\text{C}$$

Calculate the fractionation factor for oxygen in the CO_2–calcite system at 732 °C (or that of carbon from the equation proposed by Scheele and Hoefs (1992))

$$1000 \ln \alpha_{CO_2-\text{calcite}} = -3.45 + 10.01 \times 10^3/(732+273) - 1.71 \times 10^6/(732+273)^2 = 4.817172$$

$$\alpha_{CO_2-\text{calcite}} = 1.004829$$

Substitute α in Equation (9.21) and solve for F

$$F = \exp[(23.6 - 30)/(1000 \times (1.004829 - 1))] \cong 0.264$$

that means that the carbonatite melt lost 73.6% of CO_2 on the way from the partial melting zone in the upper mantle to the crustal magmatic reservoir, where the aragonite phenocrysts crystalized.

9.6.2 CARBONATITE–CO_2–H_2O SYSTEM

Apart from CO_2, carbonatite magmas may also dissolve a large amount of H_2O (Gittins, 1989). Because CO_2 is the only source of carbon for calcite, the carbon isotopic evolution follows a single-component fractionation trend. In contrast, however, oxygen is contributed by CO_2 and H_2O, thus resulting in a two-component fractionation. Ray and Ramesh (2000, 2006) derived an extended Rayleigh model of the isotope fractionation in the carbonatite–CO_2–H_2O system during the accumulation of calcite phenocrysts.

According to Equations (9.14) and (9.15), the carbon isotope evolution of the source A and the extracted calcite phenocrysts (B) is defined as

$$\delta^{13}C_A = \left(1000 + \delta^{13}C_A^0\right) F_C{}^{(\alpha-1)} - 1000$$

and

$$\delta^{13}C_B = \alpha\left(1000 + \delta^{13}C_A^0\right) F_C{}^{(\alpha-1)} - 1000$$

where α is the fractionation factor of *carbon* between *calcite and CO_2*, F_C is the fraction of remaining carbon in the source, and $\delta^{13}C_A^0$ is the initial carbon isotope ratio in the source.

The two-component oxygen isotope evolution of calcite phenocrysts is defined as

$$\delta^{18}O_B = 1000\left(\frac{\alpha_{cal-CO_2}}{a-b/F_O}-1\right)+\delta^{18}O_A\left(\frac{\alpha_{cal-CO_2}}{a-b/F_O}\right) \tag{9.35}$$

where α is the fractionation factor of *oxygen* between *calcite and* CO_2, F_O is the fraction of remaining oxygen, and a and b are parameters as defined below. The $\delta^{18}O_A$ value of the source can be derived from the equation that describes stable isotope evolution of a multicomponent reservoir

$$\delta_A-\delta_0 \approx 1000\left[\left(\frac{\alpha_{cal-CO_2}}{a}\right)\ln\left(\frac{aF_i-b}{a-b}\right)-\ln F_i\right] \tag{9.36}$$

where δ_0 and δ_A denote initial and final isotope compositions of the source, and F_i is the fraction of remaining atoms (C/O) left in the source. Note that this equation is in error in Ray and Ramesh (2000), where the $-\ln F_i$ term was not multiplied by 1000.

If the source is dominated by CO_2, complete exhausting of H_2O due to the calcite deposition occurs according to the reaction $Ca^{2+}+CO_2+H_2O \rightarrow CaCO_3+2H^+$ at

$$F_O = 1-\frac{3r_{H_2O-CO_2}}{2+r_{H_2O-CO_2}} \tag{9.37}$$

where $r_{H_2O-CO_2}$ is the molar ratio of H_2O to CO_2 in the fluid. Additional parameters of Equation (9.36) are as follows

$$a = 1/3(2+\alpha_{H_2O-CO_2});\quad b = 2/3(\alpha_{H_2O-CO_2}-1)\left(\frac{1-r_{H_2O-CO_2}}{2+r_{H_2O-CO_2}}\right) \tag{9.38}$$

where α is the fractionation factor of oxygen between *H_2O and CO_2*. To calculate carbon and oxygen isotope ratios in the calcite melt or phenocrysts, F values for both elements are related through the following equation

$$F_C = \left(\frac{1-r_{H_2O-CO_2}}{3}\right)+F_O\left(\frac{2+r_{H_2O-CO_2}}{3}\right) \tag{9.39}$$

The $r_{H_2O-CO_2}$ and X_{CO_2} values are related by the equation

$$r_{H_2O-CO_2} = 1/X_{CO_2}-1 \tag{9.40}$$

Isotope composition of calcite phenocrysts can be found by solving Equations (9.35) and (9.36), respectively, as a function of the remaining carbon/oxygen in the source. Several examples of the fractionation trends produced by the crystal deposition at various X_{CO_2} are given in Figure 9.10.

The model proposed by Ray and Ramesh (2000) represents an extension of that invented by Pineau et al. (1973). They tried to explain the trend of positively correlated $\delta^{13}C$ and $\delta^{18}O$ values observed in fresh unaltered carbonatites in terms of the fractional crystallization of calcite phenocrysts from volatile carbonatite melt, leaving behind progressively enriched CO_2 in the source. Owing to the high solubility of CO_2 in the carbonatite melt, this may happen in closed reservoirs, when the CO_2 degassing is negligible and the crystal fractionation is the dominant process. A reversal task, that means the reconstruction of isotope composition of the carbonatite melt remaining in the source after a Rayleigh devolatilization of the mixed-volatile CO_2–H_2O fluid in an open magmatic system can also be solved using

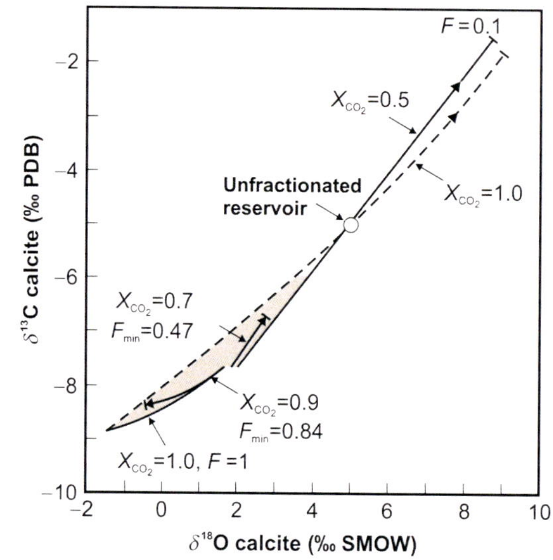

FIGURE 9.10

Evolution trends of calcite phenocrysts fractionated from a mantle-derived reservoir of carbonatite melt saturated with CO_2–H_2O fluid on cooling from 1000 to 500 °C up to 90% of crystallization. The trends were calculated using the extended Rayleigh model (Ray and Ramesh, 2000), isotope fractionation factors from Zheng (1999)—oxygen isotope fractionation in the system CO_2–calcite; Scheele and Hoefs (1992)—carbon isotope fractionation in the system CO_2–calcite; and Richet et al. (1977)—oxygen isotope fractionation in the system CO_2–H_2O. Note variable trends reflected by the δ-values of calcite deposited from the CO_2–H_2O-saturated carbonatite melt. F_{min} denotes the maximum degree of fractionation, at which all H_2O in the source was consumed by the deposited calcite assuming the reaction $Ca^{2+} + CO_2 + H_2O \rightarrow CaCO_3 + 2H^+$. Once the water is consumed, isotope ratios in the deposited calcite phenocrysts became controlled by the isotopic exchange between dissolved CO_2 and the calcite. If all CO_2 is exhausted, the isotope composition of remaining phenocrysts becomes identical with that of the residual carbonatite melt in the magmatic reservoir. However, with up to 60% fractionation, the isotope composition of the deposited calcite projects within a relatively small, shaded area, regardless of the water content in the CO_2-dominated fluid.

this model by setting the $CO_2 + H_2O$ fluid as the extract B, and the calcic carbonatite as the residuum. In this case, the reversal fractionation factors $\alpha_{CO_2-\text{calcite}}$ and $\alpha_{CO_2-H_2O}$ must be applied.

9.7 HYDROTHERMAL DEVOLATILIZATION–PRECIPITATION

9.7.1 RAYLEIGH PRECIPITATION OF CARBONATE FROM H_2O–CO_2 FLUID

At hydrothermal conditions, solubility of carbonates increases with the decreasing temperature (Rimstidt, 1997). Hence, the carbonate precipitation can also be triggered by a gradually increasing temperature of the fluid percolating through cracks. This mechanism operates during prograde metamorphism accompanied by the formation of syn-tectonic tension fractures at specific stress conditions limited by the building of a high overburden pressure (vertical load due to thrusting) and a strong horizontal stress triggered by the compression (Hurai et al., 2008a). The carbonate precipitation in such structures is identical with the Rayleigh-type fractionation of carbonate phenocrysts from volatile carbonatite melt (Ray and Ramesh, 2000) discussed in the preceding chapter, except for the fluid composition, which is now dominated by H_2O, and temperatures are considerably lower.

Stable isotope evolution in the precipitated carbonate is defined in Equation (9.35) with the fractionation factor α for the *carbon* isotope fractionation between *calcite and* CO_2, and in Equation (9.36) with the fractionation factor α for the *oxygen* isotope fractionation between *calcite and* H_2O. The a and b parameters of Equation (9.36) are now defined as

$$a = 1 + 2/3(\alpha_{CO_2-H_2O} - 1);\quad b = 2/3(\alpha_{CO_2-H_2O} - 1)\left(\frac{1 - r_{CO_2-H_2O}}{1 + 2r_{CO_2-H_2O}}\right) \tag{9.41}$$

where $r_{CO_2-H_2O}$ is the CO_2/H_2O molar ratio in the fluid ($r_{CO_2-H_2O} = 1/r_{H_2O-CO_2}$), and $\alpha_{CO_2-H_2O}$ is the isotope fractionation factor of oxygen between CO_2 and H_2O. The F values for oxygen and carbon are related by the following equation:

$$F_C = 1/3\left(1 - \frac{1}{r_{CO_2-H_2O}}\right) + 1/3F_O\left(2 + \frac{1}{r_{CO_2-H_2O}}\right) \tag{9.42}$$

whereas $r_{CO_2-H_2O}$ and X_{H_2O} values can be derived from

$$r_{CO_2-H_2O} = 1/X_{H_2O} - 1 \tag{9.43}$$

Fractionation trends produced by the carbonate deposition at fixed X_{CO_2} are shown in Figure 9.11. Note that their slopes change from moderate to steep ones with the carbonate deposition. However, if at a certain point the system is replenished with a new portion of the hydrothermal fluid, the slope of the fractionation trend reverts to a moderate one.

9.7.2 PRECIPITATION–DEVOLATILIZATION MODEL FOR CARBONATES

Open hydrothermal systems occur in extensional fractures, where the carbonate precipitation at decreasing temperature is enabled by the devolatilization of dissolved carbon-bearing components, pH change, boiling, or mixing. At low temperatures, the calculated fractionation trends are complicated by the introduction of the HCO_3^- component. A corrected model of the devolatilization-induced carbonate precipitation proposed by Zheng (1990a) is presented hereafter *in extenso* due to numerous typographic errors in the original paper (Equations 4, 5, 11, 14b, 15b, 16a).

The Rayleigh fractionation of precipitated carbonate and devolatilized phase can be expressed as

$$\delta^f_{fluid} = \delta^0_{fluid} + \ln F_1 1000 \ln \alpha^{CO_2}_{fluid} \tag{9.44}$$

where F_1 is the mole fraction of the element remaining in the degassed hydrothermal fluid, α is an isotope fractionation factor between CO_2 and the source fluid, and δ^0 and δ^f are initial and final isotope ratios of the source fluid expressed in terms of the δ-notation. The Rayleigh precipitation of calcite can be expressed in a similar way

$$\delta^f_{carb} = \delta^0_{carb} + \ln F_2 1000 \ln \alpha^{carb}_{fluid} \tag{9.45}$$

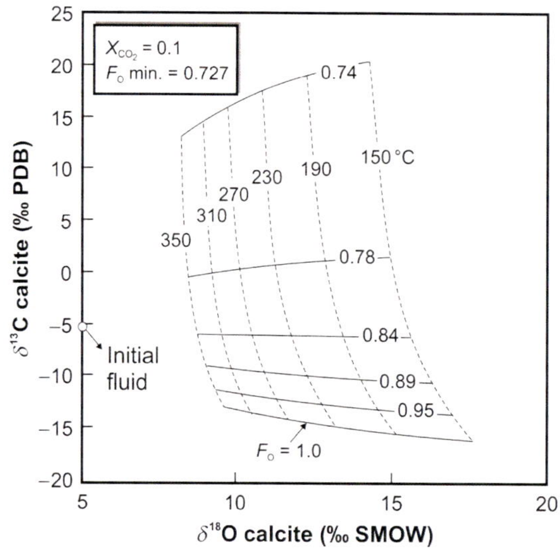

FIGURE 9.11

Isotope composition of calcite precipitated on heating from 150 to 350 °C from aqueous fluid with a mantlelike isotope signature, containing 10 mol.% CO_2. The trends were calculated using the extended Rayleigh model (Ray and Ramesh, 2000), and fractionation factors for oxygen isotopes in the calcite–H_2O (Zheng, 1999) and CO_2–H_2O (Truesdell, 1974) systems, and those for carbon isotopes in the calcite–CO_2 system (Chacko et al., 1991). $F_O$min. is the degree of oxygen fractionation, at which all CO_2 in the H_2O–CO_2 fluid is consumed by the precipitating carbonate.

where F_2 is a mole fraction of the element remaining in the fluid after precipitation of the carbonate and α is an isotope fractionation factor between precipitated carbonate and the source fluid. Because the carbonate precipitation is triggered by the devolatilization, the following approximation can be made

$$\delta^0_{\text{fluid}} = \delta^{\text{f}}_{\text{fluid}}$$

and the isotope composition of the carbonate can be obtained by the combination of Equations (9.44) and (9.45)

$$\delta_{\text{carb}} = \delta^0_{\text{fluid}} + \ln F_1 1000 \ln \alpha^{CO_2}_{\text{fluid}} + (1 + \ln F_2) 1000 \ln \alpha^{\text{carb}}_{\text{fluid}} \tag{9.46}$$

To remove one variable—the ratio of the deposited and the degassed material—the molar fraction of the carbon removed by the CO_2 devolatilization is set as equal to that bound in the deposited carbonate

$$F_1 = 1 - X_C \,(\text{devolatilization})$$

$$F_2 = F_1 - X_C = 1 - 2X_C \,(\text{precipitation})$$

where X_C is the mole fraction of carbon in the degassed carbonic compound. This variable now replaces F_1 and F_2 parameters in the preceding equations, so that for the Rayleigh model one has

$$\delta_{\text{carb}} = \delta^0_{\text{fluid}} + \ln(1 - X_i) 1000 \ln \alpha^{CO_2}_{\text{fluid}} + [1 + \ln(1 - 2X_i)] 1000 \ln \alpha^{\text{carb}}_{\text{fluid}} \tag{9.47}$$

where X_i stands for carbon or oxygen degassed in the carbonic compound. Two limiting cases apply to hydrothermal environments. H_2CO_3 is the main carbonic volatile species at a low pH and increased temperatures. Adopting the approximation $\delta^{13}C_{H_2CO_3} \approx \delta^{13}C_{CO_2}$ (Ohmoto, 1972), carbon and oxygen isotope composition of the carbonate can be now calculated as

$$\delta^{13}C_{carb} = \delta^{13}C^0_{fluid} + [1 + \ln(1 - 2X_C)]1000 \ln\alpha^{carb}_{CO2} \tag{9.48}$$

$$\delta^{18}O_{carb} = \delta^{18}O^0_{fluid} + [1 + \ln(1 - 2X_O)]1000 \ln\alpha^{carb}_{H_2O} + \ln(1 - X_{CO_2})1000 \ln\alpha^{CO_2}_{H_2O}$$

where X_O is the mole fraction of oxygen degassed in CO_2. At a higher pH and a low temperature (<100 °C), HCO_3^- is the dominant carbon-bearing species, and the Rayleigh model can be written as

$$\delta^{13}C_{carb} = \delta^{13}C^0_{fluid} + [1 + \ln(1 - 2X_C)]1000 \ln\alpha^{carb}_{HCO_3} + \ln(1 - X_C)1000 \ln\alpha^{CO_2}_{HCO_3}$$

$$\delta^{18}O_{carb} = \delta^{18}O^0_{fluid} + [1 + \ln(1 - 2X_O)]1000 \ln\alpha^{carb}_{H_2O} + \ln(1 - X_O)1000 \ln\alpha^{CO_2}_{H_2O} \tag{9.49}$$

Owing to the term $\ln(1 - 2X_i)$, equations for the Rayleigh-type fractionation are valid only for $X_i < 0.5$.

Batch fractionation of isotopes between the precipitated carbonate and the devolatilized phase can be expressed as

$$\delta^f_{fluid} = \delta^0_{fluid} - (1 - F_1)1000 \ln\alpha^{CO_2}_{fluid} \tag{9.50}$$

where F_1 is the mole fraction of the element remaining in the degassed hydrothermal fluid, α is the isotope fractionation factor between CO_2 and fluid, and δ^0 and δ^f are initial and final isotope ratios expressed in the standard δ-notation. Batch precipitation of calcite can be expressed in a similar way

$$\delta^f_{fluid} = \delta^0_{fluid} - (1 - F_2)1000 \ln\alpha^{carb}_{fluid} \tag{9.51}$$

Since $\delta^0_{fluid} = \delta^f_{fluid}$ and mole fraction of the carbon removed by CO_2 devolatilization is equal to that exhausted by the carbonate precipitation

$$\delta_{carb} = \delta^0_{fluid} - X_i 1000 \ln\alpha^{CO_2}_{fluid} + (1 - 2X_i)1000 \ln\alpha^{carb}_{fluid} \tag{9.52}$$

For a high-temperature, CO_2-degassing fluid one has

$$\delta^{13}C_{carb} = \delta^{13}C^0_{fluid} + (1 - 2X_C)1000 \ln\alpha^{carb}_{CO_2} \tag{9.53}$$

$$\delta^{18}O_{carb} = \delta^{18}O^0_{fluid} + (1 - 2X_O)1000 \ln\alpha^{carb}_{H_2O} - X_O 1000 \ln\alpha^{CO_2}_{H_2O}$$

whereas the calcite deposited from a low-temperature, HCO_3^--dominant fluid has

$$\delta^{13}C_{carb} = \delta^{13}C^0_{fluid} + (1 - 2X_C)1000 \ln\alpha^{carb}_{HCO_3} - X_C 1000 \ln\alpha^{CO_2}_{HCO_3} \tag{9.54}$$

$$\delta^{18}O_{carb} = \delta^{18}O^0_{fluid} + (1 - 2X_O)1000 \ln\alpha^{carb}_{H_2O} - X_O 1000 \ln\alpha^{CO_2}_{H_2O}$$

As shown in Figure 9.12, considerable differences do not exist between the two models up to 20 mol.% of the carbon degassed. In both cases, fractionation trends generated by the CO_2 and the HCO_3^- degassing have opposite slopes due to the contrasting carbon isotope fractionation in the carbonate–CO_2 and carbonate–HCO_3^- systems (Ohmoto and Rye, 1979; Ohmoto and Goldhaber, 1997).

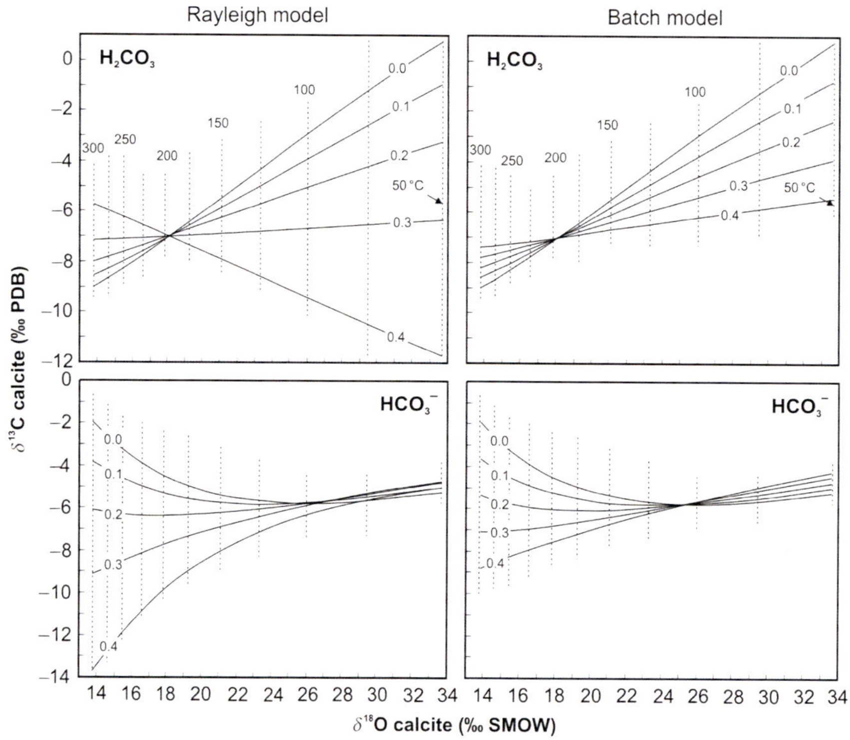

FIGURE 9.12

Fractionation trends in calcite triggered by the Rayleigh and batch devolatilization–precipitation (Zheng, 1993a) of a hydrothermal fluid with $\delta^{13}C = -7‰$ PDB and $\delta^{18}O = +8‰$ SMOW at temperatures 50–300 °C. Numbers on fractionation curves correspond to mole fractions of the carbon degassed. The amount of degassed oxygen was set to zero (water dominates in the fluid). The trends were calculated using oxygen isotope fractionation factors in the calcite–H_2O (Zheng, 1999), and CO_2–H_2O systems (Truesdell, 1974), and those of carbon in the calcite–CO_2 and HCO_3–CO_2 systems (Ohmoto and Rye, 1979).

9.8 FLUID–ROCK INTERACTION

The fluid–rock interaction is an exchange of isotopes and/or elements taking place at increasing temperatures. The interaction involves the dissolution–precipitation, chemical exchange reactions, redox reactions, diffusion, and their combinations.

The fluid–rock ratio in silicate rocks can be calculated from the modified mass-balance equation (Taylor, 1974), assuming an aqueous fluid

$$w\delta^0_{H_2O} + r\delta^0_{rock} = w\delta_{H_2O} + r\delta_{rock} \tag{9.55}$$

where superscript "0" refers to the initial δ-value, w is the atomic percentage of oxygen in the fluid, and r is the atomic percentage of the fractionated oxygen in the rock.

Rearrangement results in the equation expressing the w/r ratio in a closed system

$$\frac{w}{r} = \frac{\delta_{\text{rock}} - \delta^0_{\text{rock}}}{\delta^0_{\text{H}_2\text{O}} - \delta_{\text{H}_2\text{O}}} \tag{9.56}$$

The w/r ratio is called the effective (molar) ratio, differing from the actual ratio by the effectiveness of the exchange reaction. The w and r variables correspond to atomic% of an element in the fluid and the rock, respectively. The mass w/r ratio can be calculated from the extended equation

$$\frac{w}{r} = \left[\frac{\delta_{\text{rock}} - \delta^0_{\text{rock}}}{\delta^0_{\text{H}_2\text{O}} - \delta_{\text{H}_2\text{O}}}\right] \times \left[\frac{c^0_{\text{rock}}}{c^0_{\text{H}_2\text{O}}}\right] \tag{9.57}$$

where c is the concentration of an element in rock or fluid (mass%). The initial δ^0_{rock} value can be obtained by the analysis of fresh, unaltered rock, whereas the final δ_{rock} value corresponds to that of the altered rock. The initial $\delta^0_{\text{H}_2\text{O}}$ can be derived from the D/H ratio in hydrothermal OH-bearing minerals precipitating from meteoric water. The $\delta_{\text{H}_2\text{O}}$ value can be approximated to that of the An_{30} plagioclase, which is volumetrically dominant in most rocks and it is most susceptible to the isotopic exchange with the fluid phase. Knowing the temperature of alteration, the δ^{18}O of the fluid can be derived from the oxygen isotope fractionation factor of the plagioclase–water system.

Equation (9.56) reflects an uninterrupted circulation and equilibration of water in a closed convective system. In fact, a portion or all of the water escapes from the partially or fully open system. In the case of the latter, each infinitesimal portion of water percolates through the rock only once, and the following equation can be written as (Taylor, 1977; Gregory et al., 1989)

$$\frac{w}{r} = \ln\left[1 + \left(\frac{w}{r}\right)_{\text{closed}}\right] \tag{9.58}$$

The actual fluid–rock interaction is located somewhere between the two extremes of the closed and open systems. At a low w/r ratio, oxygen in the surrounding rock dominates over the oxygen in the fluid, thus resulting in the change of the fluid isotope ratio. In an opposite case, the rock isotope ratio becomes modified. In fact, however, both methods yield minimum w/r ratios, because an appreciable volume of water moves through a fracture in the rock without exchanging isotopes.

Silicate minerals are sensitive to the isotopic exchange so that a cryptic alteration can be revealed in apparently fresh rocks, using isotope compositions of two minerals, of which one contains the hydroxyl group. For instance, if granite is altered by hot meteoric water, the δ^{18}O values of biotite and plagioclase remain unchanged at initial stages and $w/r < 0.1$, while the δD value in the biotite rapidly drops down to -100‰ compared to the nonaltered mineral. During advanced alteration accompanied by the progressively increasing w/r ratio, the δD value remains fixed, but δ^{18}O starts to decrease rapidly.

Infiltration of a hot fluid in a cold rock triggers the change in thermodynamic equilibrium and precipitation of minerals. At a higher CO_2 partial pressure, carbonates are ubiquitous in newly formed phases. Oxygen isotope composition of the residual aqueous fluid can be derived from Equation (9.56) (Zheng and Hoefs, 1993a)

$$\delta^{18}\text{O}_{\text{fluid}} = \delta^{18}\text{O}^0_{\text{fluid}} + \frac{r'}{w'}\left(\delta^{18}\text{O}^0_{\text{rock}} - \delta^{18}\text{O}_{\text{rock}}\right) \tag{9.59}$$

and a similar equation is also valid for the carbon isotope composition

$$\delta^{13}C_{\text{fluid}} = \delta^{13}C^0_{\text{fluid}} + \frac{r}{w}\left(\delta^{13}C^0_{\text{rock}} - \delta^{13}C_{\text{rock}}\right) \tag{9.60}$$

Isotope composition of the precipitating carbonate now reads

$$\delta^{18}O_{\text{carb}} = \delta^{18}O^0_{\text{fluid}} + 1000\ln\alpha^{\text{carb}}_{H_2O} + \frac{r'}{w'}\Delta^{18}O \tag{9.61}$$

where $\Delta = \delta^0_{\text{rock}} - \delta_{\text{rock}}$ and a similar equation is also valid for the carbon isotope composition

$$\delta^{13}C_{\text{carb}} = \delta^{13}C^0_{\text{fluid}} + 1000\ln\alpha^{\text{carb}}_{CO_2/HCO_3} + \frac{r}{w}\Delta^{13}C \tag{9.62}$$

Whereas the Δ value can be easily derived from the isotope compositions of altered and unaltered rocks, the atomic ratios of carbon and oxygen are unknown. To overcome this problem, the approximation of $r'/w' = r/w$ is applied. Isotopic trends in carbonates precipitating due to the fluid–rock interaction are displayed in Figure 9.13.

9.9 HYDROTHERMAL ALTERATION

Hydrothermal alteration of carbonates is a special type of mixing process, which does not trigger the precipitation, but only alters the isotope composition of the already precipitated carbonate. Assuming an open system (Equation (9.58)), the following expression can be inferred (Zheng and Hoefs, 1993a)

$$\frac{w}{r} = \ln\left[\frac{\delta^{18}O^0_{H_2O} + \left(\Delta^{18}O^{\text{carb}}_{H_2O} - \delta^{18}O^0_{\text{carb}}\right)}{\delta^{18}O^0_{H_2O} - \left(\delta^{18}O_{\text{carb}} - \Delta^{18}O^{\text{carb}}_{H_2O}\right)}\right] \tag{9.63}$$

where $\Delta^{18}O$ is the equilibrium isotope fractionation of oxygen between carbonate and water. A similar equation is valid for carbon, however, owing to the presence of HCO_3^- and H_2CO_3, the term $X_{HCO_3^-}$ expressing the molar ratio of HCO_3^- in the solution must be introduced:

$$\frac{w}{r}X_{HCO_3^-} = \ln\left[\frac{\delta^{13}C^0_{HCO_3^-} + \left(\Delta^{13}C^{\text{carb}}_{HCO_3^-} - \delta^{13}C^0_{\text{carb}}\right)}{\delta^{13}C^0_{HCO_3^-} - \left(\delta^{13}C_{\text{carb}} - \Delta^{13}C^{\text{carb}}_{HCO_3^-}\right)}\right] \tag{9.64}$$

where $\Delta^{13}C$ is the equilibrium carbon isotope fractionation between carbonate and HCO_3^-. These equations describe the behavior of an open system (Figure 9.14), where the isotopic composition of the altered carbonate can be expressed as

$$\delta^{18}O_{\text{carb}} = \left(\delta^{18}O^0_{H_2O} + \Delta^{18}O^{\text{carb}}_{H_2O}\right) - \left(\delta^{18}O^0_{H_2O} + \Delta^{18}O^{\text{carb}}_{H_2O} - \delta^{18}O^0_{\text{carb}}\right)e^{-\frac{w}{r}} \tag{9.65}$$

$$\delta^{13}C_{\text{carb}} = \left(\delta^{13}C^0_{HCO_3^-} + \Delta^{13}C^{\text{carb}}_{HCO_3^-}\right) - \left(\delta^{13}C^0_{HCO_3^-} + \Delta^{13}C^{\text{carb}}_{HCO_3^-} - \delta^{13}C^0_{\text{carb}}\right)e^{-\frac{w}{r}X_{HCO_3^-}} \tag{9.66}$$

Isotopic exchange in a closed system differs from that in the open system by full and continuous equilibrium. This process can be modeled by adapted Equation (9.63)

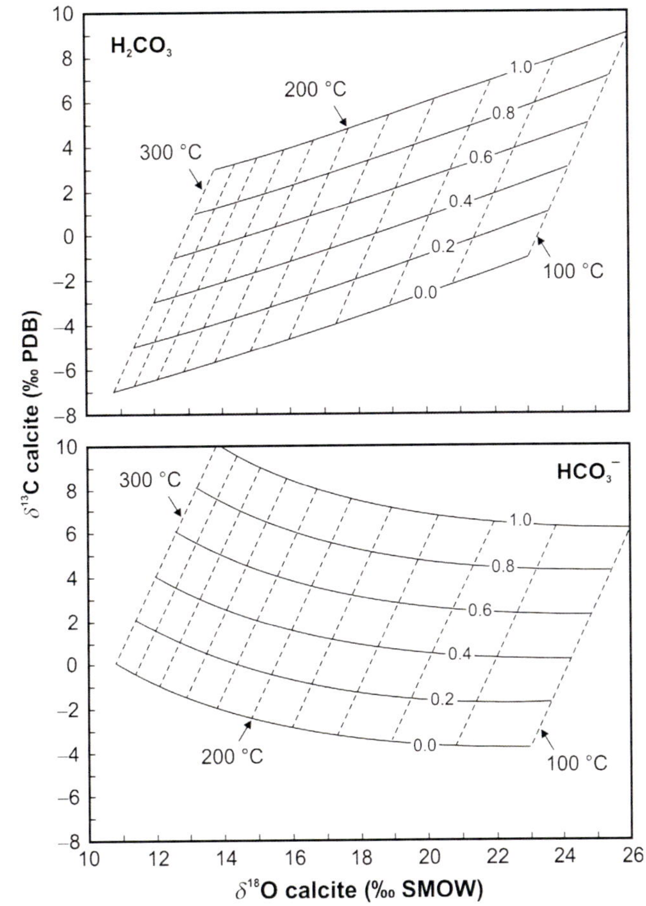

FIGURE 9.13

Covariation of isotopes in carbonates precipitating by the fluid–rock interaction (Zheng, 1993a). The marine carbonate (fresh rock: $\delta^{13}C$ 0% PDB, $\delta^{18}O$ +28% SMOW, altered rock: $\delta^{13}C$ −10% PDB, $\delta^{18}O$ +25% SMOW) interacted with a magmatic fluid with $\delta^{13}C$ −5% PDB and $\delta^{18}O$ +5% SMOW, dominated by H_2CO_3 (top) or HCO_3^- (bottom). Calculation was made using oxygen isotope fractionation factors in the calcite–H_2O system (Zheng, 1999) and those of carbon in the calcite–CO_2 and HCO_3^-–CO_2 systems (Ohmoto and Rye, 1979).

$$\delta^{18}O_{carb} = \frac{\delta^{18}O^{0}_{carb} + \frac{w}{r}\left(\delta^{18}O^{0}_{H_2O} + \Delta^{18}O^{carb}_{H_2O}\right)}{1 + \frac{w}{r}} \tag{9.67}$$

$$\delta^{13}C_{carb} = \frac{\delta^{13}C^{0}_{carb} + \frac{w}{r}X_{HCO_3^-}\left(\delta^{13}C^{0}_{HCO_3^-} + \Delta^{13}C^{carb}_{HCO_3^-}\right)}{1 + \frac{w}{r}X_{HCO_3^-}} \tag{9.68}$$

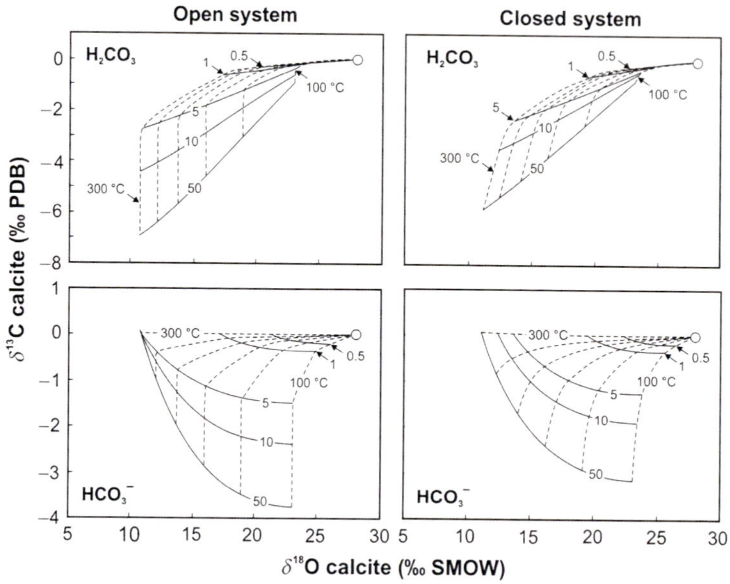

FIGURE 9.14

Alteration of marine limestone ($\delta^{13}C$ 0% PDB, $\delta^{18}O$ +28% SMOW) with magmatic aqueous fluid ($\delta^{13}C$ −5% PDB, $\delta^{18}O$ +5% SMOW) containing 10 mol.% H_2CO_2 or HCO_3^- at fluid–rock ratios between 0.5 and 50, and temperatures between 100 and 300 °C. Calculations were conducted using the model of Zheng (1993a) with oxygen isotope fractionation factors in the calcite–H_2O system (Zheng, 1999) and those of carbon in the calcite–CO_2 and HCO_3^-–CO_2 systems (Ohmoto and Rye, 1979).

9.10 FLUID MIXING AND INFILTRATION

Assuming two phases A and B mixing in a random ratio r, and $r = A/B$, isotope ratios of two elements or isotopes X and Y in the mixture M corresponds to (Zheng and Hoefs, 1993a)

$$(1+r)X_M = rX_A + X_B \tag{9.69}$$

$$(1+r)Y_M = rY_A + Y_B \tag{9.70}$$

If the concentration of an element or isotope Y in both phases is roughly equal and the concentration of the element X is defined by the parameter $p = X_B/X_A$, Equation (9.69) acquires the form

$$(p+r)X_M = rX_A + pX_B \tag{9.71}$$

Equations (9.70) and (9.71) can be combined to eliminate r, thus yielding a general mixing equation expressing isotope ratios of two elements in two phases

$$X_M = \frac{Y_M(pX_B - X_A) - pX_BY_A + X_AY_B}{Y_M(p-1) - pY_A + Y_B} \tag{9.72}$$

Note that the hyperbolic equation (9.72) reduces to linear one, if concentrations of elements are equal in the mixed phases. Hence, mixing waters with different $\delta^{18}O$ or δD values will project along a linear trend, because the oxygen and hydrogen concentrations in two mixing aqueous reservoirs are equal.

Equation (9.72) can be used for all mixing models with a variable concentration of the element X and a fixed concentration of the element Y. These variables can be replaced by isotopic ratios expressed in terms of the δ-notation or as absolute numbers. Commonly, oxygen isotope composition is set as Y, because the concentration of oxygen is roughly similar in hydrothermal fluids dominated by H_2O, as well as in silicate melts and rocks. In contrast, the variable X is replaced by isotopic ratios of elements with changing concentrations in mixing reservoirs, for instance, carbon or sulfur in aqueous solutions, strontium in rocks and melts, and the like.

For example, assuming a constant temperature, hydrothermal fluids A and B with constant oxygen but different carbon concentrations mix according to the equation

$$\begin{aligned}&\delta^{18}O_{AB}\left(\delta^{13}C_A - P\delta^{13}C_B\right) + \delta^{18}O_{AB}\delta^{13}C_{AB}(P-1) + \delta^{13}C_{AB}\left(\delta^{18}O_B - P\delta^{18}O_B\right)\\&+P\delta^{18}O_A\delta^{13}C_B - \delta^{18}O_B\delta^{13}C_A = 0\end{aligned} \tag{9.73}$$

where P denotes the ratio of carbon concentrations in B and A, the subscript "AB" designates the isotopic composition of the element in the mixture.

However, if the precipitation of carbonate occurs, this equation does not tell us anything about the isotope composition of the precipitated mineral. Moreover, the mixing fluids most likely have different temperatures. Assuming that the fluid A is dominated by HCO_3^- (high pH, low T), whereas the fluid B contains dissolved H_2CO_3 (low pH, high T), the carbon isotope composition of precipitating carbonate can be expressed as

$$\delta^{13}C_{carb} = \frac{X_A\left(\delta^{13}C_A + 1000\ln\alpha^{carb}_{HCO_3^-}\right) + P(1-X_A)\left(\delta^{13}C_B + 1000\ln\alpha^{carb}_{CO_2}\right)}{P + X_A(1-P)} \tag{9.74}$$

and that of oxygen

$$\delta^{18}O_{carb} = \delta^{18}O_B + 1000\ln\alpha^{carb}_{H_2O} + X_A\left(\delta^{18}O_A - \delta^{18}O_B\right) \tag{9.75}$$

where X_A is the molar fraction of the fluid A in the AB mixture, and α is the corresponding coefficient of the isotope fractionation. A model of the calcite precipitation triggered by mixing is in Figure 9.15.

9.11 ISOTOPE FRACTIONATION TRENDS IN NATURAL HYDROTHERMAL CARBONATES

Oxygen and carbon isotope ratios in hydrothermal carbonates from open extensional veins in silicate rocks commonly exhibit positively sloped fractionation trends indicating their precipitation from a H_2CO_3-dominated fluid at gradually decreasing temperatures triggered either by the devolatilization or by the fluid–rock interaction or by combination of both mechanisms. Alternatively, a similar trend can also be produced by the influx of meteoric water in continental crust or that of seawater in submarine hydrothermal systems (Zheng, 1990a; Zheng and Hoefs, 1993a, b). Vectors of the fractionation trends are directed toward higher δ-values, which means that precipitating carbonates become enriched in heavier isotopes with decreasing temperatures. In the case of the fluid–rock interaction precipitation

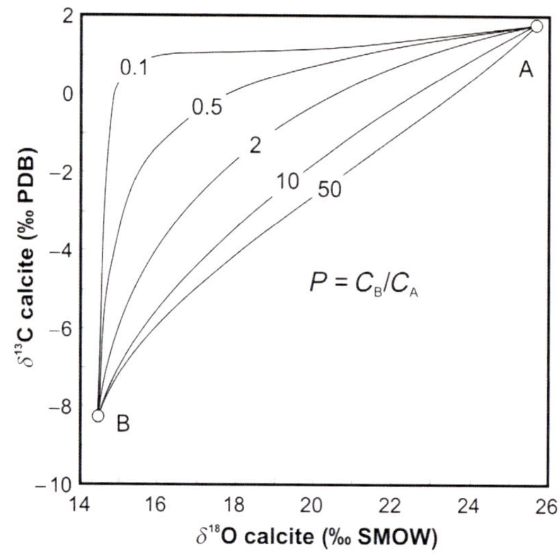

FIGURE 9.15

Carbon and oxygen isotope composition of calcite precipitated by mixing of two fluids (A: marine water 50 °C warm, with carbon dissolved as HCO_3^-, $\delta^{13}C = 0‰$, $\delta^{18}O = 0‰$, B: magmatic water 250 °C hot, with carbon dissolved as H_2CO_3, $\delta^{13}C = -7‰$, $\delta^{18}O = +7‰$). Trends were calculated using the model of Zheng and Hoefs (1993a), oxygen isotope fractionation factors in the calcite–H_2O system (Zheng, 1999), and those of carbon in the calcite–CO_2 and HCO_3^-–CO_2 systems (Ohmoto and Rye, 1979).

mechanism, the open system can be recognized according to a shallower slope of the fractionation trend compared to the steeper slope typical of the closed system (Zheng and Hoefs, 1993a). At low temperatures, the decrease in $\delta^{13}C$ values accompanied by the increasing $\delta^{18}O$ values can be observed at decreasing temperatures.

Opposite fractionation vectors, which is the depletion in both $\delta^{13}C$ and $\delta^{18}O$ values, are diagnostic of rising temperatures and the absence of mixing in closed hydrothermal systems. The closed systems prevent degassing the carbonic species. Owing to the decreasing carbonate solubility, the carbonate precipitation is triggered by increasing temperatures.

An instructive example of a closed hydrothermal system developed in siliciclastic rocks and black shales comes from the Rudňany ore field of the Western Carpathians, where siderite veins, up to several km long and up to 40 m thick, are hosted in a Carboniferous low-to-medium grade metamorphic basement and its anchimetamorphic Permian sedimentary cover (Hurai et al., 2002a). Stable isotope profiling across siderite crystals from cavities revealed a decrease in their $\delta^{18}O$ values. The $\delta^{13}C$ values remained almost constant, thereby indicating the absence of devolatilization. Stable isotope compositions of earlier massive siderites project as a continuation of the crystal fractionation trend to higher $\delta^{18}O$ values, that means to lower temperatures. It has been concluded that the precipitation of siderite occurred at increasing temperatures in a closed hydrothermal system where neither mixing nor devolatilization occurred. For this case, the devolatilization-absent Rayleigh precipitation model of Zheng (1990a) can be used to calculate the stable isotope composition of the parental fluid at temperatures constrained by fluid inclusions (Figure 9.16). According to this model, the calculated $\delta^{18}O$ value of the siderite-precipitating fluid was constant (7.7‰), whereas the $\delta^{13}C$ values varied in a narrow range

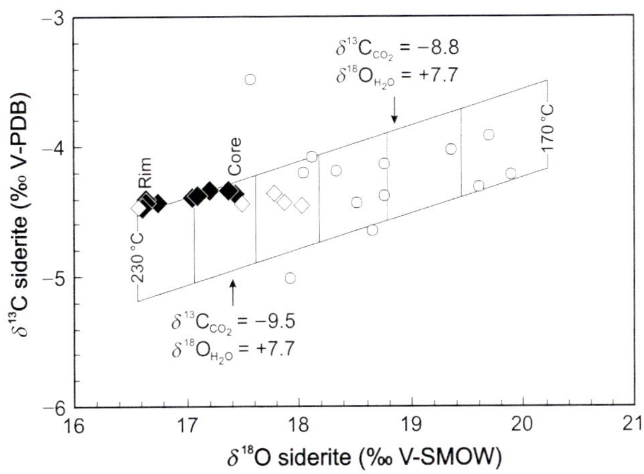

FIGURE 9.16

Stable isotope composition of siderite from the Rudňany ore field (the Gemeric tectonic unit of Western Carpathians, Slovakia) and the calculated C–O isotope composition of the siderite-forming fluid. The model used is a devolatilization-absent Rayleigh precipitation (Zheng, 1990a) from aqueous fluid with H_2CO_3 as the dominant dissolved carbonic species, with isotope fractionation factors from Rosenbaum and Sheppard (1986), Carothers et al. (1988), and Truesdell (1974). Temperatures were constrained by fluid inclusions (homogenization temperatures and K/Na ratios in leachates). Solid diamonds correspond to single siderite crystal from the 6th horizon, open diamonds are other siderite crystals from the 6th and 22nd horizons, and open circles represent massive siderites from four siderite veins several km long and up to 40 m thick (Hurai et al., 2002a).

between −8.8‰ and −9.5‰. The calculated limits for the $\delta^{18}O$ and $\delta^{13}C$ values are narrower than those estimated from the oxygen isotope fractionation in the siderite–H_2O system (6–10‰) and the carbon isotope fractionation in the siderite–CO_2 system (from −7‰ to −9.5‰, compare with Figure 9.6). The almost horizontal fractionation trend indicates that the $\delta^{13}C$ value has not been substantially modified by the precipitating siderite (Figure 9.11), hence, the fluid reservoir must have been continually replenished.

Closed hydrothermal systems hosted in carbonatic rocks display a different pattern from that of silicic rocks, because apart from oxygen, carbon isotopes are also exchanged with the surrounding rocks at low fluid–rock ratios. There may also be continuous replenishment of the system with CO_2 liberated by the dissolution of the host carbonates. For such systems, the alteration model satisfactorily reproduces fractionation trends observed in the deposited hydrothermal carbonates.

An example is presented from calcite veinlets in calcareous sandstones of the Eocene–Oligocene flysch of a back-arc basin, and Upper Cretaceous–Oligocene flysch of the accretion wedge of the Outer Carpathians (Figure 9.17). Carbonates occur here as calcite and dolomite cement of sandstones, stratiform deposits of Mn–carbonate ores (mainly Mn–calcite and rhodochrosite) in Oligocene shales of the back-arc basin (Rojkovič et al., 2008), and infillings of small-scale joints and tension gashes originated during syn-sedimentary folding, thrusting, and regional rotation of the accretion wedge (Świerczewska et al., 2000). Maximum temperatures can be set to the range of 100–250 °C, indicated by fluid inclusions in calcite and quartz from veinlets, and the vitrinite reflectance data from surrounding rocks (Hurai et al., 2002b).

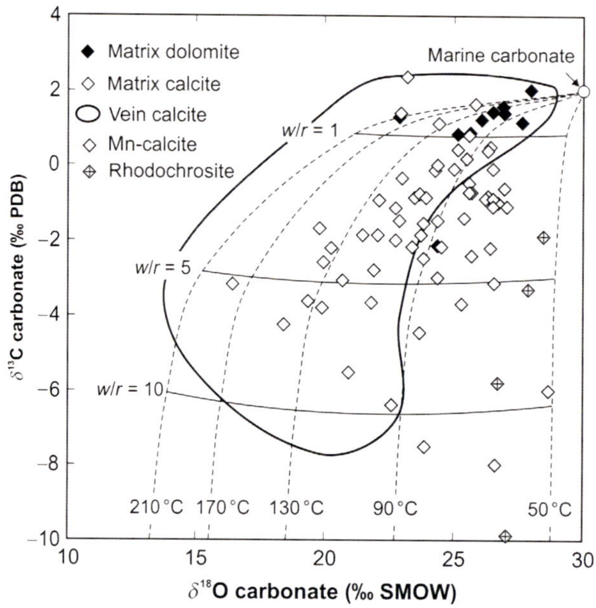

FIGURE 9.17

Stable isotope composition of carbonates from the Tertiary accretion wedge and fore-arc basin of Western Carpathians compiled from Rojkovič et al. (2008) and unpublished reports. Fractionation curves calculated using a closed-system alteration model (Zheng, 1993a) with aqueous fluid containing 5 mol.% of dissolved HCO_3^- $\delta^{13}C = -25‰$, $\delta^{18}O = +3‰$ and isotope fractionation factors from Zheng (1999) and Ohmoto and Rye (1979).

Stable isotope ratios in the stratiform carbonates and the carbonate cements are reminiscent of an isotopic exchange with marine carbonate at various *w*/*r* ratios and temperatures. The stable isotope record of the stratiform rhodochrosite is consistent with an early diagenetic origin (50–100 °C) and the influence of fluids with $\delta^{13}C < -15‰$ and $\delta^{18}O \sim +3‰$. The variable $\delta^{13}C$ values indicate widely fluctuating *w*/*r* ratios. The *w*/*r* ratios increase with the decreasing $\delta^{13}C$ value of the percolating fluid, when the dissolved carbon may have been liberated by the decomposition of organic acids, thus potentially approaching a limiting value of $-25 \pm 3‰$ PDB typical of the organogenic carbon.

The general isotope trend of the carbonate cement is also reminiscent of a hydrothermally altered marine carbonate, but the simple alteration model can be ruled out owing to differences between the dolomitic and calcitic matrix. Stable isotope ratios of the dolomite cement are limited to $\delta^{13}C$ values from $-2‰$ to $+2‰$ and the $\delta^{18}O$ values from $+23‰$ to $+28‰$, thus approaching the isotope composition of unmodified marine carbonates. This indicates very low *w*/*r* ratios and precipitation of the dolomite in a closed system. In contrast, the calcite cement is characteristic of wider $\delta^{13}C$ values from $+2.5‰$ to $-7.5‰$, and $\delta^{18}O$ values from $+16‰$ to $+26‰$, thus indicating higher temperatures and *w*/*r* ratios diagnostic of a transition to an open hydrothermal system. The calcitic cement can be genetically related to the calcite precipitating within small-scale fractures and tension gashes, because both groups overlap in the $\delta^{18}O$–$\delta^{13}C$ diagram. In this case, the precipitation triggered by the fluid–rock interaction would be the appropriate model to infer the isotope composition of the parental fluid; nevertheless, the newly calcite from veinlets maintains the trend diagnostic of a hydrothermally altered marine carbonate, though shifted to even lower $\delta^{13}C$ and $\delta^{18}O$ values. This would indicate an increased temperature of convectively (adiabatically) heated fluids in fractures contrasting with a conductive heat transfer along grain boundaries, where the calcite cement crystalized at lower temperatures. The crystallization of late-stage quartz with inclusions of thermogenic methane (Hurai et al., 2002b) and P–T data from immiscible methane-aqueous fluids (see Figure 5.26 in Chapter 5) unequivocally indicate the influx of hot, upward-moving fluids into colder sediments within a collapsing accretion wedge, thus supporting also the model of the carbonate precipitation.

The fluid isotope composition and w/r ratios can roughly be estimated from the extent of stable isotope fractionation. The influence of deeply circulating meteoric waters with negative $\delta^{18}O$ values can be excluded; otherwise, the near-vertical trend of the carbonates most enriched in ^{18}O could not be reproduced. Hence, the seawater enriched in ^{18}O by the isotopic exchange with silicate rocks (sandstones, shales) is the best candidate for the parental hydrothermal fluid circulating in the fractures. At temperatures between 100 and 250 °C, the stable existence of H_2CO_3 and/or HCO_3^- is possible, and the compounds may be identified according the fractionation trend of samples most depleted in carbon: the positive trend indicates the predominance of H_2CO_3, whereas the negatively correlated or $\delta^{18}O$-axis-parallel trends reflect the predominance of HCO_3^-. In our set (Figure 9.17), the samples most depleted in ^{13}C show a negative correlation, thus being consistent with HCO_3^- as the main dissolved carbon species. A similar trend, however, can also be obtained with H_2CO_3 at rising temperatures and progressively decreasing w/r ratios.

Hydrothermal systems with low w/r ratios ($w/r = 1$–10) could be designated as rock-buffered, whereas those with the increased w/r ratio as fluid-buffered (Gray et al., 1991; Richards et al., 2002). Hydrothermal veins in Paleozoic rocks of the Bohemian Massif of Czech Republic yield many examples of transitions from closed, rock-buffered systems to open, fluid-buffered hydrothermal systems.

Time-relative and structural relationships suggest that the veins were related either to Variscan or post-Variscan events. The extent of T_h values of aqueous inclusions in the carbonates corresponded to 70–146 °C. The homogenization temperatures were within error limits in a good agreement with the maturity of organic matter in surrounding rocks. Aqueous monophase inclusions in Devonian calcites indicated a trapping temperature lower than 70 °C, and thus the temperature of 50 °C has been used as the lowermost limit. The calcite $-HCO_3^-$ system was selected owing to the low formation temperature indicated by fluid inclusions (Figure 9.18).

The isotopic data on carbonates have been considered as two separate subsets representing the Devonian and the Silurian carbonate host-rocks and veins, inasmuch as the geological context for both subsets and lithologies was different. The best match to the observed trend has been achieved for the Devonian limestones and their veins assuming an open isotopic system. The model is consistent with dominantly brittle deformation of host rocks. In contrast, a closed isotopic system was attributed to significantly ductilely deformed Silurian host rocks. The fluid isotope composition of the Devonian hydrothermal system corresponded to $\delta^{13}C = 0‰$ V-PDB and $\delta^{18}O = 0‰$ V-SMOW, and that of the Silurian fluid system to $\delta^{13}C = -10‰$ V-PDB and $\delta^{18}O = 0‰$ V-SMOW. The concentration of 10 mol.% HCO_3^- in the fluid most satisfactorily reproduced the observed trends. The three-component isotope model overlapped the field data at temperatures between 50 and 140 °C within the subset of Devonian strata and at 70–140 °C within the subset of Silurian strata. The inferred water/rock (w/r) mass ratios varied from up to 5 for the Silurian sediments to as much as 100–1000 for the Devonian strata.

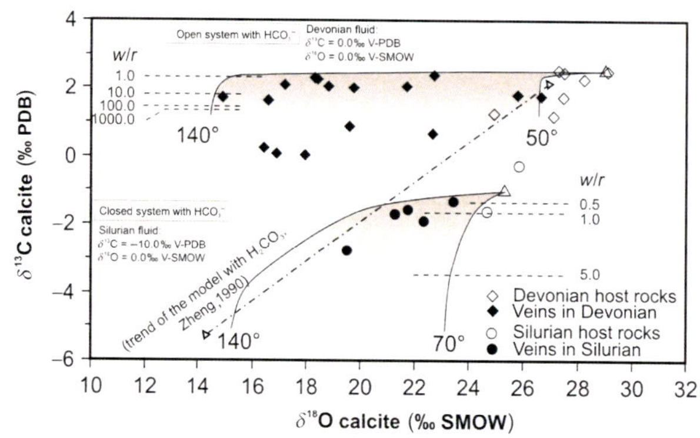

FIGURE 9.18

The alteration model (Zheng and Hoefs, 1993b) with isotope fractionation factors for oxygen in the system calcite–H_2O (Zheng, 1999) and carbon in the system calcite–HCO_3^- (Ohmoto and Rye, 1979) applied to Variscan fluids percolating in the Prague basin. Conditions and input data are as follows: unaltered Devonian rock: $\delta^{13}C = +2.5‰$ V-PDB, $\delta^{18}O = +29.0‰$ V-SMOW, fluid isotope composition: $\delta^{13}C = 0‰$ V-PDB, $\delta^{18}O = 0‰$ V-SMOW, mole fraction of gas species (HCO_3^-) in the fluid: 0.1, temperatures of alteration: 50–140 °C. Unaltered Silurian rock: $\delta^{13}C = -1.0‰$ V-PDB, $\delta^{18}O = +25.3‰$ V-SMOW, fluid isotope composition: $\delta^{13}C = -10.0‰$ V-PDB, $\delta^{18}O = 0‰$ V-SMOW, mole fraction of gas species (HCO_3^-) in the fluid: 0.1, temperature of alteration: 70–140 °C.

The estimated *w/r* ratios were in agreement with the geological context, showing that the Silurian strata with short and thin fractures were less permeable than the Devonian ones containing longer and wider fractures. The isotopic model corroborated HCO_3^- as the dominant dissolved carbonic gas species. The H_2CO_3 component could be excluded because for the same input parameters, the model exhibited a steep positive inclination inconsistent with the observed trend. The obtained fluid inclusion and stable isotope data documented that the late Variscan vein minerals in the Barrandian Palaeozoic rocks crystalized in a closed, rock-buffered fluid system, except for fault-related veins and those from the top of the Devonian lithostratigraphic sequence. Vertical fluid pathways were restricted or entirely missing, and only lateral or intralayer fluid migration over a short distance was possible (Slobodník et al., 2012).

An extremely rock-buffered fluid system has been documented within syntectonic Variscan calcite veins crosscutting Devonian limestones of the Moravian Karst (Slobodník et al., 2006). Isotopic ratios in older white, more deformed, and younger rose calcites were as follows: $\delta^{18}O$ between +17.7‰ and +26.1‰ (white), +14.8‰ and +20.7‰ SMOW (rose), $\delta^{13}C$ values ranged from +0.1‰ to +2.5‰ (white), and from −0.3‰ to +1.6‰ V-PDB (rose). Parent fluids corresponded to the H_2O–NaCl system with salinities between 0.4 and 17.3 mass% NaCl equivalents. Pressure-corrected T_h values suggested crystallization temperatures between 120 and 170 °C, and pressures of 300–880 bar corresponding to a depth of between 2.1 and 3.2 km, assuming the lithostatic load.

The distribution of isotopic data along a zero fractionation line in the plot correlating $\delta^{13}C$ values for the calcites and neghboring rocks (limestones) reflects the identical carbon isotope composition (Figure 9.19), thus suggesting the derivation of the carbon in the vein calcites from the adjacent rocks. A similar behavior was observed also for oxygen in the A- and B-groups of veins, contrasting with a different trend in the C-group vein calcites. The different oxygen isotopic composition indicates

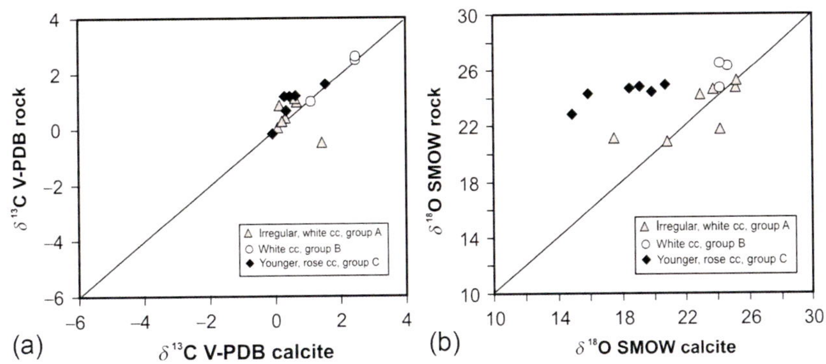

FIGURE 9.19

Stable isotope compositions of calcite veins and adjoining limestone $\delta^{13}C$ values from the Moravian Karst (Slobodník et al., 2006). (a) The $\delta^{13}C$ values for three groups of the Late Variscan calcite veins (A, B, C) are distributed along the zero fractionation line. (b) The $\delta^{18}O$ values are either distributed close to the zero fractionation line or exhibit a horizontal trend. The observed relationships indicate predominantly a rock-buffered, closed system accompanied by a restricted bed-scale fluid migration.

another isotopic reservoir. The observed relationships reflect a very restricted distance of the fluid flow—a bed-scale fluid migration.

Post-Variscan Zn–Pb–Cu dolomite veins occur in the Nízký Jeseník Mountains (NJM) and the Upper Silesian Basin (USB) within Lower and Upper Carboniferous siliciclastic deposits. Homogenization temperatures of fluid inclusions ranged between 66 and 148 °C, with generally higher temperatures in NJM than those in USB. The highest homogenization temperatures (up to 148 °C) were found near major regional faults, whereas the lowest occurred in a distant position or at higher stratigraphic levels. High salinity (16.6–28.4 mass% NaCl equivalents) aqueous solutions corresponded to the H_2O–NaCl–$CaCl_2 \pm MgCl_2$ system.

The $\delta^{18}O$ values of dolomites from NJM varied between 15.5‰ and 22.5‰ V-SMOW and the $\delta^{13}C$ values between −5.6‰ and −3.0‰ V-PDB. A narrow range of the $\delta^{18}O$ values of dolomites occurred in USB (between 17.2‰ and 21.9‰ V-SMOW), whereas the $\delta^{13}C$ values showed a range between −5.2‰ and +2.0‰ V-PDB. Several dolomites, mainly from NJM, plotted along a horizontal line with a $\delta^{13}C$ value of −4.8‰ V-PDB (Figure 9.20). Dolomite samples with enriched $\delta^{13}C$ values form negative trends. The most pronounced negative trend occurs in the USB dolomites.

Isotopic compositions of dolomites and T_h values were used to calculate the isotopic composition of the parent fluid using equilibrium isotopic fractionation equations for two-component systems from Zheng (1999), Ohmoto and Goldhaber (1997), and Ohmoto and Rye (1979). HCO_3^- was assumed to be the dominant dissolved carbon species. The calculated oxygen isotope ratios in the fluid varied between 0.9‰ and 3.0‰ V-SMOW, and those of carbon between −8.1‰ and −5.1‰ V-PDB for the NJM dolomites. The $\delta^{18}O$ values of the USB fluids ranged from between −0.9‰ and +1.9‰ V-SMOW and those of $\delta^{13}C$ between −7.4‰ and −0.2‰ V-PDB.

Isotopic modeling with a three-component system was used to explain the trends observed (Figure 9.21). The Rayleigh devolatilization–precipitation model (Zheng, 1990a) and the alteration model (Zheng and Hoefs, 1993b) have been employed for the range of temperatures between 80

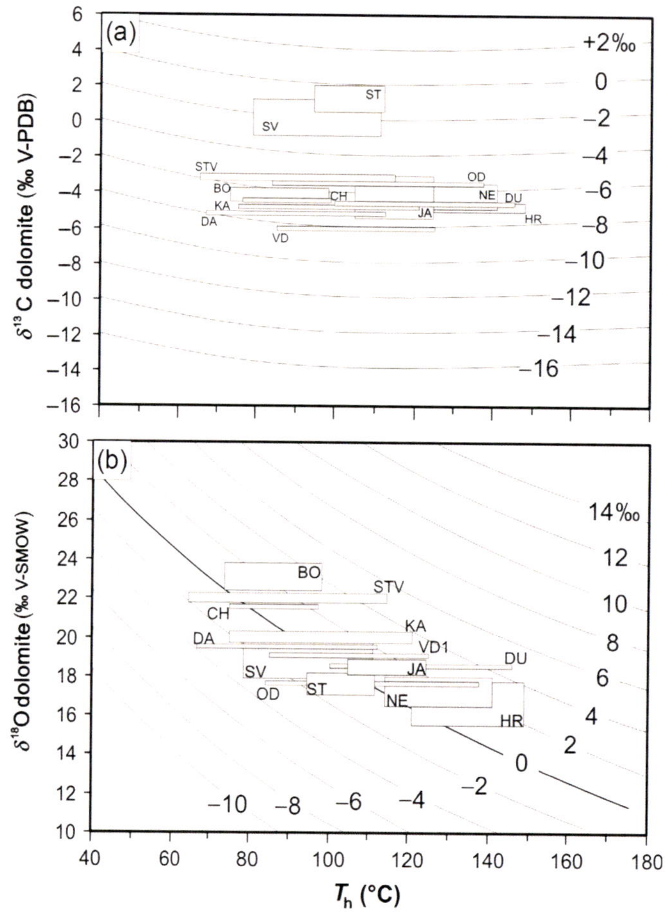

FIGURE 9.20

Isotopic composition of carbon (a) and oxygen (b) of the dolomite-forming fluids in the Moravo–Silesian unit calculated using homogenization temperatures (T_h), isotopic composition of dolomites, and equilibrium fractionation equations for the dolomite–CO_2 and dolomite–H_2O systems (Zheng, 1999; Ohmoto and Goldhaber, 1997; Ohmoto and Rye, 1979). Abbreviations for the localities in NJM: HR, Hrabůvka; NE, Nejdek; DU, Dubová; OD, Odry; JA, Jakubčovice; VD, Vendelín; KA, Kajlovec; STV, Stará Ves; BO, Bohučovice; ST, Sařič; SV, Sviadnov; DA, Darkov; CH, Chlebovice (Kučera et al., 2010).

and 160 °C indicated by fluid inclusions. The best fit for the NJM data with the $\delta^{13}C$ values of around −4.8‰ V-PDB was provided by the Rayleigh fractionation model with the dolomite crystallization in an open system (Zheng, 1990a). The model supports an assumption that HCO_3^- was the dominant dissolved carbon species. H_2CO_3 can be excluded because the corresponding trendline would have a steep positive inclination for the same conditions. The short negative trends in NJM dolomites can be explained by changes in the isotopic composition of the parental fluid: the $\delta^{13}C$ value could vary between −5.5‰ and −8.5‰ V-PDB, and the $\delta^{18}O$ between 0‰ and −1.5‰ V-SMOW.

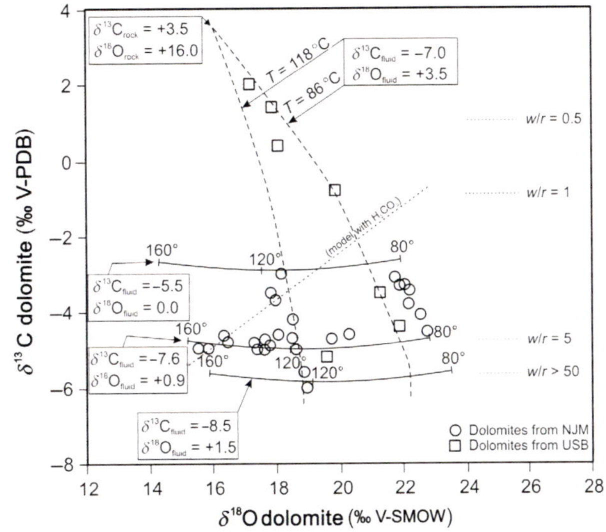

FIGURE 9.21

Stable isotope composition of dolomite-forming fluids from the NJM and the USB. Solid lines represent the Rayleigh devolatilization–precipitation model (Zheng, 1990a) with HCO_3^- as dominant carbon species dissolved in H_2O at temperatures of 80–160 °C and changing fluid isotope compositions specified in boxes. Dotted line represents the same model with H_2CO_3 as the dominant carbon-bearing species. Dashed lines correspond to alteration with dissolved HCO_3^- in an open fluid system (Zheng and Hoefs, 1993a). *w/r* is the water/rock ratio.

Because the fluid chemistry was similar in USB and NJM, the distinct negative trend in the USB dolomites could reflect a different fluid evolution and/or precipitation mechanism. The alteration model (Zheng and Hoefs, 1993b) could explain the observed trend. Datapoints particularly match the ~90 °C isotherm with the fluid isotope composition of carbon and oxygen corresponding to −7.0‰ V-PDB and −3.5‰ SMOW, respectively. Modeled *w/r* ratios varied between ≤0.5 and 5. The samples with *w/r* ratio between ≤0.5 and 1 are interpreted as deposited in rock-buffered conditions and those with *w/r* ratio ≥5 as deposited in fluid-buffered conditions. In the case of the rock-buffered system, isotopic composition of the rock corresponded to $\delta^{13}C = +3.5$‰ V-PDB and $\delta^{18}O = +16.0$‰ V-SMOW.

Both models are compatible, because they are based on open systems with HCO_3^- as the dominant carbon species. Moreover, the models fit well to the measured homogenization temperatures. The $\delta^{13}C$ value of the fluid used in the alteration model corresponded to −7.0‰ V-PDB, being thus compatible with that in the Rayleigh devolatilization–precipitation model ($\delta^{13}C$ between −8.5‰ and −5.5‰ V-PDB). The low $\delta^{13}C$ fluid values, with respect to the given geological context, could indicate the isotopic signature of carbon derived from surrounding rocks mixed with the organic carbon (Irwin et al., 1977) extracted and brought by fluids from sedimentary sequences (note the growing *w/r* along with the decreasing $\delta^{13}C$—Figure 9.21).

The negative trend in the alteration model is interpreted as reflecting the variation in the extent of the rock buffering beneath USB. The isotopic composition of the model rock reservoir ($\delta^{13}C = +3.5$‰

V-PDB and $\delta^{18}O = +16.0‰$ SMOW) could represent an average value of the limestones and siliciclastics underlying USB. Devonian and Lower Carboniferous limestones have $\delta^{13}C$ values between +0.1‰ and +3.0‰ V-PDB and the $\delta^{18}O$ values between +20.9‰ and +27.8‰ V-SMOW (Hladíková et al., 1997; Slobodník et al., 2001). Lower Carboniferous siliciclastic rocks have, however, much lower $\delta^{18}O$ values, between +11.0‰ and +11.7‰ V-SMOW. It can be concluded that post-Variscan hydrothermal dolomite crystalized in stable platform conditions. The isotopic modeling is in accord with crystallization temperatures of between 80 and 160 °C and reveals also variable *w/r* ratios in the open hydrothermal system. However, the stable isotope geochemistry and fluid compositions are largely influenced by specific lithology (limestone, shale, and organic-rich sediments) also in the veins precipitating in fluid-buffered conditions, particularly in the NJM area (Kučera et al., 2010).

9.12 INTERPRETATION OF STABLE ISOTOPES AND FLUID INCLUSIONS IN SIDERITE AND MAGNESITE DEPOSITS OF WESTERN CARPATHIANS

World-class replacement-type magnesite and siderite, as well as vein-type siderite deposits, occur in the low-grade Variscan basement of the Alpine–Carpathian orogenic belt, and the combined fluid inclusion and stable isotope study provided clue to their origins. Siderite veins of the Carpathian basement, with a maximum length of 7 km (Droždiak, Zlatník, Hrubá veins), belong to the world's largest carbonate vein systems. Almost 700 of 1300 known hydrothermal veins (Figure 9.22) have been exploited since Celtic times, and the barite exploitation persisted in small amounts until recently. The largest metasomatic siderite–ankerite deposits of the Western Carpathians (Železník, Dobšiná, Rákoš) were exhausted and abandoned between World Wars I and II, or in the second half of the twentieth century (Grecula et al., 1995). The replacement-type deposit near Nižná Slaná produced annually ~one million metric tons of siderite ore (Hanzelová, 2001), but it ceased in 2008. The only active open pit operation is at the Styrian Erzberg (Eastern Alps, Austria), annually producing ~1.4 million metric tons of siderite (Prochaska, 1997).

Apart from the vein and replacement-type siderite, the Paleozoic basement of the Western Carpathians contains numerous medium-sized deposits of coarse-grained (sparry) magnesite, which has become one of the most important mineral resources of Slovakia during the second half of the twentieth century and was exploited until recently. An average annual production of the raw magnesite attained 1.2–1.6 million metric tons (Mt) in the period 1991–2003. Potential economic reserves were recently around 350 Mt and a total of 1600 Mt of raw magnesite was produced since the beginning of its exploitation in the twentieth century (Grecula et al., 2000; Csikósová et al., 2000).

Strikes and shapes of Gemeric siderite veins largely match the cleavage structure of the Variscan basement, showing flat dips in the northern periphery, and subvertical and locally reversal dips in the southern part. The arcuate orientation of the siderite veins is coherent with the foliation fabrics of the Gemeric basement (Figures 9.22 and 9.23). Siderite veins occur mainly in Carboniferous sequences with basic volcanoclastics. They fade out in Early Permian anchimetamorphic conglomerates at the top and within Early Palaeozoic low- to medium-grade sequences at the bottom.

Mineral infilling of siderite veins is dependent on the composition of country rocks. Siderite predominates in the vein segments crosscutting Carboniferous sequences intercalated with basic volcanoclastics. Barite precipitates only in the uppermost segments penetrating Early Permian conglomerates, where Mg content of the siderite also rapidly increases (Figure 9.24). Quartz–stibnite veins occur along

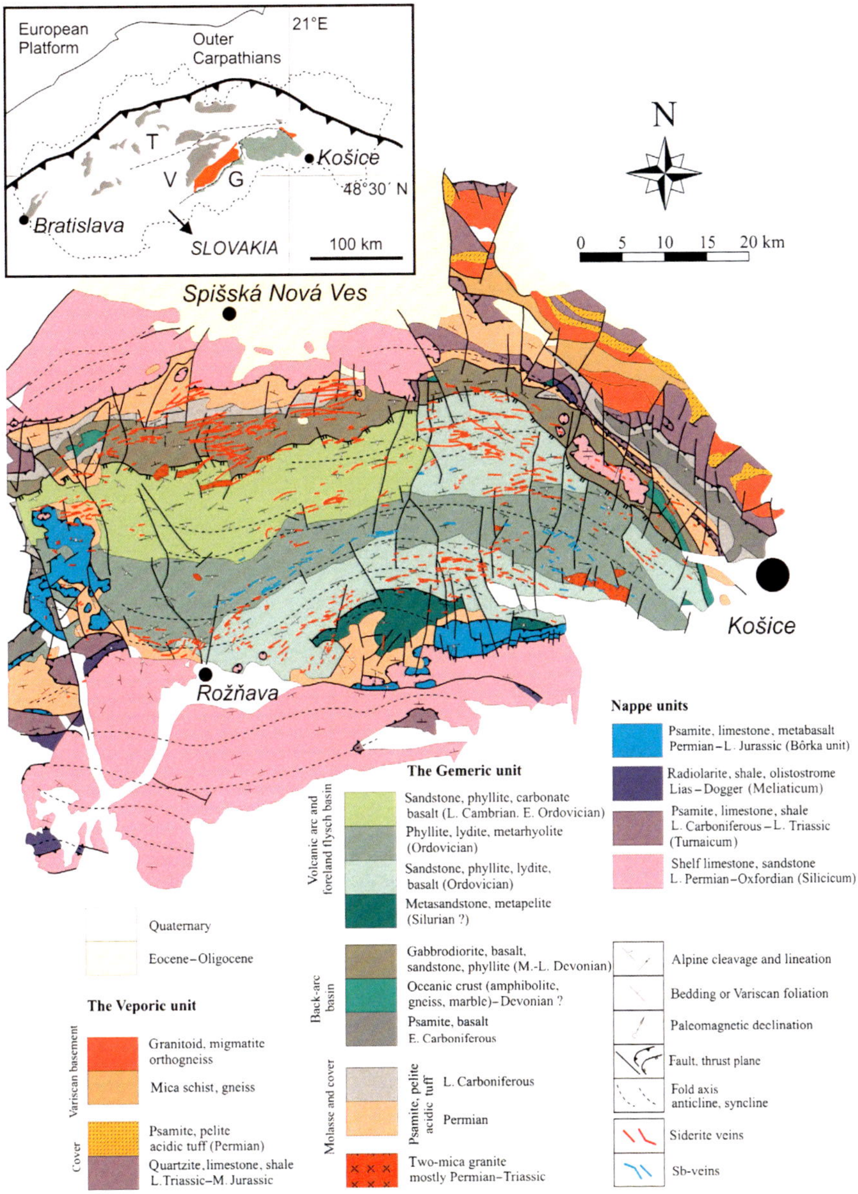

FIGURE 9.22

Distribution of hydrothermal veins in the Gemeric unit of Western Carpathians. Simplified geological map (Lexa et al., 2000, 2003) was updated according to data from Vozárová et al. (2012). The distribution of hydrothermal veins was redrawn from the metallogenetic map of Lexa et al. (2004). The inset shows the positions of Variscan basements of the Gemeric (G), Veporic (V), and Tatric (T) tectonic units in the territory of Slovakia.

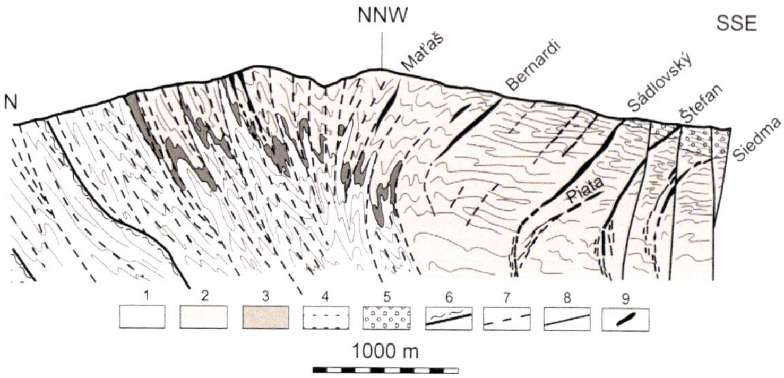

FIGURE 9.23

Geological cross section of the Rožňava-Nadabula vein system in southernmost part of the Gemeric unit (strongly modified after Grecula et al., 1995). The Early Paleozoic basement: 1—black shales; 2—metarhyolite; 3—variegated metavolcanic complex; 4—granitized metapelite. The Late Palaeozoic cover: 5—conglomerate. Other symbols: 6—thrusting planes; 7—cleavage; 8—faults; 9—ore veins.

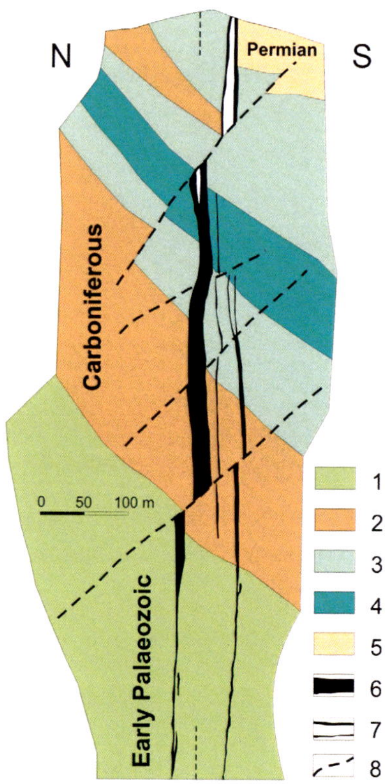

FIGURE 9.24

Transversal cross section of the Droždiak vein in the vicinity of the Poráč shaft in the northern part of the Gemeric unit (Bartalský in Kantor and Ďurkovičová, 1977) with simplified geological background. Early Paleozoic strata: 1—chlorite-sericite schist; Carboniferous strata: 2—graphitic schist; 3—metabasaltic tuff; 4—metabasalt; Permian strata: 5—conglomerate and sandstone; 6—siderite; 7—barite-siderite fill; 8—fault.

the base of a variegated Early Paleozoic complex composed of porphyroids, marbles, lydites, and black shales (Figure 9.22). The quartz-stibnite veins represent joint veins originated during brittle deformation along weakened tectonic zones intruded also by Permian granites. The relative age of the siderite and stibnite veins is unclear because they occur in different levels of the Variscan basement.

Mineralogical and paragenetic studies revealed a main siderite superimposed by a quartz–sulfide phase in all Gemeric siderite veins (Bernard, 1961; Varček, 1985; Rojkovič, 1985). The initial siderite phase was further subdivided into fuchsite, siderite, and barite stages. The early siderite and the late quartz–sulfide phases are separated by the quartz–tourmaline stage, which comprises rejuvenated siderite II, hematite–specularite, muscovite–phengite, monazite, xenotime, apatite, zircon, rutile, and subordinate Fe–As sulfides (pyrite, arsenopyrite). The quartz–sulfide phase with ankerite as the dominant carbonate associated with Cu-sulfides and sulphosalts is superimposed by cinnabar and native quicksilver.

The succession of quartz–stibnite veins resembles that of siderite veins, except for the missing fuchsite, siderite, barite, and cinnabar stages (Beňka and Caňo, 1992). The initial quartz–tourmaline stage predates sulfides. Similar to siderite veins, Cu-sulfides and sulphosalts of the sulfide stage are associated with ankerite, but they are postdated by Sb-sulfides and sulphosalts. Muscovite–phengite, monazite, xenotime, apatite, rutile, arsenopyrite, pyrite, zircon, and titanite assemblage of the quartz–tourmaline stage is similar to that in siderite veins (Urban et al., 2006).

In spite of the extensive mining, the origin of siderite and magnesite deposits remained unclear. The metamorphic–hydrothermal origin for siderite was proposed by Grecula (1982), and later extended and redefined by Žák et al. (1991) and Grecula et al. (1995). The model explained the origin of all hydrothermal mineralizations in terms of the mixing Variscan metamorphic fluids with meteoric waters along Permian rifts. Updated versions invoked an infiltration of a halite-fractionated oxidizing seawater, which replaced Carboniferous marine reef carbonates (Radvanec and Prochaska, 2001; Radvanec et al., 2004a,c) and also gave rise to siderite veins (Radvanec et al., 2004b; Grecula and Radvanec, 2005a). Grecula and Radvanec (2005b) proposed that the veins formed during Middle Permian times, and their evolution, including the final sulfidic stages, were completed during the Late Permian. The genetic model is identical with that favored for the magnesite and siderite replacement-type deposits of the Eastern Alps, where the influx of ore-forming basinal brines during the Late Permian–Early Triassic rifting (Prochaska, 1999, 2001; Ebner et al., 1999) was corroborated by the Sm–Nd dating of magnesite (Henjes-Kunst et al., 2014).

Published fluid inclusion and stable isotope data from the Carpathian magnesite deposits (Vozárová et al., 1995; Huraiová et al., 2002; Koděra and Radvanec, 2002; Radvanec et al., 2004a, b) indicated strong metamorphogenic affinity of mineral-forming fluids, the presence of variegated fluid types, including high-salinity brines and CO_2-rich low-salinity aqueous fluids. In contrast, each paragenetic stage of hydrothermal veins is typical of a specific type of fluid inclusions. High-density $N_2 \pm CO_2$ inclusions in barite (Hurai et al., 2008b) as well as superdense CO_2 inclusions in the quartz associated with stibnite and tourmaline (Urban et al., 2006) rule out their entrapment at low pressures and high thermal gradients during the Permian and Triassic rifting. The minimum formation depth of 6 km and thermal gradients between 20 and 40 km constrained by the isochores of fluid inclusions in the barite (see Figure 5.28 in Chapter 5) are reminiscent of a thick and cold continental crust. Indeed, the LA-ICPMS dating of monazite from the quartz-tourmaline stage of siderite veins in the Rožňava–Nadabula ore field (Figure 9.23) yielded the age of 139 ± 1 Ma (J.-L. Paquette, unpublished), which almost exactly overlaps the Vallanginian–Berriasian boundary and unequivocally places the superimposed barite and sulphidic assemblages in the Cretaceous period.

Fluid inclusion data from siderite veins appear at first look to be compatible with the model of the Permian infiltration of evaporated seawater. Average salinities around 20 mass% (18–26 mass% NaCl equivalents in the massive siderite, 14–35 mass% in siderite crystals from cavities) and a NaCl–KCl–$CaCl_2$–H_2O composition with minor CO_2 (Hurai et al., 2008b) were similar to typical basinal brines. A high Br content in the ore-forming fluids and Br/Cl and Na/Cl ratios plotting along the seawater fractionation trend (see Figure 8.3 in Chapter 8) were additional indicators supporting the infiltration model. However, the degree of evaporation indicated by the Br/Cl values (~40%) was controversial with the composition of evaporites in the Permian sediments of the Gemeric unit. These sediments contain only gypsum and anhydrite, thereby indicating a maximum of 4.4–11% evaporation (see Figure 8.1 in Chapter 8). Indeed, the plot correlating SO_4/Cl versus K/Na ratios documents the deviation of the fluid inclusion compositions from the seawater evaporation path (Figure 9.25), showing only the siderite veins intersecting the Permian sediments to plot close to the SET at ~10% evaporation. Other data reveal an extraneous enrichment in Br, probably due to a hydrothermal leaching of organic matter. A concomitant depletion in SO_4 tending to increase with the increasing temperature indicated by K/Na ratios was attributed to a progressive thermogenic reduction of sulfates (Hurai et al., 2008b, 2011). Interestingly, fluid inclusion leachates from magnesites exhibit even larger Br enrichment (Hurai et al., 2011), thus disproving the Br/Cl ratio as the reliable proof of the fluid origin.

Stable isotope research yielded decisive arguments for unequal origins and different ages of the siderite and magnesite deposits. Data from the Rudňany ore field are consistent with the crystallization at rising temperature without the CO_2 devolatilization, what is reflected by varying $\delta^{18}O$ and almost fixed $\delta^{13}C$ ratios of the deposited siderite (Figure 9.16). In contrast, the varying $\delta^{13}C$ and almost fixed $\delta^{18}O$ values are typical for literally all magnesite deposits (Hurai et al., 2011). A larger extent of carbon isotope fractionation compared to that of oxygen isotopes can be interpreted only in terms of the precipitation assisted by the CO_2 devolatilization at high fluid/rock ratios. An example in Figure 9.26 shows the typical trend as recorded in the Burda deposit reproduced using the alteration model (Zheng and Hoefs, 1993a). Derived *w/r* ratios are greater than 5, thus indicating an open, fluid-buffered hydrothermal system.

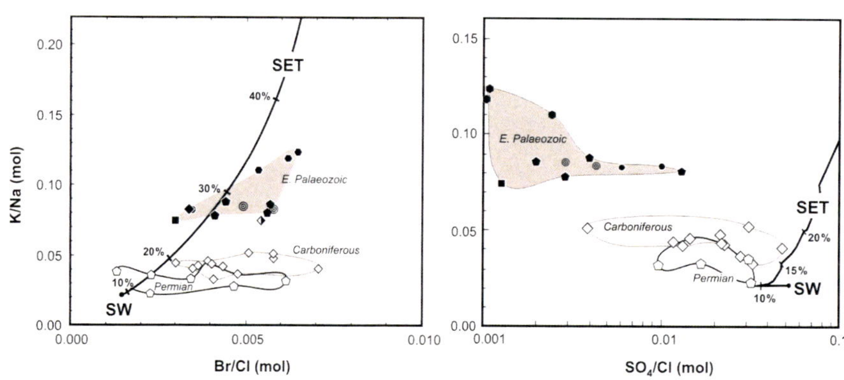

FIGURE 9.25

Covariation of K, Na, Br, Cl, and SO_4 in crush–leach analyses of siderite from Gemeric hydrothermal polymetallic veins (Hurai et al., 2008b). The seawater (SW) evaporation trend (SET) with percentages of evaporation (McCaffrey et al., 1987) is shown for comparison.

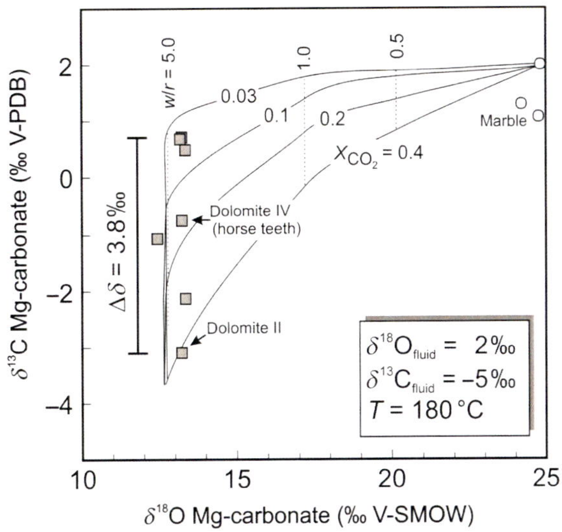

FIGURE 9.26

The alteration model (Zheng and Hoefs, 1993a) calculated for the Burda deposit of the Gemeric unit at fixed *T*, $\delta^{18}O$, and $\delta^{13}C$ values of the ore-forming fluid, with projection points of various generations (II–IV) of metasomatic dolomite and magnesite. Isotope composition of the parent fluid (data in the insert) and the fluid/rock (*w*/*r*) ratios were calculated from equations for oxygen and carbon isotope fractionations in the H_2O–calcite (O'Neil et al., 1969) and the CO_2–calcite (Ohmoto and Rye, 1979) systems, respectively. Isotope ratios in non-metasomatised marble from surrounding Carboniferous sediments were taken from Vozárová et al. (1995).

Assuming the closed system, the variations in oxygen isotope compositions are controlled only by the changing temperature. Hence, a vertical oxygen isotope gradient can be employed for constraining depths and geothermal gradients together with fluid inclusion isochores. An example is shown in Figure 9.27, which plots isochores of aqueous fluid inclusions in siderite and the temperature-dependent oxygen isotope gradient along a 0–600 m depth interval of the Droždiak siderite vein in the northern part of the Gemeric unit. The diagram is the only type of diagram that enables a simultaneous presentation of fluid inclusion and stable isotope data.

The observed fluid inclusion densities and oxygen isotope gradients are inconsistent with a shallow circulation of basinal brines under hydrostatic pressure during incipient intracontinental Permian rifting. The estimated depths of burial and thermal gradients adjusted to fluid inclusion densities, K/Na ratios, and the oxygen isotope gradients between 1.8‰/km and 2.3‰/km correspond to 4.3–6 km and 25–45 °C/km, respectively. The calculated conditions are substantially different from thermal gradients presumably in excess of 50–80 °C/km due to shallow burial (max. 2 km) and high heat flow induced by the active Permian volcanism. In these conditions, oxygen isotope gradients higher that 5‰/km are expected.

Even lower isotopic gradient, 0.9‰/km, was determined in the Rožňava–Nadabula ore field in the southern part of the Gemeric unit (Žák et al., 2005). Fluid isochores and K/Na ratios (Figure 9.28) suggest greater burial depths (9–14 km) and lower geothermal gradients (~20 °C/km) consistent with the superdense CO_2 inclusions in the neighboring quartz–stibnite deposit near Čučma.

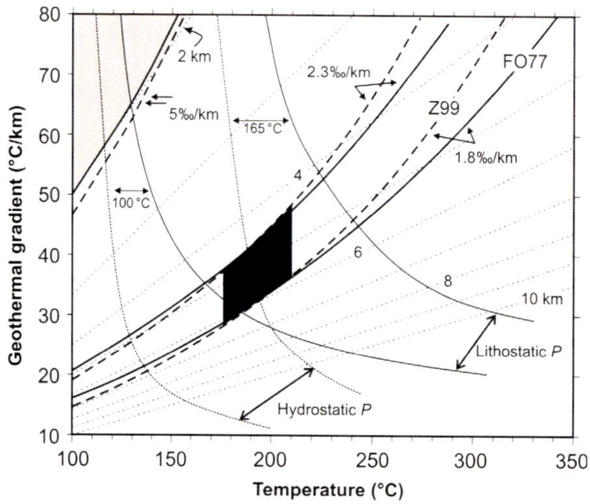

FIGURE 9.27

Formation temperatures and depths constrained by densities of primary aqueous inclusions and $\delta^{18}O$ values of siderite from the Droždiak vein in the northern part of the Gemeric tectonic unit (Hurai et al., 2008b). The strongly bent curves are isochores of 20 mass% NaCl–H_2O solutions (Gehrig, 1980) calculated for the limiting measured homogenization temperatures 100 and 165 °C and hydrostatic (dotted curves) and lithostatic (solid curves) loads, assuming the average densities of fluids and rocks corresponding to 1.0 and 2.75 g/cm^3, respectively. The diagonal, moderately bent curves are iso-gradients (‰/km) calculated from the equations of the temperature-dependent oxygen isotope equilibrium fractionation in the siderite–water system (Friedman and O'Neil, 1977; Zheng, 1999). The solid field between the 1.8‰/km and 2.3‰/km iso-gradient curves represents the range of temperatures, depths, and geothermal gradients consistent with fluid inclusion microthermometry data, the oxygen isotope composition of siderite, and K/Na cationic exchange thermometry data (Verma and Santoyo, 1997) applied to fluid inclusion leachates. The shaded triangle in the upper-left corner represents the most conservative estimate of *P–T* conditions during the intra-continental Permian–Triassic rifting.

Distinct geometry, fluid inclusion, and stable isotope proxies are consistent with the vein opening during the progressive north-vergent thrusting of the Mesozoic nappes accompanied by the formation of an arcuate deformation front in the subjacent Gemeric basement (Lexa et al., 2003). An accelerated pressure increase accompanied by the attenuated temperature due to low thermal conductivity of rocks resulted in the depression of thermal gradients with time along the prograde *P–T* path. This model accounts for the ^{18}O-depletion during growth of siderite crystals and increasing pressures and temperatures during crystallization of early siderite and the superimposed barite in the north-Gemeric unit. The maximum depth of burial (14 km) determined from fluid inclusions in barite near the base of the Permian molasse and also in the southern part of the Gemeric unit can be explained only by an integrated thickness of the autochthonous sedimentary cover and the imbricated Mesozoic accretion wedge thrusted over the Gemeric basement during the Early Cretaceous orogenic convergence. The different character of fluids reported from initial stages of the siderite and quartz–stibnite veins reflects the vein opening at different depths of the crustal-scale deformation front associated with the rapid elevation of the core of the Gemeric cleavage fan.

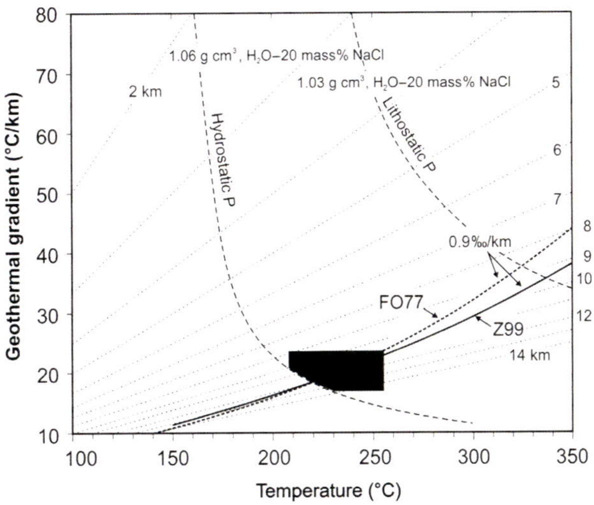

FIGURE 9.28

Formation temperatures and depths constrained by densities of primary aqueous inclusions and $\delta^{18}O$ values of siderite from the Rožňava–Nadabula ore field in the southern part of the Gemeric tectonic unit (Hurai et al., 2008b). The strongly bent curves are isochores of 20 mass% NaCl–H_2O solutions (Gehrig, 1980) calculated for hydrostatic and lithostatic loads, assuming the average densities of fluids and rocks corresponding to 1.0 and 2.75 g/cm^3, respectively. The diagonal, moderately bent curves are iso-gradients (‰/km) calculated from the equations of the temperature-dependent oxygen isotope equilibrium fractionation in the siderite–water system (Friedman and O'Neil, 1977; Zheng, 1999). The solid field between iso-gradient curves represents the range of temperatures, depth, and geothermal gradients consistent with fluid inclusion microthermometry data, the oxygen isotope composition of siderite, and K/Na thermometry data (Verma and Santoyo, 1997) applied to fluid inclusion leachates.

The syn-tectonic model of hydrothermal vein formation coupled to the Early Cretaceous deformation requires particular stress conditions, reflecting the vertical stress related to a thick overburden that acted simultaneously with the horizontal compression. In theory, the compressive stress should generate tensional failure parallel to σ_1, that is perpendicular to cleavage, assuming that the cleavage represents the xy plane of instantaneous and finite strain ellipsoids, respectively. Therefore, according to classical stress considerations, one would predict vertical N–S-trending veins perpendicular to the main Cretaceous foliation structure (Price and Cosgrove, 1990). The fact that most Gemeric veins are sub-parallel to the tectonic fabrics along the whole arcuate trend of the Cretaceous deformation front (Figure 9.22) indicates the fluid pressure was greater than the sum of the tensional strength of the rock and the plane-perpendicular stress σ_n (Cosgrove, 1997). Consequently, the opening of tensile fractures parallel to the main anisotropy—the Cretaceous foliation—can be explained in terms of a low differential stress ($\sigma_1 - \sigma_3$), corresponding to a small difference between the horizontal tectonic stress and the vertical overburden pressure, and a large difference between tensional strengths $T_p - T_n$, where T_p is the tensional strength parallel to the main anisotropy and T_n perpendicular to it. Cosgrove (1997) showed that the tensional failure occurs parallel to main anisotropy in the direction perpendicular to the main compressive stress, if $(T_p - T_n) > (\sigma_1 - \sigma_3)$. Therefore, we assume that the veining event in the Gemeric unit occurred within a narrow time interval, with specific stress conditions marked by the building of high overburden pressure (vertical load due to thrusting) and strong horizontal stress (horizontal push related to the formation of the Cretaceous fabrics). At these conditions, both vertical and horizontal stress components may have achieved similar magnitudes and the high fluid pressure associated with the large difference between the tensional strengths may have generated veins parallel to the main compressive cleavage (O. Lexa in Hurai et al., 2008a).

SUGGESTED FURTHER READING

Chacko, T., Cole, D.R., Horita, J., 2001. Equilibrium oxygen, hydrogen and carbon isotope fractionation factors applicable to geologic systems. Rev. Mineral. Geochem. 43, 1–81.

Marini, L., Moretti, R., Accornero, M., 2011. Sulfur isotopes in magmatic-hydrothermal systems, melts, and magmas. Rev. Mineral. Geochem. 73, 423–492.

Seal II, R.R., 2006. Sulfur isotope geochemistry of sulfide minerals. Rev. Mineral. Geochem. 61, 663–677.

Valley, J.W., 2001. Stable isotope thermometry at high temperatures. Rev. Mineral. Geochem. 43, 365–413.

APPENDIX

I

EUTECTIC AND PERITECTIC POINTS OF AQUEOUS SYSTEMS WITH SALTS

Table I.1 Binary Systems

System	Point	*T* (°C)	Molar Concentration	Assemblage	Source
$CaBr_2$	Te_S	−83.0		$CaBr_2 \cdot 6H_2O+I$ (ice)	2
LiCl	Te_S	−74.8		$LiCl \cdot H_2O+I$	2
$CaCl_2$	$Te_{S\text{-mod}}$	−50.4	Ca 3.993	$CaCl_2 \cdot 6H_2O+I$	3
$CaCl_2$	$Te_{S\text{-exp}}$	−49.8	Ca 3.902	$CaCl_2 \cdot 6H_2O+I$	6
$FeCl_3$	Te_S	−36.5		$FeCl_3 \cdot 10H_2O+I$	2
$FeCl_2$	Te_S	−35.5		$FeCl_2 \cdot 6H_2O+I$	2
$MgCl_2$	$Te_{S\text{-mod}}$	−33.0	Mg 2.836	$MgCl_2 \cdot 12H_2O+I$	3
NaCl	$Te_{M\text{-mod}}$	−28.0		NaCl+I	3
KF	Te_S	−21.5		$KF \cdot 4H_2O+I$	2
NaCl	Te_S	−21.2	Na 5.2	$NaCl \cdot 2H_2O+I$	1
$MgCl_2$	Tp_{mod}	−19.4		$MgCl_2 \cdot 12H_2O+MgCl_2 \cdot 6H_2O$	3
$MgCl_2$	Tp_{exp}	−19.4	Mg 5.242	$MgCl_2 \cdot 12H_2O+MgCl_2 \cdot 6H_2O$	7
$MgCl_2$	Tp_{mod}	−16.8	Mg 4.853	$MgCl_2 \cdot 12H_2O+MgCl_2 \cdot 8H_2O$	3
KCl	$Te_{S\text{-mod}}$	−10.5	K 3.295	KCl+I	3
$KHCO_3$	Te_S	−6.0		$KHCO_3+I$	2
$MgSO_4$	$Te_{M\text{-mod}}$	−5.0	Mg 2.02	$MgSO_4 \cdot 7H_2O+I$	5
$MgSO_4$	$Te_{S\text{-mod}}$	−3.6	Mg 1.74	$MgSO_4 \cdot 12H_2O+I$	5
$MgSO_4$	$Te_{M\text{-exp}}$	−3.5	Mg 1.7	$MgSO_4 \cdot 7H_2O+I$	6
NaF	Te_S	−3.5		NaF+I	2
$MgCl_2$	Tp_{exp}	−3.4	Mg 5.506	$MgCl_2 \cdot 8H_2O+MgCl_2 \cdot 6H_2O$	7
$MgCl_2$	Tp_{mod}	−3.4	Mg 5.507	$MgCl_2 \cdot 8H_2O+MgCl_2 \cdot 6H_2O$	3
Na_2SiO_3	Te_S	−2.7		$Na_2SiO_3 \cdot 9H_2O+I$	2
$NaHCO_3$	Te_S	−2.3		$NaHCO_3+I$	2
Na_2CO_3	Te_S	−2.1		$Na_2CO_3 \cdot 10H_2O+I$	2
K_2SO_4	$Te_{S\text{-exp}}$	−1.9	K 0.451	K_2SO_4+I	6
K_2SO_4	$Te_{S\text{-mod}}$	−1.6	K 0.41	K_2SO_4+I	5
Na_2SO_4	$Te_{S\text{-exp}}$	−1.2	Na 0.277	$Na_2SO_4 \cdot 10H_2O+I$	4
Na_2SO_4	$Te_{S\text{-mod}}$	−1.2	Na 0.296	$Na_2SO_4 \cdot 10H_2O+I$	5

Continued

Geofluids. http://dx.doi.org/10.1016/B978-0-12-803241-1.09985-8

Table I.1 Binary Systems—Cont'd

System	Point	*T* (°C)	Molar Concentration	Assemblage	Source
Na_2SO_4	$Te_{S\text{-}exp}$	−1.1	Na 0.296	$Na_2SO_4 \cdot 10H_2O + I$	4
$CaSO_4$	$Te_{M\text{-}mod}$	−0.1	Ca 0.013	$CaSO_4 \cdot 2H_2O + I$	5
NaCl	Tp_{exp}	−0.1	Na 6.11	$NaCl + NaCl \cdot 2H_2O + I$	6
$CaCl_2$	Tp_{exp}	30.0		$CaCl_2 \cdot 6H_2O + CaCl_2 \cdot 4H_2O$	8
$CaCl_2$	Tp_{exp}	45.0		$CaCl_2 \cdot 4H_2O + CaCl_2 \cdot 2H_2O$	8
$MgCl_2$	Tp_{mod}	115.0	Mg 9.48	$MgCl_2 \cdot 6H_2O + MgCl_2 \cdot 4H_2O$	9
$MgCl_2$	Tp_{exp}	116.7	Mg 9.02	$MgCl_2 \cdot 6H_2O + MgCl_2 \cdot 4H_2O$	6
$MgCl_2$	Tp_{exp}	181.5	Mg 13.26	$MgCl_2 \cdot 4H_2O + MgCl_2 \cdot 2H_2O$	6
$MgCl_2$	Tp_{mod}	181.5	Mg 13.37	$MgCl_2 \cdot 4H_2O + MgCl_2 \cdot 2H_2O$	9

Abbreviations: T_{eS}*—stable eutectic point;* T_{eM}*—metastable eutectic point;* T_p*—peritectic point; exp—experimental data; mod—model data. References: (1) Davis et al. (1990). (2) Borisenko (1977, 1982). (3) Spencer et al. (1990). (4) Buchshteyn et al. (1953). (5) Marion and Farren (1999). (6) Linke and Seidell (1965). (7) D´Ans (1933). (8) Schiffries (1990). (9) Pabalan and Pitzer (1987). Compiled by Milovský (2000).*

Table I.2 Ternary Systems

System	Point	*T* (°C)	Molar Concentrations	Assemblage	Source
$MgCl_2$–NaCl	$Te_{M\text{-}mod}$	−80.0		$MgCl_2 \cdot 6H_2O + NaCl + I$	1
LiCl–$CaCl_2$	Te_S	−78.0		$LiCl \cdot 5H_2O + CaCl_2 \cdot 6H_2O + I$	2
LiCl–NaCl	Te_S	−77.0		$LiCl \cdot 5H_2O + NaCl \cdot 2H_2O + I$	2
$CaCl_2$–NaCl	$Te_{M\text{-}mod}$	−70.0	Na 0.51; Ca 3.78	$CaCl_2 \cdot 4H_2O + NaCl + I$	1
$MgCl_2$–NaCl	$Te_{M\text{-}mod}$	−55.0		$MgCl_2 \cdot 8H_2O + NaCl + I$	1
$CaCl_2$–$MgCl_2$	Te_S	−52.2	Ca 3.509; Mg 0.525	$MgCl_2 \cdot 12H_2O + CaCl_2 \cdot 6H_2O + I$	3
$CaCl_2$–NaCl	$Te_{S\text{-}mod}$	−51.6	Na 0.508; Ca 3.78	$CaCl_2 \cdot 6H_2O + NaCl \cdot 2H_2O + I$	3
$CaCl_2$–KCl	Te_S	−50.5		$CaCl_2 \cdot 6H_2O + KCl \cdot 4H_2O + I$	2
$MgCl_2$–NaCl	$Te_{M\text{-}mod}$	−37.0		$MgCl_2 \cdot 12H_2O + NaCl + I$	1
Na_2CO_3–K_2CO_3	Te_S	−37.0		$(Na, K)_2CO_3 \cdot 6H_2O + K_2CO_3 \cdot 6H_2O + I$	2
NaCl–$FeCl_2$	Te_S	−37.0		$NaCl \cdot 2H_2O + FeCl_2 \cdot 6H_2O + I$	2
$MgCl_2$–NaCl	$Te_{S\text{-}mod}$	−35.0	Na 1.1; Mg 2.48	$MgCl_2 \cdot 12H_2O + NaCl \cdot 2H_2O + I$	1
KCl–$MgCl_2$	$Te_{S\text{-}exp}$	−34.3	K 0.166; Mg 2.381	$KCl + MgCl_2 \cdot 12H_2O + I$	4
KCl–$MgCl_2$	$Te_{S\text{-}mod}$	−33.7	K 0.34; Mg 2.806	$KCl + MgCl_2 \cdot 12H_2O + I$	3
$CaCl_2$–NaCl	Tp_{mod}	−26.1	Na 0.394; Ca 4.365	$CaCl_2 \cdot 6H_2O + NaCl \cdot 2H_2O + NaCl$	3
NaCl–KCl	Te_S	−23.1	Na 4.717; K 1.028	$KCl + NaCl \cdot 2H_2O + I$	3
NaCl–KCl	Te_S	−22.9	Na 4.67; K 1.05	$KCl + NaCl \cdot 2H_2O + I$	1
NaCl–$Na_2B_5O_8$	Te_S	−22.7		$NaCl \cdot 2H_2O + Na_2B_5O_8 \cdot 10H_2O + I$	2
$CaCl_2$–$MgCl_2$	*Tp*	−22.5	Ca 1.609; Mg 3.591	$MgCl_2 \cdot 12H_2O + MgCl_2 \cdot 8H_2O + CaCl_2 \cdot 6H_2O$	3

Table I.2 Ternary Systems—Cont'd

System	Point	*T* (°C)	Molar Concentrations	Assemblage	Source
$KCl–MgCl_2$	*Tp*	−22.1	K 0.312; Mg 3.632	$MgCl_2 \cdot 12H_2O + KCl + KCl \cdot MgCl_2 \cdot H_2O$	3
$MgCl_2–NaCl$	Tp_{mod}	−22.0	Na 0.63; Mg 3.52	$MgCl_2 \cdot 12H_2O + NaCl + NaCl \cdot 2H_2O$	1
$NaCl–NaHCO_3$	Te_S	−21.8		$NaCl \cdot 2H_2O + NaHCO_3 + I$	2
NaCl–KF	Te_S	−21.7		$NaCl \cdot 2H_2O + KF \cdot 4H_2O + I$	2
$NaCl–Na_2SO_4$	Te_S	−21.7		$NaCl + Na_2SO_4 \cdot 5H_2O + I$	2
$NaCl–Na_2CO_3$	Te_S	−21.4		$NaCl \cdot 2H_2O + Na_2CO_3 \cdot 10H_2O + I$	2
$NaCl–Na_2SO_4$	$Te_{S\text{-}exp}$	−21.4	Cl 5.14; SO_4 0.016	$NaCl + Na_2SO_4 \cdot 10H_2O + I$	4
$NaCl–Na_2SO_4$	$Te_{S\text{-}mod}$	−21.3	Cl 5.16; SO_4 0.014	$NaCl + Na_2SO_4 \cdot 10H_2O + I$	5
$NaCl–Na_2SO_4$	$Te_{S\text{-}exp}$	−21.3	Cl 5.26; SO_4 0.029	$NaCl + Na_2SO_4 \cdot 10H_2O + I$	4
NaCl–NaF	Te_S	−21.2		$NaCl \cdot 2H_2O + NaF + I$	2
$CaCl_2–MgCl_2$	*Tp*	−17.6	Ca 2.88; Mg 4.626	$MgCl_2 \cdot 12H_2O + MgCl_2 \cdot 8H_2O$	3
$KCl–MgCl_2$	*Tp*	−16.9	K 0.019; Mg 4.854	$MgCl_2 \cdot 12H_2O + MgCl_2 \cdot 8H_2O + I$	3
$KCl–K_2SO_4$	$Te_{S\text{-}exp}$	−10.9	Cl 3.24; SO_4 0.052	$KCl + K_2SO_4 + I$	4
$KCl–K_2SO_4$	$Te_{S\text{-}mod}$	−10.9	Cl 3.27; SO_4 0.054	$KCl + K_2SO_4 + I$	5
$CaCl_2–MgCl_2$	*Tp*	−6.6	Ca 1.337; Mg 4.51	$MgCl_2 \cdot 8H_2O + MgCl_2 \cdot 6H_2O + CaCl_2 \cdot 6H_2O$	3
$K_2SO_4–MgSO_4$	$Te_{S\text{-}exp}$	−5.2	K 0.246; Mg 1.88	$MgSO_4 \cdot 12H_2O + MgSO_4 \cdot K_2SO_4 \cdot 6H_2O + I$	4
$Na_2SO_4–MgSO_4$	$Te_{S\text{-}exp}$	−5.0	Na 0.28; Mg 1.69	$Na_2SO_4 \cdot 10H_2O + MgSO_4 \cdot 12H_2O + I$	4
$Na_2SO_4–MgSO_4$	$Te_{S\text{-}mod}$	−5.0	Na 0.329; Mg 1.66	$Na_2SO_4 \cdot 10H_2O + MgSO_4 \cdot 12H_2O + I$	5
$K_2SO_4–MgSO_4$	$Te_{S\text{-}mod}$	−4.8	K 0.29; Mg 1.78	$MgSO_4 \cdot 12H_2O + MgSO_4 \cdot K_2SO_4 \cdot 6H_2O + I$	5
$KCl–MgCl_2$	*Tp*	−3.4	K 0.01; Mg 5.508	$MgCl_2 \cdot 8H_2O + MgCl_2 \cdot 6H_2O + KCl \cdot MgCl_2 \cdot H_2O$	3
$Na_2CO_3–NaHCO_3$	Te_S	−3.3		$Na_2CO_3 \cdot 10H_2O + NaHCO_3 + I$	2
$Na_2SO_4–NaHCO_3$	Te_S	−3.1		$Na_2SO_4 \cdot 10H_2O + NaHCO_3 + I$	2
$K_2SO_4–MgSO_4$	$Te_{M\text{-}exp}$	−3.0	K 0.498; Mg 0.855	$K_2SO_4 + MgSO_4 \cdot K_2SO_4 \cdot 6H_2O + I$	4
$Na_2SO_4–K_2SO_4$	$Te_{S\text{-}exp}$	−3.0	Na 0.358; K 0.5	$Na_2SO_4 \cdot 10H_2O + K_2SO_4 + I$	4
$K_2SO_4–MgSO_4$	$Te_{M\text{-}mod}$	−2.8	K 0.532; Mg 0.802	$K_2SO_4 + MgSO_4 \cdot K_2SO_4 \cdot 6H_2O + I$	5
$Na_2SO_4–K_2SO_4$	$Te_{S\text{-}exp}$	−2.7	Na 0.344; K 0.457	$Na_2SO_4 \cdot 10H_2O + K_2SO_4 + I$	4
$Na_2SO_4–K_2SO_4$	$Te_{S\text{-}mod}$	−2.7	Na 0.384; K 0.483	$Na_2SO_4 \cdot 10H_2O + K_2SO_4 + I$	5
NaCl–KCl	*Tp*	−2.3	Na 5.47; K 1.32	$KCl + NaCl + NaCl \cdot 2H_2O$	3
NaCl–KCl	Tp_{mod}	−1.9	Na 5.504; K 1.264	$NaCl + NaCl \cdot 2H_2O + I$	3

Abbreviations: T_{eS}*—stable eutectic point;* T_{eM}*—metastable eutectic point;* T_p*—peritectic point; exp—experimental data; mod—model data. References: (1) Davis et al. (1990). (2) Borisenko (1977, 1982). (3) Spencer et al. (1990). (4) Buchshteyn et al. (1953). (5) Marion and Farren (1999). Compiled by Milovský (2000).*

APPENDIX

SELECTED EQUATIONS OF STATE II

H_2O

Relationship between T and D [g/cm^3] of the liquid along L+G curve (Haas, 1976)

$$D = \frac{1.0 + 0.1342489\ \vartheta^{1/3} - 3.946263 \times 10^{-3}\vartheta}{V_c - 0.3151548\ \vartheta^{1/3} - 1.203374 \times 10^{-3}\vartheta + 7.48908 \times 10^{-13}\vartheta^4} \tag{II.1}$$

V_c—critical relative (reciprocal) volume (3.1975 cm^3/g), $\vartheta = 647.27 - T$ [K]

Pressure P [bar] along the L+G curve (Haas, 1976)

$$\ln P = 12.50849 - \frac{4616.913}{z} + \frac{3.193455 \times 10^{-4}w}{z}\left[10^{\left(1.1965\times10^{-11}w^2\right)} - 1\right] + \left[-0.01013137 \times 10^{\left(-0.0057148y^{1.25}\right)}\right] \tag{II.2}$$

$z = T + 0.01$, $y = 647.27 - T$, $w = z^2 - 293{,}700$, T [K]. The original equations contain several errors, which are corrected here.

CO_2

Density D [g/cm^3] along the boiling and condensation curves L+G (Angus et al., 1976)

$$D = 0.4660553\left[C_0\left(1 - \frac{T}{T_c}\right)^{0.347} + \sum_{i=1}^{2} C_i\left(1 - \frac{T}{T_c}\right)^{(i+1)/3} + 1\right] \tag{II.3}$$

$C_0 = 1.9073793$, $C_1 = 0.38225012$, $C_2 = 0.42897885$ for liquid; $C_0 = -1.7988929$, $C_1 = -0.71728276$, $C_2 = 1.7739244$ for vapor, $T_c = 304.21$ K.

Pressure P [bar] along the L+G curve (Angus et al., 1976)

$$P = 73.825\ \exp\left[C_0\left(1 - \frac{T}{304.21}\right)^{1.935} + \sum_{i=1}^{4} C_i\left(\frac{304.21}{T} - 1\right)^{i}\right] \tag{II.4}$$

Geofluids. http://dx.doi.org/10.1016/B978-0-12-803241-1.09984-6

T [K], $C_0 = 11.377371$, $C_1 = -6.8849249$, $C_2 = -9.5924263$, $C_3 = 13.679755$, $C_4 = -8.6056439$.

Molar volume $\overline{V}$ [cm^3/mol] of a homogeneous fluid (Shmonov and Shmulovich, 1974)

$$\begin{aligned}\overline{V} = {} & 17.771 - 25.28856 \log(P - 1233) + 0.1479286T \\ & + \log(P - 1447)(27.40306 - 0.0354434T)\end{aligned} \tag{II.5}$$

P [bar], T [K]. The equation is valid at 100–700 °C and 2000–8000 bar, and extrapolation is possible up to 1000 °C and 10 kbar. Sterner and Pitzer (1994) published equations, permitting the extrapolation up to 10 GPa.

CH_4

Density of vapor D [g/cm^3] and temperature T [K] along the L+G curve (Kleinrahm and Wagner, 1986)

$$\ln\frac{D}{D_c} = C_1\left(1-\frac{T}{T_c}\right)^{2/6} + C_2\left(1-\frac{T}{T_c}\right)^{5/6} + C_3\left(1-\frac{T}{T_c}\right) + C_4\left(1-\frac{T}{T_c}\right)^{28/6} \tag{II.6}$$

T_c—temperature at the critical point (190.551 K), D_c—density at the critical point (0.16266 g cm^{-3}), $C_1 = 1.497557$, $C_2 = -0.7690676$, $C_3 = 0.4989926$, $C_4 = 0.01651778$.

Density of liquid D [g/cm^3] and temperature T [K] along the L+G curve (Kleinrahm and Wagner, 1986)

$$\begin{aligned}\ln\frac{D}{D_c} = {} & C_1\left(1-\frac{T}{T_c}\right)^{2/6} + C_2\left(1-\frac{T}{T_c}\right)^{3/6} + C_3\left(1-\frac{T}{T_c}\right) \\ & + C_4\left(1-\frac{T}{T_c}\right)^{13/6} + C_5\left(1-\frac{T}{T_c}\right)^{26/6} + C_6\left(1-\frac{T}{T_c}\right)^{55/6}\end{aligned} \tag{II.7}$$

$C_1 = -1.351579$, $C_2 = -1.054388$, $C_3 = -3.585975$, $C_4 = -6.313057$, $C_5 = -16.97401$, $C_6 = -57.08145$.

Pressure P [bar] and temperature T [K] along the L+G curve (Kleinrahm and Wagner, 1986)

$$\ln\frac{P}{P_c} = \frac{T_c}{T}\left(C_1\left(1-\frac{T}{T_c}\right) + C_2\left(1-\frac{T}{T_c}\right)^{1.5} + C_3\left(1-\frac{T}{T_c}\right)^{2} + C_4\left(1-\frac{T}{T_c}\right)^{4.5}\right) \tag{II.8}$$

$P_c = 45.992$ bar, $T_c = 190.551$ K, $C_1 = -6.035927$, $C_2 = 1.412132$, $C_3 = -0.4992592$, $C_4 = -1.438047$.

The MRK EOS for homogeneous CH_4 fluid (Kerrick and Jacobs, 1981)

$$P = \frac{RT(1+y+y^2-y^3)}{\overline{V}(1-y)^3} - \frac{c(T) + d(T)/\overline{V} + e(T)/\overline{V}^2}{\overline{V}(\overline{V}+b)\sqrt{T}} \tag{II.9}$$

$$\begin{aligned} y &= b/4\overline{V} \\ c(T) &= 13.403\times10^6 + 9.28\times10^4 T + 2.7T^2 \\ d(T) &= 5.216\times10^9 - 6.8\times10^6 T + 3.28\times10^3 T^2 \\ e(T) &= -2.3322\times10^{11} + 6.738\times10^8 T + 3.179\times10^5 T^2 \end{aligned}$$

$R = 83.14$ cm^3 bar mol^{-1} K^{-1}, P [bar], T [K], $\overline{V}$ [cm^3/mol], $b = 60$ cm^3/mol. The equation is valid for 50–8600 bar and 47–397 °C.

H_2O–NACL

Vapor-saturated ice liquidus (Hall et al., 1988)

$$m = 0.00 + 1.78\theta - 0.0442\theta^2 + 0.000557\theta^3 \quad \text{(II.10)}$$

$\theta = |T_m|$, T_m—temperature of melting of ice, m—NaCl concentration (mass %). The equation is valid for T_m from 0 °C to −21.21 °C.

Vapor-saturated halite liquidus (the L+G+NaCl curve, Sterner et al., 1988)

$$m = 26.242 + 0.4928\Psi + 1.42\Psi^2 - 0.223\Psi^3 + 0.04129\Psi^4 + 0.006295\Psi^5 - 0.001967\Psi^6 + 0.0001112\Psi^7 \quad \text{(II.11)}$$

$\Psi = T_m\text{NaCl}$ (°C)/100, m—NaCl concentration (mass %). The equation is valid for T_mNaCl between 0 and 801 °C.

Pressure P [bar] along the L+G+NaCl curve (Bischoff, 1991)

$$P = 41.749 - 1.2125T + 0.0136213T^2 - 7.52333 \times 10^{-5}T^3 + 2.19664 \times 10^{-7}T^4 - 2.82583 \times 10^{-10}T^5 + 1.27231 \times 10^{-13}T^6 \quad \text{(II.12)}$$

Density of liquid D [g cm^{-3}] along the L+G+NaCl curve (Bischoff, 1991)

$$D = 1.1027 + 0.0017752T - 1.5056 \times 10^{-5}T^2 + 4.0404 \times 10^{-8}T^3 - 4.1455 \times 10^{-11}T^4 + 1.5125 \times 10^{-14}T^5 \quad \text{(II.13)}$$

Equations (II.12) and (II.13) are valid for the temperatures between 100 and 801 °C.

Density D [g cm^{-3}] of the aqueous liquid along the vapor-saturated ice and halite liquidi (Bodnar, 1983)

$$D = 0.9923 - 3.0512 \times 10^{-2}A^2 - 2.1977 \times 10^{-4}A^4 + 8.6241 \times 10^{-2}B - 4.1768 \times 10^{-2}AB + 1.4825 \times 10^{-2}A^2B + 1.446 \times 10^{-3}A^3B - 3.0852 \times 10^{-9}A^8B + 1.3051 \times 10^{-2}AB^2 - 6.1402 \times 10^{-3}A^2B^2 - 1.2843 \times 10^{-3}B^3 + 3.7604 \times 10^{-4}A^2B^3 - 9.9594 \times 10^{-9}A^2B^7 \quad \text{(II.14)}$$

$A = T_h$ (°C)/100, B = mass % NaCl/10, T_h—temperature of homogenization to liquid. Standard error corresponds to ±0.0046 g/cm^3. The equation is valid to 610 °C, except for near-critical densities. The original equation contains a typographic error, which is corrected here.

A simpler and more accurate equation can be used for temperatures to 100 °C (Potter and Brown, 1977)

$$D = \frac{1000D_{H_2O} + M_{NaCl}\overline{m}D_{H_2O}}{1000 + A_0\overline{m}D_{H_2O} + B_0\overline{m}^{3/2}D_{H_2O} + C_0\overline{m}^2D_{H_2O}} \quad \text{(II.15)}$$

D—density [g cm^{-3}], $\overline{m}$—molality, D_{H_2O}—density of pure water [g cm^{-3}]. Further constants valid for selected temperatures are in Table II.1.

Temperatures along the L+G curve of the H_2O–NaCl system are higher than that of pure water at the same pressure (Haas, 1976)

$$\ln T_0 = \frac{\ln T_x}{(a + bT_x)} \quad \text{(II.16)}$$

Table II.1 Constants of Equation (II.15)

T [°C]	A_0	B_0	C_0	D_{H_2O}
0	12.43	3.07	−0.02	0.999839
25	16.62	1.773	0.098	0.997047
40	17.45	1.71	0.04	0.991640
50	18.00	1.666	0.002	0.988038
75	18.18	1.19	0.12	0.974844

$$a = 1 + 5.93582 \times 10^{-6}\overline{m} - 5.19386 \times 10^{-5}\overline{m}^2 + 1.23156 \times 10^{-5}\overline{m}^3$$

$$b = 0 + 1.1542 \times 10^{-6}\overline{m} + 1.41254 \times 10^{-7}\overline{m}^2 - 1.92476 \times 10^{-8}\overline{m}^3 - 1.70717 \times 10^{-9}\overline{m}^4 + 1.0539 \times 10^{-10}\overline{m}^5$$

T_0 is temperature of pure water, T_x is temperature of H_2O–NaCl solutions [K], $\overline{m}$ is the molality [mol NaCl/1000 g H_2O]. T_0 calculated from Equation (II.16) is used in Equation (II.2) to derive the pressure along the L+G curve of the H_2O–NaCl system. The equation is valid for temperatures up to 325 °C.

Empirical EOS for a homogeneous fluid (Zhang and Frantz, 1987; Brown and Lamb, 1989)

$$P = A_1 + A_2 T$$
$$A_1 = 6.1\mathrm{e}^{-3} + (0.2385 - a_1)T_h - (2.855\mathrm{e}^{-3} + a_2)T_h^2 - (a_3 T_h + a_4 T_h^2)\overline{m} \quad \text{(II.17)}$$
$$A_2 = a_1 + a_2 T_h + 9.888\mathrm{e}^{-6} T_h^2 + (a_3 + a_4 T_h)\overline{m}$$

T_h is the homogenization temperature (°C), $\overline{m}$ is the molality, $a_1 = 18.6102$, $a_2 = -0.00952838$, $a_3 = 1.34907$, $a_4 = -0.00767355$ for densities between 0.98 and 1.2 g/cm^3, $a_1 = 28.73$, $a_2 = -0.06477$, $a_3 = -0.2009$, $a_4 = -0.003186$ for densities < 0.98 g/cm^3. Equation (II.17) with corresponding parameters a_i listed in Table II.2 enables the calculation of isochores for the H_2O–$CaCl_2$ and H_2O–KCl systems, and for pure water. Density equations published in the original paper (Zhang and Frantz, 1987) are not recommended due to inconsistencies with experimental data.

H_2O–NACL–KCL

Salinity (mass %) along the vapor-saturated halite liquidus (Sterner et al., 1988)

$$m = a + b\Psi + cR + dR\Psi + e\Psi^2 + fR^2\Psi + gR^3\Psi^2 + hR^3\Psi^3 + iR\Psi^3 + j\Psi^3 + kR^4\Psi^2 + lR\Psi^7 + mR^2\Psi^6 + nR^2\Psi^7 + oR^3\Psi^7 + pR^6\Psi^6 + qR^7\Psi^4 + rR^9\Psi^5 + s\Psi^6 \quad \text{(II.18)}$$

Table II.2 Constants a_i for Equations (II.17) and (II.22)

	a_1	a_2	a_3	a_4
H_2O	28.57	−0.06059		
$CaCl_2$	28.48	−0.06445	−0.4159	−0.007438
KCl	28.46	−0.06403	−0.2306	−0.003166
NaCl	28.73	−0.06477	−0.2009	−0.003186

The equation is valid from −2.3 °C to 801 °C.

Salinity (mass %) along the vapor-saturated sylvite liquidus (Sterner et al., 1988)

$$m = a + b\Psi + cR + dR\Psi + e\Psi^2 + fR^2 + gR\Psi^2 + hR^3\Psi + iR\Psi^3 + j\Psi^3 + k\Psi^4 + lR\Psi + mR^8\Psi^6 + nR^9\Psi^2 + oR^9\Psi^9 + pR^9 + q\Psi^5 + r\Psi^6 \quad \text{(II.19)}$$

The equation is valid from −22.9 °C to 770 °C.

Salinity (mass %) along the vapor-saturated hydrohalite liquidus

$$m = a + b\Psi + cR \quad \text{(II.20)}$$

The equation is valid from −22.9 °C to 0.1 °C. m—salinity [mass %], $\Psi = T/100$, R—NaCl mass fraction [NaCl/(NaCl+KCl)]. Constants of Equations (II.18)–(II.20) are listed in Table II.3.

Salinity (mass %) along the vapor-saturated ice liquidus (Hall et al., 1988)

$$m = \left[(-b/2) + \left(b^2/4 + a^3/27\right)^{1/2}\right]^{1/3} + \left[(-b/2) - \left(b^2/4 + a^3/27\right)^{1/2}\right]^{1/3} \quad \text{(II.21)}$$

$a = c/e$, $b = -\theta/e$, $\theta = |T_{\mathrm{m}}|$—temperature of ice melting, $c = 0.4597 + 0.1440X$, $e = 2.227 \times 10^{-4} + 1.999 \times 10^{-4}X + 4.633 \times 10^{-5}X^2 + 1.123 \times 10^{-4}X^3$, X is the NaCl mass fraction [NaCl/(NaCl+KCl)].

Table II.3 Constants of Equations (II.18)–(II.20)

	Halite Field (Equation (II.18))	Sylvite Field (Equation (II.19))	Hydrohalite Field (Equation (II.20))
a	39.19843568	21.88644809	40.36947594
b	13.59271070	20.27990158	14.80771966
c	−12.95646989	3.25451170	−14.08238722
d	−22.39663458	−6.43202667	
e	−1.52357614	−9.60299248	
f	9.29672450	7.03770644	
g	6.55485254	5.80727278	
h	−0.66512956	6.97961319	
i	−0.39464296	−0.77994533	
j	0.83679331	4.07787875	
k	−3.61131899	−0.87240194	
l	0.00275314	0.00037040	
m	0.00822892	−0.31288876	
n	−0.00608193	−9.92786125	
o	0.00344002	0.00215197	
p	−0.00531261	17.59451286	
q	0.04128563	0.09174297	
r	0.00629464	−0.00377589	
*s**	−0.00488294		

Error in original work is corrected here

H_2O–NACL–KCL–$CACL_2$

Isochore for a homogeneous fluid (Zhang and Frantz, 1987)

$$P = P_{(ID)} + \sum_i P_i \overline{m}_i \tag{II.22}$$

$$P_{(ID)} = 0.0061 + (0.2385 - a_1)T_h - (0.002855 + a_2)T_h^2 + \left(a_1 + a_2 T_h + 9.888e - 6T_h^2\right)T$$
$$P_i = -\left(a_{3,i}T_h + a_{4,i}T_h^2\right) + (a_{3,i} + a_{4,i}T_h)T$$

i—solute (NaCl, $CaCl_2$, KCl), T_h, T [°C], P [bar], $\overline{m}_i$—molality of ith compound, constants $a_{j,i}$ are listed in Table II.2.

H_2O–CH_4

Lin and Bodnar (2010) published empirical equation that relates the slope of the iso-T_h line—$\Delta P/\Delta T$—in one phase field to the homogenization temperature to a liquid—T_h, for H_2O–CH_4 fluid inclusions with ≤4 mol. % CH_4

$$\Delta P/\Delta T = a + b\overline{m} + c\overline{m}^4 + d(T_h)^2 + e\overline{m}T_h + f\overline{m}(T_h)^4 \tag{II.23}$$

where $\overline{m}$ is the CH_4 molality and a–f are fitting parameters: $a = 19.5386$, $b = 30.49833$, $c = -0.01859$, $d = -0.00010733$, $e = -0.12478$, $f = 8.56213e^{-10}$. The equation is valid to 500 °C and 3 kbar.

H_2O–CO_2–NACL

Diamond (1992a) derived equations that allow the determination of salinity m (in mass % NaCl eq.) along univariant curves of the H_2O–NaCl–CO_2 system, provided that the inclusion pressure is fixed by a temperature of metastable partial homogenization of the CO_2-rich phase (in the metastable absence of the CO_2-clathrate). If the CO_2-phase homogenizes to liquid and $T_hCO_2 < T_m$cla (which means that the CO_2-clathrate coexists with H_2O- and CO_2-liquids along the Q2-B segment in Figure 4.33), the total salinity is expressed by the following least-squares fit

$$\begin{aligned} m = {} & 15.6151 - 0.03627T_h + 0.00164T_h^2 - 0.9491T_m - 0.00287T_h \times T_m - 0.02464T_m^2 \\ & - 0.00107T_h \times T_m^2 - 0.00222T_m^3 \end{aligned} \tag{II.24}$$

If the CO_2-rich phase homogenizes to vapor and $T_hCO_2 < T_m$cla (that means that the CO_2-clathrate coexists with the H_2O-liquid and CO_2-vapor along the Q2-Q1 segment, Figure 4.33), the total salinity is expressed by equation

$$\begin{aligned} m = {} & 15.5131 + 0.065705T_h - 0.00778T_h^2 - 1.05135T_m + 0.02687T_hT_m - 0.04717T_m^2 \\ & + 0.00138T_hT_m^2 - 0.00411T_m^3 \end{aligned} \tag{II.25}$$

where T_h is the temperature of homogenization of the CO_2-rich phase between −20 and +10 °C, T_m is the temperature of CO_2-clathrate dissociation between −5 and +13 °C (Equation (II.24)) or between −8 and +10 °C (Equation (II.25)).

The MRK EOS (Equation (3.11)) for a homogeneous H_2O–NaCl–CO_2 fluid has the following parameters

$$a = X_{CO_2}{}^2 a_{CO_2} + (1 - X_{CO_2})^2 (\Gamma + m\Lambda) \times 10^6$$
$$+ 2X_{CO_2}(1 - X_{CO_2})\left(\left(a_{CO_2}{}^0 \exp(\Theta + m\Xi + m^2\gamma + m^3\varphi)10^6\right)^{0.5} + 0.5R^2K\right) \quad \text{(II.26)}$$
$$b = X_{CO_2} b_{CO_2} + (1 - X_{CO_2})(\psi + m\Omega) \text{ (Bowers and Helgeson, 1983a)}$$

$$b_{CO_2} = 29.7, \; a_{CO_2}{}^2 = 46{,}000{,}000 \text{ (de Santis et al., 1974)}$$
$$a_{CO_2} = 73{,}030{,}000 - 71{,}400t + 21.57t^2 \text{ (Holloway, 1977)} \quad \text{(II.27)}$$

$$\ln K = -11.071 + 5953/T - 2{,}746{,}000/T^2 + 464{,}600{,}000/T^3 \text{ (de Santis et al., 1974)}$$
$$\Theta = 4.881243 + 0.1823047\text{e}^{-2}t - 0.1712269\text{e}^{-4}t^2 + 0.6479419\text{e}^{-8}t^3$$
$$\Xi = 0.02636494 - 0.536994\text{e}^{-3}t + 0.2687074\text{e}^{-5}t^2 - 0.4321741\text{e}^{-8}t^3$$
$$Y = 0.6802827\text{e}^{-2} - 0.948023\text{e}^{-4}t + 0.3770339\text{e}^{-6}t^2 - 0.5075318\text{e}^{-9}t^3$$
$$\phi = 0.5235827\text{e}^{-4} - 0.3505272\text{e}^{-7}t$$
$$\Gamma = 111.3057 + 50.70033\exp(-0.982646\text{e}^{-2}t)$$
$$\Lambda = -8.05658\exp(-0.982646\text{ e}^{-2}t)$$
$$\Psi = 14.6$$
$$\Omega = -0.04420283 \text{ (Bowers and Helgeson, 1983a)}$$
$$R = 83.11665,\; P[\text{bar}],\; T[K],\; t\text{—temperature } [°C],\; m\text{—mass\% NaCl relative to } H_2O.$$

CO_2–CO–CH_4–H_2–H_2S–N_2

The MRK EOS (Equation (3.11)) can be used for calculating isochores of homogeneous gaseous mixtures. Holloway (1977) generalized mixing rules for mixture of n-species proposed by Redlich and Kwong

$$b = \sum_j^n X_i b_i \quad \text{(II.28)}$$

$$a = \sum_i^n \sum_j^n X_i X_j a_{ij} \quad \text{(II.29)}$$

where $X_{i,j}$ is the molar fraction of ith species, and

$$a_{ij} = \sqrt{a_i a_j} \quad \text{(II.30)}$$

The a_i and b_i constants for individual gas species are independent of temperature for anhydrous compositions (Table II.4). A sample calculation is given as Problem 3.1.

Table II.4 Parameters a_i and b_i for Selected Gaseous Species (Holloway, 1981)

Species	a_i	b_i
CO_2	46,000,000	29.700
CO	16,980,000	27.380
CH_4	31,590,000	29.703
H_2	3,560,000	15.150
H_2S	87,900,000	20.000
N_2	15,382,000	26.800

III OPTICAL PROPERTIES OF FLUID INCLUSION PHASES

Phase	Name	Refractive Index	Birefringence	System	Habit	Notes
CO_2 (liquid)	Carbon dioxide	1.195				
CO_2 (solid)	Carbon dioxide	1.4		Cubic	Microcrystalline aggregates	Translucent, homogeneous aggregates or octahedrons in gaseous mixtures
H_2O	Ice	1.31	$\varepsilon 1.313 - \omega 1.309$ Negligible	Hexagonal	Rounded laths or grains	Anisotropic, but looks like isotropic, colorless
H_2O	Water	1.32–1.33				
$CO_2 \cdot 5\frac{3}{4}\ H_2O$	CO_2-clathrate	Near to liquid water		Cubic	Rounded grains, octahedrons, pentagonal dodecahedrons	Colorless, hardly discernible in low-salinity aqueous solutions
$Na_2SO_4 \cdot 10H_2O$	Mirabilite	1.39–1.40	Negligible	Monoclinic	Prismatic	
$NaCl \cdot 2H_2O$	Hydrohalite	1.416	Negligible	Monoclinic	Pseudohexagonal plates	Grainy appearance, high relief compared to ice, colorless to pale green
$Na_2CO_3 \cdot 10H_2O$	Natron	1.40–1.44	Low	Monoclinic	Elongated laths	T_m (incongruent) +32 °C
CaF_2	Fluorite	1.43		Cubic	Cubes, octahedrons	Fluorescent in UV
$MgSO_4 \cdot 6H_2O$	Hexahydrite	1.45	Low	Monoclinic	Fibrous, tabular	
$NaHCO_3$	Nahcolite	1.37–1.58	Very high	Monoclinic	Tabular, needlelike	Twinning, changes of relief during rotation in polarized light
$Na_2CO_3 \cdot H_2O$	Thermonatrite	1.42–1.52	Medium	Orthorhombic	Granular	Colorless, yellowish

Continued

Geofluids. http://dx.doi.org/10.1016/B978-0-12-803241-1.09983-4

Phase	Name	Refractive Index	Birefringence	System	Habit	Notes
Na_2SO_4	Thenardite	1.46–1.48	Low	Orthorhombic	Prismatic, tabular	Fluorescent in UV
KCl	Sylvite	1.49		Cubic	Cubes	Rounded edges, colorless to pale yellow
$MgCl_2{\cdot}6H_2O$	Pseudo-bischofite	1.49–1.50	Low to medium	Monoclinic	Fibrous	Colorless, translucent
$MgCl_2{\cdot}12H_2O$	Hydro-bischofite	1.50–1.53	Negligible	Monoclinic	Tiny grains	Colorless
$CaSO_4{\cdot}2H_2O$	Gypsum	1.52–1.53	Very low	Monoclinic	Prismatic, fibrous	
NaCl	Halite	1.544		Cubic	Cubes, rarely octahedrons	Colorless, refraction index equal to quartz
KCl	Sylvite	1.49		Cubic	Cubes, octahedrons	Rounded edges, dissolves more rapidly than halite
$FeCl_n$	Fe-chlorides	Variable	Medium to high	Trigonal	Tabular, rhombic, pseudohexagonal	Usually light green
SiO_2	Quartz	1.54–1.55	Low	Trigonal		
$CaCl_2{\cdot}6H_2O$	Antarcticite	1.49–1.55 ε1.393 ω1.417	Low-negligible	Trigonal	Similar to hydrohalite, sometimes rounded	colorless
$CaSO_4$	Anhydrite	1.57–1.61	Low	Orthorhombic	Prismatic	Associated with gypsum
$CaCO_3$	Aragonite	1.53–1.69	High	Orthorhombic	Prismatic, needlelike	
(Ca, Mg) CO_3	Carbonates	1.49–1.66	Very high	Trigonal	Rhombohedral	High relief, changing during rotation in polarized light
$BaSO_4$	Barite	1.64–1.65	Low	Orthorhombic	Tabular	

Reprinted from van den Kerkhof, A.M., Hein, U., 2001. Fluid inclusion petrography, Lithos 55, 27–47, Copyright 2001, with permission from Elsevier.

APPENDIX

STABLE ISOTOPE FRACTIONATION FACTORS

IV

Geofluids. http://dx.doi.org/10.1016/B978-0-12-803241-1.09982-2

Table IV.1 Oxygen

	Quartz–Mineral			Mineral–H_2O			Calcite–Mineral			Mineral–OH			CO_2–Mineral			Min. *T*	Max. *T*		
	*C**	*B*	*A*	*C*	*B*	*A*	*C*	*B*	*A*	*C*	*B*	*A*	*C*	*B*	*A*	(°C)	(°C)	Method	Refs.**
Acmite	0.30	2.38	−0.99	4.18	−7.14	2.17	−0.17	2.45	−0.99							0	1200	Increment	1
Actinolite	0.52	3.48	−1.45	3.96	−8.25	2.37	0.05	3.58	1.45							0	1200	Increment	2
Albite	0.15	1.39	−0.57	4.33	−6.15	1.98	−0.32	1.49	−0.57							0	1200	Increment	1
Albite				2.91		−3.41										350	800	Experim.	3
Albite				3.13		−3.70										500	800	Spectr.	4
Albite				2.39		−2.51										400	500	Experim.	5
Albite				1.59		−1.16										500	800	Experim.	5
Albite				2.79	−0.32	−2.83										200	800	Increment	6
Albite				4.13	−6.30	0.27										0	1200	Theory/ exper.	7
Albite				4.32	−6.27	2.00										0	1200	Increment	1
Albite				0.94												600	1300	Experim.	8
Albite				0.97												500	800	Emp./ theory	9
Albite				0.5												400	800	Experim.	10
Almandine	0.72	4.26	−1.79	3.76	−9.02	2.52	0.25	4.36	−1.79							0	1200	Increment	1
Amesite	0.53	3.53	−1.47	3.95	−8.30	2.38	0.06	3.63	−1.47							0	1200	Increment	2
Analcime				3.48		−4.62										0	400	Experim.	11
Andalusite	0.28	2.26	−0.94	4.2	−7.02	2.15	−0.19	2.36	−0.94							0	1200	Increment	1
Andalusite	0.28	2.30	−0.95													0	1200	Increment	12
Andradite	0.72	4.29	−1.80	3.76	−9.05	2.52	0.26	4.38	−1.80							0	1200	Increment	1
Anglesite	0.54	0.78	−0.36	3.94	−5.54	1.87										0	1200	Increment	13
Anhydrite				3.88		−2.90										100	600	Experim.	14
Anhydrite				3.21		−4.72										100	550	Experim.	15
Anhydrite	0.29	−0.17	0.05	4.19	−4.59	1.71										0	1200	Increment	13
Ankerite	0.36	−0.15	0.03	4.12	−4.62	1.71							−1.82	9.96	−3.45	0	1200	Increment	13
Annite	0.59	3.79	−1.59	3.89	−8.56	2.43	0.13	3.89	1.59							0	1200	Increment	2
Annite				3.89	−8.56	2.43										0	1000	Increment	2
Anorthite	0.36	2.73	−1.14	4.12	−7.5	2.24	−0.11	2.83	−1.14							0	1200	Increment	1
Anorthite				3.13		−3.70										500	800	Theory	4
Anorthite				1.99												600	1300	Experim.	8
Anthophyllite	0.52	3.49	−1.46	3.96	−8.26	2.38	0.06	3.60	−1.46							0	1200	Increment	2
Aragonite	0.57	1.03	−0.47	3.91	−5.79	1.92							−1.61	11.14	−3.66	0	1200	Increment	13
Axinite	0.28	2.26	−0.94	4.20	−7.02	2.15	−0.19	2.36	−0.94							0	1200	Increment	2
Barite				3.01		−7.3										110	350	Experim.	16

Barite – 1 m $NaCl–H_2O$				2.64		−5.3										110	350	Experim.	16
Barite	0.54	0.71	−0.33	3.94	−5.47	1.86										0	1200	Increment	13
Beryl	0.50	3.4	−1.42	3.98	−8.16	2.34	−0.03	3.5	−1.42							0	1200	Increment	1
Biotite				3.69		−0.60										500	800	Emp./ theory	9
Biotite	0.64	3.99	−1.67	3.84	−8.76	2.46	0.18	4.10	1.67	1.14	4.89	−2.08				0	1200	Increment	2
Biotite				0.67	−1.06	−1.95										200	800	Increment	6
Bi–phosphate	0.19	−0.21	0.07	4.29	−4.55	1.72	−0.28	−0.11	0.07							0	1200	Increment	17
Calcite				2.78		−2.89										0	500	Exp./ theory	18
Calcite				3.77	−3.64	−1.07										0	1200	Theory/ exper.	7
Calcite				0.38												600	1300	Experim.	8
Calcite	0.47	−0.10		4.01	−4.66	1.71							−1.71	10.01	−3.45	0	1200	Increment	13
Cassiterite	0.56	5.80	−3.04				3.68	−9.58	2.00							0	1200	Increment	19
Celestite	0.52	0.52	−0.25	3.96	−5.28	1.82										0	1200	Increment	13
Cerianite	1.67	9.32	−4.97				2.57	−13.09	3.93							0	1200	Increment	19
Cerussite	0.60	1.34	−0.59	3.88	−6.10	1.98							−1.58	11.45	−3.72	0	1200	Increment	13
Chamosite	0.42	3.04	−1.27	4.06	−7.81	2.3	−0.04	3.14	−1.27							0	1200	Increment	2
Chlorapatite	0.73	2.25	−0.97	3.75	−7.02	2.15	0.27	2.35	−0.97							0	1200	Increment	17
Chlorite										0.69	4.30	−1.80				0	1200	Increment	2
Chlorite				1.56		−4.70													20
Chlorite				4.81	−8.38														21
Chlorite				1.19	6.78	−13.68													22
Chlorite				0.81	−1.01	−2.01													6
Chlorite				5.44		−1.63										500	800	Theory	32
Chloritoid	0.49	3.35	−1.40	3.99	−8.11	2.30	0.02	3.45	−1.4							0	1200	Increment	2
Chondrodite	0.74	4.34	−1.82	3.73	−9.11	2.52	0.27	4.45	−1.82							0	1200	Increment	2
Chromite	1.63	9.22	−4.91				2.62	−12..99	3.87							0	1200	Increment	19
Clinochlore				0.51	3.43	−1.43										0	1000	Increment	2
Clinochlore	0.51	3.43	−1.43	3.97	−8.19	2.36	0.04	3.53	−1.43							0	1200	Increment	2
Clinohumite	0.70	4.19	−1.75	3.78	−8.95	2.5	0.23	4.29	−1.75							0	1200	Increment	2
CO_2 gas				4.68	0.82	−2.11										100	1300	Theory	23
Coesite	0.02	0.17	−0.07													0	1200	Increment	12
Cordierite	0.38	2.85	−1.19	4.1	−7.62	2.26	−0.08	2.95	−1.19							0	1200	Increment	1
Corundum	2.00	9.94	−5.32				2.24	−13.71	4.28							0	1200	Increment	19
Cristobalite	−0.08	−0.03	0.02													0	1200	Increment	12
Cummingtonite	0.52	3.48	−1.45	3.96	−8.25	2.37	0.05	3.58	−1.45							0	1200	Increment	2
Datolite	0.15	1.44	−0.6	4.33	−6.21	1.99	−0.31	1.54	−0.6							0	1200	Increment	2
Diopside				2.75												600	1300	Experim.	8

Continued

Table IV.1 Oxygen—Cont'd

	Quartz–Mineral			Mineral–H_2O			Calcite–Mineral			Mineral–OH			CO_2–Mineral			Min. *T*	Max. *T*		
	*C**	*B*	*A*	*C*	*B*	*A*	*C*	*B*	*A*	*C*	*B*	*A*	*C*	*B*	*A*	(°C)	(°C)	Method	Refs.**
Diopside	0.56	3.67	−1.53	3.92	−8.43	2.40	0.10	3.78	−1.53							0	1200	Increment	1
Dolomite				3.2		−1.50										300	510	Experim.	33
Dolomite	0.42	−0.12	0.01	4.06	−4.65	1.71							−1.76	9.99	−3.45	0	1200	Increment	13
Enstatite	0.51	3.45	−1.44	3.97	−8.22	2.37	0.05	3.55	−1.44							0	1200	Increment	1
Epidote	0.42	3.05	−1.27	4.05	−7.81	2.29	−0.04	3.15	−1.27							0	1200	Increment	2
Epidote				0.85	−0.99	−2.03										200	800	Increment	6
Fayalite	0.84	4.69	−1.97	3.64	−9.46	2.59	0.38	4.79	−1.97							0	1200	Increment	1
Ferrosilite	0.48	3.33	−1.39	3.99	−8.09	2.35	0.02	3.43	−1.39							0	1200	Increment	1
Fluorapatite	0.67	1.87	−0.81	3.81	−6.63	2.08	0.21	1.97	−0.81							0	1200	Increment	17
Forsterite				3.67												600	1300	Experim.	8
Forsterite	0.93	4.95	−2.09	3.55	−9.72	2.64	0.46	5.05	−2.09							0	1200	Increment	1
Gedrite	0.63	3.96	−1.66	3.85	−8.72	2.46	0.17	4.06	1.66							0	1200	Increment	2
Geikielite	1.46	8.85	−4.7				2.78	−12.63	3.66							0	1200	Increment	19
Glauconite	0.49	3.34	−1.39	3.99	−8.11	2.34	0.02	3.44	1.39							0	1200	Increment	2
Glaucophane	0.45	3.18	−1.32	4.03	−7.94	2.31	−0.02	3.28	−1.32							0	1200	Increment	2
Garnet				2.88												500	800	Emp./theory	9
Grossular	0.74	4.35	−1.82	3.74	−9.11	2.52	0.27	4.45	−1.82							0	1200	Increment	1
Hedenbergite	0.55	3.61	−1.51	3.93	−8.37	2.40	0.08	3.71	−1.51							0	1200	Increment	1
Hematite	1.55	9.05	−4.82				2.69	−12.82	3.78							0	1200	Increment	19
Hercynite	1.92	9.80	−5.24				2.32	−13.58	4.2							0	1200	Increment	19
Hornblende	0.59	3.80	−1.59	3.89	−8.56	2.43	0.13	3.9	1.59	1.22	5.21	−2.22				0	1200	Increment	2
Hornblende				3.15		−0.30										500	800	Emp./theory	9
Humite	0.71	4.24	−1.78	3.77	−9.01	2.51	0.24	4.34	−1.78							0	1200	Increment	2
Hydroxyapatite	0.65	1.71	−0.75	3.83	−6.48	2.05	0.19	1.81	−0.75							0	1200	Increment	17
Illite	0.34	2.60	−1.08	4.14	−7.36	2.21	−0.13	2.70	1.08	0.93	5.91	−2.47				0	1200	Increment	2
Ilmenite	1.36	8.61	−4.57				2.88	−12.38	3.53							0	1200	Increment	19
Jacobsite	1.23	8.24	−4.37				3.02	−12.02	3.33							0	1200	Increment	19
Jadeite	0.31	2.42	−1.01	4.17	−7.19	2.18	−0.16	2.52	−1.01							0	1200	Increment	1
Kaolinite	0.19	1.68	−0.69	4.29	−6.44	2.03	−0.28	1.78	−0.69	2.71	2.64	−1.3				0	1200	Increment	2
K–Feldspar	0.16	1.50	−0.62	4.32	−6.27	2.00	−0.3	1.60	−0.62							0	1200	Increment	1
Kyanite	0.32	2.52	−1.05	4.16	−7.29	2.20	−0.14	2.63	−1.05							0	1200	Increment	1
Kyanite	0.33	2.55	−1.06													0	1200	Increment	12
Lepidolite	0.38	2.81	−1.17	4.10	−7.58	2.25	−0.09	2.91	1.17							0	1200	Increment	2

Leucite	0.22	1.91	−0.79	4.26	−6.67	2.08	−0.24	2.01	−0.79							0	1200	Increment	1
Lizardite	0.51	3.42	−1.43	3.97	−8.19	2.36	0.04	3.52	−1.43							0	1200	Increment	2
Ilvaite	1.05	5.32	−2.24	3.43	−10.08	2.69	0.58	5.42	−2.24							0	1200	Increment	2
Magnesiochromite	1.76	9.49	−5.07				2.48	−13.27	4.03							0	1200	Increment	19
Magnesioferrite	1.29	8.43	−4.47				2.95	−12.2	3.43							0	1200	Increment	19
Magnesite				3.36		−2.96													24
Magnesite	0.41	−0.13	0.01	4.07	−4.64	1.72							−1.77	9.98	−3.46	0	1200	Increment	13
Magnetite				5.57												500	800	Emp./ theory	9
Magnetite				6.29												600	1300	Experim.	8
Magnetite				6.11												400	800	Experim.	10
Magnetite	1.22	8.22	−4.35				3.02	−12.00	3.31							0	1200	Increment	19
Magnetite (Spinel-type)	1.99	7.30	−3.13	2.49	−12.1	3.00	1.52	7.41	−3.13							0	1200	Increment	34
Magnetite (inv. Spinel-type)	1.60	6.59	−2.81	2.88	−11.4	2.89	1.13	6.70	−2.81							0	1200	Increment	34
Malayaite	0.60	3.83	−1.60	3.88	−8.60	2.43	0.14	3.94	−1.60							0	1200	Increment	1
Margarite	0.38	2.84	−1.18	4.10	−7.61	2.25	−0.08	2.94	1.18							0	1200	Increment	2
Monazite	0.26	−0.18	0.05	4.22	−4.58	72.00	−0.20	−0.08	0.05							0	1200	Increment	17
Muscovite	0.38	2.84	−1.18	4.10	−7.61	2.25	−0.08	2.97	1.18	0.97	5.73	−2.4				0	1200	Increment	2
Muscovite				2.25		−0.60										500	800	Emp./ theory	9
Muscovite				2.38		−3.89										400	650	Experim.	3
Muscovite				1.90		−3.1										500	800	Spectr.	4
Muscovite				1.46	−0.78	−2.28										200	800	Increment	6
Muscovite				4.10	−7.61	2.25										0	1200	Increment	2
Natrite				3.83	−6.48	2.05										0	1200	Increment	25
Nepheline	0.37	2.77	−1.15	4.11	−7.53	2.24	−0.10	2.89	−1.15							0	1200	Increment	1
Norbergite	0.86	4.75	−2.00	3.62	−9.51	2.60	0.39	4.85	−2.00							0	1200	Increment	2
Norsethite				3.99	−5.01	1.77										0	1200	Increment	25
Otavite					2.76	−3.96										0	500	Experim.	35
Otavite	0.41	−0.12	0.01	4.07	−4.64	1.71							−1.77	9.97	−3.45	0	1200	Increment	13
Paragonite	0.37	2.77	−1.15	4.11	−7.54	2.24	−0.10	2.87	1.15							0	1200	Increment	2
Pargasite	0.71	4.22	−1.77	3.77	−8.99	2.51	0.24	4.32	1.77							0	1200	Increment	2
Pectolite	0.41	2.95	−1.23	4.08	−7.71	2.28	−0.06	3.05	−1.23							0	1200	Increment	2
Pennine	0.51	3.43	−1.43	3.97	−8.19	2.36	0.04	3.53	−1.43							0	1200	Increment	2
Phenakite	1.06	5.34	−2.26	3.42	−10.11	2.70	0.60	5.45	−2.26							0	1200	Increment	1
Phengite	0.35	2.64	−1.10	4.13	−7.41	2.22	−0.12	2.74	1.10							0	1200	Increment	2
Phlogopite	0.62	3.92	−1.64	3.86	−8.68	2.45	0.16	4.02	1.64	1.13	4.95	−2.11				0	1200	Increment	2
Phlogopite				3.86	−8.68	2.45										0	1000	Increment	2
Plattnerite	0.56	5.80	−3.04				3.68	−9.58	2.00							0	1200	Increment	19

Continued

Table IV.1 Oxygen—Cont'd

	Quartz–Mineral			Mineral–H_2O			Calcite–Mineral			Mineral–OH			CO_2–Mineral			Min. *T*	Max. *T*		
	*C**	*B*	*A*	*C*	*B*	*A*	*C*	*B*	*A*	*C*	*B*	*A*	*C*	*B*	*A*	(°C)	(°C)	Method	Refs.**
Prehnite	0.30	2.40	−1.00	4.18	−7.17	2.18	0.17	2.5	−1.00							0	1200	Increment	2
Pyrolusite	0.64	6.16	−3.23				3.61	−9.94	2.19							0	1200	Increment	19
Pyrope	0.73	4.30	−1.80	3.75	−9.07	2.52	0.26	4.40	−1.80							0	1200	Increment	1
Pyrophyllite	0.08	0.86	−0.35	4.4	−5.62	1.87	−0.38	0.96	−0.35							0	1200	Increment	2
Pyroxene				2.67												500	800	Emp./theory	9
Pyroxene				2.08												400	800	Experim.	10
Pyroxene	2.75															600	1300	Experim.	8
Quartz				4.48	−4.77	1.71	−0.47	0.10								0	1200	Increment	1
Quartz				3.38		−3.40										200	500	Experim.	26
Quartz				2.51		−1.96										500	750	Experim.	26
Quartz				3.55		−2.57										195	573	Spectr.	27
Quartz				3.23		−2.94										573	1000	Spectr.	27
Quartz				4.10		−3.70										500	800	Theory/emp.	4
Quartz				3.19	0.42	−3.50										200	727	Spectr./experim.	28
Quartz				3.28		−5.53										0	100	Theory	29
Quartz				3.34		−3.31										250	500	Experim.	5
Quartz				3.13		−2.94										400	500	Experim.	5
Quartz				2.05		−1.14										500	800	Experim.	5
Quartz				5.89	−7.99	2.85										250	600	3 isotopes	10
Quartz				3.31		2.71										180	550		30
Quartz				3.70		−3.21										200	800	Increment	6
Quartz				4.24	−3.77	−1.04										0	1200	Theory/exper.	7
Quartz, amorph.				3.52		−4.35										34	93	Emp./exp.	36
Rhodochrosite	0.29	−0.17	0.04	4.19	−4.59	1.72							−1.89	9.94	−3.46	0	1200	Increment	13
Rhodonite	0.63	3.94	−1.65	3.85	−8.71	2.46	0.16	4.04	−1.65							0	1200	Increment	1
Riebeckite	0.43	3.08	−1.28	4.05	−7.85	2.3	−0.04	3.18	1.28							0	1200	Increment	2
Rutile	0.79	6.82	−3.59				3.45	−10.6	2.55							0	1200	Increment	19
Scheelite	0.25	3.85	−2.01	3.99	−7.63	0.96	−0.22	3.99	−2.02							0	1200	Increment	31
Serpentine	0.49	3.35	−1.40	3.99	−8.12	2.35	0.02	3.45	−1.40							0	1200	Increment	2
Siderite				3.13		−3.50										33	197	Experim.	37
Siderite	0.25	−0.19	0.05	4.23	−4.58	1.73							−1.93	9.92	−3.47	0	1200	Increment	13

Sillimanite	0.22	1.89	−0.78	4.26	−6.65	2.07	−0.25	1.99	−0.78							0	1200	Increment	1
Sillimanite	0.23	1.94	−0.80													0	1200	Increment	12
Smithsonite	0.21	−0.2	0.06	4.27	−4.56	1.73							−1.97	9.90	−3.47	0	1200	Increment	13
Spessartine	0.73	4.31	−1.81	3.75	−9.07	2.52	0.26	4.41	−1.81							0	1200	Increment	1
Spinel	2.06	10.04	−5.38				2.18	−13.81	4.34							0	1200	Increment	19
Staurolite	0.39	2.9	−1.21	4.09	−7.66	2.27	−0.07	3.00	−1.21							0	1200	Increment	2
Stishovite	0.22	1.88	−0.78	4.26	−6.64	2.08	−0.25	1.98	−0.78							0	1200	Increment	1
Stishovite	0.22	1.91	−0.79													0	1200	Increment	12
Strontianite				2.69		−3.24										0	500	Exp./ theory	38
Strontianite	0.58	1.16	−0.52	3.90	−5.93	1.95							−1.6	11.27	−3.69	0	1200	Increment	13
Talc	0.28	2.27	−0.94	4.20	−7.04	2.14	−0.19	2.38	−0.94	0.87	6.14	−2.56				0	1200	Increment	2
Tephroite	0.86	4.75	−2.00	3.62	−9.51	2.60	0.39	4.85	−2.00							0	1200	Increment	1
Thorianite	1.66	9.30	−4.96				2.58	−13.07	3.92							0	1200	Increment	19
Thorite	0.79	4.50	−1.89	3.69	−9.27	2.55	0.32	4.61	−1.89							0	1200	Increment	1
Thuringite	0.58	3.73	−1.56	3.90	−8.50	2.42	0.11	3.84	−1.56							0	1200	Increment	2
Titanite	0.67	4.11	−1.72	3.81	−8.87	2.49	0.21	4.21	−1.72							0	1200	Increment	1
Topaz	0.30	2.41	−1.00	4.18	−7.18	2.17	−0.16	2.52	−1.00							0	1200	Increment	2
Tourmaline	0.27	2.22	−0.92	4.21	−6.99	2.14	−0.20	2.32	−0.92	0.92	6.28	−2.62				0	1200	Increment	2
Tremolite	0.53	3.52	−1.47	3.90	−8.28	2.38	0.06	3.62	1.47	1.19	5.45	−2.31				0	1200	Increment	2
Tridymite	−0.43	−0.15	0.09													0	1200	Increment	12
Ulvöspinel	1.43	8.78	−4.67				2.81	−12.56	3.63							0	1200	Increment	19
Uraninite	1.61	9.19	−4.90				2.63	−12.97	3.86							0	1200	Increment	19
Uvarovite	0.72	4.28	−1.80	3.76	−9.05	2.52	0.26	4.39	−1.8							0	1200	Increment	1
Vesuvianite	0.93	4.97	−2.09	3.55	−9.74	2.63	0.47	5.07	−2.09							0	1200	Increment	2
Willemite	0.69	4.18	−1.75	3.79	−8.94	2.50	0.23	4.28	−1.75							0	1200	Increment	1
Witherite				2.57		−4.23										0	500	Exp./ theory	38
Witherite	0.61	1.38	−0.61	3.87	−6.15	1.99							−1.57	11.49	−3.73	0	1200	Increment	13
Wolframite	0.38	4.76	−2.48	3.86	−8.54	1.44	−0.1	4.90	−2.50							0	1200	Increment	31
Wollastonite	0.67	4.11	−1.72	3.81	−8.87	2.49	0.21	4.21	−1.72							0	1200	Increment	1
Xenotime	−0.13	−0.32	0.14	4.60	−4.45	1.75	−0.59	−0.22	0.14							0	1200	Increment	17
Zircon	0.72	4.26	−1.79	3.76	−9.03	2.52	0.25	4.37	−1.79							0	1200	Increment	1
Zoisite	0.43	3.07	−1.28	4.05	−7.84	2.30	−0.04	3.18	−1.28	1.13	5.78	−2.45				0	1200	Increment	2

Coefficients of the polynomial isotope fractionation equation: $1000\ ln\alpha = A + B \times 10^3/T + C \times 10^6/T^2$ (T is temperature in K).

**References: (1) Zheng (1993b). (2) Zheng (1993a). (3) O'Neil and Taylor (1967). (4) Bottinga and Javoy (1973). (5) Matsuhisa et al. (1978). (6) Richter and Hoernes (1988). (7) Clayton and Kieffer (1991). (8) Chiba et al. (1989). (9) Bottinga and Javoy (1975). (10) Matthews et al. (1983). (11) Karlsson and Clayton (1990). (12) Zheng (1993c). (13) Zheng (1999). (14) Lloyd (1968). (15) Chiba et al. (1981). Note: The equation typed in the paper reads $3.21 \times 10e^3/T^2$–4.72, however it should read $3.21 \times 10e^6/T^2$–4.72 to be consistent with their Figure 1. (16) Kusakabe and Robinson (1977). (17) Zheng (1996). (18) O'Neil et al. (1969). Note: The equations are recalculated by Friedman and O'Neil (1977) to account for a different value of the H_2O–CO_2 fractionation. (19) Zheng (1991). (20) Werner and Taylor (1971). (21) Cole (1985). (22) Savin and Lee (1988). (23) Richet et al. (1977). (24) Steele (1998). (25) Zheng (2011). (26) Clayton et al. (1972). (27) Shiro and Sakai (1972). (28) Becker and Clayton (1976). (29) Kawabe (1978). (30) Truesdell (1974). (31) Zheng (1992). (32) Javoy (1977). (33) Northrop and Clayton (1966). (34) Zheng (1995). (35) Kim and O'Neil (1997). (36) Kita et al. (1985). (37) Carothers et al. (1988). (38) Friedman and O'Neil (1977).*

Table IV.2 Sulfur

Compound–H_2S	*A**	*B*	*C*	Range (°C)	Method	References**
Ag_2S			−0.8 ± 0.1		Experim.	1
Ag_2S			−0.62	280–700	Unknown	2
Bi_2S_3			−0.67	150–600	Unknown	2
CaS, SrS, BaS			0.60 ± 0.1		Experim.	1
CdS			0.18		Increment	3
CdS			−0.4 ± 0.1		Experim.	1
COS	−1.15	0.43	0.67	>25	Experim.	1
CoS, NiS, MnS			0.10 ± 0.05		Experim.	1
Cu_2S			−0.06	510–630	Unknown	2
Cu_2S			−0.75 ± 0.1		Experim.	1
Cu_5FeS_4			−0.25 ± 0.1		Experim.	1
Cu_5FeS_4			−0.07		Increment	3
$CuFe_2S_3$			0.05 ± 0.05		Experim.	1
$CuFe_2S_3$			0.04		Increment	3
$CuFeS_2$			−0.05 ± 0.08	200–600	Experim.	1
$CuFeS_2$			0.05		Increment	3
CuS			0.04	280–490	Unknown	2
CuS			−0.4 ± 0.1		Experim.	1
$FeNi_2S_4$			−0.27		Increment	3
FeS			0.10 ± 0.05	200–600	Experim.	1
FeS			0.25		Increment	3
FeS_2			0.40 ± 0.08	200–700	Experim.	1
HgS			−0.7 ± 0.1		Experim.	1
HS^-	−0.6		−0.06 ± 0.15	50–350	Experim.	1
MgS			0.50 ± 0.1		Experim.	1
MoS_2			0.45 ± 0.1		Experim.	1
MnS_2			0.40 ± 0.10		Experim.	1
PbS			−0.63 ± 0.05		Experim.	1
S, S_8	± 0.5		−0.16	200–400	Experim.	1
S^{2-}	−1.23	−1.23	−0.21		Experim.	1
Sb_2S_3			−0.75 ± 0.1		Experim.	1
SnS			−0.45 ± 0.1		Experim.	1
SO_2	−0.5 ± 0.5		4.70	350–1050	Experim.	1
Sulphate	0.56 ± 0.5		6.463	200–400	Experim.	4
Sulphate	6.0 ± 0.5		5.26	200–350	Experim.	1
Sulphate	±1.0		8.0	>400	Experim.	1
Sulphate			6.5 ± 0.3	600–1000	Experim.	5
Sulphite	−5.0	5.82	4.12	>25	Experim.	1
ZnS			0.10 ± 0.05	50–705	Experim.	1

*Coefficients of the polynomial isotope fractionation equation: $1000\ \ln\alpha = A + B \times 10^3/T + C \times 10^6/T^2$, T is temperature in K

**References: (1) Ohmoto and Rye (1979), (2) Seal (2006), (3) Li and Liu (2006), (4) Ohmoto and Lasaga (1982), (5) Marini et al. (2011).

Table IV.3 Carbon

Compound–CO_2	*A**	*B*	*C*	*D*	Range (°C)	References**
C	−3.04	0	−4.53		600–1200	7
C	9.32	−22.89	6.921	−0.6637	<700	2
Calcite	2.962	−11.346	5.538	−0.388	<372	3
Calcite	8.27	−18.11	8.557	−0.8914	<600	2
Calcite	−2.72	9.58	−3.46	0	500–1200	7
CO	9.1	−17.56	−2.84	0	<700	2
CO_3^{2-}	−22.24	8.885	0	0	<100	6
CO_3^{2-}	6.14	−17.66	8.196	−0.8361	<330	2
Dolomite	3.132	−11.346	5.538	−0.388	<600	1
Dolomite	8.44	−18.11	8.737	−0.8914	<600	2
HCO_3^-	−35.7	20.16	−2.16	0	<290	2
CH_4	4.363	−8.933	−5.21	0.4194	<700	1
Magnesite	−2.73	9.86	−3.12	0	500–700	4
Siderite	0.82	0	0.861	0	33–197	5
Siderite	0	1.75	0	−4.57		8

**Coefficients of the polynomial isotope fractionation equation:* $1000 \ln\alpha = A + B \times 10^3/T + C \times 10^6/T^2 + D \times 10^9/T^{-3}$*, T is temperature in K.*
***References: (1) Ohmoto and Goldhaber (1997). (2) Ohmoto and Rye (1979). (3) Chacko et al. (1991). (4) Steele (1998). (5) Carothers et al. (1988). (6) Thode et al. (1965). (7) Scheele and Hoefs (1992). (8) Golyshev et al. (1981).*

Table IV.4 Hydrogen

Compound–H_2O	*A**	*B*	*C*	*D*	Range (°C)	References**
Brucite		−27.9	8.8	−2.24	27–627	1
Brucite	14.5	−38.6	8.72		144–510	10
Chlorite	−28				500–700	11
Clinozoisite	−138.8		29.2		150–300	2
Epidote	−61.9		9.3		300–600	9
Hornblende	7.9		−23.9		400–700	4
Kaolinite	19.36		−4.53		<230	6
Kaolinite	−7.7		−2.2		0–330	7,12
Mg–biotite	−2.8		−21.3		400–850	4
Muscovite	19.1		−22.1		450–750	4
Phlogopite	27.1		−22.4		400–850	4
Smectite/illite	94.7	−45.3			0–150	8
Tourmaline	2.3		−27.9		450–800	3
Tremolite	14.9		−31		650–850	5
Zoisite	−27.3		−15.07		280–650	2

**Coefficients of the polynomial isotope fractionation equation:* $1000 \ln\alpha = A + B \times 10^3/T + C \times 10^6/T^2 + D \times 10^9/T^{-3}$*, T is temperature in K.*
***References: (1) Reynard and Caracas (2009). (2) Graham et al. (1980). (3) Jibao and Yaqian (1997). (4) Suzuoki and Epstein (1976). (5) Graham et al. (1984). (6) Lambert and Epstein (1980). (7) Gilg and Sheppard (1996). (8) Capuano (1992). (9) Chacko et al. (1999). (10) Satake and Matsuo (1984). (11) Graham et al. (1987). (12) Sheppard and Gilg (1996).*

APPENDIX

RAMAN BANDS SORTED BY VIBRATIONS

V

cm^{-1}	Formula	Name	All Bands	Source
74	$Cu_2PO_4(OH)$	Libethenite	**74**, 92, 113, 140, **160**, 227, 270, **301**, 319, 371, 392, 430, 461, 561, 590, 944, **979**, 1008, **1022**, **3475**	57
80	$Cu_4(OH)_6SO_4$	Brochantite2	**80**, **91**, 132, 138, 142, 157, 188, 199, 242, 319, 387, 426, 450, 484, 509, 595, 609, 910, **973**, 1097, **3401**, **3566**, **3588**	13
87	HBO_2	Metaboric acid III	**87**, **119**, 404, 419, 469, 478, **599**, 638, **810**, 1100, 1343, 1460, 3195, 3295, 3365	19
91	$Cu_4(OH)_6SO_4$	Brochantite2	**80**, **91**, 132, 138, 142, 157, 188, 199, 242, 319, 387, 426, 450, 484, 509, 595, 609, 910, **973**, 1097, **3401**, **3566**, **3588**	13
100	ZnO	Zincite	**100**, 331, **438**, 489	30
102	$PbCl(OH)$	Paralaurionite	81, **102**, **132**, 220, 273, **330**, 598, 733	19
105	$Na_2B_4O_7{\cdot}4H_2O$	Kernite	**105**, 175, 256, 388, 426, 456, **496**, 554, 579, **737**, 802, 848, **856**, **929**, 978, 1009, 1034, 1089, 1139	19
106	$Cu_2(OH)_3Cl$	Atacamite2	66, **106**, **120**, **139**, **150**, 450, **514**, 586, 820, 845, **908**, 975, **3351**, **3453**	13
109	$Cu_3(PO_4)(OH)_3$	Cornetite	**109**, 131, 141, 158, 174, 297, 313, 363, **446**, **477**, 515, 583, 639, 945, **971**, **996**, 1016, **1083**, 1137, **3414**	57
110	$Ca_5(PO_4)_2(OH)_4$	Pseudomalachite	88, **110**, 131, 176, 368, 450, **480**, 537, 609, 972, 1000, **1086**	30
110	$Cu_5(PO_4)_2(OH)_4$	Reichenbachite	8989, **110**, 135, 175, 188, 214, 255, 298, 365, 412, 453, **480**, 515, 540, 572, 607, 636, 951, 971, 998, **1083**, **3380**, **3428**	57

Continued

Geofluids. http://dx.doi.org/10.1016/B978-0-12-803241-1.09973-1

cm^{-1}	Formula	Name	All Bands	Source
117	$Cu_2(OH)_3Cl$	Paratacamite	95, **117**, **140**, 165, 364, 420, **511**, 576, 799, 868, 892, 928, 970, **3311**, **3355**, **3441**	30
118	$CuCl_2$	Tolbachite	95, 106, **118**, **140**, 149, 213, 362, 416, 477, **511**, 575, 799, 819, 845, 869, 893, 910, 928, 971	30
119	HBO_2	Metaboric acid III	**87**, **119**, 404, 419, 469, 478, **599**, 638, **810**, 1100, 1343, 1460, 3195, 3295, 3365	19
120	$Cu_2(OH)_3Cl$	Atacamite2	66, **106**, **120**, **139**, **150**, 450, **514**, 586, 820, 845, **908**, 975, **3351**, **3453**	13
121	$Cu_2(OH)_3Cl$	Atacamite	106, **121**, 361, 419, 446, 475, **511**, 818, 846, **912**, **975**	30
129	$Fe_4Be_3(SiO_4)_3S$	Danalite	**129**, 155, 173, 237, 361, 443, 531, 566, 681, **891**, 943, 1023	19
131	$Mg_6[Cl_2/B_{14}O_{26}]$	Boracite	**131**, 602, **660**, **840**, 1142	19
131	$Mn_4Be_3(SiO_4)_3S$	Helvite	**131**, 158, 176, 238, 288, 362, 444, 527, 565, 678, **981**, 942, 1020	19
131	$YNbO_4$	Fergusonite monoclinic	**131**, **324**, **336**, 658, 675, 698, **811**	19
132	$PbCl(OH)$	Paralaurionite	81, **102**, **132**, 220, 273, **330**, 598, 733	19
132	$CaLiAl_2(AlBeSi_2)O_{10}(OH)_2$	Bityite	104, **132**, 365, **478**, 684, **898**, 1049	19
136	PbS	Galena	**136**, 270	10
137	$NaMgF_3$	Neighborite	102, 127, **137**, **182**, 208, **226**	19
137	$Pb_3O_2Cl_2$	Mendipite	**137**, 268, 290, 335, 432, 471	19
139	$Cu_2(OH)_3Cl$	Atacamite2	66, **106**, **120**, **139**, **150**, 450, **514**, 586, 820, 845, **908**, 975, **3351**, **3453**	13
140	$Cu_2(OH)_3Cl$	Paratacamite	95, **117**, **140**, 165, 364, 420, **511**, 576, 799, 868, 892, 928, 970, **3311**, **3355**, **3441**	30
140	$CuCl_2$	Tolbachite	95, 106, **118**, **140**, 149, 213, 362, 416, 477, **511**, 575, 799, 819, 845, 869, 893, 910, 928, 971	30
143	$LiAlSi_4O_{10}$	Petalite	**143**, **281**, 318, 357, 383, **491**, **793**	19
143	TiO_2	Anatase	**143**, 195, 395, 514, **638**	10
146	K_4PbF_6	K–Pb hexafluoride	63, 89, **146**, 375	19
147	$Mg_2(OH)_2(CO_3)\cdot 3H_2O$	Artinite	**147**, 173, 472, 704, **1094**, 3030, 3229, **3593**	10

cm^{-1}	Formula	Name	All Bands	Source
147	$Cu_4Cl[Cl_{0.29}(OH)_{0.71}](OH)_6$	Claringbullite	119, 136, **147**, 163, 231, 260, 356, 389, 447, **511**, 579, **815**, **906**, **970**, 1049, 3211, 3331, **3351**, **3433**, 3458	31
148	$Be_2BO_3(OH,F)$	Hambergite	**148**, **158**, 992	22
150	HBO_2	Metaboric acid I cubic	104, **150**, **231**, 271, **385**, 434, 454, 478, **511**, **618**, **759**, **780**, **820**, 970, 1025, 1240, 1980, 2380	10
150	TiO_2	Brookite	127, **150**, 247, 318, 366, 412, **645**	10
150	$Cu_2(OH)_3Cl$	Atacamite2	66, **106**, **120**, **139**, **150**, 450, **514**, 586, 820, 845, **908**, 975, **3351**, **3453**	13
152	$Pb_2CO_3Cl_2$	Phosgenite	**152**, 176, **182**, 282, 667, 760, 833, 1058, 1061, 1293, 1304, 1327, 1384, 1518, 1672	12
152	$Pb_5(PO_4, AsO_4)_3Cl$	Pyromorphite–As	**152**, 177, 186, 206, 318, 339, 344, 377, 391, 409, 414, 768, 776, **815**, 818, 825, 917, **944**, 1014, 3325, **3378**, 3444	34
154	$Cu_2(OH)_2(CO_3)$	Malachite	**154**, **178**, 272, **434**, 537, 721, **1098**, 1492, 3386, **3468**	10
156	$Na_2CO_3{\cdot}H2O$	Thermonatrite	**156**, 185, 230, **1062**, 1432	10
156	$SnCl_2{\cdot}2H_2O$	Clathrate	78, 123, **156**, 187	19
157	S_8	Sulphur	**157**, 187, **219–222**, 246, 437, 462, **473**	10
158	$Be_2BO_3(OH,F)$	Hambergite	**148**, **158**, 992	22
159	$PbCl_2$	Cotunnite	**159**, **179**, 417	19
160	$Cu_5(PO4)_2(OH)_4$	Ludjibaite	**160**, 190, 226, 263, **301**, 368, 387, 402, 449, 557, 586, 633, 736, 784, 815, 855, 925, **981**, 1019, 1046, 1072, 1115, **3470**	57
160	$Cu_2PO_4(OH)$	Libethenite	**74**, 92, 113, 140, **160**, 227, 270, **301**, 319, 371, 392, 430, 461, 561, 590, 944, **979**, 1008, **1022**, **3475**	57
161	$(Ce, La)CO_3F$	Bastnäsite	**161**, 175, 181, 234, 261, 286, 288, 352, 402, 566, 601, 606, 719, 735, **1085**, **1096**, 1432, 1445, 1504, 1737	11
161	$(Mg, Fe, Mn)Al_3(BO_3)(SiO_4)O_2$	Grandidierite	**161**, 228, 242, 269, **343**, 363, 376, **425**, 442, 489, **577**, 659, 864, **949**, 979, 1042	8
164	$Na_2Ca(CO_3)_2{\cdot}5H_2O$	Gaylussite	**164**, 265, 723, **1071**, **2944**, 3334	10

Continued

cm^{-1}	Formula	Name	All Bands	Source
166	$PbMoO_4$	Wulfenite	**166**, **315**, 348, 742, 765, **868**, **1736**, 3290, **3350**, **3377**, 3533, 3566	19
168	K_2SnCl_6	Potassium–tin chloride	**168**, 239, **318**	19
169	Rb_2SnCl_6	Rubidium–tin chloride	**169**, 239, **318**, 549	19
171	$CaCl_2{\cdot}2H_2O$	Sinjarite	124, **171**, 194, 297, 535	19
173	$Mg_2(OH)_2(CO_3){\cdot}3H_2O$	Artinite	**147**, 173, 472, 704, **1094**, 3030, 3229, **3593**	10
177	$(Ca, Na)(Al, Si)AlSi_3O_8$	Oligoclase	**177**, 275, 329, 403, 453, 474, **511**, 747, 808, 1007, 1114	19
178	$Cu_2(OH)_2(CO_3)$	Malachite	**154**, **178**, 272, **434**, 537, 721, **1098**, 1492, 3386, **3468**	10
179	$PbCl_2$	Cotunnite	**159**, **179**, 417	19
181	$CaMgB_6O_{11}{\cdot}6H_2O$	Hydroboracite	**181**, 257, 322, 383, 398, 524, 548, 564, 606, **753**, 837, 876, **3595**	19
182	$NaMgF_3$	Neighborite	102, 127, **137**, **182**, 208, **226**	19
182	$Pb_2CO_3Cl_2$	Phosgenite	**152**, 176, **182**, 282, 667, 760, 833, 1058, 1061, 1293, 1304, 1327, 1384, 1518, 1672	12
185	$SrCl_2$	Strontium chloride	**185**	19
186	$BaCl_2$	Barium chloride	**186**	19
186	$Cu_8Si_8O_{22}(OH)_4{\cdot}H_2O$	Planchéite	104, 122, **139**, 148, **186**, 227, 302, 328, 403, 421, 476, 505, 699, 828, 906, 1048, **1081**, 1355, **1510**, **3250**, **3479**	32
188	$KLi_2AlSi_4O_{10}(OH, F)_2$	Polylithionite2	89, 99, **188**, **245**, 268, 409, 452, 552, **715**, 765, 791, **1157**	19
191	$NaAl(CO_3)(OH)_2$	Dawsonite2	152, **191**, 389, 519, **590**, 824, 897, **1091**, **1506**, 3252, **3283**	19
192	$NaAl(CO_3)(OH)_2$	Dawsonite	**192**, 260, 590, 728, 1065, **1091**, 1506, 3250, **3282**	10
198	$FeAs_2$	Löllingite	**198**, **250**	52
202	$Al_{13}Si_5O_{20}(OH, F)_{18}Cl$	Zunyite	168, **202**, 270, 317, 360, 383, **449**, **610**, 954, 992	8
205	$NH_4Al(CO_3)(OH)_2$	Dawsonite–NH_4	79, 93, 110, 120, 139, **205**, 292, 306, 367, 572, 585, 620, 645, 686, 733, **1107**, 3388, 3460	19
206	$Al_7(BO_3)(SiO_4)_3O_3$	Dumortierite	169, **206**, 293, 353, 403, 418, **508**, 810, 842, **946**, 1076, 1112	8
206	$Be_4Si_2O_7(OH)_2$	Bertrandite	183, **206**, 231, 330, 358, 579, 694, 711, 928, 990, 1073, **3551**, **3587**	19
206	CuCl	Nantokite	127, 155, **206**, 262, 356, 393, **463**	31
207	$Na_2Mg(CO_3)_2$	Eitelite	**207**, **263**, 710, **1105**	54

cm^{-1}	Formula	Name	All Bands	Source
207	$KAl(CO_3)(OH)_2$	Dawsonite–K	77, 169, **207**, 282, 363, 503, 580, 679, 730, 832, **1102**, 1486, 3406, 3432	19
210	$LiOH{\cdot}H_2O$	Li hydroxide monohydrate	**210**, **243**, 362, 388, **510**, 834	19
214	FeS	Pyrrhotite hex.	**214**, **274**, 385, 586	13
216	$CuCl_2{\cdot}2H_2O$	Eriochalcite	117, **216**, 235, 247, 390, **407**, 672, 692, 1620, 3176, 3317, **3367**, 3462	31
217	As	Arsenic	**217**, 281, 393, 607, 659	19
219	S_8	Sulphur	**157**, 187, **219–222**, 246, 437, 462, 473	10
222	$NaAlSi_3O_8$	Kumdykolite	155, **222**, 265, 407, 464, **492**	61
224	$Ca(Ce, Y)(CO_3)_2F$	Synchysite–Y	**224**, 244, **290**, 374, 402, 479, 517, 603, 655, 993, 1054, **1080**, **1095**	19
225	$H_6Si_2O_7$	Pyrosilicic acid	**225**, 426, **631**, 916, **3414**	19
226	$NaMgF_3$	Neighborite	102, 127, **137**, **182**, 208, **226**	19
226		Rubellite	**226**, 246, 276, 332, **379**, 410, 517, 645, **716**, 733, 753, 775, 883, 982, 1036, **1070**	19
227	$MgCl_2{\cdot}6H_2O$	Bischofite	206, **227**, 348, 355, 1554, **1586, 1601, 1604**, 1803, 3356	19
231	HBO_2	Metaboric acid I cubic	104, **150**, **231**, 271, **385**, 434, 454, 478, **511, 618, 759, 780, 820**, 970, 1025, 1240, 1980, 2380	19
235	Al_2SiO_5	Sillimanite	142, **235**, 310, 456, 597, 708, 874, **907**, 964, 1127	10
241	α-FeOOH	Goethite2	**241, 298,** 396, 478, 547, 678, 991	19
243	$LiOH{\cdot}H_2O$	Li hydroxide monohydrate	**210**, **243**, 362, 388, **510**, 834	19
243	$Sr(Ce, Ca, La)(CO_3)_2(OH)_2{\cdot}H_2O$	Ancylite–Ce	125, 189, **243**, 371, 699, 725, **1083**	8, 58
245	$CaB_2(SiO_4)_2$	Danburite	166, 182, **245**, 348, 480, **611**, 628, 974, 1107, 1664, 3079, 3912	19
245	$KLi_2AlSi_4O_{10}(OH, F)_2$	Polylithionite2	89, 99, **188**, **245**, 268, 409, 452, 552, **715**, 765, 791, **1157**	19
250	$FeAs_2$	Löllingite	**198**, **250**	52
253	As	Arsenolamprite	220, 225, **253**	22
255	$C_{10}H_{22}$	Decane saturated	**255**, **843**, 1132, **1302**, 1435, 1451, **2848**, 2858, **2872**, 2897, 2917, 2934	46
256	PbF_4	Lead fluoride	**256**	19

Continued

cm^{-1}	Formula	Name	All Bands	Source
259	$BeAlSiO_4(OH)$	Euclase	145, 180, 237, **259**, 290, **397**, 444, 452, 574, **880**, **908**, 975, 1023, 1059, 1120, **3575**, 3587	19
261	$KAl_4Si_5Al_2O_{20}(OH, F)_4$	Muscovite	178, 197, 216, **261**, 385, 407, 639, **702**, 754, 914, 957, 1117, **3627**	10
262	H_2WO_4	Tungstic acid	188, **262**, **685**, 804, 942	19
262	SrF_2	Strontium fluoride	**262**	19
263	$Na_2Mg(CO_3)_2$	Eitelite	**207, 263,** 710, **1105**	54
264	$KLi_{1.5}Al_{1.5}AlSi_3O_{10}F_2$	Trilithionite	182, 197, 215, **264**, 405, 480, 561, 633, **710**, 751, 782	8
266	$Al_2SiO_4(OH, F)_2$	Topaz2	238, **266**, 285, 402, 456, 856, 928, **3637**	19
267	$Ba_3Ce_2(CO_3)_5F_2$	Cebaite	177, 213, **267**, 621, 701, 715, 730, 1072, 1084, **1091**	8
268	$Al_2SiO_4(OH, F)_2$	Topaz	155, 166, 240, **268**, 286, 318, 333, 403, 457, 856, 928	13
268	$Al_2Si_2O_5(OH)_4$	Dickite	120, 132, 178, 198, 245, **268**, **336**, 420, 434, **463**, 748, 792, 916, 935, 1050, 1104, **3623**, 3648, **3683**, **3700**	56
268	$Al_2Si_2O_5(OH)_4$	Nacrite	120, 132, 140, 197, 238, **268**, **332**, 418, 433, **462**, 748, 798, 916, 935, 1085, 1102, **3618**, 3628, 3650, **3698**	56
272	$Al_2Si_2O_5(OH)_4$	Kaolinite	120, 132, 200, 248, **272**, **338**, 420, 434, **461**, 749, 791, 916, 935, 1048, 1105, **3620**, 3650, 3668, 3682, **3692**	56
272	$PbCl(OH)$	Laurionite	131, 142, 174, **272**, **328**, 594, 662, 729, **3506**	12
274	$3NaAlSi_3O_8{\cdot}Na_2CO_3$	Cancrinite	117, **274**, 349, 422, 462, 503, 684, **774**, 957, **1057**	19
274	FeS	Pyrrhotite hex.	**214**, **274**, 385, 586	13
280	BeF_2	Beryllium fluoride	**280**, 420, **500**, 560, 630	19
281	$LiAlSi_4O_{10}$	Petalite	**143**, **281**, 318, 357, 383, **491**, **793**	19
282	BeF_2	Beryllium fluoride in melt	230, **282**, 333, 385, 490, 750, 805	19
283	$CaCO_3$	Calcite–(Mg)	157, **283**, 714, 1069, **1088**, 1438	13
285	Cs_2ZnCl_4	Cs–Zn tetrachloride	275, **285**, **296**	19
286	Pb_2Cl_3OH	Penfieldite	167, 200, 230, **286**, **303**, 440, 615, 772, 978	19
289	$NaAlF_4$	Sodium fluoroaluminate	174, 193, **289**, 304, 315, **352**, **622**, 641, **812**, 893	19
290	$Ca(Ce, Y)(CO_3)_2F$	Synchysite–Y	**224**, 244, **290**, 374, 402, 479, 517, 603, 655, 993, 1054, **1080**, **1095**	19

cm^{-1}	Formula	Name	All Bands	Source
290	Fe_2O_3	Hematite	223, **290**, **409**, 498, 609, 1313	10
291	K_2ZnCl_4	K–Zn tetrachloride	277, **291**, **293**, 317	19
291	KCl	Sylvite	213, **291**	10
291	Rb_2ZnCl_4	Rb–Zn tetrachloride	284, **291**, 300	19
292	Al_2SiO_5	Andalusite2	**292**, 320, 358, 551, **918**, **959**, 1019, 1064, 1110	19
292	$NaAlSi_3O_8$	Albite	183, 210, **292**, 457, 480, **508**, 764, 816, 977, 1032, 1098	10
293	Al_2SiO_5	Andalusite	**293**, 323, 361, 453, 553, 719, 834, **920**, 992, 1065, 1111	10
293	$CuFeS_2$	Chalcopyrite	**293**, 322, 352, 378	10
293	K_2ZnCl_4	K–Zn tetrachloride	277, **291**, **293**, 317	19
296	$BeCl_2$	Beryllium chloride2	175, 205, 243, **296**, 327, 478, 574, 620	19
296	Cs_2ZnCl_4	Cs–Zn tetrachloride	275, **285**, **296**	19
298	α-FeOOH	Goethite2	**241, 298,** 396, 478, 547, 678, 991	19
301	$Cu_2PO_4(OH)$	Libethenite	**74**, 92, 113, 140, **160**, 227, 270, **301**, 319, 371, 392, 430, 461, 561, 590, 944, **979**, 1008, **1022**, **3475**	57
301	$Cu_5(PO4)_2(OH)_4$	Ludjibaite	**160**, 190, 226, 263, **301**, 368, 387, 402, 449, 557, 586, 633, 736, 784, 815, 855, 925, **981**, 1019, 1046, 1072, 1115, **3470**	57
303	Pb_2Cl_3OH	Penfieldite	167, 200, 230, **286**, **303**, 440, 615, 772, 978	19
304	SiO_2	Tridymite	210, **304**, **349**, **431**, 799, 1073	19
309	C_6H_6	Benzene3	**309**, **999**, 1580, **3064**	69
310	C_7H_{16}	Heptane saturated	**310**, 834, 899, 1302, **1452**, **2849**, 2859, **2872**, 2895, 2916, 2936, 2962	46
311	Bi	Bismuth	79, **311**, 465, 621	19
313	$Pb_5(AsO_4)_3Cl$	Mimetite	98, 105, 145, 175, **313**, 340, 371, 385, 408, 422, 765, **788**, 806, **813**, 837, 854, 860, **3266**, **3420**, 3548	34
315	$PbMoO_4$	Wulfenite	**166**, **315**, 348, 742, 765, **868**, **1736**, 3290, **3350**, **3377**, 3533, 3566	19
317	CaF_2	Fluorite2	**317**	19
318	K_2SnCl_6	Potassium–tin chloride	**168**, 239, **318**	19
318	Rb_2SnCl_6	Rubidium–tin chloride	**169**, 239, **318**, 549	19
319	$Pb_3Ca_2(AsO_4)_3Cl$	Hedyphane	141, 151, 162, 177, 192, 201, **319**, 334, 349, 372, 394, 432, 452, 770, 787, **811**, **819**, 834, 848, 3352, 3409, **3458**	34

Continued

cm^{-1}	Formula	Name	All Bands	Source
322	CaF_2	Fluorite	**322**, 641	10
324	FeS_2	Marcasite	**324**, 387	10
324	$YNbO_4$	Fergusonite monoclinic	**131**, **324**, **336**, 658, 675, 698, **811**	19
326	K_2NaAlF_6	Elpasolite	135, **326**, 379, 387, 446, **559**, 668, 873, 989, 1009, 1055	19
327	$SnCl_4{\cdot}5H_2O$	Sn–chloridepentahydr. in aq.	137, 157, 198, **327**, 452	19
328	$Al_6(BO_3)_5(OH, F)_3$	Jeremejevite	301, **328**, **372**, 417, 512, 608, 710, 749, **962**, 1067, 3624, **3672**	19
328	$PbCl(OH)$	Laurionite	131, 142, 174, **272**, **328**, 594, 662, 729, **3506**	12
329	$CaWO_4$	Scheelite	55, 79, 111, 207, **329**, 395, **793**, 834, **909**, 1020, 1120, 1625, 3267, 3285, **3351**, **3379**, 3413, **3545**	19
330	$PbCl(OH)$	Paralaurionite	81, **102**, **132**, 220, 273, **330**, 598, 733	19
331	$SnCl_4{\cdot}5H_2O$	Sn–chloridepentahydr. in aq.	154, 236, **331**, 3421, 3542	19
332	$Al_2Si_2O_5(OH)_4$	Nacrite	120, 132, 140, 197, 238, **268**, **332**, 418, 433, **462**, 748, 798, 916, 935, 1085, 1102, **3618**, 3628, 3650, **3698**	56
334	$FeCl_3$	Ferric iron chloride	**334**, 418	19
336	$YNbO_4$	Fergusonite monoclinic	**131**, **324**, **336**, 658, 675, 698, **811**	19
336	$Al_2Si_2O_5(OH)_4$	Dickite	120, 132, 178, 198, 245, **268**, **336**, 420, 434, **463**, 748, 792, 916, 935, 1050, 1104, **3623**, 3648, **3683**, **3700**	56
338	$Al_2Si_2O_5(OH)_4$	Kaolinite	120, 132, 200, 248, **272**, **338**, 420, 434, **461**, 749, 791, 916, 935, 1048, 1105, **3620**, 3650, 3668, 3682, **3692**	56
340		Tourmaline green	178, 204, 225, 240, 305, **340**, **377**, 413, 444, 675, **726**, 763, 819, 840, 975, 2025, 1044, **1060**, 1070	19
343	$(Mg, Fe, Mn)Al_3(BO_3)(SiO_4)O_2$	Grandidierite	**161**, 228, 242, 269, **343**, 363, 376, **425**, 442, 489, **577**, 659, 864, **949**, 979, 1042	8
347	$Be(OH)_2$	Behoite	133, 207, 276, **347**, **398**, 444, 456, 599, 684, 757, 1055	19
349	SiO_2	Tridymite	210, **304**, **349**, **431**, 799, 1073	19
349	ZnF_2	Zinc fluoride	69, 249, **349**, 520	19

cm^{-1}	Formula	Name	All Bands	Source
350	ZnS	Sphalerite	218, 274, 300, 310, **349**, **350**, 419, 639, 669	28
351	β-$Be(OH)_2$	Behoite-β	148, 286, **351**, **404**, 450, 459, 465, 553, 622, 717, 1069, 2544, 2551, 2566, **3413**, 3464, 3476	19
352	$NaAlF_4$	Sodium fluoroaluminate	174, 193, **289**, 304, 315, **352**, **622**, 641, **812**, 893	19
356	K_2ZnF_4	K–Zn tetrafluoride	167, 180, **356**	19
356	$ZrSiO_4$	Zircon	202, 212, 225, **356**, 392, 438, **972–974**, **1005–1008**	10
358	$NaCl$	Halite	**358**	10
361	$PbCrO_4$	Crocoite	141, 339, **361**, 378, 405, **842**, 864	30
369	NaF	Villiaumite	217, 263, **369**, 580	19
370	As_2O_3	Arsenolite	85, 185, 269, **370**, 471, 560, **781**	8
372	$Al_6(BO_3)_5(OH, F)_3$	Jeremejevite	301, **328**, **372**, 417, 512, 608, 710, 749, **962**, 1067, 3624, **3672**	19
374	$MgB_6O_{10}{\cdot}6H_2O$	Magnesium hexaborate	**374**, **634**, **638**, **641**, **852**, **953**, **964**	19
376	$Be_2(PO_4)(OH){\cdot}4H_2O$	Moraesite	124, 200, **376**, 415, **584**, 696, **1029**, 1080	8
378	SnF_2	Tin fluoride	80, 131, 199, **378**, 418	19
379		Rubellite	**226**, 246, 276, 332, **379**, 410, 517, 645, **716**, 733, 753, 775, 883, 982, 1036, **1070**	19
380	$KAlSi_3O_8{\cdot}H_2O$	K–cymrite	114, **380**, **833**, 3541, 3623	53
380	$Al(OH)_3$	Gibbsite	242, 255, 322, **380**, 538, 569, 979	10
385	HBO_2	Metaboric acid I cubic	104, **150**, **231**, 271, **385**, 434, 454, 478, **511**, **618**, **759**, **780**, **820**, 970, 1025, 1240, 1980, 2380	19
386	$Na_2Mg_3Al_2Si_8O_{22}(OH)_2$	Glaucophane	210, 257, 339, **386**, 411, 445, 525, 562, **670**, 778, 896, 1000, **1047**, 1108, 3615, 3631, 3647, 3660	19
389	α-$FeOOH$	Goethite	242, 299, **389**, **547**, 681	10
390	$KAlSi_3O_8$	Kokchetavite	109, **390**, **836**	53
393	$FeNb_2O_6$	Columbite	132, 203, 275, **393**, 525, 634, **869**	19
396	$Be_3Al_2Si_6O_{18}$	Beryl	206, 289, 322, **396**, 422, 444, 528, 581, **685**, 770, 915, 1012, 1068, 1243	8

Continued

cm^{-1}	Formula	Name	All Bands	Source
397	$BeAlSiO_4(OH)$	Euclase	145, 180, 237, **259**, 290, **397**, 444, 452, 574, **880**, **908**, 975, 1023, 1059, 1120, **3575**, 3587	19
397	$Cu_3(OH)_2(CO_3)_2$	Azurite	170, 246, 279, **397**, **1095**, 1457, **3453**	10
398	$Be(OH)_2$	Behoite	133, 207, 276, **347**, **398**, 444, 456, 599, 684, 757, 1055	19
399	MgF_2	Sellaite	286, **399**, 816	19
399	$NaAlSiO_4$	Nepheline	220, 264, 350, **399**, **415**, 467, 920, 967, **989**, 1086	19
400	$Ca_2Mn_5Be_4(PO_4)_6(OH)_4{\cdot}6H_2O$	Roscherite	173, 210, 259, **400**, 559, 579, 599, 631, 771, **979**, 1066	19
402	$BaCl_2{\cdot}2H_2O$	Clathrate	193, 256, **402**, 490, 550, **3331**, 3451	19
404	β-$Be(OH)_2$	Behoite-β	148, 286, **351**, **404**, 450, 459, 465, 553, 622, 717, 1069, 2544, 2551, 2566, **3413**, 3464, 3476	19
407	$CuCl_2{\cdot}2H_2O$	Eriochalcite	117, **216**, 235, 247, 390, **407**, 672, 692, 1620, 3176, 3317, **3367**, 3462	31
408	$MgAl_2O_4$	Spinel	313, **408**, 666, 768	10
409	Fe_2O_3	Hematite	223, **290**, **409**, 498, 609, 1313	10
415	$NaAlSiO_4$	Nepheline	220, 264, 350, **399**, **415**, 467, 920, 967, **989**, 1086	19
417	Al_2O_3	Corundum	378, **417**, 644, 750	10
418	SiO_2	Cristobalite2	112, 230, **418**, 780, 1075, 1405	19
419	$Al_{4.5}Si_{1.5}O_{9.75}$	Mullite	161, 264, 308, **419**, 521, 608, 886, **964**, 1133	19
420	SiO_2	Cristobalite	114, 230, 273, 286, **420**, 792, 1075	10
425	$(Mg, Fe, Mn)Al_3(BO_3)(SiO_4)O_2$	Grandidierite	**161**, 228, 242, 269, **343**, 363, 376, **425**, 442, 489, **577**, 659, 864, **949**, 979, 1042	8
428	FeS_2	Pyrite	342, 377, **428**	10
431	SiO_2	Tridymite	210, **304**, **349**, **431**, 799, 1073	19
432	$(K, Cs)Al_4Be_4(B, Be)_{12}O_{28}$	Rhodizite	298, **432**, **473**, 648, 804, 865	19
434	$Cu_2(OH)_2(CO_3)$	Malachite	**154**, **178**, 272, **434**, 537, 721, **1098**, 1492, 3386, **3468**	10
434	$Mg_5Si_6O_{20}(OH)_4$	Talc	113, 196, 295, 335, 366, **434**, **678**, 786, 793, 1018, 1055, **3677**	10
438	ZnO	Zincite	**100**, 331, **438**, 489	30
438	$ZrSiO_4$	Zircon	202, 212, 225, **356**, 392, 438, **972–974**, **1005–1008**	10
440	$KMgClSO_4{\cdot}2.75H_2O$	Kainite	**440**, **630**, **1022**, **1039**, 1157, 1176, 1197, 1207	19

cm^{-1}	Formula	Name	All Bands	Source
443	$LiFe^{3+}(PO_4)(OH)$	Tavorite	157, 201, 443, 576, 589, 629, **951**, 1001, 1069	19
443	$Na_2LiBe_2F_7$	Na–Li fluoroberyllate	352, 370, **443**, **520**, 568, 779, 890	19
444	TiO_2	Rutile	139, 238, **444**, **609**, 695, 920	10
446	$Al_3Si_2O_7(OH)_3$	Piezotite	**446**, **1336**, **1607**, 2932	19
446	$Li_2B_4O_7 \cdot 5H_2O$	Diomignite	391, **446**, 493, **543**, **772**, 845, 896, **1028**, 1097, 1352	22
446	$Cu_3(PO_4)(OH)_3$	Cornetite	**109**, 131, 141, 158, 174, 297, 313, 363, **446**, **477**, 515, 583, 639, 945, **971**, **996**, 1016, **1083**, 1137, **3414**	57
447	$Be_2(PO_4)(OH) \cdot 4H_2O$	Moraesite2	148, 172, 230, 347, 401, **447**, **503**, 631, 676, 1004, 1097	19
447	$Pb_4Fe_3O_8(OH, Cl)$	Hematophanite	236, 263, **447**, 664	19
448	$AlOOH$	Diaspore	331, **448**	10
449	$Al_{13}Si_5O_{20}(OH, F)_{18}Cl$	Zunyite	168, **202**, 270, 317, 360, 383, **449**, **610**, 954, 992	8
449	CeO_2	Cerianite–(Ce)	184, **449–465**, 571	17
450	$S_2O_3^{2-}$	Thiosulphate ions in aq.	**450**, **1000**	59
452	$AlOOH$	Diaspore2	159, 337, **452**, **501**, 558, **667**, 793, 1050, 1190	19
452	$BaSO_4$	Barite2	74, 89, 97, 152, 189, **452**, **462**, 616, 629, 647, 965, **988**, 1104, **1144**, 1166	18
453	$K_2Mg(SO_4)_2 \cdot 4H_2O$	Leonite	**453**, **632**, **1025**, **1041**, 1158, 1196, 3428	19
453	$K_3Na(SO_4)_2$	Glaserite	**453**, **619**, **678**, **990**	19
456	$K_2Mg_2(SO_4)_3$	Langbeinite	**456**, **466**, 607, 626, 640, **1053**	19
461	$Al_2Si_2O_5(OH)_4$	Kaolinite	120, 132, 200, 248, **272**, **338**, 420, 434, **461**, 749, 791, 916, 935, 1048, 1105, **3620**, 3650, 3668, 3682, **3692**	56
461	$BaSO_4$	Barite	**461**, 617, **988**, 1143	10
462	$(Fe, Mn)WO_4$	Wolframite	205, 236, 260, 353, 401, **462**, 484, 508, 695, 790, 806, **881**, 1061, 1079, **1156**	19
462	$Na_4BeAlSi_4O_{12}Cl$	Tugtupite	160, 206, 269, 326, 439, **462**, 510, 553, 631, **1052**, 1103, 1130	19
462	$BaSO_4$	Barite2	74, 89, 97, 152, 189, **452**, **462**, 616, 629, 647, 965, **988**, 1104, **1144**, 1166	18
462	$Al_2Si_2O_5(OH)_4$	Nacrite	120, 132, 140, 197, 238, **268**, **332**, 418, 433, **462**, 748, 798, 916, 935, 1085, 1102, **3618**, 3628, 3650, **3698**	56

Continued

cm^{-1}	Formula	Name	All Bands	Source
463	CuCl	Nantokite	127, 155, **206**, 262, 356, 393, **463**	31
463	$Al_2Si_2O_5(OH)_4$	Dickite	120, 132, 178, 198, 245, **268**, **336**, 420, 434, **463**, 748, 792, 916, 935, 1050, 1104, **3623**, 3648, **3683**, **3700**	56
464	α-SiO_2	Quartz	128, 206, 265, 356, 402, **464**, 485, 520, 608, 698, 807, 1066, 1161	10
464	$Ca_2B_6O_{11}\cdot 13H_2O$	Inyoite	352, 388, **464**, **615**, 733, 808, 912, 956, 1096, 3223, 3334, **3443**, **3531**	19
464	$LiAlSi_2O_6$	Spodumene	282, 320, 382, **464**, **492**, 355, 697, **707**, 776, 794, 808, **977**, 1019, 1062, **1074**, 1135, 1157	19
465	$Na_8(Al_6Si_6O_{24})Cl_2$	Sodalite	114, 167, 263, 297, 411, **465**, **988**, 1066	13
465	CeO_2	Cerianite–(Ce)	184, **449–465**, 571	17
466	$K_2Mg_2(SO_4)_3$	Langbeinite	**456**, **466**, 607, 626, 640, **1053**	19
471	CuS	Covellite	263, **471**	10
473	$(K, Cs)Al_4Be_4(B, Be)_{12}O_{28}$	Rhodizite	298, **432**, **473**, 648, 804, 865	19
473	S_8	Sulphur	**157**, 187, **219–222**, 246, 437, 462, **473**	10
475	$NaAlSi_3O_8$	Albite2	157, 286, **475**, **503**, 1090	19
476	$Cs_2Al_2Si_4O_{12}\cdot H_2O$	Pollucite	298, 390, **476**, 1132	19
476	$Na_4B_{10}O_{17}\cdot 7H_2O$	Ezcurrite	132, 226, 286, 315, 336, **476**, **513**, **570**, **759**, 881, **954**, 1040, 1059, **1349**, **1388**, **2859**, 3335, 3375	19
477	$BeAl_2O_4$	Chrysoberyl	218, 242, 353, 422, 459, **477**, **518**, 546, **639**, 658, 679, 711, 778, 816, **931**	19
477	$Cu_3(PO_4)(OH)_3$	Cornetite	**109**, 131, 141, 158, 174, 297, 313, 363, **446**, **477**, 515, 583, 639, 945, **971**, **996**, 1016, **1083**, 1137, **3414**	57
477	$Fe_5(SO_4)_6(OH)_2\cdot 20H_2O$	Ferricopiapite	268, 310, 451, **477**, **996**, **1019**, 1148	8, 58
478	$CaLiAl_2(AlBeSi_2)O_{10}(OH)_2$	Bityite	104, **132**, 365, **478**, 684, **898**, 1049	19
480	$Cu_5(PO_4)_2(OH)_4$	Reichenbachite	89, **110**, 135, 175, 188, 214, 255, 298, 365, 412, 453, **480**, 515, 540, 572, 607, 636, 951, 971, 998, **1083**, **3380**, **3428**	57
480	$Ca_5(PO_4)_2(OH)_4$	Pseudomalachite	88, **110**, 131, 176, 368, 450, **480**, 537, 609, 972, 1000, **1086**	30
481	BPO_4	Boron orthophosphate	239, 460, **481**, 541, 1067, **1116**, 1261	19

cm^{-1}	Formula	Name	All Bands	Source
484	$MgAlBO_4$	Sinhalite	**484**, 608, 744, 860, 1045	19
491	$LiAlSi_4O_{10}$	Petalite	**143**, **281**, 318, 357, 383, **491**, **793**	19
492	$CaSO_4{\cdot}2H_2O$	Gypsum	414, **492–494**, 621–623, 671, **1006–1008**, 1142, 3250, **3405**, 3491, 3500	4
492	$LiAlSi_2O_6$	Spodumene	282, 320, 382, **464**, **492**, 355, 697, **707**, 776, 794, 808, **977**, 1019, 1062, **1074**, 1135, 1157	19
492	$NaAlSi_3O_8$	Kumdykolite	155, **222**, 265, 407, 464, **492**	61
494	$CaSO_4{\cdot}2H_2O$	Gypsum	414, **492–494**, 621–623, 671, **1006–1008**, 1142, 3250, **3405**, 3491, 3500	10
496	$Na_2B_4O_7{\cdot}4H_2O$	Kernite	**105**, 175, 256, 388, 426, 456, **496**, 554, 579, **737**, 802, 848, **856**, **929**, 978, 1009, 1034, 1089, 1139	19
500	BeF_2	Beryllium fluoride	**280**, 420, **500**, 560, 630	19
500	H_3BO_3	Sassolite	187, 211, **500**, 880, 1168, 3165, **3245–3247**	22
501	AlOOH	Diaspore2	159, 337, **452**, **501**, 558, **667**, 793, 1050, 1190	19
501	SiO_2	Moganite	129, 141, 220, 265, 317, 370, 377, 398, 432, 449, 463, **501**, 693, 792, 833, 1058, 1084, 1171, 1177	68
502	$Na_2(Si, Al, Be)_7(OH, F)_{14}$	Leifite	388, 472, **502**, 532, 698, 836, 965, 1032, 1128	19
503	$Be_2(PO_4)(OH){\cdot}4H_2O$	Moraesite2	148, 172, 230, 347, 401, **447**, **503**, 631, 676, 1004, 1097	19
503	$NaAlSi_3O_8$	Albite2	157, 286, **475**, **503**, 1090	19
508	$Al_7(BO_3)(SiO_4)_3O_3$	Dumortierite	169, **206**, 293, 353, 403, 418, **508**, 810, 842, **946**, 1076, 1112	8
508	$C_{10}H_8$	Naphthalene	**508**, **759**, 1017, 1166, 1240, 1274, **1375**, **1459**, **1572**, 1626, 3054	45
508	$NaAlSi_3O_8$	Albite	183, 210, **292**, 457, 480, **508**, 764, 816, 977, 1032, 1098	10
510	$LiOH{\cdot}H_2O$	Li hydroxide monohydrate	**210**, **243**, 362, 388, **510**, 834	19
511	$Cu_2(OH)_3Cl$	Atacamite	106, **121**, 361, 419, 446, 475, **511**, 818, 846, **912**, **975**	30
511	$Cu_4Cl[Cl_{0.29}(OH)_{0.71}](OH)_6$	Claringbullite	119, 136, **147**, 163, 231, 260, 356, 389, 447, **511**, 579, **815**, **906**, **970**, 1049, 3211, 3331, **3351**, **3433**, 3458	31
511	$(Ca, Na)(Al, Si)AlSi_3O_8$	Oligoclase	**177**, 275, 329, 403, 453, 474, **511**, 747, 808, 1007, 1114	19

Continued

cm^{-1}	Formula	Name	All Bands	Source
511	HBO_2	Metaboric acid I cubic	104, **150**, **231**, 271, **385**, 434, 454, 478, **511**, **618**, **759**, **780**, **820**, 970, 1025, 1240, 1980, 2380	19
511	$Cu_2(OH)_3Cl$	Paratacamite	95, **117**, **140**, 165, 364, 420, **511**, 576, 799, 868, 892, 928, 970, **3311**, **3355**, **3441**	30
511	$CuCl_2$	Tolbachite	95, 106, **118**, **140**, 149, 213, 362, 416, 477, **511**, 575, 799, 819, 845, 869, 893, 910, 928, 971	30
513	$KAlSi_3O_8$	Orthoclase2	282, 402, 404, 458, 474, **513**, 761, 805, 992, 1128	19
513	$Na_4B_{10}O_{17}{\cdot}7H_2O$	Ezcurrite	132, 226, 286, 315, 336, **476**, **513**, **570**, **759**, 881, **954**, 1040, 1059, **1349**, **1388**, **2859**, 3335, 3375	19
514	$Cu_2(OH)_3Cl$	Atacamite2	66, **106**, **120**, **139**, **150**, 450, **514**, 586, 820, 845, **908**, 975, **3351**, **3453**	13
514	$KAlSi_3O_8$	Microcline	159, 178, 199, 263, 267, 286, 455, 475, **514**, 651, 749, 813, 1007, 1128, 1142	10
514	$KAlSi_3O_8$	Sanidine	163, 284, 462, 475, **514**, 767, 813, 1123	10
514	$KAlSi_3O_8$	Orthoclase	157, 177, 197, 284, 458, 477, **514**, 583, 751, 814, 967, 1035, 1062, 1137	10
515	$(NH_4)AlSi_3O_8{\cdot}0.5H_2O$	Buddingtonite	135–138, 288, 296, 448, 478, **515–517**, 579, 805, 1097, 1120, 1145, 3223	19
515	$RbAlSi_3O_8$	Rubicline	90, 114, 142, 154, 286, 447, 478, **515**, 638, 809, 821, 838, 1013, 1040, 1114, 1185	19
517	$(NH_4)AlSi_3O_8{\cdot}0.5H_2O$	Buddingtonite	135–138, 288, 296, 448, 478, **515–517**, 579, 805, 1097, 1120, 1145, 3223	19
518	$BeAl_2O_4$	Chrysoberyl	218, 242, 353, 422, 459, **477**, **518**, 546, **639**, 658, 679, 711, 778, 816, **931**	19
520	$Be_2F_7^{2-}$	Fluoroberyllate ion in melt	**520**, 954	19
520	$Na_2LiBe_2F_7$	Na–Li fluoroberyllate	352, 370, **443**, **520**, 568, 779, 890	19
520	Si	Silica	**520**	19
521	SiO_2	Coesite	116, 151, 176, 204, 269, 326, 355, 427, 466, **521**, 785, 815, 837, 1036, 1065, 1144, 1164	10
522	$Be_2F_7^{3-}$	Fluoroberyllate ion in melt	**522**	19
523	$Al_2(PO_4)F_2OH{\cdot}7H_2O$	Fluellite	280, 410, **523**, 583, 649, 1037, 1119	19

cm^{-1}	Formula	Name	All Bands	Source
526	LiCl	Lithium chloride	**526**, **532**	19
529	$LiB_5O_8{\cdot}5H_2O$	Lithium pentaborate	92, 115, 124, 490, **529**, 569, 776, 796, 924, 3105, 3210, 3317, 3416	19
529	$NaB_5O_8{\cdot}5H_2O$	Sborgite	121, **529**, **856**	66
530	S_3^-	Trisulphur ion	**530–540**, 1070, 1600, 2130, 2650	59
531	$Na_5Al_3F_{14}$	Chiolite	250, 318, 356, **531**, 553, 1088	19
532	LiCl	Lithium chloride	**526**, **532**	19
540	S_3^-	Trisulphur ion	**530–540**, 1070, 1600, 2130, 2650	59
543	$Li_2B_4O_7{\cdot}5H_2O$	Diomignite	391, **446**, 493, **543**, **772**, 845, 896, **1028**, 1097, 1352	22
545	$Al(OH)_3$	Gibbsite2	117, 243, 321, 383, **545**, **568**, 896, 980	19
547	α-FeOOH	Goethite	242, 299, **389**, **547**, 681	10
547	BeF_2	Beryllium fluoride in melt	255, 385, **547**, 800	19
547	BeF_4^{2-}	Fluoroberyllate ion in melt	**547**, **550**	19
547	$Na_3Ca(Al_3Si_3O_{12})S$	Lazurite	257, 288, **547**, 801, 1094, 1639	13
548	$CsB_5O_8{\cdot}4H_2O$	Ramanite–Cs	94, 98, 293, **548**, 907, 3119, **3396**	24
551	$BeCl_2$	Beryllium chloride	263, **551**	19
551	NH_4BeF_4	NH_4 fluoroberyllate	264, 385, **551**	19
552	$NaCaB_5O_9{\cdot}5H_2O$	Probertite	157, 240, 337, 432, 452, 487, **552**, **577**, **726**, 747, 821, 920, 1032, **3358**, 3492, 3549, **3616**	19
554	$RbB_5O_8{\cdot}4H_2O$	Ramanite–Rb	101, **554**, 914, **3374**, 3426	24
555	Na_3AlF_6	Cryolite	485, **555**, **620**	10
556	$KB_5O_8{\cdot}4H_2O$	Santite	107, **556**, 765, 918	66
558	$KB_5O_8{\cdot}4H_2O$	Potassium pentaborate	372, 459, 509, **558**, 768, 788, 856, 919	19
559	K_2BeF_4	K fluoroberyllate	268, 387, 392, **559**, **562**, 669	19
559	K_2NaAlF_6	Elpasolite	135, **326**, 379, 387, 446, **559**, 668, 873, 989, 1009, 1055	19
559	$CuC_2O_4{\cdot}0{\cdot}5H_2O$	Moolooite	210, 299, **559**, 586, 611, 832, 932, 1487, **1516**, 1616	30
560	As_2O_3	Arsenolite	85, 185, 269, **370**, 471, 560, **781**	8
560	BeF_4^{2-}	Tetrafluoroberyllate in melt	**560**	19
562	K_2BeF_4	K fluoroberyllate	268, 387, 392, **559**, **562**, 669	19
563	$Li_3Na_3Al_2F_{12}$	Cryolithionite	340, 364, 378, 413, **563**	19
566	$Na_3Li_3Al_2F_{12}$	Cryolithionite	353, 390, **566**	8

Continued

cm^{-1}	Formula	Name	All Bands	Source
568	$Al(OH)_3$	Gibbsite2	117, 243, 321, 383, **545**, **568**, 896, 980	19
568	Na_2BeF_4	Na–Be tetrafluoride	**568**	19
570	$Na_4B_{10}O_{17}\cdot 7H_2O$	Ezcurrite	132, 226, 286, 315, 336, **476**, **513**, **570**, **759**, 881, **954**, 1040, 1059, **1349**, **1388**, **2859**, 3335, 3375	19
571	$K_2B_4O_7\cdot 4H_2O$	Potassium tetraborate	397, 456, **571**, 767, **864**, 979	19
571	$Na_2B_4O_5(OH)_4\cdot 8H_2O$	Borax	344, 405, 463, **571**, 776, 943, 997	10
571	$Na_2Si_2O_5$	Sodium disilicate glass	**571**, 772, 943, **1099**	19
576	$Na_2B_4O_7\cdot 10H_2O$	sodium tetraborate	385, 461, **576**, 756, 852, 948, 1036	22
577	$Na_2B_4O_5(OH)_4\cdot 3H_2O$	Tinclaconite	107, 156, 320, 344, 389, 406, 462, **577**, 465, 943, 993	19
577	$NaCaB_5O_9\cdot 5H_2O$	Probertite	157, 240, 337, 432, 452, 487, **552**, **577**, **726**, 747, 821, 920, 1032, **3358**, 3492, 3549, **3616**	19
577	(Mg, Fe, Mn)$Al_3(BO_3)(SiO_4)O_2$	Grandidierite	**161**, 228, 242, 269, **343**, 363, 376, **425**, 442, 489, **577**, 659, 864, **949**, 979, 1042	8
578	$Na_2B_4O_5(OH)_4\cdot 8H_2O$	Borax2	344, 405, **578**, 767, 943, 997, **3351**, 3499	19
579	KF	Potassium fluoride	203, 261, **579**	19
583	$Mg_2B_4O_7\cdot 9H_2O$	Magnesium tetraborate	408, 450, **583**, 787, 845, 949, 1045	19
584	$Be_2(PO_4)(OH)\cdot 4H_2O$	Moraesite	124, 200, **376**, 415, **584**, 696, **1029**, 1080	8
584	$CaBe(PO_4)(F, OH)$	Herderite	166, 194, 424, 445, 516, 529, **584**, **594**, **983**, **1005**, 1128, 1140, **3564**, **3573**, 3620	19
589	$NaCaB_5O_9\cdot 8H_2O$	Ulexite	228, 263, 300, 330, 392, 440, 485, 531, **589**, **592**, **595**, **747**, **862**, 1013, 1316, 3559, **3570**, 3588	19
592	$NaCaB_5O_9\cdot 8H_2O$	Ulexite	228, 263, 300, 330, 392, 440, 485, 531, **589**, **592**, **595**, **747**, **862**, 1013, 1316, 3559, **3570**, 3588	19
594	$CaBe(PO_4)(F, OH)$	Herderite	166, 194, 424, 445, 516, 529, **584**, **594**, **983**, **1005**, 1128, 1140, **3564**, **3573**, 3620	19
595	$B(OH)_2^-$	Boron dihydrate in aq.	**595**	19
595	HBO_2	Metaboric acid	401, 415, 475, **595**, 809	10
595	$Na_3B_3F_6O_3$	Na hexafluoroboroxine	350, 378, **595**, 730, 1060	19

cm^{-1}	Formula	Name	All Bands	Source
595	$NaCaB_5O_9{\cdot}8H_2O$	Ulexite	228, 263, 300, 330, 392, 440, 485, 531, **589**, **592**, **595**, **747**, **862**, 1013, 1316, 3559, **3570**, 3588	19
599	HBO_2	Metaboric acid III	**87**, **119**, 404, 419, 469, 478, **599**, 638, **810**, 1100, 1343, 1460, 3195, 3295, 3365	19
608	$Ca[B_2BO_4(OH)_3]{\cdot}H_2O$	Colemanite	306, 383, 531, **608**, 666, 742, 3514, **3588**	19
608	$CaB_3O_3(OH)_5{\cdot}H_2O$	Meyerhofferite	163, 334, 380, 433, 472, 491, **608**, 725, 879, 1000	19
609	TiO_2	Rutile	139, 238, **444**, **609**, 695, 920	10
610	$Al_{13}Si_5O_{20}(OH, F)_{18}Cl$	Zunyite	168, **202**, 270, 317, 360, 383, **449**, **610**, 954, 992	8
611	$CaB_2(SiO_4)_2$	Danburite	166, 182, **245**, 348, 480, **611**, 628, 974, 1107, 1664, 3079, 3912	19
612	$Ca_2Be_4(Fe, Mn)_5(PO_4)_6(OH)_4{\cdot}6H_2O$	Greifensteinite	164, 216, 305, 407, 435, 481, **612**, **628**, **993**, **1056**, 1133, 1176	19
615	$Ca_2B_6O_{11}{\cdot}13H_2O$	Inyoite	352, 388, **464**, **615**, 733, 808, 912, 956, 1096, 3223, 3334, **3443**, **3531**	19
615	$Fe_8Si_6O_{15}[(OH_6)Cl_4]$	Ferropyrosmalite	**615**, 1022, **3551**, **3579**, **3637**	16
616	$Na_2Mg(SO_4)_2{\cdot}4H_2O$	Astrakanite	454, **616**, **991**, 1067, 1115, 1153	19
618	HBO_2	Metaboric acid I cubic	104, **150**, **231**, 271, **385**, 434, 454, 478, **511**, **618**, **759**, **780**, **820**, 970, 1025, 1240, 1980, 2380	19
619	$K_3Na(SO_4)_2$	Glaserite	**453**, **619**, **678**, **990**	19
620	Na_3AlF_6	Cryolite	485, **555**, **620**	10
620	$NaB_3O_3(OH)_4$	Ameghnite	135, 143, 171, 199, 348, 363, 465, 503, 528, **620**, **755**, 883	19
622	$LiAlF_4$	Li Tetrafluoraluminate	210, 320, **622**, 760	19
622	$NaAlF_4$	Sodium fluoroaluminate	174, 193, **289**, 304, 315, **352**, **622**, 641, **812**, 893	19
623	$NaAl(PO_4)F$	Lacroixite	319, 503, 609, **623**, **1001**, 1043, 1119	19
626	$Na_7Ca_3(OH)(SO_4)_6{\cdot}0.8H_2O$	Cesanite	448, 474, 626, 647, **1004**, 1104	10
628	$Ca_2B_6O_{11}{\cdot}9H_2O$	Calcium triborate	472, **628**, 740, 866	19
628	$Ca_2Be_4(Fe, Mn)_5(PO_4)_6(OH)_4{\cdot}6H_2O$	Greifensteinite	164, 216, 305, 407, 435, 481, **612**, **628**, **993**, **1056**, 1133, 1176	19

Continued

cm^{-1}	Formula	Name	All Bands	Source
630	$KMgClSO_4 \cdot 2.75H_2O$	Kainite	**440**, **630**, **1022**, **1039**, 1157, 1176, 1197, 1207	19
631	$H_6Si_2O_7$	Pyrosilicic acid	**225**, 426, **631**, 916, **3414**	19
631	$Mg_2B_6O_{11} \cdot 15H_2O$	Inderite	232, **631**, 3630, 3557	19
632	$K_2Mg(SO_4)_2 \cdot 4H_2O$	Leonite	**453**, **632**, **1025**, **1041**, 1158, 1196, 3428	19
633	$CaSiO_3$	Wollastonite	315, **633**, **967**, 1037	19
633	SnO_2	Cassiterite2	474, **633**, 774	19
634	$MgB_6O_{10} \cdot 6H_2O$	Magnesium hexaborate	**374**, **634**, **638**, **641**, **852**, **953**, **964**	19
634	SnO_2	Cassiterite	230, 283, **634**	8
637	$Al_2(PO_4)(OH)_3$	Augelite	124, 163, 226, 252, 368, 410, 464, 526, 565, **637**, 859, 896, 1014, 1108, 1162, 3428, **3469**, **3537**	19
638	$MgB_6O_{10} \cdot 6H_2O$	Magnesium hexaborate	**374**, **634**, **638**, **641**, **852**, **953**, **964**	19
638	TiO_2	Anatase	**143**, 195, 395, 514, **638**	10
639	$BeAl_2O_4$	Chrysoberyl	218, 242, 353, 422, 459, **477**, **518**, 546, **639**, 658, 679, 711, 778, 816, **931**	19
641	$MgB_6O_{10} \cdot 6H_2O$	Magnesium hexaborate	**374**, **634**, **638**, **641**, **852**, **953**, **964**	19
645	TiO_2	Brookite	127, **150**, 247, 318, 366, 412, **645**	10
646	$LiAl(PO_4)(OH, F)$	Montebrasite	138, 170, 196, 298, 428, 484, 602, 630, **646**, 801, 965, **1013**, **1046**, **1058**, 1107	19
647	$LiAl(PO_4)(F, OH)$	Amblygonite	130, 303, 426, 488, 603, **647**, **1012**, **1048**, 1064, **3369**, 3408	19
652	$NaCa_2Si_3O_8(OH)$	Pectolite	150, 202, 316, 357, 414, **652**, 974, 1026	19
655	K_2SiF_6	Hieratite cubic	405, 476, 540, **655**, 839	18
656	$SrSO_4$	Celestine	452, 627, **656**, **1000**, 1156, 1190	10
660	$Mg_6[Cl_2/B_{14}O_{26}]$	Boracite	**131**, 602, **660**, **840**, 1142	19
660	$WO_3 \cdot H_2O$	Tungstite	215, 236, 257, 269, 325, 379, **660–66**3, 684, **805**, **958**	19
661	Na_2SiF_6	Na hexafluorosilicate	476, **661**, **674**	19
663	$WO_3 \cdot H_2O$	Tungstite	215, 236, 257, 269, 325, 379, **660–66**3, 684, **805**, **958**	19
664	$MgSiO_3$	Enstatite	237, 343, 382, 414, **664**, **684**, 1011	10
665	$CaMgSi_2O_6$	Diopside	325, 393, **665**, 1010	14
665	$MnSiO_3$	Rhodonite	257, 414, **665**, 711, 737, **998**	19
665	$NaCa_2Mg_4Al_3Si_6O_{22}(OH)_2$	Pargasite	229, **665**, **1017**	10

cm^{-1}	Formula	Name	All Bands	Source
667	AlOOH	Diaspore2	159, 337, **452**, **501**, 558, **667**, 793, 1050, 1190	19
667	$Ca_2[Mg_4(Al, Fe)]Si_7AlO_{22}(OH)_2$	Hornblende	224, **667**, **1040**	10
668	$FeFe_2O_4$	Magnetite	193, 306, 538, **668**	10
670	$Na_2Mg_3Al_2Si_8O_{22}(OH)_2$	Glaucophane	210, 257, 339, **386**, 411, 445, 525, 562, **670**, 778, 896, 1000, **1047**, 1108, 3615, 3631, 3647, 3660	19
672	Fe_2SiO_4	Fe–spinel	334, **672**, 779, **845**, 924	19
673	$(Cu, Al)_2H_2Si_2O_5(OH)_4{\cdot}nH_2O$	Chrysocolla	109, 146, 183, 213, 245, 336, 424, 485, **673**, 792, **933**, **1058**, 1602, 3295, 3504, 3583, **3624**, 3652, 3708	32
674	Na_2SiF_6	Na hexafluorosilicate	476, **661**, **674**	19
678	BeO	Bromellite	388, **678**, **684**, 722, 1081, 1097	22
678	$K_3Na(SO_4)_2$	Glaserite	**453**, **619**, **678**, **990**	19
678	$Mg_5Si_6O_{20}(OH)_4$	Talc	113, 196, 295, 335, 366, **434**, **678**, 786, 793, 1018, 1055, **3677**	10
679	$(Mg, Fe)_5Al(OH)_6(AlSi_3O_{10})$	Clinochlore	104, 198, 358, 548, **679**, 3477, 3605, 3647, 3679	10
679	$K(Mg, Fe)_3AlSi_3O_{10}(OH, F)_2$	Biotite	178, 549, **679**, 717, 767, 3658, 3680	10
680	$FeTiO_3$	Ilmenite2	162, 228, 255, 369, 453, **680**	19
680	$K_2(Mg, Fe)_6Si_5Al_2O_{20}(OH, F)_4$	Phlogopite	192, 279, 331, 372, **680**, 792, 1038, 1096, **3673**, **3715**	10
683	Nb_2O_5	Niobium oxide	238, 308, 463, 551, 635, **683**, 815, 897, **993**	19
684	BeO	Bromellite	388, **678**, **684**, 722, 1081, 1097	22
684	$MgSiO_3$	Enstatite	237, 343, 382, 414, **664**, **684**, 1011	10
685	$Be_3Al_2Si_6O_{18}$	Beryl	206, 289, 322, **396**, 422, 444, 528, 581, **685**, 770, 915, 1012, 1068, 1243	8
685	$FeTiO_3$	Ilmenite	232, 373, **685**	10
685	H_2WO_4	Tungstic acid	188, **262**, **685**, 804, 942	19
686	$Be_3Al_2Si_6O_{18}$	Beryl2	324, 401, **686**, 1006, **1068**, 1112, **3593**, 3604	19
695	$LiKCO_3$	Li–K carbonate	**695**, 729, 870, **1046**, **1050**, 1388, 1418, 1483, 1517, 1753	19
702	$CaCO_3$	Aragonite–(Sr)	144, 154, 181, 191, 207, 274, **702**, 707, 1060, 1084, **1086**, 1463	13

Continued

cm^{-1}	Formula	Name	All Bands	Source
702	$KAl_4Si_5Al_2O_{20}(OH, F)_4$	Muscovite	178, 197, 216, **261**, 385, 407, 639, **702**, 754, 914, 957, 1117, **3627**	10
703	$Pb_{1-x}O_{3-2x}(UO_2)_4(OH)_{4+2x}\cdot 4H_2O$	Fourmarierite	373, 490, **703**, **822**	8
703	C_8H_{18}	Cyclooctane	**703**, 1467, **2855**, 2925	70
707	$LiAlSi_2O_6$	Spodumene	282, 320, 382, **464**, **492**, 355, 697, **707**, 776, 794, 808, **977**, 1019, 1062, **1074**, 1135, 1157	19
707	$ZnAl_2O_4$	Gahnite	416, 625, 658, **707**	19
710	$Ca_2(Mn, Fe)Al_2[OH/BO_3/Si_4O_{12}]$	Axinite	104, 271, 414, 482, **710**, 897, **977**, 3354	19
710	$KLi_{1.5}Al_{1.5}AlSi_3O_{10}F_2$	Trilithionite	182, 197, 215, **264**, 405, 480, 561, 633, **710**, 751, 782	8
713	$LiBO_2$	Lithium metaborate	228, 290, 403, 463, 630, 662, **713**, 1081, 1128, **1459**	19
714	$CaCO_3$	Mg–calcite	157, 284, **714**, **1087**	4
715	$KLi_2AlSi_4O_{10}(OH, F)_2$	Polylithionite2	89, 99, **188**, **245**, 268, 409, 452, 552, **715**, 765, 791, **1157**	19
716		Rubellite	**226**, 246, 276, 332, **379**, 410, 517, 645, **716**, 733, 753, 775, 883, 982, 1036, **1070**	19
716	$KLi_2AlSi_4O_{10}(OH, F)_2$	Polylithionite	181, 245, 409, 553, **716**, 781, 872	19
717	$KAl_4Si_5B_2O_{20}(OH, F)_4$	Boromuscovite	247, 411, 476, 555, **717**, 784, **3614**	19
718	WO_3	Tungsten trioxide	272, 325, **718**, **807**	19
725	$Li_2B_4O_7$	Lithium tetraborate	158, 197, 222, 256, 351, 493, 531, 562, **725**, **781**, 980, 1034, 1060, 1166	19
726	$NaCaB_5O_9\cdot 5H_2O$	Probertite	157, 240, 337, 432, 452, 487, **552**, **577**, **726**, 747, 821, 920, 1032, **3358**, 3492, 3549, **3616**	19
731	$HBF_2(OH)_2$	Hydroxyfluoroboric acid	283, 349, 509, 611, **731**, 758, **871**, 3536	19
737	$Na_2B_4O_7\cdot 4H_2O$	Kernite	**105**, 175, 256, 388, 426, 456, **496**, 554, 579, **737**, 802, 848, **856**, **929**, 978, 1009, 1034, 1089, 1139	19
738	$NaBO_2\cdot 2H_2O$	Sodium monoborate2	520, 553, 568, **738**, 832, 952, 1053, 1274, 1293	19
740	$Mg(BO_2)_2\cdot 3H_2O$	Pinnoite	**740**	19
741	$NaBO_2\cdot 2H_2O$	Sodium monoborate	**741**, **746**, 1132, 1276	19
744	$B(OH)_4^-$	Boron tetrahydrate in aq.	**744**	21
745	$MgB_2O_4\cdot 3H_2O$	Magnesium diborate	390, 479, 508, 629, **745**, 860, 900, 1156, 1301	19
746	$LiBO_2\cdot 8H_2O$	Li metaborate hydrous	390, **746**, **852**	19

cm^{-1}	Formula	Name	All Bands	Source
746	$NaBO_2{\cdot}2H_2O$	Sodium monoborate	**741**, **746**, 1132, 1276	19
747	$NaCaB_5O_9{\cdot}8H_2O$	Ulexite	228, 263, 300, 330, 392, 440, 485, 531, **589**, **592**, **595**, **747**, **862**, 1013, 1316, 3559, **3570**, 3588	19
753	$BF_2(OH)_2^-$	Boronfluoridedihydrate in aq.	**753**	19
753	$CaMgB_6O_{11}{\cdot}6H_2O$	Hydroboracite	**181**, 257, 322, 383, 398, 524, 548, 564, 606, **753**, 837, 876, **3595**	19
755	$CaB_2O_4{\cdot}6H_2O$	Calcium metaborate II	**755**, **859**	19
755	$NaB_3O_3(OH)_4$	Ameghnite	135, 143, 171, 199, 348, 363, 465, 503, 528, **620**, **755**, 883	19
755	$Pb_7(UO_2)_{14}O_{19}(OH)_4{\cdot}12H_2O$	Wolsendorfite	348, 510, **755**	8
758	$CaB_2O_4{\cdot}4H_2O$	Calcium metaborate I	390, 547, **758**, 854	19
759	HBO_2	Metaboric acid I cubic	104, **150**, **231**, 271, **385**, 434, 454, 478, **511**, **618**, **759**, **780**, **820**, 970, 1025, 1240, 1980, 2380	19
759	$C_{10}H_8$	Naphthalene	**508**, **759**, 1017, 1166, 1240, 1274, **1375**, **1459**, **1572**, 1626, 3054	45
759	$Na_4B_{10}O_{17}{\cdot}7H_2O$	Ezcurrite	132, 226, 286, 315, 336, **476**, **513**, **570**, **759**, 881, **954**, 1040, 1059, **1349**, **1388**, **2859**, 3335, 3375	19
763	BF_3OH^-	Fluoroboric acid in aq.	**763**	19
763	$CaB_6O_{10}{\cdot}4H_2O$	Calcium hexaborate	383, 391, 460, 482, 572, 636, 745, **763**, 813, **855**, **861**, **895**, 952, 996	19
764	$BF_3{\cdot}2H_2O$	Boron fluoride	280, 350, 510, 730, **764**, 3530	19
767	$B_5O_6(OH)_4^-$	Pentaboratetetrahydrate in aq.	**767**	21
767	NH_4BF_4	NH_4 tetrafluoroborate	354, 527, **767**	19
768	$CsBF_4$	Cesium tetrafluoroborate	353, 356, 523, 527, 530, **768**	19
769	$NaBF_4$	Na–tetrafluoroborate in aq.	**773**, **769**	19
770	BF_4^-	Tetrafluoroborate ion in aq.	**770**	19
772	KBF_4	K–tetrafluoroborate	358, 360, 365, 527, 530, 533, 535, 538, **772**	19
772	$Li_2B_4O_7{\cdot}5H_2O$	Diomignite	391, **446**, 493, **543**, **772**, 845, 896, **1028**, 1097, 1352	22
772	$RbBF_4$	Rb tetrafluoroborate	526, 531, 535, **772**	19
773	$LiBF_4$	Li–tetrafluoroborate in aq.	**773**	19
773	$NaBF_4$	Na–tetrafluoroborate in aq.	**773**, **769**	19
773	$Si(OH)_4$	$Si(OH)_4$ monomer	**773**	9

Continued

cm^{-1}	Formula	Name	All Bands	Source
774	$3NaAlSi_3O_8{\cdot}Na_2CO_3$	Cancrinite	117, **274**, 349, 422, 462, 503, 684, **774**, 957, **1057**	19
774	KBF_4	Avogadrite	359, 535, **774**	8
780	HBO_2	Metaboric acid I cubic	104, **150**, **231**, 271, **385**, 434, 454, 478, **511**, **618**, **759**, **780**, **820**, 970, 1025, 1240, 1980, 2380	19
781	$Li_2B_4O_7$	Lithium tetraborate	158, 197, 222, 256, 351, 493, 531, 562, **725**, **781**, 980, 1034, 1060, 1166	19
782	HBO_2	Metaboric acid	428, 475, 518, 533, **782**	10
783	H_4SiO_4	Orthosilicic acid	**783**, **3413**	19
785	HBO_2	Metaboric acid II	77, 94, 108, 119, 131, 147, 178, 198, 477, 522, 536, **785**, **3130**	19
785	$NaBF_4$	Sodium tetrafluoroborate	347, 364, 531, 547, **785**	19
785	C_6H_{12}	Cyclohexane2	**785**, **1028**, 1266, 1440, 2664, **2855**, 2895, 2923, 2938	69
788	$Pb_5(AsO_4)_3Cl$	Mimetite	98, 105, 145, 175, **313**, 340, 371, 385, 408, 422, 765, **788**, 806, **813**, 837, 854, 860, **3266**, **3420**, 3548	34
788	SiC	Moissanite	143, 149, 767, **788**, 797, **966**, 1514, 1712	19
790	$Cu(UO_2)_2(SiO_3OH)_2{\cdot}6H_2O$	Cuprosklodowskite	207, 275, 302, **790**, **973**	8
793	$CaWO_4$	Scheelite	55, 79, 111, 207, **329**, 395, **793**, 834, **909**, 1020, 1120, 1625, 3267, 3285, **3351**, **3379**, 3413, **3545**	19
793	$LiAlSi_4O_{10}$	Petalite	**143**, **281**, 318, 357, 383, **491**, **793**	19
795	$(Ca, Ba, Pb, K, Na)U_2O_7{\cdot}5H_2O$	Calciouranoite	207, **795**, 964	8
798	$LiBF_4$	Lithium tetrafluoroborate	348, 523, **798**	19
804	C_6H_{12}	Cyclohexane	**804**, 2855, 2941	49
805	$WO_3{\cdot}H_2O$	Tungstite	215, 236, 257, 269, 325, 379, **660–663**, 684, **805**, **958**	19
805	Y(Nb, Ta)O_4	Fergusonite (Norway)	285, 667, **805**, **1419**, 1725, 2219, 3124, 3846	19
805	$CuUO_2(OH)_4$	Vandenbrandeite	182, 249, 279, 467, **805**, 887	8
806	$MnFe_2O_4$	Jacobsite	**806**, **839**	19
806	$SiO_2{\cdot}n(CH_4, C_2H_6, C_3H_8,$ i-$C_4H_{10})$	Isobutene in chibaite	**806**	35
807	WO_3	Tungsten trioxide	272, 325, **718**, **807**	19
810	HBO_2	Metaboric acid III	**87**, **119**, 404, 419, 469, 478, **599**, 638, **810**, 1100, 1343, 1460, 3195, 3295, 3365	19
811	$YNbO_4$	Fergusonite monoclinic	**131**, **324**, **336**, 658, 675, 698, **811**	19

cm^{-1}	Formula	Name	All Bands	Source
811	$Pb_3Ca_2(AsO_4)_3Cl$	Hedyphane	141, 151, 162, 177, 192, 201, **319**, 334, 349, 372, 394, 432, 452, 770, 787, **811**, **819**, 834, 848, 3352, 3409, **3458**	34
812	$NaAlF_4$	Sodium fluoroaluminate	174, 193, **289**, 304, 315, **352**, **622**, 641, **812**, 893	19
813	$Pb_5(AsO_4)_3Cl$	Mimetite	98, 105, 145, 175, **313**, 340, 371, 385, 408, 422, 765, **788**, 806, **813**, 837, 854, 860, **3266**, **3420**, 3548	34
814	$Y(Nb, Ta)O_4$	Fergusonite	**814**, **1412**, 1694, **2216**, 2425, 3143, 3839	19
815	$Cu_4Cl[Cl_{0.29}(OH)_{0.71}](OH)_6$	Claringbullite	119, 136, **147**, 163, 231, 260, 356, 389, 447, **511**, 579, **815**, **906**, **970**, 1049, 3211, 3331, **3351**, **3433**, 3458	31
815	$Pb_5(PO_4,AsO_4)_3Cl$	Pyromorphite–As	**152**, 177, 186, 206, 318, 339, 344, 377, 391, 409, 414, 768, 776, **815**, 818, 825, 917, **944**, 1014, 3325, **3378**, 3444	34
816	$Ca_3(Fe, Ti)_2Si_3O_{12}$	Andradite	174, 236, 325, 370, 452, 494, 516, 574, **816**, **842**, **874**, 995	10
819	$Pb_3Ca_2(AsO_4)_3Cl$	Hedyphane	141, 151, 162, 177, 192, 201, **319**, 334, 349, 372, 394, 432, 452, 770, 787, **811**, **819**, 834, 848, 3352, 3409, **3458**	34
820	HBO_2	Metaboric acid I cubic	104, **150**, **231**, 271, **385**, 434, 454, 478, **511**, **618**, **759**, **780**, **820**, 970, 1025, 1240, 1980, 2380	19
822	$Pb_{1-x}O_{3-2x}(UO_2)_4(OH)_{4+2x}\cdot 4H_2O$	Fourmarierite	373, 490, **703**, **822**	8
824	Mg_2SiO_4	Forsterite	227, 303, 423, 548, 608, **824**, **856**, 882, 921, **964**	10
827	$Ca_3Al_2Si_3O_{12}$	Grossular	181, 247, 280, 373, 420, 550, **827**, 848, 1007	10
828	$Ca_3Cr_2Si_3O_{12}$	Uvarovite	176, 242, 272, 370, 509, 526, 590, **828**, **894**	10
830	$Pb_{3+x}[(UO_2)_4O_{4+x}(OH)_{3-x}]_2\cdot 2H_2O$	Curite	162, 221, 302, 456, 786, **830**, 1024	8
830	$Ba(UO_2)_6O_4(OH)_6\cdot 8H_2O$	Billietite	803, **830**	58
832	$4[Mg_2SiO_4]\cdot[Mg, Fe, Ti](OH, F)_2$	Hydroxylclinohumite	110, 186, 212, 233, 306, 321, 348, 549, 590, 609, 756, 786, **832**, **845**, **861**, 931, 968, 3393, 3410, **3560**, **3570**, **3578**	13
833	$KAlSi_3O_8\cdot H_2O$	K–cymrite	114, **380**, **833**, 3541, 3623	53

Continued

cm^{-1}	Formula	Name	All Bands	Source
834	$Cu(UO_2)_2(SO_4)_2(OH)_2{\cdot}8H_2O$	Johannite	201, 244, **834**, 1043, 1090	8,58
835	$MgHBO_3$	Ascharite	324, **835**, **3555**	19
836	$Mg(UO_2)_2(PO_4)_2{\cdot}10H_2O$	Saléeite	193, **836**, 913, **989**	8
836	$KAlSi_3O_8$	Kokchetavite	109, **390**, **836**	53
839	$MnFe_2O_4$	Jacobsite	**806**, **839**	19
840	$Mg_6[Cl_2/B_{14}O_{26}]$	Boracite	**131**, 602, **660**, **840**, 1142	19
842	$PbCrO_4$	Crocoite	141, 339, **361**, 378, 405, **842**, 864	30
842	$Ca_3(Fe, Ti)_2Si_3O_{12}$	Andradite	174, 236, 325, 370, 452, 494, 516, 574, **816**, **842**, **874**, 995	10
843	$C_{10}H_{22}$	Decane saturated	**255**, **843**, 1132, **1302**, 1435, 1451, **2848**, 2858, **2872**, 2897, 2917, 2934	46
845	Fe_2SiO_4	Fe–spinel	334, **672**, 779, **845**, 924	19
845	$4[Mg_2SiO_4]{\cdot}[Mg, Fe, Ti](OH, F)_2$	Hydroxylclinohumite	110, 186, 212, 233, 306, 321, 348, 549, 590, 609, 756, 786, **832**, **845**, **861**, 931, 968, 3393, 3410, **3560**, **3570**, **3578**	13
849	$Mn_3Al_2Si_3O_{12}$	Spessartine	175, 221, 321, 350, 500, 552, 630, **849**, **879**, **905**, 1029	10
850	$UO_2Cl_5{}^{3-}$	Uranyl chloride Cl^{3-} in aq.	**850**	51
850	$MgB_3O_3(OH)_5{\cdot}5H_2O$	Kurnakovite	391, 422, 466, 627, **850**, 944	19
852	$LiBO_2{\cdot}8H_2O$	Lithium metaborate	390, **746**, **852**	19
852	$MgB_6O_{10}{\cdot}6H_2O$	Magnesium hexaborate	**374**, **634**, **638**, **641**, **852**, **953**, **964**	19
854	$Mg_2B_6O11{\cdot}17H_2O$	Magnesium triborate	390, 412, 491, 653, 743, **854**, 1060	19
854	$UO_2Cl_4{}^{2-}$	Uranyl chloride Cl^{2-} in aq.	**854**	51
855	$CaB_6O_{10}{\cdot}4H_2O$	Calcium hexaborate	383, 391, 460, 482, 572, 636, 745, **763**, 813, **855**, **861**, **895**, 952, 996	19
856	Mg_2SiO_4	Forsterite	227, 303, 423, 548, 608, **824**, **856**, 882, 921, **964**	10
856	$Na_2B_4O_7{\cdot}4H_2O$	Kernite	**105**, 175, 256, 388, 426, 456, **496**, 554, 579, **737**, 802, 848, **856**, **929**, 978, 1009, 1034, 1089, 1139	19
856	$NaB_5O_8{\cdot}5H_2O$	Sodium pentaborate	386, 464, 493, **529**, 773, **856**, 922	19
856	$NaB_5O_8{\cdot}5H_2O$	Sborgite	121, **529**, **856**	66
857	COS	Carbonyl sulphide	**856**	4
858	$UO_2Cl_3{}^{-}$	Uranyl chloride Cl^{-} in aq.	**858**	51
858	$Mg_2B_6O_{11}{\cdot}15H_2O$	Inderite synthetic	396, 424, 493, 584, 637, 745, **858**, 954, 1101	19
859	$CaB_2O_4{\cdot}6H_2O$	Calcium metaborate II	**755**, **859**	19
861	$4[Mg_2SiO_4]{\cdot}[Mg, Fe, i](OH, F)_2$	Hydroxylclinohumite	110, 186, 212, 233, 306, 321, 348, 549, 590, 609, 756, 786,	13

cm^{-1}	Formula	Name	All Bands	Source
			832, **845**, **861**, 931, 968, 3393, 3410, **3560**, **3570**, **3578**	
861	$CaB_6O_{10}\cdot 4H_2O$	Calcium hexaborate	383, 391, 460, 482, 572, 636, 745, **763**, 813, **855**, **861**, **895**, 952, 996	19
862	$NaCaB_5O_9\cdot 8H_2O$	Ulexite	228, 263, 300, 330, 392, 440, 485, 531, **589**, **592**, **595**, **747**, **862**, 1013, 1316, 3559, **3570**, 3588	19
862	UO_2Cl_2	Uranyl chloride Cl^0 in aq.	**862**	51
863	$Fe_3Al_2Si_3O_{12}$	Almandine	170, 216, 323, 342, 370, 500, 556, **863**, **897**, **916**, 1038	10
864	$K_2B_4O_7\cdot 4H_2O$	Potassium tetraborate	397, 456, **571**, 767, **864**, 979	19
866	UO_2Cl^+	Uranyl chloride Cl^+ in aq.	**866**	51
868	$PbMoO_4$	Wulfenite	**166**, **315**, 348, 742, 765, **868**, **1736**, 3290, **3350**, **3377**, 3533, 3566	19
869	$FeNb_2O_6$	Columbite	132, 203, 275, **393**, 525, 634, **869**	19
869	Zn_2SiO_4	Willemite	171, 385, **869**, 905, 944	19
869	C_5H_{12}	Pentane saturated	**869**, 2849–2861, 2912–2929, 2965–2969	70
870	UO_2^{2+}	Uranyl in aq.	**870**	50
870	$(Fe,Mn)Nb_2O_6$	Ferrocolumbite	134, 390, 529, **870–874**	19
871	$HBF_2(OH)_2$	Hydroxyfluoroboric acid	283, 349, 509, 611, **731**, 758, **871**, 3536	19
871	$Mg_3Al_2Si_3O_{12}$	Pyrope	56, 211, 364, 650, **871**, **902**, **928**, 1066	10
874	$Ca_3(Fe, Ti)_2Si_3O_{12}$	Andradite	174, 236, 325, 370, 452, 494, 516, 574, **816**, **842**, **874**, 995	10
874	$(Fe,Mn)Nb_2O_6$	Ferrocolumbite	134, 390, 529, **870–874**	19
874	$SiO_2\cdot n(CH_4, C_2H_6, C_3H_8, i\text{-}C_4H_{10})$	Propane in chibaite	**874**	35
875	H_3BO_3	Boric acid in aq.	495, **875**	23
875	$(Mn,Fe)Nb_2O_6$	Manganocolumbite	133, 390, 530, **875**	19
877	Be_2SiO_4	Phenakite	447, 527, 601, 618, 778, **877**, 916, 924, 952, 999, 1020	19
879	$Mn_3Al_2Si_3O_{12}$	Spessartine	175, 221, 321, 350, 500, 552, 630, **849**, **879**, **905**, 1029	10
880	$BeAlSiO_4(OH)$	Euclase	145, 180, 237, **259**, 290, **397**, 444, 452, 574, **880**, **908**, 975, 1023, 1059, 1120, **3575**, 3587	19
880	H_3BO_3	Sassolite	187, 211, 500, **880**, 1168, 3165, **3245–3247**	22

Continued

cm^{-1}	Formula	Name	All Bands	Source
880	$NaH_4(CO_3)(BO_3){\cdot}2H_2O$	Qilianshanite	187, 204, 355, 498, 539, 687, 858, **880**, 1012, 1029, **1038**, 1112	19
881	(Fe, Mn)WO_4	Wolframite	205, 236, 260, 353, 401, **462**, 484, 508, 695, 790, 806, **881**, 1061, 1079, **1156**	19
881		Manganotantalite	125, 395, 538, 613, **881**	19
891	$Fe_4Be_3(SiO_4)_3S$	Danalite	**129**, 155, 173, 237, 361, 443, 531, 566, 681, **891**, 943, 1023	19
894	$Ca_3Cr_2Si_3O_{12}$	Uvarovite	176, 242, 272, 370, 509, 526, 590, **828**, **894**	10
895	$CaB_6O_{10}{\cdot}4H_2O$	Calcium hexaborate	383, 391, 460, 482, 572, 636, 745, **763**, 813, **855**, **861**, **895**, 952, 996	19
897	$Fe_3Al_2Si_3O_{12}$	Almandine	170, 216, 323, 342, 370, 500, 556, **863**, **897**, **916**, 1038	10
898	$CaLiAl_2(AlBeSi_2)O_{10}(OH)_2$	Bityite	104, **132**, 365, **478**, 684, **898**, 1049	19
899	C_8H_{18}	Octane saturated	**899**, 2849–2861, 2912–2929, 2965–2969	70
902	$Mg_3Al_2Si_3O_{12}$	Pyrope	56, 211, 364, 650, **871**, **902**, **928**, 1066	10
905	$Mn_3Al_2Si_3O_{12}$	Spessartine	175, 221, 321, 350, 500, 552, 630, **849**, **879**, **905**, 1029	10
906	$Cu_4Cl[Cl_{0.29}(OH)_{0.71}](OH)_6$	Claringbullite	119, 136, **147**, 163, 231, 260, 356, 389, 447, **511**, 579, **815**, **906**, **970**, 1049, 3211, 3331, **3351**, **3433**, 3458	31
907	Al_2SiO_5	Sillimanite	142, **235**, 310, 456, 597, 708, 874, **907**, 964, 1127	10
908	$Cu_2(OH)_3Cl$	Atacamite2	66, **106**, **120**, **139**, **150**, 450, **514**, 586, 820, 845, **908**, 975, **3351**, **3453**	13
908	$BeAlSiO_4(OH)$	Euclase	145, 180, 237, **259**, 290, **397**, 444, 452, 574, **880**, **908**, 975, 1023, 1059, 1120, **3575**, 3587	19
908	NaH_2PO_4	Na–hydrogen phosphate	389, 535, **908**, 970, 1173, 3441, 3517	19
909	$CaWO_4$	Scheelite	55, 79, 111, 207, **329**, 395, **793**, 834, **909**, 1020, 1120, 1625, 3267, 3285, **3351**, **3379**, 3413, **3545**	19
910	Na_3PO_4	Sodium phosphate	391, 482, 524, 544, **910**, 942, 993, 3308, 3402, **3572**	19
912	$Cu_2(OH)_3Cl$	Atacamite	106, **121**, 361, 419, 446, 475, **511**, 818, 846, **912**, **975**	30
916	$Fe_3Al_2Si_3O_{12}$	Almandine	170, 216, 323, 342, 370, 500, 556, **863**, **897**, **916**, 1038	10

cm^{-1}	Formula	Name	All Bands	Source
918	Al_2SiO_5	Andalusite2	**292**, 320, 358, 551, **918**, **959**, 1019, 1064, 1110	19
920	Al_2SiO_5	Andalusite	**293**, 323, 361, 453, 553, 719, 834, **920**, 992, 1065, 1111	10
928	ClO_4^-	Perchlorate ion in aq.	460, 645, **928**	10
928	$Mg_3Al_2Si_3O_{12}$	Pyrope	56, 211, 364, 650, **871**, **902**, **928**, 1066	10
929	$Na_2B_4O_7{\cdot}4H_2O$	Kernite	**105**, 175, 256, 388, 426, 456, **496**, 554, 579, **737**, 802, 848, **856**, **929**, 978, 1009, 1034, 1089, 1139	19
931	$BeAl_2O_4$	Chrysoberyl	218, 242, 353, 422, 459, **477**, **518**, 546, **639**, 658, 679, 711, 778, 816, **931**	19
931	$Na_2WO_4{\cdot}2H_2O$	Sodium wolframate	336, 842, **931**, 3309, **3463**	19
933	$(Cu, Al)_2H_2Si_2O_5(OH)_4{\cdot}nH_2O$	Chrysocolla	109, 146, 183, 213, 245, 336, 424, 485, **673**, 792, **933**, **1058**, 1602, 3295, 3504, 3583, **3624**, 3652, 3708	32
944	$Pb_5(PO_4,AsO_4)_3Cl$	Pyromorphite–As	**152**, 177, 186, 206, 318, 339, 344, 377, 391, 409, 414, 768, 776, **815**, 818, 825, 917, **944**, 1014, 3325, **3378**, 3444	34
946	$Al_7(BO_3)(SiO_4)_3O_3$	Dumortierite	169, **206**, 293, 353, 403, 418, **508**, 810, 842, **946**, 1076, 1112	8
947	$Ca_3(PO_4)_2$	Tricalcium phosphate	407, **947**, **970**	27
949	$(Mg, Fe, Mn)Al_3(BO_3)(SiO_4)O_2$	Grandidierite	**161**, 228, 242, 269, **343**, 363, 376, **425**, 442, 489, **577**, 659, 864, **949**, 979, 1042	8
951	$LiFe^{3+}(PO_4)(OH)$	Tavorite	157, 201, 443, 576, 589, 629, **951**, 1001, 1069	19
951	$LiFePO_4$	Triphylite	152, 251, 444, 572, 594, 628, **951**, 1002, 1068, 1082	19
952	Al_2SiO_5	Kyanite	302, 325, 360, 386, 405, 419, 437, 486, 562, 669, **952**	10
953	$Mg_3BO_3(OH, F)_3$	Fluoborite	241, **953**, 1341, 3645, 3656, **3669**, 3686	19
953	$MgB_6O_{10}{\cdot}6H_2O$	Magnesium hexaborate	**374**, **634**, **638**, **641**, **852**, **953**, **964**	19
955	$CaMgPO_4F$	Isokite	273, 425, 457, 607, **955**, 1021	19
958	Rb_2SO_4	Rubidium sulphate	449, 617, **958**, **1099**, **1130**	19
958	$WO_3{\cdot}H_2O$	Tungstite	215, 236, 257, 269, 325, 379, **660–663**, 684, **805**, **958**	19
959	$(Ca, Sr)_5(PO_4)_3(OH)$	Apatite–Sr	442, 575, 586, **959**, 1044, 1071, 3295, 3353	19

Continued

cm^{-1}	Formula	Name	All Bands	Source
959	Al_2SiO_5	Andalusite2	**292**, 320, 358, 551, **918**, **959**, 1019, 1064, 1110	19
962	$Al_6(BO_3)_5(OH, F)_3$	Jeremejevite	301, **328**, **372**, 417, 512, 608, 710, 749, **962**, 1067, 3624, **3672**	19
962	$Ca_5(PO_4)_3OH$	Hydroxylapatite	431, 591, **962**, **3572**	27
963	$Ca_5(PO_4)_3Cl$	Apatite–Cl	430, 581, **963**, 1039, 1127	10
964	$Al_{4.5}Si_{1.5}O_{9.75}$	Mullite	161, 264, 308, **419**, 521, 608, 886, **964**, 1133	19
964	Mg_2SiO_4	Forsterite	227, 303, 423, 548, 608, **824**, **856**, 882, 921, **964**	10
964	$MgB_6O_{10}{\cdot}6H_2O$	Magnesium hexaborate	**374**, **634**, **638**, **641**, **852**, **953**, **964**	19
965	(Ce, La)CO_3(OH, F)	Hydroxylbastnäsite	112, 155, 436, 591, **965**, 1053, 1330, **3317**, **3352**, 3430, 3535, 3714	11
965	$Ca_5(PO_4)_3F$	Apatite–F	432, 449, 581, 592, 608, **965**, 1042, 1053, 1081	10
966	SiC	Moissanite	143, 149, 767, **788**, 797, **966**, 1514, 1712	19
967	$CaSiO_3$	Wollastonite	315, **633**, **967**, 1037	19
967	$Na_6Mg_2(SO_4)(CO_3)_4$	Tychite	212, 246, 300, 462, 629, 718, **967**, **1108**, 1135	19
970	$Cu_4Cl[Cl_{0.29}(OH)_{0.71}](OH)_6$	Claringbullite	119, 136, **147**, 163, 231, 260, 356, 389, 447, **511**, 579, **815**, **906**, **970**, 1049, 3211, 3331, **3351**, **3433**, 3458	31
970	$Ca_3(PO_4)_2$	Tricalcium phosphate	407, **947**, **970**	27
971	$Cu_3(PO_4)(OH)_3$	Cornetite	**109**, 131, 141, 158, 174, 297, 313, 363, **446**, **477**, 515, 583, 639, 945, **971**, **996**, 1016, **1083**, 1137, **3414**	57
973	$Cu(UO_2)_2(SiO_3OH)_2{\cdot}6H_2O$	Cuprosklodowskite	207, 275, 302, **790**, **973**	8
973	$Cu_4(OH)_6SO_4$	Brochantite	90, 123, 139, 196, 390, 421, 450, 481, 595, 609, **973**, **3371**, **3399**, **3564**, **3587**	30
973	$Cu_4(OH)_6SO_4$	Brochantite2	**80**, **91**, 132, 138, 142, 157, 188, 199, 242, 319, 387, 426, 450, 484, 509, 595, 609, 910, **973**, 1097, **3401**, **3566**, **3588**	13
974	$KAl(SO_4)_2{\cdot}12H_2O$	Alum–K	442, 455, 614, **974**, **989**, 1104, 1130, 3072, **3396**	10
975	$Cu_2(OH)_3Cl$	Atacamite	106, **121**, 361, 419, 446, 475, **511**, 818, 846, **912**, **975**	30
977	Ca_2(Mn, Fe)$Al_2[OH/BO_3/Si_4O_{12}]$	Axinite	104, 271, 414, 482, **710**, 897, **977**, 3354	19
977	$LiAlSi_2O_6$	Spodumene	282, 320, 382, **464**, **492**, 355, 697, **707**, 776, 794, 808, **977**, 1019, 1062, **1074**, 1135, 1157	19

cm^{-1}	Formula	Name	All Bands	Source
978	$PbSO_4$	Anglesite	438, 450, 608, 641, **978**, 1068, 1160	10
978	SO_4^{2-}	Sulphate ion in aq.	450, 620, **978–983**	9
979	$Cu_2PO_4(OH)$	Libethenite	**74**, 92, 113, 140, **160**, 227, 270, **301**, 319, 371, 392, 430, 461, 561, 590, 944, **979**, 1008, **1022**, **3475**	57
979	$Ca_2Mn_5Be_4(PO_4)_6(OH)_4{\cdot}6H_2O$	Roscherite	173, 210, 259, **400**, 559, 579, 599, 631, 771, **979**, 1066	19
980	$(Mn, Fe, Mg, Ca)_2(PO_4)(F, OH)$	Triplite	425, 610, **980**, 1034	10
981	$Cu_5(PO4)_2(OH)_4$	Ludjibaite	**160**, 190, 226, 263, **301**, 368, 387, 402, 449, 557, 586, 633, 736, 784, 815, 855, 925, **981**, 1019, 1046, 1072, 1115, **3470**	57
981	$Mn_4Be_3(SiO_4)_3S$	Helvite	**131**, 158, 176, 238, 288, 362, 444, 527, 565, 678, **981**, 942, 1020	19
983	$CaBe(PO_4)(F, OH)$	Herderite	166, 194, 424, 445, 516, 529, **584**, **594**, **983**, **1005**, 1128, 1140, **3564**, **3573**, 3620	19
983	$CuSO_4{\cdot}5H_2O$	Copper sulphate	259, 333, 368, 423, 450, 611, **983**, 1144	19
983	$K_2Ca(SO_4)_2{\cdot}H_2O$	Syngenite	474, 494, 642, 662, **983**, **1007**, 1142, 1168, **3301**, **3378**	10
983	K_2SO_4	Arcanite	457, 622, **983**, 1093, 1109	10
983	SO_4^{2-}	Sulphate ion in aq.	450, 620, **978–983**	4
984	$(K, Na)_3Na(SO_4)_2$	Aphthitalite	447, 457, 619, **984**, 1093, 1104	10
984	$Be_2(BO_3)(OH, F){\cdot}H_2O$	Berborite	78, 379, 394, **984**, **1377**, **3504**	19
984	$MgSO_4{\cdot}6H_2O$	Hexahydrite	445, 466, 610, **984**, 1085, 1146, 3258, **3428**	10
984	$MgSO_4{\cdot}7H_2O$	Epsomite	447, 612, **984**, 1061, 1095, 1134, **3303**, **3425**	10
985	$Ca_3Al_2(SO_4)(OH, F)_{10}{\cdot}2H_2O$	Creedite	441, 481, 602, 628, 658, **985**, 1080, 1132, 1182	19
985	$CaBe(PO_4)(OH)$	Hydroxylherderite	170, 197, 427, 448, 531, 579, 596, **985**, **1004**, 1129, 1141	19
985	$CaHPO_4{\cdot}2H_2O$	Brushite	408, 521, 875, **985**, 1056, 1132, 3160, 3261, **3469**, **3536**	27
986	$CsAl(SO_4)_2{\cdot}12H_2O$	Cs–Al sulphate	191, 447, 624, **986**, 1091	19
987	$CePO_4$	Monazite	466, 620, **987**, 1054	10
987	$K_2MgCa(SO_4)_4{\cdot}2H_2O$	Polyhalite	625, **987**, **1013**	19
988	$Na_8(Al_6Si_6O_{24})Cl_2$	Sodalite	114, 167, 263, 297, 411, **465**, **988**, 1066	13

Continued

cm^{-1}	Formula	Name	All Bands	Source
988	$Be_2BO_3(OH, F)$	Hambergite2	280, 324, 398, 583, 623, 739, **988**, **3419**, **3484**, 3562, **3600**, 3662	19
988	$Al_2(SO_4)_3{\cdot}18H_2O$	Aluminium sulphate	617, 972, **988**, 1127	19
988	$BaSO_4$	Barite	**461**, 617, **988**, 1143	10
988	$BaSO_4$	Barite2	74, 89, 97, 152, 189, **452**, **462**, 616, 629, 647, 965, **988**, 1104, **1144**, 1166	18
988	C_6H_6	Benzene2	**988**, **3058**	49
989	$KAl(SO_4)_2{\cdot}12H_2O$	Alum–K	442, 455, 614, **974**, **989**, 1104, 1130, 3072, **3396**	10
989	$Na_2SO_4{\cdot}10H_2O$	Mirabilite	458, 627, **989**, 1129, 3340, **3506**	10
989	$NaAlSiO_4$	Nepheline	220, 264, 350, **399**, **415**, 467, 920, 967, **989**, 1086	19
990	$K_3Na(SO_4)_2$	Glaserite (aphthitalite)	**453**, **619**, **678**, **990**	19
991	$Na_2Mg(SO_4)_2{\cdot}4H_2O$	Astrakanite (blodite)	454, **616**, **991**, 1067, 1115, 1153	19
992	$Na_2Mg(SO_4)_2{\cdot}4H_2O$	Blodite (astrakanite)	454, 618, **992**, 1188	8
993	$Ca_2Be_4(Fe, Mn)_5(PO_4)_6(OH)_4{\cdot}6H_2O$	Greifensteinite	164, 216, 305, 407, 435, 481, **612**, **628**, **993**, **1056**, 1133, 1176	19
993	Nb_2O_5	Niobium oxide	238, 308, 463, 551, 635, **683**, 815, 897, **993**	19
994	C_6H_6	Benzene	608, **994**, 1179, 1587, 1607, 2927, 2950, **3063**	44
994	$BeSO_4{\cdot}4H_2O$	Berylium sulphate	181, 281, 415, 495, 588, 630, **994**, 1080, 1112	19
994	Na_2SO_4	Thenardite	452, 469, 621, 632, **994**, 1103, 1153	10
994	$Na_6(CO_3)(SO_4)_2$	Burkeite	451, 474, 620, 633, 644, **994**, 1065	10
995	$CsHSO_4$	Cesium hydrogen sulphate	419, 473, 570, 587, 610, 857, **995**, 1180, 1249	19
996	$Cu_3(PO_4)(OH)_3$	Cornetite	**109**, 131, 141, 158, 174, 297, 313, 363, **446**, **477**, 515, 583, 639, 945, **971**, **996**, 1016, **1083**, 1137, **3414**	57
996	$Fe_5(SO_4)_6(OH)_2{\cdot}20H_2O$	Ferricopiapite	268, 310, 451, **477**, **996**, **1019**, 1148	8, 58
998	$MnSiO_3$	Rhodonite	257, 414, **665**, 711, 737, **998**	19
998	YPO_4	Xenotime	485, 642, **998**, 1056, 1394	10
999	C_6H_6	Benzene3	**309**, **999**, 1580, **3064**	69
1000	C_2H_6	Ethane2	**1000**, 1440, **2895**, **2955**	64
1000	$S_2O_3^{2-}$	Thiosulphate ions in aq.	**450**, **1000**	59
1000	$MgSO_4{\cdot}4H_2O$	Starkeyite	116, 401, 462, 565, 616, 664, **1000**, 1086, 1156, 1186, 3331, **3427**, **3481**, 3558	10
1000	$SrSO_4$	Celestine	452, 627, **656**, **1000**, 1156, 1190	10

cm^{-1}	Formula	Name	All Bands	Source
1001	$K_3NaFeCl_6$	Rinneite	287, 524, **1001**	19
1001	$NaAl(PO_4)F$	Lacroixite	319, 503, 609, **623**, **1001**, 1043, 1119	19
1002	$Na_2Ca(SO_4)_2$	Glauberite	485, 618, 644, **1002**, 1106, 1140	10
1002	$Na_6(F, Cl)(SO_4)_2$	Sulfohalite	471, 634, **1002**, 1125	10
1004	$CaBe(PO_4)(OH)$	Hydroxylherderite	170, 197, 427, 448, 531, 579, 596, **985**, **1004**, 1129, 1141	19
1004	$Na_7Ca_3(OH)(SO_4)_6{\cdot}0.8H_2O$	Cesanite	448, 474, 626, 647, **1004**, 1104	10
1005	$CaBe(PO_4)(F, OH)$	Herderite	166, 194, 424, 445, 516, 529, **584**, **594**, **983**, **1005**, 1128, 1140, **3564**, **3573**, 3620	19
1005	$K_2Ca_5(SO_4)_6{\cdot}H_2O$	Goergeyite	433, 440, 480, 595, 602, 631, **1005**, **1013**, 1085, 1115, 1162, 3525, **3580**	10
1005	$MgSO_4{\cdot}5H_2O$	Pentahydrite	371, 447, 602, **1005**, 1106, 1159, 3289, **3343**, **3391**, 3494, 3553	10
1005	$ZrSiO_4$	Zircon	202, 212, 225, **356**, 392, 438, **972–974**, **1005–1008**	13
1006	$CaSO_4{\cdot}2H_2O$	Gypsum	414, **492–494**, 621–623, 671, **1006–1008**, 1142, 3250, **3405**, 3491, 3500	4
1006	Li_2SO_4	Lithium sulphate	173, 208, 393, 480, 633, **1006**, 1099, 1149, 1178	19
1007	$K_2Ca(SO_4)_2{\cdot}H_2O$	Syngenite	474, 494, 642, 662, **983**, **1007**, 1142, 1168, **3301**, **3378**	10
1007	$KFeSO_4$	Potassium–iron sulphate	162, 219, 441, 475, 591, **1007**, 1024, **1068**, 1281	19
1007	$Li_2SO_4{\cdot}H_2O$	Lithium sulphate hydrate	484, 635, **1007**, 1102, 1149, 1176	19
1007	$MnBe(PO_4)(OH)$	Väyrynenite	96, 125, 186, 235, 332, 381, 462, 508, 599, 619, 741, 800, 986, **1007**, 1075, 1126	19
1008	$CaSO_4{\cdot}2H_2O$	Gypsum	414, **492–494**, 621–623, 671, **1006–1008**, 1142, 3250, **3405**, 3491, 3500	10
1008	$ZrSiO_4$	Zircon	202, 212, 225, **356**, 392, 438, 972–974, **1005–1008**	10
1011	$Mg_3B_2(SO_4)(OH)_8(OH, F)_2$	Sulfoborite	275, 363, 460, 621, 657, 778, 901, 934, **1011**, 1081, 1134, 3242, 3374, 3561, 3573, **3592**, 3618	19
1012	$Be_3(PO_4)_2$	Beryllium phosphate	351, 466, 540, **1012**, **1051**	19
1012	$LiAl(PO_4)(F, OH)$	Amblygonite	130, 303, 426, 488, 603, **647**, **1012**, **1048**, 1064, **3369**, 3408	19
1012	$NaBePO_4$	Beryllonite	139, 161, 433, 548, **1012**, **1056**	26

Continued

cm^{-1}	Formula	Name	All Bands	Source
1013	$CaMgSi_2O_6$	Diopside2	324, 360, 391, 667, 855, **1013**, 1046	19
1013	$K_2Ca_5(SO_4)_6{\cdot}H_2O$	Goergeyite	433, 440, 480, 595, 602, 631, **1005**, **1013**, 1085, 1115, 1162, 3525, **3580**	10
1013	$K_2MgCa(SO_4)_4{\cdot}2H_2O$	Polyhalite	625, **987**, **1013**	19
1013	$LiAl(PO_4)(OH, F)$	Montebrasite	138, 170, 196, 298, 428, 484, 602, 630, **646**, 801, 965, **1013**, **1046**, **1058**, 1107	19
1015	$CaSO_4$	Anhydrite	430, 500, 515, 611, 629, 674–676, **1015–1018**, 1131	4
1015	$KAl_3(SO_4)_2(OH)_6$	Jarosite	442, 560, 625, **1015**, 1112, 1161	19
1015	HCO_3^-	Hydrocarbonate ion in aq.2	630, 641, 673, 842, **1015**, **1312**, **1364**, 1628, 1682, 2619	43
1017	$H_{4-n}SiO_4^{n-}$	Deprotonated monomers	**1017**	9
1017	HCO_3^-	Hydrocarbonate ion in aq.	**1017**, 1091, 1360	9
1017	$NaCa_2Mg_4Al_3Si_6O_{22}(OH)_2$	Pargasite	229, **665**, **1017**	10
1018	$CaSO_4$	Anhydrite	430, 500, 515, 611, 629, 674–676, **1015–1018**, 1131	10
1018	$HfSiO_4$	Hafnon	148, 159, 215, 350, 403, 448, 984, **1018**	19
1019	$Fe_5(SO_4)_6(OH)_2{\cdot}20H_2O$	Ferricopiapite	268, 310, 451, **477**, **996**, **1019**, 1148	8, 58
1021	$ZnSO_4{\cdot}6H_2O$	Zinc sulphate hexahydrate	277, 420, 503, 624, 664, **1021**, 1083, 1188	19
1022	$KMgClSO_4{\cdot}2.75H_2O$	Kainite	**440**, **630**, **1022**, **1039**, 1157, 1176, 1197, 1207	19
1022	$Cu_2PO_4(OH)$	Libethenite	**74**, 92, 113, 140, **160**, 227, 270, **301**, 319, 371, 392, 430, 461, 561, 590, 944, **979**, 1008, **1022**, **3475**	57
1023	$MgSO_4$	Magnesium sulphate	451, 475, 499, 608, 681, 697, **1023**, 1053, 1136, 1220, 1256	10
1025	$K_2Mg(SO_4)_2{\cdot}4H_2O$	Leonite	**453**, **632**, **1025**, **1041**, 1158, 1196, 3428	19
1026	$KAl_3[(OH)_3(SO_4)]_2$	Alunite	485, 509, 654, **1026**, 1079, 1190, **3482**, **3509**	10
1027	$KAl_3[(OH)_3(SO_4)]_2$	Alunite2	236, 347, 384, 486, 564, 655, **1027**, 1081, 1190, **3480**, **3508**	19
1028	$AlPO_4{\cdot}2H_2O$	Variscite	225, 435, 612, **1028**, 1076, 1160, **3573**	19
1028	$KHCO_3$	Kalicinite	635, 673–675, **1028**, **1271**, 1277	25
1028	$Li_2B_4O_7{\cdot}5H_2O$	Diomignite	391, **446**, 493, **543**, **772**, 845, 896, **1028**, 1097, 1352	22
1028	C_6H_{12}	Cyclohexane2	**785**, **1028**, 1266, 1440, 2664, **2855**, 2895, 2923, 2938	69

cm^{-1}	Formula	Name	All Bands	Source
1029	$KHCO_3$	Kalicinite2	79, 106, 184, 636, 676, 832, 960, **1029**, **1282**, 1448, 1682, **2567**	43
1029	$Be_2(PO_4)(OH)\cdot 4H_2O$	Moraesite	124, 200, **376**, 415, **584**, 696, **1029**, 1080	8
1034	$MgSO_4\cdot 2H_2O$	Sanderite	447, 492, 597, 630, **1034**, 1164, **3446**, 3539	10
1036	$Fe_2(SO_4)_3\cdot 9H_2O$	Ferric iron sulphate hydrate	251, 455, 631, **1036**, 1049, 1086, 1189	19
1038	$NaH_4(CO_3)(BO_3)\cdot 2H_2O$	Qilianshanite	187, 204, 355, 498, 539, 687, 858, **880**, 1012, 1029, **1038**, 1112	19
1039	$KMgClSO_4\cdot 2.75H_2O$	Kainite	**440**, **630**, **1022**, **1039**, 1157, 1176, 1197, 1207	19
1040	$Ca_2[Mg_4(Al, Fe)]Si_7AlO_{22}(OH)_2$	Hornblende	224, **667**, **1040**	10
1040	Cs_2CO_3	Cesium carbonate	227, 662, 672, 679, **1040**, **1048**	25
1041	$K_2Mg(SO_4)_2\cdot 4H_2O$	Leonite	**453**, **632**, **1025**, **1041**, 1158, 1196, 3428	19
1046	$LiAl(PO_4)(OH, F)$	Montebrasite	138, 170, 196, 298, 428, 484, 602, 630, **646**, 801, 965, **1013**, **1046**, **1058**, 1107	19
1046	$LiKCO_3$	Li–K carbonate	**695**, 729, 870, **1046**, **1050**, 1388, 1418, 1483, 1517, 1753	19
1046	$MgSO_4\cdot H_2O$	Kieserite	436, 481, 629, **1046**, 1117, 1215, **3297**	10
1046	$NaHCO_3$	Nahcolite	684, **1046**, 1266, 1432	4
1047	$Na_2Mg_3Al_2Si_8O_{22}(OH)_2$	Glaucophane	210, 257, 339, **386**, 411, 445, 525, 562, **670**, 778, 896, 1000, **1047**, 1108, 3615, 3631, 3647, 3660	19
1048	Cs_2CO_3	Cesium carbonate	227, 662, 672, 679, **1040**, **1048**	25
1048	$LiAl(PO_4)(F, OH)$	Amblygonite	130, 303, 426, 488, 603, **647**, **1012**, **1048**, 1064, **3369**, 3408	19
1049	NO_3^-	Nitrate ion in aq.	690, **1049**, 1355	10
1050	HSO_4^-	Hydrogen sulphate in aq.	890, **1050**	4
1050	$LiKCO_3$	Li–K carbonate	**695**, 729, 870, **1046**, **1050**, 1388, 1418, 1483, 1517, 1753	19
1051	$2PbCO_3\cdot Pb(OH)_2$	Hydrocerussite	415, 681, **1051**, 1055	30
1051	$Be_3(PO_4)_2$	Beryllium phosphate	351, 466, 540, **1012**, **1051**	19
1052	$Na_4BeAlSi_4O_{12}Cl$	Tugtupite	160, 206, 269, 326, 439, **462**, 510, 553, 631, **1052**, 1103, 1130	19
1053	$K_2Mg_2(SO_4)_3$	Langbeinite	**456**, **466**, 607, 626, 640, **1053**	19

Continued

cm^{-1}	Formula	Name	All Bands	Source
1055	BN	Boron nitride cubic	**1055**, **1305**	19
1055	$2PbCO_3 \cdot Pb(OH)_2$	Hydrocerussite	415, 681, 1051, 1055	30
1056	$Ca_2Be_4(Fe, Mn)_5(PO_4)_6(OH)_4 \cdot 6H_2O$	Greifensteinite	164, 216, 305, 407, 435, 481, **612**, **628**, **993**, **1056**, 1133, 1176	19
1056	$NaBePO_4$	Beryllonite	139, 161, 433, 548, **1012**, **1056**	26
1056	$PbCO_3$	Cerussite	150, 180, 215, 682, 839, **1056**, 1378	10
1057	$3NaAlSi_3O_8 \cdot Na_2CO_3$	Cancrinite	117, **274**, 349, 422, 462, 503, 684, **774**, 957, **1057**	19
1058	$(Cu, Al)_2H_2Si_2O_5(OH)_4 \cdot nH_2O$	Chrysocolla	109, 146, 183, 213, 245, 336, 424, 485, **673**, 792, **933**, **1058**, 1602, 3295, 3504, 3583, **3624**, 3652, 3708	32
1058	$LiAl(PO_4)(OH, F)$	Montebrasite	138, 170, 196, 298, 428, 484, 602, 630, **646**, 801, 965, **1013**, **1046**, **1058**, 1107	19
1059	$(Mg, Fe)Al_2(PO_4)_2(OH)_2$	Lazulite	253, 344, 611, 630, 741, **1059**, 1100, 1136, 3384, 3400	19
1059	$BaCO_3$	Witherite	136, 152, 692, **1059**, 1420	10
1060	K_2CO_3	Gregoryite	141, 192, 234, 677, 686, 697, 702, 1043, **1060–1064**, 1374, 1405, 1426	25
1060	$Na_3H(CO_3)_2 \cdot 2H_2O$	Trona	140, 185, 225, **1060**, 1430	10
1062	$Na_2CO_3 \cdot H_2O$	Thermonatrite	**156**, 185, 230, **1062**, 1432	10
1062	Rb_2CO_3	Rubidium carbonate	221, 685, **1062**	25
1062	CO_3^{2-}	Carbonate ion in aq.	684, **1062–1072**, **1376–1382**, 1438	7
1064	$Al_4(PO_4)_3(OH)_3$	Trolleite	269, 428, 621, 690, 790, 978, **1064**, 1100, 1120, **3428**, 3510	19
1064	K_2CO_3	Gregoryite	141, 192, 234, 677, 686, 697, 702, 1043, **1060–1064**, 1374, 1405, 1426	10
1065	CO_3^{2-}	Carbonate ion in aq.2	684, 885, **1065**, 1376, 1421, 1764	43
1068	$Be_3Al_2Si_6O_{18}$	Beryl2	324, 401, **686**, 1006, **1068**, 1112, **3593**, 3604	19
1068	$KFeSO_4$	Potassium–iron sulphate	162, 219, 441, 475, 591, **1007**, 1024, **1068**, 1281	19
1070	$Na_2Ca_2(CO_3)_3$	Shortite	139, 154, 201, 266, 695, 710, 730, **1070**, **1090**, 1402, 1470	19
1070	Na_2SO_4	Na–sulphate in aq.	491, 682, **1070**, 1198, 1694	19
1070		Rubellite	**226**, 246, 276, 332, **379**, 410, 517, 645, **716**, 733, 753, 775, 883, 982, 1036, **1070**	19
1071	$Na_2Ca(CO_3)_2 \cdot 5H_2O$	Gaylussite	**164**, 265, 723, **1071**, **2944**, 3334	10

cm^{-1}	Formula	Name	All Bands	Source
1072	CO_3^{2-}	Carbonate ion in aq.	684, **1062–1072**, **1376–1382**, 1438	7
1073	$SrCO_3$	Strontianite	149, 183, 250, 700, 1057, **1073**, 1450	10
1074	$LiAlSi_2O_6$	Spodumene	282, 320, 382, **464**, **492**, 355, 697, **707**, 776, 794, 808, **977**, 1019, 1062, **1074**, 1135, 1157	19
1076	$CaB(SiO_4)OH$	Datolite	149, 275, 707, **1076**, 1428, 1742	19
1078	Na_2CO_3	Natrite	698, **1078**, 1428	25
1079	$(Fe, Mn)WO_4$	Wolframite	205, 236, 260, 353, 401, **462**, 484, 508, 695, 790, 806, **881**, 1061, 1079, **1156**	19
1079	$Ca(Ce, La)(CO_3)_2F$	Synchysite–Ce	247, 354, 742, 864, **1079**, **1093**	19
1080	$Ca(Ce, Y)(CO_3)_2F$	Synchysite–Y	**224**, 244, **290**, 374, 402, 479, 517, 603, 655, 993, 1054, **1080**, **1095**	19
1080	$(Ce, La)CO_3(OH, F)$	Hydroxylbastnäsite2	**1080**, **1087**, **1098**, **3235**, **3493**, **3568**, **3638**	60
1081	$Ca(Ce, La)_2(CO_3)_2F_2$	Parisite–Ce	166, 255, 264, 278, 354, 394, 598, 739, 867, 957, **1081**, **1092**, **1099**	19
1081	$Cu_8Si_8O_{22}(OH)_4{\cdot}H_2O$	Planchéite	104, 122, **139**, 148, **186**, 227, 302, 328, 403, 421, 476, 505, 699, 828, 906, 1048, **1081**, 1355, **1510**, **3250**, **3479**	32
1083	$Cu_3(PO_4)(OH)_3$	Cornetite	**109**, 131, 141, 158, 174, 297, 313, 363, **446**, **477**, 515, 583, 639, 945, **971**, **996**, 1016, **1083**, 1137, **3414**	57
1083	$Cu_5(PO_4)_2(OH)_4$	Reichenbachite	89, **110**, 135, 175, 188, 214, 255, 298, 365, 412, 453, **480**, 515, 540, 572, 607, 636, 951, 971, 998, **1083**, **3380**, **3428**	57
1083	$Sr(Ce, Ca, La)(CO_3)_2(OH)_2{\cdot}H_2O$	Ancylite–Ce	125, 189, **243**, 371, 699, 725, **1083**	8, 58
1085	$(Ce, La)CO_3F$	Bastnäsite	**161**, 175, 181, 234, 261, 286, 288, 352, 402, 566, 601, 606, 719, 735, **1085**, **1096**, 1432, 1445, 1504, 1737	11
1085	$CaCO_3$	Calcite	156, 283, 711, **1085**	4
1086	$Ca_5(PO_4)_2(OH)_4$	Pseudomalachite	88, **110**, 131, 176, 368, 450, **480**, 537, 609, 972, 1000, **1086**	30
1086	$CaCO_3$	Aragonite–(Sr)	144, 154, 181, 191, 207, 274, **702**, 707, 1060, 1084, **1086**, 1463	13

Continued

cm^{-1}	Formula	Name	All Bands	Source
1087	$CaCO_3$	Mg–calcite	157, 284, **714**, **1087**	4
1087	$MnCO_3$	Rhodochrosite	185, 289, 718, **1087**, 1416	10
1087	$Na_2Ca(CO_3)_2$	Nyerereite	677, 711, 1071, **1087**	29
1087	(Ce, La)CO_3(OH, F)	Hydroxylbastnäsite2	**1080**, **1087**, **1098**, **3235**, **3493**, **3568**, **3638**	60
1088	$(Ce_2,Ca)(CO_3)_2F_2$	Parisite	152, 161, 258, 263, 353, 601, 742, **1088**, 1420, 1735	11
1088	$CaCO_3$	Calcite–(Mg)	157, **283**, 714, 1069, **1088**, 1438	13
1090	(Fe, Mg)CO_3	Siderite	194, 301, 738, **1090**, 1442	10
1090	$CaCO_3$	Vaterite	118, 301, 740, 750, 1074, **1090**, 1465	10
1090	$Na_2Ca_2(CO_3)_3$	Shortite	139, 154, 201, 266, 695, 710, 730, **1070**, **1090**, 1402, 1470	19
1091	$Ba_3Ce_2(CO_3)_5F_2$	Cebaite	177, 213, **267**, 621, 701, 715, 730, 1072, 1084, **1091**	8
1091	Li_2CO_3	Zabuyelite	95, 712, 746, **1091**, 1456–1459	1
1091	$NaAl(CO_3)(OH)_2$	Dawsonite	**192**, 260, 590, 728, 1065, **1091**, 1506, 3250, **3282**	10
1091	$NaAl(CO_3)(OH)_2$	Dawsonite2	152, **191**, 389, 519, **590**, 824, 897, **1091**, **1506**, 3252, **3283**	19
1092	Ca(Ce, La)$_2(CO_3)_2F_2$	Parisite–Ce	166, 255, 264, 278, 354, 394, 598, 739, 867, 957, **1081**, **1092**, **1099**	19
1093	Ca(Ce, La)$(CO_3)_2$F	Synchysite–Ce	247, 354, 742, 864, **1079**, **1093**	19
1093	$ZnCO_3$	Smithsonite	196, 303, 731, **1093**, 1408	10
1094	$Mg_2(OH)_2(CO_3)\cdot 3H_2O$	Artinite	**147**, 173, 472, 704, **1094**, 3030, 3229, **3593**	10
1094	$MgCO_3$	Magnesite	329, 738, **1094**	4
1095	Al_2SiO_4(OH, F)$_3$	Hydroxyltopaz	229, 377, 439, 457, 897, 996, **1095**, 1122, 1353, 3288, 3366	19
1095	Ca(Ce, Y)$(CO_3)_2$F	Synchysite–Y	**224**, 244, **290**, 374, 402, 479, 517, 603, 655, 993, 1054, **1080**, **1095**	19
1095	$Cu_3(OH)_2(CO_3)_2$	Azurite	170, 246, 279, **397**, **1095**, 1457, **3453**	10
1096	(Ce, La)CO_3F	Bastnäsite	**161**, 175, 181, 234, 261, 286, 288, 352, 402, 566, 601, 606, 719, 735, **1085**, **1096**, 1432, 1445, 1504, 1737	11
1097	(Ca, Mg)CO_3	Dolomite	176, 299, 725, **1097**	4
1098	$Cu_2(OH)_2(CO_3)$	Malachite	**154**, **178**, 272, **434**, 537, 721, **1098**, 1492, 3386, **3468**	10
1098	(Ce, La)CO_3(OH, F)	Hydroxylbastnäsite2	**1080**, **1087**, **1098**, **3235**, **3493**, **3568**, **3638**	60
1099	Ca(Ce, La)$_2(CO_3)_2F_2$	Parisite–Ce	166, 255, 264, 278, 354, 394, 598, 739, 867, 957, **1081**, **1092**, **1099**	19

cm^{-1}	Formula	Name	All Bands	Source
1099	Li_2SO_4	Lithium sulphate	173, 208, 393, 480, 633, **1006**, 1099, 1149, 1178	19
1099	$Na_2Si_2O_5$	Sodium disilicate glass	**571**, 772, 943, **1099**	19
1099	Rb_2SO_4	Rubidium sulphate	449, 617, **958**, **1099**, **1130**	19
1102	$KAl(CO_3)(OH)_2$	Dawsonite–K	77, 169, **207**, 282, 363, 503, 580, 679, 730, 832, **1102**, 1486, 3406, 3432	19
1104	$AlPO_4$	Berlinite	119, 163, 196, 221, 418, 439, 462, 524, **1104**, **1111**, 1230	6
1105	$Na_2Mg(CO_3)_2$	Eitelite	**207**, **263**, 710, **1105**	54
1107	$NH_4Al(CO_3)(OH)_2$	Dawsonite–NH_4	79, 93, 110, 120, 139, **205**, 292, 306, 367, 572, 585, 620, 645, 686, 733, **1107**, 3388, 3460	19
1108	$Na_6Mg_2(SO_4)(CO_3)_4$	Tychite	212, 246, 300, 462, 629, 718, **967**, **1108**, 1135	19
1111	$AlPO_4$	Berlinite	119, 163, 196, 221, 418, 439, 462, 524, **1104**, **1111**, 1230	6
1115	$Na_3Mg(CO_3)_2Cl$	Northupite	120, 148, 180, 213, 250, 714, 1107, **1115**, 1554	11
1116	BPO_4	Boron orthophosphate	239, 460, **481**, 541, 1067, **1116**, 1261	19
1119	$Mg_5(CO_3)_4(OH)_2{\cdot}4H_2O$	Hydromagnesite	184, 202, 232, **1119**, 1487	10
1122	$Mg_5[(OH)(CO_3)_2]_2{\cdot}5H_2O$	Dypingite	203, 249, 311, 434, 727, 1092, **1122**, 1447, 3421, 3515, **3648**	10
1124	$Be(HCO_3)_3$	Be-hydrocarbonate	484, 766, **1124**	19
1130	Rb_2SO_4	Rubidium sulphate	449, 617, **958**, **1099**, **1130**	19
1144	$BaSO_4$	Barite2	74, 89, 97, 152, 189, **452**, **462**, 616, 629, 647, 965, **988**, 1104, **1144**, 1166	18
1151	SO_2	Sulphur dioxide gaseous	524, **1151**	4
1156	$(Fe, Mn)WO_4$	Wolframite	205, 236, 260, 353, 401, **462**, 484, 508, 695, 790, 806, **881**, 1061, 1079, **1156**	19
1157	$KLi_2AlSi_4O_{10}(OH, F)_2$	Polylithionite2	89, 99, **188**, **245**, 268, 409, 452, 552, **715**, 765, 791, **1157**	19
1245	CO_2	Carbon dioxide gas $^{13}C^{16}O^{18}O$	1235, **1245**, **1343**, 1356	39
1259	CO_2	Carbon dioxide gas $^{12}C^{16}O^{18}O$	1250, **1259**, **1366**, 1377	39
1262	CO_2	Carbon dioxide $^{13}CO_2$	**1262**, **1369**	38
1267	CO_2	Carbon dioxide gas $^{13}C^{16}O_2$	1249, **1267**, **1370**, 1389	39
1271	$KHCO_3$	Kalicinite	635, 673–675, **1028**, **1271**, 1277	10
1276	CO_2	Carbon dioxide gas in aq.	**1276**, **1384**	7

Continued

cm^{-1}	Formula	Name	All Bands	Source
1282	$KHCO_3$	Kalicinite2	79, 106, 184, 636, 676, 832, 960, **1029**, **1282**, 1448, 1682, **2567**	43
1285	CO_2	Carbon dioxide gas	**1285**, **1388**, 1409	4
1286	CO_2	Carbon dioxide gas $^{12}C^{16}O_2$	1265, **1286**, **1388**, 1410	39
1302	$C_{10}H_{22}$	Decane saturated	**255**, **843**, 1132, **1302**, 1435, 1451, **2848**, 2858, **2872**, 2897, 2917, 2934	46
1305	BN	Boron nitride cubic	**1055**, **1305**	19
1312	HCO_3^-	Hydrocarbonate ion in aq.2	630, 641, 673, 842, **1015**, **1312**, **1364**, 1628, 1682, 2619	43
1315	C	Diamond ^{13}C	**1315**	20
1332	C	Diamond	**1332**	10
1332	C	Diamond 6H	1208, 1210, 1270, 1298, 1332	15
1332	C	Diamond 8H	405, 546, 732, 1112, 1304, **1332**, 1353, **1357**	15
1336	$Al_3Si_2O_7(OH)_3$	Piezotite	**446**, **1336**, **1607**, 2932	19
1341	$C_{18}H_{12}$	Triphenylene	264, 282, 408, 420, 544, 620, 699, 777, 1062, 1164, 1171, 1181, 1230, 1247, 1299, 1318, **1341**, 1396, 1605, 1616, **3085**	44
1343	CO_2	Carbon dioxide $^{13}C^{16}O^{18}O$	1235, **1245**, **1343**, 1356	39
1349	$Na_4B_{10}O_{17}{\cdot}7H_2O$	Ezcurrite	132, 226, 286, 315, 336, **476**, **513**, **570**, **759**, 881, **954**, 1040, 1059, **1349**, **1388**, **2859**, 3335, 3375	19
1351	$C_{24}H_{12}$	Coronene	370, 484, 1027, **1351**, **1369**, 1451, 1597, 1618, 1629, 1807, 3015, 3041, 3052, 3073	44
1355	C	Graphite disordered	1150, **1355**, 1500, 1622	3
1357	C	Diamond 8H	405, 546, 732, 1112, 1304, **1332**, 1353, **1357**	15
1364	HCO_3^-	Hydrocarbonate ion in aq.2	630, 641, 673, 842, **1015**, **1312**, **1364**, 1628, 1682, 2619	43
1366	CO_2	Carbon dioxide gas $^{12}C^{16}O^{18}O$	1250, **1259**, **1366**, 1377	39
1369	CO_2	Carbon dioxide $^{13}CO_2$	**1262**, **1369**	38
1369	$C_{24}H_{12}$	Coronene	370, 484, 1027, **1351**, **1369**, 1451, 1597, 1618, 1629, 1807, 3015, 3041, 3052, 3073	44
1370	CO_2	Carbon dioxide gas $^{13}CO_2$	**1370**	4
1370	CO_2	Carbon dioxide gas $^{13}C^{16}O_2$	1249, **1267**, **1370**, 1389	39
1371	$C_{22}H_{14}$	Pentacene	184, 266, 446, 501, 751, 996, 1158, 1177, 1353, **1371**, 1410, 1458, 1533, **1598**, 1619, 3032, 3043	44

cm^{-1}	Formula	Name	All Bands	Source
1374	$C_{20}H_{12}$	Perylene	193, 419, 451, 534, 550, 796, 846, 953, 980, 1047, 1102, 1142, 1224, 1299, **1374**, 1399, 1570, **1622**, 3011, 3052, 3063	44
1375	$C_{10}H_8$	Naphthalene	**508**, **759**, 1017, 1166, 1240, 1274, **1375**, **1459**, **1572**, 1626, 3054	45
1376	CO_3^{2-}	Carbonate ion in aq.	684, **1062–1072**, **1376–1382**, 1438	7
1377	$Be_2(BO_3)(OH, F)\cdot H_2O$	Berborite	78, 379, 394, **984**, **1377**, **3504**	19
1382	CO_3^{2-}	Carbonate ion in aq.	684, **1062–1072**, **1376–1382**, 1438	7
1383	$C_{18}H_{12}$	Chrysene	295, 681, 771, 1019, 1366, **1383**, 1434, 1576, 1603, 1618, 3063	44
1384	CO_2	Carbon dioxide gas in aq.	**1276**, **1384**	7
1388	CO_2	Carbon dioxide gas	**1285**, **1388**, 1409	4
1388	CO_2	Carbon dioxide gas $^{12}C^{16}O_2$	1265, **1286**, **1388**, 1410	39
1388	$Na_4B_{10}O_{17}\cdot 7H_2O$	Ezcurrite	132, 226, 286, 315, 336, **476**, **513**, **570**, **759**, 881, **954**, 1040, 1059, **1349**, **1388**, **2859**, 3335, 3375	19
1412	$Y(Nb, Ta)O_4$	Fergusonite	**814**, **1412**, 1694, **2216**, 2425, 3143, 3839	19
1419	$Y(Nb, Ta)O_4$	Fergusonite (Norway)	285, 667, **805**, **1419**, 1725, 2219, 3124, 3846	19
1451	C_6H_{14}	Hexane saturated	370, 866, 889, **1451**, 2849, 2859, **2872**, 2893, 2915, **2936**, 2961	46
1452	C_7H_{16}	Heptane saturated	**310**, 834, 899, 1302, **1452**, **2849**, 2859, **2872**, 2895, 2916, 2936, 2962	46
1459	$LiBO_2$	Lithium metaborate	228, 290, 403, 463, 630, 662, **713**, 1081, 1128, **1459**	19
1459	$C_{10}H_8$	Naphthalene	**508**, **759**, 1017, 1166, 1240, 1274, **1375**, **1459**, **1572**, 1626, 3054	45
1506	$NaAl(CO_3)(OH)_2$	Dawsonite2	152, **191**, 389, 519, **590**, 824, 897, **1091**, **1506**, 3252, **3283**	19
1510	$Cu_8Si_8O_{22}(OH)_4\cdot H_2O$	Planchéite	104, 122, **139**, 148, **186**, 227, 302, 328, 403, 421, 476, 505, 699, 828, 906, 1048, **1081**, 1355, **1510**, **3250**, **3479**	32
1511	O_2	Oxygen gas $^{16}O–^{18}O$	**1511**	40

Continued

cm^{-1}	Formula	Name	All Bands	Source
1516	$CuC_2O_4.½H_2O$	Moolooite	210, 299, **559**, 586, 611, 832, 932, 1487, **1516**, 1616	30
1553	O_2	Oxygen gas $^{16}O_2$	**1553**	40
1572	$C_{10}H_8$	Naphthalene	**508**, **759**, 1017, 1166, 1240, 1274, **1375**, **1459**, **1572**, 1626, 3054	45
1582	C	Graphite ordered	**1582**	3
1586	$MgCl_2{\cdot}6H_2O$	Bischofite	206, **227**, 348, 355, 1554, **1586**, **1601**, **1604**, 1803, 3356	19
1598	$C_{22}H_{14}$	Pentacene	184, 266, 446, 501, 751, 996, 1158, 1177, 1353, **1371**, 1410, 1458, 1533, **1598**, 1619, 3032, 3043	44
1601	$MgCl_2{\cdot}6H_2O$	Bischofite	206, **227**, 348, 355, 1554, **1586**, **1601**, **1604**, 1803, 3356	19
1604	$MgCl_2{\cdot}6H_2O$	Bischofite	206, **227**, 348, 355, 1554, **1586**, **1601**, **1604**, 1803, 3356	19
1607	$Al_3Si_2O_7(OH)_3$	Piezotite	**446**, **1336**, **1607**, 2932	19
1622	$C_{20}H_{12}$	Perylene	193, 419, 451, 534, 550, 796, 846, 953, 980, 1047, 1102, 1142, 1224, 1299, **1374**, 1399, 1570, **1622**, 3011, 3052, 3063	44
1635	C_6H_{12}	Hexene	1294, **1635**, 2996	49
1644	$KMgCl_3{\cdot}6H_2O$	Carnallite	**1644**, **3442**	19
1650	R-C(O)OH	Carboxylic acid	**1650**	69
1736	$PbMoO_4$	Wulfenite	**166**, **315**, 348, 742, 765, **868**, **1736**, 3290, **3350**, **3377**, 3533, 3566	19
2090	CO	Carbon monoxide gas ^{13}CO	**2090**	38
2143	CO	Carbon monoxide gas	**2143**	4
2216	Y(Nb, Ta)O_4	Fergusonite	**814**, **1412**, 1694, **2216**, 2425, 3143, 3839	19
2290	N_2	Nitrogen gas $^{14}N{-}^{15}N$	**2290**	13
2331	N_2	Nitrogen gas	**2331**	4
2567	$KHCO_3$	Kalicinite2	79, 106, 184, 636, 676, 832, 960, **1029**, **1282**, 1448, 1682, **2567**	43
2574	HS^-	Hydrosulphide ion in aq.	**2574**	4
2578	CH_4	Methane	**2578**, **2917**, 3006, 3070	62
2580	H_2S	Hydrogen sulphide liquid	**2580**	4
2590	H_2S	Hydrogen sulphide in aq·	**2590**	4
2611	H_2S	Hydrogen sulphide gas	**2611**	4
2750	H_2O	Water liquid	1630, **2750–3900**	10
2848	$C_{10}H_{22}$	Decane saturated	**255**, **843**, 1132, **1302**, 1435, 1451, **2848**, 2858, **2872**, 2897, 2917, 2934	46

cm^{-1}	Formula	Name	All Bands	Source
2849	C_7H_{16}	Heptane saturated	**310**, 834, 899, 1302, **1452**, **2849**, 2859, **2872**, 2895, 2916, 2936, 2962	46
2855	C_6H_{12}	Cyclohexane2	**785**, **1028**, 1266, 1440, 2664, **2855**, 2895, 2923, 2938	69
2855	C_8H_{18}	Cyclooctane	**703**, 1467, **2855**, 2925	70
2859	$Na_4B_{10}O_{17}{\cdot}7H_2O$	Ezcurrite	132, 226, 286, 315, 336, **476**, **513**, **570**, **759**, 881, **954**, 1040, 1059, **1349**, **1388**, **2859**, 3335, 3375	19
2872	C_7H_{16}	Heptane saturated	**310**, 834, 899, 1302, **1452**, **2849**, 2859, **2872**, 2895, 2916, 2936, 2962	46
2872	C_6H_{14}	Hexane saturated	370, 866, 889, **1451**, 2849, 2859, **2872**, 2893, 2915, **2936**, 2961	46
2872	$C_{10}H_{22}$	Decane saturated	**255**, **843**, 1132, **1302**, 1435, 1451, **2848**, 2858, **2872**, 2897, 2917, 2934	46
2882	C_3H_8	Propane2	**2882**	63
2883	C_3H_8	Propane3	**2883**, 2937, 2973	64
2885	$SiO_2{\cdot}n(CH_4, C_2H_6, C_3H_8, i\text{-}C_4H_{10})$	Ethane in chibaite	991, **2885**, 2939	35
2885	C_4H_{10}	Butane	837, 2865, **2885**, 2927, 2960, 2987	64,70
2890	C_3H_8	Propane	**2890**	4
2895	C_2H_6	Ethane2	**1000**, 1440, **2895**, **2955**	64
2909	CH_4	Methane in aq.	**2909**	18
2911	$SiO_2{\cdot}n(CH_4, C_2H_6, C_3H_8, i\text{-}C_4H_{10})$	Methane in chibaite	2903, **2911**	35
2911	$:CH_2$	Methylene	**2911–2913**	49
2917	CH_4	Methane	1535, **2917**, 3020, 3070	36,37
2917	CH_4	Methane	**2578**, **2917**, 3006, 3070	62
2919	CH_4	Methane3	1532, **2919**, 3030, 3090	64
2936	C_6H_{14}	Hexane saturated	370, 866, 889, **1451**, 2849, 2859, **2872**, 2893, 2915, **2936**, 2961	46
2944	$Na_2Ca(CO_3)_2{\cdot}5H_2O$	Gaylussite	**164**, 265, 723, **1071**, **2944**, 3334	10
2954	C_2H_6	Ethane	993, 2914, **2954**	37
2955	C_2H_6	Ethane2	**1000**, 1440, **2895**, **2955**	64
3040	NH_4^+	Ammonia in aq.	2870, **3040**	10
3058	C_6H_6	Benzene2	**988**, **3058**	49

Continued

cm^{-1}	Formula	Name	All Bands	Source
3063	C_6H_6	Benzene	608, **994**, 1179, 1587, 1607, 2927, 2950, **3063**	44
3064	C_6H_6	Benzene3	**309**, **999**, 1580, **3064**	69
3085	$C_{18}H_{12}$	Triphenylene	264, 282, 408, 420, 544, 620, 699, 777, 1062, 1164, 1171, 1181, 1230, 1247, 1299, 1318, **1341**, 1396, 1605, 1616, **3085**	44
3090	H_2O	Ice	**3090**, 3218	2
3102	H_2O	Ice2	**3102**, 3225	42
3130	HBO_2	Metaboric acid II	77, 94, 108, 119, 131, 147, 178, 198, 477, 522, 536, **785**, **3130**	19
3165	H_3BO_3	Sassolite	187, 211, 500, 880, 1168, 3165, **3245–3247**	22
3235	(Ce, La)CO_3(OH, F)	Hydroxylbastnäsite2	**1080**, **1087**, **1098**, **3235**, **3493**, **3568**, **3638**	60
3245	H_3BO_3	Sassolite	187, 211, 500, 880, 1168, 3165, **3245–3247**	22
3247	H_3BO_3	Sassolite	187, 211, 500, 880, 1168, 3165, **3245–3247**	21
3250	$Cu_8Si_8O_{22}(OH)_4 \cdot H_2O$	Planchéite	104, 122, **139**, 148, **186**, 227, 302, 328, 403, 421, 476, 505, 699, 828, 906, 1048, **1081**, 1355, **1510**, **3250**, **3479**	32
3266	$Pb_5(AsO_4)_3Cl$	Mimetite	98, 105, 145, 175, **313**, 340, 371, 385, 408, 422, 765, **788**, 806, **813**, 837, 854, 860, **3266**, **3420**, 3548	34
3282	$NaAl(CO_3)(OH)_2$	Dawsonite	**192**, 260, 590, 728, 1065, **1091**, 1506, 3250, **3282**	10
3283	$NaAl(CO_3)(OH)_2$	Dawsonite2	152, **191**, 389, 519, **590**, 824, 897, **1091**, **1506**, 3252, **3283**	19
3297	$MgSO_4 \cdot H_2O$	Kieserite	436, 481, 629, **1046**, 1117, 1215, **3297**	10
3301	$K_2Ca(SO_4)_2 \cdot H_2O$	Syngenite	474, 494, 642, 662, **983**, **1007**, 1142, 1168, **3301**, **3378**	10
3303	$MgSO_4 \cdot 7H_2O$	Epsomite	447, 612, **984**, 1061, 1095, 1134, **3303**, **3425**	10
3311	$Cu_2(OH)_3Cl$	Paratacamite	95, **117**, **140**, 165, 364, 420, **511**, 576, 799, 868, 892, 928, 970, **3311**, **3355**, **3441**	30
3312	NH_3	Ammonia in water	3250, **3312**, 3400	47
3317	(Ce, La)CO_3(OH, F)	Hydroxylbastnäsite	112, 155, 436, 591, **965**, 1053, 1330, **3317**, **3352**, 3430, 3535, 3714	11
3331	$BaCl_2 \cdot 2H_2O$	Ba–chloride dihydrate	193, 256, **402**, 490, 550, **3331**, 3451	19
3336	NH_3	Ammonia	**3336**	4

cm^{-1}	Formula	Name	All Bands	Source
3343	$MgSO_4{\cdot}5H_2O$	Pentahydrite	371, 447, 602, **1005**, 1106, 1159, 3289, **3343**, **3391**, 3494, 3553	10
3350	$PbMoO_4$	Wulfenite	**166**, **315**, 348, 742, 765, **868**, **1736**, 3290, **3350**, **3377**, 3533, 3566	19
3351	$Cu_2(OH)_3Cl$	Atacamite2	66, **106**, **120**, **139**, **150**, 450, **514**, 586, 820, 845, **908**, 975, **3351**, **3453**	13
3351	$Cu_4Cl[Cl_{0.29}(OH)_{0.71}](OH)_6$	Claringbullite	119, 136, **147**, 163, 231, 260, 356, 389, 447, **511**, 579, **815**, **906**, **970**, 1049, 3211, 3331, **3351**, **3433**, 3458	31
3351	$CaWO_4$	Scheelite	55, 79, 111, 207, **329**, 395, **793**, 834, **909**, 1020, 1120, 1625, 3267, 3285, **3351**, **3379**, 3413, **3545**	19
3351	$Na_2B_4O_5(OH)_4{\cdot}8H_2O$	Borax2	344, 405, **578**, 767, 943, 997, **3351**, 3499	19
3352	$(Ce, La)CO_3(OH, F)$	Hydroxylbastnäsite	112, 155, 436, 591, **965**, 1053, 1330, **3317**, **3352**, 3430, 3535, 3714	11
3355	$Cu_2(OH)_3Cl$	Paratacamite	95, **117**, **140**, 165, 364, 420, **511**, 576, 799, 868, 892, 928, 970, **3311**, **3355**, **3441**	30
3356	$LiCl{\cdot}XH_2O$	Li–chloride hydrate	**3356**, **3376**	2
3358	$NaCaB_5O_9{\cdot}5H_2O$	Probertite	157, 240, 337, 432, 452, 487, **552**, **577**, **726**, 747, 821, 920, 1032, **3358**, 3492, 3549, **3616**	19
3362	$\alpha\text{-}CaCl_2{\cdot}4H_2O$	Li–chloride hydrate	**3362**, **3423**, **3444**, 3475	2
3367	$CuCl_2{\cdot}2H_2O$	Eriochalcite	117, **216**, 235, 247, 390, **407**, 672, 692, 1620, 3176, 3317, **3367**, 3462	31
3369	$\alpha\text{-}CaCl_2{\cdot}4H_2O$	α-tetrahydrate	3197, 3215, **3369**, 3425, **3446**, 3472, 3519	41
3369	$LiAl(PO_4)(F, OH)$	Amblygonite	130, 303, 426, 488, 603, **647**, **1012**, **1048**, 1064, **3369**, 3408	19
3371	$Cu_4(OH)_6SO_4$	Brochantite	90, 123, 139, 196, 390, 421, 450, 481, 595, 609, **973**, **3371**, **3399**, **3564**, **3587**	30
3374	$RbB_5O_8{\cdot}4H_2O$	Ramanite–Rb	101, **554**, 914, **3374**, 3426	24
3376	$LiCl{\cdot}XH_2O$	Li–chloride hydrate	**3356**, **3376**	2
3377	$PbMoO_4$	Wulfenite	**166**, **315**, 348, 742, 765, **868**, **1736**, 3290, **3350**, **3377**, 3533, 3566	19

Continued

cm^{-1}	Formula	Name	All Bands	Source
3378	$Pb_5(PO_4,AsO_4)_3Cl$	Pyromorphite–As	**152**, 177, 186, 206, 318, 339, 344, 377, 391, 409, 414, 768, 776, **815**, 818, 825, 917, **944**, 1014, 3325, **3378**, 3444	34
3378	$K_2Ca(SO_4)_2{\cdot}H_2O$	Syngenite	474, 494, 642, 662, **983**, **1007**, 1142, 1168, **3301**, **3378**	10
3379	$Cu_4Cl[Cl_{0.29}(OH)_{0.71}](OH)_6$	Claringbullite	119, 136, **147**, 163, 231, 260, 356, 389, 447, **511**, 579, **815**, **906**, **970**, 1049, 3211, 3331, **3351**, **3433**, 3458	31
3379	$CaWO_4$	Scheelite	55, 79, 111, 207, **329**, 395, **793**, 834, **909**, 1020, 1120, 1625, 3267, 3285, **3351**, **3379**, 3413, **3545**	19
3380	$Cu_5(PO_4)_2(OH)_4$	Reichenbachite	89, **110**, 135, 175, 188, 214, 255, 298, 365, 412, 453, **480**, 515, 540, 572, 607, 636, 951, 971, 998, **1083**, **3380**, **3428**	57
3386	$FeCl_2{\cdot}4H_2O$	Fe–chloride tetrahydrate	3170, **3386**, **3404**, 3520	2
3389	$FeCl_2{\cdot}6H_2O$	Fe–chloride hexahydrate	**3389**, **3411**, 3426	2
3391	$MgSO_4{\cdot}5H_2O$	Pentahydrite	371, 447, 602, **1005**, 1106, 1159, 3289, **3343**, **3391**, 3494, 3553	10
3396	$CsB_5O_8{\cdot}4H_2O$	Ramanite–Cs	94, 98, 293, **548**, 907, 3119, **3396**	24
3396	$KAl(SO_4)_2{\cdot}12H_2O$	Alum–K	442, 455, 614, **974**, **989**, 1104, 1130, 3072, **3396**	10
3399	$Cu_4(OH)_6SO_4$	Brochantite	90, 123, 139, 196, 390, 421, 450, 481, 595, 609, **973**, **3371**, **3399**, **3564**, **3587**	30
3401	$Cu_4(OH)_6SO_4$	Brochantite2	**80**, **91**, 132, 138, 142, 157, 188, 199, 242, 319, 387, 426, 450, 484, 509, 595, 609, 910, **973**, 1097, **3401**, **3566**, **3588**	13
3402	$MgCl_2{\cdot}12H_2O$	Mg–chloride dodecahydrate	3188, 3324, 3397, **3402**, 3422, 3457, 3480, **3510**	2
3402	$NaCl{\cdot}2H_2O$	Hydrohalite	**3402**, **3419**, **3432**, 3536	2
3404	$CaCl_2{\cdot}6H_2O$	Antarcticite	3242, **3404**, 3410, **3430**	2
3404	$FeCl_2{\cdot}4H_2O$	Fe–chloride tetrahydrate	3170, **3386**, **3404**, 3520	2
3405	$CaSO_4{\cdot}2H_2O$	Gypsum	414, **492–494**, 621–623, 671, **1006–1008**, 1142, 3250, **3405**, 3491, 3500	10
3411	$FeCl_2{\cdot}6H_2O$	Fe–chloride hexahydrate	**3389**, **3411**, 3426	2
3413	β-$Be(OH)_2$	Behoite-β	148, 286, **351**, **404**, 450, 459, 465, 553, 622, 717, 1069, 2544, 2551, 2566, **3413**, 3464, 3476	19
3413	H_4SiO_4	Orthosilicic acid	**783**, **3413**	19

cm^{-1}	Formula	Name	All Bands	Source
3414	$Cu_3(PO_4)(OH)_3$	Cornetite	**109**, 131, 141, 158, 174, 297, 313, 363, **446**, **477**, 515, 583, 639, 945, **971**, **996**, 1016, **1083**, 1137, **3414**	57
3414	$H_6Si_2O_7$	Pyrosilicic acid	**225**, 426, **631**, 916, **3414**	19
3419	$Be_2BO_3(OH, F)$	Hambergite2	280, 324, 398, 583, 623, 739, **988**, **3419**, **3484**, 3562, **3600**, 3662	19
3419	$NaCl{\cdot}2H_2O$	Hydrohalite	**3402**, **3419**, **3432**, 3536	2
3420	$Pb_5(AsO_4)_3Cl$	Mimetite	98, 105, 145, 175, **313**, 340, 371, 385, 408, 422, 765, **788**, 806, **813**, 837, 854, 860, **3266**, **3420**, 3548	34
3423	α-$CaCl_2{\cdot}4H_2O$	Ca–chloride tetrahydrate	**3362**, **3423**, **3444**, 3475	2
3423	$CaCl_2{\cdot}2H_2O$	Sinjarite	3376, 3403, **3423**, 3462, 3471	2
3425	$MgSO_4{\cdot}7H_2O$	Epsomite	447, 612, **984**, 1061, 1095, 1134, **3303**, **3425**	10
3427	$MgSO_4{\cdot}4H_2O$	Starkeyite	116, 401, 462, 565, 616, 664, **1000**, 1086, 1156, 1186, 3331, **3427**, **3481**, 3558	10
3428	$Cu_5(PO_4)_2(OH)_4$	Reichenbachite	89, **110**, 135, 175, 188, 214, 255, 298, 365, 412, 453, **480**, 515, 540, 572, 607, 636, 951, 971, 998, **1083**, **3380**, **3428**	57
3428	$Al_4(PO_4)_3(OH)_3$	Trolleite	269, 428, 621, 690, 790, 978, **1064**, 1100, 1120, **3428**, 3510	19
3428	$MgSO_4{\cdot}6H_2O$	Hexahydrite	445, 466, 610, **984**, 1085, 1146, 3258, **3428**	10
3428	$LiCl{\cdot}5H_2O$	Li–chloride pentahydrate	3288, 3339, **3428**, **3449**, 3523	65
3430	$CaCl_2{\cdot}6H_2O$	Antarcticite	3242, **3404**, 3410, **3430**	2
3432	$NaCl{\cdot}2H_2O$	Hydrohalite	**3402**, **3419**, **3432**, 3536	2
3434	γ-$CaCl_2{\cdot}4H_2O$	Ca–chloride tetrahydrate	**3434**	2
3435	γ-$CaCl_2{\cdot}4H_2O$	γ-tetrahydrate	3392, **3435**, 3463, 3575	41
3441	$Cu_2(OH)_3Cl$	Paratacamite	95, **117**, **140**, 165, 364, 420, **511**, 576, 799, 868, 892, 928, 970, **3311**, **3355**, **3441**	30
3443	$Ca_2B_6O_{11}{\cdot}13H_2O$	Inyoite	352, 388, **464**, **615**, 733, 808, 912, 956, 1096, 3223, 3334, **3443**, **3531**	19
3444	α-$CaCl_2{\cdot}4H_2O$	Ca–chloride tetrahydrate	**3362**, **3423**, **3444**, 3475	2
3446	α-$CaCl_2{\cdot}4H_2O$	α-tetrahydrate	3197, 3215, **3369**, 3425, **3446**, 3472, 3519	41
3446	$MgSO_4{\cdot}2H_2O$	Sanderite	447, 492, 597, 630, **1034**, 1164, **3446**, 3539	10
3449	$LiCl{\cdot}5H_2O$	Li–chloride pentahydrate	3288, 3339, **3428**, **3449**, 3523	65

Continued

cm^{-1}	Formula	Name	All Bands	Source
3453	$Cu_2(OH)_3Cl$	Atacamite2	66, **106**, **120**, **139**, **150**, 450, **514**, 586, 820, 845, **908**, 975, **3351**, **3453**	13
3453	$Cu_3(OH)_2(CO_3)_2$	Azurite	170, 246, 279, **397**, **1095**, 1457, **3453**	10
3458	$Pb_3Ca_2(AsO_4)_3Cl$	Hedyphane	141, 151, 162, 177, 192, 201, **319**, 334, 349, 372, 394, 432, 452, 770, 787, **811**, **819**, 834, 848, 3352, 3409, **3458**	34
3463	$Na_2WO_4{\cdot}2H_2O$	Sodium wolframate	336, 842, **931**, 3309, **3463**	19
3468	$Cu_2(OH)_2(CO_3)$	Malachite	**154**, **178**, 272, **434**, 537, 721, **1098**, 1492, 3386, **3468**	10
3469	$Al_2(PO_4)(OH)_3$	Augelite	124, 163, 226, 252, 368, 410, 464, 526, 565, **637**, 859, 896, 1014, 1108, 1162, 3428, **3469**, **3537**	19
3469	$CaHPO_4{\cdot}2H_2O$	Brushite	408, 521, 875, **985**, 1056, 1132, 3160, 3261, **3469**, **3536**	27
3470	$Cu_5(PO4)_2(OH)_4$	Ludjibaite	**160**, 190, 226, 263, **301**, 368, 387, 402, 449, 557, 586, 633, 736, 784, 815, 855, 925, **981**, 1019, 1046, 1072, 1115, **3470**	57
3475	$Cu_2PO_4(OH)$	Libethenite	**74**, 92, 113, 140, **160**, 227, 270, **301**, 319, 371, 392, 430, 461, 561, 590, 944, **979**, 1008, **1022**, **3475**	57
3479	$Cu_8Si_8O_{22}(OH)_4{\cdot}H_2O$	Planchéite	104, 122, **139**, 148, **186**, 227, 302, 328, 403, 421, 476, 505, 699, 828, 906, 1048, **1081**, 1355, **1510**, **3250**, **3479**	32
3480	$KAl_3[(OH)_3(SO_4)]_2$	Alunite2	236, 347, 384, 486, 564, 655, **1027**, 1081, 1190, **3480**, **3508**	19
3481	$MgSO_4{\cdot}4H_2O$	Starkeyite	116, 401, 462, 565, 616, 664, **1000**, 1086, 1156, 1186, 3331, **3427**, **3481**, 3558	10
3482	$KAl_3[(OH)_3(SO_4)]_2$	Alunite	485, 509, 654, **1026**, 1079, 1190, **3482**, **3509**	10
3484	$Be_2BO_3(OH, F)$	Hambergite2	280, 324, 398, 583, 623, 739, **988**, **3419**, **3484**, 3562, **3600**, 3662	19
3493	(Ce, La)CO_3(OH, F)	Hydroxylbastnäsite2	**1080**, **1087**, **1098**, **3235**, **3493**, **3568**, **3638**	60
3504	$Be_2(BO_3)(OH, F){\cdot}H_2O$	Berborite	78, 379, 394, **984**, **1377**, **3504**	19
3506	PbCl(OH)	Laurionite	131, 142, 174, **272**, **328**, 594, 662, 729, **3506**	12
3506	$Na_2SO_4{\cdot}10H_2O$	Mirabilite	458, 627, **989**, 1129, 3340, **3506**	10
3508	$KAl_3[(OH)_3(SO_4)]_2$	Alunite2	236, 347, 384, 486, 564, 655, **1027**, 1081, 1190, **3480**, **3508**	19

cm^{-1}	Formula	Name	All Bands	Source
3509	$KAl_3[(OH)_3(SO_4)]_2$	Alunite	485, 509, 654, **1026**, 1079, 1190, **3482**, **3509**	10
3510	$MgCl_2{\cdot}12H_2O$	Mg–chloride dodecahydrate	3188, 3324, 3397, **3402**, 3422, 3457, 3480, **3510**	2
3525	$K_2Ca_5(SO_4)_6{\cdot}H_2O$	Goergeyite	433, 440, 480, 595, 602, 631, **1005**, **1013**, 1085, 1115, 1162, 3525, **3580**	10
3531	$Ca_2B_6O_{11}{\cdot}13H_2O$	Inyoite	352, 388, **464**, **615**, 733, 808, 912, 956, 1096, 3223, 3334, **3443**, **3531**	19
3536	$CaHPO_4{\cdot}2H_2O$	Brushite	408, 521, 875, **985**, 1056, 1132, 3160, 3261, **3469**, **3536**	27
3537	$Al_2(PO_4)(OH)_3$	Augelite	124, 163, 226, 252, 368, 410, 464, 526, 565, **637**, 859, 896, 1014, 1108, 1162, 3428, **3469**, **3537**	19
3545	$CaWO_4$	Scheelite	55, 79, 111, 207, **329**, 395, **793**, 834, **909**, 1020, 1120, 1625, 3267, 3285, **3351**, **3379**, 3413, **3545**	19
3551	$Be_4Si_2O_7(OH)_2$	Bertrandite	183, **206**, 231, 330, 358, 579, 694, 711, 928, 990, 1073, **3551**, **3587**	19
3551	$Fe_8Si_6O_{15}[(OH_6)Cl_4]$	Ferropyrosmalite	**615**, 1022, **3551**, **3579**, **3637**	16
3555	$MgHBO_3$	Ascharite	324, **835**, **3555**	19
3560	$4[Mg_2SiO_4]{\cdot}[Mg, Fe, Ti](OH, F)_2]$	Hydroxylclinohumite	110, 186, 212, 233, 306, 321, 348, 549, 590, 609, 756, 786, **832**, **845**, **861**, 931, 968, 3393, 3410, **3560**, **3570**, **3578**	13
3562	$LiOH{\cdot}H_2O$	Li–hydroxide monohydrate	**3562**, **3664**	67
3564	$Cu_4(OH)_6SO_4$	Brochantite	90, 123, 139, 196, 390, 421, 450, 481, 595, 609, **973**, **3371**, **3399**, **3564**, **3587**	30
3564	$CaBe(PO_4)(F, OH)$	Herderite	166, 194, 424, 445, 516, 529, **584**, **594**, **983**, **1005**, 1128, 1140, **3564**, **3573**, 3620	19
3566	$Cu_{\mathbf{4}}(OH)_6SO_4$	Brochantite2	**80**, **91**, 132, 138, 142, 157, 188, 199, 242, 319, 387, 426, 450, 484, 509, 595, 609, 910, **973**, 1097, **3401**, **3566**, **3588**	13
3568	$(Ce, La)CO_3(OH, F)$	Hydroxylbastnäsite2	**1080**, **1087**, **1098**, **3235**, **3493**, **3568**, **3638**	60
3570	$NaCaB_5O_9{\cdot}8H_2O$	Ulexite	228, 263, 300, 330, 392, 440, 485, 531, **589**, **592**, **595**, **747**, **862**, 1013, 1316, 3559, **3570**, 3588	19

Continued

cm^{-1}	Formula	Name	All Bands	Source
3570	$4[Mg_2SiO_4]\cdot[Mg, Fe, Ti](OH, F)_2]$	Hydroxylclinohumite	110, 186, 212, 233, 306, 321, 348, 549, 590, 609, 756, 786, **832**, **845**, **861**, 931, 968, 3393, 3410, **3560**, **3570**, **3578**	13
3572	$Ca_5(PO_4)_3OH$	Hydroxylapatite	431, 591, **962**, **3572**	27
3572	Na_3PO_4	Sodium phosphate	391, 482, 524, 544, **910**, 942, 993, 3308, 3402, **3572**	19
3573	$AlPO_4\cdot 2H_2O$	Variscite	225, 435, 612, **1028**, 1076, 1160, **3573**	19
3573	$CaBe(PO_4)(F, OH)$	Herderite	166, 194, 424, 445, 516, 529, **584**, **594**, **983**, **1005**, 1128, 1140, **3564**, **3573**, 3620	19
3575	$BeAlSiO_4(OH)$	Euclase	145, 180, 237, **259**, 290, **397**, 444, 452, 574, **880**, **908**, 975, 1023, 1059, 1120, **3575**, 3587	19
3578	$4[Mg_2SiO_4]\cdot[Mg, Fe, Ti](OH, F)_2$	Hydroxylclinohumite	110, 186, 212, 233, 306, 321, 348, 549, 590, 609, 756, 786, **832**, **845**, **861**, 931, 968, 3393, 3410, **3560**, **3570**, **3578**	13
3579	$Fe_8Si_6O_{15}[(OH_6)Cl_4]$	Ferropyrosmalite	**615**, 1022, **3551**, **3579**, **3637**	16
3587	$Cu_4(OH)_6SO_4$	Brochantite	90, 123, 139, 196, 390, 421, 450, 481, 595, 609, **973**, **3371**, **3399**, **3564**, **3587**	30
3587	$Be_4Si_2O_7(OH)_2$	Bertrandite	183, **206**, 231, 330, 358, 579, 694, 711, 928, 990, 1073, **3551**, **3587**	19
3588	$Cu_4(OH)_6SO_4$	Brochantite2	**80**, **91**, 132, 138, 142, 157, 188, 199, 242, 319, 387, 426, 450, 484, 509, 595, 609, 910, **973**, 1097, **3401**, **3566**, **3588**	13
3588	$Ca[B_2BO_4(OH)_3]\cdot H_2O$	Colemanite	306, 383, 531, **608**, 666, 742, 3514, **3588**	19
3592	$Mg_3B_2(SO_4)(OH)_8(OH, F)_2$	Sulfoborite	275, 363, 460, 621, 657, 778, 901, 934, **1011**, 1081, 1134, 3242, 3374, 3561, 3573, **3592**, 3618	19
3593	$Be_3Al_2Si_6O_{18}$	Beryl2	324, 401, **686**, 1006, **1068**, 1112, **3593**, 3604	19
3593	$Mg_2(OH)_2(CO_3)\cdot 3H_2O$	Artinite	**147**, 173, 472, 704, **1094**, 3030, 3229, **3593**	10
3595	$CaMgB_6O_{11}\cdot 6H_2O$	Hydroboracite	**181**, 257, 322, 383, 398, 524, 548, 564, 606, **753**, 837, 876, **3595**	19
3600	$Be_2BO_3(OH, F)$	Hambergite2	280, 324, 398, 583, 623, 739, **988**, **3419**, **3484**, 3562, **3600**, 3662	19
3614	$KAl_4Si_5B_2O_{20}(OH, F)_4$	Boromuscovite	247, 411, 476, 555, **717**, 784, **3614**	19

cm^{-1}	Formula	Name	All Bands	Source
3616	$NaCaB_5O_9 \cdot 5H_2O$	Probertite	157, 240, 337, 432, 452, 487, **552**, **577**, **726**, 747, 821, 920, 1032, **3358**, 3492, 3549, **3616**	19
3618	$Al_2Si_2O_5(OH)_4$	Nacrite	120, 132, 140, 197, 238, **268**, **332**, 418, 433, **462**, 748, 798, 916, 935, 1085, 1102, **3618**, 3628, 3650, **3698**	56
3620	$Al_2Si_2O_5(OH)_4$	Kaolinite	120, 132, 200, 248, **272**, **338**, 420, 434, **461**, 749, 791, 916, 935, 1048, 1105, **3620**, 3650, 3668, 3682, **3692**	56
3620	$NaOH$	Sodium hydroxide in melt	**3620**	55
3624	$(Cu, Al)_2H_2Si_2O_5(OH)_4 \cdot nH_2O$	Chrysocolla	109, 146, 183, 213, 245, 336, 424, 485, **673**, 792, **933**, **1058**, 1602, 3295, 3504, 3583, **3624**, 3652, 3708	32
3630	H_2O	Water in hydrocarbons	**3630**	48
3623	$Al_2Si_2O_5(OH)_4$	Dickite	120, 132, 178, 198, 245, **268**, **336**, 420, 434, **463**, 748, 792, 916, 935, 1050, 1104, **3623**, 3648, **3683**, **3700**	56
3627	$KAl_4Si_5Al_2O_{20}(OH, F)_4$	Muscovite	178, 197, 216, **261**, 385, 407, 639, **702**, 754, 914, 957, 1117, **3627**	10
3637	$Al_2SiO_4(OH)_2$	Topaz2	238, **266**, 285, 402, 456, 856, 928, **3637**	19
3637	$Fe_8Si_6O_{15}[(OH_6)Cl_4]$	Ferropyrosmalite	**615**, 1022, **3551**, **3579**, **3637**	16
3638	$(Ce, La)CO_3(OH, F)$	Hydroxylbastnäsite2	**1080**, **1087**, **1098**, **3235**, **3493**, **3568**, **3638**	60
3648	$Mg_5[(OH)(CO_3)_2]_2 \cdot 5H_2O$	Dypingite	203, 249, 311, 434, 727, 1092, **1122**, 1447, 3421, 3515, **3648**	10
3657	H_2O	Water vapor	**3657–3756**	10
3664	$LiOH \cdot H_2O$	Li–hydroxide monohydrate	**3562**, **3664**	67
3669	$Mg_3BO_3(OH, F)_3$	Fluoborite	241, **953**, 1341, 3645, 3656, **3669**, 3686	19
3672	$Al_6(BO_3)_5(OH, F)_3$	Jeremejevite	301, **328**, **372**, 417, 512, 608, 710, 749, **962**, 1067, 3624, **3672**	19
3673	$K_2(Mg, Fe)_6Si_5Al_2O_{20}(OH, F)_4$	Phlogopite	192, 279, 331, 372, **680**, 792, 1038, 1096, **3673**, **3715**	10
3677	$Mg_5Si_6O_{20}(OH)_4$	Talc	113, 196, 295, 335, 366, **434**, **678**, 786, 793, 1018, 1055, **3677**	10
3683	$Al_2Si_2O_5(OH)_4$	Dickite	120, 132, 178, 198, 245, **268**, **336**, 420, 434, **463**, 748, 792, 916, 935, 1050, 1104, **3623**, 3648, **3683**, **3700**	56

Continued

cm^{-1}	Formula	Name	All Bands	Source
3692	$Al_2Si_2O_5(OH)_4$	Kaolinite	120, 132, 200, 248, **272**, **338**, 420, 434, **461**, 749, 791, 916, 935, 1048, 1105, **3620**, 3650, 3668, 3682, **3692**	56
3698	$Al_2Si_2O_5(OH)_4$	Nacrite	120, 132, 140, 197, 238, **268**, **332**, 418, 433, **462**, 748, 798, 916, 935, 1085, 1102, **3618**, 3628, 3650, **3698**	56
3700	$Al_2Si_2O_5(OH)_4$	Dickite	120, 132, 178, 198, 245, **268**, **336**, 420, 434, **463**, 748, 792, 916, 935, 1050, 1104, **3623**, 3648, **3683**, **3700**	56
3715	$K_2(Mg, Fe)_6Si_5Al_2O_{20}(OH, F)_4$	Phlogopite	192, 279, 331, 372, **680**, 792, 1038, 1096, **3673**, **3715**	10
3756	H_2O	Water vapor	**3657–3756**	10
3900	H_2O	Aqueous liquid	1630, **2750–3900**	10
4156	H_2	Hydrogen gaseous	354, 586, 813, 1032, 4126, 4143, **4156**, 4161	10

Sources: (1) Anderson et al. (2001). (2) Baumgartner and Bakker (2010). (3) Beyssac et al. (2002a). (4) Burke (1995). (5) Buzgar and Apopei (2009). (6) Cinta and Onac (2009). (7) Davis and Oliver (1972). (8) Downs (2006). (9) Frezzotti et al. (2011). (10) Frezzotti et al. (2012b). (11) Frost and Dickfos (2007). (12) Frost and Williams (2004). (13) Geological Institute SAS, Banská Bystrica, Slovakia, unpubl. (14) Hurai et al. (2007). (15) Karczemska (2010). (16) Koděra et al. (2003). (17) Zaitsev et al. (2011). (18) Prinsloo et al. (1997). (19) R. Thomas, personal comm. (20) Solopova et al. (2012). (21) Thomas (2002a). (22) Thomas and Davidson (2010). (23) Thomas et al. (2009). (24) Thomas et al. (2008a). (25) Thomas et al. (2011a). (26) Thomas et al. (2011b). (27) Wopenka and Pasteris (2005). (28) Yang et al. (2009). (29) Zaitsev (2010). (30) Burgio and Clark (2001). (31) Frost et al. (2003). (32) Frost and Xi (2012). (33) Frost and Xi (2013). (34) Frost et al. (2007). (35) Momma et al. (2011). (36) Brunsgaard et al. (2002). (37) Kiefer et al. (2008). (38) Morizet et al. (2009). (39) Irmer and Graupner (2002). (40) Champagnon et al. (1997). (41) Baumgartner and Bakker (2009a). (42) Bakker (2004). (43) Rudolph et al. (2006). (44) Colangelli et al. (1992). (45) Srivastava and Singh (2007). (46) Gorelik et al. (2000). (47) Simonelli and Shultz (2001). (48) Pironon et al. (2000). (49) Zhang et al. (2007). (50) Maya and Begun (1981). (51) Nguyen-Trung et al. (1992). (52) Mikhno et al. (2013). (53) Kanzaki et al. (2012). (54) Sharygin et al. (2013). (55) Thomas et al. (2012a). (56) Wiewiora et al. (1979). (57) Kharbish et al. (2014). (58) Borisenko et al. (2011). (59) Jacquemet et al. (2014). (60) Yang et al. (2008). (61) Kotková et al. (2014). (62) Mernagh and Bastrakov (2013). (63) Arai et al. (2012). (64) Atamas et al. (2004). (65) Dubois et al. (2010). (66) Li et al. (1995). (67) Walrafen and Douglas (2006). (68) Kingma and Hemley (1994). (69) Orange et al. (1996). (70) Dollish et al. (1974).
Bold highlights main bands.

RAMAN BANDS SORTED BY NAME

VI

Name	Formula	Vibration Bands (cm^{-1})	Source
α-Tetrahydrate	α-$CaCl_2 \cdot 4H_2O$	3197, 3215, **3369**, 3425, **3446**, 3472, 3519	41
γ-Tetrahydrate	γ-$CaCl_2 \cdot 4H_2O$	3392, **3435**, 3463, 3375	41
Albite	$NaAlSi_3O_8$	183, 210, **292**, 457, 480, **508**, 764, 816, 977, 1032, 1098	10
Albite2	$NaAlSi_3O_8$	157, 286, **475**, **503**, 1090	19
Almandine	$Fe_3Al_2Si_3O_{12}$	170, 216, 323, 342, 370, 500, 556, **863**, **897**, **916**, 1038	10
Aluminium sulphate	$Al_2(SO_4)_3 \cdot 18H_2O$	617, 972, **988**, 1127	19
Alum-K	$KAl(SO_4)_2 \cdot 12H_2O$	442, 455, 614, **974**, **989**, 1104, 1130, 3072, **3396**	10
Alunite	$KAl_3[(OH)_3(SO_4)]_2$	485, 509, 654, **1026**, 1079, 1190, **3482**, **3509**	10
Alunite2	$KAl_3[(OH)_3(SO_4)]_2$	236, 347, 384, 486, 564, 655, **1027**, 1081, 1190, **3480**, **3508**	19
Amblygonite	$LiAl(PO_4)(F,OH)$	130, 303, 426, 488, 603, **647**, **1012**, **1048**, 1064, **3369**, 3408	19
Ameghnite	$NaB_3O_3(OH)_4$	135, 143, 171, 199, 348, 363, 465, 503, 528, **620**, **755**, 883	19
Ammonia	NH_3	**3336**	4
Ammonia in aqueous solution	NH_4^+	2870, **3040**	10
Ammonia in water	NH_3	3250, **3312**, 3400	47
Anatase	TiO_2	**143**, 195, 395, 514, **638**	10
Ancylite-Ce	$Sr(Ce,Ca,La)(CO_3)_2(OH)_2 \cdot H_2O$	125, 189, **243**, 371, 699, 725, **1083**	8, 58
Andalusite	Al_2SiO_5	**293**, 323, 361, 453, 553, 719, 834, **920**, 992, 1065, 1111	10
Andalusite2	Al_2SiO_5	**292**, 320, 358, 551, **918**, **959**, 1019, 1064, 1110	19
Andradite	$Ca_3(Fe,Ti)_2Si_3O_{12}$	174, 236, 325, 370, 452, 494, 516, 574, **816**, **842**, **874**, 995	10
Anglesite	$PbSO_4$	438, 450, 608, 641, **978**, 1068, 1160	10

Continued

Geofluids. http://dx.doi.org/10.1016/B978-0-12-803241-1.09972-X

Name	Formula	Vibration Bands (cm^{-1})	Source
Anhydrite	$CaSO_4$	430, 500, 515, 611, 629, 674–676, **1015–1018**, 1131	10
Antarcticite	$CaCl_2 \cdot 6H_2O$	3242, **3404**, 3410, **3430**	2
Apatite-Cl	$Ca_5(PO_4)_3Cl$	430, 581, **963**, 1039, 1127	10
Apatite-F	$Ca_5(PO_4)_3F$	432, 449, 581, 592, 608, **965**, 1042, 1053, 1081	10
Apatite-Sr	$(Ca,Sr)_5(PO_4)_3(OH)$	442, 575, 586, **959**, 1044, 1071, 3295, 3353	19
Aphthitalite (glaserite)	$(K,Na)_3Na(SO_4)_2$	447, 457, 619, **984**, 1093, 1104	10
Aragonite-(Sr)	$CaCO_3$	144, 154, 181, 191, 207, 274, **702**, 707, 1060, 1084, **1086**, 1463	13
Arcanite	K_2SO_4	457, 622, **983**, 1093, 1109	10
Arsenic	As	**217**, 281, 393, 607, 659	19
Arsenolamprite	As	220, 225, **253**	22
Arsenolite	As_2O_3	85, 185, 269, **370**, 471, 560, **781**	8
Artinite	$Mg_2(OH)_2(CO_3)\cdot 3H_2O$	**147**, 173, 472, 704, **1094**, 3030, 3229, **3593**	10
Ascharite	$MgHBO_3$	324, **835**, **3555**	19
Astrakanite (blodite)	$Na_2Mg(SO_4)_2\cdot 4H_2O$	454, **616**, **991**, 1067, 1115, 1153	19
Atacamite	$Cu_2(OH)_3Cl$	106, **121**, 361, 419, 446, 475, **511**, 818, 846, **912**, **975**	30
Atacamite2	$Cu_2(OH)_3Cl$	66, **106**, **120**, **139**, **150**, 450, **514**, 586, 820, 845, **908**, 975, **3351**, **3453**	13
Augelite	$Al_2(PO_4)(OH)_3$	124, 163, 226, 252, 368, 410, 464, 526, 565, **637**, 859, 896, 1014, 1108, 1162, 3428, **3469**, **3537**	19
Avogadrite	KBF_4	359, 535, **774**	8
Axinite	$Ca_2(Mn,Fe)Al_2[OH/BO_3/Si_4O_{12}]$	104, 271, 414, 482, **710**, 897, **977**, 3354	19
Azurite	$Cu_3(OH)_2(CO_3)_2$	170, 246, 279, **397**, **1095**, 1457, **3453**	10
Ba-chloride dihydrate	$BaCl_2\cdot 2H_2O$	193, 256, **402**, 490, 550, **3331**, 3451	19
Barite	$BaSO_4$	**461**, 617, **988**, 1143	10
Barite2	$BaSO_4$	74, 89, 97, 152, 189, **452**, **462**, 616, 629, 647, 965, **988**, 1104, **1144**, 1166	18
Barium chloride	$BaCl_2$	**186**	19
Bastnäsite	$(Ce,La)CO_3F$	**161**, 175, 181, 234, 261, 286, 288, 352, 402, 566, 601, 606, 719, 735, **1085**, **1096**, 1432, 1445, 1504, 1737	11
Be-hydrocarbonate	$Be(HCO_3)_3$	484, 766, **1124**	19
Behoite	$Be(OH)_2$	133, 207, 276, **347**, **398**, 444, 456, 599, 684, 757, 1055	19
Behoite-β	β-$Be(OH)_2$	148, 286, **351**, **404**, 450, 459, 465, 553, 622, 717, 1069, 2544, 2551, 2566, **3413**, 3464, 3476	19
Benzene	C_6H_6	608, **994**, 1179, 1587, 1607, 2927, 2950, **3063**	44

Name	Formula	Vibration Bands (cm^{-1})	Source
Benzene2	C_6H_6	**988**, **3058**	49
Benzene3	C_6H_6	**309**, **999**, 1580, **3064**	69
Berborite	$Be_2(BO_3)(OH,F)\cdot H_2O$	78, 379, 394, **984**, **1377**, **3504**	19
Berlinite	$AlPO_4$	119, 163, 196, 221, 418, 439, 462, 524, **1104**, **1111**, 1230	6
Bertrandite	$Be_4Si_2O_7(OH)_2$	183, **206**, 231, 330, 358, 579, 694, 711, 928, 990, 1073, **3551**, **3587**	19
Beryl	$Be_3Al_2Si_6O_{18}$	206, 289, 322, **396**, 422, 444, 528, 581, **685**, 770, 915, 1012, 1068, 1243	8
Beryl2	$Be_3Al_2Si_6O_{18}$	324, 401, **686**, 1006, **1068**, 1112, **3593**, 3604	19
Berylium sulphate	$BeSO_4\cdot 4H_2O$	181, 281, 415, 495, 588, 630, **994**, 1080, 1112	19
Beryllium fluoride	BeF_2	**280**, 420, **500**, 560, 630	19
Beryllium fluoride in melt	BeF_2	255, 385, **547**, 800	19
Beryllium chloride	$BeCl_2$	263, **551**	19
Beryllium chloride2	$BeCl_2$	175, 205, 243, **296**, 327, 478, 574, 620	19
Beryllium phosphate	$Be_3(PO_4)_2$	351, 466, 540, **1012**, **1051**	19
Beryllonite	$NaBePO_4$	139, 161, 433, 548, **1012**, **1056**	26
Billietite	$Ba(UO_2)_6O_4(OH)_6\cdot 8H_2O$	803, **830**	58
Biotite	$K(Mg,Fe)_3AlSi_3O_{10}(OH,F)_2$	178, 549, **679**, 717, 767, 3658, 3680	10
Bischofite	$MgCl_2\cdot 6H_2O$	206, **227**, 348, 355, 1554, **1586**, **1601**, **1604**, 1803, 3356	19
Bismuth	Bi	79, **311**, 465, 621	19
Bityite	$CaLiAl_2(AlBeSi_2)O_{10}(OH)_2$	104, **132**, 365, **478**, 684, **898**, 1049	19
Blodite (astrakanite)	$Na_2Mg(SO_4)_2\cdot 4H_2O$	454, 618, **992**, 1188	8
Boracite	$Mg_6[Cl_2/B_{14}O_{26}]$	**131**, 602, **660**, **840**, 1142	19
Borax	$Na_2B_4O_5(OH)_4\cdot 8H_2O$	344, 405, 463, **571**, 776, 943, 997	10
Borax2	$Na_2B_4O_5(OH)_4\cdot 8H_2O$	344, 405, **578**, 767, 943, 997, **3351**, 3499	19
Boric acid in aq.	H_3BO_3	495, **875**	23
Boromuscovite	$KAl_4Si_5B_2O_{20}(OH,F)_4$	247, 411, 476, 555, **717**, 784, **3614**	19
Boron dihydrate in aq.	$B(OH)_2^-$	**595**	19
Boron fluoride	$BF_3\cdot 2H_2O$	280, 350, 510, 730, **764**, 3530	19
Boron nitride cubic	BN	**1055**, **1305**	19
Boron orthophosphate	BPO_4	239, 460, **481**, 541, 1067, **1116**, 1261	19
Boron tetrahydrate in aq.	$B(OH)_4^-$	**744**	21
Boronfluoride dihydrate in aq.	$BF_2(OH)_2^-$	**753**	19
Brochantite	$Cu_4(OH)_6SO_4$	90, 123, 139, 196, 390, 421, 450, 481, 595, 609, **973**, **3371**, **3399**, **3564**, **3587**	30
Brochantite2	$Cu_4(OH)_6SO_4$	**80**, **91**, 132, 138, 142, 157, 188, 199, 242, 319, 387, 426, 450, 484, 509, 595, 609, 910, **973**, 1097, **3401**, **3566**, **3588**	13

Continued

Name	Formula	Vibration Bands (cm^{-1})	Source
Bromellite	BeO	388, **678**, **684**, 722, 1081, 1097	22
Brookite	TiO_2	127, **150**, 247, 318, 366, 412, **645**	10
Brushite	$CaHPO_4 \cdot 2H_2O$	408, 521, 875, **985**, 1056, 1132, 3160, 3261, **3469**, **3536**	27
Buddingtonite	$(NH_4)AlSi_3O_8 \cdot 0.5H_2O$	135–138, 288, 296, 448, 478, **515–517**, 579, 805, 1097, 1120, 1145, 3223	19
Burkeite	$Na_6(CO_3)(SO_4)_2$	451, 474, 620, 633, 644, **994**, 1065	10
Butane	C_4H_{10}	837, 2865, **2885**, 2927, 2960, 2987	64, 70
Ca-chloride tetrahydrate	α-$CaCl_2 \cdot 4H_2O$	**3362**, **3423**, **3444**, 3475	2
Ca-chloride tetrahydrate	γ-$CaCl_2 \cdot 4H_2O$	**3434**	2
Calcite	$CaCO_3$	156, 283, 711, **1085**	4
Calcite-(Mg)	$CaCO_3$	157, **283**, 714, 1069, **1088**, 1438	13
Calciouranoite	$(Ca,Ba,Pb,K,Na)U_2O_7 \cdot 5H_2O$	207, **795**, 964	8
Calcium hexaborate	$CaB_6O_{10} \cdot 4H_2O$	383, 391, 460, 482, 572, 636, 745, **763**, 813, **855**, **861**, **895**, 952, 996	19
Calcium metaborate I	$CaB_2O_4 \cdot 4H_2O$	390, 547, **758**, 854	19
Calcium metaborate II	$CaB_2O_4 \cdot 6H_2O$	**755**, **859**	19
Calcium triborate	$Ca_2B_6O_{11} \cdot 9H_2O$	472, **628**, 740, 866	19
Cancrinite	$3NaAlSi_3O_8 \cdot Na_2CO_3$	117, **274**, 349, 422, 462, 503, 684, **774**, 957, **1057**	19
Carbon dioxide gas $^{13}CO_2$	CO_2	1262, **1369**	38
Carbon dioxide gas in aq.	CO_2	**1276**, **1384**	7
Carbon dioxide gas	CO_2	**1285**, **1388**, 1409	4
Carbon dioxide gas $^{12}C^{16}O_2$	CO_2	1265, **1286**, **1388**, 1410	39
Carbon dioxide gas $^{13}C^{16}O_2$	CO_2	1249, **1267**, **1370**, 1389	39
Carbon dioxide gas $^{12}C^{16}O^{18}O$	CO_2	1250, **1259**, **1366**, 1377	39
Carbon dioxide gas $^{13}C^{16}O^{18}O$	CO_2	1235, **1245**, **1343**, 1356	39
Carbon monoxide gas	CO	**2137**	38
Carbon monoxide gas ^{13}CO	CO	**2090**	38
Carbonate ion in aq.	CO_3^{2-}	684, **1062–1072**, **1376–1382**, 1438	7
Carbonate ion in aq.	CO_3^{2-}	684, 885, **1065**, 1376, 1421, 1764	43
Carbonyl sulphide	COS	**857**	4
Carboxylic acid	R-C(O)OH	**1650**	69
Carnallite	$KMgCl_3 \cdot 6H_2O$	**1644**, **3442**	19
Cassiterite	SnO_2	230, 283, **634**	8
Cassiterite2	SnO_2	474, **633**, 774	19

Name	Formula	Vibration Bands (cm^{-1})	Source
Cebaite	$Ba_3Ce_2(CO_3)_5F_2$	177, 213, **267**, 621, 701, 715, 730, 1072, 1084, **1091**	8
Celestite	$SrSO_4$	452, 627, **656**, **1000**, 1156, 1190	10
Cerianite	CeO_2	184, **449–465**, 571	17
Cerussite	$PbCO_3$	150, 180, 215, 682, 839, **1056**, 1378	10
Cesanite	$Na_7Ca_3(OH)(SO_4)_6{\cdot}0.8H_2O$	448, 474, 626, 647, **1004**, 1104	10
Cesium carbonate	Cs_2CO_3	227, 662, 672, 679, **1040**, **1048**	25
Cesium hydrogen sulphate	$CsHSO_4$	419, 473, 570, 587, 610, 857, **995**, 1180, 1249	19
Cesium tetrafluoroborate	$CsBF_4$	353, 356, 523, 527, 530, **768**	19
Ce–Zn tetrachloride	Cs_2ZnCl_4	275, **285**, **296**	19
Claringbullite	$Cu_4Cl[Cl_{0.29}(OH)_{0.71}](OH)_6$	119, 136, **147**, 163, 231, 260, 356, 389, 447, **511**, 579, **815**, **906**, **970**, 1049, 3211, 3331, **3351**, **3433**, 3458	31
Clathrate	$SnCl_2{\cdot}2H_2O$	78, 123, **156**, 187	19
Clathrate	$BaCl_2{\cdot}2H_2O$	193, 256, **402**, 490, 550, **3331**, 3451	19
Clinochlore	$(Mg,Fe)_5Al(OH)_6(AlSi_3O_{10})$	104, 198, 358, 548, **679**, 3477, 3605, 3647, 3679	10
Coesite	SiO_2	116, 151, 176, 204, 269, 326, 355, 427, 466, **521**, 785, 815, 837, 1036, 1065, 1144, 1164	10
Colemanite	$Ca[B_2BO_4(OH)_3]{\cdot}H_2O$	306, 383, 531, **608**, 666, 742, 3514, **3588**	19
Columbite	$FeNb_2O_6$	132, 203, 275, **393**, 525, 634, **869**	19
Copper sulphate	$CuSO_4{\cdot}5H_2O$	259, 333, 368, 423, 450, 611, **983**, 1144	19
Cornetite	$Cu_3(PO_4)(OH)_3$	**109**, 131, 141, 158, 174, 297, 313, 363, **446**, **477**, 515, 583, 639, 945, **971**, **996**, 1016, **1083**, 1137, **3414**	57
Coronene	$C_{24}H_{12}$	370, 484, 1027, **1351**, **1369**, 1451, 1597, 1618, 1629, 1807, 3015, 3041, 3052, 3073	44
Corundum	Al_2O_3	378, **417**, 644, 750	10
Cotunnite	$PbCl_2$	**159**, **179**, 417	19
Covellite	CuS	263, **471**	10
Creedite	$Ca_3Al_2(SO_4)(OH,F)_{10}{\cdot}2H_2O$	441, 481, 602, 628, 658, **985**, 1080, 1132, 1182	19
Cristobalite	SiO_2	114, 230, 273, 286, **420**, 792, 1075	10
Cristobalite2	SiO_2	112, 230, **418**, 780, 1075, 1405	19
Crocoite	$PbCrO_4$	141, 339, **361**, 378, 405, **842**, 864	30
Cryolite	Na_3AlF_6	485, **555**, **620**	10
Cryolithionite	$Li_3Na_3Al_2F_{12}$	340, 364, 378, 413, **563**	19
Cs–Al sulphate	$CsAl(SO_4)_2{\cdot}12H_2O$	191, 447, 624, **986**, 1091	19
Cs–Zn tetrachloride	Cs_2ZnCl_4	275, **285**, **296**	19
Cuprosklodowskite	$Cu(UO_2)_2(SiO_3OH)_2{\cdot}6H_2O$	207, 275, 302, **790**, **973**	8
Curite	$Pb_{3+x}[(UO_2)_4O_{4+x}(OH)_{3-x}]_2{\cdot}2H_2O$	162, 221, 302, 456, 786, **830**, 1024	8

Continued

Name	Formula	Vibration Bands (cm^{-1})	Source
Cyclohexane	C_6H_{12}	**804**, 2855, 2941	49
Cyclohexane2	C_6H_{12}	**785**, **1028**, 1266, 1440, 2664, **2855**, 2895, 2923, 2938	69
Cyclooctane	C_8H_{18}	**703**, 1467, **2855**, 2925	70
Danalite	$Fe_4Be_3(SiO_4)_3S$	**129**, 155, 173, 237, 361, 443, 531, 566, 681, **891**, 943, 1023	19
Danburite	$CaB_2(SiO_4)_2$	166, 182, **245**, 348, 480, **611**, 628, 974, 1107, 1664, 3079, 3912	19
Datolite	$CaB(SiO_4)OH$	149, 275, 707, **1076**, 1428, 1742	19
Dawsonite	$NaAl(CO_3)(OH)_2$	**192**, 260, 590, 728, 1065, **1091**, 1506, 3250, **3282**	10
Dawsonite2	$NaAl(CO_3)(OH)_2$	152, **191**, 389, 519, **590**, 824, 897, **1091**, **1506**, 3252, **3283**	19
Dawsonite-K	$KAl(CO_3)(OH)_2$	77, 169, **207**, 282, 363, 503, 580, 679, 730, 832, **1102**, 1486, 3406, 3432	19
Dawsonite-NH_4	$NH_4Al(CO_3)(OH)_2$	79, 93, 110, 120, 139, **205**, 292, 306, 367, 572, 585, 620, 645, 686, 733, **1107**, 3388, 3460	19
Decane saturated	$C_{10}H_{22}$	**255**, **843**, 1132, **1302**, 1435, 1451, **2848**, 2858, **2872**, 2897, 2917, 2934	46
Deprotonated monomers	$H_{4-n}SiO_4^{n-}$	**1017**	9
Diamond	C	**1332**	10
Diamond ^{13}C	C	**1315**	20
Diamond 6H	C	1208, 1210, 1270, 1298, 1332	15
Diamond 8H	C	405, 546, 732, 1112, 1304, **1332**, 1353, **1357**	15
Diaspore	AlOOH	331, **448**	10
Diaspore2	AlOOH	159, 337, **452**, **501**, 558, **667**, 793, 1050, 1190	19
Dickite	$Al_2Si_2O_5(OH)_4$	120, 132, 178, 198, 245, **268**, **336**, 420, 434, **463**, 748, 792, 916, 935, 1050, 1104, **3623**, 3648, **3683**, **3700**	56
Diomignite	$Li_2B_4O_7 \cdot 5H_2O$	391, **446**, 493, **543**, **772**, 845, 896, **1028**, 1097, 1352	22
Diopside	$CaMgSi_2O_6$	325, 393, **665**, 1010	14
Diopside2	$CaMgSi_2O_6$	324, 360, 391, 667, 855, **1013**, 1046	19
Dolomite	$(Ca,Mg)CO_3$	176, 299, 725, **1097**	4
Dumortierite	$Al_7(BO_3)(SiO_4)_3O_3$	169, **206**, 293, 353, 403, 418, **508**, 810, 842, **946**, 1076, 1112	8
Dypingite	$Mg_5[(OH)(CO_3)_2]_2 \cdot 5H_2O$	203, 249, 311, 434, 727, 1092, **1122**, 1447, 3421, 3515, **3648**	10
Eitelite	$Na_2Mg(CO_3)_2$	**207**, **263**, 710, **1105**	54
Elpassolite	K_2NaAlF_6	135, **326**, 379, 387, 446, **559**, 668, 873, 989, 1009, 1055	19

Name	Formula	Vibration Bands (cm^{-1})	Source
Enstatite	$MgSiO_3$	237, 343, 382, 414, **664**, **684**, 1011	10
Epsomite	$MgSO_4 \cdot 7H_2O$	447, 612, **984**, 1061, 1095, 1134, **3303**, **3425**	10
Eriochalcite	$CuCl_2 \cdot 2H_2O$	117, **216**, 235, 247, 390, **407**, 672, 692, 1620, 3176, 3317, **3367**, 3462	31
Ethane	C_2H_6	993, 2914, **2954**	37
Ethane2	C_2H_6	**1000**, 1440, **2895**, **2955**	64
Ethane in chibaite	$SiO_2 \cdot n(CH_4, C_2H_6, C_3H_8, i\text{-}C_4H_{10})$	991, **2885**, 2939	35
Euclase	$BeAlSiO_4(OH)$	145, 180, 237, **259**, 290, **397**, 444, 452, 574, **880**, **908**, 975, 1023, 1059, 1120, **3575**, 3587	19
Ezcurrite	$Na_4B_{10}O_{17} \cdot 7H_2O$	132, 226, 286, 315, 336, **476**, **513**, **570**, **759**, 881, **954**, 1040, 1059, **1349**, **1388**, **2859**, 3335, 3375	19
Fe-chloride hexahydrate	$FeCl_2 \cdot 6H_2O$	**3389**, **3411**, 3426	2
Fe-chloride tetrahydrate	$FeCl_2 \cdot 4H_2O$	3170, **3386**, **3404**, 3520	2
Fergusonite	$Y(Nb,Ta)O_4$	**814**, **1412**, 1694, **2216**, 2425, 3143, 3839	19
Fergusonite (Norway)	$Y(Nb,Ta)O_4$	285, 667, **805**, **1419**, 1725, 2219, 3124, 3846	19
Fergusonite monoclinic	$YNbO_4$	**131**, **324**, **336**, 658, 675, 698, **811**	19
Ferricopiapite	$Fe_5(SO_4)_6(OH)_2 \cdot 20H_2O$	268, 310, 451, **477**, **996**, **1019**, 1148	8, 58
Ferric iron chloride	$FeCl_3$	**334**, 418	19
Ferric iron sulphate hydrate	$Fe_2(SO_4)_3 \cdot 9H_2O$	251, 455, 631, **1036**, 1049, 1086, 1189	19
Ferrocolumbite		134, 390, 529, **870–874**	19
Ferropyrosmalite	$Fe_8Si_6O_{15}[(OH_6)Cl_4]$	**615**, 1022, **3551**, **3579**, **3637**	16
Fe-spinel	Fe_2SiO_4	334, **672**, 779, **845**, 924	19
Fluellite	$Al_2(PO_4)F_2OH \cdot 7H_2O$	280, 410, **523**, 583, 649, 1037, 1119	19
Fluoborite	$Mg_3BO_3(OH,F)_3$	241, **953**, 1341, 3645, 3656, **3669**, 3686	19
Fluorite	CaF_2	**322**, 641	10
Fluorite2	CaF_2	**317**	19
Fluoroberyllate ion in melt	$Be_2F_7^{2-}$	**520**, 954	19
Fluoroberyllate ion in melt	$Be_2F_7^{3-}$	**522**	19
Fluoroberyllate ion in melt	BeF_4^{2-}	**547**, **550**	19
Fluoroberyllate ion in melt	BeF_4^{2-}	**560**	19
Fluoroboric acid in aq.	BF_3OH^-	**763**	19
Forsterite	Mg_2SiO_4	227, 303, 423, 548, 608, **824**, **856**, 882, 921, **964**	10

Continued

Name	**Formula**	**Vibration Bands (cm^{-1})**	**Source**
Fourmarierite	$Pb_{1-x}O_{3-2x}(UO_2)_4(OH)_{4+2x}\cdot 4H_2O$	373, 490, **703**, **822**	8
Gahnite	$ZnAl_2O_4$	416, 625, 658, **707**	19
Galena	PbS	**136**, 270	10
Gaylussite	$Na_2Ca(CO_3)_2\cdot 5H_2O$	**164**, 265, 723, **1071**, **2944**, 3334	10
Gibbsite	$Al(OH)_3$	242, 255, 322, **380**, 538, 569, 979	10
Gibbsite2	$Al(OH)_3$	117, 243, 321, 383, **545**, **568**, 896, 980	19
Glaserite	$K_3Na(SO_4)_2$	**453**, **619**, **678**, **990**	19
Glauberite	$Na_2Ca(SO_4)_2$	485, 618, 644, **1002**, 1106, 1140	10
Glaucophane	$Na_2Mg_3Al_2Si_8O_{22}(OH)_2$	210, 257, 339, **386**, 411, 445, 525, 562, **670**, 778, 896, 1000, **1047**, 1108, 3615, 3631, 3647, 3660	19
Goergeyite	$K_2Ca_5(SO_4)_6\cdot H_2O$	433, 440, 480, 595, 602, 631, **1005**, **1013**, 1085, 1115, 1162, 3525, **3580**	10
Goethite	α-FeOOH	242, 299, **389**, **547**, 681	10
Goethite2	α-FeOOH	**241**, **298**, 396, 478, 547, 678, 991	19
Grandidierite	$(Mg,Fe,Mn)Al_3(BO_3)(SiO_4)O_2$	**161**, 228, 242, 269, **343**, 363, 376, **425**, 442, 489, **577**, 659, 864, **949**, 979, 1042	8
Graphite disordered	C	1150, **1355**, 1500, 1622, 2680, 2714, 3240	3
Graphite ordered	C	**1582**	3
Gregoryite	K_2CO_3	141, 192, 234, 677, 686, 697, 702, 1043, **1060–1064**, 1374, 1405, 1426	25
Greifensteinite	$Ca_2Be_4(Fe,Mn)_5(PO_4)_6(OH)_4\cdot 6H_2O$	164, 216, 305, 407, 435, 481, **612**, **628**, **993**, **1056**, 1133, 1176	19
Grossular	$Ca_3Al_2Si_3O_{12}$	181, 247, 280, 373, 420, 550, **827**, 848, 1007	10
Gypsum	$CaSO_4\cdot 2H_2O$	414, **492–494**, 621–623, 671, **1006–1008**, 1142, 3250, **3405**, 3491, 3500	4
Hafnon	$HfSiO_4$	148, 159, 215, 350, 403, 448, 984, **1018**	19
Halite	NaCl	**358**	10
Hambergite	$Be_2BO_3(OH,F)$	**148**, **158**, 992	22
Hambergite2	$Be_2BO_3(OH,F)$	280, 324, 398, 583, 623, 739, **988**, **3419**, **3484**, 3562, **3600**, 3662	19
Hedyphane	$Pb_3Ca_2(AsO_4)_3Cl$	141, 151, 162, 177, 192, 201, **319**, 334, 349, 372, 394, 432, 452, 770, 787, **811**, **819**, 834, 848, 3352, 3409, **3458**	34
Helvite	$Mn_4Be_3(SiO_4)_3S$	**131**, 158, 176, 238, 288, 362, 444, 527, 565, 678, **981**, 942, 1020	19
Hematite	Fe_2O_3	223, **290**, **409**, 498, 609, 1313	10
Hematophanite	$Pb_4Fe_3O_8(OH,Cl)$	236, 263, **447**, 664	19
Heptane saturated	C_7H_{16}	**310**, 834, 899, 1302, **1452**, **2849**, 2859, **2872**, 2895, 2916, 2936, 2962	46
Herderite	$CaBe(PO_4)(F,OH)$	166, 194, 424, 445, 516, 529, **584**, **594**, **983**, **1005**, 1128, 1140, **3564**, **3573**, 3620	19

Name	Formula	Vibration Bands (cm^{-1})	Source
Hexahydrite	$MgSO_4 \cdot 6H_2O$	445, 466, 610, **984**, 1085, 1146, 3258, **3428**	10
Hexane saturated	C_6H_{14}	370, 866, 889, **1451**, 2849, 2859, **2872**, 2893, 2915, **2936**, 2961	46
Hexene	C_6H_{12}	1294, **1635**, 2996	49
Hieratite cubic	K_2SiF_6	405, 476, 540, **655**, 839	18
Hornblende	$Ca_2[Mg_4(Al,Fe)]Si_7AlO_{22}(OH)_2$	224, **667**, **1040**	10
Hydroboracite	$CaMgB_6O_{11} \cdot 6H_2O$	**181**, 257, 322, 383, 398, 524, 548, 564, 606, **753**, 837, 876, **3595**	19
Hydrocarbonate ion in aq.	HCO_3^-	**1017**, 1091, 1360	9
Hydrocarbonate ion in aq.2	HCO_3^-	630, 641, 673, 842, **1015**, **1312**, **1364**, 1628, 1682, 2619	43
Hydrogen gaseous	H_2	354, 586, 813, 1032, 4126, 4143, **4156**, 4161	10
Hydrogen sulphate in aq.	HSO_4^-	890, **1050**	4
Hydrogen sulphide in aq.	H_2S	**2590**	4
Hydrogen sulphide gas	H_2S	**2611**	4
Hydrogen sulphide liquid	H_2S	**2580**	4
Hydrocerussite	$2PbCO_3 \cdot Pb(OH)_2$	415, 681, **1051**, **1055**	30
Hydrohalite	$NaCl \cdot 2H_2O$	**3402**, **3419**, **3432**, 3536	2
Hydromagnesite	$Mg_5(CO_3)_4(OH)_2 \cdot 4H_2O$	184, 202, 232, **1119**, 1487	10
Hydrosulphide ion in aq.	HS^-	**2574**	4
Hydroxyfluoroboric acid	$HBF_2(OH)_2$	283, 349, 509, 611, **731**, 758, **871**, 3536	19
Hydroxylapatite	$Ca_5(PO_4)_3OH$	431, 591, **962**, **3572**	27
Hydroxylbastnäsite	$(Ce,La)CO_3(OH,F)$	112, 155, 436, 591, **965**, 1053, 1330, **3317**, **3352**, 3430, 3535, 3714	11
Hydroxylbastnäsite2	$(Ce,La)CO_3(OH,F)$	**1080**, **1087**, **1098**, **3235**, **3493**, **3568**, **3638**	60
Hydroxylclinohumite	$4[Mg_2SiO_4] \cdot [Mg,Fe,Ti](OH,F)_2$	110, 186, 212, 233, 306, 321, 348, 549, 590, 609, 756, 786, **832**, **845**, **861**, 931, 968, 3393, 3410, **3560**, **3570**, **3578**	13
Hydroxylherderite	$CaBe(PO_4)(OH)$	170, 197, 427, 448, 531, 579, 596, **985**, **1004**, 1129, 1141	19
Hydroxyltopaz	$Al_2SiO_4(OH,F)_3$	229, 377, 439, 457, 897, 996, **1095**, 1122, 1353, 3288, 3366	19
Chalcopyrite	$CuFeS_2$	**293**, 322, 352, 378	10
Chiolite	$Na_5Al_3F_{14}$	250, 318, 356, **531**, 553, 1088	19
Chrysene	$C_{18}H_{12}$	295, 681, 771, 1019, 1366, **1383**, 1434, 1576, 1603, 1618, 3063	44
Chrysoberyl	$BeAl_2O_4$	218, 242, 353, 422, 459, **477**, **518**, 546, **639**, 658, 679, 711, 778, 816, **931**	19

Continued

Name	Formula	Vibration Bands (cm^{-1})	Source
Crysocolla	$(Cu,Al)_2H_2Si_2O_5(OH)_4\cdot nH_2O$	109, 146, 183, 213, 245, 336, 424, 485, **673**, 792, **933**, **1058**, 1602, 3295, 3504, 3583, **3624**, 3652, 3708	32
Ice	H_2O	**3090**, 3218	2
Ice2	H_2O	**3102**, 3225	42
Ilmenite	$FeTiO_3$	232, 373, **685**	10
Ilmenite2	$FeTiO_3$	162, 228, 255, 369, 453, **680**	19
Inderite	$Mg_2B_6O_{11}\cdot 15H_2O$	232, **631**, 3630, 3557	19
Inderite synthetic	$Mg_2B_6O_{11}\cdot 15H_2O$	396, 424, 493, 584, 637, 745, **858**, 954, 1101	19
Inyoite	$Ca_2B_6O_{11}\cdot 13H_2O$	352, 388, **464**, **615**, 733, 808, 912, 956, 1096, 3223, 3334, **3443**, **3531**	19
Isobutene in chibaite	$SiO_2\cdot n(CH_4, C_2H_6, C_3H_8, i\text{-}C_4H_{10})$	**806**	35
Isokite	$CaMgPO_4F$	273, 425, 457, 607, **955**, 1021	19
Jacobsite	$MnFe_2O_4$	**806**, **839**	19
Jarosite	$KAl_3(SO_4)_2(OH)_6$	442, 560, 625, **1015**, 1112, 1161	19
Jeremejevite	$Al_6(BO_3)_5(OH,F)_3$	301, **328**, **372**, 417, 512, 608, 710, 749, **962**, 1067, 3624, **3672**	19
Johannite	$Cu(UO_2)_2(SO_4)_2(OH)_2\cdot 8H_2O$	201, 244, **834**, 1043, 1090	8, 58
K-cymrite	$KAlSi_3O_8\cdot H_2O$	114, **380**, **833**, 3541, 3623	53
K-fluoroberyllate	K_2BeF_4	268, 387, 392, **559**, **562**, 669	19
Kainite	$KMgClSO_4\cdot 2.75H_2O$	**440**, **630**, **1022**, **1039**, 1157, 1176, 1197, 1207	19
Kalicinite	$KHCO_3$	635, 673–675, **1028**, **1271**, 1277	25
Kalicinite2	$KHCO_3$	79, 106, 184, 636, 676, 832, 960, **1029**, **1282**, 1448, 1682, **2567**	43
Kaolinite	$Al_2Si_2O_5(OH)_4$	120, 132, 200, 248, **272**, **338**, 420, 434, **461**, 749, 791, 916, 935, 1048, 1105, **3620**, 3650, 3668, 3682, **3692**	56
Kernite	$Na_2B_4O_7\cdot 4H_2O$	**105**, 175, 256, 388, 426, 456, **496**, 554, 579, **737**, 802, 848, **856**, **929**, 978, 1009, 1034, 1089, 1139	19
Kieserite	$MgSO_4\cdot H_2O$	436, 481, 629, **1046**, 1117, 1215, **3297**	10
Kokchetavite	$KAlSi_3O_8$	109, **390**, **836**	53
K–Pb hexafluoride	K_4PbF_6	63, 89, **146**, 375	19
K-tetrafluoroborate	KBF_4	358, 360, 365, 527, 530, 533, 535, 538, **772**	19
Kumdykolite	$NaAlSi_3O_8$	155, **222**, 265, 407, 464, **492**	61
Kurnakovite	$MgB_3O_3(OH)_5\cdot 5H_2O$	391, 422, 466, 627, **850**, 944	19
Kyanite	Al_2SiO_5	302, 325, 360, 386, 405, 419, 437, 486, 562, 669, **952**	10
K–Zn tetrafluoride	K_2ZnF_4	167, 180, **356**	19
K–Zn tetrachloride	K_2ZnCl_4	277, **291**, **293**, 317	19
Lacroixite	$NaAl(PO_4)F$	319, 503, 609, **623**, **1001**, 1043, 1119	19

Name	Formula	Vibration Bands (cm^{-1})	Source
Langbeinite	$K_2Mg_2(SO_4)_3$	**456**, **466**, 607, 626, 640, **1053**	19
Laurionite	PbCl(OH)	131, 142, 174, **272**, **328**, 594, 662, 729, **3506**	12
Lazulite	$(Mg,Fe)Al_2(PO_4)_2(OH)_2$	253, 344, 611, 630, 741, **1059**, 1100, 1136, 3384, 3400	19
Lazurite	$Na_3Ca(Al_3Si_3O_{12})S$	257, 288, **547**, 801, 1094, 1639	13
Lead fluoride	PbF_4	**256**	19
Leifite	$Na_2(Si,Al,Be)_7(OH,F)_{14}$	388, 472, **502**, 532, 698, 836, 965, 1032, 1128	19
Leonite	$K_2Mg(SO_4)_2 \cdot 4H_2O$	**453**, **632**, **1025**, **1041**, 1158, 1196, 3428	19
Li-hydroxide monohydrate	$LiOH \cdot H_2O$	**3562**, **3664**	67
Li-hydroxide monohydrate	$LiOH \cdot H_2O$	**210**, **243**, 362, 388, **510**, 834	19
Li-chloride hydrate	α-$CaCl_2 \cdot 4H_2O$	**3362**, **3423**, **3444**, 3475	2
Li-chloride hydrate	$LiCl \cdot XH_2O$	**3356**, **3376**	2
Li-chloride pentahydrate	$LiCl \cdot 5H_2O$	3288, 3339, **3428**, **3449**, 3523	65
Li–K carbonate	$LiKCO_3$	**695**, 729, 870, **1046**, **1050**, 1388, 1418, 1483, 1517, 1753	19
Li-metaborate hydrous	$LiBO_2 \cdot 8H_2O$	390, **746**, **852**	19
Li-tetrafluorborate in aq.	$LiBF_4$	**773**	19
Li-tetrafluoraluminate	$LiAlF_4$	210, 320, **622**, 760	19
Libethenite	$Cu_2PO_4(OH)$	**74**, 92, 113, 140, **160**, 227, 270, **301**, 319, 371, 392, 430, 461, 561, 590, 944, **979**, 1008, **1022**, **3475**	57
Lithium chloride	LiCl	**526**, **532**	19
Lithium metaborate	$LiBO_2$	228, 290, 403, 463, 630, 662, **713**, 1081, 1128, **1459**	19
Lithium metaborate hydrous	$LiBO_2 \cdot 8H_2O$	390, **746**, **852**	19
Lithium pentaborate hydrous	$LiB_5O_8 \cdot 5H_2O$	92, 115, 124, 490, **529**, 569, 776, 796, 924, 3105, 3210, 3317, 3416	19
Lithium sulphate	Li_2SO_4	173, 208, 393, 480, 633, **1006**, 1099, 1149, 1178	19
Lithium sulphate hydrate	$Li_2SO_4 \cdot H_2O$	484, 635, **1007**, 1102, 1149, 1176	19
Lithium tetraborate	$Li_2B_4O_7$	158, 197, 222, 256, 351, 493, 531, 562, **725**, **781**, 980, 1034, 1060, 1166	19
Lithium tetrafluorborate	$LiBF_4$	348, 523, **798**	19
Löllingite	$FeAs_2$	**198**, **250**	52
Ludjibaite	$Cu_5(PO4)_2(OH)_4$	**160**, 190, 226, 263, **301**, 368, 387, 402, 449, 557, 586, 633, 736, 784, 815, 855, 925, **981**, 1019, 1046, 1072, 1115, **3470**	57
Magnesite	$MgCO_3$	329, 738, **1094**	4

Continued

Name	Formula	Vibration Bands (cm^{-1})	Source
Magnesium diborate	$MgB_2O_4\cdot 3H_2O$	390, 479, 508, 629, **745**, 860, 900, 1156, 1301	19
Magnesium hexaborate	$MgB_6O_{10}\cdot 6H_2O$	**374**, **634**, **638**, **641**, **852**, **953**, **964**	19
Magnesium sulphate	$MgSO_4$	451, 475, 499, 608, 681, 697, **1023**, 1053, 1136, 1220, 1256	10
Magnesium tetraborate	$Mg_2B_4O_7\cdot 9H_2O$	408, 450, **583**, 787, 845, 949, 1045	19
Magnesium triborate	$Mg_2B_6O_{11}\cdot 17H_2O$	390, 412, 491, 653, 743, **854**, 1060	19
Magnetite	$FeFe_2O_4$	193, 306, 538, **668**	10
Malachite	$Cu_2(OH)_2(CO_3)$	**154**, **178**, 272, **434**, 537, 721, **1098**, 1492, 3386, **3468**	10
Manganocolumbite	$(Mn\cdot Fe)Nb_2O_5$	133, 390, 530, **875**	19
Manganotantalite	$(Mn\cdot Fe)Ta_2O_5$	125, 395, 538, 613, **881**	19
Marcasite	FeS_2	**324**, 387	10
Mendipite	$Pb_3O_2Cl_2$	**137**, 268, 290, 335, 432, 471	19
Metaboric acid	HBO_2	428, 475, 518, 533, **782**	10
Metaboric acid 2	HBO_2	401, 415, 475, **595**, 809	10
Metaboric acid I cubic	HBO_2	104, **150**, **231**, 271, **385**, 434, 454, 478, **511**, **618**, **759**, **780**, **820**, 970, 1025, 1240, 1980, 2380	19
Metaboric acid II	HBO_2	77, 94, 108, 119, 131, 147, 178, 198, 477, 522, 536, **785**, **3130**	19
Metaboric acid III	HBO_2	**87**, **119**, 404, 419, 469, 478, **599**, 638, **810**, 1100, 1343, 1460, 3195, 3295, 3365	19
Methane	CH_4	1535, **2917**, 3020, 3070	36, 37
Methane2	CH_4	**2578**, **2917**, 3006, 3070	62
Methane3	CH_4	1532, **2919**, 3030, 3090	64
Methane in chibaite	$SiO_2\cdot n(CH_4, C_2H_6, C_3H_8, i\text{-}C_4H_{10})$	2903, **2911**	35
Methane in aq.	CH_4	**2909**	18
Methylene	$:CH_2$	**2911–2913**	49
Meyerhofferite	$CaB_3O_3(OH)_5\cdot H_2O$	163, 334, 380, 433, 472, 491, **608**, 725, 879, 1000	19
Mg-calcite	$CaCO_3$	157, 284, **714**, **1087**	4
Mg-chloride dodecahydrate	$MgCl_2\cdot 12H_2O$	3188, 3324, 3397, **3402**, 3422, 3457, 3480, **3510**	2
Microcline	$KAlSi_3O_8$	159, 178, 199, 263, 267, 286, 455, 475, **514**, 651, 749, 813, 1007, 1128, 1142	10
Mimetite	$Pb_5(AsO_4)_3Cl$	98, 105, 145, 175, **313**, 340, 371, 385, 408, 422, 765, **788**, 806, **813**, 837, 854, 860, **3266**, **3420**, 3548	34
Mirabilite	$Na_2SO_4\cdot 10H_2O$	458, 627, **989**, 1129, 3340, **3506**	10
Moissanite	SiC	143, 149, 767, **788**, 797, **966**, 1514, 1712	19
Moganite	SiO_2	129, 141, 220, 265, 317, 370, 377, 398, 432, 449, 463, **501**, 693, 792, 833, 1058, 1084, 1171, 1177	68

Name	Formula	Vibration Bands (cm^{-1})	Source
Monazite	$CePO_4$	466, 620, **987**, 1054	10
Montebrasite	$LiAl(PO_4)(OH,F)$	138, 170, 196, 298, 428, 484, 602, 630, **646**, 801, 965, **1013**, **1046**, **1058**, 1107	19
Moolooite	$CuC_2O_4{\cdot}0.5H_2O$	210, 299, **559**, 586, 611, 832, 932, 1487, **1516**, 1616	30
Moraesite	$Be_2(PO_4)(OH){\cdot}4H_2O$	124, 200, **376**, 415, **584**, 696, **1029**, 1080	8
Moraesite2	$Be_2(PO_4)(OH){\cdot}4H_2O$	148, 172, 230, 347, 401, **447**, **503**, 631, 676, 1004, 1097	19
Mullite	$Al_{4.5}Si_{1.5}O_{9.75}$	161, 264, 308, **419**, 521, 608, 886, **964**, 1133	19
Muscovite	$KAl_4Si_5Al_2O_{20}(OH,F)_4$	178, 197, 216, **261**, 385, 407, 639, **702**, 754, 914, 957, 1117, **3627**	10
Na–Be tetrafluoride	Na_2BeF_4	**568**	19
Nahcolite	$NaHCO_3$	684, **1046**, 1266, 1432	4
Na-hexafluoroboroxine	$Na_3B_3F_6O_3$	350, 378, **595**, 730, 1060	19
Na-hexafluorosilicate	Na_2SiF_6	476, **661**, **674**	19
Na-hexafluorosilicate	Na_2SiF_6	476, **661**, **674**	19
Na-hydrogen phosphate	NaH_2PO_4	389, 535, **908**, 970, 1173, 3441, 3517	19
Na–Li fluoroberyllate	$Na_2LiBe_2F_7$	352, 370, **443**, **520**, 568, 779, 890	19
Nacrite	$Al_2Si_2O_5(OH)_4$	120, 132, 140, 197, 238, **268**, **332**, 418, 433, **462**, 748, 798, 916, 935, 1085, 1102, **3618**, 3628, 3628, 3650, **3698**	56
Nantokite	$CuCl$	127, 155, **206**, 262, 356, 393, **463**	31
Naphthalene	$C_{10}H_8$	**508**, **759**, 1017, 1166, 1240, 1274, **1375**, **1459**, **1572**, 1626, 3054	45
Na-sulphate in aq.	Na_2SO_4	491, 682, **1070**, 1198, 1694	19
Na-tetrafluoroborate in aq.	$NaBF_4$	**773**, **769**	19
Natrite	Na_2CO_3	698, **1078**, 1428	25
Neighborite	$NaMgF_3$	102, 127, **137**, **182**, 208, **226**	19
Nepheline	$NaAlSiO_4$	220, 264, 350, **399**, **415**, 467, 920, 967, **989**, 1086	19
NH_4-fluoroberyllate	NH_4BeF_4	264, 385, **551**	19
NH_4-tetrafluorborate	NH_4BF_4	354, 527, **767**	19
Niobium oxide	Nb_2O_5	238, 308, 463, 551, 635, **683**, 815, 897, **993**	19
Nitrate ion in aq.	NO_3^-	690, **1049**, 1355	10
Nitrogen gas	N_2	**2331**	4
Nitrogen gas ^{14}N–^{15}N	N_2	**2290**	13
Northupite	$Na_3Mg(CO_3)_2Cl$	120, 148, 180, 213, 250, 714, 1107, **1115**, 1554	11
Nyerereite	$Na_2Ca(CO_3)_2$	677, 711, 1071, **1087**	29
Octane saturated	C_8H_{18}	**899**, 2849–2861, 2912–2929, 2965–2969	70

Continued

Name	Formula	Vibration Bands (cm^{-1})	Source
Oligoclase	$(Ca,Na)(Al,Si)AlSi_3O_8$	**177**, 275, 329, 403, 453, 474, **511**, 747, 808, 1007, 1114	19
Orthoclase	$KAlSi_3O_8$	157, 177, 197, 284, 458, 477, **514**, 583, 751, 814, 967, 1035, 1062, 1137	10
Orthoclase2	$KAlSi_3O_8$	282, 402, 404, 458, 474, **513**, 761, 805, 992, 1128	19
Orthosilicic acid	H_4SiO_4	**783**, **3413**	19
Oxygen gas $^{16}O_2$	O_2	**1553**	40
Oxygen gas $^{16}O–^{18}O$	O_2	**1511**	40
Paralaurionite	$PbCl(OH)$	81, **102**, **132**, 220, 273, **330**, 598, 733	19
Paratacamite	$Cu_2(OH)_3Cl$	95, **117**, **140**, 165, 364, 420, **511**, 576, 799, 868, 892, 928, 970, **3311**, **3355**, **3441**	30
Pargasite	$NaCa_2Mg_4Al_3Si_6O_{22}(OH)_2$	229, **665**, **1017**	10
Parisite	$(Ce_2,Ca)(CO_3)_2F_2$	152, 161, 258, 263, 353, 601, 742, **1088**, 1420, 1735	11
Parisite-Ce	$Ca(Ce,La)_2(CO_3)_2F_2$	166, 255, 264, 278, 354, 394, 598, 739, 867, 957, **1081**, **1092**, **1099**	19
Pectolite	$NaCa_2Si_3O_8(OH)$	150, 202, 316, 357, 414, **652**, 974, 1026	19
Penfieldite	Pb_2Cl_3OH	167, 200, 230, **286**, **303**, 440, 615, 772, 978	19
Pentaboratetetrahydrate in aq.	$B_5O_6(OH)_4^-$	**767**	21
Pentane	C_5H_{12}	**869**, 2849–2861, 2912–2929, 2965–2969	70
Pentacene	$C_{22}H_{14}$	184, 266, 446, 501, 751, 996, 1158, 1177, 1353, **1371**, 1410, 1458, 1533, **1598**, 1619, 3032, 3043	44
Pentahydrite	$MgSO_4 \cdot 5H_2O$	371, 447, 602, **1005**, 1106, 1159, 3289, **3343**, **3391**, 3494, 3553	10
Perchlorate ion in aq.	ClO_4^-	460, 645, **928**	10
Perylene	$C_{20}H_{12}$	193, 419, 451, 534, 550, 796, 846, 953, 980, 1047, 1102, 1142, 1224, 1299, **1374**, 1399, 1570, **1622**, 3011, 3052, 3063	44
Petalite	$LiAlSi_4O_{10}$	**143**, **281**, 318, 357, 383, **491**, **793**	19
Phenakite	Be_2SiO_4	447, 527, 601, 618, 778, **877**, 916, 924, 952, 999, 1020	19
Phlogopite	$K_2(Mg,Fe)_6Si_5Al_2O_{20}(OH,F)_4$	192, 279, 331, 372, **680**, 792, 1038, 1096, **3673**, **3715**	10
Phosgenite	$Pb_2CO_3Cl_2$	**152**, 176, **182**, 282, 667, 760, 833, 1058, 1061, 1293, 1304, 1327, 1384, 1518, 1672	12
Piezotite	$Al_3Si_2O_7(OH)_3$	**446**, **1336**, **1607**, 2932	19
Pinnoite	$Mg(BO_2)_2 \cdot 3H_2O$	**740**	19
Planchéite	$Cu_8Si_8O_{22}(OH)_4 \cdot H_2O$	104, 122, **139**, 148, **186**, 227, 302, 328, 403, 421, 476, 505, 699, 828, 906, 1048, **1081**, 1355, **1510**, **3250**, **3479**	32
Pollucite	$Cs_2Al_2Si_4O_{12} \cdot H_2O$	298, 390, **476**, 1132	19
Polyhalite	$K_2MgCa(SO_4)_4 \cdot 2H_2O$	625, **987**, **1013**	19

Name	Formula	Vibration Bands (cm^{-1})	Source
Polylithionite	$KLi_2AlSi_4O_{10}(OH,F)_2$	181, 245, 409, 553, **716**, 781, 872	19
Polylithionite2	$KLi_2AlSi_4O_{10}(OH,F)_2$	89, 99, **188**, **245**, 268, 409, 452, 552, **715**, 765, 791, **1157**	19
Potassium fluoride	KF	203, 261, **579**	19
Potassium pentaborate	$KB_5O_8 \cdot 4H_2O$	372, 459, 509, **558**, 768, 788, 856, 919	19
Potassium tetraborate	$K_2B_4O_7 \cdot 4H_2O$	397, 456, **571**, 767, **864**, 979	19
Potassium tetraborate	$K_2B_4O_7 \cdot 4H_2O$	397, 456, **571**, 767, **864**, 979	19
Potassium-iron sulphate	$KFeSO_4$	162, 219, 441, 475, 591, **1007**, 1024, **1068**, 1281	19
Potassium-tin chloride	K_2SnCl_6	**168**, 239, **318**	19
Probertite	$NaCaB_5O_9 \cdot 5H_2O$	157, 240, 337, 432, 452, 487, **552**, **577**, **726**, 747, 821, 920, 1032, **3358**, 3492, 3549, **3616**	19
Propane	C_3H_8	**2890**	4
Propane2	C_3H_8	**2882**	63
Propane3	C_3H_8	**2883**, 2937, 2973	64
Propane in chibaite	$SiO_2 \cdot n(CH_4, C_2H_6, C_3H_8, i\text{-}C_4H_{10})$	**874**	35
Pseudomalachite	$Cu_5(PO_4)_2(OH)_4$	88, **110**, 131, 176, 368, 450, **480**, 537, 609, 972, 1000, **1086**	30
Pyrite	FeS_2	342, 377, **428**	10
Pyromorphite-As	$Pb_5(PO_4,AsO_4)_3Cl$	**152**, 177, 186, 206, 318, 339, 344, 377, 391, 409, 414, 768, 776, **815**, 818, 825, 917, **944**, 1014, 3325, **3378**, 3444	34
Pyrope	$Mg_3Al_2Si_3O_{12}$	56, 211, 364, 650, **871**, **902**, **928**, 1066	10
Pyrrhotite hex.	FeS	**214**, **274**, 385, 586	13
Pyrosilicic acid	$H_6Si_2O_7$	**225**, 426, **631**, 916, **3414**	19
Qilianshanite	$NaH_4(CO_3)(BO_3) \cdot 2H_2O$	187, 204, 355, 498, 539, 687, 858, **880**, 1012, 1029, **1038**, 1112	19
Quartz	$\alpha\text{-}SiO_2$	128, 206, 265, 356, 402, **464**, 485, 520, 608, 698, 807, 1066, 1161	10
Ramanite-Cs	$CsB_5O_8 \cdot 4H_2O$	94, 98, 293, **548**, 907, 3119, **3396**	24
Ramanite-Rb	$RbB_5O_8 \cdot 4H_2O$	101, **554**, 914, **3374**, 3426	24
Rb-tetrafluoroborate	$RbBF_4$	526, 531, 535, **772**	19
Rb–Zn tetrachloride	Rb_2ZnCl_4	284, **291**, 300	19
Reichenbachite	$Cu_5(PO_4)_2(OH)_4$	89, **110**, 135, 175, 188, 214, 255, 298, 365, 412, 453, **480**, 515, 540, 572, 607, 636, 951, 971, 998, **1083**, **3428**, **3380**	57
Rhodizite	$(K,Cs)Al_4Be_4(B,Be)_{12}O_{28}$	298, **432**, **473**, 648, 804, 865	19
Rhodochrosite	$MnCO_3$	185, 289, 718, **1087**, 1416	10
Rhodonite	$MnSiO_3$	257, 414, **665**, 711, 737, **998**	19
Rinneite	$K_3NaFeCl_6$	287, 524, **1001**	19

Continued

Name	Formula	Vibration Bands (cm^{-1})	Source
Roscherite	$Ca_2Mn_5Be_4(PO_4)_6(OH)_4 \cdot 6H_2O$	173, 210, 259, **400**, 559, 579, 599, 631, 771, **979**, 1066	19
Rubellite		**226**, 246, 276, 332, **379**, 410, 517, 645, **716**, 733, 753, 775, 883, 982, 1036, **1070**	19
Rubicline	$RbAlSi_3O_8$	90, 114, 142, 154, 286, 447, 478, **515**, 638, 809, 821, 838, 1013, 1040, 1114, 1185	19
Rubidium carbonate	Rb_2CO_3	221, 685, **1062**	25
Rubidium sulphate	Rb_2SO_4	449, 617, **958**, **1099**, **1130**	19
Rubidium-tin chloride	Rb_2SnCl_6	**169**, 239, **318**, 549	19
Rutile	TiO_2	139, 238, **444**, **609**, 695, 920	10
Saléeite	$Mg(UO_2)_2(PO_4)_2 \cdot 10H_2O$	193, **836**, 913, **989**	8
Sanderite	$MgSO_4 \cdot 2H_2O$	447, 492, 597, 630, **1034**, 1164, **3446**, 3539	10
Sanidine	$KAlSi_3O_8$	163, 284, 462, 475, **514**, 767, 813, 1123	10
Sassolite	H_3BO_3	187, 211, 500, **880**, 1168, 3165, **3245-3247**	21
Santite	$KB_5O_8 \cdot 4H_2O$	107, **556**, 765, 918	66
Sborgite	$NaB_5O_8 \cdot 5H_2O$	121, **529**, **856**	66
Sellaite	MgF_2	286, **399**, 816	19
Shortite	$Na_2Ca_2(CO_3)_3$	139, 154, 201, 266, 695, 710, 730, **1070**, **1090**, 1402, 1470	19
Scheelite	$CaWO_4$	55, 79, 111, 207, **329**, 395, **793**, 834, **909**, 1020, 1120, 1625, 3267, 3285, **3351**, **3379**, 3413, **3545**	19
$Si(OH)_4$ monomer	$Si(OH)_4$	**773**	9
Siderite	$(Fe,Mg)CO_3$	194, 301, 738, **1090**, 1442	10
Silica	Si	**520**	19
Sillimanite	Al_2SiO_5	142, **235**, 310, 456, 597, 708, 874, **907**, 964, 1127	10
Sinhalite	$MgAlBO_4$	**484**, 608, 744, 860, 1045	19
Sinjarite	$CaCl_2 \cdot 2H_2O$	124, **171**, 194, 297, 535	19
Sinjarite	$CaCl_2 \cdot 2H_2O$	3376, 3403, **3423**, 3462, 3471	2
Smithsonite	$ZnCO_3$	196, 303, 731, **1093**, 1408	10
Sn-chloridepentahydr. in aq.	$SnCl_4 \cdot 5H_2O$	137, 157, 198, **327**, 452	19
Sn-chloridepentahydr. in aq.2	$SnCl_4 \cdot 5H_2O$	154, 236, **331**, 3421, 3542	19
Sodalite	$Na_8(Al_6Si_6O_{24})Cl_2$	114, 167, 263, 297, 411, **465**, **988**, 1066	13
Sodium disilicate glass	$Na_2Si_2O_5$	**571**, 772, 943, **1099**	19
Sodium fluoroaluminate	$NaAlF_4$	174, 193, **289**, 304, 315, **352**, **622**, 641, **812**, 893	19
Sodium fluoroborate in water	$NaBF_4$	**773**, **769**	19
Sodium hydroxide in melt	NaOH	**3620**	55

Name	Formula	Vibration Bands (cm^{-1})	Source
Sodium monoborate	$NaBO_2{\cdot}2H_2O$	**741**, **746**, 1132, 1276	19
Sodium monoborate2	$NaBO_2{\cdot}2H_2O$	520, 553, 568, **738**, 832, 952, 1053, 1274, 1293	19
Sodium pentaborate	$NaB_5O_8{\cdot}5H_2O$	386, 464, 493, **529**, 773, **856**, 922	19
Sodium phosphate	Na_3PO_4	391, 482, 524, 544, **910**, 942, 993, 3308, 3402, **3572**	19
Sodium tetraborate	$Na_2B_4O_7{\cdot}10H_2O$	385, 461, **576**, 756, 852, 948, 1036	22
Sodium tetrafluorborate	$NaBF_4$	347, 364, 531, 547, **785**	19
Sodium wolframate	$Na_2WO_4{\cdot}2H_2O$	336, 842, **931**, 3309, **3463**	19
Spessartine	$Mn_3Al_2Si_3O_{12}$	175, 221, 321, 350, 500, 552, 630, **849**, **879**, **905**, 1029	10
Sphalerite	ZnS	218, 274, 300, 310, **349–350**, 419, 639, 669	28
Spinel	$MgAl_2O_4$	313, **408**, 666, 768	10
Spodumene	$LiAlSi_2O_6$	282, 320, 382, **464**, **492**, 355, 697, **707**, 776, 794, 808, **977**, 1019, 1062, **1074**, 1135, 1157	19
Starkeyite	$MgSO_4{\cdot}4H_2O$	116, 401, 462, 565, 616, 664, **1000**, 1086, 1156, 1186, 3331, **3427**, **3481**, 3558	10
Strontianite	$SrCO_3$	149, 183, 250, 700, 1057, **1073**, 1450	10
Strontium fluoride	SrF_2	**262**	19
Strontium chloride	$SrCl_2$	**185**	19
Sulfoborite	$Mg_3B_2(SO_4)(OH)_8(OH,F)_2$	275, 363, 460, 621, 657, 778, 901, 934, **1011**, 1081, 1134, 3242, 3374, 3561, 3573, **3592**, 3618	19
Sulfohalite	$Na_6(F,Cl)(SO_4)_2$	471, 634, **1002**, 1125	10
Sulphate ion in aq.	SO_4^{2-}	450, 620, **978–983**	9
Sulphur	S_8	**157**, 187, **219–222**, 246, 437, 462, **473**	19
Sulphur dioxide gas	SO_2	524, **1151**	4
Sylvite	KCl	213, **291**	10
Syngenite	$K_2Ca(SO_4)_2{\cdot}H_2O$	474, 494, 642, 662, **983**, **1007**, 1142, 1168, **3301**, **3378**	10
Synchysite-Ce	$Ca(Ce,La)(CO_3)_2F$	247, 354, 742, 864, **1079**, **1093**	19
Synchysite-Y	$Ca(Ce,Y)(CO_3)_2F$	**224**, 244, **290**, 374, 402, 479, 517, 603, 655, 993, 1054, **1080**, **1095**	19
Talc	$Mg_5Si_6O_{20}(OH)_4$	113, 196, 295, 335, 366, **434**, **678**, 786, 793, 1018, 1055, **3677**	10
Tavorite	$LiFe^{3+}(PO_4)(OH)$	157, 201, 443, 576, 589, 629, **951**, 1001, 1069	19
Tetrafluoroborate ion in aq.	BF_4^-	**770**	19
Thenardite	Na_2SO_4	452, 469, 621, 632, **994**, 1103, 1153	10
Thermonatrite	$Na_2CO_3{\cdot}H_2O$	**156**, 185, 230, **1062**, 1432	10
Thiosulphate ions in aq.	$S_2O_3^{2-}$	**450**, **1000**	59
Tin fluoride	SnF_2	80, 131, 199, **378**, 418	19

Continued

Name	Formula	Vibration Bands (cm^{-1})	Source
Tinclaconite	$Na_2B_4O_5(OH)_4 \cdot 3H_2O$	107, 156, 320, 344, 389, 406, 462, **577**, 465, 943, 993	19
Tolbachite	$CuCl_2$	95, 106, **118**, **140**, 149, 213, 362, 416, 477, **511**, 575, 799, 819, 845, 869, 893, 910, 928, 971	30
Topaz	$Al_2SiO_4(OH,F)_2$	155, 166, 240, **268**, 286, 318, 333, 403, 457, 856, 928	13
Topaz2	$Al_2SiO_4(OH,F)_2$	238, **266**, 285, 402, 456, 856, 928, **3637**	19
Tourmaline green		178, 204, 225, 240, 305, **340**, **377**, 413, 444, 675, **726**, 763, 819, 840, 975, 2025, 1044, **1060**, 1070	19
Tricalcium phosphate	$Ca_3(PO_4)_2$	407, **947**, **970**	27
Tridymite	SiO_2	210, **304**, **349**, **431**, 799, 1073	19
Trilithionite	$KLi_{1.5}Al_{1.5}AlSi_3O_{10}F_2$	182, 197, 215, **264**, 405, 480, 561, 633, **710**, 751, 782	8
Triphenylene	$C_{18}H_{12}$	264, 282, 408, 420, 544, 620, 699, 777, 1062, 1164, 1171, 1181, 1230, 1247, 1299, 1318, **1341**, 1396, 1605, 1616, **3085**	44
Triphylite	$LiFePO_4$	152, 251, 444, 572, 594, 628, **951**, 1002, 1068, 1082	19
Triplite	$(Mn,Fe,Mg,Ca)_2(PO_4)(F,OH)$	425, 610, **980**, 1034	10
Trisulphur ion	S_3^-	**530–540**, 1070, 1600, 2130, 2650	59
Trolleite	$Al_4(PO_4)_3(OH)_3$	269, 428, 621, 690, 790, 978, **1064**, 1100, 1120, **3428**, 3510	19
Trona	$Na_3H(CO_3)_2 \cdot 2H_2O$	140, 185, 225, **1060**, 1430	10
Tugtupite	$Na_4BeAlSi_4O_{12}Cl$	160, 206, 269, 326, 439, **462**, 510, 553, 631, **1052**, 1103, 1130	19
Tungsten trioxide	WO_3	272, 325, **718**, **807**	19
Tungstic acid	H_2WO_4	188, **262**, **685**, 804, 942	19
Tungstite	$WO_3 \cdot H_2O$	215, 236, 257, 269, 325, 379, **660–66**3, 684, **805**, **958**	19
Tychite	$Na_6Mg_2(SO_4)(CO_3)_4$	212, 246, 300, 462, 629, 718, **967**, **1108**, 1135	19
Ulexite	$NaCaB_5O_9 \cdot 8H_2O$	228, 263, 300, 330, 392, 440, 485, 531, **589**, **592**, **595**, **747**, **862**, 1013, 1316, 3559, **3570**, 3588	19
Uranyl in water	UO_2^{2+}	**870**	50
Uranyl chloride Cl^+ in water	UO_2Cl^+	**866**	51
Uranyl chloride Cl^0 in water	UO_2Cl_2	**862**	51
Uranyl chloride Cl^- in water	$UO_2Cl_3^-$	**858**	51
Uranyl chloride Cl^{2-} in water	$UO_2Cl_4^{2-}$	**854**	51

Name	Formula	Vibration Bands (cm^{-1})	Source
Uranyl chloride Cl^{3-} in water	$UO_2Cl_5^{3-}$	**850**	51
Uvarovite	$Ca_3Cr_2Si_3O_{12}$	176, 242, 272, 370, 509, 526, 590, **828**, **894**	10
Vandenbrandeite	$CuUO_2(OH)_4$	182, 249, 279, 467, **805**, 887	8
Variscite	$AlPO_4 \cdot 2H_2O$	225, 435, 612, **1028**, 1076, 1160, **3573**	19
Vaterite	$CaCO_3$	118, 301, 740, 750, 1074, **1090**, 1465	10
Väyrynenite	$MnBe(PO_4)(OH)$	96, 125, 186, 235, 332, 381, 462, 508, 599, 619, 741, 800, 986, **1007**, 1075, 1126	19
Villiaumite	NaF	217, 263, **369**, 580	19
Water liquid	H_2O	1630, **2750–3900**	10
Water vapor	H_2O	**3657–3756**	10
Water in hydrocarbons	H_2O	**3630**	48
Willemite	Zn_2SiO_4	171, 385, **869**, 905, 944	19
Witherite	$BaCO_3$	136, 152, 692, **1059**, 1420	10
Wolframite	$(Fe,Mn)WO_4$	205, 236, 260, 353, 401, **462**, 484, 508, 695, 790, 806, **881**, 1061, 1079, **1156**	19
Wollastonite	$CaSiO_3$	315, **633**, **967**, 1037	19
Wolsendorfite	$Pb_7(UO_2)_{14}O_{19}(OH)_4 \cdot 12H_2O$	348, 510, **755**	8
Wulfenite	$PbMoO_4$	**166**, **315**, 348, 742, 765, **868**, **1736**, 3290, **3350**, **3377**, 3533, 3566	19
Xenotime	YPO_4	485, 642, **998**, 1056, 1394	10
Zabuyelite	Li_2CO_3	95, 712, 746, **1091**, 1456–1459	1
Zinc fluoride	ZnF_2	69, 249, **349**, 520	19
Zinc sulphate hexahydrate	$ZnSO_4 \cdot 6H_2O$	277, 420, 503, 624, 664, **1021**, 1083, 1188	19
Zincite	ZnO	**100**, 331, **438**, 489	30
Zircon	$ZrSiO_4$	202, 212, 225, **356**, 392, 438, **972–974**, **1005–1008**	10
Zunyite	$Al_{13}Si_5O_{20}(OH,F)_{18}Cl$	168, **202**, 270, 317, 360, 383, **449**, **610**, 954, 992	8

Sources: (1) Anderson et al. (2001). (2) Baumgartner and Bakker (2010). (3) Beyssac et al. (2002a). (4) Burke (1995). (5) Buzgar and Apopei (2009). (6) Cinta Pinzaru and Onac (2009). (7) Davis and Oliver (1972). (8) Downs (2006). (9) Frezzotti et al. (2011). (10) Frezzotti et al. (2012b). (11) Frost and Dickfos (2007). (12) Frost and Williams (2004). (13) Geological Institute SAS, Banská Bystrica, Slovakia, unpubl. (14) Hurai et al. (2007). (15) Karczemska (2010). (16) Koděra et al. (2003). (17) Zaitsev et al. (2011). (18) Prinsloo et al. (1997). (19) R. Thomas, personal comm. (20) Solopova et al. (2012). (21) Thomas (2002a). (22) Thomas and Davidson (2010). (23) Thomas et al. (2009). (24) Thomas et al. (2008a). (25) Thomas et al. (2011a). (26) Thomas et al. (2011b). (27) Wopenka and Pasteris (2005). (28) Yang et al. (2009). (29) Zaitsev (2010). (30) Burgio and Clark (2001). (31) Frost et al. (2003). (32) Frost and Xi (2012). (33) Frost and Xi (2013). (34) Frost et al. (2007). (35) Momma et al. (2011). (36) Brunsgaard Hansen et al. (2002). (37) Kiefer et al. (2008). (38) Morizet et al. (2009). (39) Irmer and Graupner (2002). (40) Champagnon et al. (1997). (41) Baumgartner and Bakker (2009a). (42) Bakker (2004). (43) Rudolph et al. (2006). (44) Colangelli et al. (1992). (45) Srivastava and Singh (2007). (46) Gorelik et al. (2000). (47) Simonelli and Shultz (2001). (48) Pironon et al. (2000). (49) Zhang et al. (2007). (50) Maya and Begun (1981). (51) Ngueyn-Trung et al. (1992). (52) Mikhno et al. (2013). (53) Kanzaki et al. (2012). (54) Sharygin et al. (2013). (55) Thomas et al. (2012a). (56) Wiewiora et al. (1979). (57) Kharbish et al. (2014). (58) Borisenko et al. (2011). (59) Jacquemet et al. (2014). (60) Yang et al. (2008). (61) Kotková et al. (2014). (62) Mernagh and Bastrakov (2013). (63) Arai et al. (2012). (64) Atamas et al. (2004). (65) Dubois et al. (2010). (66) Li et al. (1995). (67) Walrafen and Douglas (2006). (68) Kingma and Hemley (1994). (69) Orange et al. (1996). (70) Dollish et al. (1974).

APPENDIX

DATA SOURCES, COMPUTER PROGRAMS, WEBSITES VII

Table VII.1 Computer Programs for Geofluid Data Reduction and Interpretation

Name	Systems	References
CLATHRATES	H_2O–CO_2–CH_4–N_2–C_2H_6–NaCl–KCl–$CaCl_2$	Bakker (1997)
DENFIND	H_2O–CO_2–NaCl	Bowers and Helgeson (1985)
FLINCOR	H_2O–NaCl–KCl–CO_2, gas mixtures	Brown (1989)
FLINCS, FLINCALC		Cavarretta and Tecce (1995)
FLUIDS	H_2O–NaCl–KCl–$CaCl_2$–$MgCl_2$–CO_2–CO–CH_4–N_2–C_2H_6–H_2S–NH_3–H_2–O_2	Bakker (2003, 2009, 2012), Bakker and Brown (2003)
GFluid	H_2O, CO_2, CH_4, H_2, O_2, CO, C_2H_6	Zhang and Duan (2010)
No name Basic	H_2O–NaCl	Hurai (1989)
No name Fortran	H_2O–CO_2–NaCl	Nicholls and Crawford (1985)
HokieFlincs_H_2O-NaCl	H_2O–NaCl	Steele-MacInnis et al. (2012)
MACFLINCOR	H_2O–NaCl–CO_2, H_2O–NaCl–KCl, gaseous mixtures, and subsystems	Brown and Hagemann (1995) Bakker and Brown (2003)
PIT (FIT-Oil)	Oil and condensate	Thiéry et al. (2000, 2002)
SolGeo	Cation-exchange thermometry	Verma et al. (2008)
VTFLINC	Oil and condensate	Aplin et al. (1999)
SALTY	H_2O–NaCl–KCl	Bodnar et al. (1989b)

Other stand-alone programs and web calculators useful for the geofluid research:
http://gcmodel.kl-edi.ac.cn/archives/programs.htm—Zhenhao Duan's research group
http://models.kl-edi.ac.cn/fluidinc.htm—specific programs of Zhenhao Duan's research group tailored to fluid inclusion research
http://fluids.unileoben.ac.at/Home.html—Ronald Bakker's research group
http://www2.ggl.ulaval.ca/cgi-bin/alphadelta/alphadelta.cgi—on-line calculation of stable isotope fractionation factors (Beaudoin and Therrien, 2004, 2009)
http://webbook.nist.gov/chemistry/fluid/—on-line calculation of state properties of pure fluid species (National Institute of Standards and Technology)
http://wwwobs.univ-bpclermont.fr/lmv/perso/Thiery_Regis/fitoil/index.html—FIT-Oil program for calculating properties of oil inclusions

Geofluids. http://dx.doi.org/10.1016/B978-0-12-803241-1.09971-8

References

Abers, G.A., 2000. Hydrated subducted crust at 100–250 km depth. Earth Planet. Sci. Lett. 176, 323–330.

Affolter, S., Fleitmann, D., Leuenberger, M., 2014. New on-line method for water isotope analysis of speleothem fluid inclusions using laser absorption spectroscopy (WS-CRDS). Clim. Past Discuss. 10, 429–467.

Akopov, E.K., 1963. Solubility polytherms of the ternary system LiCl–NaCl–H_2O. Zh. Prikl. Khim. 36, 1916–1919 (in Russian).

Allamagny, P., 1976. Gas Encyclopaedia. Elsevier, Amsterdam, 1164 p.

Allan, M.M., Yardley, B.W.D., Forbes, L.J., Shmulovich, K.I., Banks, D.A., Shepherd, T.J., 2005. Validation of LA-ICPMS fluid inclusion analysis with synthetic fluid inclusions. Am. Mineral. 90, 1767–1775.

Allégre, C.J., 2008. Isotope Geology. Cambridge University Press, Cambridge UK, 512 p.

Anderko, A., Pitzer, K.S., 1993. Equation-of-state representation of phase equilibria and volumetric properties of the system NaCl–H_2O above 573 K. Geochim. Cosmochim. Acta 57, 1657–1680.

Andersen, T., O'Reilly, S.Y., Griffin, W.L., 1984. The trapped fluid phase in upper mantle xernoliths from Victoria, Australia: implications for mantle metasomatism. Contrib. Mineral. Petrol. 88, 72–85.

Andersen, T., Burke, E.A.J., Austrheim, H., 1989. Nitrogen-bearing, aqueous fluid inclusions in some eclogites from the Western Gneiss Region of the Norwegian Caledonides. Contrib. Mineral. Petrol. 103, 153–165.

Andersen, T., Burke, E.A.J., Neumann, E.R., 1995. Nitrogen-rich fluid in the upper mantle: fluid inclusions in spinel dunite from Lanzarote, Canary Islands. Contrib. Mineral. Petrol. 120, 20–28.

Anderson, A.J., Clark, A.H., Gray, S., 2001. The occurrence and origin of zabuyelite (Li_2CO_3) in spodumene-hosted fluid inclusions: Implications for the internal evolution of rare-element granitic pegmatites. Can. Mineral. 39, 1513–1527.

Anderson, A.T., 1974. Evidence for a picritic, volatile-rich magma beneath Mt. Shasta, California. J. Petrol. 15, 243–267.

Anderson, A.T., Newman, S., Williams, S.N., Druitt, T.H., Skirius, C., Stolper, E., 1989. H_2O, CO_2, Cl and gas in Plinian and ash-flow Bishop rhyolite. Geology 17, 221–225.

Anderson, G.M., 2005. Thermodynamics of Natural Systems, second ed. Cambridge University Press, Cambridge, 648 p.

Angus, S., Armstrong, B., de Reuck, M.K., Altunin, V.V., Gadetski, O.G., Chapela, G.A., Rowlinson, J.S., 1976. International Thermodynamic Tables of the Fluid State—Carbon Dioxide. Pergamon, Oxford.

Angus, S., Armstrong, B., de Reuck, M.K., 1979. International Thermodynamic Tables of the Fluid State. Nitrogen, vol. 6. Pergamon, Oxford.

Anovitz, L.M., Essene, E.J., 1987. Phase equilibria in the system $CaCO_3$–$MgCO_3$–$FeCO_3$. J. Petrol. 28, 389–414.

Aplin, A.C., McLeod, G., Larter, S., Pedersen, S.R., Sorensen, H., Booth, T., 1999. Combined use of confocal laser microscopy and PVT simulation for estimating the composition and physical properties of petroleum in fluid inclusions. Mar. Petrol. Geol. 16, 97–110.

Araguás-Araguás, L., Froehlich, K., Rozanski, K., 2000. Deuterium and oxygen-18 isotope composition of precipitation and atmospheric moisture. Hydrol. Process. 14, 1341–1355.

Arai, S., Ishimaru, S., Mizukami, T., 2012. Methane and propane microinclusions in olivine in titanoclinohumite-bearing dunites from the Sanbagawa high-P metamorphic belt, Japan: Hydrocarbon activity in a subduction zone and Ti mobility. Earth Planet. Sci. Lett. 353–354, 1–11.

Arai, Y., Kaminishi, G., Saito, S., 1971. The experimental determination of the PVTX relations for the carbon dioxide–nitrogen and the carbon dioxide–methane systems. J. Chem. Eng. Jpn. 4, 113–122.

Arienzo, M.M., Swart, P.K., Vonhof, H.B., 2013. Measurement of $\delta^{18}O$ and $\delta^{2}H$ values of fluid inclusion water in speleothems using cavity ring-down spectroscopy compared with isotope ratio mass spectrometry. Rapid Commun. Mass Spectrom. 27, 2616–2624.

Arnórsson, S., 1983. Chemical equilibria in Icelandic geothermal systems—Implications for chemical geothermometry investigations. Geothermics 12, 119–128.

Atamas, N.A., Yaremko, A.M., Seeger, T., Leipertz, A., Bienko, A., Latajka, Z., Ratajczak, H., Barnes, A.J., 2004. A study of the Raman spectra of alkanes in the Fermi-resonance region. J. Mol. Struct. 708, 189–195.

Audétat, A., Keppler, H., 2004. Viscosity of fluids in subduction zones. Science 303, 513–516.

Audétat, A., Lowenstern, J.B., 2014. Melt inclusions, second ed. In: Holland, H.D., Turekian, K.K. (Eds.), Treatise on Geochemistry, vol. 13. Elsevier, Oxford, pp. 143–173.

Bailly, L., Bouchot, V., Bény, C., Milési, J.-P., 2000. Fluid inclusion study of stibnite using infrared microscopy: an example from the Brouzils antimony deposit (Vendée, Armorican Massif, France). Econ. Geol. 95, 221–226.

Bailly, L., Grancea, L., Kouzmanov, K., 2002. Infrared microthermometry and chemistry of wolframite from the Baia Sprie epithermal deposit, Romania. Econ. Geol. 97, 415–423.

Baker, T., Lang, J.R., 2003. Reconciling fluid inclusion types, fluid processes, and fluid sources in skarns: an example from the Bismark deposit, Mexico. Miner. Deposita 38, 474–495.

Bakker, R.J., 1997. CLATHRATES: computer programs to calculate fluid inclusion V–X properties using clathrate melting temperatures. Comput. Geosci. 23, 1–18.

Bakker, R.J., 1999. Adaptation of the Bowers and Helgeson (1983) equation of state to the H_2O–CO_2–CH_4–N_2–NaCl system. Chem. Geol. 154, 225–236.

Bakker, R.J., 2003. Package FLUIDS 1. Computer programs for analysis of fluid inclusion data and for modelling bulk fluid properties. Chem. Geol. 194, 3–23.

Bakker, R.J., 2004. Raman spectra of fluid and crystal mixtures in the systems H_2O, H_2O–NaCl and H_2O–$MgCl_2$ at low temperatures: applications to fluid–inclusion research. Can. Mineral. 42, 1283–1314.

Bakker, R.J., 2009. Package FLUIDS. Part 3: correlations between equations of state, thermodynamics and fluid inclusions. Geofluids 9, 63–74.

Bakker, R.J., 2011. Accuracy, precision and uncertainty of peak position estimations of CO_2 and CH_4 in Raman spectra. In: European Current Research on Fluid Inclusions (ECROFI–XXI), Montanuniversität Leoben, Austria, 9–11 August, 2011, Abstracts, p. 32.

Bakker, R.J., 2012. Package FLUIDS. Part 4: thermodynamic modelling and purely empirical equations for H_2O–NaCl–KCl solutions. Mineral. Petrol. 105, 1–29.

Bakker, R.J., Brown, P.E., 2003. Computer modelling in fluid inclusion research. In: Samson, I., Anderson, A., Marshall, D. (Eds.), Fluid Inclusions: Analysis and Interpretation. In: Min. Assoc. Can. Short Course Ser., vol. 32, pp. 175–212.

Bakker, R.J., Diamond, L.W., 2000. Determination of the composition and molar volume of H_2O–CO_2 fluid inclusions by microthermometry. Geochim. Cosmochim. Acta 64, 1753–1764.

Bakker, R.J., Diamond, L.W., 2006. Estimation of volume fractions of liquid and vapour phases in fluid inclusions, and definition of inclusion shapes. Am. Mineral. 91, 635–657.

Bakker, R.J., Elburg, M.A., 2006. A magmatic–hydrothermal transition in Arkaroola (northern Flinders Ranges, South Australia): from diopside–titanite pegmatites to hematite–quartz growth. Contrib. Mineral. Petrol. 152, 541–569.

Bakker, R.J., Jansen, J.B.H., 1991. Experimental post-entrapment water loss from synthetic CO_2–H_2O inclusions in natural quartz. Geochim. Cosmochim. Acta 55, 2215–2230.

Bakker, R.J., Jansen, J.B.H., 1994. A mechanism for preferential H_2O leakage from fluid inclusions in quartz, based on TEM observations. Contrib. Mineral. Petrol. 116, 7–20.

Banks, D.A., Giuliani, G., Yardley, B.W.D., Cheilletz, A., 2000. Emerald mineralization in Colombia: fluid chemistry and the role of brine mixing. Miner. Deposita 35, 699–713.

Banks, D.A., Boyce, A.J., Samson, I.M., 2002. Constraints on the origins of fluids forming Irish Zn–Pb–Ba deposits: evidence from the composition of fluid inclusions. Econ. Geol. 97, 471–480.

Banks, D.A., Yardley, B.W.D., 1992. Crush-leach analysis of fluid inclusions in small natural and synthetic samples. Geochim. Cosmochim. Acta 56, 245–248.

Barker, C.E., 1983. Influence of time on metamorphism of sedimentary organic matter in liquid-dominated geothermal systems, western North America. Geology 11, 384–388.

Barker, C.E., Goldstein, R.H., 1990. Fluid-inclusion technique for determining maximum temperature in calcite and its comparison to the vitrinite reflectance geothermometer. Geology 18, 1003–1006.

Barker, C.E., Pawlewicz, M.J., 1986. The correlation of vitrinite reflectance with maximum temperature in humic organic matter. In: Buntebarth, G., Stegena, L. (Eds.), Paleogeothermics. In: Lecture Notes in Earth Sciences, vol. 5. Springer, Berlin, pp. 79–93.

Bartoli, O., Cesare, B., Poli, S., Acosta-Vigil, A., Esposito, R., Turina, A., Bodnar, R.J., Angel, R.J., Hunter, J., 2013. Nanogranite inclusions in migmatitic garnet: behavior during piston-cylinder remelting experiments. Geofluids 13, 405–420.

Barton, P.B., Jr., 1970. Sulfide petrology. Min. Soc. Am. Spec. Paper 3, 187–798.

Barton, P.B., Jr., Skinner, B.J., 1979. Sulfide mineral stabilities. In: Barnes, H.L. (Ed.), Geochemistry of Hydrothermal Ore Deposits, second ed. John Wiley and Sons, New York, pp. 278–403.

Bassham, J.A., Benson, A.A., Calvin, M., 1950. The path of carbon in photosynthesis. J. Biol. Chem. 185, 781–787.

Baumgartner, L.P., Valley, J.W., 2001. Stable isotope transport and contact metamorphic fluid flow. In: Valley, J.W., Cole, D.R. (Eds.), Stable Isotope Geochemistry. In: Min. Soc. Am. Rev. Mineral. Geochem., vol. 43, pp. 415–467.

Baumgartner, M., Bakker, R.J., 2009a. $CaCl_2$-hydrate nucleation in synthetic fluid inclusions. Chem. Geol. 265, 335–344.

Baumgartner, M., Bakker, R.J., 2009b. Raman spectroscopy of pure H_2O and NaCl–H_2O containing synthetic fluid inclusions in quartz—a study of polarization effects. Mineral. Petrol. 95, 1–15.

Baumgartner, M., Bakker, R.J., 2010. Raman spectra of ice and salt hydrates in synthetic fluid inclusions. Chem. Geol. 275, 58–66.

Bayram, A.F., 2001. Application of an artificial neural network model to a Na–K geothermometer. J. Volcanol. Geotherm. Res. 12, 75–81.

Beane, R.E., Bodnar, R.J., 1995. Hydrothermal fluids and hydrothermal alteration in porphyry copper deposits. In: Pierce, F.W., Bohm, J.G. (Eds.), Porphyry Copper Deposits of the American Cordillera. In: Arizona Geol. Soc. Digest, vol. 20, pp. 83–93.

Beaudoin, G., Therrien, P., 2004. The web stable isotope fractionation calculator. In: De Groot, P.A. (Ed.), Handbook of Stable Isotope Analytical Techniques, vol. I. Elsevier, Amsterdam, pp. 1045–1047.

Beaudoin, G., Therrien, P., 2009. The updated web stable isotope fractionation calculator. In: De Groot, P.A. (Ed.), Handbook of Stable Isotope Analytical Techniques, vol. II. Elsevier, Amsterdam, pp. 1120–1122.

Becker, R.H., Clayton, R.N., 1976. Oxygen isotope study of a Precambrian banded iron-formation, Hamersley Range, Western Australia. Geochim. Cosmochim. Acta 40, 1153–1165.

Befus, K.S., Gardner, J.E., Zinke, R.W., 2012. Analysing water contents in unexposed glass inclusions in quartz crystals. Am. Mineral. 97, 1898–1904.

Behr, H.-J., Fyfe, W.S., Hoefs, J., Johannes, W., Touret, J.L.R., 1989. Die geologische Aktivität von Krustenfluide. Nds. Akad. Geowiss. Veröffentl. 1, 4–75.

Behrens, H., Roux, J., Neuville, D.R., Siemann, M., 2006. Quantification of dissolved H_2O in silicate glasses using confocal microRaman spectroscopy. Chem. Geol. 229, 96–112.

Belonoshko, A.B., Saxena, S., 1992. A unified equation of state for fluids of C–H–O–N–S–Ar composition and their mixtures up to very high temperatures and pressures. Geochim. Cosmochim. Acta 56, 3611–3626.

Belonoshko, A.B., Shi, P., Saxena, S., 1992. SUPERFLUID: A FORTRAN-77 program for calculation of Gibbs free energy and volume of C–H–O–S–N–Ar mixtures. Comput. Geosci. 18, 1267–1269.

Bendtsen, J., 2001. High-resolution Fourier transform Raman spectra of the fundamental bands of $^{14}N^{15}N$ and $^{15}N_2$. J. Raman Spectrosc. 32, 989–995.

Benešová, Z., Ďurišová, J., 1989. Gas-fluid Inclusions and Their Significance for Geology. Ústřední Ústav Geol, Praha, 89 p. (in Czech).

Beňka, J., Caňo, F., 1992. Mineralogy, paragenesis and geochemistry of stibnite veins in the region of Betliar – Čučma – Volovec. Záp. Karp. Sér. Mineral. Petrogr. Geochém. Metalog. 15, 61–91 (in Slovak).

Bergman, S.C., Dubessy, J., 1984. CO_2-CO fluid inclusions in a composite peridotite xenolith: implications for upper mantle oxygen fugacity. Contrib. Mineral. Petrol. 85, 1–13.

Berman, R.G., 1988. Internally consistent thermodynamic data for minerals in the system $Na_2O-K_2O-CaO-MgO-FeO-Fe_2O_3-Al_2O_3-SiO_2-TiO_2-H_2O-CO_2$. J. Petrol. 29, 445–522.

Berman, R.G., Brown, T.H., 1985. Heat capacity of minerals in the system $Na_2O–K_2O–CaO–MgO–FeO–Fe_2O_3–Al_2O_3–SiO_2–TiO_2–CO_2$: representation, estimation, and high temperature extrapolation. Contrib. Mineral. Petrol. 89, 168–183.

Berman, R.G., Brown, T.H., Perkins, E.H., 1987. GEO-CALC: software for calculation and display of P–T–X phase diagrams. Am. Mineral. 72, 861–862.

Bernard, J.H., 1961. Mineralogie und Geochemie der Siderit-Schwerspatgänge mit Sulphiden im Gebiet von Rudňany (Tschechoslowakei). Geol. Práce – Zošit 58, 5–222.

Bertrand, P., Pittion, J.L., Bernaud, C., 1986. Fluorescence of sedimentary organic matter in relation to its chemical composition. Org. Geochem. 10, 641–647.

Beyssac, O., Lazzeri, M., 2012. Application of Raman spectroscopy to the study of graphitic carbon in the Earth sciences. In: Dubessy, J., Caumon, M.-C., Rull, F. (Eds.), Raman Spectroscopy Applied to Earth Sciences and Cultural Heritage. In: EMU Notes in Mineralogy, vol. 12, pp. 415–454.

Beyssac, O., Goffé, B., Chopin, C., Rouzaud, J.N., 2002a. Raman spectra of carbonaceous material in metasediments: a new geothermometer. J. Metamorph. Geol. 20, 859–871.

Beyssac, O., Rouzaud, J.N., Goffe, B., Brunet, F., Chopin, C., 2002b. Graphitization in a high-pressure, low-temperature metamorphic gradient: a Raman microspectroscopy and HRTEM study. Contrib. Mineral. Petrol. 143, 19–31.

Bhattacharya, A., 2012. New model to estimate water in the Earth's mantle. EOS Trans. Am. Geophys. Union 93, 316.

Bigeleisen, J., Mayer, M.G., 1947. Calculation of equilibrium constants for isotopic exchange reactions. J. Chem. Phys. 15, 261–267.

Birch, F., 1966. Compressibility: elastic constants. In: Clark, S.P. Jr., (Ed.), Handbook of Physical Constants. In: Geol. Soc. Am. Memoir, vol. 97, pp. 97–173.

Bischoff, J.L., 1991. Densities of liquids and vapors in boiling $NaCl–H_2O$ solutions: A PVTX summary from 300° to 500 °C. Am. J. Sci. 291, 309–338.

Bischoff, J.L., Pitzer, K.S., 1985. Phase relations and adiabats in boiling seafloor geothermal systems. Earth Planet. Sci. Lett. 75, 327–338.

Bischoff, J.L., Rosenbauer, R.J., Fournier, R.O., 1996. The generation of HCl in the system $CaCl_2-H_2O$: Vapor–liquid relations from 380–500 °C. Geochim. Cosmochim. Acta 60, 7–16.

Blamey, N.J.F., 2012. Composition and evolution of crustal, geothermal and hydrothermal fluids interpreted using quantitative fluid inclusion gas analysis. J. Geochem. Explor. 116–117, 17–27.

Blencoe, J.G., Naney, M.T., Anowitz, L.M., 2001. The $CO_2–H_2O$ system: III. A new experimental method for determining liquid–vapor equilibria at high subcritical temperatures. Am. Mineral. 86, 1100–1111.

Bodnar, R.J., 1983. A method of calculating fluid inclusion volumes based on vapor bubble diameters and P–V–T–X properties of inclusion fluids. Econ. Geol. 78, 535–542.

Bodnar, R.J., 1990. Petroleum migration in the Miocene Monterey Formation, CA, USA: constraints from fluid–inclusion studies. Mineral. Mag. 54, 295–304.

Bodnar, R.J., 1992. Can we recognize magmatic fluid inclusions in fossil systems based on room-temperature phase relations and microthermometric behavior? Rept. Geol. Surv. Jpn. 279, 26–30.

Bodnar, R.J., 1993. Revised equation and table for determining the freezing point depression of H_2O–NaCl solutions. Geochim. Cosmochim. Acta 57, 683–684.

Bodnar, R.J., 1994. Synthetic fluid inclusions: XII. The system H_2O–NaCl. Experimental determination of the halite liquidus and isochores for a 40 wt.% NaCl solution. Geochim. Cosmochim. Acta 58, 1053–1063.

Bodnar, R.J., 1995. Fluid inclusion evidence for a magmatic source for metals in porphyry copper deposits. In: Thompson, J.F.H. (Ed.), Magmas, Fluids and Ore Deposits. In: Min. Assoc. Can. Short Course, vol. 23, pp. 139–152.

Bodnar, R.J., 2003a. Introduction to fluid inclusions. In: Samson, I., Anderson, A., Marshall, D. (Eds.), Fluid Inclusions: Analysis and Interpretation. In: Min. Assoc. Can. Short Course, vol. 32, pp. 1–8.

Bodnar, R.J., 2003b. Reequilibration of fluid inclusions. In: Samson, I., Anderson, A., Marshall, D. (Eds.), Fluid Inclusions: Analysis and Interpretation. In: Min. Assoc. Can. Short Course, vol. 32, pp. 213–230.

Bodnar, R.J., 2003c. Introduction to aqueous-electrolyte fluid inclusions. In: Samson, I., Anderson, A., Marshall, D. (Eds.), Fluid Inclusions: Analysis and Interpretation. In: Min. Assoc. Can. Short Course, vol. 32, pp. 81–100.

Bodnar, R.J., Beane, R.E., 1980. Fluid inclusion study of the porphyry copper prospect at Red Mountain, Arizona. Econ. Geol. 75, 876–893.

Bodnar, R.J., Bethke, P.M., 1984. Systematics of stretching of fluid inclusions. I. Fluorite and sphalerite at 1 atmosphere confining pressure. Econ. Geol. 79, 141–161.

Bodnar, R.J., Vityk, M., 1995. Interpretation of microthermometric data for H_2O–NaCl fluid inclusions. In: De Vivo, B., Frezzotti, M.L. (Eds.), Fluid Inclusions in Minerals: Methods and Applications. Short Course of the Working Group IMA Inclusions in Minerals. Virginia Tech., USA, pp. 117–130.

Bodnar, R.J., Burnham, C.W., Sterner, S.M., 1985a. Synthetic fluid inclusions in natural quartz. III. Determination of phase equilibrium properties in the system H_2O–NaCl to 1000 °C and 1500 bars. Geochim. Cosmochim. Acta 49, 1861–1873.

Bodnar, R.J., Reynolds, T.J., Kuehn, C.A., 1985b. Fluid inclusion systematics in epithermal systems. In: Berger, B. R., Bethke, P.M. (Eds.), Geology and Geochemistry of Epithermal Systems. In: Rev. Econ. Geol, vol. 2, pp. 73–97.

Bodnar, R.J., Binns, P.R., Hall, D.L., 1989a. Synthetic fluid inclusions – VI. Quantitative evaluation of the decrepitation behaviour of fluid inclusions in quartz at one atmosphere confining pressure. J. Metamorph. Geol. 7, 229–242.

Bodnar, R.J., Sterner, S.M., Hall, D.L., 1989b. SALTY: a FORTRAN program to calculate compositions of fluid inclusions in the system NaCl–KCl–H_2O. Comput. Geosci. 15, 19–41.

Bodnar, R.J., Mernagh, T.P., Samson, I.M., Manning, C.E., 2013. Introduction to thematic issue on fluid and melt inclusions. Geofluids 13, 395–397.

Boettcher, A.L., O'Neil, J.R., 1980. Stable isotope, chemical and petrographic studies of high-pressure amphiboles and micas: evidence for metasomatism in the mantle source regions of alkali basalt and kimberlites. Am. J. Sci. 280–A, 549–621.

Bohlen, S.R., Lindsley, D.H., 1987. Thermometry and barometry of igneous and metamorphic rocks. Ann. Rev. Earth Planet. Sci. 15, 397–420.

Borisenko, A.S., 1977. Study of salt composition of fluid inclusions in minerals using cryometric technique. Geol. Geofiz. 8, 16–27 (in Russian).

Borisenko, A.S., 1982. Analysis of salt content of solutions in gas-liquid inclusions in minerals using cryometric methods. In: Laverov, N.P. (Ed.), Ispolzovaniye metodov termobarogeokhimiyi pri poiskach i izucheniyi rudnych mestorozhdeniy. Nedra, Moskva, pp. 37–46 (in Russian).

Borisenko, A.S., Borovikov, A.A., Vasyukova, E.A., Pavlova, G.G., Ragozin, A.L., Prokop'ev, I.R., Vladykin, N.V., 2011. Oxidized magmatogene fluids: metal-bearing capacity and role in ore formation. Russ. Geol. Geophys. 52, 144–164.

Borisova, A.Y., Thomas, R., Salvi, S., Candaudap, F., Lanzanova, A., Chmeleff, J., 2012. Tin and associated metal and metalloid chemistry by femtosecond LA-ICP-QMS microanalysis of pegmatite-leucogranite melt and fluid inclusions: new evidence for melt–melt–fluid immiscibility. Mineral. Mag. 76, 91–113.

Bottinga, Y., 1968. Calculation of fractionation factors for carbon and oxygen isotopic exchange in the system calcite–carbon dioxide–water. J. Phys. Chem. 72, 800–808.

Bottinga, Y., 1969. Calculated fractionation factors for carbon and hydrogen isotope exchange in the system calcite–carbon dioxide–graphite–methane–hydrogen–water vapor. Geochim. Cosmochim. Acta 33, 49–64.

Bottinga, Y., Javoy, M., 1973. Comments on oxygen isotope geothermometry. Earth Planet. Sci. Lett. 20, 250–265.

Bottinga, Y., Javoy, M., 1975. Oxygen isotope partitioning among the minerals in igneous and metamorphic rocks. Rev. Geophys. Space Phys. 13, 401–418.

Bottinga, Y., Weill, D.F., 1972. The viscosity of magmatic silicate liquids: a model for calculation. Am. J. Sci. 272, 438–475.

Bottrell, S.H., Carr, L.P., Dubessy, J., 1988a. A nitrogen-rich metamorphic fluid and coexisting minerals in slates from North Wales. Mineral. Mag. 52, 451–457.

Bottrell, S.H., Yardley, B., Buckley, F., 1988b. A modified crush-leach method for the analysis of fluid inclusion electrolytes. Bull. Minéral. 111, 279–290.

Boullier, A.-M., 1999. Fluid inclusions: tectonic indicators. J. Struct. Geol. 21, 1229–1235.

Boullier, A.-M., Robert, F., 1992. Paleoseismic events recorded in Archean gold–quartz vein networks, Val d'Or, Abitibi, Québec, Canada. J. Struct. Geol. 14, 161–179.

Bourdet, J., Burruss, R.C., Bodnar, R.J., Eadington, P.J., 2011. Assessment of UV-Raman for analysis of petroleum inclusions. In: ECROFI–XXI (Montanuniversität Leoben, Austria, 9–11 August 2011), Abstract Volume, pp. 50–51.

Bourdet, J., Pironon, J., 2008. Strain response and re-equilibration of CH_4-rich synthetic aqueous fluid inclusions in calcite during pressure drops. Geochim. Cosmochim. Acta 72, 2946–2959.

Bourdet, J., Pironon, J., Levresse, G., Tritlla, J., 2008. Petroleum type determination through homogenization temperature and vapour volume fraction measurements in fluid inclusions. Geofluids 8, 46–59.

Bowers, T.S., Helgeson, H.C., 1983a. Calculation of the thermodynamic and geochemical consequences of nonideal mixing in the system H_2O-CO_2-NaCl on phase relations in geological systems: equation of state for H_2O-CO_2-NaCl fluids at high pressures and temperatures. Geochim. Cosmochim. Acta 47, 1247–1275.

Bowers, T.S., Helgeson, H.C., 1983b. Calculation of the thermodynamic and geochemical consequences of nonideal mixing in the system H_2O-CO_2-NaCl on phase relations in geological systems: metamorphic equilibria at high pressures and temperatures. Am. Mineral. 68, 1059–1075.

Bowers, T.S., Helgeson, H.C., 1985. Fortran programs for generating fluid inclusion isochores and fugacity coefficients for the system H_2O–CO_2–NaCl at high pressures and temperatures. Comput. Geosci. 11, 203–213.

Boyle, R., 1672. Essay About the Origin and Virtues of Gems. William Godbid, London, 185 p.

Boyle, R., 1673. Specimen de Gemmarum origine et virtutibus. Spec. Gew. 2, 2/3.

Bozzo, A.T., Chen, J.R., Barduhn, A.J., 1973. The properties of hydrates of chlorine and carbon dioxide. In: Delyannis, A., Delyannis, E. (Eds.), 4th International Symposium on Fresh Water from the Sea, vol. 3, pp. 437–451.

Bray, C.J., Spooner, E.T.C., Thomas, A.V., 1991. Fluid inclusion volatile analysis by heated crushing, on-line gas chromatography: applications to Archean fluids. J. Geochem. Explor. 42, 167–193.

Brewster, D., 1823. On the existence of two new fluids in the cavities of minerals, which are immiscible, and which possess remarkable properties. Edinburgh Phil. J. 9, 94–107.

Brewster, D., 1845. On the existence of crystals with different primitive forms and physical properties in the cavities of minerals; with additional observations on the new fluids in which they occur. Royal Soc. Edinburgh Trans. 16, 11–22.

Brown, P.E., 1989. FLINCOR: a new microcomputer program for the reduction and investigation of fluid inclusion data. Am. Mineral. 74, 1390–1393.

Brown, P.E., Hagemann, S.G., 1995. Macflincor and its application to fluids in Archean lode-gold deposits. Geochim. Cosmochim. Acta 59, 3943–3952.

Brown, P.E., Lamb, W.M., 1989. P−V−T−X properties of fluids in the system H_2O-CO_2-NaCl: A new graphical presentations and implications for fluid inclusion studies. Geochim. Cosmochim. Acta 53, 1209–1221.

Brunsgaard Hansen S., Berg, R.W., Stenby, E.H., 2002. How to determine the pressure of a methane-containing gas mixture by means of two weak Raman bands, ν_3 and $2\nu_2$. J. Raman Spectrosc. 33, 160–164.

Buchshteyn, V.M., Val'yashko, M.G., Pel'sh, A.D., 1953. Handbook of Solubility of Salt Systems. I+II. Goschimizdat, Moskva/Leningrad, 1270 p. (in Russian).

Buddington, A.F., Lindsley, D.H., 1964. Iron-titanium oxide minerals and synthetic equivalents. J. Petrol. 5, 310–357.

Burgio, L., Clark, R.J.H., 2001. Library of FT-Raman spectra of pigments, minerals, pigment media and varnishes, and supplement to existing library of Raman spectra of pigments with visible excitation. Spectrochim. Acta A 57, 1491–1521.

Burke, E.A.J., 1995. Raman microspectrometry of fluid inclusions: The daily practice. In: de Vivo, B., Frezzotti, M. L. (Eds.), Fluid Inclusions in Minerals: Methods and Applications, Short Course of the Working Group IMA. Virginia Polytechnic Institute and State University, USA, pp. 25–44.

Burke, E.A.J., 2001. Raman microspectrometry of fluid inclusions. Lithos 55, 139–158.

Burlinson, K., 1991. Decrepitation in gold exploration—a case history from the Cotan Prospect, NT. J. Geochem. Explor. 42, 143–156.

Burnham, A.K., Sweeney, J.J., 1989. A chemical kinetic model of vitrinite maturation and reflectance. Geochim. Cosmochim. Acta 53, 2649–2657.

Burnham, C.W., 1975. Water and magmas: a mixing model. Geochim. Cosmochim. Acta 39, 1077–1084.

Burnham, C.W., 1979. Magmas and hydrothermal fluids. In: Barnes, H.L. (Ed.), Geochemistry of Hydrothermal Ore Deposits, second ed. John Wiley and Sons, New York, pp. 71–136.

Burnham, C.W., Ohmoto, H., 1980. Late-stage processes in felsic magmas. Min. Geol. Spec. Issue 8, 1–11.

Burruss, R.C., 1981a. Hydrocarbon fluid inclusions in studies of sedimentary diagenesis. In: Hollister, L.S., Crawford, M.L. (Eds.), Fluid Inclusions—Applications to Petrology. In: Min. Assoc. Can. Short Course Notes, vol. 6, pp. 128–156.

Burruss, R.C., 1981b. Analysis of fluid inclusions: phase equilibria at constant volume. Am. J. Sci. 281, 1104–1126.

Burruss, R.C., 1981c. Analysis of phase equilibria in C-O-H-S fluid inclusions. In: Hollister, L.S., Crawford, M.L. (Eds.), Fluid Inclusions—Applications to Petrology. In: Min. Assoc. Can. Short Course Notes, vol. 6, pp. 39–74.

Burruss, R.C., 1991. Practical aspects of fluorescence microscopy of petroleum fluid inclusions. In: Barker, C.E., Kopp, O.C. (Eds.), Luminescence Microscopy and Spectroscopy: Qualitative and Quantitative Applications. In: SEPM Short Course, vol. 25, pp. 1–7.

Burruss, R.C., 1992. Phase behavior in petroleum−water (brine) systems applied to fluid inclusion studies. In: McKibben, M.A., Montanez, I.P., Hall, D.L. (Eds.), PACROFI IV, Pan-American Conference on Research on Fluid Inclusions, Program and Abstracts. Dept. of Earth Sciences, University of California, Riverside, pp. 116–118.

Burruss, R.C., 2003. Raman spectroscopy of fluid inclusions. In: Samson, I., Anderson, A., Marshall, D. (Eds.), Fluid Inclusions: Analysis and Interpretation. In: Min. Assoc. Can. Short Course, vol. 32, pp. 279–289.

Buzgar, N., Apopei, A.I., 2009. The Raman study of certain carbonates. Analele Stiintifice ale Universitati Al. I. Cuza, Iasi Geologie 55, 97–112.

Campbell, A.R., Panter, K.S., 1990. Comparison of fluid inclusions in coexisting (cogenetic?) wolframite, cassiterite and quartz from St. Michels Mount and Cligga Head, Cornwall, England. Geochim. Cosmochim. Acta 54, 673–681.

Campbell, A.R., Robinson-Cook, S., 1987. Infrared fluid inclusion microthermometry on coexisting wolframite and quartz. Econ. Geol. 82, 1640–1645.

Campbell, A., Robinson-Cook, S., Amindyas, C., 1988. Observation of fluid inclusions in wolframite from Panasqueira, Portugal. Bull. Minéral. 111, 251–256.

Capuano, R.M., 1992. The temperature dependence of hydrogen isotope fractionation between clay minerals and water: Evidence from a geopressured system. Geochim. Cosmochim. Acta 56, 2547–2554.

Carmichael, I.S.E., Nicholls, J., 1967. Iron–titanium oxides and oxygen fugacities in volcanic rocks. J. Geophys. Res. 72, 4665–4687.

Carothers, W.W., Adami, L.H., Rosenbauer, R.J., 1988. Experimental oxygen isotope fractionation between siderite–water and phosphoric acid-liberated CO_2–siderite. Geochim. Cosmochim. Acta 52, 2445–2450.

Castiglioni, C., Negri, F., Rigolio, M., Zerbi, G., 2001. Raman activation in disordered graphites of the A(1)(′) symmetry forbidden k not equal 0 phonon: The origin of the D line. J. Chem. Phys. 115, 3769–3778.

Caumon, M.-C., Dubessy, J., Robert, P., Tarantola, A., 2013. Fused-silica capillary capsules (FSCCs) as reference synthetic aqueous fluid inclusions to determine chlorinity by Raman spectroscopy. Eur. J. Mineral. 25, 755–763.

Caumon, M.-C., Robert, P., Laverret, E., Tarantola, A., Randi, A., Pironon, J., Dubessy, J., Girard, J.-P., 2014. Determination of methane content in NaCl–H_2O fluid inclusions by Raman spectroscopy. Calibration and application to the external part of the Central Alps (Switzerland). Chem. Geol. 378–379, 52–61.

Cavarretta, G., Tecce, F., 1995. FLINCS and FLINCALC: PC programs for fluid inclusion calculations. Comput. Geosci. 21, 715–717.

Cemič, L., 2005. Thermodynamics in Mineral Sciences. Springer Verlag, Berlin/Heidelberg, 386 p.

Cesare, B., 1995. Graphite precipitation in C–O–H fluid inclusions: closed system compositional and density changes, and thermobarometric implications. Contrib. Mineral. Petrol. 122, 25–33.

Chabiron, A., Pironon, J., Massare, D., 2004. Characterization of water in synthetic rhyolitic glasses and natural melt inclusions by Raman spectroscopy. Contrib. Mineral. Petrol. 146, 485–492.

Chacko, T., Cole, D.R., Horita, J., 2001. Equilibrium oxygen, hydrogen and carbon isotope fractionation factors applicable to geologic systems. In: Valley, J.W., Cole, D.R. (Eds.), Stable Isotope Geochemistry. In: Rev. Mineral. Geochem, vol. 43, pp. 1–81.

Chacko, T., Mayeda, T.K., Clayton, R.N., Goldsmith, J.R., 1991. Oxygen and carbon isotope fractionations between CO_2 and calcite. Geochim. Cosmochim. Acta 55, 2867–2882.

Chacko, T., Riciputi, L.R., Cole, D.R., Horita, J., 1999. A new technique for determining equilibrium hydrogen isotope fractionation factors using the ion microprobe: application to the epidote-water system. Geochim. Cosmochim. Acta 63, 1–10.

Champagnon, B., Panczer, G., Chazallon, B., Arnaud, L., Duval, P., Lipenkov, V., 1997. Nitrogen and oxygen guest molecules in clathrate hydrates: Different sites revealed by Raman spectroscopy. J. Raman Spectrosc. 28, 711–715.

Chang, R., 1991. Chemistry, fourth ed. McGraw-Hill, New York/St. Louis/San Francisco, 1049 p.

Chaplin, M.F., 2000. A proposal for the structuring of water. Biophys. Chem. 83, 211–221.

Charlier, B., Grove, T.L., 2012. Experiments on liquid immiscibility along tholeiitic liquid lines of descent. Contrib. Mineral. Petrol. 164, 27–44.

Chatterjee, N.D., 1991. Applied Mineralogical Thermodynamics. Springer, Berlin/Heidelberg, 321 p.

Chen, Y., Provost, A., Schiano, P., Cluzel, N., 2013. Magma ascent rate and initial water concentration inferred from diffusive water loss from olivine-hosted melt inclusions. Contrib. Mineral. Petrol. 165, 525–541.

Chiba, H., Chacko, T., Clayton, R.N., Goldsmith, J.R., 1989. Oxygen isotope fractionations involving diopside, forsterite, magnetite, and calcite: Applications to geothermometry. Geochim. Cosmochim. Acta 53, 2985–2995.

Chiba, H., Kusakabe, M., Hirano, S., Matsuo, S., Somiya, S., 1981. Oxygen isotope fraction factors between anhydrite and water from 100 °C and 550 °C. Earth Planet. Sci. Lett. 53, 55–62.

Chou, I.-M., 1987. Phase equilibria in the system $NaCl-KCl-H_2O$. III. Solubilities of halite in vapour saturated liquids above 445 °C and determination of phase equilibrium properties in the system $NaCl-H_2O$ to 1000 °C and 1500 bars. Geochim. Cosmochim. Acta 51, 1965–1975.
Chovan, M., Lüders, V., Hurai, V., 1999. Fluid inclusion and C, O-isotope constraints on the origin of granodiorite-hosted Sb–As–Au–W deposit at Dubrava (Nizke Tatry Mts., Western Carpathians). Terra Nostra 99 (6), 173–175.
Christoforakos, M., 1985. Überkritische Systeme unter hohem Druck. Ph.D. Thesis, Univ. Karlsruhe.
Christoforakos, M., Franck, E.U., 1986. An equation of state for binary fluid mixtures to high temperatures and high pressures. Ber. Bunsenges. Phys. Chem. 90, 780–789.
Churakov, S.V., Gottschalk, M., 2003. Perturbation theory based equation of state for polar molecular fluids: II. Fluid mixtures. Geochim. Cosmochim. Acta 67, 2415–2425.
Ciais, P., White, J.W.C., Jouzel, J., Petit, J.R., 1995. The origin of present-day Antarctic precipitation from surface snow deuterium excess data. J. Geophys. Res. 100, 18917–18927.
Cinta Pinzaru S., Onac, B.P., 2009. Raman study of natural berlinite from a geological phosphate deposit. Vib. Spectrosc. 49, 97–100.
Clayton, R.N., Kieffer, S.W., 1991. Oxygen isotopic thermometer calibrations. In: Taylor, H.P., Jr., O'Neil, J.R., Kaplan, I.R. (Eds.), Stable Isotope Geochemistry. In: Geochem. Soc. Spec. Publ, vol. 3, pp. 3–10.
Clayton, R.N., O'Neil, J.R., Mayeda, T.K., 1972. Oxygen isotope exchange between quartz and water. J. Geophys. Res. 77, 3057–3067.
Cline, J.S., Bodnar, R.J., 1991. Can economic porphyry copper mineralization be generated by a typical calc-alkaline melt? J. Geophys. Res. 96, 8113–8126.
Clocchiatti, R., 1975. Les inclusions vitreuses des cristaux de quartz. Étude optique, thermo-optique et chimique. Mém. Soc. Géol. France 122, 1–96.
Coates, J., 2000. Interpretation of infrared spectra, a practical approach. In: Meyers, R.A. (Ed.), Encyclopedia of Analytical Chemistry. John Wiley & Sons, Chichester, pp. 10815–10837.
Colangelli, L., Mennella, V., Baratta, G.A., Bussoletti, E., Strazzulla, G., 1992. Raman and infrared spectra of polycyclic aromatic hydrocarbon molecules of possible astrophysical interest. Astrophys. J. 396, 369–377.
Cole, D.R., 1985. A preliminary evaluation of oxygen isotopic exchange between chlorite and water. Geol. Soc. Am. Abstr. Progr. 17, 550.
Collins, P.L.F., 1979. Gas hydrates in CO_2-bearing fluid inclusions and the use of freezing data for estimation of salinity. Econ. Geol. 74, 1435–1444.
Colthup, N.B., Daly, L.H., Wiberley, S.E., 1990. Introduction to Infrared and Raman Spectroscopy, third ed. Academic Press, London/San Diego, 547 p.
Connolly, J.A.D., 1990. Multivariable phase diagrams: an algorithm based on generalized thermodynamics. Am. J. Sci. 290, 666–718.
Connolly, J.A.D., 1992. Elementary phase diagrams: principles and methods. In: Pressure and Temperature Evolution of Orogenic Belts. 5th Summer School, Geologia e Petrologia dei Basamenti Cristallini. University of Siena and Italian National Research Council, Italy, pp. 203–220.
Connolly, J.A.D., 1995. Phase diagram methods for graphitic rocks and application to the system $C-O-H-FeO-TiO_2-SiO_2$. Contrib. Mineral. Petrol. 119, 94–116.
Connolly, J.A.D., Cesare, B., 1993. $C-O-H-S$ fluid composition and oxygen fugacity in graphitic metapelites. J. Metamorph. Geol. 11, 379–388.
Connolly, J.A.D., Kerrick, D.M., 1987. An algorithm and computer program for calculating composition diagrams. CALPHAD 11, 1–54.
Cosgrove, J.W., 1997. Hydraulic fractures and their implications regarding the state of stress in a sedimentary sequence during burial. In: Sengupta, S. (Ed.), Evolution of Geological Structures in Micro- and Macro-scales. Chapman & Hall, London, pp. 11–25.

Craig, H., 1961. Isotopic variations in meteoric waters. Science 133, 1702–1703.

Craw, D., 1992. Fluid evolution, fluid immiscibility and gold deposition during Cretaceous-Recent tectonics and uplift of the Otago and Alpine Schists, New Zealand. Chem. Geol. 98, 221–236.

Craw, D., Teagle, D.A.H., Belocky, R., 1993. Fluid immiscibility in late-Alpine gold-bearing veins. Eastern and Northwestern European Alps. Miner. Deposita 28, 28–36.

Crawford, M.L., 1992. Fluid inclusions—What can we learn? Earth Sci. Rev. 32, 137–139.

Csikósová, A., Novek, Z., Kameníková, K., 2000. Position of the Slovak magnesite on world market. Miner. Slov. 32, 530–532.

D'Ans, J., 1933. Die Lösungsgleichgewichte der Systeme der Salze ozeanischer Salzablagerungen. Kali-Forschungsanstalt, Berlin.

Darling, R.S., 1991. An extended equation to calculate NaCl contents from final clathrate melting temperatures in H_2O-CO_2-NaCl fluid inclusions: implications for P–T isochore location. Geochim. Cosmochim. Acta 55, 3869–3871.

Darling, W.G., 1998. Hydrothermal hydrocarbon gases: 1. Genesis and geothermometry. Appl. Geochem. 13, 815–824.

Davidson, P., Kamenetsky, V., 2001. Immiscibility and continuous felsic melt–fluid evolution within the Rio Blanco porphyry system, Chile: Evidence from inclusions in magmatic quartz. Econ. Geol. 96, 1921–1929.

Davis, A.R., Oliver, B.G., 1972. A vibrational-spectroscopic study of the species present in the CO_2-H_2O system. J. Solut. Chem. 1, 329–338.

Davis, D.W., Lowenstein, T.K., Spencer, R.J., 1990. Melting behavior of fluid inclusions in laboratory-grown halite crystals in the system $NaCl-H_2O$, $NaCl-KCl-H_2O$, $NaCl-MgCl_2-H_2O$ and $NaCl-CaCl_2-H_2O$. Geochim. Cosmochim. Acta 54, 591–601.

Davisson, M.L., Criss, R.E., 1996. Na–Ca–Cl relations in basinal fluids. Geochim. Cosmochim. Acta 60, 2743–2752.

Davy, H., 1822. On the state of water and aeriform matter in cavities found in certain crystals. Royal Soc. London Phil. Trans. 2, 367–376.

De Capitani, C., Brown, T.H., 1987. The computation of chemical equilibria in complex systems containing non-ideal solutions. Geochim. Cosmochim. Acta 51, 2639–2652.

Dele-Dubois, M.-L., Dhamelincourt, P., Schubnel, H.-J., 1980. Étude par spectrométrie Raman d´inclusions dans les diamants, saphirs et émeraudes. Rev. Gemmol. A.F.G. 63, 11–14.

De Santis, R., Breedveld, G.J.F., Prausnitz, J.M., 1974. Thermodynamic properties of aqueous gas mixtures at advanced pressures. Ind. Eng. Chem. Proc. Des. Dev. 13, 374–377.

De Vivo, B., Bodnar, R.J. (Eds.), 2003. Melt Inclusions in Volcanic Systems. Methods, Applications and Problems, first ed. Elsevier, Amsterdam/Boston, 274 p.

Debenedetti, P.G., 2003. Supercooled and glassy water. J. Phys. Condens. Matter 15, R1669–R1726.

Debenedetti, P.G., Stanley, H.E., 2003. Supercooled and glassy water. Phys. Today Online 56 (6), 40–46.

Delhaye, M., Dhamelincourt, P., 1975. Raman microprobe and microscope with laser excitation. J. Raman Spectrosc. 3, 33.

Dennis, P.F., Rowe, P.J., Atkinson, T.C., 2001. The recovery and isotopic measurement of water from fluid inclusions in speleothems. Geochim. Cosmochim. Acta 65, 871–884.

Dereppe, J.M., Pironon, J., Moreaux, C., 1994. Characterization of the composition of fluid inclusions in minerals by ^1H NMR. Am. Mineral. 79, 712–718.

Derome, D., Cathelineau, M., Lhomme, T., Cuney, M., 2003. Fluid inclusion evidence of the differential migration of H_2 and O_2 in the McArthur River unconformity-type uranium deposit (Saskatchewan, Canada). Possible role of post-ore modification of the host rocks. J. Geochem. Explor. 78–79, 525–530.

Dhamelincourt, P., Beny, J.M., Dubessy, J., Poty, B., 1979. Analyse d'inclusions fluides á la microsonde MOLE á effect Raman. Bull. Soc. Fr. Minéral. 102, 600–610.

Di Muro, A., Giordano, D., Villemant, B., Montagnac, G., Scaillet, B., Romano, C., 2006. Influence of composition and thermal history of volcanic glasses on water content as determined by micro-Raman spectrometry. Appl. Geochem. 21, 802–812.

Di Muro, A., Villemant, B., Montagnac, G., Scaillet, B., Reynard, B., 2005. Quantification of water content and speciation in natural silicic glasses (phonolites, dacites, rhyolites) by confocal microRaman spectrometry. Geochim. Cosmochim. Acta 70, 2868–2884.

Diamond, L.W., 1990. Fluid inclusion evidence for P–V–T–X evolution of hydrothermal solutions in late-Alpine gold–quartz veins at Brusson, Val d'Ayas, northwest Italia. Am. J. Sci. 290, 912–958.

Diamond, L.W., 1992a. Stability of CO_2 clathrate hydrate + CO_2 liquid + CO_2 vapour + aqueous KCl–NaCl solutions: experimental determination and application to salinity estimates of fluid inclusions. Geochim. Cosmochim. Acta 56, 273–280.

Diamond, L.W., 1992b. Gas–clathrate hydrates in fluid inclusions: phase equilibria and applications. In: McKibben, M.A., Montanez, I.P., Hall, D.L. (Eds.), PACROFI IV, Pan-American Conference on Research on Fluid Inclusions, Program and Abstracts. Dept. of Earth Sciences, University of California, Riverside, pp. 119–122.

Diamond, L.W., 1994. Salinity of multivolatile fluid inclusions determined from clathrate hydrate stability. Geochim. Cosmochim. Acta 58, 19–41.

Diamond, L.W., 1995. Introduction to phase relations of CO_2–H_2O fluid inclusions. In: De Vivo, B., Frezzotti, M.-L. (Eds.), Fluid Inclusions in Minerals: Methods and Applications. Short Course of the Working Group IMA Inclusions in Minerals. Virginia Tech., USA, pp. 131–158.

Diamond, L.W., 2001. Review of the systematics of CO_2–H_2O fluid inclusions. Lithos 55, 69–99.

Diamond, L.W., 2003. Introduction to gas-bearing, aqueous fluid systems. In: Samson, I., Anderson, A., Marshall, D. (Eds.), Fluid Inclusions: Analysis and Interpretation. In: Min. Assoc. Can. Short Course, vol. 32, pp. 101–158.

Diamond, L.W., Akinfiev, N.N., 2003. Solubility of CO_2 in water from −1.5 to 100 °C and from 0.1 to 100 MPa: evaluation of literature data and thermodynamic modelling. Fluid Phase Equilib. 208, 265–290.

Díaz-Gonzáles, L., Santoyo, E., 2008. A new precise calibration of the Na/K geothermometer using a world database of geothermal fluids and improved geochemometric techniques. Geochim. Cosmochim. Acta 72, A215.

Ding, T., Valkiers, S., Kipphardt, H., De Bièvre, P., Taylor, P.D.P., Gonfiantini, R., Krouse, R., 2001. Calibrated sulfur isotope abundance ratios of three IAEA sulfur isotope reference materials and V-CDT with a reassessment of the atomic weight of sulfur. Geochim. Cosmochim. Acta 65, 2433–2437.

Dingwell, D.B., 1996. Volcanic dilemma—flow or blow? Science 273 (5278), 1054–1055.

Dingwell, D.B., Hess, K.U., 1998. Melt viscosities in the system Na–Fe–Si–O–F–Cl: contrasting effects of F and Cl in alkaline melts. Am. Mineral. 83, 1016–1021.

Dingwell, D.B., Bagdassarov, N.S., Bussod, G.Y., Webb, S.L., 1993. Magma rheology. In: Luth, R.W. (Ed.), Experiments at High Pressure and Their Application to the Earth. In: Min. Assoc. Can. Short Course Handbook, vol. 21, pp. 131–196.

Dolgov, J.A., Bazarov, L.Sh., Bakumenko, I.T., 1968. Method of determination of fluid inclusion pressure with help of contemporaneous application of homogenization and cryometry. In: Mineralogicheskaya Termo- i Barometriya, II. Nauka, Moskva, pp. 7–11 (in Russian).

Dollish, F.R., Fateley, W.G., Bentley, F.F., 1974. Characteristic Raman Frequencies of Organic Compounds. John Wiley and Sons, New York.

Downs, R.T., 2006. The RRUFF Project. http:/rruff.geo.arizona.edu/.

Duan, Z., Møller, N., Greenberg, J.H., Weare, J., 1992a. The prediction of methane solubility in natural waters to high ionic strength from 0 °C to 250 °C and from 0 to 1600 bar. Geochim. Cosmochim. Acta 56, 1451–1460.

Duan, Z., Møller, N., Weare, J., 1992b. An equation of state for CH_4–CO_2–H_2O system: II. Mixtures from 50 to 1000 °C and 0 to 1000 bar. Geochim. Cosmochim. Acta 56, 2619–2631.

Duan, Z., Møller, N., Weare, J., 1995. Equation of state for NaCl–H_2O–CO_2 system: prediction of phase equilibria and volumetric properties. Geochim. Cosmochim. Acta 59, 2869–2882.

Dubessy, J., Audeoud, D., Wilkins, R., Kosztolanyi, C., 1982. The use of the Raman microprobe in the determination of the electrolytes dissolved in the aqueous phase of fluid inclusions. Chem. Geol. 37, 137–150.

Dubessy, J., Pagel, M., Beny, J.M., Christensen, H., Hickel, B., Kosztolanyi, Ch., Poty, B., 1988. Radiolysis evidenced by H_2–O_2 and H_2-bearing fluid inclusions in three uranium deposits. Geochim. Cosmochim. Acta 52, 1155–1167.

Dubessy, J., Poty, B., Ramboz, C., 1989. Advances in C–O–H–N–S fluid geochemistry based on micro-Raman spectrometric analysis of fluid inclusions. Eur. J. Mineral. 1, 517–534.

Dubessy, J., Buschaert, S., Lamb, W., Pironon, J., Thiéry, R., 2001. Methane-bearing aqueous fluid inclusions: Raman analysis, thermodynamic modelling and application to petroleum basins. Chem. Geol. 173, 193–205.

Dubessy, J., Guillaume, D., Lhomme, T., Pironon, J., Tarantola, A., Teinturie, S., 2002a. Advancement in the analysis of fluid inclusions using both micro-Raman spectrometry and thermodynamic modelling. Acta Univ. Carol. Geol. 46, 16–18.

Dubessy, J., Lhomme, T., Boiron, M.-C., Rull, F., 2002b. Determination of chlorinity in aqueous fluids using Raman spectroscopy of the stretching band of water at room temperature: application to fluid inclusions. Appl. Spectrosc. 56, 99–106.

Dubessy, J., Derome, D., Sausse, J., 2003. Numerical modelling of fluid mixings in the H_2O–NaCl system: application to the north Caramal U prospect (Australia). Chem. Geol. 194, 25–39.

Dubessy, J., Caumon, M.-C., Rull, F., 2012. Raman Spectroscopy Applied to Earth Sciences and Cultural Heritage. EMU Notes in Mineralogy, vol. 12, 504 p.

Dublyansky, Y.V., Spötl, C., 2009. Hydrogen and oxygen isotopes of water from inclusions in minerals: design of a new crushing system and on-line continuous-flow isotope ratio mass spectrometric analysis. Rapid Commun. Mass Spectrom. 23, 2605–2613.

Dubois, M., Marignac, C., 1997. The H_2O–NaCl–$MgCl_2$ ternary phase diagram with special application to fluid inclusion studies. Econ. Geol. 92, 114–119.

Dubois, M., Monin, C., Castelain, T., Coquinot, Y., Gouy, S., Gauthier, A., Goffé, B., 2010. Investigation of the H_2O–NaCl–LiCl system: a synthetic fluid inclusion study and thermodynamic modelling from $-50°$ to $+100$ °C and up to 12 mol/kg. Econ. Geol. 105, 329–338.

Dunaeva, A.N., Antsyshkin, D.V., Kuskov, O.L., 2010. Phase diagram of H_2O: thermodynamic functions of the phase transitions of high-pressure ices. Sol. Syst. Res. 44, 202–222.

Dunbar, N.W., Hervig, R.L., Kyle, P.R., 1989. Determination of pre-eruptive H_2O, F, and Cl contents of silicic magmas using melt inclusions: examples from Taupo volcanic center, New Zealand. Bull. Volcanol. 51, 177–184.

Dunham, J.L., 1932. The energy levels of a rotating vibrator. Phys. Rev. 41, 721–731.

Dunn, S.R., Valley, J.W., 2007. Calcite–graphite isotope thermometry: a test for polymetamorphism in marble, Tudor gabbro aureole, Ontario, Canada. J. Metamorph. Geol. 10, 487–501.

Ďurkovič, T., Koráb, T., Rudinec, R., Gašpariková, V., Snopková, P., Köhler, E., Zakovič, M., 1982. Deep Structural Drillhole ZBOJ-1. Geologický Ústav D. Štúra, Bratislava, 76 p. (in Slovak).

Ebner, F., Cerny, I., Eichhorn, R., Götzinger, M., Paar, W., Prochaska, W., Weber, L., 1999. Mineral resources in the Eastern Alps and adjoining areas. Mitt. Österr. Geol. Ges. 92, 157–184.

Economou, I.G., Peters, C.J., de Swaan, Arons J., 1995. Water–salt phase equilibria at elevated temperatures and pressures: model development and mixture predictions. J. Phys. Chem. 99, 6182–6193.

Edwards, H.G.M., Farwell, D.W., Jorge Villar, S.E., 2007. Raman microspectrometric studies of amber resins with insect inclusions. Spectrochim. Acta A 68, 1089–1095.

Eggler, D.H., Baker, D.R., 1982. Reduced volatiles in the system C–O–H: implications to mantle melting, fluid formation and genesis. In: Manghnani, M.H., Akimoto, S. (Eds.), High-Pressure Research in Geophysics. Springer, Berlin, pp. 237–250.

Einaudi, M.T., Hedenquist, J.W., Inan, E.E., 2003. Sulfidation state of fluids in active and extinct hydrothermal systems: transitions from porphyry to epithermal environments. In: Simmons, S.F., Graham, I. (Eds.), Volcanic, Geothermal, and Ore-Forming Fluids: Rulers and Witnesses of Processes within the Earth. In: Soc. Econ. Geol. Spec. Publ., vol. 10, pp. 285–313.

El-Diasty, F., 2011. Coherent anti-stokes Raman scattering: spectroscopy and microscopy. Vib. Spectrosc. 55, 1–37.

Ellis, A.J., Golding, R.M., 1963. The solubility of carbon dioxide above 200 °C in water and in sodium chloride solutions. Am. J. Sci. 261, 47–60.

Elsner, N., Fisher, S., Klinger, J., 1982. Thermophysikalische Stoffeigenschaften von Wasser. VEB Deutscher Verlag f. Grundstoffindustrie, Leipzig.

Ely, J.F., Haynes, W.M., Bain, B.C., 1989. Isochoric (P, V_m, T) measurements on CO_2 and on ($0.982CO_2$+$0.018N_2$) from 250 to 330 K at pressures to 35 MPa. J. Chem. Thermodyn. 21, 879–894.

Epstein, S., Buchsbaum, R., Lowenstam, H., Urey, H.C., 1951. Carbonate-water isotopic temperature scale. J. Geol. 62, 417–426.

Eriksson, G., Hack, K., 1990. Chemsage—a computer program for the calculation of complex chemical equilibria. Metall. Trans. B 21B, 1013–1023.

Esposito, R., Klebesz, R., Bartoli, O., Klyukin, Y.I., Moncada, D., Doherty, A.L., Bodnar, R.J., 2012. Application of the Linkam TS1400XY heating stage to melt inclusion studies. Cent. Eur. J. Geosci. 4, 208–218.

Essene, E.J., 1982. Geologic thermometry and barometry. Rev. Mineral. Geochem. 10, 153–206.

Essene, E.J., 1989. The current status of thermobarometry in metamorphic rocks. In: Daly, J.S., Cliff, R.A., Yardley, B.W.D. (Eds.), The Evolution of Metamorphic Belts. Blackwell Scientific, London, pp. 1–44.

Eugster, H., 1957. Heterogeneous reactions involving oxidation and reduction at high pressures and temperatures. J. Chem. Phys. 26, 1760–1761.

Eugster, H., Wones, D.R., 1962. Stability relations of the ferruginous biotite, annite. J. Petrol. 3, 82–125.

Evans, M.A., 1995. Fluid inclusions in veins from the Middle Devonian shales: A record of deformation conditions and fluid evolution in the Appalachian Plateau. GSA Bull. 107, 327–339.

Evans, M.A., Battles, D.A., 1999. Fluid inclusion and stable isotope analyses of veins from the central Appalachian Valley and Ridge province: implications for regional synorogenic hydrologic structure and fluid migration. GSA Bull. 111, 1841–1860.

Fabre, C., Boiron, M.-C., Dubessy, J., Cathelineau, M., Banks, D., 2002. Palaeofluid chemistry of a single fluid event: a bulk and in-situ multi-technique analysis (LIBS, Raman spectroscopy) of an Alpine fluid (Mont-Blanc). Chem. Geol. 182, 249–264.

Fabrichnaya, O.B., 1995. Thermodynamic data for phases in the $FeO-MgO-SiO_2$ system and phase relations in the mantle transition zone. Phys. Chem. Miner. 22, 323–332.

Fan, H.-R., Xie, Y.-H., Wang, K.-Y., Tao, K.-J., Wilde, S.A., 2004. REE daughter minerals trapped in fluid inclusions in the giant Bayan-Obo REE-Nb-Fe deposit, Inner Mongolia, China. Int. Geol. Rev. 46, 638–645.

Ferraro, J.R., Nakamoto, K., Brown, C.W., 2003. Introductory Raman Spectroscopy. Elsevier, Amsterdam, 434 p.

Fletcher, P., 1993. Chemical Thermodynamics for Earth Scientists. Longman Scientific and Technical/John Wiley & Sons, Harlow, England/New York, 464 p.

Fontes, J.C., Matray, J.M., 1993. Geochemistry and origin of formation brines from the Paris Basin, France. 1. Brines associated with Triassic salts. Chem. Geol. 109, 149–175.

Fournier, R.O., 1979. A revised equation for the Na/K geothermometer. Geotherm. Resour. Council Trans. 3, 221–224.

Fournier, R.O., Potter, R.W., II, 1979. Magnesium correction to the Na–K–Ca chemical geothermometer. Geochim. Cosmochim. Acta 43, 1543–1550.

Fournier, R.O., Truesdell, A.H., 1973. An empirical Na–K–Ca geothermometer for natural waters. Geochim. Cosmochim. Acta 37, 1255–1275.

Frezzotti, M.L., 2001. Silicate–melt inclusions in magmatic rocks: applications to petrology. Lithos 55, 273–299.

Frezzotti, M.L., Burke, E.A.J., de Vivo, B., Stefanini, B., Villa, I.M., 1992. Mantle fluids in pyroxenite nodules from Salt Lake Crater (Oahu, Hawaii). Eur. J. Mineral. 4, 1137–1153.

Frezzotti, M.L., Selverstone, J., Sharp, Z.D., Compagnoni, R., 2011. Carbonate dissolution during subduction revealed by diamond-bearing rocks from the Alps. Nat. Geosci. 4, 703–706.

Frezzotti, M.L., Ferrando, S., Tecce, F., Castelli, D., 2012a. Water content and nature of solutes in shallow-mantle fluids from fluid inclusions. Earth Planet. Sci. Lett. 351–352, 70–83.

Frezzotti, M.L., Tecce, F., Casagli, A., 2012b. Raman spectroscopy for fluid inclusion analysis. J. Geochem. Explor. 112, 1–20.

Friedman, I., O'Neil, J.R., 1977. Compilation of stable isotope fractionation factors of geochemical interest. US Geol. Survey Prof. Pap., 440–K, 49 p.

Fritz, S.J., 1994. A survey of charge-balance errors on published analyses of potable ground and surface waters. Groundwater 32, 539–546.

Frost, B.R., 1991. Introduction to oxygen fugacity and its petrologic importance. In: Lindsley, D.H. (Ed.), Oxide Minerals: Petrologic and Magmatic Significance. In: Min. Soc. Am. Rev. Mineral., vol. 25, pp. 1–9.

Frost, R.L., Dickfos, M.J., 2007. Raman spectroscopy of halogen containing carbonates. J. Raman Spectrosc. 38, 1516–1522.

Frost, R.L., Williams, P., 2004. Raman spectroscocopy of some basic chloride containing minerals of lead and copper. Spectrochim. Acta A Mol. Biomol. Spectrosc. 60, 2071–2077.

Frost, R.L., Xi, Y., 2012. A vibrational spectroscopic study of planchéite $Cu_8Si_8O_{22}(OH)_4 \cdot H_2O$. Spectrochim. Acta A 91, 314–318.

Frost, R.L., Xi, Y., 2013. Is chrysocolla $(Cu, Al)_2H_2Si_2O_5(OH)_4 \cdot nH_2O$ related to spertiniite $Cu(OH)_2$?—A vibrational spectroscopic study. Vib. Spectrosc. 64, 33–38.

Frost, R.L., Williams, P., Kloprogge, J.T., Martens, W.N., 2003. Raman spectroscopy of the copper chloride minerals nantokite, eriochalcite and claringbullite—implications for copper corrosion. N. Jhrb. Mineral. Monatshefte 10, 433–445.

Frost, R.L., Bouzaid, J.M., Palmer, S.J., 2007. The structure of mimetite, arsenian pyromorphite and hedyphane: a Raman spectroscopic study. Polyhedron 26, 2964–2970.

Fyfe, W.S., Price, N.J., Thompson, A.B., 1978. Fluids in the Earth's Crust. Elsevier, Amsterdam, 373 p.

Gaarenstroom, R.A., Tromp, R.A.J., deJong, M.C., Brandenberg, A.M., 1993. Pressures in the Central North Sea: implications for trap integrity and drilling safety. In: Parker, J.R. (Ed.), Petroleum Geology of Northwest Europe. Proceedings of the 4th Conference, pp. 1305–1313.

Gaboury, D., Daigneault, R., 2000. Flat vein formation in a transitional crustal setting by self-induced fluid pressure equilibrium—an example from the Géant Dormant gold mine, Canada. Ore Geol. Rev. 17, 155–178.

Galeener, F.L., Mikkelsen, J.C., 1981. Raman diffusion profilometry: OH in vitreous SiO_2. Appl. Phys. Lett. 38, 336–338.

Garnier, V., Giuliani, G., Maluski, H., Ohnenstetter, D., Trong, T.P., Quang, V.H., Van, L.P., Van, T.V., Schwarz, D., 2002. Ar–Ar ages in phlogopites from marble-hosted ruby deposits in northern Vietnam: evidence for Cenozoic ruby formation. Chem. Geol. 188, 33–49.

Gehrig, M., 1980. Phasengleichgewichte und pVT–daten ternärer Mischungen aus Wasser, Kohlendioxid und Natriumchlorid bis 3 kbar und 550 °C. Ph.D. Thesis, University Karlsruhe, Hochschulverlag, Freiburg.

Gehrig, M., Lentz, H., Franck, E.U., 1986. The system water–carbon dioxide–sodium chloride to 773 K and 300 MPa. Ber. Bunsenges. Phys. Chem. 90, 525–533.

George, S.C., Krieger, F.W., Eadington, P.J., Quezada, R.A., Greenwood, P.F., Eisenberg, L.I., Hamilton, P.J., Wilson, M., 1997. Geochemical comparison of oil-bearing fluid inclusions and produced oil from the Toro sandstone, Papua New Guinea. Org. Geochem. 26, 155–173.

Germann, K., Lüders, V., Banks, D.A., Simon, K., Hoefs, J., 2003. Late Hercynian polymetallic vein-type base-metal mineralization in the Iberian Pyrite Belt: fluid–inclusion and stable–isotope geochemistry (S–O–H–Cl). Miner. Deposita 38, 953–967.

Ghiorso, M.S., 1990. Thermodynamic properties of hematite–ilmenite–geikielite solid solutions. Contrib. Mineral. Petrol. 104, 645–667.

Ghiorso, M.S., Sack, R.O., 1991. Fe–Ti oxide geothermometry: thermodynamic formulation and the estimation of intensive variables in silicic magmas. Contrib. Mineral. Petrol. 108, 485–510.

Ghiorso, M.S., Sack, R.O., 1995. Chemical mass transfer in magmatic processes. IV. A revised and internally consistent thermodynamic model for the interpolation and extrapolation of liquid-solid equilibria in magmatic systems at elevated temperatures and pressures. Contrib. Mineral. Petrol. 119, 197–212.

Giggenbach, W.F., 1980. Geothermal gas equilibria. Geochim. Cosmochim. Acta 44, 549–556.

Giggenbach, W.F., 1988. Geothermal solute equilibria. Derivation of Na–K–Mg–Ca geoindicators. Geochim. Cosmochim. Acta 52, 2749–2765.

Gilg, H.A., Sheppard, S.M.F., 1996. Hydrogen isotope fractionation between kaolinite and water revisited. Geochim. Cosmochim. Acta 60, 529–533.

Gionfantini, R., Stichler, W., Rozanski, K., 1995. Standards and intercomparison materials distributed by the International Atomic Energy Agency for stable isotope measurements. In: Reference and Intercomparison Materials for Stable Isotopes of Light Elements. Proceedings of a Consultants Meeting, Vienna, 1–3 December 1993, pp. 13–29.

Giordano, D., Romano, C., Papale, P., Dingwell, D.B., 2004a. The viscosity of trachytes, and comparison with basalts, phonolites, and rhyolites. Chem. Geol. 213, 49–61.

Giordano, D., Romano, C., Poe, B., Dingwell, D.B., Behrens, H., 2004b. The combined effects of water and fluorine on the viscosity of silicic magmas. Geochim. Cosmochim. Acta 68, 5159–5168.

Giordano, D., Russell, J.K., Dingwell, D.B., 2008. Viscosity of magmatic liquids: A model. Earth Planet. Sci. Lett. 271, 123–134.

Gittins, J., 1989. The origin and evolution of carbonate magmas. In: Bell, K. (Ed.), Carbonatites: Genesis and Evolution. Unwin Hyman, London, pp. 580–599.

Giuliani, G., Dubessy, J., Banks, D., Quang, V.H., Lhomme, T., Pironon, J., Garnier, V., Trong, T.P., Van, L.P., Ohnenstetter, D., Schwarz, D., 2003. CO_2–H_2S–COS–S_8–AlO(OH)-bearing fluid inclusions in ruby from marble-hosted deposits in Luc Yen area, north Vietnam. Chem. Geol. 194, 167–185.

Goldstein, R.H., 1986. Re-equilibration of fluid inclusions in low-temperature calcium–carbonate cement. Geology 14, 792–795.

Goldstein, R.H., 2001. Fluid inclusions in sedimentary and diagenetic systems. Lithos 55, 159–193.

Goldstein, R.H., Reynolds, T.J., 1994. Systematics of Fluid Inclusions in Diagenetic Minerals. SEPM Short Course, vol. 31, 199 p.

Golyshev, S.I., Padalko, N.L., Pechenkin, S.A., 1981. Fractionation of stable oxygen and carbon isotopes in carbonate systems. Geokhymiya 10, 1427–1441.

Gorelik, V.S., Chervyakov, A.V., Kol'tsova, L.V., Veryaskin, S.S., 2000. Raman spectra of saturated hydrocarbons and gasolines. J. Russ. Laser Res. 21, 323–334.

Gottschalk, M., 1997. Internally consistent thermodynamic data for rock-forming minerals in the system SiO_2–TiO_2–Al_2O_3–Fe_2O_3–CaO–MgO–FeO–K_2O–Na_2O–H_2O–CO_2. Eur. J. Mineral. 9, 175–223.

Graham, C.M., Sheppard, S.M.F., Heaton, T.H.E., 1980. Experimental hydrogen isotope studies: Hydrogen isotope fractionation in the systems epidote–H_2O, zoisite–H_2O and AlO(OH)–H_2O. Geochim. Cosmochim. Acta 44, 353–364.

Graham, C.M., Harmon, R.S., Sheppard, S.M.F., 1984. Experimental hydrogen isotope studies: Hydrogen isotope exchange between amphibole and water. Am. Mineral. 69, 128–138.

Graham, C.M., Viglino, J.A., Harmon, R.S., 1980. Experimental study of hydrogen-isotope exchange between aluminous chlorite and water and of hydrogen diffusion in chlorite. Am. Mineral. 72, 566–579.

Graney, J.R., Kesler, S.E., Jones, H.D., 1991. Application of gas-analysis of jasperoid inclusion fluids to exploration for micron gold deposits. J. Geochem. Explor. 42, 91–106.

Gratier, J.P., Jenatton, L., 1984. Deformation by solution–deposition, and re-equilibration of fluid inclusions in crystals depending on temperature, internal pressure and stress. J. Struct. Geol. 6, 189–200.

Gray, D.R., Gregory, R.T., Durney, D.W., 1991. Rock-buffered fluid–rock interaction in deformed quartz-rich turbidite sequences, Eastern Australia. J. Geophys. Res. 96, 19681–19704.

Grecula, P., 1982. The Gemericum—Segment of Palaeotethynian Riftogeneous Basin. Alfa, Bratislava, 263 p. (in Slovak).

Grecula, P., Radvanec, M., 2005a. Age of siderite–sulphide mineralization of the Gemericum: Variscan or Alpine? Miner. Slov. 37, 343–345 (in Slovak).

Grecula, P., Radvanec, M., 2005b. Geotectonic model of the Gemericum. Miner. Slov. 37, 193–198 (in Slovak).

Grecula, P., Abonyi, A., Abonyiová, M., Antaš, J., Bartalský, B., Bartalský, J., Dianiška, I., Drnzík, E., Ďuďa, R., Garguľák, M., Gazdačko, Ľ., Hudáček, J., Kobulský, J., Lörincz, L., Macko, J., Návesňák, D., Németh, Z., Novotný, L., Radvanec, M., Rojkovič, I., Rozložník, L., Rozložník, O., Varček, C., Zlocha, J., 1995. Mineral Deposits of the Slovak Ore Mountains. Geocomplex, Bratislava, 834 p.

Grecula, P., Radvanec, M., Németh, Z., 2000. Magnesite and talc mineralization in Slovakia. Miner. Slov. 32, 533–542.

Gregory, R.T., Criss, R.E., Taylor, H.P., Jr., 1989. Oxygen isotope exchange kinetics of mineral pairs in closed and open systems: applications to problems of hydrothermal alteration of igneous rocks and Precambrian iron formations. Chem. Geol. 75, 1–42.

Griffiths, J., 2008. A brief history of mass spectrometry. Anal. Chem. 80, 5678–5683.

Grimmer, J.O.W., Pironon, J., Teinturier, S., Mutterer, J., 2003. Recognition and differentiation of gas condensates and other oil types using microthermometry of petroleum inclusions. J. Geochem. Explor. 78–79, 367–371.

Guilhaumou, N., Jouaffre, D., Velde, B., Bény, C., 1988. Raman microprobe analysis on gaseous inclusions from diagenetically altered Terres Noires (S–E France). Bull. Minéral. 111, 577–585.

Guilhaumou, N., Szydlowski, N., Pradier, B., 1990. Characterization of hydrocarbon fluid inclusions by infra-red and fluorescence microspectrometry. Mineral. Mag. 54, 311–324.

Guillaume, D., Teinturier, S., Dubessy, J., Pironon, J., 2003. Calibration of methane analysis by Raman spectrometry in H_2O–NaCl–CH_4 fluid inclusions. Chem. Geol. 194, 41–49.

Gurenko, A.A., Belousov, A.B., Trumbull, R.B., Sobolev, A.V., 2005. Explosive basaltic volcanism of the Chikurachki Volcano (Kuril arc, Russia): insights on pre-eruptive magmatic conditions and volatile budget revealed from phenocryst-hosted melt inclusions and groundmass glasses. J. Volcanol. Geotherm. Res. 147, 203–232.

Günther, D., Frischknecht, R., Heinrich, C.A., Kahlert, H.J., 1997. Capabilities of an argon fluoride 193 nm excimer laser for laser ablation inductively coupled plasma mass spectrometry microanalysis of geological materials. J. Anal. At. Spectrom. 12, 939–944.

Haar, L., Gallagher, J.S., Kell, G.S., 1984. NBS/NRC Steam Tables. Thermodynamic and Transport Properties and Computer Programs for Vapour and Liquid State of Water in SI units. Hempshire Publ. Corp., McGraw Hill, New York.

Haas, J.L., Jr., 1976. Physical properties of the coexisting phases and the thermochemical properties of the H_2O component in boiling NaCl solutions. US Geol. Surv. Bull. 1421–A, 1–73.

Haas, J.L., Fisher, J.R., 1976. Simultaneous evaluation and correlation of thermodynamic data. Am. J. Sci. 276, 525–545.

Hagemann, S.G., Lüders, V., 2003. P–T–X conditions of hydrothermal fluids and precipitation mechanism of stibnite–gold mineralization at the Wiluna lode-gold deposits, Western Australia: conventional and infrared microthermometric constraints. Miner. Deposita 38, 936–952.

Halbach, H., Chatterjee, N.D., 1982. An empirical Redlich–Kwong type equation of state for water to 1000 °C and 200 kbar. Contrib. Mineral. Petrol. 79, 337–345.

Hall, D.L., Sterner, S.M., Bodnar, R.J., 1988. Freezing point depression of NaCl−KCl−H_2O solutions. Econ. Geol. 83, 197–202.

Hall, D.L., Bodnar, R.J., Craig, J.R., 1991. Evidence for postentrapment diffusion of hydrogen into peak metamorphic fluid inclusions from the massive sulfide deposits at Ducktown, Tennessee. Am. Mineral. 76, 1344–1355.

Hall, W.E., Friedman, I., 1963. Composition of fluid inclusions, Cave in Rock fluorite district, Illinois and Upper Mississippi Valley zinc lead district. Econ. Geol. 58, 886–911.

Halter, W.E., Pettke, T., Heinrich, C.A., Rothen-Rutishauser, B., 2002. Major to trace element analysis of melt inclusions by laser-ablation ICP-MS: methods of quantification. Chem. Geol. 183, 63–86.

Hammerli, J., Rusk, B., Spandler, C., Emsbo, P., Oliver, N.H.S., 2013. In situ quantification of Br and Cl in minerals and fluid inclusions by LA-ICP-MS: A powerful tool to identify fluid sources. Chem. Geol. 337–338, 75–87.

Hansteen, T.H., Burke, E.A.J., 1994. Aphthitalite in high-temperature fluid inclusions in quartz from the Eikeren-Skrim granite complex, the Oslo paleorift. Norsk Geol. Tiddskr. 74, 238–240.

Hanzelová, M., 2001. Material flow of production process. Acta Montan. Slov. 5, 92–96 (in Slovak).

Harvey, A.H., Friend, D.G., 2004. Physical properties of water. In: Palmer, D.A., Fernández-Prini, R., Harvez, A. H. (Eds.), Aqueous Systems at Elevated Temperature and Pressure: Physical Chemistry in Water, Steam and Hydrothermal Solutions. Elsevier, Amsterdam/San Diego/Oxford/London, pp. 1–27.

Hawes, G.W., 1881. On liquid carbon dioxide in smoky quartz. Am. J. Sci. 21, 209–216.

Hedenquist, J.W., Henley, R.W., 1985. The importance of CO_2 on freezing point measurements of fluid inclusions: evidence from active geothermal systems and implications for epithermal ore deposition. Econ. Geol. 80, 1379–1406.

Hedenquist, J.W., Lowenstern, J.B., 1994. The role of magmas in the formation of hydrothermal ore deposits. Nature 370, 519–527.

Heijlen, W., Muchez, P., Banks, D.A., 2001. Origin and evolution of high-salinity, Zn–Pb mineralizing fluids in the variscides of Belgium. Miner. Deposita 36, 165–176.

Heilig, M., Franck, E.U., 1989. Calculation of thermodynamic properties of binary fluid mixtures to high temperatures and high pressures. Ber. Bunsenges. Phys. Chem. 93, 898–905.

Hein, U.F., 1990. Kompaktkurs—Einführung in die Mikrothermometrie. Inst. f. Angewandte Geophysik, Petrologie und Lagerstättenforschung, TU Berlin, 103 p.

Heinrich, C.A., Pettke, T., Halter, W.E., Aigner-Torres, M., Audétat, A., Günther, D., Hattendorf, B., Bleiner, D., Guillong, M., Horn, I., 2003. Quantitative multi-element analysis of minerals, fluid and melt inclusions by laser-ablation inductively-coupled-plasma mass-spectrometry. Geochim. Cosmochim. Acta 67, 3473–3496.

Heinrich, C.A., Ryan, C.G., Mernagh, T.P., Eadington, P.J., 1992. Segregation of ore metals between magmatic brine and vapor: a fluid inclusion study using PIXE microanalysis. Econ. Geol. 87, 1566–1583.

Helgeson, H.C., Delany, J.M., Nesbitt, H.W., Bird, D.K., 1978. Summary and critique of the thermodynamic properties of rock-forming minerals. Am. J. Sci. 278-A, 1–229.

Heinrich, W., 2007. Fluid immiscibility in metamorphic rocks. Min. Soc. Am. Rev. Mineral. Geochem. 65, 389–430.

Hendel, E.M., Hollister, L.S., 1981. An empirical solvus for CO_2−H_2O−2.6 wt. % salt. Geochim. Cosmochim. Acta 45, 225–228.

Henjes-Kunst, F., Prochaska, W., Niedermayr, A., Sullivan, N., Baxter, E., 2014. Sm–Nd dating of hydrothermal carbonate formation: an example from the Breitenau magnesite deposit (Styria, Austria). Chem. Geol. 387, 184–201.

Hidden, W.E., 1882. A phenomenal find of fluid-bearing quartz crystals. N. Y. Acad. Sci. Trans. 1, 131–136.

Hiza, M.J., Kidnay, A.J., Miller, R.C., 1982. Equilibrium Properties of Fluid Mixtures. 2. A Bibliography of Experimental Data on Selected Fluids. IFI/Plenum Press, New York/Washington/London, 246 p.

Hladíková, J., Hladil, J., Zusková, J., 1997. Discrimination between facies and global controls in isotope composition of carbonates: carbon and oxygen isotopes at the Devonian reef margin in Moravia (HV−105 Křtiny borehole). J. Czech Geol. Soc. 42, 1–16.

Hoefs, J., 1997. Stable Isotope Geochemistry. Springer, Berlin, 201 p.

Holbrook, P., 1999. A simple closed form force balanced solution for pore pressure, overburden and the principal effective stresses in the Earth. Mar. Petrol. Geol. 16, 303–319.

Holland, T.J.B., 1981. Thermodynamics of simple mineral systems. In: Newton, R., Navrotsky, A., Wood, B.J. (Eds.), Thermodynamics of Minerals and Melts. Springer Verlag, Berlin/Heidelberg/New York, pp. 19–34.

Holland, T.J.B., Powell, R., 1990. An enlarged and updated internally consistent thermodynamic dataset with uncertainties and correlations: the system K_2O–Na_2O–CaO–MgO–MnO–FeO–Fe_2O_3–Al_2O_3–TiO_2–SiO_2–C–H_2–O_2. J. Metamorph. Geol. 8, 89–124.

Holland, T.J.B., Powell, R., 1998. An internally consistent thermodynamic data set for phases of petrological interest. J. Metamorph. Geol. 16, 309–343.

Holland, T.J.B., Powell, R., 2011. An improved and extended internally consistent thermodynamic dataset for phases of petrological interest, involving a new equation of state for solids. J. Metamorph. Geol. 29, 333–383.

Hollister, L.S., Burruss, R.C., 1976. Phase equilibria in fluid inclusions from the Khtada Lake metamorphic complex. Geochim. Cosmochim. Acta 40, 163–175.

Hollister, L.S., Crawford, M.L. (Eds.), 1981. Fluid Inclusions. Applications in Petrology. Min. Assoc. Can. Short Course Handbook, vol. 6, 304 p.

Holloway, J.R., 1977. Fugacity and activity of molecular species in supercritical fluids. In: Fraser, D.G. (Ed.), Thermodynamics in Geology. Reidel Publishing Company, Dordrecht, pp. 161–182.

Holloway, J.R., 1981. Compositions and volumes of supercritical fluids in the earth's crust. In: Hollister, L.S., Crawford, M.L. (Eds.), Short Course in Fluid Inclusions: Application to Petrology. In: Min. Assoc. Can. Short Course Handbook, vol. 6, pp. 13–35.

Holloway, J.R., 1987. Igneous fluids. In: Carmichael, I.S.E., Eugster, H.P. (Eds.), Thermodynamic Modelling of Geological Materials: Minerals, Fluids and Melts. In: Rev. Mineral, vol. 17, pp. 211–233.

Holloway, J.R., Reese, R.L., 1974. The generation of N_2−CO_2−H_2O fluids for use in hydrothermal experimentation: I. Experimental method and equilibrium calculations in the C−O−H−N system. Am. Mineral. 59, 587–597.

Holtz, F., Johannes, W., 1994. Maximum and minimum water contents of granitic melts: implications for chemical and physical properties of ascending magmas. Lithos 32, 149–159.

Holtz, F., Johannes, W., Tamic, N., Behrens, H., 2001. Maximum and minimum water contents of granitic melts generated in the crust: a reevaluation and implications. Lithos 56, 1–14.

Horita, J., Clayton, R.N., 2007. Comment on the studies of the oxygen isotope fractionation between calcium carbonates and water at low temperatures by Zhou and Zheng (2003; 2005). Geochim. Cosmochim. Acta 71, 3131–3135.

Horita, J., Friedman, T.J., Lazar, B., Holland, H.D., 1991. The composition of Permian seawater. Geochim. Cosmochim. Acta 55, 417–432.

Huff, T.A., Nabelek, P.I., 2007. Production of carbonic fluids during metamorphism of graphitic pelites in a collisional orogen—an assessment from fluid inclusions. Geochim. Cosmochim. Acta 71, 4997–5015.

Hui, H., Zhang, Y., 2007. Toward a general viscosity equation for natural anhydrous and hydrous silicate melts. Geochim. Cosmochim. Acta 71, 403–416.

Huizenga, J.-M., 2005. COH, an Excel spreadsheet for composition calculations in the C−O−H fluid system. Comput. Geosci. 31, 797–800.

Hurai, V., 1989. BASIC program for interpretation of microthermometric data from H_2O and H_2O–NaCl fluid inclusions. Comput. Geosci. 15, 135–142.

Hurai, V., 1992. Immiscibility in the system H_2O−CO_2−NaCl: applications to fluid inclusion thermobarometry. N. Jhrb. Mineral. Abh. 165, 5–17.

Hurai, V., 2010. Fluid inclusion geobarometry: pressure corrections for immiscible H_2O–CH_4 and H_2O–CO_2 fluids. Chem. Geol. 278, 201–211.

Hurai, V., Horn, E.-E., 1992. A boundary layer-induced immiscibility in naturally re-equilibrated H_2O–CO_2–NaCl inclusions from metamorphic quartz (Western Carpathians, Czechoslovakia). Contrib. Mineral. Petrol. 112, 414–427.

Hurai, V., Huraiová, M., 2011. Origin of ferroan alabandite and manganoan sphalerite from the Tisovec skarn, Slovakia. N. Jhrb. Mineral. Abh. 188, 119–134.

Hurai, V., Simon, K., Bezák, V., 1997. Contrasting chemistry and H, O, C isotope composition of greenschist-facies, Hercynian and Alpine metamorphic fluids (Western Carpathians). Chem. Geol. 136, 281–293.

Hurai, V., Simon, K., Wiechert, U., Hoefs, J., Konečný, P., Huraiová, M., Pironon, J., Lipka, J., 1998. Immiscible separation of metalliferous Fe/Ti-oxide melts from fractionating alkali basalt: P–T–f_{O2} conditions and two-liquid elemental partitioning. Contrib. Mineral. Petrol. 133, 12–29.

Hurai, V., Janák, M., Ludhová, L., Horn, E.E., Thomas, R., Majzlan, J., 2000. Nitrogen-bearing fluids, brines and carbonate liquids in Variscan migmatites of the Tatra Mountains, Western Carpathians—heritage of high pressure metamorphism. Eur. J. Mineral. 12, 1283–1300.

Hurai, V., Harčová, E., Huraiová, M., Ozdín, D., Prochaska, W., Wiegerová, V., 2002. Origin of siderite veins in the Western Carpathians: I. $P-T-X-{}^{13}C-\delta^{18}O$ relations in ore-forming brines of the Rudňany deposit. Ore Geol. Rev. 21, 67–101.

Hurai, V., Kihle, J., Kotulová, J., Marko, F., Świerczewska, A., 2002. Origin of methane in quartz crystals from the Tertiary accretionary wedge and fore-arc basin of the Western Carpathians. Appl. Geochem. 17, 1259–1271.

Hurai, V., Marko, F., Tokarski, A.K., Świerczewska, A., Kotulová, J., Biroň, A., 2006. Fluid inclusion evidence for deep burial of the Tertiary accretionary wedge of the Carpathians. Terra Nova 18, 440–446.

Hurai, V., Huraiová, M., Konečný, P., Thomas, R., 2007. Mineral–melt–fluid composition of carbonate-bearing cumulate xenoliths in Tertiary alkali basalts of southern Slovakia. Mineral. Mag. 71, 63–79.

Hurai, V., Lexa, O., Schulmann, K., Montigny, R., Prochaska, W., Frank, W., Konečný, P., Král', J., Thomas, R., Chovan, M., 2008. Mobilization of ore fluids during Alpine metamorphism: evidence from hydrothermal veins in the Variscan basement of Western Carpathians, Slovakia. Geofluids 8, 181–207.

Hurai, V., Prochaska, W., Lexa, O., Schulmann, K., Thomas, R., Ivan, P., 2008. High-density nitrogen inclusions in barite from a giant siderite vein: implications for Alpine evolution of the Variscan basement of Western Carpathians, Slovakia. J. Metamorph. Geol. 26, 487–498.

Hurai, V., Janák, M., Thomas, R., 2010. Fluid-assisted retrogression of garnet and P–T history of metapelites from HP/UHP metamorphic terrane (Pohorje Mountains, Eastern Alps). Contrib. Mineral. Petrol. 160, 203–218.

Hurai, V., Huraiová, M., Koděra, P., Prochaska, W., Vozárová, A., Dianiška, I., 2011. Fluid inclusion and stable C–O isotope constraints on the origin of metasomatic magnesite deposits of the Western Carpathians, Slovakia. Russ. Geol. Geophys. 52, 1474–1490.

Hurai, V., Huraiová, M., Milovský, R., Luptáková, J., Konečný, P., 2013. High-pressure aragonite phenocrysts in carbonatite and carbonated syenite xenoliths within an alkali basalt. Am. Mineral. 98, 1074–1077.

Hurai, V., Wierzbicka-Wieczorek, M., Pentrák, M., Huraiová, M., Thomas, R., Świerczewska, A., Luptáková, J., 2014. X-ray diffraction and vibrational spectroscopic characteristics of hydroxylclinohumite from ruby-bearing marbles (Luc Yen District, Vietnam). Intl. J. Mineral., Article ID 648530, 11 p.

Huraiová, M., Dubessy, J., Konečný, P., 1991. Composition and oxidation state of upper mantle fluids from Southern Slovakia. Plinius 5, 110–111.

Huraiová, M., Vozárová, A., Repčok, I., 2002. Fluid inclusion and stable isotope constraints on the origin of magnesite at Burda, Ochtiná, Lubeník, and Ploské deposits (Slovakia, Western Carpathians). Geol. Carpath. Spec. Issue 53, 98–100.

Huraiová, M., Dubessy, J., Konečný, P., Simon, K., Král', J., Zielinski, G., Lipka, J., Hurai, V., 2005. Glassy orthopyroxene granodiorites of the Pannonian Basin: tracers of ultra-high-temperature deep-crustal anatexis triggered by Tertiary basaltic volcanism. Contrib. Mineral. Petrol. 148, 615–633.

Ihinger, P.D., Hervig, R.L., McMillan, P.F., 1994. Analytical methods for volatiles in glasses. In: Carroll, M.R., Holloway, J.R. (Eds.), Volatiles in Magmas. In: Rev. Mineral., vol. 30, pp. 67–121.

Irmer, G., Graupner, T., 2002. Isotopes of C and O in CO_2: a Raman study using gas standards and natural fluid inclusions. Acta Univ. Carol. Geol. 46, 35–36.

Irwin, H., Curtis, C., Coleman, M., 1977. Isotopic evidence for source of diagenetic carbonates formed during burial of organic-rich sediments. Nature 269, 209–213.

Irwin, J.J., Roedder, E., 1995. Diverse origins of fluid in magmatic inclusions at Bingham (Utah, USA), Butte (Montana, USA), St. Austell (Cornwall, UK), and Ascension Island (Mid-Atlantic, UK), indicated by laser microprobe analysis of Cl, K, I, Ba, Te, Ar, Kr, and Xe. Geochim. Cosmochim. Acta 59, 295–312.

Jacobs, G.K., Kerrick, D.M., 1981. Methane: An equation of state with application to the ternary system H_2O–CO_2–CH_4. Geochim. Cosmochim. Acta 45, 607–614.

Jacobsen, R.T., Stewart, R.B., Jahangiri, M., 1986. Thermodynamic properties of nitrogen from the freezing line to 2000 K at pressures to 1000 MPa. J. Phys. Chem. Ref. Data 15, 735–909.

Jacquemet, N., Guillaume, D., Zwick, A., Pokrovski, G.S., 2014. In situ Raman spectroscopy identification of the S^{3-} ion in S-rich hydrothermal fluids from synthetic fluid inclusions. Am. Mineral. 99, 1109–1118.

Jakobsen, J.K., Veksler, I.V., Tegner, C., Brooks, C.K., 2005. Immiscible iron- and silica-rich melts in basalt petrogenesis documented in the Skaergaard intrusion. Geology 33, 885–888.

Janák, M., O'Brien, P., Hurai, V., Reutel, Ch., 1996. Metamorphic evolution and fluid composition of garnet–clinopyroxene amphibolites from the Tatra Mts., Western Carpathians. Lithos 39, 57–79.

Janák, M., Hurai, V., Ludhová, L., O'Brien, P.J., Horn, E.-E., 1999. Dehydration-melting and devolatilization during exhumation of the high-grade metapelites: the Tatra Mountains, Western Carpathians. J. Metamorph. Geol. 17, 379–395.

Javoy, M., 1977. Stable isotopes and geothermometry. J. Geol. Soc. London 133, 609–636.

Jehlička, J., Bény, C., Rouzaud, J.-N., 1997. Raman microspectrometry of accumulated non-graphitized solid bitumens. J. Raman Spectrosc. 28, 717–724.

Jia, Y., Li, X., Kerrich, R., 2000. A fluid inclusion study of Au-bearing quartz vein systems in the Central and North Deborah Deposits of the Bendigo Gold Field, Central Victoria, Australia. Econ. Geol. 95, 467–494.

Jiang, S., Pitzer, K.S., 1996. Phase equilibria and volumetric properties of aqueous $CaCl_2$ by an equation of state. AIChE J. 42, 585–594.

Jibao, G., Yaqian, Q., 1997. Hydrogen isotope fractionation and hydrogen diffusion in the tourmaline-water system. Geochim. Cosmochim. Acta 61, 4679–4688.

Johnson, J., Oelkers, E.H., Helgeson, H.C., 1992. SUPCRT92: a software package for calculating the standard molal thermodynamic properties of minerals, gases, aqueous species, and reactions from 1 to 5000 bar and 0 to 1000 °C. Comput. Geosci. 18, 899–947.

Kagi, H., Lu, R., Davidson, P., Goncharov, A.F., Mao, H.K., Hemley, R.J., 2000. Evidence for ice VI as an inclusion in cuboid diamonds from high P–T near infrared spectroscopy. Mineral. Mag. 64, 1089–1097.

Kamenetsky, V.S., Kamenetsky, M.B., 2011. Magmatic fluids immiscible with silicate melts: examples from inclusions in phenocrysts and glasses, and implications for magma evolution and metal transport. In: Yardley, B., Manning, C., Garven, G. (Eds.), Frontiers in Geofluids. Wiley–Blackwell, Chichester, pp. 293–311.

Kamenetsky, V.S., Crawford, A.J., Eggins, S.M., Mühe, R., 1997. Phenocrysts and melt inclusion chemistry of near-axis seamounts, Valu Fa Ridge, Lau Basin: insight into mantle wedge melting and the addition of subduction components. Earth Planet. Sci. Lett. 151, 205–223.

Kamenetsky, V.S., De Vivo, B., Naumov, V.B., Kamenetsky, M.B., Mernagh, T.P., Van Achterbergh, E., Ryan, Ch.G., Davidson, P., 2003. Magmatic inclusions in the search for natural silicate–salt melt immiscibility:

methodology and examples. In: De Vivo, B., Bodnar, R.J. (Eds.), Melt Inclusions in Volcanic Systems. Methods, Applications and Problems, first ed. Elsevier, Amsterdam/Boston, pp. 65–82.

Kamenetsky, V.S., Naumov, V.B., Davidson, P., van Achterbergh, E., Ryan, C.G., 2004. Immiscibility between silicate magmas and aqueous fluids: a melt inclusion pursuit into the magmatic–hydrothermal transition in the Omsukchan Granite (NE Russia). Chem. Geol. 210, 73–90.

Kamenetsky, V.S., Kamenetsky, M.B., Sharygin, V.V., Faure, K., Golovin, A.V., 2007. Chloride and carbonate immiscible liquids at the closure of the kimberlite magma evolution (Udachnaya-East kimberlite, Siberia). Chem. Geol. 237, 384–400.

Kantor, J., Ďurkovičová, J., 1977. Sulfur isotopes in selected barite deposits of Western Carpathians. Unpublished report. Geofond, Bratislava, 244 p. (in Slovak).

Kanzaki, M., Xue, X., Amalberti, J., Zhang, Q., 2012. Raman and NMR spectroscopic characterization of high-pressure K–cymrite ($KAlSi_3O_8.H_2O$) and its anhydrous form (kokchetavite). J. Mineral. Petrol. Sci. 107, 114–119.

Karczemska, A.T., 2010. Diamonds in meteorites—Raman mapping and cathodoluminescence studies. J. Achiev. Mater. Manuf. Eng. 43, 94–107.

Karlsson, H.R., Clayton, R.N., 1990. Oxygen isotope fractionation between analcime and water: an experimental study. Geochim. Cosmochim. Acta 54, 1359–1368.

Karpinskiy, A.P., 1880. On the occurrence of inclusions of liquid carbon dioxide in mineral substances. Gornyj Zh. 2, 96–117 (in Russian).

Katsura, T., Ito, E., 1992. The system $MgO–SiO_2–CO_2–H_2O$ at high pressure: a preliminary investigation of CO_2 concentration in mantle fluids. In: Syono, Y., Manghnani, M.H. (Eds.), High-Pressure Research: Application to Earth and Planetary Sciences. Terra Scientific, Tokyo, pp. 275–281.

Kawabe, I., 1978. Calculation of oxygen isotope fractionation in quartz–water system with special reference to the low temperature fractionation. Geochim. Cosmochim. Acta 42, 613–621.

Kawakami, Y., Yamamoto, J., Kagi, H., 2003. Micro-Raman densimeter for CO_2 inclusions in mantle-derived minerals. Appl. Spectrosc. 57, 1333–1339.

Kazahaya, K., Matsuo, S., 1984. A new ball-milling method for extraction of fluid inclusions from minerals. Geochem. J. 19, 45–54.

Keenan, J.H., Keyes, F.G., Hill, P.G., Moore, J.G., 1978. Steam Tables: Thermodynamic Properties of Water, Including Vapor, Liquid and Solid Phases (SI Units). Wiley, New York.

Kendall, C., Caldwell, E.A., 1998. Fundamentals of isotope geochemistry. In: Kendall, C., McDonnell, J.J. (Eds.), Isotope Tracers in Catchment Hydrology. Elsevier, Amsterdam, pp. 51–86.

Kerrick, D.M., Jacobs, G.K., 1981. A modified Redlich–Kwong equation for H_2O, CO_2, and $H_2O–CO_2$ mixtures. Am. J. Sci. 281, 735–767.

Kesler, S.E., Haynes, P.S., Creech, M.Z., Gorman, J.A., 1986. Applications of fluids inclusion and rock-gas analysis in mineral exploration. J. Geochem. Explor. 25, 201–215.

Kesler, S.E., Bodnar, R.J., Mernagh, T.P., 2013. Role of fluid and melt inclusion studies in geologic research. Geofluids 13, 398–404.

Kharbish, S., Andráš, P., Luptáková, J., Milovská, S., 2014. Raman spectra of oriented and non-oriented Cu hydroxy-phosphate minerals: libethenite, cornetite, pseudomalachite, reichenbachite and ludjibaite. Spectrochim. Acta A Mol. Biomol. Spectrosc. 130, 152–163.

Khavari Khorasani G., 1987. Novel development in fluorescence microscopy of complex organic mixtures: application in petroleum geochemistry. Org. Geochem. 11, 157–168.

Kiefer, J., Seeger, T., Steuer, S., Schorsch, S., Weikl, M.C., Leipertz, A., 2008. Design and characterization of a Raman-scattering-based sensor system for temporally resolved gas analysis and its application in a gas turbine power plant. Meas. Sci. Technol. 19, 1–9.

Kihle, J., Hurum, J.H., Liebe, L., 2012. Preliminary results on liquid petroleum occurring as fluid inclusions in intracellular mineral precipitates in the vertebrate of Pliosaurus funkei. Norsk Geol. Tidsskr. 92, 341–352.

Kim, S.-T., O'Neil, J.R., 1997. Equilibrium and non-equilibrium oxygen isotope effects in synthetic carbonates. Geochim. Cosmochim. Acta 61, 3461–3475.

Kingma, K.J., Hemley, R., 1994. Raman spectroscopy study of microcrystalline silica. Am. Mineral. 79, 269–273.

Kita, I., Taguchi, S., Matsubaya, O., 1985. Oxygen isotope fractionation between amorphous silica and water at 34–93 °C. Nature 314, 63–64.

Kleinrahm, R., Wagner, W., 1986. Measurement and correlation of the equilibrium liquid and vapour densities and the vapour pressure along the coexistence curve of methane. J. Chem. Thermodyn. 18, 739–760.

Klemme, S., 2004. Evidence for fluoride melts in Earth's mantle formed by liquid immiscibility. Geology 32, 441–444.

Knight, C.L., Bodnar, R.J., 1989. Synthetic fluid inclusions: IX. Critical PVTX properties of $NaCl-H_2O$ solutions. Geochim. Cosmochim. Acta 53, 3–8.

Kodĕra, P., Murphy, P.J., Rankin, A.H., 2003. Retrograde mineral reactions in saline fluid inclusions: the transformation ferropyrosmalite ↔ clinopyroxene. Am. Mineral. 88, 151–158.

Kodĕra, P., Radvanec, M., 2002. Comparative mineralogical and fluid inclusion study of the Hnúšt'a–Mútnik talc–magnesite and Miková–Jedl'ová magnesite deposit (Western Carpathians, Slovakia). Bol. Parana. Geociênc. 50, 131–150.

Kolker, A., Lindsley, D.H., 1989. Geochemical evolution of the Maloin pluton, Laramie Anorthosite Complex, Wyoming: petrology and mixing relations. Am. Mineral. 74, 307–324.

Königsberger, J., 1901. Die Minerallagerstätten im Biotitprotogrin des Aarmassivs. N. Jhrb. Mineral. Abh. 14, 43–119.

Königsberger, J., Müller, W.J., 1906. Über die Flüssigkeitseinschlüsse im Quarz alpiner Mineralklüfte. Zentralbl. Mineral. Geol. Paläont. 72–77.

Konnerup-Madsen, J., 2001. A review of the composition and evolution of hydrocarbon gases during solidification of the Ilímaussaq alkaline complex, South Greenland. Geol. Greenland Surv. Bull. 190, 159–166.

Konnerup-Madsen, J., Dubessy, J., Rose-Hansen, J., 1985. Combined Raman microprobe spectrometry and microthermometry of fluid inclusions in minerals from igneous rocks of the Gardar Province (south Greenland). Lithos 18, 271–280.

Kotková, J., Škoda, R., Machovič, V., 2014. Kumdykolite from the ultrahigh-pressure granulite of the Bohemian Massif. Am. Mineral. 99, 1798–1801.

Kouzmanov, K., Bailly, L., Ramboz, C., Rouer, O., Bény, J.-M., 2002. Morphology, origin and infrared microthermometry of fluid inclusions in pyrite from the Radka epithermal copper deposit, Srednogorie zone, Bulgaria. Miner. Deposita 37, 599–613.

Kozlowski, A., 1985. Do melt inclusions lose volatiles during homogenization studies? In: VIIIth ECROFI Meeting, Abstracts, Göttingen, pp. 88–89.

Krader, T., Franck, E.U., 1987. The ternary systems H_2O-CH_4-NaCl and $H_2O-CH_4-CaCl_2$ to 800 K and 250 MPa. Ber. Bunsenges. Phys. Chem. 91, 627–634.

Kretschmar, U., Scott, S.D., 1976. Phase relations involving arsenopyrite in the system Fe–As–S and their application. Can. Mineral. 14, 364–386.

Krumrei, T.V., Pernicka, E., Kaliwoda, M., Markl, G., 2007. Volatiles in a peralkaline system: abiogenic hydrocarbons and F–Cl–Br systematics in the naujaite of the Ilímaussaq intrusion, South Greenland. Lithos 95, 298–314.

Kučera, J., Muchez, Ph., Slobodník, M., Prochaska, W., 2010. Geochemistry of highly saline fluids in the Moravo-Silesian Palaeozoic siliciclastic sequences: genetic implications. Int. J. Earth Sci. (Geol. Rundsch.) 99, 269–284.

Kumagai, Y., Kawamoto, T., Yamamoti, J., 2014. Evolution of carbon dioxide-bearing saline fluids in the mantle wedge beneath the Northeast Japan arc. Contrib. Mineral. Petrol. 168, 1056.

Kusakabe, M., Robinson, B.W., 1977. Oxygen and sulfur isotope equilibria in the $BaSO_4-HSO_4^--H_2O$ system from 110 to 350 °C and applications. Geochim. Cosmochim. Acta 41, 1033–1040.

Kyser, T.K., 1986. Stable isotope variations in the mantle. In: Valley, J.W., Taylor, H.P., Jr., (Eds.), Stable Isotopes in High–Temperature Geological Processes. In: Min. Soc. Am. Rev. Mineral, vol. 16, pp. 141–164.

Kyser, T.K., O'Neil, J.R., 1984. Hydrogen isotope systematics of submarine basalts. Geochim. Cosmochim. Acta 48, 2123–2133.

Lagache, M., Weisbrod, A., 1977. The system: two alkali feldspars–KCl–NaCl–H_2O at moderate to high temperatures and low pressures. Contrib. Mineral. Petrol. 62, 77–101.

Lahfid, A., Beyssac, O., Deville, E., Negro, F., Chopin, C., Goffé, B., 2010. Evolution of the Raman spectrum of carbonaceous material in low-grade metasediments of the Glarus Alps (Switzerland). Terra Nova 22, 354–360.

Lambert, S.J., Epstein, S., 1980. Stable isotope investigations of an active geothermal system in Valles Caldera, Jemez Mountains, New Mexico. J. Volcanol. Geotherm. Res. 8, 111–129.

Landenburger, S., Marks, M., Markl, G., 2012. Analysis of crush leach solutions from hydrothermal ore deposits by combining ion chromatography (IC) and total reflection X-ray fluorescence spectroscopy (TXRF). In: European Mineralogical Conference Abstract, vol. 1.

Landis, G.P., Hofstra, A.H., 1991. Fluid inclusion gas chemistry as a potential minerals exploration tool – case studies from Creede, CO, Jerrit Canyon, NV, Coeur-dalene District, ID and MT, southern Alaska mesothermal veins, and midcontinent MVT's. J. Geochem. Explor. 42, 25–59.

Landsberg, G., Mandelstam, L., 1928. Eine neue Erscheinung bei der Lichtzerstreuung in Krystalen. Naturwissenschaften 16, 557–558.

Larkin, P.J., 2011. IR and Raman spectroscopy. Principles and Spectral Interpretation. Elsevier, Waltham/San Diego/Oxford/Amsterdam, 228 p.

Larsen, R.B., 1993. Geofluid—a FORTRAN–77 program to compute chemical properties of gas species in C–O–H fluids. Comput. Geosci. 19, 1295–1320.

Le Losq, C., Neuville, D.R., Moretti, R., Roux, J., 2012. Determination of water content in silicate glasses using Raman spectrometry: implications for the study of explosive volcanism. Am. Mineral. 97, 779–790.

Lecumberri-Sanchez, P., Steele-MacInnis, M., Bodnar, R.J., 2012. A numerical model to estimate trapping conditions of fluid inclusions that homogenize by halite disappearance. Geochim. Cosmochim. Acta 92, 14–22.

Lecumberri-Sanchez, P., Steele-MacInnis, M., Bodnar, R.J., 2014. Synthetic fluid inclusions XIX. Experimental determination of the vapour-saturated liquidus of the system H_2O–NaCl–$FeCl_2$. Geochim. Cosmochim Acta 148, 34–49.

Lecumberri-Sanchez, P., Steele-MacInnis, M., Bodnar, R.J., 2014. Synthetic fluid inclusions XIX. Experimental determination of the vapour-saturated liquidus of the system H_2O–NaCl–$FeCl_2$. Geochim. Cosmochim. Acta (in press).

Lecuyer, C., O'Neil, J.R., 1994. Stable isotope compositions of fluid inclusions in biogenic carbonates. Geochim. Cosmochim. Acta 58, 353–363.

Leeder, O., Thomas, R., Klemm, W., 1987. Einschlüsse in Mineralen. VEB Deutscher Verlag f. Grundstoffindustrie, Leipzig, 180 p.

Leisen, M., Boiron, M.-C., Richard, A., Dubessy, J., 2012a. Determination of Cl and Br concentrations in individual fluid inclusions by combining microthermometry and LA-ICPMS analysis: implications for the origin of salinity in crustal fluids. Chem. Geol. 330–331, 197–206.

Leisen, M., Dubessy, J., Boiron, M.-C., Lach, P., 2012b. Improvement of the determination of element concentrations in quartz-hosted fluid inclusions and Pitzer thermodynamic modelling of ice melting temperature. Geochim. Cosmochim. Acta 90, 110–125.

Lemmlein, G.G., 1929. Sekundäre Flüssigkeitseinschlüsse in Mineralen. Z. Kristallogr. 71, 237–256.

Lemmlein, G.G., 1950. Al-Biruni's Mineralogical Information. Sbornik Biruni, Moskva/Leningrad, 106–127.

Lemmlein, G.G., 1973. Morphology and Genesis of the Crystals. Nauka, Moskva, 327 p. (in Russian).

Leroy, J., 1979. Contribution á l´étalonnage de la presion interne des inclusions fluides lors de leur décrépitation. Bull. Minéral. 102, 584–593.

Leschik, M., Heide, G., Frischat, G.H., Behrens, H., Wiedenbeck, M., Wagner, N., Heide, K., Geißler, H., Reinholz, U., 2004. Determination of H_2O and D_2O contents in rhyolitic glasses. Phys. Chem. Glasses 45, 238–251.

Leško, B., Babák, B., Borkovcová, D., Cesnek, V., Ďurkovič, T., Fáber, P., Fil'ková, V., Gašpariková, V., Harča, V., Hradil, F., 1985. The Supporting Well Hanušovce-1, 6003 m. Geologický Ústav D. Štúra, Bratislava, 207 p. (in Slovak).

Lewis, G.N., 1908. The osmotic pressure of concentrated solutions, and the laws of the perfect solution. J. Am. Chem. Soc. 30, 668–683.

Lewis, G.N., Randall, M., 1923. Thermodynamics and the Free Energy of Chemical Substances, first ed. McGraw-Hill Book Company Inc., New York, 653 p.

Lewis, J.C., Byrne, T.B., Pasteris, J.D., London, D., Morgan, G.B., IV, 2000. Early Tertiary fluid flow and pressure–temperature conditions in the Shimanto accretionary complex of south-west Japan: constraints from fluid inclusions. J. Metamorph. Geol. 18, 319–333.

Lexa, J., Bezák, V., Elečko, M., Mello, J., Polák, M., Potfaj, M., Vozár, J., Schnabel, G.W., Pálenský, P., Császár, G., Rylko, W., Mackiv, B., 2000. Geological Map of Western Carpathians and Adjacent Areas 1:500 000. Geological Survey of Slovakia, Bratislava.

Lexa, J., Bačo, P., Chovan, M., Petro, M., Rojkovič, I., Tréger, M., 2004. Metallogenetic Map of Slovakia. Ministry of Environment and Geological Survey of Slovakia, Bratislava.

Lexa, O., Schulmann, K., Ježek, J., 2003. Cretaceous collision and indentation in the West Carpathians: view based on structural analysis and numerical modelling. Tectonics 22, 1066.

Li, J., Li, S., 2014. Application of hydrothermal diamond anvil cell to homogenization experiments of silicate melt inclusions. Acta Geol. Sin. 88, 854–864.

Li, J., Xia, S., Gao, S., 1995. FT–IR and Raman spectroscopic study of hydrated borates. Spectrochim. Acta 51A, 519–532.

Li, Y.B., Liu, J.M., 2006. Calculation of sulfur isotope fractionation in sulfides. Geochim. Cosmochim. Acta 70, 1789–1795.

Lin, F., Bodnar, R.J., 2010. Synthetic fluid inclusions XVIII: experimental determination of the PVTX properties of H_2O–CH_4 to 500 °C, 3 kbar and $XCH_4 \leq 4$ mol. %. Geochim. Cosmochim. Acta 74, 3260–3273.

Lin, F., Bodnar, R.J., Becker, S.P., 2007a. Experimental determination of the Raman CH_4 symmetric stretching (ν_1) band position from 1–650 bar and 0.3–22 °C: application to fluid inclusion studies. Geochim. Cosmochim. Acta 71, 3746–3756.

Lin, F., Bodnar, R.J., Becker, S.P., 2007b. Correlation of methane Raman ν_1 band position with fluid density and interactions at the molecular level. J. Raman Spectrosc. 38, 1510–1515.

Lindaas, S.E., Kulis, J., Campbell, A.R., 2002. Near–infrared observation and microthermometry of pyrite-hosted fluid inclusions. Econ. Geol. 97, 603–618.

Linke, W.F., Seidell, A., 1965. Solubilities of Inorganic and Metal Organic Compounds, fourth ed. Am. Chem. Soc., Washington, DC

Lin-Vien, D., Colthup, N.B., Fateley, W.G., Grasselli, J.G., 1991. The Handbook of Infrared and Raman Characteristic Frequencies of Organic Molecules. Academic Press, San Diego, 503 p.

Liu, D.H., Xiao, X.M., Tian, H., Min, Y.S., Zhou, Q., Cheng, P., Shen, J.G., 2013a. Sample maturation calculated using Raman spectroscopic parameters for solid organics: methodology and geological applications. Chin. Sci. Bull. 58, 1258–1298.

Liu, W.B., Li, Y.G., Lu, J.F., 1999. A new equation of state for real aqueous ionic fluids based on electrolyte perturbation theory, mean spherical approximation and statistical associating fluid theory. Fluid Phase Equilib. 158–160, 595–606.

Liu, Y.S., Hu, Z.C., Li, M., Gao, S., 2013b. Applications of LA-ICP-MS in the elemental analyses of geological samples. Chin. Sci. Bull. 58, 3863–3878.

Lloyd, R.M., 1968. Oxygen isotope behavior in the sulfate–water system. J. Geophys. Res. 73, 6099–6110.

Lockner, D.A., 1995. Rock failure. In: Rock Physics and Phase Relations. Handbook of Physical Constants. In: AGU Reference Shelf, vol. 3, pp. 127–147.
Loppnow, G.R., Shoute, L., Schmidt, K.J., Savage, A., Hall, R.H., Bulmer, J.T., 2004. UV Raman spectroscopy of hydrocarbons. Royal Soc. London Phil. Trans. A 362, 2461–2476.
Louck, R.R., 2000. Precise geothermometry on fluid inclusion populations that trapped mixtures of immiscible fluids. Am. J. Sci. 300, 23–59.
Lowell, J.D., Guilbert, J.M., 1970. Lateral and vertical alteration–mineralization zoning in porphyry ore deposits. Econ. Geol. 65, 373–408.
Lowenstern, J.B., 2003. Melt inclusions come on age: volatiles, volcanoes, and Sorby's legacy. In: De Vivo, B., Bodnar, R.J. (Eds.), Melt Inclusions in Volcanic Systems. Methods, Applications and Problems, first ed. Elsevier, Amsterdam/Boston, pp. 1–21.
Lu, W.J., Chou, I.-M., Burruss, R.C., Song, Y.C., 2007. A unified equation for calculating methane vapour pressures in CH_4–H_2O system with measured Raman shift. Geochim. Cosmochim. Acta 71, 3969–3978.
Lu, W.J., Chou, I.-M., Burruss, R.C., 2008. Determination of methane concentrations in water in equilibrium with methane hydrate in the absence of a vapour phase by in situ Raman spectroscopy. Geochim. Cosmochim. Acta 72, 412–422.
Lüders, V., 1996. Contribution of infrared microscopy to fluid inclusion studies in some opaque minerals (wolframite, stibnite, bournonite): Metallogenetic implications. Econ. Geol. 91, 1462–1468.
Lüders, V., Gutzmer, J., Beukes, N.J., 1999. Fluid inclusion studies in cogenetic hematite, hausmannite, and gangue minerals from high-grade manganese ores in the Kalahari Manganese Field, South Africa. Econ. Geol. 94, 589–596.
Lüders, V., Romer, R.L., Cabral, A.R., Schmidt, Ch., Banks, D.A., Schneider, J., 2005. Genesis of itabirite-hosted Au–Pd–Pt-bearing hematite–(quartz) veins, Quadrilátero Ferrífero, Minais Gerais, Brazil: constraints from fluid inclusion infrared microthermometry, bulk crush–leach analysis and U–Pb systematics. Miner. Deposita 40, 289–306.
Lüders, V., Ziemann, M., 1999. Possibilities and limits of infrared light microthermometry applied to studies of pyrite-hosted fluid inclusions. Chem. Geol. 154, 169–178.
Ludwig, R., 2001. Water: from clusters to the bulk. Angew. Chem. Int. Ed. 40, 1808–1827.
Luque, F.J., Crespo-Feo, E., Barrenechea, J.F., Ortega, L., 2012. Carbon isotopes of graphite: implications on fluid history. Geosci. Front. 3, 197–207.
Maillard, A., 1936. Content of deuterium in light hydrocarbons of various petroleums. C. R. Acad. Sci. 203, 804–806.
Mallamace, F., Corsaro, C., Stanley, H.E., 2013. Possible relation of water structural relaxation to water anomalies. PNAS 110, 4899–4904.
Mancano, D.P., Campbell, A.R., 1995. Microthermometry of enargite-hosted fluid inclusions from Lepanto, Philippines, high-sulfidation Cu–Au deposit. Geochim. Cosmochim. Acta 59, 3909–3916.
Mango, F.D., Hightower, J., 1997. The catalytic decomposition of petroleum into natural gas. Geochim. Cosmochim. Acta 61, 5347–5350.
Marini, L., Moretti, R., Accornero, M., 2011. Sulfur isotopes in magmatic-hydrothermal systems, melts and magmas. Min. Soc. Am. Rev. Mineral. Geochem. 73, 423–492.
Marion, G.M., Farren, R.E., 1999. Mineral solubilities in the Na–K–Mg–Ca–Cl–SO_4–H_2O system: a re-evaluation of the sulfate chemistry in the Spencer–Møller–Weare model. Geochim. Cosmochim. Acta 63, 1305–1318.
Marko, F., Hurai, V., Dyda, M., Almeida, G., Prochaska, W., Thomas, R., 2006. Tectonic and fluid inclusion constraints on the origin of quartz veins with giant crystals in the Tocantins structural province (Cristalândia, central Brazil). J. South Am. Earth Sci. 21, 241–253.
Mason, B.J., 1971. The Physics of Clouds, second ed. Oxford University Press, London/New York, 671 p.
Matsuhisa, Y., Goldsmith, J.R., Clayton, R.N., 1978. Mechanism of hydrothermal crystallization of quartz at 250 °C and 15 kbar. Geochim. Cosmochim. Acta 42, 173–182.

Matsumo, T., Honda, M., McDougall, I., Yatsevich, I., O'Reilly, S.Y., 1996. Plume-like neon in a metasomatic apatite from Australian lithospheric mantle. Nature 288, 162–164.

Matsumo, T., Honda, M., McDougall, I., Yatsevich, I., O'Reilly, S.Y., 1998. Noble gases in anhydrous lherzolites from the Newer Volcanics, southern Australia: a MORB-like reservoir in the subcontinental mantle. Geochim. Cosmochim. Acta 62, 2521–2533.

Matsumura, M., Hashimoto, Y., Kimura, G., Ohmori-Ikehara, K., Enjohji, M., Ikesawa, E., 2003. Depth of oceanic-crust underplating in a subduction zone: inferences from fluid-inclusion analyses of crack-seal veins. Geology 31, 1005–1008.

Matthews, A., Goldsmidth, J.R., Clayton, R.N., 1983. Oxygen isotope fractionation involving pyroxenes: the calibration of mineral-pair geothermometers. Geochim. Cosmochim. Acta 47, 631–644.

Matthews, A., Ayalon, A., Bar-Matthews, M., 2000. D/H ratios of fluid inclusions of Soreq Cave (Israel) speleothems as a guide to the eastern Mediterranean Metoric Line relationship in the last 120 ky. Chem. Geol. 166, 183–191.

Mavrogenes, J.A., Bodnar, R.J., 1994. Hydrogen movement into and out of fluid inclusions in quartz: experimental evidence and geologic implications. Geochim. Cosmochim. Acta 58, 141–148.

Maya, L., Begun, G.M., 1981. A Raman spectroscopic study of hydroxo and carbonato species of the uranyl (VI) ion. J. Inorg. Nucl. Chem. 43, 2827–2832.

Mc Caffrey, M.A., Lazar, B., Holland, H.D., 1987. The evaporation path of seawater and the coprecipitation of Br^- and K^+ with halite. J. Sediment. Petrol. 57, 928–937.

McCaig, A.M., Tritlla, J., Banks, D.A., 2000. Fluid mixing and recycling during Pyrenean thrusting: evidence from fluid inclusion halogen ratios. Geochim. Cosmochim. Acta 64, 3395–3412.

McCain, W.D., Jr., 1994. Heavy components control reservoir fluid behavior. J. Petrol. Technol. 46, 746–750.

McCrea, J.M., 1950. On the isotopic chemistry of carbonates and a paleotemperature scale. J. Chem. Phys. 18, 849–857.

McGarry, S., Bar-Matthews, M., Vaks, A., Ayalon, A., 2004. Constraints on hydrological and paleotemperature variations in the eastern Mediterranean region in the last 140 ka given by the δD values of speleothem fluid inclusions. Quat. Sci. Rev. 23, 919–934.

McKinney, C.R., McCrea, J.M., Epstein, S., Allen, H.A., Urey, H.C., 1950. Improvements in mass spectrometers for the measurement of small differences in isotope abundance ratios. Rev. Sci. Instrum. 21, 724–730.

Mercier, M., Di Muro, A., Metrich, N., Giordano, D., Belhadj, O., Mandeville, C.W., 2010. Spectroscopic analysis (FTIR, Raman) of water in mafic and intermediate glasses and glass inclusions. Geochim. Cosmochim. Acta 74, 5641–5656.

Mernagh, T.P., Bastrakov, E.N., 2013. An evaluation of hydrogen sulfide in anorogenic gold fluids and the uncertainties associated with vapor-rich inclusions. Geofluids 13, 494–505.

Mernagh, T.P., Wilde, A.R., 1989. The use of Raman microprobe for the determination of salinity in fluid inclusions. Geochim. Cosmochim. Acta 53, 765–771.

Michard, G., Fouillac, C., 1981. Sodium/lithium ratio in water applied to geothermometry of geothermal reservoirs. Geothermics 10, 55–70.

Mikhno, A.O., Schmidt, U., Korsakov, A.V., 2013. Origin of K–cymrite and kokchetavite in the polyphase mineral inclusions from Kokchetav UHP calc-silicate rocks: evidence from confocal Raman imaging. Eur. J. Mineral. 25, 807–816.

Milata, V., Segl'a, P., Brezová, V., Gatial, A., Kováčik, V., Miglierini, M., Stankovský, Š., Šíma, J., 2008. Applied Molecular Spectroscopy. Slovenská Technická Univerzita, Bratislava, 600 p. (in Slovak).

Milovský, R., 2000. Overthrust Mechanisms of Superficial Nappes of Central Western Carpathians: The Role of Fluid Overpressure on their Bases. Ph.D. Thesis, GÚ SAV, Bratislava, 110 p. (in Slovak).

Milovský, R., Hurai, V., Plašienka, D., 2002. Hydrotectonic regime at soles of overthrust sheets: textural and fluid inclusion evidence from basal cataclasites of the Muráň Nappe (Western Carpathians, Slovakia). Geodin. Acta 16, 1–20.

Milovský, R., van den Kerkhof, A., Hoefs, J., Hurai, V., Prochaska, W., 2012. Cathodoluminescence, fluid inclusion and stable C–O isotope study of tectonic breccias from thrusting plane of a thin-skinned calcareous nappe. Int. J. Earth. Sci. (Geol. Rundsch.) 101, 535–554.

Misra, K.C., 2012. Introduction to Geochemistry. Principles and Applications, Wiley–Blackwell, 438 p.

Momma, K., Ikeda, T., Nishikuboa, K., Takahashi, N., Honma, Ch., Takada, M., Furukawa, Y., Nagase, T., Kudoh, Y., 2011. New silica clathrate minerals that are isostructural with natural gas hydrates. Nat. Commun. 2, 1–7, Art. 196.

Montel, F., 1993. Phase equilibria needs for petroleum exploration and production industry. Fluid Phase Equilib. 84, 343–367.

Moon, K.J., 1991. Application of fluid inclusions in mineral exploration. J. Geochem. Explor. 42, 205–221.

Moore, W.J., 1979. Physical Chemistry. SNTL – Nakladatelství technické literatury, Praha (in Slovak).

Morizet, Y., Nichols, A.R.L., Kohn, S.C., Brooker, R.A., Dingwell, D.B., 2007. The influence of H_2O and CO_2 on the glass transition temperature: insights into the effects of volatiles on magma viscosity. Eur. J. Mineral. 19, 657–669.

Morizet, Y., Paris, M., Gaillard, F., Scaillet, B., 2009. Raman quantification factor calibration for CO–CO_2 gas mixture in synthetic fluid inclusions: application to oxygen fugacity calculation in magmatic systems. Chem. Geol. 264, 58–70.

Morrison, G., 1981. Effect of water upon critical points of carbon dioxide and ethane. J. Phys. Chem. 85, 759–761.

Mullis, J., 1979. The system methane–water as a geologic thermometer and barometer from the external parts of the Central Alps. Bull. Minéral. 102, 526–536.

Mullis, J., 1987. Fluid inclusion studies during very low-grade metamorphism. In: Frey, M. (Ed.), Low Grade Metamorphism. Blackie, Glasgow, pp. 162–189.

Mullis, J., Dubessy, J., Poty, B., O'Neil, J., 1994. Fluid regimes during late stages of a continental collision: physical, chemical, and stable isotope measurements of fluid inclusions in fissure quartz from a geotraverse through the Central Alps, Switzerland. Geochim. Cosmochim. Acta 58, 2239–2267.

Munz, I.A., 2001. Petroleum inclusions in sedimentary basins: systematics, analytical methods and applications. Lithos 55, 195–212.

Munz, I.A., Iden, K., Johansen, H., Vagle, K., 1998. The fluid regime during fracturing of the Embla field, Central Trough, North Sea. Mar. Petrol. Geol. 15, 751–768.

Munz, I.A., Johansen, H., Holm, K., Lacharpagne, J.-C., 1999. The petroleum characteristics of the Frøy field and the Rind discovery, Norwegian North Sea. Mar. Petrol. Geol. 16, 633–651.

Murck, B.W., Burruss, R.C., Hollister, L.S., 1978. Phase equilibria in fluid inclusions in ultramafic xenoliths. Am. Mineral. 63, 40–46.

Mysen, B.O., 1988. Structure and Properties of Silicate Melts. Elsevier, Amsterdam, 354 p.

Mysen, B.O., 1991. Relations between structure, redox equilibria of iron, and properties of magmatic liquids. In: Perchuk, L.L., Kushiro, I. (Eds.), Physical Chemistry of Magmas. Springer, Berlin/Heidelberg/New York, pp. 41–98.

Mysen, B.O., Virgo, D., 1989. Redox equilibria, structure, and properties of Fe-bearing aluminosilicate melts: relationships among temperature, composition, and oxygen fugacity in the system Na_2O–Al_2O_3–SiO_2–Fe–O. Am. Mineral. 74, 58–76.

Nacken, R., 1921. Welche Folderungen ergeben sich aus dem Auftreten von Flüssigkeitseinschlüssen in Mineralien? Zentralbl. Mineral. 12, 35–43.

Naden, J., Kilias, S.P., Leng, M.J., Cheliotis, I., Shepherd, T.J., 2003. Do fluid inclusions preserve $\delta^{18}O$ values of hydrothermal fluids in epithermal systems over geological time? Evidence from paleo- and modern geothermal systems, Milos Island, Aegean Sea. Chem. Geol. 197, 143–159.

Nasdala, L., Smith, D.C., Kaindl, R., Gaft, M., Ziemann, M.A., 2004. Raman spectroscopy: analytical perspectives in mineralogical research. In: Beran, A., Libowitzky, E. (Eds.), Spectroscopic Methods in Mineralogy. In: EMU Notes in Mineralogy, vol. 6. Eötvös University Press, Budapest, pp. 281–343.

Nasdala, L., Grötzschel, R., Probst, S., Bleisteiner, B., 2010. Irradiation damage in monazite–(Ce): an example to establish the limits of Raman confocality and depth resolution. Can. Mineral. 48, 351–359.

Naumov, V.B., Balickiy, V.C., Khetchikov, L.N., 1966. About relationship between temperatures of formation, homogenization and decrepitation of gas-liquid inclusions. Dokl. Akad. Nauk SSSR 171, 146–148 (in Russian).

Naumov, V.B., Solovova, I.P., Kovalenker, V.I., 1992. Juvenile water under pressures of 15–17 kbar and its concentration in magma: first data on inclusions in plagioclases of andesites. Dokl. Akad. Nauk 324, 654–658.

Naumov, V.B., Tolstykh, M.L., Kovalenker, V.A., Kononkova, N.N., 1996. Fluid overpressure in andesite melts from Central Slovakia: evidence from inclusions in minerals. Petrology 4, 265–276.

Naumov, V.B., Kamenetsky, V.S., Thomas, R., Kononkova, N.N., Ryzhenko, B.N., 2008. Inclusions of silicate and sulphate melts in chrome diopside from the Inagli deposit, Yakutia, Russia. Geochem. Int. 46, 554–564.

Nedkvitne, T., Karlsen, D.A., Bjørlykke, K., Larter, S., 1993. Relationship between reservoir diagenetic evolution and petroleum emplacement in the Ula field, North Sea. Mar. Petrol. Geol. 10, 255–270.

Nguyen-Trung, C., Begun, G.M., Palmer, D.A., 1992. Aqueous uranium complexes. 2. Raman spectroscopic study of the complex formation of the dioxouranium (VI) ion with a variety of inorganic and organic ligands. Inorg. Chem. 31, 5280–5287.

Nicholls, J., Crawford, M.L., 1985. FORTRAN programs for calculation of fluid properties from microthermometric data on fluid inclusions. Comput. Geosci. 11, 619–645.

Nicholson, K., 1993. Geothermal Fluids: Chemistry and Exploration Techniques. Springer, Berlin/Heidelberg, Germany, 253 p.

Nier, A.O., 1947. A mass spectrometer for isotope and gas analysis. Rev. Sci. Instrum. 18, 398–411.

Nieva, D., Nieva, R., 1987. Developments in geothermal energy in Mexico—part twelve: a cationic geothermometer for prospecting of geothermal resources. Heat Rec. Syst. 7, 243–258.

Niiranen, T., Poutiainen, M., Manttari, I., 2007. Geology, geochemistry, fluid inclusion characteristics, and U–Pb age studies on iron oxide Cu–Au deposits in the Kolari region, northern Finland. Ore Geol. Rev. 30, 75–105.

Norman, D.I., Ting, W., Putnam, B.R., Smith, R.W., 1985. Mineralization of the Hansonburg Mississippi-Valley-type deposit. Can. Mineral. 23, 353–368.

Norman, D.I., Sawkins, F.J., 1987. Analysis of volatiles in fluid inclusions by mass spectrometry. Chem. Geol. 61, 1–10.

Norman, D.I., Harrison, R.W., Andres, C.B., 1991. Geology and geochemical analysis of mineralizing fluids at the St-Cloud and United-States Mines, Chloride mining district, New Mexico. J. Geochem. Explor. 42, 61–89.

Northrop, D.A., Clayton, R.N., 1966. Oxygen–isotope fractionations in systems containing dolomite. J. Geol. 74, 174–196.

O'Leary, M.H., 1981. Carbon isotope fractionation in plants. Phytochemistry 20, 553–567.

O'Leary, M.H., 1988. Carbon isotopes in photosynthesis. Bioscience 38, 328–336.

O'Neil, J.R., Epstein, S., 1966. Oxygen isotope fractionation in the system dolomite−calcite−carbon dioxide. Science 152, 198–201.

O'Neil, J.R., Taylor, H.P., Jr., 1967. The oxygen isotope and cation exchange chemistry of feldspars. Am. Mineral. 52, 1414–1437.

O'Neil, J.R., Clayton, R.N., Mayeda, T.K., 1969. Oxygen isotope fractionation in divalent metal carbonates. J. Chem. Phys. 51, 5547–5558.

O'Reilly, C., Parnell, J., 1999. Fluid flow and thermal histories for Cambrian-Ordovician platform deposits, New York: evidence from fluid inclusion studies. Geol. Soc. Am. Bull. 111, 1844–1896.

Oakes, C.S., Bodnar, R.J., Simonson, J.M., 1990. The system $NaCl-CaCl_2-H_2O$. I: the ice liquidus at 1 atm total pressure. Geochim. Cosmochim. Acta 54, 603–610.

Ohba, T., Kazahaya, K., Matsuo, S., 1995. Diffusional ^{18}O loss from inclusion water in a natural hydrothermal quartz from the Kaneuchi tungsten deposit, Japan. Geochim. Cosmochim. Acta 59, 3039–3047.

Ohba, T., Matsuo, S., 1988. Precise determination of hydrogen and oxygen isotope ratios of water in fluid inclusions of quartz and halite. Geochem. J. 22, 55–68.

Ohmoto, H., 1972. Systematics of sulfur and carbon isotopes in hydrothermal ore deposits. Econ. Geol. 67, 551–578.

Ohmoto, H., Goldhaber, M.B., 1997. Sulfur and carbon isotopes. In: Barnes, H.L. (Ed.), Geochemistry of Hydrothermal Ore Deposits. John Wiley and Sons, New York/Chichester/Brisbane/Toronto, pp. 517–611.

Ohmoto, H., Kerrick, D.M., 1977. Devolatilization equilibria in graphitic systems. Am. J. Sci. 277, 1013–1044.

Ohmoto, H., Lasaga, A.C., 1982. Kinetics of reactions between aqueous sulfates and sulfides in hydrothermal systems. Geochim. Cosmochim. Acta 46, 1717–1745.

Ohmoto, H., Rye, R.O., 1979. Isotopes of sulfur and carbon. In: Barnes, H.L. (Ed.), Geochemistry of Hydrothermal Ore Deposits. John Wiley and Sons, New York/Chichester/Brisbane/Toronto, pp. 509–563.

Orange, D., Knittle, E., Farber, D., Williams, Q., 1996. Raman spectroscopy of crude oils and hydrocarbon fluid inclusions: a feasibility study. In: Dyar, M.D., McCammon, C., Schaefer, M.W. (Eds.), Mineral Spectroscopy: A Tribute to Roger G. Burns. In: Geochem. Soc. Spec. Publ., vol. 5, pp. 65–81.

Oružinský, V., Hurai, V., 1985. Fluid inclusions of secondary quartzites of the Štiavnica stratovolcano. Miner. Slov. 17, 415–424 (in Slovak).

Orvošová, M., Hurai, V., Simon, K., Wiegerová, V., 2004. Fluid inclusion and stable isotopic evidence for early hydrothermal karstification in vadose caves of the Nízke Tatry Mountains (Western Carpathians). Geol. Carpath. 55, 421–429.

Osadchii, E.G., Chareev, D.A., 2006. Thermodynamic studies of pyrrhotite–pyrite equilibria in the Ag–Fe–S system by solid–state galvanic cell technique at 518–723 K and total pressure of 1 atm. Geochim. Cosmochim. Acta 70, 5617–5633.

Osborne, M., Swarbrick, R.E., 1997. Mechanisms for generating overpressure in sedimentary basins: a reevaluation. AAPG Bull. 81, 1023–1041.

Pabalan, R.T., Pitzer, K.S., 1987. Thermodynamics of concentrated electrolyte mixtures and the prediction of mineral solubilities to high temperatures for mixtures in the system $Na-K-Mg-Cl-SO_4-OH-H_2O$. Geochim. Cosmochim. Acta 51, 2429–2443.

Pang, L.S.K., George, S.C., Quezada, R.A., 1998. A study of the gross composition of oil-bearing fluid inclusions using high-performance liquid chromatography. Org. Geochem. 29, 1149–1161.

Pasteris, J.D., 1987. Fluid inclusions in mantle xenoliths. In: Newton, P.H. (Ed.), Mantle Xenoliths. John Wiley and Sons, New York, pp. 691–707.

Pasteris, J.D., 1989. In situ analysis in geological thin-sections by Laser Raman microprobe microspectroscopy: a cautionary note. Appl. Spectrosc. 43, 567–570.

Pasteris, J.D., Wanamaker, B.J., 1988. Laser Raman microprobe analysis of experimentally re-equilibrated fluid inclusions in olivine: some implications for mantle fluids. Am. Mineral. 73, 1074–1088.

Peng, D.-Y., Robinson, D.B., 1976. A new two–constant equation of state. Ind. Eng. Chem. Fundam. 15, 59–64.

Peretti, A., Dubessy, J., Mullis, J., Prost, R., Trommsdorff, V., 1992. Highly reducing conditions during alpine metamorphism of the Maleno peridotite (Sondrio, Northern Italy) indicated by mineral paragenesis and H_2 in fluid inclusions. Contrib. Mineral. Petrol. 112, 329–340.

Peretyazhko, I.S., Prokofiev, V.Y., Zagorskiy, V.Y., Smirnov, S.Z., 2000. Role of boric acids in the formation of pegmatite and hydrothermal minerals: petrologic consequences of sassolite (H_3BO_3) discovery in fluid inclusions. Petrology 8, 214–237.

Pettke, T., 2008. Analytical protocols for element concentration and isotope ratio measurements in fluid inclusions by LA-(MC)-ICP-MS. In: Sylvester, P. (Ed.), Laser Ablation ICP-MS in the Earth Sciences: Current Practices and Outstanding Issues. In: Min. Assoc. Can. Short Course Series, vol. 40, pp. 189–218.

Pettke, T., Diamond, L.W., 1995. Rb/Sr isotopic analysis of fluid inclusions in quartz: evaluation of bulk extraction procedures and geochronometer systematics using synthetic fluid inclusions. Geochim. Cosmochim. Acta 59, 4009–4027.

Pettke, T., Halter, W.E., Webster, J.D., Aigner-Torres, M., Heinrich, C.A., 2004. Accurate quantification of melt inclusion chemistry by LA-ICPMS: a comparison with EMP and SIMS and advantages and possible limitations of these methods. Lithos 78, 333–361.

Pettke, T., Oberli, F., Audétat, A., Guillong, M., Simon, A.C., Hanley, J.J., Klemm, L.M., 2012. Recent developments in element concentration and isotope ratio analysis of individual fluid inclusions by laser ablation single and multiple collector ICP-MS. Ore Geol. Rev. 44, 10–38.

Phillip, W.J., 1972. Hydraulic fracturing and mineralization. J. Geol. Soc. London 12, 337–359.

Phillips, W., 1875. The rocks of the mining districts of Cornwall and their relation to metalliferous deposits. Geol. Soc. London Quart. J. 31, 319–345.

Philpotts, A.R., Ague, J.J., 2009. Principles of Igneous and Metamorphic Petrology. Cambridge University Press, Cambridge, 667 p.

Pichavant, M., Ramboz, C., Weisbrod, A., 1982. Fluid immiscibility in natural processes: use and misuse of fluid inclusion data. Chem. Geol. 37, 1–27.

Pineau, F., Javoy, M., Allégre, C., 1973. Étude systématique des isotopes de l'oxygéne, du carbone et du strontium dans les carbonatites. Geochim. Cosmochim. Acta 37, 2363–2377.

Ping, H., Thiéry, R., Chen, H., 2011. Thermodynamic modelling of petroleum inclusions: the prediction of the saturation pressure of crude oils. Geofluids 11, 328–340.

Pironon, J., Canals, M., Dubessy, J., Walgenwitz, F., Laplace-Builhe, C., 1998. Volumetric reconstruction of individual oil inclusions by confocal scanning laser microscopy. Eur. J. Mineral. 10, 1143–1150.

Pironon, J., Thiéry, R., Teinturier, S., Walgenwitz, F., 2000. Water in petroleum inclusions: evidence from Raman and FT–IR measurements, *PVT* consequences. J. Geochem. Explor. 69–70, 663–668.

Pironon, J., Grimmer, J.O.W., Teinturier, S., Guillaume, D., Dubessy, J., 2003. Dissolved methane in water: temperature effect on Raman quantification in fluid inclusions. J. Geochem. Explor. 78–79, 111–115.

Pitzer, K.S., 1973. Thermodynamics of electrolytes. I. Theoretical basis and general equations. J. Phys. Chem. 77, 268–277.

Pitzer, K.S., Pabalan, R.T., 1986. Thermodynamics of NaCl in steam. Geochim. Cosmochim. Acta 50, 1445–1454.

Placzek, G., 1934. Rayleigh–Streuung und Raman–Effekt. In: Marx, E. (Ed.), Handbuch der Radiologie, vol. 6. Akad. Verlagsges., Leipzig, pp. 205–374

Porcelli, D.L., O'Nions, R.K., O'Reilly, S.Y., 1986. Helium and strontiom isotopes in ultramafic xenoliths. Chem. Geol. 54, 237–249.

Potter, J., Konnerup-Madsen, J., 2003. A review of the occurrence and origin of abiogenic hydrocarbons in igneous rocks. Geol. Soc. London Spec. Publ. 214, 151–173.

Potter, J., Rankin, A.H., Treolar, P.J., 2004. Abiogenic Fischer–Tropsch synthesis of hydrocarbons in alkaline igneous rocks; fluid inclusion, textural and isotopic evidence from the Lovozero complex, N.W. Russia. Lithos 75, 311–330.

Potter, R.W., Brown, D.L., 1977. The volumetric properties of aqueous sodium chloride solutions from 0° to 500 °C and pressures up to 2000 bars based on a regression of available data in the literature. US Geol. Surv. Bull. 1421–C, 36 p.

Potter, R.W., Babcock, R.S., Brown, D.L., 1977. A new method for determining the solubility of salts in aqueous solutions at elevated temperatures. J. Res. US Geol. Surv. 5, 389–395.

Poty, B.P., Stalder, H.A., Weisbrod, A.M., 1974. Fluid inclusion studies in quartz from fissures of Western and Central Alps. Schweiz. Mineral. Petrogr. Mitt. 54, 719–752.

Powell, R., Holland, T.J.B., 1988. An internally consistent thermodynamic dataset with uncertainties and correlations: 3. Application to geobarometry, worked examples and a computer program. J. Metamorph. Geol. 6, 173–204.

Powell, R., Holland, T., Worley, B., 1998. Calculating phase diagrams involving solid solutions via non-linear equations, with examples using THERMOCALC. J. Metamorph. Geol. 16, 577–588.

Preston-Thomas, H., 1990. The international temperature scale of 1990 (ITS–90). Metrologia 27, 3–10.

Prezbindowski, D.R., Larese, R.E., 1987. Experimental stretching of fluid inclusions in calcite—implications for diagenetic studies. Geology 15, 333–336.

Price, N.J., Cosgrove, J.W., 1990. Analysis of Geological Structures. Cambridge Univ. Press, Cambridge, 494 p.

Prinsloo, L.C., Heyns, A.M., Ehrl, R., Range, K.-J., 1997. The vibrational spectra of potassium hexafluorosilicate, K_2SiF_6. Eur. J. Solid State Inorg. Chem. 34, 881–893.

Prochaska, W., 1997. Formation of different siderite provinces during the Alpine tectono-metamorphic event in the Eastern Alps of Austria. In: Papunen, H. (Ed.), Mineral Deposits. Balkema, Rotterdam/Brookfield, pp. 845–848.

Prochaska, W., 1999. Die Bedeutung der chemischen Zusammensetzung von Einschlussfluiden und laugbaren Salzen für die Genese von hydrothermalen und sedimentären Karbonatgesteinen in den Ostalpen. Mitt. Österr. Geol. Ges. 90, 175–183.

Prochaska, W., 2001. Magnesite mineralization of the Eastern Alps and the Carpathians. In: Piestrzyński, A. (Ed.), Mineral Deposits at the Beginning of the 21st Century. Balkema, Lisse, pp. 1017–1019.

Radvanec, M., Prochaska, W., 2001. Successive replacement of Upper Carboniferous calcite to dolomite and magnesite in Dúbrava magnesite deposit (Western Carpathians, Slovakia). Miner. Slov. 33, 517–525.

Radvanec, M., Grecula, P., Žák, K., 2004a. Siderite mineralization of the Gemericum superunit (Western Carpathians, Slovakia): review and a revised genetic model. Ore Geol. Rev. 24, 267–298.

Radvanec, M., Koděra, P., Prochaska, W., 2004b. Mineralogy, fluid inclusion and C, O, Sr isotope study of the Košice-Medvedia magnesite deposit, Western Carpathians, Slovakia. Acta Petrol. Sin. 20, 855–876.

Radvanec, M., Koděra, P., Prochaska, W., 2004c. Mg replacement at the Gemerská Poloma talc-magnesite deposit, Western Carpathians, Slovakia. Acta Petrol. Sin. 20, 773–790.

Rahl, J.M., Anderson, K.M., Brandon, M.T., Fassoulas, C., 2005. Raman spectroscopic carbonaceous material thermometry of low-grade metamorphic rocks: calibration and application to tectonic exhumation in Crete, Greece. Earth Planet. Sci. Lett. 240, 339–354.

Raman, C.V., Krishnan, K.S., 1928. A new type of secondary radiation. Nature 121, 501–502.

Ramboz, C., Pichavant, M., Weisbrod, A., 1982. Fluid immiscibility in natural processes: use and misuse of fluid inclusion data II. Interpretation of fluid inclusion data in terms of immiscibility. Chem. Geol. 37, 29–48.

Ramsay, J.G., 1980. The crack-seal mechanism of rock deformation. Nature 284, 135–139.

Ray, J., Ramesh, R., 2000. Rayleigh fractionation of stable isotopes from a multicomponent source. Geochim. Cosmochim. Acta 64, 299–306.

Ray, J., Ramesh, R., 2006. Stable carbon and oxygen isotopic compositions of Indian carbonatites. Int. Geol. Rev. 48, 17–46.

Rayleigh, J.W.S., 1896. Theoretical considerations respecting the separation of gases by diffusion and similar processes. Phil. Mag. 42, 493.

Rayleigh, J.W.S., 1902. On the distillation of binary mixtures. Phil. Mag. Ser. 6, 521–537.

Ražnjevic, 1975. Thermodynamic Tables. Alfa, Bratislava (in Slovak).

Redlich, O., Kwong, J.N.S., 1949. On the thermodynamics of solutions. V. An equation of state. Fugacities of gaseous solutions. Chem. Rev. 44, 233–244.

Reese, C.L., 1898. Petroleum inclusions in quartz crystals. Am. Chem. Soc. J. 20, 795–797.

Reynard, B., Caracas, R., 2009. D/H isotopic fractionation between brucite $Mg(OH)_2$ and water from first-principles vibrational modelling. Chem. Geol. 262, 159–168.

Reynolds, T.J., Beane, R.E., 1985. Evolution of hydrothermal fluid characteristics at the Santa Rita, New Mexico, porphyry copper deposit. Econ. Geol. 80, 1328–1347.

Richards, I.J., Connelly, J.B., Gregory, R.T., Gray, D.R., 2002. The importance of diffusion, advection, and host-rock lithology on vein formation: a stable isotope study from the Paleozoic Ouachita orogenic belt, Arkansas and Oklahoma. Geol. Soc. Am. Bull. 114, 1343–1355.

Richards, J.P., Kerrich, R., 1993. Observations of zoning and fluid inclusions in pyrite using a transmitted infrared light microscope ($\lambda \leq 1.9$ μm). Econ. Geol. 88, 716–723.

Richet, P., Bottinga, Y., 1995. Rheology and configurational entropy of silicate melts. In: Stebbins, J., McMillan, P.F., Dingwell, D.B. (Eds.), Structure, Dynamics and Properties of Silicate Melts. In: Min. Soc. Am. Rev. Mineral., vol. 32, pp. 67–93.

Richet, P., Bottinga, Y., Javoy, M., 1977. A review of hydrogen, carbon, nitrogen, oxygen, sulphur, and chlorine stable isotope fractionation among gaseous molecules. Ann. Rev. Earth Planet. Sci. 5, 65–110.

Richter, R., Hoernes, S., 1988. The application of the increment method in comparison with experimentally derived and calculated O-isotope fractionations. Chem. Erde 48, 1–18.

Rickers, K., Thomas, R., Heinrich, W., 2004. Trace-element analysis of individual synthetic and natural fluid inclusions with synchrotron radiation XRF using Monte Carlo simulations for quantification. Eur. J. Mineral. 16, 23–35.

Rickers, K., Thomas, R., Heinrich, W., 2006. The behaviour of trace elements during the chemical evolution of the H_2O-, B-, and F-rich granite–pegmatite–hydrothermal system at Ehrenfriedersdorf, Germany: a SXRF study of melt and fluid inclusions. Miner. Deposita 41, 229–245.

Rimstidt, J.D., 1997. Gangue mineral transport and deposition. In: Barnes, H.L. (Ed.), Geochemistry of Hydrothermal Deposits. John Wiley and Sons, New York, pp. 487–515.

Robie, R.A., Hemingway, B.S., 1995. Thermodynamic properties of minerals and related substances at 298.15 K and 1 bar (10^5 Pascals) pressure and at higher temperatures. US Geol. Surv. Bull. 2131, 461 p.

Robie, R.A., Hemingway, B.S., Fisher, J.R., 1979. Thermodynamic properties of minerals and related substances at 298.15 K and 1 bar (10^5 Pascals) pressure and at higher temperatures. US Geol. Surv. Bull. 1452, 456 p.

Rocholl, A., Heusser, E., Kirsten, T., Oehm, J., Richter, H., 1996. A noble gas profile across a Hawaiian mantle xenolith: coexisting accidental and cognate noble gases derived from the lithospheric and astenospheric mantle beneath Oahu. Geochim. Cosmochim. Acta 60, 4773–4783.

Rocholl, A., Simon, K., Jochum, K.P., Bruhn, F., Gehann, R., Kramar, U., Luecke, W., Molzahn, M., Pernicka, E., Seufert, M., Spettel, B., Stummeier, J., 1997. Chemical characterization of NIST silicate glass certified reference material SRM 610 by ICP-MS, TIMS, LIMS, SSMS, INAA, AAS and PIXE. Geostand. Newslett. 21, 101–114.

Roedder, E., 1958. Technique for the extraction and partial chemical analysis of fluid-filled inclusions in minerals. Econ. Geol. 3, 235–269.

Roedder, E., 1965. Liquid CO_2 inclusions in olivine-bearing nodules and phenocrysts from basalts. Am. Mineral. 50, 1746–1782.

Roedder, E., 1967. Metastable "superheated" ice in liquid-water inclusions under high negative pressure. Science 155, 1411–1417.

Roedder, E., 1979. Origin and significance of magmatic inclusions. Bull. Minéral. 102, 487–510.

Roedder, E., 1984. Fluid Inclusions. Min. Soc. Am. Rev. Mineral., vol. 12. 644 p.

Roedder, E., Bodnar, R.J., 1980. Geologic pressure determinations from fluid inclusion studies. Ann. Rev. Earth Planet. Sci. 8, 263–301.

Roedder, E., Coombs, D.S., 1967. Immiscibility in granitic melts, indicated by fluid inclusions in ejected granitic blocks from Ascension Island. J. Petrol. 8, 417–451.

Roedder, E., Weiblen, P.W., 1970. Lunar petrology of silicate melt inclusions, Apollo 11 rocks. Proc. Apollo 11 Lunar Sci. Conf. Geochim. Cosmochim. Acta Suppl. 1, 801–837.

Roeske, C.A., O'Leary, M.H., 1984. Carbon isotope effects on the enzyme-catalyzed carboxylation of ribulose biphosphate. Biochemistry 23, 6275–6284.

Rojkovič, I., 1985. Paragenesis and mineral succession. In: Cambel, B., Jarkovský, J. (Eds.), The Rudňany Ore Field—Geochemical and Metalogenetic Characteristics. Veda, Bratislava, pp. 183–193 (in Slovak).

Rojkovič, I., Soták, J., Konečný, P., Čech, P., 2008. Stratiform manganese mineralization in the Paleogene and Jurassic shale formations of the Western Carpathians: mineralogy, geochemistry and ore-forming processes. Geol. Carpath. 59, 503–514.

Rosasco, G.J., Roedder, E., 1975. Application of a new microprobe spectrometer to nondestructive analysis of sulfate and other ions in individual phases in fluid inclusions in minerals. Geochim. Cosmochim. Acta 43, 1907–1916.

Rosasco, G.J., Roeder, E., Simmons, J.H., 1975. Laser-excited Raman spectroscopy for nondestructive partial analysis of individual phases in fluid inclusions in minerals. Science 190, 557–560.

Rosenbaum, J., Sheppard, S.M., 1986. An isotopic study of siderites, dolomites and ankerites at high temperatures. Geochim. Cosmochim Acta 50, 1147–1150.

Rossano, S., Mysen, B., 2012. Raman spectroscopy of silicate glasses and melts in geological systems. In: Dubessy, J., Caumon, M.-C., Rull, F. (Eds.), Raman Spectroscopy Applied to Earth Sciences and Cultural Heritage. In: EMU Notes in Mineralogy, vol. 12, pp. 321–366.

Rosso, K.M., Bodnar, R.J., 1995. Microthermometric and Raman spectroscopic detection limits of CO_2 in fluid inclusions and the Raman spectroscopic characterization of CO_2. Geochim. Cosmochim. Acta 59, 3961–3975.

Rudolph, W.W., Fischer, D., Irmer, G., 2006. Vibrational spectroscopic studies and density functional theory calculations of speciation in the CO_2–water system. Appl. Spectrosc. 60, 130–144.

Rull, F., 2002. Structural investigation of water and aqueous inclusions by Raman spectroscopy. Pure Appl. Chem. 74, 1859–1870.

Rye, R.O., O'Neil, J.R., 1968. The O^{18} content of water in primary fluid inclusions from Providencia, North Central Mexico. Econ. Geol. 63, 232–238.

Sack, R.O., Ghiorso, M.S., 1989. Importance of considerations of mixing properties in establishing an internally consistent thermodynamic database: thermochemistry of minerals in the system $Mg_2SiO_4–Fe_2SiO_4–SiO_2$. Contrib. Mineral. Petrol. 102, 41–68.

Sahagian, D.L., Proussevitch, A.A., 1992. Bubbles in volcanic systems. Nature 359, 485.

Salvi, S., Williams-Jones, A.E., 1997. Fluid–inclusion volatile analysis by gas chromatography: application of a wide–bore porous–polymer capillary column to the separation of organic and inorganic compounds. Can. Mineral. 35, 1391–1414.

Samson, I., Anderson, A., Marshall, D. (Eds.), 2003. Fluid Inclusions. Analysis and Interpretation. In: Min. Assoc. Can. Short Course Series, vol. 32, 374 p.

Sanjuan, B., Millot, R., Brach, M., Asmundsson, R., Giroud, N., 2010. Use of a new sodium/lithium (Na/Li) geothermometric relationship for high-temperature dilute geothermal fluids from Iceland. In: Proc. World Geotherm. Congress, Bali, Indonesia, 25–29 April 2010.

Satake, H., Matsuo, S., 1984. Hydrogen isotopic fractionation factor between brucite and water in the temperature range from 100° to 510 °C. Contrib. Mineral. Petrol. 86, 19–24.

Sauerzapf, U., Lattard, D., Bouchard, M., Engelmann, R., 2008. The titanomagnetite–ilmenite equilibrium: new experimental data and thermo–oxybarometric applications to the crystallization of basic to intermediate rocks. J. Petrol. 49, 1161–1185.

Saul, A., Wagner, W., 1989. A fundamental equation for water covering the range from the melting line to 1273 K at pressures up to 25000 MPa. J. Phys. Chem. Ref. Data 18, 1537–1564.

Savin, S.M., Lee, M., 1988. Isotopic studies of phyllosilicates. In: Bailey, S.W. (Ed.), Hydrous Phyllosilicates. In: Min. Soc. Am. Rev. Mineral., vol. 19. pp. 189–223.

Saxena, S.K., Fei, Y., 1987. Fluids at crustal pressures and temperatures. Contrib. Mineral. Petrol. 95, 370–375.

Scambelluri, M., Piccardo, G.B., Philippot, P., Robbiano, A., Negretti, L., 1997. High-salinity fluid inclusions formed from recycled seawater in deeply subducted alpine serpentinite. Earth Planet. Sci. Lett. 148, 485–499.

Scheele, N., Hoefs, J., 1992. Carbon isotope fractionation between calcite, graphite and CO_2: an experimental study. Contrib. Mineral. Petrol. 112, 35–45.

Scherfel, A., 1878. Einschlüsse in Zinkblende. Mitt. Berg- u. Hüttenm. Zeitung 37, 49.

Scheuchzer, J.J., 1723. Helveticus sive Itinera per Helvetiae alpinas regiones. Leiden, 216 p.

Schiano, P., 2003. Primitive mantle magmas recorded as silicate melt inclusions in igneous minerals. Earth Sci. Rev. 63, 121–144.

Schiffries, C.M., 1990. Liquid-absent aqueous fluid inclusions and phase equilibria in the system $CaCl_2-NaCl-H_2O$. Geochim. Cosmochim. Acta 54, 611–619.

Schrötter, H.W., Klöckner, H.W., 1979. Raman scattering cross sections in gases and liquids. In: Weber, A. (Ed.), Raman Spectroscopy of Gases and Liquids. In: Topics in Current Physics 11, Springer-Verlag, Berlin/Heidelberg/New York, pp. 123–166.

Schütze, H., 1980. The isotopic index—a method for approximate calculation of isotopic exchange equilibriums between crystalline substances by means of increments. Chem. Erde 39, 321–334.

Seal, R.R., II, 2006. Sulfur isotope geochemistry of sulfide minerals. Min. Soc. Am. Rev. Mineral. Geochem. 61, 633–677.

Seifert, W., Rhede, D., Förster, H.-J., Thomas, R., 2009. Accessory minerals as fingerprints for the thermal history and geochronology of the Caledonian Rumburk granite. N. Jhrb. Miner. Abh. 186, 215–233.

Seifert, W., Thomas, R., Rhede, D., Förster, H.-J., 2010. Origin of coexisting wüstite, Mg–Fe and REE phosphate minerals in graphite-bearing fluorapatite from the Rumburk granite. Eur. J. Mineral. 22, 495–507.

Seifert, W., Rhede, D., Thomas, R., Förster, H.-J., Lucassen, F., Dulski, P., Wirth, R., 2011. Distinctive properties of rock-forming blue quartz: inferences from a multi-analytical study of submicron mineral inclusions. Mineral. Mag. 75, 2519–2534.

Seitz, J.C., Pasteris, J.D., Chou, I.-M., 1996. Raman spectroscopic characterization of gas mixtures: II. Quantitative composition and pressure determination of the CO_2–CH_4 system. Am. J. Sci. 296, 577–600.

Seo, J.H., Guillong, M., Aerts, M., Zajacz, Z., Heinrich, C.A., 2011. Microanalysis of S, Cl, and Br in fluid inclusions by LA-ICP-MS. Chem. Geol. 284, 35–44.

Sereffidin, F., Schwarcz, H.P., Ford, D.C., 2002. Palaeotemperature reconstruction using isotopic variations in speleothem fluid inclusion water. Geochim. Cosmochim. Acta 66, 697.

Setzmann, U., Wagner, W., 1991. A new equation of state and tables of thermodynamic properties for methane covering the range from the melting line to 625 K at pressures up to 1000 MPa. J. Phys. Chem. Ref. Data 20, 1061–1155.

Severs, M.J., Azbej, T., Thomas, J.B., Mandeville, C.W., Bodnar, R.J., 2007. Experimental determination of H_2O loss from melt inclusions during laboratory heating. Chem. Geol. 237, 358–371.

Sharp, Z., 2007. Principles of Stable Isotope Geochemistry. Pearson Prentice Hall, Upper Saddle River, NJ, 344 p.

Sharp, Z.D., Kirschner, D.L., 1994. Quartz–calcite oxygen isotope thermometry: a calibration based on natural isotopic variations. Geochim. Cosmochim. Acta 58, 4491–4501.

Sharygin, I.S., Golovin, A.V., Korsakov, A.V., Pokhilenko, N.P., 2013. Eitelite in sheared peridotite xenoliths from Udachnaya–East kimberlite pipe (Russia)—a new locality and host rock type. Eur. J. Mineral. 25, 825–834.

Shaw, H.R., 1972. Viscosities of magmatic silicate liquids: an empirical method of prediction. Am. J. Sci. 272, 870–893.

Shepherd, T.J., Chenery, S.R., 1995. Laser ablation ICP-MS elemental analysis of individual fluid inclusions: an evaluation study. Geochim. Cosmochim. Acta 59, 3997–4007.

Shepherd, T.J., Rankin, A.H., Alderton, D.H.M., 1985. A Practical Guide to Fluid Inclusion Studies. Blackie, Glasgow, 239 p.

Shepherd, T.J., Bottrell, S.H., Miller, M.F., 1991. Fluid inclusion volatiles as an exploration guide to black shale-hosted gold deposits, Dlgellau gold belt, North Wales, UK. J. Geochem. Explor. 42, 5–24.

Sheppard, S.M.F., Epstein, S., 1970. D/H and 18O/16O ratios of minerals of possible mantle or lower crustal origin. Earth Planet. Sci. Lett. 9, 232–239.

Sheppard, S.M.F., Gilg, H.A., 1996. Stable isotope geochemistry of clay minerals. Clay Miner. 31, 1–24.

Sheppard, S.M.F., Schwarcz, H.P., 1970. Fractionation of carbon and oxygen isotopes and magnesium between coexisting metamorphic calcite and dolomite. Contrib. Mineral. Petrol. 26, 161–198.

Shi, P., Saxena, S.K., 1992. Thermodynamic modelling of the C−H−O−S fluid system. Am. Mineral. 77, 1038–1049.

Shiro, Y., Sakai, H., 1972. Calculation of the reduced partition function ratios of alpha–beta quartz and calcite. Jpn. Chem. Soc. Bull. 45, 2355–2359.

Shmonov, V.M., Shmulovich, K.I., 1974. Molar volumes and equation of state for CO_2 in the intervals 100–1000 °C and 2000–10000 bar. Dokl. Akad. Nauk SSSR 217, 935–938 (in Russian).

Shmulovich, K.I., Mercury, L., Thiéry, R., Ramboz, C., El Mekki, M., 2009. Experimental superheating of water and aqueous solutions. Geochim. Cosmochim. Acta 73, 2457–2470.

Shoute, L.C.T., Schmidt, K.J., Hall, R.H., Webb, M.A., Rifai, S., Abel, P., Arboleda, P.H., Savage, A., Bulmer, J. T., Loppnow, G.R., 2002. UV Raman spectroscopy of oilsands-derived bitumen and commercial petroleum products. Appl. Spectrosc. 56, 1308–1313.

Shvarov Yu, V., 1999. Algorithmization of the numeric equilibrium modelling of dynamic geochemical processes. Geochem. Int. 37, 571–576.

Sibson, R.H., 1987. Earthquake rupturing as a mineralizing agent in hydrothermal systems. Geology 15, 701–704.

Sibson, R.H., 1990. Conditions for fault–valve behaviour. In: Knippe, R.J., Rutter, E.H. (Eds.), Deformation Mechanism, Rheology and Tectonics. In: Geol. Soc. Spec. Publ, vol. 54, pp. 15–28.

Sibson, R.H., 1992. Fluid flow accompanying faulting: field evidence and models. In: Simpson, D.W., Richards, P. G. (Eds.), Earthquake Predictions: An International Review. In: Am. Geophys. Union Monogr., Maurice Ewing Series, vol. 4, pp. 593–603.

Sibson, R.H., McMoore, J., Rankin, A.H., 1975. Seismic pumping—a hydrothermal fluid transport mechanism. J. Geol. Soc. London 131, 653–659.

Sibson, R.H., Robert, F., Poulsen, K.H., 1988. High-angle reverse faults, fluid pressure cycling and mesothermal gold–quartz deposits. Geology 16, 551–555.

Sillitoe, R.H., Hedenquist, J.W., 2003. Linkages between volcanotectonic settings, ore-fluid compositions, and epithermal precious metal deposits. In: Simmons, S.F., Graham, I. (Eds.), Volcanic, Geothermal, and Ore-Forming Fluids: Rulers and Witnesses of Processes within the Earth. In: Soc. Econ. Geol. Spec. Publ, vol. 10, pp. 315–343.

Simon, K., 2001. Does δD from fluid inclusion in quartz reflect the original hydrothermal fluid? Chem. Geol. 177, 483–495.

Simonelli, D., Shultz, M.J., 2001. Temperature dependence for the relative Raman cross section of the ammonia/water complex. J. Mol. Spectrosc. 205, 221–226.

Sirbescu, M.-L.C., Krukowski, E.G., Schmidt, C., Thomas, R., Samson, I.M., Bodnar, R.J., 2013. Analysis of boron in fluid inclusions by microthermometry, laser ablation ICP-MS, and Raman spectroscopy: application to the Cryo-Genie pegmatite, San Diego County, California, USA. Chem. Geol. 342, 138–150.

Skaare, B.B., Kihle, J., Torsvik, T., 2011. Biodegradation of crude oil as potential source of organic acid in produced water. In: Lee, K., Neff, J. (Eds.), Produced Water: Environmental Risks and Advances in Mitigation Technologies. Springer, New York, pp. 89–113.

Skinner, B.J., 1966. Thermal expansion. In: Clark, S.P. (Ed.), Handbook of Physical Constants. In: Geol. Soc. Am. Memoir, vol. 97, pp. 78–96.

Skippen, G.B., Marshall, D.D., 1991. The metamorphism of granulites and devolatilization of the lithosphere. Can. Mineral. 29, 693–705.

Slack, C.R., Hatch, M.D., 1967. Comparative studies on the activity of carboxylases and other enzymes in relation to the new pathway of photosynthetic carbon dioxide fixation in tropical grasses. Biochem. J. 103, 660–665.

Slobodník, M., Muchez, Ph., Dolníček, Z., 2001. Zn−Pb−Cu mineralization in reef carbonates beneath the Upper Silesian Basin, Czech Republic: possible extremity of the Silesian-Cracow MVT district. In: Piestrzyński, A. (Ed.), Mineral Deposits at the Beginning of the 21st Century. Balkema, Rotterdam, pp. 209–212.

Slobodník, M., Muchez, Ph., Král, J., Keppens, E., 2006. Variscan veins: record of fluid circulation and Variscan tectonothermal events in Upper Palaeozoic limestones of the Moravian Karst, Czech Republic. Geol. Mag. 143, 491–508.

Slobodník, M., Melichar, R., Hurai, V., Bakker, R., 2012. Litho-stratigraphic effect on Variscan fluid flow within the Prague synform, Barrandian: evidence based on C, O, Sr isotopes and fluid inclusions. Mar. Petrol. Geol. 35, 128–138.

Smekal, A., 1923. Zur Quantentheorie der Dispersion. Naturwissenschaften 11, 873–875.

Soave, G., 1972. Equilibrium constants from a modified Redlich–Kwong equation of state. Chem. Eng. Sci. 27, 1197–1203.

Sobolev, V.S., Kostyuk, V.P., Bazarova, T.Yu., Bazarov, L.Sh., 1967. Melt inclusions in phenocrysts of nepheline basalts. Dokl. Akad. Nauk SSSR, Earth Sci. Sect. 173, 1760–1763.

Solopova, N., Spivak, A., Kupenko, I., Yu, Litvin, Dubrovinsky, L., 2012. High-pressure ^{13}C–diamond-forming reaction of decomposition of $Na_2{}^{13}CO_3$ melt at 20–60 GPa. Geophys. Res. Abstr, 14.

Solovova, I.P., Naumov, V.B., Kovalenko, V.I., Girnis, A.V., Guzhkova, A.V., 1990. The history of spinel lherzolite formation (Draiser Weiher, West Germany) according to microinclusion data. Geokhimiya 10, 1400–1411 (in Russian).

Song, K.Y., Kobayashi, R., 1987. Water content of CO_2 in equilibrium with liquid water and/or hydrates. Soc. Petrol. Eng. Form. Eval. 2, 500–508.

Sorby, H.C., 1858. On the microscopic structure of crystals, indicating the origin of minerals and rocks. Geol. Soc. London Quart. J. 14, 453–500.

Sourirajan, S., Kennedy, G.C., 1962. The system $H_2O-NaCl$ at elevated temperatures and pressures. Am. J. Sci. 260, 115–141.

Span, R., Lemmon, E.W., Jacobsen, R.T., Wagner, W., Yokozeki, A., 2000. A reference equation of state for the thermodynamic properties of nitrogen for temperatures from 63.151 to 1000 K and pressures up to 2200 MPa. J. Phys. Chem. Ref. Data 29, 1361–1433.

Span, R., Wagner, W., 1996. A new equation of state for carbon dioxide covering the fluid region from the triple-point temperature to 1100 K at pressures up to 800 MPa. J. Phys. Chem. Ref. Data 25, 1509–1596.

Spear, F.S., 1993. Metamorphic Phase Equilibria and Pressure-Temperature-Time paths. Min. Soc. Am., Washington, DC, 799 p.

Spencer, R.J., Møller, N., Weare, J.H., 1990. The prediction of mineral solubilities in natural waters: a chemical equilibrium model for the $Na-K-Ca-Mg-Cl-SO_4-H_2O$ systems at the temperatures below 25 °C. Geochim. Cosmochim. Acta 54, 575–590.

Srivastava, A., Singh, V.B., 2007. Theoretical and experimental studies of vibrational spectra of naphthalene and its cation. Ind. J. Pure Appl. Phys. 45, 714–720.

Stasiuk, L.D., Snowdon, L.R., 1997. Fluorescence micro-spectrometry of synthetic and natural hydrocarbon fluid inclusions: crude oil chemistry, density and application to petroleum migration. Appl. Geochem. 12, 229–241.

Steele, D.F., 1998. Theoretical Carbon and Oxygen Isotope Fractionation Factors in Calcium Magnesium Carbonates. M.Phil. Thesis, University of Liverpool, Liverpool.

Steele-MacInnis, M., Lecumberri-Sanchez, P., Bodnar, R.J., 2012. HokieFlincs_H2O-NaCl: a Microsoft Excel spreadsheet for interpreting microthermometric data from fluid inclusions based on the *PVTX* properties of H_2O–NaCl. Comput. Geosci. 49, 334–337.

Steele-MacInnis, M., Lecumberri-Sanchez, P., Bodnar, R.J., 2015. Synthetic fluid inclusions XX. Critical PTx properties of H2O–FeCl2 fluids. Geochim. Cosmochim. Acta. 148, 50–61.

Steele-MacInnis, M., Lecumberri-Sanchez, P., Bodnar, R.J., 2015. Synthetic fluid inclusions XX. Critical *PTx* properties of H_2O–$FeCl_2$ fluids. Geochim. Cosmochim. Acta (in press).

Sterner, S.M., 1992. Homogenization of fluid inclusions to the vapor phase—the apparent homogenization phenomenon. Econ. Geol. 87, 1616–1623.

Sterner, S.M., Bodnar, R.J., 1991. Synthetic fluid inclusions X.: Experimental determination of P−V−T−X properties in the CO_2–H_2O system to 6 kb and 700 °C. Am. J. Sci. 291, 1–54.

Sterner, S.M., Hall, D.L., Bodnar, R.J., 1988. Synthetic fluid inclusions. V. Solubility relations in the system $NaCl-KCl-H_2O$ under vapour-saturated conditions. Geochim. Cosmochim. Acta 52, 989–1005.

Sterner, S.M., Hall, D.L., Keppler, H., 1995. Compositional re-equilibration of fluid inclusions in quartz. Contrib. Mineral. Petrol. 119, 1–15.

Sterner, S.M., Pitzer, K.S., 1994. An equation of state for carbon dioxide valid from zero to extreme pressures. Contrib. Mineral. Petrol. 117, 362–374.

Stevens, G., Clemens, J.D., 1993. Fluid-absent melting and the role of fluids in the lithosphere: a slanted summary? Chem. Geol. 108, 1–17.

Stolper, E., 1982. The speciation of water in silicate melts. Geochim. Cosmochim. Acta 46, 2609–2620.

Stormer, J.C., 1983. The effects of recalculations on estimates of temperature and oxygen fugacity from analyses of multicomponent iron–titanium oxides. Am. Mineral. 68, 586–594.

Student, J.J., Bodnar, R.J., 1996. Melt inclusion microthermometry: petrologic constraints from the H_2O–saturated haplogranite system. Petrology 4, 291–306.

Student, J.J., Bodnar, R.J., 1999. Synthetic fluid inclusions XIV: Coexisting silicate melt and aqueous fluid inclusions in the haplogranite–H_2O–NaCl–KCl system. J. Petrol. 40, 1509–1525.

Student, J.J., Bodnar, R.J., 2004. Silicate melt inclusions in porphyry copper deposits: identification and homogenization behaviour. Can. Mineral. 42, 1583–1599.

Sun, Q., Qin, C., 2011. Raman OH stretching band of water as an internal standard to determine carbonate concentrations. Chem. Geol. 283, 274–278.

Sun, Q., Zhao, L., Li, N., Liu, J., 2010. Raman spectroscopic study for the determination of Cl concentration (molarity scale) in aqueous solutions: applications to fluid inclusions. Chem. Geol. 272, 55–61.

Suzuoki, T., Epstein, S., 1976. Hydrogen isotope fractionation between OH-bearing minerals and water. Geochim. Cosmochim. Acta 40, 1229–1240.

Swanenberg, H.E.C., 1979. Phase equilibria in carbonic systems and their application to freezing studies of fluid inclusions. Contrib. Mineral. Petrol. 68, 303–306.

Swanenberg, H.E.C., 1980. Fluid inclusions in high-grade metamorphic rocks from S.W. Norway. Geol. Ultraiect. 25, 1–146.

Šulgan, M., Hurai, V., 1989. Some ore-bearing factors of intrusions in relation to Cu–Mo porphyry deposits. Miner. Slov. 21, 123–130 (in Slovak).

Świerczewska, A., Tokarski, A.K., Hurai, V., 2000. Joints and mineral veins during structural evolution: case study from the Outer Carpathians (Poland). Geol. Quart. 44, 333–339.

Takenouchi, S., 1991. Fluid inclusion gas–composition of some mineral deposits and a geothermal area. J. Geochem. Explor. 42, 107–132.

Takenouchi, S., Kennedy, G.C., 1964. The binary system H_2O-CO_2 at high temperatures and pressures. Am. J. Sci. 262, 1055–1074.

Takenouchi, S., Kennedy, G.C., 1965. The solubility of carbon dioxide in NaCl solutions at high temperatures and pressures. Am. J. Sci. 263, 445–454.

Tarantola, A., Mullis, J., Vennemann, T., Dubessy, J., de Capitani, Ch., 2007. Oxidation of methane at the CH_4/$H_2O-(CO_2)$ transition zone in the external part of the Central Alps, Switzerland: evidence from stable isotope investigations. Chem. Geol. 237, 329–357.

Tarantola, A., Diamond, L.W., Stünitz, H., 2010a. Modification of fluid inclusions in quartz by deviatoric stress I: experimentally induced changes in inclusion shapes and microstructures. Contrib. Mineral. Petrol. 160, 825–843.

Tarantola, A., Diamond, L.W., Stünitz, H., 2010b. Modification of fluid inclusions in quartz by deviatoric stress II: experimentally induced changes in inclusion volume and composition. Contrib. Mineral. Petrol. 160, 845–864.

Tarantola, A., Diamond, L.W., Stünitz, H., Thust, A., Pec, M., 2012. Modification of fluid inclusions in quartz by deviatoric stress. III: influence of principal stresses on inclusion density and orientation. Contrib. Mineral. Petrol. 164, 537–550.

Taylor, H.P., Jr., 1974. The application of oxygen and hydrogen isotope studies to problems of hydrothermal alteration and ore deposition. Econ. Geol. 69, 843–883.

Taylor, H.P., Jr., 1977. Water/rock interactions and the origin of H_2O in granitic batholiths. J. Geol. Soc. London 133, 509–558.

Taylor, R.P., Jackson, S.E., Longerich, H.P., Webster, J.D., 1997. In situ trace-element analysis of individual silicate melt inclusions by laser ablation microprobe inductively coupled plasma-mass spectrometry (LAM-ICP-MS). Geochim. Cosmochim. Acta 61, 2559–2567.

Thiéry, R., van den Kerkhof, A.M., Dubessy, J., 1994a. $v-X$ properties of CH_4-CO_2 and CO_2-N_2 fluid inclusions: modelling for T<31 °C and P<400 bar. Eur. J. Mineral. 6, 753–771.

Thiéry, R., Vidal, J., Dubessy, J., 1994b. Phase equilibria modelling applied to fluid inclusions. Liquid-vapor equilibria and calculation of the molar volume in the CO_2–CH_4–N_2 system. Geochim. Cosmochim Acta 58, 1073–1082.

Thiéry, R., Pironon, J., Walgenwitz, F., Montel, F., 2000. PIT (Petroleum Inclusion Thermodynamic): a new modelling tool for the characterization of hydrocarbon fluid inclusions from volumetric and microthermometric measurements. J. Geochem. Explor. 69–70, 701–704.

Thiéry, R., Pironon, J., Walgenwitz, F., Montel, F., 2002. Individual characterization of petroleum fluid inclusions (composition and *P–T* trapping conditions) by microthermometry and confocal laser scanning microscopy: inferences from applied thermodynamics of oil. Mar. Petrol. Geol. 19, 847–859.

Thode, H.G., Shima, M., Rees, C.E., Krishnamurty, K.V., 1965. Carbon-13 isotope effects in systems containing carbon dioxide, bicarbonate, carbonate, and metal ions. Can. J. Chem. 43, 582–595.

Thomas, R., 1982. Ergebnisse der thermobarogeochemischen Untersuchungen an Flüssigkeitseinschlüssen in Mineralen der postmagmatischen Zinn-Wolfram-Mineralisation des Erzgebirges. Freiberg. Forschungsh. C 370, 85 p.

Thomas, R., 1994. Estimation of the viscosity and the water content of silicate melts from melt inclusion data. Eur. J. Mineral. 6, 511–535.

Thomas, R., 2000. Determination of water contents in granite melt inclusions by confocal laser Raman microprobe spectroscopy. Am. Mineral. 85, 868–872.

Thomas, R., 2002a. Determination of the H_3BO_3 concentration in fluid and melt inclusions in granite pegmatites by laser Raman microprobe spectrometry. Am. Mineral. 87, 56–68.

Thomas, R., 2002b. Determination of water contents in melt inclusions by laser Raman spectroscopy. In: De Vivo, B., Bodnar, R.J. (Eds.), Melt Inclusions: Methods, Applications and Problems, Short Course Proceedings. September 26–30th 2002, Napoli, Italy, pp. 211–216.

Thomas, R., Davidson, P., 2006. Progress in the determination of water in glasses and melt inclusions with Raman spectroscopy: a short review. Zeitschr. Geol. Wiss. 34, 159–163.

Thomas, R., Davidson, P., 2007. Progress in the determination of water in glasses and melt inclusions with Raman spectroscopy: a short review. Acta Petrol. 22, 15–20.

Thomas, R., Davidson, P., 2010. Hambergite-rich melt inclusions in morganite crystals from the Muiane pegmatite, Mozambique and some remarks on the paragenesis of hambergite. Mineral. Petrol. 100, 227–239.

Thomas, R., Davidson, P., 2012a. Evidence of a water-rich silica gel state during the formation of a simple pegmatite. Mineral. Mag. 76, 2785–2801.

Thomas, R., Davidson, P., 2012b. Water in granite and pegmatite-forming melts. Ore Geol. Rev. 46, 32–46.

Thomas, R., Davidson, P., 2012c. The application of Raman spectroscopy in the study of fluid and melt inclusions. Zeitschr. Dt. Ges. Geowiss. 163/2, 113–126.

Thomas, R., Davidson, P., 2013. The missing link between granites and granitic pegmatites. J. Geosci. 58, 77–94.

Thomas, R., Davidson, P., 2015. Comment on "A Petrological Assessment of Internal Zonation in Granitic Pegmatites" by David London (2014). Lithos 212–215, 462–468.

Thomas, R., Klemm, W., 1997. Microthermometric study of silicate melt inclusions in Variscan granites from SE Germany: volatile contents and entrapment conditions. J. Petrol. 38, 1753–1765.

Thomas, R., Webster, J.D., 2000. Strong tin enrichment in a pegmatite-forming melt. Miner. Deposita 35, 570–582.
Thomas, R., Webster, J.D., Heinrich, W., 2000. Melt inclusions in pegmatite quartz: complete miscibility between silicate melts and hydrous fluids at low pressure. Contrib. Mineral. Petrol. 139, 394–401.
Thomas, R., Förster, H.-J., Rickers, K., Webster, J.D., 2005. Formation of extremely F-rich hydrous melt fractions and hydrothermal fluids during differentiation of highly evolved tin-granite magmas: a melt/fluid-inclusion study. Contrib. Mineral. Petrol. 148, 582–601.
Thomas, R., Kamenetsky, V.S., Davidson, P., 2006a. Laser Raman spectroscopic measurements of water in unexposed glass inclusions. Am. Mineral. 91, 467–470.
Thomas, R., Webster, J.D., Rhede, D., Seifert, W., Rickers, K., Förster, H.-J., Heinrich, W., Davidson, P., 2006b. The transition from peraluminous to peralkaline granitic melts: evidence from melt inclusions and accessory minerals. Lithos 91, 137–149.
Thomas, R., Davidson, P., Hahn, A., 2008a. Ramanite-(Cs) and ramanite-(Rb): new cesium and rubidium pentaborate tetrahydrate minerals identified with Raman spectrsocopy. Am. Mineral. 93, 1034–1042.
Thomas, R., Davidson, P., Badanina, E., 2009. A melt and fluid inclusion assemblage in beryl from pegmatite in the Orlovka amazonite granite, East Transbaikalia, Russia: implications for pegmatite-forming melt systems. Mineral. Petrol. 96, 129–140.
Thomas, R., Davidson, P., Schmidt, Ch., 2011a. Extreme alkali bicarbonate- and carbonate-rich fluid inclusions in granite pegmatite from Precambrian Ronne granite, Bornholm Island, Denmark. Contrib. Mineral. Petrol. 161, 315–329.
Thomas, R., Webster, J.D., Davidson, P., 2011b. Be-daughter minerals in fluid and melt inclusions: implications for the enrichment of Be in granite–pegmatite systems. Contrib. Mineral. Petrol. 161, 483–495.
Thomas, R., Davidson, P., Badanina, E., 2012a. Water- and boron-rich melt inclusions in quartz from the Malkhan pegmatite, Transbaikalia, Russia. Minerals 2, 435–458.
Thomas, R., Davidson, P., Beurlen, H., 2012b. The competing models for the origin and internal evolution of granitic pegmatites in the light of melt and fluid inclusion research. Mineral. Petrol. 106, 55–73.
Thomas, S.-M., Thomas, R., Davidson, P., Reichart, P., Koch-Müller, M., Dollinger, G., 2008b. Application of Raman spectroscopy to quantify trace water concentrations in glasses and garnets. Am. Mineral. 93, 1550–1557.
Tissot, B.P., Welte, D.H., 1984. Petroleum Formation and Occurrence. Springer Verlag, Berlin/Heidelberg/New York, 377 p.
Tkachenko, S.I., Shmulovich, K., 1992. Liquid–vapour equilibria in water–salt systems (NaCl, KCl, $CaCl_2$, $MgCl_2$) at 400–600 °C. Dokl. Akad. Nauk SSSR 326, 1055–1059 (in Russian).
Tobin, R.C., Claxton, B.L., 2000. Multidisciplinary thermal maturity studies using vitrinite reflectance and fluid inclusion microthermometry: a new calibration of old techniques. AAPG Bull. 84, 1647–1665.
Tödheide, K., 1982. Hydrothermal solutions. Ber. Bunsenges. Phys. Chem. 86, 1005–1016.
Tödheide, K., Franck, E.U., 1963. Das Zweiphasegebiet und die kritische Kurve im Kohlendioxid–Wasser bis zu Drucken von 3500 bar. Z. Phys. Chem. 37, 387–401.
Tonani, F., 1980. Some remarks on the application of geochemical techniques in geothermal exploration. In: Proc. Advances in Eur. Geotherm. Res., 2nd Symposium, Strasbourg, pp. 428–443.
Toulmin, P., Barton, P.B., 1964. A thermodynamic study of pyrite and pyrrhotite. Geochim. Cosmochim. Acta 28, 641–671.
Touret, J.L.R., 1971a. Le faciès granulite en Norvège méridionale: I Les associations minéralogiques. Lithos 4, 239–249.
Touret, J.L.R., 1971b. Le faciès granulite en Norvège méridionale: II Les inclusions fluides. Lithos 4, 423–436.
Touret, J.L.R., 1981. Fluid inclusions in high-grade metamorphic rocks. In: Hollister, L.S., Crawford, M.L. (Eds.), Fluid Inclusions. Applications to Petrology. In: Min. Assoc. Can. Short Course Handbook, vol. 6, pp. 182–208.
Trommsdorf, V., Skippen, G., 1986. Vapor loss ("boiling") as a mechanism for fluid evolution in metamorphic rocks. Contrib. Mineral. Petrol. 94, 317–322.

Truesdell, A.H., 1974. Oxygen isotope activities and concentrations in aqueous salt solutions at elevated temperatures—consequences for isotope geochemistry. Earth Planet. Sci. Lett. 23, 387–396.

Truesdell, A.H., 1976. Geochemical techniques in exploration. In: Proceedings of the 2nd United Nations Symposium on the Development and Use of Geothermal Resources, San Francisco, CA, USA, pp. 53–79.

Tugarinov, A.I., Naumov, V.B., 1970. Dependence of temperature of decrepitation on composition of gas–liquid inclusions in minerals. Dokl. Akad. Nauk SSSR 195, 182–184 (in Russian).

Tuinstra, F., Koenig, J.L., 1970. Raman spectrum of graphite. J. Chem. Phys. 53, 1126–1130.

Ujiie, K., Yamaguchi, A., Taguchi, S., 2008. Stretching of fluid inclusions in calcite as an indicator of frictional heating on faults. Geology 36, 111–114.

Ulrich, M.R., Bodnar, R.J., 1988. Systematics of stretching of fluid inclusions. II. Barite at one atmosphere confining pressure. Econ. Geol. 83, 1037–1046.

Urban, M., Thomas, R., Hurai, V., Konečný, P., Chovan, M., 2006. Superdense CO_2 inclusions in Cretaceous quartz–stibnite veins hosted in low-grade Variscan basement of the Western Carpathians, Slovakia. Miner. Deposita 40, 867–873.

Urey, H.C., 1947. The thermodynamic properties of isotopic substances. J. Chem. Soc. London 562–581.

Urey, H.C., Brickwedde, F.G., Murphy, G.M., 1932. A hydrogen isotope of mass 2 and its concentration. Phys. Rev. 40, 1.

Urey, H.C., Epstein, S., McKinney, C., McCrea, J., 1948. Method for measurement of paleotemperatures. Bull. Geol. Soc. Am. 59, 1359–1360.

Urusova, M.A., 1975. Volumetric properties of natrium chloride aqueous solutions at increased temperatures and pressures. Zh. Neorg. Khim. 22, 3103–3110 (in Russian).

Valley, J.W., 1986. Stable isotope geochemistry of metamorphic rocks. Min. Soc. Am. Rev. Mineral. 16, 445–489.

Valley, J.W., 2001. Stable isotope thermometry at high temperatures. In: Valley, J.W., Cole, D.R. (Eds.), Stable Isotope Geochemistry. In: Min. Soc. Am. Rev. Mineral. Geochem., vol. 43, pp. 365–413.

Valley, J.W., O'Neil J.R., J.R., 1981. $^{13}C/^{12}C$ exchange between calcite and graphite: a possible thermometer in Grenville marbles. Geochim. Cosmochim. Acta 45, 411–419.

van den Kerkhof, A.M., 1988. The system CO_2–CH_4–N_2 in fluid inclusions: Theoretical modelling and geological applications. Ph.D. Thesis, Free University Press, Amsterdam.

van den Kerkhof, A.M., 1990. Isochoric phase diagrams in the systems CO_2–CH_4 and CO_2–N_2: Application to fluid inclusions. Geochim. Cosmochim. Acta 54, 621–629.

van den Kerkhof, A.M., Hein, U., 1992. H_2S–bearing systems. In: McKibben, M.A., Montanez, I.P., Hall, D.L. (Eds.), PACROFI IV, Pan-American Conference on Research on Fluid Inclusions, Program and Abstracts. Dept. of Earth Sciences, University of California, Riverside, pp. 164–169.

van den Kerkhof, A.M., Thiéry, R., 1994. Phase transitions and density calculation in the CO_2–CH_4–N_2 system. In: de Vivo, B., Frezzotti, M.-L. (Eds.), Fluid Inclusions in Minerals: Methods and Applications. Virginia Tech., USA, pp. 171–190.

van den Kerkhof, A.M., Hein, U., 2001. Fluid inclusion petrography. Lithos 55, 27–47.

van den Kerkhof, A.M., Olsen, S., 1990. A natural example of superdense CO_2 inclusions: microthermometry and Raman analysis. Geochim. Cosmochim. Acta 54, 895–901.

van den Kerkhof, A.M., Thiéry, R., 2001. Carbonic inclusions. Lithos 55, 49–68.

van den Kerkhof, A.M., Touret, J.L.R., Maijer, C., Jansen, J.B.H., 1991. Retrograde methane-dominated fluid inclusions from high-temperature granulites of Rogaland, southwestern Norway. Geochim. Cosmochim. Acta 55, 2533–2544.

van den Kerkhof, A.M., Kronz, A., Simon, K., Scherer, T., 2004. Fluid-controlled quartz recovery in granulite as revealed by cathodoluminescence and trace element analysis (Bamble Sector, Norway). Contrib. Mineral. Petrol. 146, 637–652.

van den Kerkhof, A.M., Kronz, A., Simon, K., 2014. Deciphering fluid inclusions in high-grade rocks. Geosci. Front. 5, 683–695.

Vanko, D.A., Bodnar, R.J., Sterner, S.M., 1988. Synthetic fluid inclusions: VIII. Vapor-saturated halite solubility in part of the system $NaCl-CaCl_2-H_2O$, with application to fluid inclusions from oceanic hydrothermal systems. Geochim. Cosmochim. Acta 52, 2451–2456.

Varček, C., 1985. Metalogenetic characteristics of the Spišsko–gemerské rudohorie Mountains and the position of the Rudňany ore field. In: Cambel, B., Jarkovský, J. (Eds.), The Rudňany Ore Field—Geochemical and Metallogenetic Characteristics. Veda, Bratislava, pp. 61–77 (in Slovak).

Veksler, I.V., 2004. Liquid immiscibility and its role at the magmatic–hydrothermal transition: a summary of experimental studies. Chem. Geol. 210, 7–31.

Veksler, I.V., 2009. Extreme iron enrichment and liquid immiscibility in mafic intrusions: experimental evidence revisited. Lithos 111, 72–82.

Veksler, I.V., Thomas, R., 2002. An experimental study of B-, P- and F-rich synthetic granite pegmatite at 0.1 and 0.2 GPa. Contrib. Mineral. Petrol. 143, 673–683.

Verma, S., Santoyo, E., 1997. New improved equations for K/Na, Na/Li and SiO_2 geothermometers by outlier detection and rejection. J. Volcanol. Geotherm. Res. 79, 9–23.

Verma, S.P., Pandarinath, K., Santoyo, E., 2008. SolGeo: a new computer program for solute geothermometers and its application to Mexican geothermal fields. Geothermics 37, 597–621.

Villa, I.M., Frezotti, M.L., De Vivo, B., Stefanini, B., Burke, E.A.J., 1991. Ar geochemistry and fluid inclusion thermobarometry of pyroxenite nodules from Oahu. Fluid Inc. Res. 24, 215.

Vityk, M.O., Bodnar, R.J., 1995. Textural evolution of synthetic fluid inclusions in quartz during reequilibration, with applications to tectonic reconstruction. Contrib. Mineral. Petrol. 121, 309–323.

Vityk, M.O., Bodnar, R.J., Schmidt, C.S., 1994. Fluid inclusions as tectonothermobarometers: relation between pressure-temperature history and reequilibration morphology during crustal thickening. Geology 22, 731–734.

Vityk, M.O., Bodnar, R.J., Dudok, I.V., 1996. Fluid inclusions in "Marmarosh Diamonds": evidence for tectonic history of the Folded Carpathian Mountains, Ukraine. Tectonophysics 255, 163–174.

Vohnhof, H.B., van Breukelen, M.R., Postma, O., Rowe, P.J., Atkinson, T.C., Kroon, D., 2006. A continuous-flow crushing device for on-line δ^2H analysis of fluid inclusion water in speleothems. Rapid Commun. Mass Spectrom. 20, 2553–2558.

Volk, H., Fuentes, D., Fuerbach, A., Miese, Ch., Koehler, W., Bärsch, N., Barcikowski, S., 2010. First on-line analysis of petroleum from single inclusion using ultrafast laser ablation. Org. Chem. 41, 74–77.

Vozárová, A., Ďurkovičová, J., Repčok, I., 1995. Data on chemical and isotope composition of Carboniferous and Mesozoic carbonates of Inner Western Carpathians. Slovak Geol. Mag. 1, 127–152.

Vozárová, A., Šarinová, K., Rodionov, N., Laurinc, D., Paderin, I., Sergeev, S., Lepekhina, E., 2012. U–Pb ages of detrital zircons from Paleozoic metasandstones of the Gelnica Terrane (Southern Gemeric Unit, Western Carpathians, Slovakia): evidence for Avalonian-Amazonian provenance. Int. J. Earth Sci. (Geol. Rundsch.) 101, 919–936.

Voznyak, D.K., 2007. Microinclusions for the Reconstruction of the Conditions of Endogeneous Mineral Formation. Naukova Dumka, Kiiv, 279 p. (in Ukrainian).

Wagner, W., Pruss, A., 2002. The IAPWS formulation 1995 for the thermodynamic properties of ordinary water substance for general and scientific use. J. Phys. Chem. Ref. Data 31, 387–535.

Wagner, W., Saul, A., Pruss, A., 1994. International equations for the pressure along the melting and along the sublimation curve of ordinary water substance. J. Pharm. Sci. 23, 515–527.

Walder, I.F., Norman, D.I., Segalstad, T.V., 1991. Fluid inclusion gas analysis of hydrothermal vein molybdenum deposits associated with the Drammen granite in the Permian Oslo Rift, Norway. J. Geochem. Explor. 42, 195–203.

Waldmann, L., 1943. Zur Theorie der Isotopentrennung durch Austauschreaktionen. Naturwissenschaften 31, 205–206.

Walrafen, G.E., Douglas, R.T.W., 2006. Raman spectra from very concentrated aqueous NaOH and from wet and dry, solid, and anhydrous molten, LiOH, NaOH, and KOH. J. Chem. Phys. 124, 114504-1–114504-14.

Wang, A., Dhamelincourt, P., Dubessy, J., Guerard, D., Landais, P., Lelaurain, M., 1989. Characterization of graphite alteration in an uranium deposit by micro-Raman spectroscopy, X-ray diffraction, transmission electron microscopy and scanning electron microscopy. Carbon 27, 209–218.

Wang, X., Chou, I.-M., Hu, W., Burruss, R.C., Sun, Q., Song, Y., 2011. Raman spectroscopic measurements of CO_2 density: experimental calibration with high-pressure optical cell (HPOC) and fused capillary capsule (FSCC) with application to fluid inclusion observations. Geochim. Cosmochim. Acta 75, 4080–4093.

Wark, D.A., Watson, E.B., 2006. TitaniQ: a titanium-in-quartz geothermometer. Contrib. Mineral. Petrol. 152, 743–754.

Weast, R.D., 1979. Handbook of Chemistry and Physics, 60th ed. CRC Press, Cleveland, OH, 2385 p.

Webster, J.D. (Ed.), 2006. Melt Inclusions in Plutonic Rocks. Min. Assoc. Can. Short Course, vol. 36, p. 248.

Webster, J.D., Thomas, R., Rhede, D., Förster, H.-J., Seltmann, R., 1997. Melt inclusions in quartz from an evolved peraluminous pegmatite: geochemical evidence for strong tin enrichment in fluorine-rich and phosphorus-rich residual liquids. Geochim. Cosmochim. Acta 61, 2589–2604.

Webster, J.D., Kinzler, R.J., Mathez, E.A., 1999. Chloride and water solubility in basalt and andesite melts and implications for magmatic degassing. Geochim. Cosmochim. Acta 63, 729–738.

Welsch, H., 1973. Die Systeme Xenon–Wasser und Methan–Wasser bei hohen Drücken und Temperaturen. Ph.D. Thesis, Univ. Karlsruhe.

Wendland, M., Hasse, H., Maurer, G., 1999. Experimental pressure–temperature data on three- and four-phase equilibria of fluid, hydrate and ice phases in the system carbon dioxide–water. J. Chem. Eng. Data 44, 901–906.

Werner, D.B., Taylor, H.P., Jr., 1971. Temperatures of serpentinization of ultramafic rocks based on $^{18}O/^{16}O$ fractionation between coexisting serpentine and magnetite. Contrib. Mineral. Petrol. 32, 165–185.

White, N.C., Hedenquist, J.W., 1990. Epithermal environments and styles of mineralization: variations and their causes, and guidelines for exploration. J. Geochem. Explor. 36, 445–474.

Whitney, D.L., Evans, B.W., 2010. Abbreviations for names of rock-forming minerals. Am. Mineral. 95, 185–187.

Wiewiora, A., Wieckowski, T., Sokolowska, A., 1979. The Raman spectra of kaolinite sub-group minerals and of pyrophyllite. Arch. Mineral. 35, 5–14.

Wilkins, R.W.T., Barkas, J.P., 1978. Fluid inclusions, deformation and recrystallization in granite tectonits. Contrib. Mineral. Petrol. 65, 293–299.

Wilkinson, J.J., 2001. Fluid inclusions in hydrothermal ore deposits. Lithos 55, 229–272.

Wilkinson, J.J., Rankin, A.H., Mulshaw, S.C., Nolan, J., Ramsey, M.H., 1994. Laser ablation-ICP-AES for the determination of metals in fluid inclusions: an application to the study of magmatic fluids. Geochim. Cosmochim. Acta 58, 1133–1146.

Wilkinson, J.J., Lonergan, L., Fairs, T., Herrington, R.J., 1998. Fluid inclusion constraints on conditions and timing of hydrocarbon migration and quartz cementation in Brent Group reservoir sandstones, Columba Terrace, northern North Sea. In: Parnell, J. (Ed.), Dating and Duration of Fluid Flow and Fluid-Rock Interaction. In: Geol. Soc. Spec. Publ. London, vol. 144, pp. 69–89.

Williams-Jones, A.E., Samson, I.M., 1990. Theoretical estimation of halite solubility in the system $NaCl-CaCl_2-H_2O$: applications to fluid inclusions. Can. Mineral. 28, 299–304.

Winter, J.D., 2010. An Introduction to Igneous and Metamorphic Petrology, second ed. Prentice Hall, New York, 702 p.

Wones, D.R., 1981. Mafic silicates as indicators of intensive variables in granitic magmas. Min. Geol. 31, 191–212.

Wood, S.A., Spera, F.J., 1984. Adiabatic decompression of aqueous solutions: applications to hydrothermal fluid migration in the crust. Geology 12, 707–710.

Wopenka, B., Pasteris, J.D., 2005. A mineralogical perspective on the apatite in bone. Mater. Sci. Eng. C25, 131–143.

Worden, R.H., 1996. Controls on halogen concentrations in sedimentary formation waters. Mineral. Mag. 60, 259–279.

Wunder, D., Koráb, T., Antonov, P., Dubiniuk, P., Sůrová, E., Golovackij, J., Širáňová, V., Michalko, J., Kuba, J., 1991. Prognostic evaluation of natural hydrocarbon resources in the Zborov anticlinorium—the geological assessment of the Zborov-1 drillhole. Open-file Report, Geofond, Bratislava, 45 p. (in Slovak).

Xiao, Y., Hoefs, J., van den Kerkhof, A.M., Fiebig, J., Zheng, Y.-F., 2000. Fluid history of UHP metamorphism in Dabie Shan, China: a fluid inclusion and oxygen isotope study on the coesite-bearing eclogite from Bixiling. Contrib. Mineral. Petrol. 139, 1–16.

Yanatieva, O.K., 1946. Solubility polytherms in the system $CaCl_2-MgCl_2-H_2O$ and $CaCl_2-NaCl-H_2O$. Zh. Prikl. Khim. 19, 709–722 (in Russian).

Yamamoto, J., Kagi, H., 2006. Extended micro-Raman densimeter for CO_2 applicable to mantle-originated fluid inclusions. Chem. Lett. 35, 610–611.

Yamamoto, J., Kagi, H., Kawakami, Y., Hirano, N., Nakamura, M., 2007. Paleo-Moho depth determined from the pressure of CO_2 fluid inclusions: Raman spectroscopic barometry of mantle- and crust-derived rocks. Earth Planet. Sci. Lett. 3–4, 369–377.

Yang, H., Dembowski, R.F., Conrad, P., Downs, R.T., 2008. Crystal structure and Raman spectrum of hydroxyl-bastnäsite-(Ce), $CeCO_3(OH)$. Am. Mineral. 93, 698–701.

Yang, Y., Zheng, H., Zhang, L., 2009. Determining pressure with daughter minerals in fluid inclusions by Raman spectroscopy: Sphalerite as an example. Acta Geol. Sin. 83, 628–632.

Yardley, B.W.D., Banks, D.A., Barnicoat, A.C., 2000. The chemistry of crustal brines: tracking their origins. In: Porter, T.M. (Ed.), Hydrothermal Iron Oxide Copper Gold and Related Deposits: A Global Perspective. Australian Mineral Foundation, Adelaide, pp. 61–70.

Yermakov, N.P., 1950. Investigations of Mineral-Forming Solutions. Izd. Kharkovskogo Gosudarstvennogo Universiteta, Kharkov (in Russian).

Yermakov, N.P., 1965. Research on the Nature of Mineral Forming Solutions. In: Intl. Ser. Monogr. Earth Sci., vol. 22, Pergamon Press, Oxford.

Yermakov, N.P., Dolgov, J.A., 1979. Thermobarogeochemistry. Nedra, Moskva (in Russian).

Zakharchenko, A.I., 1950. Research on liquid inclusions in quartz. L'vov. Sborn. Min. Geol. Obshch. 4, 167–187 (in Russian).

Zaitsev, A.N., 2010. Nyerereite from calcite carbonatite at the Kerimasi Volcano, Northern Tanzania. Geol. Ore Depos. 52, 630–640.

Zaitsev, A.N., Chakhmouradian, A.R., Siidra, O.I., Spratt, J., Williams, C.T., Stanley, C.J., Petrov, S.V., Britvin, S. N., Polyakova, E.A., 2011. Fluorine-, yttrium- and lanthanide-rich cerianite-(Ce) from carbonatitic rocks of the Kerimasi volcano and surrounding explosion craters, Gregory Rift, northern Tanzania. Mineral. Mag. 75, 2813–2822.

Zajacz, Z., Halter, W., Malfait, W.J., Bachmann, O., Bodnar, R.J., Hirschmann, M.M., Mandeville, C.W., Morizet, Y., Muntener, O., Ulmer, P., Webster, J.D., 2005. A composition-independent quantitative determination of the water content in silicate glasses and silicate melt inclusions by confocal Raman spectroscopy. Contrib. Mineral. Petrol. 150, 631–642.

Zajacz, Z., Halter, W., Pettke, T., Guillong, M., 2008. Determination of fluid/melt partition coefficients by LA-ICPMS analysis of coexisting fluid and silicate melt inclusions: controls on element partitioning. Geochim. Cosmochim. Acta 72, 2169–2197.

Zhang, C., Duan, Z.H., 2009. A model for C–O–H fluid in the Earth's mantle. Geochim. Cosmochim. Acta 73, 2089–2102.

Zhang, C., Duan, Z.H., 2010. GFluid: an Excel spreadsheet for investigating C–O–H fluid composition under high temperatures and pressures. Comput. Geosci. 36, 569–572.

Zhang, L., Liu, J.X., Zhou, H., Chen, Z.S., 1989. Oxygen isotope fractionation in the quartz–water–salt system. Econ. Geol. 84, 1643–1650.

Zhang, L., Ren, Z.-Y., Nichols, A.R.L., Zhang, Y.-H., Zhang, Y., Qian, S.-P., Liu, J.-Q., 2014. Lead isotope analysis of melt inclusions by LA-MC-ICP-MS. J. Anal. Atom. Spectrom. 29, 1393–1405.

Zhang, N., Tian, Z.J., Leng, Y.Y., Wang, H.T., Song, F.Q., Meng, J.H., 2007. Raman characteristics of hydrocarbon and hydrocarbon inclusions. Sci. China Ser. D-Earth Sci. 50, 1171–1178.

Zhang, Y.G., Frantz, J.D., 1987. Determination of the homogenization temperatures and densities of supercritical fluids in the system $NaCl-KCl-CaCl_2-H_2O$ using synthetic fluid inclusions. Chem. Geol. 64, 225–350.

Zhao, D., Essene, E.J., Zhang, Y., 1999. An oxygen barometer for rutile–ilmenite assemblages: oxidation state of metasomatic agents in the mantle. Earth Planet. Sci. Lett. 166, 127–137.

Zheng, Y.-F., 1990a. Carbon–oxygen isotopic covariation in hydrothermal calcite during degassing of CO_2. A quantitative evaluation and application to the Kushikino gold mining area in Japan. Miner. Deposita 25, 246–250.

Zheng, Y.-F., 1990b. The effect of Rayleigh degassing of magma on sulphur isotope composition: a quantitative evaluation. Terra Nova 2, 74–78.

Zheng, Y.-F., 1991. Calculation of oxygen isotope fractionation in metal oxides. Geochim. Cosmochim. Acta 55, 2299–2307.

Zheng, Y.-F., 1992. Oxygen isotope fractionation in wolframite. Eur. J. Mineral. 4, 1331–1335.

Zheng, Y.-F., 1993a. Calculation of oxygen isotope fractionation in hydroxyl-bearing silicates. Earth Planet. Sci. Lett. 120, 247–263.

Zheng, Y.-F., 1993b. Calculation of oxygen isotope fractionation in anhydrous silicate minerals. Geochim. Cosmochim. Acta 57, 1079–1091.

Zheng, Y.-F., 1993c. Oxygen isotope fractionation in SiO_2 and Al_2SiO_5 polymorphs: effects of crystal structure. Eur. J. Mineral. 5, 651–658.

Zheng, Y.-F, 1995. Oxygen isotope fractionation in magnetites: structural effect and oxygen inheritance. Chem. Geol. 121, 309–316.

Zheng, Y.-F., 1996. Oxygen isotope fractionations involving apatites: application to paleotemperature determination. Chem. Geol. 127, 177–187.

Zheng, Y.-F., 1999. Oxygen isotope fractionation in carbonate and sulphate minerals. Geochem. J. 33, 109–126.

Zheng, Y.-F., 2011. On the theoretical calculations of oxygen isotope fractionation factors for carbonate–water systems. Geochem. J. 45, 341–354.

Zheng, Y.-F., Hoefs, J., 1993a. Carbon and oxygen isotopic covariations in hydrothermal calcites. Theoretical modelling on mixing processes and application to Pb–Zn deposits in the Harz Mountains, Germany. Miner. Deposita 28, 79–89.

Zheng, Y.-F., Hoefs, J., 1993b. Stable isotope geochemistry of hydrothermal mineralizations in the Harz Mountains: I. Carbon and oxygen isotopes of carbonates and implications for the origin of hydrothermal fluids. Monograph Ser. Mineral. Dep. 30, 169–187.

Zheng, Y.-F., Wang, Z.-R., Li, S.-G., Zhao, Z.-F., 2002. Oxygen isotope equilibrium between eclogite minerals and its constraints on mineral Sm–Nd chronometer. Geochim. Cosmochim. Acta 66, 625–634.

Zimova, M., Webb, S.L., 2006. The effect of chlorine on the viscosity of $Na_2O–Fe_2O_3–Al_2O_3–SiO_2$ melts. Am. Mineral. 91, 344–352.

Zirkel, F., 1873. Die mikroskopische Beschaffenheit der Mineralien und Gesteine. Wilhelm Engelmann, Leipzig, 502 p.

Žák, K., Radvanec, M., Grecula, P., Bartalský, B., 1991. Sr, S, C, O-isotopes and metamorphic–hydrothermal model of vein mineralization in the Gemericum. Miner. Slov. 23, 95–108 (in Slovak).

Žák, K., Radvanec, M., Grecula, P., 2005. Siderite mineralization of the Gemericum superunit (Western Carpathians, Slovakia): review and a revised genetic model [Ore Geology Reviews 24, 267-298] – a replay. Ore Geol. Rev. 26, 173–180.

Index

Note: Page numbers followed by *f* indicate figures, *b* indicate boxes and *t* indicate tables.

MIX

Papier | Fördert gute Waldnutzung

FSC® C083411

Zeitfracht Medien GmbH
Ferdinand-Jühlke-Straße 7
99095 Erfurt, Deutschland
produktsicherheit@kolibri360.de